Organic Photovoltaics

Edited by
Christoph Brabec, Vladimir Dyakonov,
and Ullrich Scherf

Related Titles

Scheer, R., Schock, H.-W.

Chalcogenide Photovoltaics

Physics, Technologies, and Thin Film Devices

2009

ISBN: 978-3-527-31459-1

Ronda, C. (ed.).

Luminescence

From Theory to Applications

2007

ISBN: 978-3-527-31402-7

Hadziioannou, G., Malliaras, G. G. (eds.)

Semiconducting Polymers

Chemistry, Physics and Engineering

2007

ISBN: 978-3-527-31271-9

Klauk, H. (ed.)

Organic Electronics

Materials, Manufacturing and Applications

2006

ISBN: 978-3-527-31264-1

Müllen, K., Scherf, U. (eds.)

Organic Light Emitting Devices

Synthesis, Properties and Applications

2006

ISBN: 978-3-527-31218-4

Brütting, W. (ed.)

Physics of Organic Semiconductors

2005

ISBN: 978-3-527-40550-3

Würfel, P.

Physics of Solar Cells

From Principles to New Concepts

2005

ISBN: 978-3-527-40428-5

Roth, S., Carroll, D.

One-Dimensional Metals

Conjugated Polymers, Organic Crystals, Carbon Nanotubes

2004

ISBN: 978-3-527-30749-4

Castaner, L., Silvestre, S.

Modelling Photovoltaic Systems Using PSpice

2003

ISBN: 978-0-470-84527-1

Organic Photovoltaics

Materials, Device Physics, and
Manufacturing Technologies

Edited by
Christoph Brabec, Vladimir Dyakonov, and
Ullrich Scherf

WILEY-VCH Verlag GmbH & Co. KGaA

The Editors

Dr. Christoph Brabec
Konarka Austria
Altenbergerstr. 69
4040 Linz
Austria

Prof. Dr. Vladimir Dyakonov
Universität Würzburg
Experimentalphysik VI
Fakultät für Physik + Astronomie
Am Hubland
97074 Würzburg
Germany

Prof. Dr. Ullrich Scherf
Institut für Makromolekulare Chemie
Bergische Universität Wuppertal
Gauss-Straße 20
42097 Wuppertal
Germany

All books published by **Wiley-VCH** are carefully produced. Nevertheless, authors, editors, and publisher do not warrant the information contained in these books, including this book, to be free of errors. Readers are advised to keep in mind that statements, data, illustrations, procedural details or other items may inadvertently be inaccurate.

Library of Congress Card No.:
applied for

British Library Cataloguing-in-Publication Data
A catalogue record for this book is available from the British Library.

Bibliographic information published by the Deutsche Nationalbibliothek
Die Deutsche Nationalbibliothek lists this publication in the Deutsche Nationalbibliografie; detailed bibliographic data are available on the Internet at http://dnb.d-nb.de.

© 2008 WILEY-VCH Verlag GmbH & Co. KGaA, Weinheim

All rights reserved (including those of translation into other languages). No part of this book may be reproduced in any form – by photoprinting, microfilm, or any other means – nor transmitted or translated into a machine language without written permission from the publishers. Registered names, trademarks, etc. used in this book, even when not specifically marked as such, are not to be considered unprotected by law.

Composition   Thomson Digital, Noida, India
Printing   betz-druck GmbH, Darmstadt
Bookbinding   Litges & Dopf GmbH, Heppenheim
Cover Design   Schulz Grafik Design, Fußgönheim

Printed in the Federal Republic of Germany
Printed on acid-free paper

ISBN: 978-3-527-31675-5

Contents

Preface *XV*
List of Contributors *XVII*

I	**Materials** *1*
A	**Donors** *1*

1 Regioregular Polythiophene Solar Cells: Material Properties and Performance *3*
Paul C. Ewbank, Darin Laird, and Richard D. McCullough

1.1	Introduction *3*
1.1.1	Overview of Nomenclature and Synthesis *3*
1.1.2	Advantages of the HT Architecture *5*
1.2	Assembly and Morphology *6*
1.2.1	Conformation *7*
1.2.2	Aggregation *7*
1.2.3	Solid Deposition *8*
1.2.4	Solid-State Crystalline Order *9*
1.2.5	Solid-State Phase Behavior and Thermal Analysis *10*
1.2.6	Anisotropy *12*
1.3	Characterization of Impurities *14*
1.3.1	Fractionation and Effects of M_w *14*
1.3.2	Inorganic Impurities *15*
1.4	Optical and Electronic Properties of PAT *16*
1.4.1	Optical Properties: Intermolecular Excitons *16*
1.4.2	HT-PT Electron Transport: Conductivity and Mobility *17*
1.5	Benefits of HT-Regioregular Polythiophenes in Solar Cells *17*
1.6	Bulk Heterojunctions: Focus on HT-PAT/PCBM Blends *18*
1.6.1	Homogeneous PCBM Assembly *18*
1.6.2	HT-PAT/PCBM Blends: Component Ratio *19*
1.6.3	HT-PAT/PCBM Blends: Annealing *20*

Organic Photovoltaics: Materials, Device Physics, and Manufacturing Technologies.
Edited by Christoph Brabec, Vladimir Dyakonov, and Ullrich Scherf
Copyright © 2008 WILEY-VCH Verlag GmbH & Co. KGaA, Weinheim
ISBN: 978-3-527-31675-5

1.6.3.1	PCBM Phase Separation and Assembly	20
1.6.3.2	Polymer Phase Separation and Assembly	21
1.6.3.3	Evolution of Open-Circuit Voltage	21
1.6.3.4	Evolution of Short-Circuit Current	23
1.6.3.5	Evolution of Fill Factor, Power Conversion Efficiency (η)	23
1.6.4	HT-PAT/PCBM Blend: Layer Thickness	24
1.6.5	Summary	24
1.7	HT-PT in Other Blends	25
1.7.1	C_{60} and Non-PCBM Fulleroids	25
1.7.2	Carbon Nanotubes and Other Organics	25
1.7.3	Hybrid Organic/Inorganic Nanocomposites	26
1.8	Surface Analysis of HTPT Films	26
1.8.1	AFM and STM	26
1.8.2	XPS/ESCA and Auger Spectroscopy	27
1.8.3	Rutherford Backscattering Spectrometry	28
1.8.4	X-Ray	29
1.8.5	Other Techniques (SIMS, UPS)	29
1.9	Summary and Future Directions	30
	Appendix 1.A. Survey of Photovoltaic Cells Incorporating Regioregular Polythiophenes (2001–2006)	30
	References	41
2	**Fluorene-Containing Polymers for Solar Cell Applications**	**57**
	David Jones	
2.1	Introduction	57
2.1.1	Bulk Heterojunctions	59
2.2	Fluorene-Containing Materials	61
2.2.1	Polyfluorene-Containing Photovoltaics	61
2.2.2	Polyfluorene Copolymers	63
2.2.2.1	Electron Transport Materials	63
2.2.2.2	Hole Transport Materials	64
2.2.3	Devices	65
2.2.3.1	Bulk Heterojunctions	65
2.2.3.2	Dye-Sensitized Solar Cells	66
2.3	Bulk Heterojunction Device Performance	68
2.3.1	Morphology	68
2.3.1.1	Techniques for Probing Thin-Film Morphology	69
2.3.1.2	Relating Film Morphology to Device Performance	71
2.3.1.3	Inkjet Printing	75
2.3.1.4	Microemulsions of Blends	76
2.4	Low-Bandgap Materials	76
2.4.1	New Low-Bandgap Materials	76
2.4.2	Alternative Structures	82
2.4.2.1	Carbazoles	82
2.4.2.2	Fluorenones	84

2.5	Future Directions 84
2.5.1	Controlled Morphology 84
2.6	Conclusions 86
	References 86

3	**Carbazole-Based Conjugated Polymers as Donor Material for Photovoltaic Devices** 93
	Wojciech Pisula, Ashok K. Mishra, Jiaoli Li, Martin Baumgarten, and Klaus Müllen
3.1	Introduction 93
3.2	Synthesis of Carbazole-Based Polymers 96
3.3	Supramolecular Order of Carbazole-Based Polymers 111
3.4	Photovoltaic Devices 116
3.4.1	Polycarbazole 116
3.4.2	Ladder-Type Polymers Based on 2,7-Carbazole 121
3.5	Conclusions 125
	References 126

4	**New Construction of Low-Bandgap Conducting Polymers** 129
	Zhengguo Zhu, David Waller, and Christoph J. Brabec
4.1	Introduction 129
4.2	Low-Bandgap Polymers Containing 4,7-Di-2-Thienyl-2,1,3-Benzothiadiazole Moieties 130
4.3	Low-Bandgap Polymers Containing 4,8-Di-2-Thienyl-Benzo[1,2-c:4,5-c']bis[1,2,5]thiadiazole Segments 136
4.4	Low-Bandgap Polymers Containing 4,9-Di-2-Thienyl[1,2,5]thiadiazolo[3,4-g]quinoxalines 137
4.5	Low-Bandgap Polymers Containing Thieno[3,4-b]pyrazines 138
4.6	Arylene Vinylene Based Low-Bandgap Polymers 140
4.7	Low-Bandgap Polymers Containing 4H-Cyclopenta[2,1-b;3,4-b']dithiophene or Its Analogues 142
4.8	Low-Bandgap Polymers Based on Other Types of Building Blocks 146
	References 148

| B | **Acceptors** 153 |

5	**Fullerene-Based Acceptor Materials** 155
	David F. Kronholm and Jan C. Hummelen
5.1	Introduction and Overview 155
5.2	Fullerenes as n-Type Semiconductors 158
5.2.1	Electron Accepting and Transport 158
5.2.2	Other Electronic Properties 159
5.3	[60]PCBM 162
5.4	Variations in Fullerene Derivative and Effect on OPV Device 165

5.4.1	Morphology Considerations – Solubility and Miscibility of the Fullerene Derivative 165
5.4.2	Solubility and Supersaturation in the Donor/Acceptor Blend 166
5.4.3	Miscibility 168
5.4.4	Morphology Fixation and Insoluble Fullerene Layers 169
5.4.5	Optical Absorption of the Fullerene Derivative 169
5.4.6	More Strongly Absorbing Fullerene Derivatives: [70]PCBM and [84]PCBM 170
5.4.7	LUMO Variation 170
5.4.8	Deuterated PCBM 172
5.5	Practical Considerations and Potential in Commercial Devices 172
5.5.1	Powder Morphology and Dissolution 172
5.5.2	Stability of the Fullerene Derivative and the Device Film 173
5.5.3	Impurities 174
5.5.4	Commercial-Scale Application 174
	References 175

6 Hybrid Polymer/Nanocrystal Photovoltaic Devices 179
Neil C. Greenham

6.1	Introduction 179
6.2	Classes of Polymer/Nanocrystal Device 181
6.2.1	Devices Based on CdSe Nanoparticles 181
6.2.1.1	Synthesis of CdSe Nanoparticles 181
6.2.1.2	Devices Using CdSe Nanoparticles 186
6.2.2	Devices Based on Metal Oxide Nanoparticles 189
6.2.2.1	Synthesis of ZnO Nanoparticles 190
6.2.2.2	Devices Based on ZnO Nanoparticles 190
6.2.3	Devices Based on Low-Bandgap Nanoparticles 192
6.2.4	Polymer Brush Devices 195
6.2.5	All-Nanoparticle Devices 195
6.3	Physical Processes in Polymer/Nanoparticle Devices 196
6.3.1	Absorption and Exciton Transport 197
6.3.2	Charge Transfer 198
6.3.3	Charge Separation and Recombination 202
6.3.4	Charge Transport 204
6.3.5	Electrical Characteristics and Morphology 206
6.4	Conclusions 207
	References 208

C Transport Layers 211

7 PEDOT-Type Materials in Organic Solar Cells 213
Andreas Elschner and Stephan Kirchmeyer

7.1	Introduction 213
7.2	Chemical Structure and Impact on Electronic Properties 214

7.2.1	Chemical Structure of PEDOT-Type Materials	214
7.2.2	Polymerization	215
7.2.3	Morphology: π–π Stacking and Crystallization	216
7.2.4	Redox States of PEDOT	217
7.3	PEDOT-Type Materials in Organic Solar Cells	218
7.3.1	Preparation of PEDOT Layers	218
7.4	High-Conductive PEDOT:PSS as TCO-Substitution in OSCs	220
7.4.1	Conductivity of PEDOT:PSS	221
7.4.2	Morphology Impact on Conductivity	222
7.4.3	Optical Properties of PEDOT:PSS	226
7.4.4	Long-Term Stability	228
7.5	PEDOT-Type Materials as Hole-extracting Layers in OSCs	229
7.5.1	PEDOT:PSS as Buffer Layer in Solar Cells	229
7.5.2	Electronic Effects at the PEDOT:PSS–Semiconductor Interface	231
7.6	Conclusions	233
	References	234

8	**The Dispersion Approach for Buffer Layers and for the Active Light Absorption Layer**	**243**
	Bjoern Zeysing and Bernhard Weßling	
8.1	Introduction	243
8.2	Photovoltaic Devices	243
8.3	Conductive Polymers	245
8.3.1	Polyaniline	247
8.4	Polymers in Photovoltaic Devices	250
8.4.1	ITO Replacement	251
8.4.2	Polymer Photovoltaic Devices	252
8.5	The Dispersion Approach as a Productive Tool for Photoactive Layer Deposition	255
8.6	Discussion of the Influence of Polymer Morphology on Device Performance	257
8.7	Summary	257
	References	258

II	**Device Physics**	**261**
A	**Overview of the State-of-the-Art**	**261**

9	**Titanium Oxide Films as Multifunctional Components in Bulk Heterojunction "Plastic" Solar Cells**	**263**
	Kwanghee Lee, Jin Young Kim, and Alan J. Heeger	
9.1	Introduction	263
9.2	Sol–Gel Processed Titanium Oxide as an Optical Spacer in Polymer Solar Cells	263
9.3	Air-Stable Bulk Heterojunction Polymer Solar Cells	269
9.4	Efficient Polymer Solar Cells in the Tandem Architecture	272

x | Contents

9.5	Conclusions	277
	References	279

B	**Bulk Heterojunction Solar Cells**	**281**

10	**Performance Improvement of Polymer: Fullerene Solar Cells Due to Balanced Charge Transport**	**283**
	L. Jan Anton Koster, Valentin D. Mihailetchi, Martijn Lenes, and Paul W.M. Blom	
10.1	Introduction	283
10.2	MDMO-PPV:PCBM-Based Solar Cells	287
10.3	Annealed P3HT:PCBM-Based Solar Cells	291
10.4	Slowly Dried P3HT:PCBM Solar Cells	294
10.5	Conclusions	295
	References	296

11	**Morphology of Bulk Heterojunction Solar Cells**	**299**
	Joachim Loos, Svetlana van Bavel, and Xiaoniu Yang	
11.1	Introduction	299
11.2	The Bulk Heterojunction of a Polymer Solar Cell	299
11.3	Our Characterization Toolbox	302
11.3.1	Microscopy	302
11.3.2	Characterization of Nanoscale Electrical Properties	303
11.4	Morphology Determining Factors	307
11.4.1	Molecular Architecture	307
11.4.2	Solvents and Preparation Methods	309
11.4.3	Annealing	311
11.4.3.1	Probing the Morphology Stability	311
11.4.3.2	Morphology Control via Annealing	315
11.4.4	Confinements	315
11.5	The P3HT/PCBM System: Nanoscale Morphology of an Efficient Bulk Heterojunction	319
11.6	Summary	322
	References	324

C	**Hybrid Solar Cells**	**327**

12	**TiO$_2$ Template/Polymer Solar Cells**	**329**
	Vignesh Gowrishankar, Brian E. Hardin, and Michael D. McGehee	
12.1	Introduction	329
12.2	Basic Operation	330
12.3	General Device Structure and Material Choices	331
12.3.1	General Device Structure	331

12.3.2	Transparent Conducting Oxide (Cathode)	*331*
12.3.3	Titania	*332*
12.3.4	Semiconducting Polymer	*332*
12.3.5	Anode	*332*
12.4	Device Structures	*332*
12.4.1	TiO_2/Polymer Bilayers	*334*
12.4.2	TiO_2 Nanoparticles/Polymer	*335*
12.4.3	Mesoporous Titania/Polymer	*337*
12.4.4	Ideal Nanostructures	*339*
12.5	Pore Filling	*341*
12.5.1	Spin Casting	*341*
12.5.2	Melt Infiltration and Dip Coating	*342*
12.5.3	*In Situ* Polymerization	*343*
12.6	Effects of Pore Filling on Polymer Mobility and Exciton Harvesting	*344*
12.6.1	Ordered Versus Disordered Polymers	*344*
12.6.2	Measurement Techniques	*344*
12.6.3	Pore Size Effects for Semicrystalline Polymers	*345*
12.6.4	Significance of Polymer Mobility in OCPVs	*346*
12.6.5	Pore Filling and Exciton Harvesting	*347*
12.7	Organic Composite Photovoltaic Modeling	*347*
12.8	Future Outlook	*348*
12.8.1	Low-Bandgap Polymers	*349*
12.8.2	Polymer Engineering	*349*
12.8.3	Increasing Exciton Diffusion Lengths via Energy Transfer	*350*
12.8.4	Interface Modification	*351*
12.8.5	Conclusion	*352*
	References	*352*
13	**Metal Oxide–Polymer Bulk Heterojunction Solar Cells**	***357***
	Waldo J.E. Beek, Martijn M. Wienk, and René A.J. Janssen	
13.1	Introduction	*357*
13.2	Planar Metal Oxide–Polymer Bilayer Cells	*363*
13.2.1	Metal Oxide–Poly(p-Phenylene Vinylene)	*363*
13.2.2	Metal Oxide–Polythiophene	*365*
13.3	Filling Nanoporous and Nanostructured Metal Oxides with Conjugated Polymers	*368*
13.3.1	Polymers in Nanoporous TiO_2	*368*
13.3.2	Filling Structured Inorganic Semiconductors with Polymers	*371*
13.3.2.1	Structured Porous TiO_2	*371*
13.3.2.2	Oriented Nanorods	*372*
13.4	Nanoparticle–Polymer Hybrid Solar Cells	*375*
13.4.1	TiO_2 Nanoparticles	*376*
13.4.2	ZnO Nanoparticles	*377*
13.4.2.1	Photophysics of Nanocrystalline ZnO–Polymer Blends	*378*
13.4.2.2	Photovoltaic Properties of *nc*-ZnO–Polymer Blends	*380*

13.4.2.3	Morphology of nc-ZnO:Polymer Blends	383
13.5	Metal Oxide Networks and Conjugated Polymers	385
13.5.1	In situ Blends Based on TiO_x	386
13.5.2	In situ Blends Based on ZnO	388
13.6	Conclusions and Outlook	392
	References	393

III	**Technology** 399	
A	**Electrodes** 399	

14	**High-Performance Electrodes for Organic Photovoltaics**	**401**
	Cecilia Guillén and José Herrero	
14.1	Introduction	401
14.2	Metal Electrodes	403
14.2.1	Metal Properties	403
14.2.2	Metal/Organic Semiconductor Interactions	406
14.3	Metal Oxide Electrodes	409
14.3.1	Metal Oxide Properties	409
14.3.2	Metal Oxide/Organic Semiconductor Interactions	411
14.4	Conducting Polymer Electrodes	413
14.4.1	Conducting Polymer Properties	413
14.4.2	Conducting Polymer/Organic Semiconductor Interactions	416
14.5	Multilayer Electrodes	417
14.6	Conclusions	419
	References	419

15	**Reel-to-Reel Processing of Highly Conductive Metal Oxides**	**425**
	Matthias Fahland	
15.1	Introduction	425
15.2	Materials	427
15.3	Deposition Technology	429
15.4	Equipment	431
15.4.1	Vacuum System	432
15.4.2	Winding System	433
15.4.3	Inline Measurement System	434
15.5	Alternative Approaches	435
	References	437

16	**Novel Electrode Structures for Organic Photovoltaic Devices**	**441**
	Michael Niggemann and Andreas Gombert	
16.1	Introduction	441
16.2	Buried Nanoelectrodes	442
16.2.1	Experimental	443
16.3	Organic Photovoltaic Devices on Functional Microprism Substrates	447
16.3.1	Optical Simulations	448

16.3.2	Dimensioning of the Microstructure *452*	
16.3.3	Experimental *454*	
16.4	Anode Wrap-Through Organic Solar Cell *457*	
16.4.1	Organic Solar Cell with Inverted Layer Sequence *458*	
16.4.2	Calculation of Optimal Device Geometry for the Wrap-Through Device *459*	
16.4.3	Performance of Wrap-Through Devices *461*	
16.5	Summary *463*	
	References *465*	

B	**Packaging** *469*	

17	**Flexible Substrates Requirements for Organic Photovoltaics** *471*	
	William A. MacDonald	
17.1	Introduction *471*	
17.2	Polyester Substrates *471*	
17.3	Properties of Base Substrates *473*	
17.3.1	Optical Properties *473*	
17.3.2	Thermal Properties *474*	
17.3.3	Solvent Resistance *475*	
17.3.4	Surface Quality *478*	
17.3.5	Mechanical Properties *479*	
17.3.6	UV Stability *481*	
17.3.7	Barrier *482*	
17.3.8	Summary of the Key Properties of Base Substrates *485*	
17.4	Concluding Remarks *487*	
	References *487*	

18	**Barrier Films for Photovoltaics Applications** *491*	
	Lorenza Moro and Robert Jan Visser	
18.1	Introduction *491*	
18.2	Requirements for OPV Environmental Barriers *492*	
18.3	Degradation Mechanisms of OPV Cells *494*	
18.4	Current Approaches to Oxygen and Moisture Barriers *496*	
18.5	Barix Multilayer Technology *498*	
18.6	Conclusions *506*	
	References *507*	

C	**Production** *511*	

19	**Roll-to-Roll Processing of Thin-Film Organic Semiconductors** *513*	
	Arved C. Hübler and Heiko Kempa	
19.1	Introduction *513*	
19.2	Coating *514*	

19.3	Patterning	*516*
19.4	Roll-to-Roll Processes	*522*
19.5	OPVC Fabrication	*526*
19.6	Conclusions	*527*
	References	*528*

20 Socio-Economic Impact of Low-Cost PV Technologies *531*
Gilles Dennler and Christoph J. Brabec

20.1	Introduction	*531*
20.1.1	The Energy Supply	*531*
20.1.2	The Oil Shortage	*534*
20.1.3	The Global Warming	*537*
20.1.4	Renewable Energies	*539*
20.2	Photovoltaic Energy	*541*
20.2.1	World Market	*541*
20.2.2	Technologies	*543*
20.2.3	Political Incentives	*544*
20.2.4	Potential of PV	*546*
20.3	Organic Photovoltaics and its Potential as a Low-Cost PV Technology	*549*
20.3.1	The Costs of PV	*549*
20.3.1.1	Conclusion	*554*
20.3.2	The Costs of OPV: BOM and BOS	*554*
20.3.2.1	BOM of OPV	*555*
20.3.3	Cost Model for OPV: Representative for any Low-Cost and Low-Performance Technology	*557*
20.3.4	Summary	*563*
	References	*564*

Index *567*

Preface

In 2007 the Norwegian Nobel prize committee gave for the first time the Nobel Prize for Peace to a large scientific consortium, which aims at investigating and proving the rapid environmental and economical change in the earth's climate, also called global warming. Stabilizing the global climate requires an in depth modification of mankind's energy supply habits, and, this will require again a lot of energy. Solar radiation is the renewable energy source with practically unlimited access, attracting equal interest from both politicians and scientists. Today, large expectations are set in photovoltaics to become a major energy supplying technology before 2030.

The main attraction of organic solar cells is their compatibility to conventional printing and coating technologies. Among all photovoltaic technologies, organic solar cells are unique as they will be fabricated by printing or coating processes resulting in a true low cost technology.

World wide research in organic solar cells just started 10 years ago, and since then the number of publications is growing exponentially. Over the last 8 years, the number of papers on all aspects of organic solar cells has increased by about 65% per annum and in 2006, already 10% of the scientific publications in the field of photovoltaics reported on organic solar cells. The performance of organic solar cells was evolving constantly over the last few years, from approx. 1% in 2000 to >5% in 2007. Mastering the next big challenge, organic solar cells with up to 10% efficiency, requires the intensified exchange of knowledge and experiences between all the different scientific subdisciplines in a truly interdisciplinary approach involving materials chemistry, materials characterization, device physics, as well as device, process and production technology.

This book aims to contribute to this very important interdisciplinary information exchange and reviews latest developments in organic photovoltaics in the fields of materials, device physics/technology and production aspects. Despite this book mainly reports on materials and components suitable for solution (wet) processing,

Organic Photovoltaics: Materials, Device Physics, and Manufacturing Technologies.
Edited by Christoph Brabec, Vladimir Dyakonov, and Ullrich Scherf
Copyright © 2008 WILEY-VCH Verlag GmbH & Co. KGaA, Weinheim
ISBN: 978-3-527-31675-5

there is, of course, a clear commitment from the editors to the undoubtly high importance of dye sensitized and vacuum processed organic photovoltaic devices. The presentation and discussion of these and further promising concepts will be covered in a future edition of this book.

April 2008

Christoph J. Brabec
Vladimir Dyakonov
Ullrich Scherf

List of Contributors

Martin Baumgarten
Max Planck Institute for Polymer
Research
Ackermannweg 10
55128 Mainz
Germany

Waldo J. E. Beek
Eindhoven University of Technology
Molecular Materials and Nanosystems
P.O. Box 513
5600 MB Eindhoven
The Netherlands

Paul W. M. Blom
University of Groningen
Zernike Institute for Advanced
Materials
Nijenborgh 4
9747 AG Groningen
The Netherlands

Christoph J. Brabec
Konarka Austria
Altenbergerstrasse 69
4040 Linz
Austria

Svetlana van Bavel
Eindhoven University of Technology
Department of Chemical Engineering
and Chemistry
Laboratory of Materials and Interface
Chemistry
P.O. Box 513
5600 MB Eindhoven
The Netherlands

Gilles Dennler
Konarka Austria
Altenbergerstrasse 69
4040 Linz
Austria

Andreas Elschner
H.C. Starck GmbH
Central Research & Development
Chempark, Building B202
51368 Leverkusen
Germany

Paul C. Ewbank
Carnegie Mellon University
Mellon College of Sciences
Department of Chemistry
4400 Fifth Avenue
Pittsburgh, PA 15213
USA

Organic Photovoltaics: Materials, Device Physics, and Manufacturing Technologies.
Edited by Christoph Brabec, Vladimir Dyakonov, and Ullrich Scherf
Copyright © 2008 WILEY-VCH Verlag GmbH & Co. KGaA, Weinheim
ISBN: 978-3-527-31675-5

Matthias Fahland
Fraunhofer Institute for Electron Beam
and Plasma Technology
Winterbergstr. 28
01277 Dresden
Germany

Andreas Gombert
Fraunhofer Institute for
Solar Energy Systems
Heidenhofstr. 2
79110 Freiburg
Germany

Vignesh Gowrishankar
Stanford University
Department of Materials Science and
Engineering
215 McCullough
416 Escondido Mall, Bldg. 550
Stanford, CA 94305-2205
USA

Neil C. Greenham
University of Cambridge
Department of Physics
Optoelectronics Group
Cavendish Laboratory
J. J. Thomson Avenue
Cambridge CB3 0HE
UK

Cecilia Guillén
Complutense University
Departamento de Energía (CIEMAT)
Avda. Complutense 22
Madrid 28040
Spain

Brian E. Hardin
Stanford University
Department of Materials Science and
Engineering
215 McCullough
416 Escondido Mall, Bldg. 550
Stanford, CA 94305-2205
USA

Alan J. Heeger
University of California, Santa Barbara
Center for Polymers and Organic Solids
Santa Barbara, CA 93106-5090
USA

Gwangju Institute of Science and
Technology
Heeger Center for Advanced Materials
Gwangju 500-712
Korea

José Herrero
Complutense University
Departamento de Energía (CIEMAT)
Avda. Complutense 22
Madrid 28040
Spain

A. C. Hübler
Chemnitz University of Technology
Institute for Print and Media
Technology
Reichenhainer Str. 70
09126 Chemnitz
Germany

Jan C. Hummelen
University of Groningen
Zernike Institute for Advanced
Materials and Stratingh Institute of
Chemistry
Molecular Electronics
Nijenborgh 4
9747 AG Groningen
The Netherlands

List of Contributors

René A. J. Janssen
Eindhoven University of Technology
Molecular Materials and Nanosystems
P.O. Box 513
5600 MB Eindhoven
The Netherlands

David Jones
University of Melbourne
Bio21 Institute, School of Chemistry
Building 102 (Level 4)
30 Flemington Road, Parkville
Melbourne, Victoria 3010
Australia

H. Kempa
Chemnitz University of Technology
Institute for Print and Media
Technology
Reichenhainer Str. 70
09126 Chemnitz
Germany

Jin Young Kim
Gwangju Institute of Science and
Technology
Department of Materials Science
and Engineering
Gwangju 500-712
Korea

Stephan Kirchmeyer
H.C. Starck GmbH
Central Research & Development
Chempark, Building B202
51368 Leverkusen
Germany

L. Jan Anton Koster
University of Cambridge
Optoelectronics Group
Department of Physics
J.J. Thompson Avenue
Cambridge CB3 OHE
UK

David F. Kronholm
Solenne BV
Zernikepark 12
9747 AN Groningen
The Netherlands

Darin Laird
Plextronics, Inc.
Pittsburgh, PA
USA

Kwanghee Lee
Gwangju Institute of Science and
Technology
Department of Materials
Science and Engineering
Gwangju 500-712
Korea

Martijn Lenes
University of Groningen
Zernike Institute for Advanced
Materials
Nijenborgh
9747 AG Groningen
The Netherlands

Jiaoli Li
Max Planck Institute for
Polymer Research
Ackermannweg 10
55128 Mainz
Germany

Joachim Loos
Eindhoven University of Technology
Department of Chemical Engineering
and Chemistry
P.O. Box 513
5600 MB Eindhoven
The Netherlands

William A. MacDonald
DuPont Teijin Films (UK) Limited
P.O. Box 2002
Wilton, Middlesbrough TS90 8JF
UK

Richard D. McCullough
Carnegie Mellon University
Mellon College of Sciences
Department of Chemistry
4400 Fifth Avenue
Pittsburgh, PA 15213
USA

Michael D. McGehee
Stanford University
Department of Materials Science and Engineering
215 McCullough
416 Escondido Mall, Bldg. 550
Stanford, CA 94305-2205
USA

Valentin D. Mihailetchi
ECN Solar Energy
P.O. Box 1
1755 ZG Petten
The Netherlands

Ashok K. Mishra
Max Planck Institute for
Polymer Research
Ackermannweg 10
55128 Mainz
Germany

Lorenza Moro
SRI International
Material Research Laboratory
333 Ravenswood Avenue
Menlo Park, CA 94025
USA

Klaus Müllen
Max Planck Institute for
Polymer Research
Ackermannweg 10
55128 Mainz
Germany

Michael Niggemann
Fraunhofer Institute for
Solar Energy Systems (ISE)
Department Materials Research and Applied Optics
Heidenhofstraße 2
79110 Freiburg
Germany

Wojciech Pisula
Evonik Degussa GmbH
Process Technology & Engineering
Process Technology – New Processes
Rodenbacher Chaussee 4
63457 Hanau-Wolfgang
Germany

Max Planck Institute for
Polymer Research
Ackermannweg 10
Mainz 55128
Germany

Nicolas Schiller
Fraunhofer Institut für
Elektronenstrahl- und Plasmatechnik (FEP)
Winterbergstrasse 28
01277 Dresden
Germany

Robert Jan Visser
Vitex Systems, Inc.
2184 Bering Drive
San Jose, CA 95131
USA

David Waller
Konarka Technologies Inc.
Boott Mill South
3rd Floor, 116 John Street, Suite 12
Lowell, MA 01852
USA

Bernhard Weßling
Ormecon GmbH
Ferdinand-Harten-Strasse 7
22949 Ammersbek
Germany

Martijn M. Wienk
Eindhoven University of Technology
Molecular Materials and Nanosystems
P.O. Box 513
5600 MB Eindhoven
The Netherlands

Xiaoniu Yang
Chinese Academy of Sciences
Changchun Institute of
Applied Chemistry
State Key Laboratory of
Polymer Physics and Chemistry
Renmin Street No. 5625
Changchun 130022
P.R. China

Bjoern Zeysing
Ormecon GmbH
Ferdinand-Harten-Strasse 7
22949 Ammersbek
Germany

Zhengguo Zhu
Konarka Technologies Inc.
Boott Mill South
3rd Floor, 116 John Street, Suite 12
Lowell, MA 01852
USA

I
Materials

A
Donors

1
Regioregular Polythiophene Solar Cells: Material Properties and Performance

Paul C. Ewbank, Darin Laird, and Richard D. McCullough

1.1
Introduction

Structural optimization of polymeric semiconductor devices is a fundamental problem for organic electronic materials. Transport properties necessitate controlling molecular orientation to maximize electronic connectivity and properties in a device. For example, electrodes in solar cells are orthogonal to those in field effect transistors, requiring efficient transport in different dimensions of a thin film, and may optimize differently. In addition, semiconductive polymers are blended with acceptors molecules that add yet another parameter in the structural optimization in organic photovoltaics or solar cells. In this context, organic photovoltaics (OPV) are studied to understand materials properties, optimize device parameters, and minimize cost. We focus on head-to-tail (HT) regioregular poly(3-alkylthiophene)s (rr-PAT or HT-PAT). Earlier reviews have focused on polymer synthesis and structure [2–4] or conjugated polymer based devices [5–9]. Herein, we seek to merge current understanding of HT-PAT structure and morphology with the material specific properties exploited in OPV devices. It is important to understand the details of rr-PT structure, so that optimization of donor/acceptor blends can be optimized. Tracing the path from single molecules through aggregation in solution to deposition of solid phases, we try to understand the morphological effects of purification, characterization, deposition, annealing, and other process modifications. Performance of HT-PT in OPVs is surveyed in the context of this assembly model.

1.1.1
Overview of Nomenclature and Synthesis

Facile synthesis, ease of modification, and performance make polythiophenes the most versatile and extensively studied conjugated polymer in electronic devices. Irregular placement of side chain functional groups bestows solubility and tractability but at a cost: steric interactions arise, limiting backbone conformation and modes

1 Regioregular Polythiophene Solar Cells: Material Properties and Performance

Possible couplings

HT (Head-to-Tail)   HH (Head-to-Head)   TT (Tail-to-Tail)

Chemically distinct regioisomers

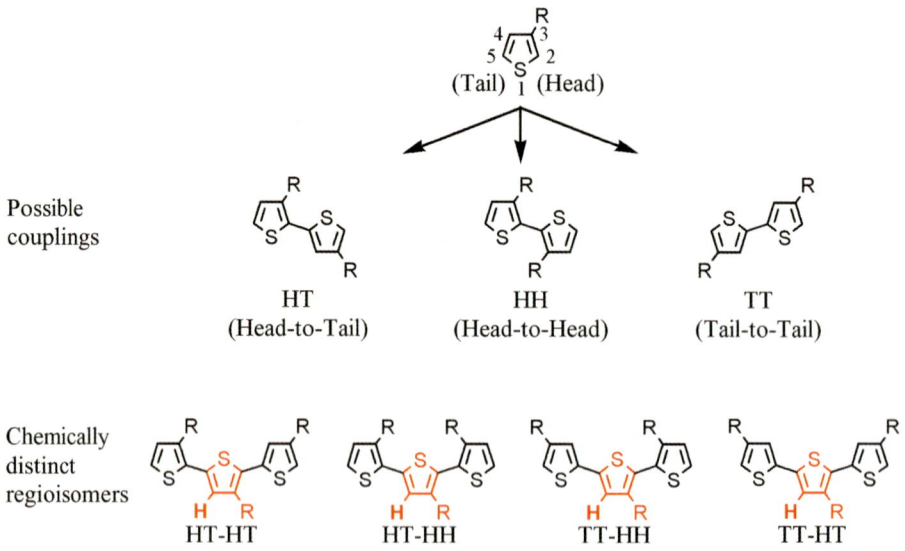

HT-HT   HT-HH   TT-HH   TT-HT

Figure 1.1 Coupling regiochemistry.

of molecular association. At the extreme, twisting of the backbone of the conjugated polymer limits effective conjugation length, widening the HOMO (highest occupied molecular orbital)–LUMO (lowest unoccupied molecular orbital) gap while forming nonplanar conformers that cannot associate with efficient intermolecular orbital overlap. Studies on 3-functional thiophenes found that controlling regiochemistry during synthesis minimizes deconjugation. Nomenclature of polythiophenes has been developed to distinguish coupling isomers. By convention, the sterically crowded 2-position of the asymmetric monomer is designated the "head" (H) and the less hindered 5-position the "tail" (T) (Figure 1.1). Joining two rings can yield three isomeric dimers, HT, HH, or TT, with different steric environments. Regiochemistry is commonly quantified as the proportion of couplings or coupling sequences enchained. For example, a 51-mer with 49 HT couplings and 1 TT defect would be 98% HT. Note that the proportion of a single defect changes with the degree of polymerization and may be undetectable in large chains. Other common prefixes seen in the literature include "rr" or "rR" for any regioregular polythiophene with HT content >90%, and "ir" (irregular), "rIR" (regioirregular), or "ran" (random) for <90%. A statistically randomly coupled PT has been reported for only one protocol [10]. To avoid confusion, we limit our use of "ran" to this product. The smallest chemically distinct sequences within a chain, a thiophene ring and its two neighbors, are denoted by the coupling regiochemistry in the triad (Figure 1.1). The desired HT–HT regioisomer dominates in all PAT except ran-PT.

For alkyl derivatives the HH isomer is severely congested, requiring significant energy (>5 kcal) to force planarity [2–4]. In the HH isomer case, properties dependent on efficient π-stacking, including conduction and charge mobility, are disrupted. Random incorporation leads to variable sequence length distributions [11], limiting the size of polymer segments capable of π-stacking and their oxidation potential. In contrast, HT and TT isomers can easily access planarity or can adopt slight twists in the solid state, since up to 20° twist at the HT or TT junction is found in a shallow potential well (>1 kcal) in energy surface calculations. Defect-free HT architectures can stack at all points along the polymer chain. Domains develop from long, low oxidation potential segments with reorganization energy, the "cost" of structural changes stabilizing oxidation or reduction of the π-system, rendered negligible [12].

The synthesis of PTs falls into two distinct classes: oxidative coupling (chemically or electrochemically) and metal-assisted cross-coupling reactions. Oxidative couplings lead to HT enriched sequences, with $FeCl_3$ polymerization typically yielding ∼70% HT–HT, not a statistically random distribution of the four triads. Control is possible exploiting electronic effects that localize charge, directing reactivity of the monomer, but is severely limited to select functionality. Metal-assisted cross-couplings are more versatile with selectivity derived from the metal-monomer adduct. Functional group tolerance may be altered by choice of coupling protocol. Coupling selectivity depends on the size of the metal and bulk of its ligands [10]. It can be tuned from a statistically random to a quasi "living" process, giving HT-regiospecificity for all but one coupling (HT_n–TT) along with narrow polydispersity and targeted molecular weight [13–15]. Several highly selective and efficient metal-assisted cross-coupling polymerizations produce HT-regioregular 2,5-coupled architectures from asymmetric monomers and are reviewed elsewhere in detail [2–4,16].

1.1.2
Advantages of the HT Architecture

The benefits of the predominantly HT architecture are immediately apparent in PATs. Easily accessible planar conformations add new dimension to structural organization, with aggregation and stacking extending electronic connections across molecular ensembles. Self-assembly is inherent, preventing disorganized agglomeration under common processing conditions. Plasticization of alkyl chains by suitable solvent allows reorganization and segregation of the aromatic backbone from aliphatic substituents, forming π-stacks of linearly extended molecules – even in solution. The drive to self-assemble is a specific function of the HT architecture. However, in films cast quickly from solvent, trapping chains in a nonequilibrium state can be found, leading to large internal tensile strain [17]. Irregular PT neither develops strain nor spontaneously forms ordered structures. Slow casting deposits strain-free films. Processing conditions can bias π-stack orientation parallel to a surface, with linear [18] or radial [19] symmetry, or perpendicular to a surface [20]. Nanorods or fibers can be formed with domains oriented parallel [21] or perpendicular [22] to the long axis. Crystalline needles several hundred micrometers long have been reported [23,24]. Such well-defined organization of organic semiconductors is

1.2
Assembly and Morphology

Aggregation and assembly of solid-state structures can be rationalized from a simple associative model (Figure 1.2). Molecular conformation, determined by solvation, prefigures association [30]. Growing aggregates may either be plastic, reorganizing for optimal molecular contact or kinetically trapped in a durable but poorly organized state. If the melting temperature of the side chains is below observation temperature, a 2D ordered liquid crystalline phase is observed [25]. Slow drying of films allows reorganization, while solvent remains as a plasticizer, and can yield a 3D ordered crystalline solid [21]. Films commonly dry too quickly for this process to finish, trapping an intermediate, partially crystalline structure within amorphous material

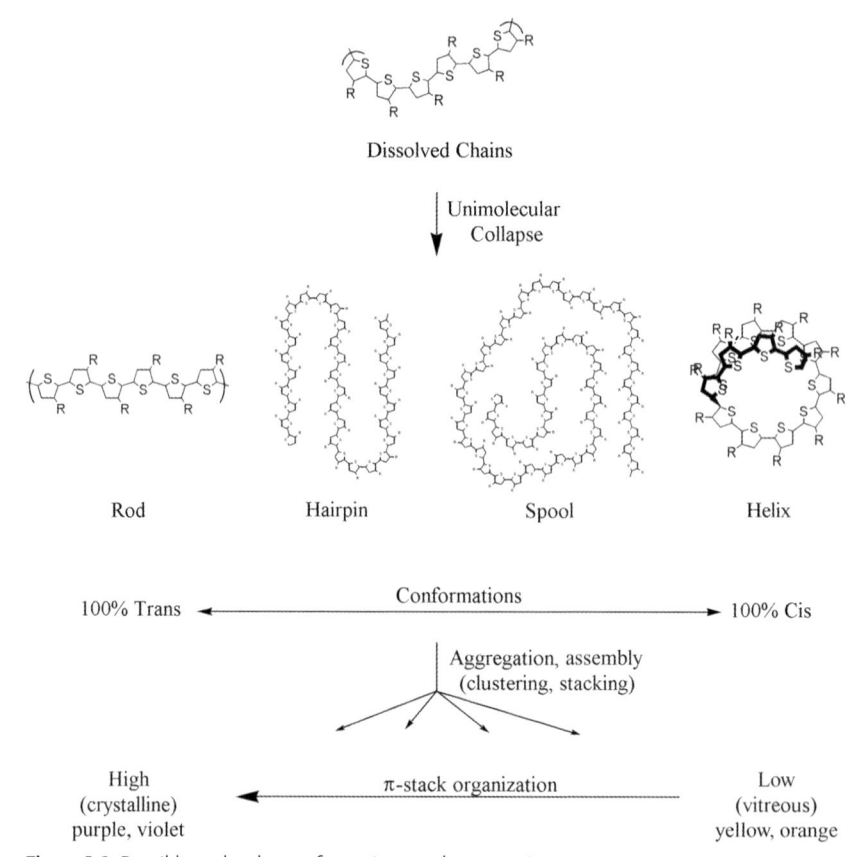

Figure 1.2 Possible molecular conformations and aggregation.

[26]. It is important to realize that in blends of rr-PT with other molecules in PV devices, that is, C_{60} blends, the structural preferences for rr-PT do not go away and in most cases the rr-PT can dominate the structure and morphology, with the major change being only the domain size.

1.2.1
Conformation

The molecular conformation of ir-PT and HT-PT in a good solvent is intermediate between ideal extremes of a random coil and rigid rod. The molecular conformation of rr-PT is also a mixture, with the rigid rodlike conformers dominating. The extent of enchained defects determines conformer bias in ir-PT. Sterically hindered HH couplings prevent coplanar orientation of rings, whereas HT and TT couplings allow both *cis*- and *trans*-coplanar orientation. A highly twisted ir-PT matches the volume and shape of polystyrene GPC standards more closely, giving comparable M_w by light scattering and GPC [27], and exists as isolated chains in dilute THF [28]. A controversial point is whether eliminating defects gives rodlike conformers. Kiriy *et al.* note several structures allowing quasi-planar orientation of conjoined rings [29,30]. These can be imagined as a continuum from a linearly extended all-*trans*-planar (rod), through intermediates with increasing *cis* orientation, for example, folded ("hairpin") or coiled ("spool"), to an all-*cis* (helix) conformer. Recent STM studies on single molecules [31] and thin films of HT-PAT [32] verified both *trans*-planar (linear) and *cis*-planar (coiled) conformers are common in the solid state. Deviation of shape from a random coil causes GPC analysis, referenced to more spherical polystyrene standards, to consistently overestimate molecular weight by a factor of 1.8–2.0 [33].

1.2.2
Aggregation

HT-regioregular polythiophenes aggregate even in good solvents. For example, HT-PDDT does not exist as dissolved, single chains even in dilute $CHCl_3$ solution near the boiling point. Aggregate shapes range from a disklike nematic liquid crystalline phase to needlelike structures with little two-dimensional order [25]. Stable colloidal suspensions in a solvent good for the alkyl chain but poor for the backbone (hexanes) have also been studied. This example of incomplete solvation caused unimolecular collapse followed by 1D aggregation into nanorods. Films made from solutions of rr-PTs in hexanes, spin cast on quartz, had an electronic spectrum comparable to the solution, indicating that the structure was conserved on deposition. Notably, the solution spectrum differed from the spectra of thin films spin cast from $CHCl_3$, confirming that different structures form under each protocol [30]. Rapid spin coating permits study of nonequilibrium structures by AFM at various stages of assembly [29,30]. Slow evaporation of extremely dilute solutions in a good solvent ($CHCl_3$) gave domains consistent with 1D assembly of linearly extended (all-*trans*) conformers. High concentration gave a lamellar network. In both, red-shifted optical absorption and vibronic structure indicate intermolecular interactions. The fluidity and amorphous nature of the alkyl side chains allows for reorientation from random coils in solution to linear rods in the solid state, and this mechanism adequately explains anisotropic assembly of crystalline domains. For example, very slow deposition from CHCl$_3$ allows the formation of well-defined "crystalline" nanorods with large (1 µm × 1 µm × 500 µm) dimensions [21].

In contrast, a nonsolvent (MeOH) drives unimolecular folding into poorly ordered flat structures at low concentrations [29,30]. Chain collapse was verified in a related PT derivative [34]. Concentration-independent vibronic structure indicating a molecular property is observed; higher concentrations give spherical agglomeration but no change in optical structure. Solvophobic interactions are minimized in a poor solvent by unimolecular collapse into planar, disklike objects that may further aggregate by cofacial stacking [29,30]. A good solvent for the alkyl chains seems to plasticize the molecules, allowing reorganization of the aggregate into domains with rodlike morphology. Colloidal solutions prepared by dissolving HT-PHT in CHCl$_3$ develop vibronic structure as solvent quality decreases [26]. Filtration removes this structure from the spectrum, leaving spectral features identical to polymer in good solvent and proving that aggregates are responsible.

1.2.3
Solid Deposition

The deposition rate often determines to what extent molecules can organize. Spin casting deposits films with thicknesses correlating with initial solution concentration and spin speed [35], but varying rates of drying influence molecular assembly (Figure 1.3). Drop cast films can show little internal strain, whereas PHT spin cast

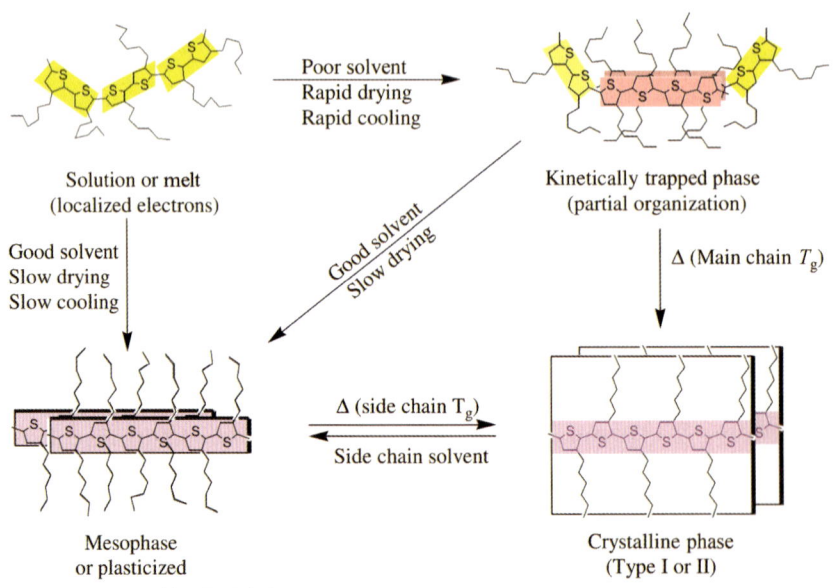

Figure 1.3 Assembly of solid phases.

from xylenes or trichlorobenzene show strain increasing with spin speed and drying rate [17]. For the solvents studied, near maximal strain was reached at a spin rate of 3000 rpm. Importantly, no strain was seen in irregular samples. Energy can be released in a dramatic fashion: cutting the film with an AFM probe allows the edges to pull away, leaving trenches as wide as 2.3 µm. Slow drying promotes molecular assembly and relieves strain. Very slow deposition from a good solvent allows formation of well-defined "crystalline" nanorods [21] or, for narrow polydispersities, nanofibrils with domain width matching the length of the linearly extended molecule [12,24]. "Crystals" of PBT examined by SEM develop complex lamellar structure with layer thickness ranging from 15.6 to 104 nm, comparable to the 7 to 130 nm length distribution estimated from GPC measurements. Fractionation during a gradient crystallization process was proposed with each layer comprised of similar M_w chains [24]. Clearly, the solution behavior at high concentration, for example, as solvent evaporates, is critical for obtaining well-defined solid-state domains.

1.2.4
Solid-State Crystalline Order

When crystallinity in PAT is observed, chain association forms discrete lamellae of π-stacks (Figure 1.4). Microcrystalline domains are dispersed in a matrix of amorphous material. For 70% HT-PAT, with short HT–HT sequences [11], broad stacking distribution centers around 3.8 Å [36]. X-ray contour maps show weak diffraction in the small-angle region (1 0 0 peak). In the large-angle region, backbone scattering from segments with differing stacking distances generates a broad, amorphous halo (0 1 0 peak). In contrast, 98% HT-PT show a very narrow, well-defined ring at wide angles, indicating a well-organized narrow distribution around 3.81 Å [37]. The stacked structure is thought to be analogous in both materials. However, the entire HT-regioregular molecule can stack, eliminating poorly conductive, amorphous domains. Electrochemical doping of HT-PT can further compress stacks to 3.6 Å with compensating expansion along the 1 0 0 dimension to accommodate counter ions [38]. Packing disruption by HH linkage is seen in a 0% HT-PAT example (TT-HH). Large 4.4 Å d-spacing is too far to allow significant intermolecular π-orbital overlap, and lamellae of stacked molecules are not distinctly observed. Backbone association is very weak, suggesting that alkyl association drives order.

Two HT-PAT polymorphs have been characterized [39]. The most commonly observed and thermodynamically stable structure (Type I) develops with superposed π-systems 3.8 Å apart (Figure 1.4). A second polymorph (Type II), seen as a contaminant during solution casting or by crystallization near the side chain melting temperature [40], packs with 4.3 Å separation along the π-stacking axis and fully interdigitated alkyl chains. One morph seems to arise from the organizational preference of the backbone, and the other from the preference of the alkyl chain. Winokur's discussion of nanoscale structure–property relationships for conjugated polymers is a recommended overview [41].

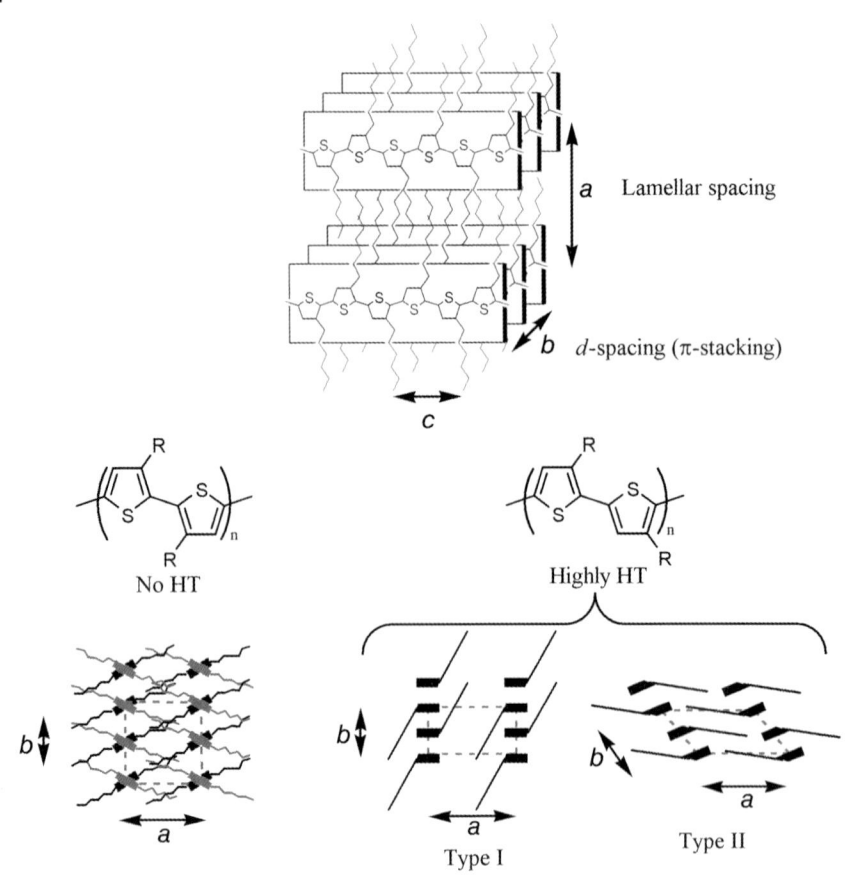

Figure 1.4 Crystal morphology in polythiophenes [39,232].

1.2.5
Solid-State Phase Behavior and Thermal Analysis

Thermal behavior in rr-PTs is important, since thermal annealing of OPV films is common in the optimization. Processing conditions can kinetically trap material in one or more phases and varied domain sizes, giving variational thermal properties to the rr-PT films. Polythiophenes have added complexity from segregated aliphatic and aromatic components, and distinct transitions are possible for each. Representative data for several HT-PATs are summarized as a qualitative guide (Table 1.1). Glass transitions (T_g) are not commonly reported but seem to indicate both low- and high-temperature processes. Dividing data this way allows roughly linear correlation of changes in transition temperature versus alkyl chain length (Graph 1.1). For the sake of discussion, we refer to the low-temperature transitions as "side chain" melting and the high temperature as "backbone" melting for the reasons explained below. The most thoroughly characterized T_g is for HT-PBT and correlates with the

1.2 Assembly and Morphology | 11

Table 1.1 Thermal characterization of poly(3-alkyl)thiophenes.

				Side chain		Main chain		
Substituent	%HT	M_n	PDI	T_{g1} (°C)	T_{m1} (°C)	T_{g2} (°C)	T_{m2} (°C)	Reference
CH_3	ir-PT	—	—	—	—	145	—	[164]
$(CH_2)_3CH_3$	ir-PT	—	—	75.4	—	—	—	[45]
	>97	—	—	—	—	—	243	[40]
	HT^a	—	—	67	—	—	272	[42]
$(CH_2)_5CH_3$	ir-PT	—	—	12	—	—	178	[165]
	92	$M_w = 8.7k$	—	20.3	—	—	222	[166]
		—	—	—	—	110	230	[167]
	>98.5	$M_w = 50k$	—	—	50	125	215	[114]
$(CD_2)_5CD_3$	HT^a	2582	1.40	—	—	—	46.3^b	[168]
		7196	1.28	—	—	—	162.7^b	[168]
		12 728	1.35	—	—	—	209.2^b	[168]
		27 051	1.32	—	—	—	231.9^b	[168]
$(CH_2)_7CH_3$	ir-PT	—	—	11.2	—	—	—	[169]
	89	$M_w = 14.2k$	—	−9.2	—	—	222	[166]
	>98	11 900	1.34	—	—	—	175	[170]
$(CH_2)_9CH_3$	>99	3400	1.20	—	54, 68^c	—	—	[171]
$(CH_2)_9CH_3$	>99	14 100	1.64	—	44, 76	—	166	[171]
$(CH_2)_{11}CH_3$	ir-PT	—	—	5.6	56.1	—	116.3	[45]
	>98.5	$M_w = 162k$	—	—	64	—	155	[40]
	>98.5	—	—	−19	—	—	—	[172]
$(CH_2)_{15}CH_3$	98	14 000	1.61	—	93	—	145	[170]
$(CH_2)_{21}CH_3$	>95	7900	3.16	—	53.3, 70.6	—	103.8	[173]

^{a} Not reported.
^{b} Second heating.
^{c} First heating.

Alkyl length versus transition temperature

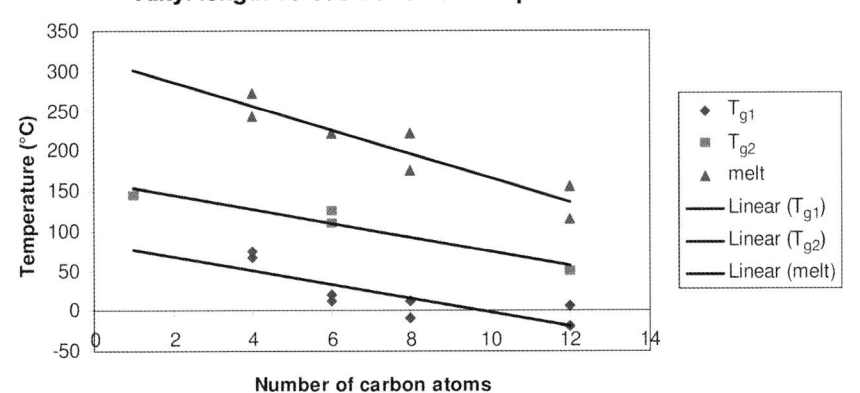

Graph 1.1 Variation in transition temperature with alkyl chain length.

onset of backbone ring twisting [42]. Although occurring at high temperature, this "twist glass" transition seems best correlated with distal alkyl chain melting reported for other derivatives. In this series the slightly longer hexyl derivative has T_g near room temperature (25 °C) and longer chains occurred below. A second, higher temperature glass transition has been reported in only a few cases. Though the data set is small, it seems correlated with T_g reported for the methyl derivative. The latter cannot have two substituent-dependent processes and may be considered the structural archetype for thermal processes occurring only in the backbone. For PHT different studies reported either low or high T_g. No accounts describe both in the same sample; thus, it is unclear whether they arise from features of a common structure or from commingled phases. However, distinct glass transitions for (aliphatic) chain ends and the (aromatic) crystalline core of the mesophase are conceivable.

Melting (T_m) processes distinct for main chain and backbone have been reported. Side chain melting was not commonly observed for alkyl shorter than decyl; thus, they are excluded from the graph. When observed, it was at lower temperature than the backbone process and decreased with increasing alkyl length. The melting process may be imagined as melting of chain ends, with the π-stacked core remaining intact, giving a liquid crystalline mesophase [41]. Further heating expands the stacking distance in the core but does not eliminate crystallinity until melting [43]. A consequence of the two-phase model is that structural changes deriving from both regimes may be noted. In one example, graphs of OPV properties versus temperature show inflections near both transition temperatures [44].

Studies on ir-PT find that $FeCl_3$ doping raises T_g, possibly increasing organization in the quinoid form, but eliminates distinct melting [45]. Phase separation in blends was studied as a function of alkyl length and %HT [46–48]. Mixtures of HT-PAT with alkyl length differing by C_4H_8 clearly segregate. Smaller (C_2H_4) differences give seemingly stable blends until the proportion of the low-melting component becomes large, then phases separate. PHT ranging from 75 to 92% HT ($\Delta HT = 17$ mol%) cocrystallize, with melting temperature gradually increasing with HT content. Plasticizing additives are infrequently studied. Parity blends of HT-PHT or HT-POT with propylene carbonate melt at 80–120 °C lower temperature [49]. Polythiophene-based plasticizers designed with high HH content have been studied morphologically but not thermally characterized [50].

1.2.6
Anisotropy

Mechanical and physical properties of polymers are determined largely by orientation. Efficient molecular alignment can maximize properties in one dimension by depleting it in others, to the extreme of being undetectable. For example, vibronic coupling, the intermolecular exciton, and intensity of the absorption spectrum are intense when measured parallel to the direction of polarization, along the long axis of the molecule. Measured orthogonal, the absorption is weak and unstructured, with the absorption maximum shifted to higher energy [20,51–55]. Extreme polarization affects all optoelectronic properties including conductivity [56], electron spin resonance (ESR) [57], vacuum UV [58], UV–Vis [20,26,51,59], spectroscopic ellipsometry [52–54], emission [29,54], IR, and Raman spectra [60]. Devices have been described with polarized electroluminescence [51], photoconductivity [55], photocarrier spin [61], and charge carrier mobility [62,63]. As domain organization improves, reducing isotropy, orientation control over large areas is needed to maximize properties including photon harvesting and charge transport. This permits optimization for specific device geometries. For example, conduction between laterally spaced source and drain electrodes in an FET requires π-stacking parallel to the (bottom) gate electrode, whereas solar cells need orthogonal alignment to connect vertically separated (top and bottom) electrodes. With respect to a surface, edge-on versus face-on orientation of π-orbitals in stacks tunes electronic contact at the interface from negligible to high, influencing charge transfer across the junction [64]. Ideally, orientation will be selected during processing and device assembly.

Bulk HT-polythiophenes can be modeled as crystalline domains imbedded in amorphous matrix. The nanocrystallites preferentially orient with one molecular surface in contact with the substrate, being isotropic in two rather than three dimensions in thin films. Slow deposition by drop casting orients molecules edge-on, with the π-stacking (b-axis) parallel to the substrate plane, in all samples studied including ir-PT. Depositing material faster, for example, by spin casting, skews molecular orientation from edge-on at low speeds to face-on at high speed [65]. Extremely fast spin casting precludes assembly, preventing significant reorganization of solvent-plasticized chains before solvent is lost, depositing poorly organized agglomerates. At temperatures below the side chain melt, a larger proportion of amorphous material is kinetically trapped. Observed polarization with deposition rate depends upon molecular substitution and regioregular structure, both critical parameters controlling assembly, and correlates with developing crystallinity [18]. Samples with >91% HT content and low molecular weight maintain the substrate parallel stacking motif. In contrast, samples with low regioregularity (81% HT) and high molecular weight orient the b-axis perpendicular to the surface. The effect clearly arises from HT content and M_w rather than film thickness [1,66]. Epitaxial order persists only for a few monolayers, however. It is unclear whether solution aggregation alone or film organization at the meniscus of evaporating solvent is responsible [67]. Electrochemically deposited HT-PT films have the preferred orientation, with b-axis parallel to substrate, despite modifications of the unit cell including ∼0.2 Å compression of the π-stack (b-axis) and ∼1 Å expansion of the lamellar spacing (a-axis) [38,68,69].

Several methods have been studied to increase order in polythiophenes. Anisotropy can be increased locally, on the scale of one domain, through self-assembly on a surface to form nanowires. The crystallographic b-axis orients in the longitudinal dimension of the rod [21]. Solution crystallization forming filterable nanowires has also been proposed [70]. Alternatively, directional crystallization from a crystalline organic template orients π-stacks orthogonal to the long rod axis, with amorphous regions separating domains [22]. It is unclear whether the same assembly model applies in this case.

Macroscopic single domains have not been realized for rr-HTPT. Orientation of cast films over a large area by mechanical techniques can create significant polarity. In a simple example, physically robust, electrochemically prepared films may be physically stretched [59]. Highly regioregular polythiophenes are generally too brittle for this treatment and require painting onto a substrate that is then stretched [26]. Rubbing is another scalable, easily implemented technique. Microscopic areas (10 µm × 10 µm) can be polarized by rubbing with an AFM tip [71]. On similar scale, stress created by drying and shrinkage of a superposed polymer creates a radial distribution of linearly polarized domains [19]. Larger areas of rr-PT rubbed with velvet become anisotropic, but ir-PT films do not [51,72]. A dichroic ratio of about 6 is observed in the photoluminescence. Shear and surface compression combine to orient 85% of crystalline domains along the rubbing direction, with the stacking axis orthogonal to the substrate [71]. Processing films in the plastic state, or while plasticized by evaporating solvent, provides additional control. Rubbing rr-PHT films with the edge of a quartz slide during casting from a high-boiling solvent (chlorobenzene, CB) aligns the π-stacking b-axis parallel to the substrate, and the crystallographic c-axis along the rubbing vector [55]. A method dubbed "friction transfer" draws a pellet of HT-PT across a substrate heated above the side chain melting temperature, depositing films >100 nm thick. Surface normal orientation of stacks with record dichroic ratio of 10–100 is accessible [20]. Application of ordering techniques in OPV blends has not been investigated to our knowledge.

1.3
Characterization of Impurities

1.3.1
Fractionation and Effects of M_w

Soxhlet extraction by increasingly "good" solvents provides a convenient means to separate impurities and narrowed M_w fractions from broad polydisperse samples as has been shown by Pron. Solubility of HTPT increases in the following order: $CH_3OH <$ acetone $<$ hexane $< CH_2Cl_2 <$ THF $<$ xylene $<$ chloroform [73]. Successive fractions show increased M_n, M_w, and regioregularity, visible absorption λ_{max} red shifting accordingly [73]. An ir-PHT prepared via $FeCl_3$ oxidation afforded fractions ranging from $M_n = 5k$ (75% HT; PDI $= 1.6$; $\lambda_{max} = 437$ nm) to $M_n = 34k$ (85% HT; PDI $= 2.6$; $\lambda_{max} = 443$ nm). Similarly, a commercial HT-PHT from Aldrich (Rieke) afforded fractions ranging from $M_n = 3k$ (81% HT; PDI $= 1.3$; $\lambda_{max} = 433$ nm) to $M_n = 33k$ (98% HT; PDI $= 1.8$; $\lambda_{max} = 452$ nm). Changing selectivity of the coupling reaction has been proposed to explain the trend toward higher HT content [26]. However, studies on metal-assisted cross couplings indicate no change in mechanism or selectivity is necessary to explain regioregular synthesis [13–15]. Chains initiated by a single TT defect subsequently grow by HT coupling. Defects are constant, at one per chain, becoming a steadily decreasing proportion of enchained couplings as M_w increases.

A systematic study of bulk properties versus M_w summarizes the changes [74]. Thin-film CV shows an oxidation at 0.74 V in all samples, decreasing with increasing M_n. Additional oxidations at 0.58–0.60 and 0.97–1.00 V, absent from ir-PHT and the acetone fraction (DP = 14), appear in the hexanes fraction (DP = 26) and intensify with M_n. Increased conjugation was proposed in agreement with red shifting of the solution UV–Vis λ_{max} with increasing M_n. Solid-state spectra trend similarly but develop vibronic structure. The ~2.09 eV interchain exciton proposed by Brown and coworkers [75] appears in thin films of the hexanes fraction, intensifying with increasing DP. This parallels improving nanofibril width and order reported for increasing M_n [12].

1.3.2
Inorganic Impurities

Commercial availability of clean, well-characterized HT-PAT obviates the need for synthesis. However, all commercial rr-PTs are not the same and care must be taken to know the PDIs, the level of impurities, and the degree of regioregularity. Samples from Plextronics or Plexcore from Aldrich are examples of a very high grade of rr-PT. Molecular characterization has been reviewed [2–4]. Subsequent analysis by MALDI-MS, yielding correction factors (1.8–2.0) for molecular weight determinations by GPC, should be noted [33]. The characterization of inorganic impurities is less well addressed, leaving their effect uncertain. Some report devices made with commercial HT-PT, as received, can be as good or better than purified samples, though the device lifetime is not mentioned [76]. Others find that impurities can limit charge carrier mobility [77,78] or give additional peaks in the DSC thermogram [76]. Adsorption into nanoporous TiO_2 is electrostatically hindered, requiring fractionation for maximal HT-PT infiltration [79]. The crude product of polymerization contains phosphine ligands, from the catalyst, and inorganic salts. By-product zinc bromide (from some synthetic methods) salts modify the absorption spectrum and partially quench fluorescence. Some commercially obtained samples with this problem contained high residual Zn (6.1×10^{-6}%) and Br (6.4×10^{-6}%) levels [80]. Soxhlet extraction with methanol (24 h) then hexanes (24 h) reduced Zn to negligible concentration, and only bromine associated with chain ends remained. Others note similar problems with ash as in the elemental analysis could not be simply resolved by precipitating from a good solvent ($CHCl_3$) into a poor solvent (MeOH) unless dilute solutions and long times were used [81]. Chelation seems an effective general method for removing Pd, Cu, Ni, and metallic residues [82,83].

Molecular oxygen commonly contaminates by forming charge transfer complexes with polythiophene [84]. Exposing a 100 nm HT-PHT film to air for a few seconds raises conductivity four orders of magnitude to its equilibrium value [85]. Oxygen-doping effects include a broad, low energy absorption in the visible spectrum ($\lambda_{max} \sim 1.97$ eV; 629 nm) and radicals detected by EPR [84]. In diodes exposed to oxygen in the dark mild doping causes a shift of the flat band voltage [86]. Light exposure increases doping efficiency with greatest effect near the absorption maximum of the charge transfer complex. However, charge carrier density is unstable, continually increasing during the ~1 h experiment. Photobleaching is possible in organic solvents (not in OPVs due to fast electron transfer to C_{60}) under certain conditions, but chain cleavage requires metal contaminants [87]. Several methods have been proposed to remove oxygen from films. Annealing above 100 °C in inert atmosphere reduced oxygen content below limits detectable by Rutherford backscattering spectrometry (RBS) [88,89]. Heating HT-PHT in vacuum (147 °C) reduced conductivity by four orders of magnitude but several hours were necessary to render conductivity negligible [85]. Chemical dedoping with tetrakisdimethylaminoethylene neutralized charge carriers, reducing conductivity from $\sigma = 3 \times 10^{-5}$ to $\sigma = 5 \times 10^{-10}\,\mathrm{S\,cm^{-1}}$ [90]. Deoxygenation of prepared FET structures by applying a negative gate bias was also reported [91].

1.4
Optical and Electronic Properties of PAT

1.4.1
Optical Properties: Intermolecular Excitons

Optical absorption of HT-PT depends upon substituent inductive effects, chain twisting, and chain association. This was reviewed [2–4] before the discovery of the intermolecular exciton band [75]. Samples with varying %HT show intensified absorption at 2.09 eV as defects decrease. This feature was present in all ir-PT and HT-PT, at constant energy, and was not completely bleached by heating or changing solvent quality. A neutral (Frenkel type) interchain exciton was proposed. Intensity of this absorption is correlated with the microstructural order in the polymer and diminished as HH defect content increased. The improved interchain interaction in HT-regioregular P3AT demonstrates transformation from classical unimolecular, one-dimensional properties to quasi-two-dimensional behavior.

The conjugation length independent exciton absorption implies that HOMO–LUMO gap estimation from UV–Vis absorption band edge can be misleading, identifying a low energy process accessible to only a fraction of the bulk material. This interchain process is the lowest energy absorption present in ir-PT but is much more intense in self-assembled HT-PT; the bandwidth is larger, implying higher mobility. This parameter seems useful for comparing disparate samples with different (or unknown) thermal histories. In a recent study of properties versus chain length, fractionated HT-PHT was spin coated from $CHCl_3$ onto glass [92]. Retroanalysis of thin-film absorption spectra indicates varying intensities of the 2.07 eV (593 nm) interchain exciton, seeming to be correlated with the degree of crystallinity, in all but the lowest M_w fraction. However, X-ray studies of the films showed a well-resolved 1 0 0 lamellar spacing only in the high M_w fraction. Annealing at high temperature developed the 1 0 0 peak in all fractions [92]. The initial difference may reflect solid-state structure derived from preorganized aggregates, detected by the 2.09 eV absorption. Either the processing or the structure of the high M_w fraction promotes formation of nanocrystalline domains in solution that may template further organization on drying.

1.4.2
HT-PT Electron Transport: Conductivity and Mobility

Conductivity in organic semiconductors depends on generating efficient pathways through the bulk. Irregular PATs have the disadvantage in this regard, with conductivities typically on the order of $\sigma = 0.1$–$1.0\,\text{S cm}^{-1}$ when doped with iodine. In contrast, HT-PAT conductivities typically are $\sigma = 100$–$200\,\text{S cm}^{-1}$ with some samples as high as $1000\,\text{S cm}^{-1}$ [2–4]. Balanced TOF carrier mobility for carefully cleaned HT-PHT, spin cast from $CHCl_3$ onto ITO, shows hole mobility $\mu_h = 3 \times 10^{-4}\,\text{cm}^2\,\text{V}^{-1}\,\text{s}^{-1}$ and electron mobility $\mu_e = 1.5 \times 10^{-4}\,\text{cm}^2\,\text{V}^{-1}\,\text{s}^{-1}$ [77]. Controlling donor and acceptor mobility is important for optimization of the OPV device, since the charge separation and generated current will depend on a balance.

1.5
Benefits of HT-Regioregular Polythiophenes in Solar Cells

Contemporary photocurrent generation and device design are addressed in detail elsewhere in this book. The following discussion of photocurrent generation considers four processes: (i) light absorption, generating an exciton; (ii) exciton diffusion (to an interface); (iii) charge separation into independent holes and electrons; and (iv) charge conduction to an electrode (Figure 1.5). Competing processes, depicted in blue, include exciton recombination and electron back transfer from the acceptor to the donor. In this model, there are several advantages of HT-regioregular polythiophenes over other p-type organic semiconductors. Polythiophenes are strong

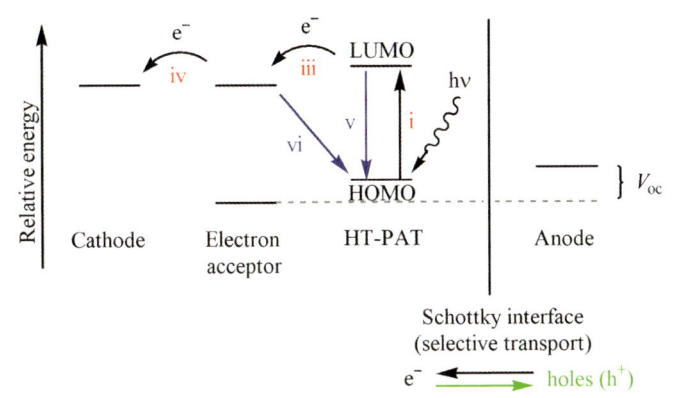

Figure 1.5 Simplified photocurrent generation mechanism: (i) absorption, (ii) exciton diffusion (not shown), (iii) charge separation, and (iv) conduction; competing processes include (v) recombination/relaxation and (vi) electron back transfer.

absorbers with a broad absorbance profile. Domains with extended π-stacking support intermolecular excitons, responsible for much of the low energy absorption, with absorption most intense in self-assembled HT-PT [75]. The spectral window is effectively widened. Blending with n-type semiconductors expands this further by introducing additional molecular and charge-transfer absorptions. For example, blending 1 : 1 with phenyl-C_{61}-butyric acid methyl ester (PCBM) increases absorption into the UV and infrared [93]. Spontaneous polymer assembly, occurring even under mild heating, generates domains with high charge carrier mobility – aiding charge separation and conduction to an electrode. Consequent anisotropy suggests polarization effects (e.g., directionally optimized transport) become important in phase-separated heterojunctions.

1.6
Bulk Heterojunctions: Focus on HT-PAT/PCBM Blends

The discovery of fast, efficient photoinduced electron transfer from electron-rich conjugated polymers to fullerenes inspired the creation of polymer-based solar cells [94]. Studies were initially hampered by low solubility and a tendency for crystallization, but solubilizing substituents largely reduced these difficulties, albeit while adding the prospect of polymorphism effecting device performance [95]. Successful implementation in photovoltaic devices requires processing conditions controlling phase and domain (crystallite) size. In the context of the above structural propensities of rr-PT, the assembly of HT-PT into nanocrystalline domains with its partner fullerene is critical for efficient OPVs. Hierarchical assembly by fullerene derivatives and resultant structural motifs invites more extensive discussion than given here and more importantly more structural work on rr-PT/fullerene blends. Recent developments of liquid fullerene derivatives, one-, two-, and three-dimensional ordered solids, are recommended starting points for further study [227–229]. This section focuses on PCBM and blends with HT-PAT as a currently popular research target to understand processing parameters optimizing OPV performance. Properties of photovoltaic devices incorporating HT-regioregular polythiophenes and reported in the past 5 years are surveyed in Appendix 1.A.

1.6.1
Homogeneous PCBM Assembly

It is useful to understand the innate behavior of PCBM, before studying its behavior in blends, as the archetype of the segregated phase in heterojunction devices. As received, PCBM powder diffracts well [96] but thin films spin cast from $CHCl_3$ are amorphous, even after annealing (12 h, 100 °C) [96,97]. Crystalline organization has been studied by Yang et al. [98]. Thin films spin cast from toluene (Tol) or chlorobenzene, formed by homogeneously distributed small molecular aggregates, were mechanically stable and separable from the substrate. Substantial variations in thickness were evident in the Tol cast film, suggesting that CB is a poorer solvent from

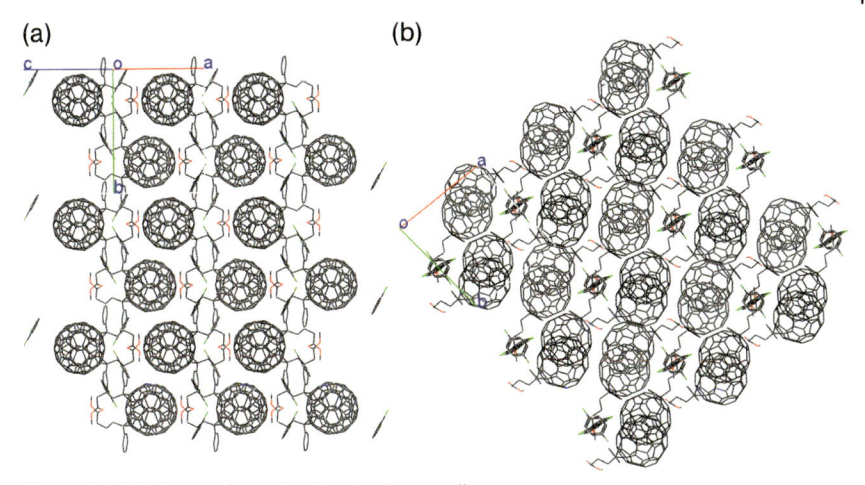

Figure 1.6 PCBM crystal packing, 3 × 3 × 3 unit cells.
(a) Dichlorobenzene solvate viewed along [1 0 −1] plane;
(b) chlorobenzene solvate viewed along [1 1 0] plane.

which crystals form. Reducing the rate of solvent evaporation by drop-casting films prevented discrete film formation. Large aggregates with darker (more dense) clusters developed instead. Slow solvent evaporation under saturated atmosphere allowed crystal growth to several micrometers in size. Small crystals, on the order of 0.5 μm long, were grown by annealing spin-cast samples (50 h/130 °C). Selected area electron diffraction (SAED) analysis of each product showed concentric rings from tiny crystallites in spin-cast films that evolved to a distinct pattern of well-aligned spots as the crystallite size increased. Three major Debye–Scherrer diffraction rings, with d-spacing of 4.6, 3.1, and 2.1 Å, resolved into several reflections with similar d-spacing. The authors found no concurrence with published crystal structures, but concluded that the same crystal phase was present in all samples, even upon annealing, with crystallite size changing as a function of processing conditions. Rispens *et al.* reported that distinct crystal structures can arise from different solvents [95]. Structurally, both sample phases separate into wavy, fullerene-rich domains and its methano-ester/solvent-rich domains (Figure 1.6). Samples crystallized from DCB are clearly lamellar, implying poor electronic connection between fullerene layers. In contrast, samples crystallized from CB periodically touch and improve connectivity. Solar cells with active layers spin cast from CB showed higher I_{sc} and efficiency than those cast from DCB or xylenes, implying that PCBM organization was responsible.

1.6.2
HT-PAT/PCBM Blends: Component Ratio

Optical absorption spectra of spin-cast 50 wt% HT-PDDT/fullerene mixtures are a simple composite of the component spectra, indicating negligible ground-state charge transfer [99]. Blending hinders self-assembly during deposition, reducing absorption attributable to the nanocrystalline phase. Absorption intensities of PCBM/HT-PHT blended films, spin cast from DCB, decrease and absorption maxima blue shift, relative to pristine polymer, because of PT disordering. For >67% PCBM blends, nonphotoinduced charge transfer complexes become important [100]. Charge carrier generation efficiency in blends changes little in comparison to pristine polymer. Rather, the primary effect in >50% PCBM is the increased charge carrier lifetime. In PHT-rich phases (80–100%) the majority of charge carriers are short lived (10–15 ps); in 50% and lower blends new, long-lived (>500 ps) carriers appear, attributable to interfacial charge separated species [101]. The longer lifetime increases, the distance carriers can diffuse before quenching, increasing the probability of separated charge reaching an electrode. Charge transport is dispersive (non-Gaussian) for extreme compositions where short lifetimes dominate. Nondispersive (Gaussian) transport is observed in the narrow composition range (44% < PHT 77%), giving the highest photocurrent lifetimes. Qualitatively, nondispersive photocurrent profiles correlate with better structural order, while dispersive transport is observed for extreme phase segregation [102].

Current density is generally highest in 1:1 blends [100,103–105]. Decreasing PCBM content from 75 to 41 wt% is reported to improve diode characteristics, though V_{oc} remains stable [106]. There are conflicting reports of fill factor (FF) evolution versus component ratio under identical thermal processing. One study finds a maximum at 47–50% PCBM with poor performance of higher ratios attributed to fullerene crystallization [106]. Another study of thickness-optimized films reports gradual increase in efficiency as PCBM content increases from 30 to 60% [105].

1.6.3
HT-PAT/PCBM Blends: Annealing

1.6.3.1 PCBM Phase Separation and Assembly

High mobilities reported for electrons in PCBM ($\mu_e = 2 \times 10^{-3}$ cm^2 V^{-1} s^{-1} [107]) and holes in assembled HT-PHT fibrils ($\mu_h = 1 \times 10^{-2}$ cm^2 V^{-1} s^{-1} [12]) indicate suitably complimentary charge carrier materials, provided phase assembly yields effective pathways. Annealing blends below the polymer T_{g2} limits diffusion and assembly of the fullerene phase. Devices with 1:2 PHT:PCBM films heated to 90 °C seem to reach a stable morphology after ~20 min, with little change in efficiency during 15 min additional anneal. Heating above T_{g2} (150 °C) causes a sharp efficiency spike followed by collapse to nearly zero [108]. For 1:1 PHT:PCBM films above T_{g2} (130 °C, 30 min) 9 μm PCBM crystals form [109]. This temperature-dependent behavior, seen in different compositions, reflects sharply increased fullerene mobility through the flexible polymer composite. Crystal growth in CHCl$_3$ cast films has been thoroughly studied, indicating that temperature, rather than time, determines the ultimate size of PCBM crystallites [110]. Brief annealing (20 s/130 °C) of 50–60% PCBM yielded tiny, isolated crystallites, but for 67–75% large grains formed [106]. Slightly longer annealing of 1:1 films (130 °C/5 min) generated >10 μm PCBM crystals and 10–30 nm dark spots, thought to be PCBM-rich clusters, homogeneously distributed in the sample [111]. For 33% PCBM mixtures crystallite dimensions up to 200 μm are reported (5 min, 175 °C). A contemporaneous study confirmed that triclinic PCBM crystals grew in films cast from CB, with length and connectivity increasing with anneal temperature and composition ratio [112]. Crystal growth leeched PCBM from the vicinity. Following this process by SAED found morphology stabilized when PCBM was expelled from the polymer matrix. Very long annealing (12 h, 100 °C) irreversibly transformed morphology to PCBM crystallites imbedded in a matrix of essentially pure, semicrystalline PHT. Long annealing can reliably give small fullerene crystallites segregated by crystalline polymer domains, which limit growth [113].

Relevant studies on MDMO-PPV (poly(2-methoxy-5-(3′,7′-dimethyl-octyloxy))-*p*-phenylene vinylene):PCBM blends indicate that PCBM crystallite size can be controlled by choice of casting solvent [230]. Drying behavior differs with solvent, as the relative solubility of blended components changes in shrinking microdroplets, with morphologies ranging from homogeneous to oblate spheroids with a PCBM-rich core surrounded by a thin polymer-rich shell [231]. The polymer remains dissolved while much of the PCBM crystallizes, giving extreme segregation and low interface area.

1.6.3.2 Polymer Phase Separation and Assembly

Polythiophene demixing and assembly occurs at mild temperatures above the side chain melt. Structural reorganization commences, improving charge carrier mobility [114]. Films (50% PHT:PCBM) develop a structured absorption spectrum and further heating only increases optical density [110]. Polymer assembly above T_{m1} but below T_{g2} (80 °C) includes intensification of the TEM electron diffraction signal for π-stacked polythiophene (1 peak, 3.8 Å) [115], shifts electrochemical oxidation by 0.2 V to lower potential, and enhances hole mobility by an order of magnitude [110]. Bright field TEM imaging of fibril formation and SAED confirmed crystallization of PHT *without* PCBM crystal growth. Long annealing (1 h, 120 °C) lengthens fibrils but does not widen them significantly [113]. Fibrils bend, appearing isotropic in three dimensions – not merely at the electrode surface. Heating above T_{g2} allows migration of entrapped PCBM, assembling the second phase. Study of the process by solid-state NMR indicates that the fullerene is initially buried in and strongly interacts with a matrix of polymer alkyl chains. Heating above T_{g2} (150 °C, 30 min) segregates phases, the fullerene then interacting most strongly with its own chain [116]. Annealing HT-PHT near T_{g2} (125 °C) gives maximum crystallinity in X-ray studies [117].

1.6.3.3 Evolution of Open-Circuit Voltage

Controlling energy states of the donor, semiconductive polymer and acceptor is critical for better understanding of p-/n-interfaces, including mechanisms of exciton dissociation and open-circuit voltage (V_{oc}) generation. It is well recognized that the critical processes of exciton dissociation leading to charge separation in bulk heterojunction solar cells occur at the junction between the absorber and the electron-transport layer. In this regard, the V_{oc} scales with respect to the offset between the HOMO level in the p-type absorber and the LUMO of the electron-transporting n-type layer. The maximum attainable voltage derives from the energy difference between the donor oxidation potential (e.g., HOMO) and acceptor reduction potential (e.g., LUMO) [118]. Thermally promoted phase separation and assembly of extended, low-energy hole and electron transport domains are expected to increase this parameter relative to an amorphous state. Experimentally, some authors note decreasing V_{oc} attributed to lower oxidation potential of the self-assembled polymer [119,120]; others find only weak dependence of V_{oc} on changes in the polymer HOMO [121]. The latter implies that morphological changes in the PCBM phase can be very important. Polymer flexibility, however, remains important only by allowing PCBM to migrate. For example, annealing PBT below T_{g1} has little effect on solar cell properties. Above the glass transition V_{oc} decreases and performance changes [120]. Evolving properties are compared for several 1:1 PHT:PCBM OPVs in Table 1.2 to identify changes. In most examples, absolute V_{oc} magnitude initially increases then declines during heating. This optimal, nonequilibrium structure is most commonly reported. Intriguingly, there appears to be a solvent effect: films cast from DCB increase V_{oc} on short annealing, whereas those cast from CB decrease. This indicates electronically distinct morphologies and is consistent

Table 1.2 Evolution of properties due to annealing 1:1 PHT:PCBM solar cells.

PHT:PCBM thickness (nm)	M_n/PDI	Casting solvent	Anneal temperature/ time	FF	V_{oc}	J_{sc} (I_{sc})	PCE (η)	Reference
100–200	28k/1.57	CB	50 °C/4 min	116[a]	93[a]	136[a]	157[a]	[103]
100–120[b]	—/—	DCB	75 °C/4 min	143	167	(300)	625	[174]
100–120[b]	—/—	DCB	75 °C[c]/4 min	150	183	(340)	875	[174]
100–200	28k/1.57	CB	80 °C/4 min	117[a]	71[a]	156[a]	185[a]	[103]
70–80	—/—	CB	100 °C/? min	157	92.0	334	480	[124]
100–200	—/—	CHCl$_3$	100 °C/10 min	123	105	(176)	227	[175]
100–200	28k/1.57	CB	110 °C/4 min	119[a]	64[a]	181[a]	143[a]	[103]
(Backbone) T_{g2} = 110–125 °C								
110	21.8k/2.39	CB	120 °C/4 min	133	84.3	222	247	[76]
60–70	12.8k/1.4	CB	120 °C/30 min	161	81	296	380	[120]
60–70	14.8k/2.1	CB	120 °C/30 min	175	95	379	566	[120]
60–70	33.4k/1.1	CB	120 °C/30 min	133	83	186	200	[120]
110	21.8k/2.39	CB	120 °C/90 min	137	85.9	202	237	[76]
60–70	280k/1.1	CB	140 °C/30 min	160	81	186	227	[120]
100–200	28k/1.57	CB	140 °C/4 min	143[a]	61[a]	207[a]	200[a]	[103]
100–200	28k/1.57	CB	170 °C/4 min	143[a]	61[a]	167[a]	157[a]	[103]
100–200	28k/1.57	CB	200 °C/4 min	92[a]	89[a]	56[a]	43[a]	[103]
(Backbone) T_m = 215–230 °C								
100–200	28k/1.57	CB	230 °C/4 min	98[a]	84[a]	48[a]	36[a]	[103]

[a] Interpolated from graph.
[b] Ratio uncertain.
[c] Concurrent with 2.7 V forward bias.

with the solvent effect observed in pristine films cast from DCB or CHCl$_3$, attributed to different crystal morphs [95]. One study observed V_{oc} decreasing for temperatures up to the vicinity of T_{g2}, minimizing at 140 °C, and increasing again at higher temperatures [103]. These changes seem to be related to demixing of the two components rather than changes in the polymer alone, as mobility in bulk HT-PHT increases steadily up to 140 °C [122]. Molecular weight dependent polymer phase separation may become important. Crystalline, low M_n PAT shows nanorod assembly accompanied by steep declines in V_{oc} whereas high M_n, containing more amorphous material, experiences only moderate decreases [120]. A larger proportion of amorphous material remains because longer molecules with poorly mobile, entangled chains do not assemble as quickly. Chains at least $M_n = 12\,800$ (~77 rings) were necessary to maintain solar cell performance.

1.6.3.4 Evolution of Short-Circuit Current

Short-circuit current density/current (J_{sc}, I_{sc}) depends on the efficiency of charge separation, scaling with the area of the donor/acceptor interface where this process occurs [123]. However, a reliable estimation of interface area remains elusive. Optical interference complicates understanding, causing nonlinear variations with film thickness [105]. With this caveat, we present some of the noted trends in the literature. Comparing solar cells with varying composition, the largest J_{sc} correlates with highest polymer content [106,124,125]. Improvement is noted for 1 : 1 mixtures at all temperatures up to the melt (Table 1.2), suggesting that polymer reorganization is critical. Optimization of the polythiophene phase shows molecular weight dependence related to crystallization. Samples with $M_n < 10k$, shown to crystallize in linearly extended conformation [12], maximize and then decrease, the maxima shifting to higher temperature as chains lengthen [120]. For $M_n > 10k$ no maximum is reached during annealing on the timescale observed. This trend suggests that short rr-P3HT chains reach thermodynamic equilibrium more quickly than large ones, as would be expected during crystallization. Previous rationalization of the poor performance by low M_n PT proposes minimal long-range interconnection of crystallites. In contrast, higher M_n fibrils tend to branch and interconnect or have chains bridging nanocrystalline domains [120].

The assembly model presented here is consistent with these observations. Different M_w samples equilibrate at different rates, and the above mentioned are being compared at different degrees of organization *en route* to a crystalline phase-separated anisotropic extreme. High order is reached fastest by small molecules with π-stacking parallel to the bottom electrode, polarizing current flow orthogonal to the electrodes and minimizing current produced by the solar cell. Larger molecules develop highly conductive low-resistance domains, but crystalline anisotropy does not develop unless special conditions facilitate reorganization [21,23].

1.6.3.5 Evolution of Fill Factor, Power Conversion Efficiency (η)

Fill factor measures the shape of the current–voltage curve, and is strongly dependent on carrier recombination strength, buildup of space charge, shunt, and series resistance. Efficient confinement of charge to different phases limits photocurrent by the slowest charge carriers, allowing defective assembly to limit current in each phase [126]. Annealing improves this parameter relative to the amorphous state, it is thought, by improving morphology. High-M_w species may require higher temperatures for optimization because chains reorganize more slowly [120]. Studies on PBT find that FF does not change when annealed below the glass transition but increases when heated above T_{g1}. Long annealing (30 min) of PHT samples maximizes FF for low M_w samples (<10 000) below T_{g2}; for high M_w species the maxima seems to occur in the vicinity of T_{g2}. Note that transition temperatures were not measured on these samples but are inferred from the literature.

One research goal is to reliably optimize device properties, maximizing power conversion efficiency. This is expected to depend upon controlled demixing and assembly of each phase. Evolution of η most closely parallels the increase in I_{sc} in samples annealed for 30 min [44,120]. Low M_w samples peak below 100 °C then decline; high M_w show some variability, but generally increase up to 150 °C. Higher temperatures degrade performance. Thermal analysis of component materials is needed to clarify the relationship.

Appendix 1.A provides a survey of polymeric solar cells and their reported efficiencies, fill factors, and short-circuit current and current density – noting important processing conditions and sample preparation.

1.6.4
HT-PAT/PCBM Blend: Layer Thickness

Active layer thickness influences absorption and charge separation events in opposite ways. Optical density is higher for thicker cells, maximizing photon capture. But thin films, with close electrode spacing, have shorter diffusion distance for separated charges and capture more efficiently the total photogenerated charge. Additionally, optical interference causes wavelength dependent reduction of J_{sc} and η within a device, giving thickness-dependent maxima [105,127]. Modeling suggests that a combination of the total absorbed light with optical intensity distribution is necessary for explanation. Total light in active layer is not the principal cause, rather the distribution of the optical electrical field throughout the transparent stack is most important. Modifying composition can offset these limitations, indicating that concurrent optimization is necessary.

1.6.5
Summary

Photovoltaic properties arise from several interdependent factors, complicating correlation and optimization. Studies of polymer and fullerene samples give clear insight into the solid-state structure of each material. Though not directly applicable to intimate blends, thermally driven demixing develops structure consistent with the parent phase. Morphological investigations are just beginning to probe the nature and extent of interface modification on annealing. Physical and electronic changes seem to be well correlated with side chain and backbone glass or melting transitions in HT-PT, altering structure in the polymer phase and controlling diffusion of the fullerene phase. Under experimental conditions PCBM does not melt but may dissolve to some extent in the polymer matrix. From this perspective, it is clear that intrinsic polymer properties are primarily responsible for reorganization in active layer blends.

1.7
HT-PT in Other Blends

Properties of photovoltaic devices incorporating HT-regioregular polythiophenes reported in the past 5 years are surveyed in Appendix 1.A. These are summarized by composition, noting dependence on HT-PT thermal properties when possible.

1.7.1
C_{60} and Non-PCBM Fulleroids

The low solubility of C_{60} hampers its use, in part, because reliable, large interface area formation is difficult to achieve. Extreme phase separation by fullerene crystallization has limited performance, with spin casting only recently yielding high-efficiency cells [128]. Effective alternatives to solution casting may facilitate use. One approach deposits fullerene films *en vacuo*, then topically applies polythiophene in a solvent that is good for both materials. Corrosion of the fullerene layer during drying forms an interface with a graduated distribution of components [129].

Annealing behavior of a C_{60}/PCBM composite depends largely on polythiophene thermal properties: spin-cast films (1:1 PHT:C_{60}) do not change when heated to 100 °C, but at 125 °C the V_{oc} declines and I_{sc} goes up. Annealing at this temperature increases the cell efficiency for films of 1:1 composition, but fullerene-rich films decrease. Solvent mixtures have been explored to control morphology during deposition. For example, spin casting from $CHCl_3$/toluene mixtures greatly increases the interface area and reduces the fullerene crystallite size relative to $CHCl_3$ alone, as determined by cross-sectional SEM (scanning electron microscopy) [130].

1.7.2
Carbon Nanotubes and Other Organics

Solar cells based on carbon nanotubes or single-walled nanotubes (SWNT) have recently been reviewed [131]. SWNT are attractive organic n-channel materials, in part, because each has a lengthy dimension of efficient intramolecular electron transport. But inseparable, metallic nanotube impurities can limit useful blends to low concentration. Native polythiophene has low conductivity (10^{-8} S m^{-1}), but blending 20% with purified nanotubes increases this by six orders of magnitude (10^{-2} S m^{-1}) [132]. Blending as little as 0.1% with PHT quenches most of the fluorescence, demonstrating efficient charge transfer. The ~2.06 eV absorption intensifies in solution and solid state, confirming that π-stacked thiophene structures form [133]. STM studies of 5% SWNT/PHT drop cast from CHCl$_3$ indicate the structure is a polymer sheath 5–10 nm thick with π-stacks orthogonal to the nanotube [134]. Application of this binary mixture in photovoltaic devices thus seems limited by morphology. The large surface area of a nanotube mat has also been employed as an electrode, with the potential for interfacial charge separation. Thin transparent films have been proposed as flexible alternative to ITO, with device efficiencies reaching 2.5% [135].

Other organic materials are less frequently studied, in part due to the comparative scarcity of efficient n-channel semiconductors. Prominent candidates have recently been reviewed in the context of transistor devices [136]. An attractive aspect of alternative conjugated architectures is the prospect for targeting a region of the spectrum to optimize photon harvesting, much like dye-sensitized solar cells. The maximum 1.5% efficiency observed in representative devices remains far below the benchmark of ~5% PHT:PCBM blends (see Appendix 1.A). Phase separation and morphology in nanotube blends are poorly understood.

1.7.3
Hybrid Organic/Inorganic Nanocomposites

Attempts to merge inorganic with organic photovoltaic technology are often undermined by solubility and processing difficulties. Two techniques are being explored as possible solutions. The first deposits porous inorganic architectures, subsequently intercalating polythiophene. The second method deposits a suspension of stabilized inorganic nanoparticles or nanorods in polymer solution. Efficiencies in both cases were generally low, though optimizing polymer processing to form nanofibrils obtained $\eta = 2.6\%$.

1.8
Surface Analysis of HTPT Films

Surface characterization techniques are integral to modern materials science. Many techniques have been developed to study surface chemistry and physics, but the relative novelty of regioregular polythiophenes limits the number of reports.

1.8.1
AFM and STM

By far the most commonly used techniques, in the context of HTPT, AFM and STM have been used to characterized surface morphology and molecular organization. Regioregular polythiophenes were initially imagined as being rather rigid with transoid ring orientation giving rodlike character. STM studies of monolayers formed on HOPG revealed that chains were surprisingly flexible. Dilute solutions drop-cast from CHCl₃ onto HOPG form 2D nanocrystalline domains with π-system orthogonal to the surface. Linearly associated domains were connected by folded sequences of approximately eight rings in cisoid orientation [32,137]. Polymers 70–80 rings long could contain several folds, even adjacent forming "S" shapes. Epitaxial effects restricted the majority of folds to 60° or 120° to match the symmetry of the underlying lattice [138].

Controlled deposition of patterned polymers is an emerging area of scrutiny. A promising method to deposit ordered arrays of polymer onto a substrate by controlling the movement of a meniscus (e.g., zone casting [139]) generates arrays of discrete dots, well-separated lines, and ladderlike structures of cross-connected lines. AFM images of HT-PHT deposited in the ladder structure show very regular fiber widths and spacing, with only a small amount of secondary branching [140].

1.8.2
XPS/ESCA and Auger Spectroscopy

X-ray photoelectron spectroscopy (XPS), also known as electron spectroscopy for chemical analysis (ESCA), is a sensitive method to study composition and chemical bonding at or near the surface (>10 nm) [141,142]. This process is more efficient and thus more sensitive for light elements. Compositional profiles of thick films are obtained by alternating analysis with ion-beam milling to remove successive surface layers. For example, XPS study of thin (15–20 nm) 70% HT-PHT films spin cast from toluene onto silica indicate sulfur atoms are not exposed, presumably because of screening by alkyl chains [143]. After removal of the surface layer, the intensity of the sulfur signal quickly intensifies by a factor of three and then saturates through the remainder of the film. Similar analysis of a phase separating 2% siloxane/ir-PT blend found it was possible to adjust segregation and thus surface coverage by siloxane between 25 and 100% [144].

XPS has been used to study ionization in polythiophenes. Chemical doping of ir-PT showed distinct neutral and charged species varying with dopant concentration [145]. Electrochemical reduction of ir-poly(3(fluorophenyl)thiophene) similarly generated a mixture of thiophene anions. Shifting orbital energies suggests that much of the charge is localized on sulfur [146]. Degradation of HT-PHT films by a 180 eV electron beam caused slight shift in C 1s and S 2p binding energies, but this was not definitive for a change in structure [147].

Complex structures can be analyzed, simultaneously monitoring several nuclei. For example, intercalation of HT-PHT into mesoporous titania on ITO was verified by depth profiling and XPS monitoring the C 1s, S 2p, O 1s, Ti 2p, and In 3d orbitals [148]. Other studies have focused on fullerene/conjugated polymer heterojunctions. Under the conditions of deposition, some diffusion of C_{60} into POT has been observed [149]. Heating bilayers above the side chain melting temperature to allow interdiffusion has been explored to generate bulk heterojunctions [150]. Auger depth profiling of 160 nm HT-POT/80 nm C_{60} bilayer annealed for 5 min showed that the C_{60} layer reduced to 35 nm and a C_{60}/PT gradient over the next 115 nm before reaching the homogeneous polymer surface layer. Heating above the polymer melt degraded photovoltaic performance, but the product was not profiled.

Mobile ionic impurities have been profiled. HT-PHT was spin cast from $CHCl_3$ over Au electrodes patterned on alkali free, common borosilicate, and alkali-rich glass slides (<0.02, 6, and 12% Na_2O, respectively) [151]. After brief heating (150 °C) to remove residual O_2, the samples were comparably conductive. Charging the films to 200 V overnight led to degraded performance in the films on alkali-rich glass and longer stress resulted in breakdown. XPS depth profiling of the films revealed sodium contamination of the HT-PHT by up to 1 atom per thiophene ring distributed uniformly through the layer. Auger analysis of the sodium indicated principally Na^+.

XPS was used to follow film formation as successive layers of HT-PHT were deposited on gold [152] or HOPG [153]. Band bending, from equilibration of the Fermi energy levels at the interface of both materials, was noted by slight shifting of the C 1s and S 2p orbitals to higher binding energy. Films studied were too thin to ensure that the flat band regime was reached, but orbital alignment was clearly different than for long oligothiophenes. Hole injection barriers (K_h) were estimated at 0.59 eV for Au and 0.29 eV for HOPG. Electron injection barriers (K_e) were estimated at 1.11 eV for Au and 1.41 eV for HOPG. Ionization energies of HT-PHT on Au (4.89 eV) [152] and HOPG (4.85 eV) [153] have been determined by UPS. Band alignment with gold contacts varies with π-stacking orientation [154]. The interface dipole and hole injection barrier are larger when π-stacks orient parallel to the surface, insulated by an alkyl substituent barrier, and smaller when stacks are vertical, orienting π-molecular orbitals toward the interface.

Grazing emission X-ray fluorescence (GEXRF) has been used to detect residual Mg, P, and Ni and determine M_w by comparing (terminal) halogen content to sulfur content in poly(thienylenevinylene)s [155]. This technique detects light elements ($Z = 4$–13) that conventional XRF cannot.

1.8.3
Rutherford Backscattering Spectrometry

This ion beam technique is an inherently standardless nuclear analytical method, attractive for high sensitivity and depth sensitive measurements on films a few nanometers to a few microns thick. Sensitivity ranges from several percent for light nuclei to ppm for heavy nuclei. Oxygen contamination of HT-PT has been profiled after handling in air to partially dope with O_2 and adsorb H_2O and thermal cycling to remedy the problem. The oxygen signal visible in pristine samples was reduced below detectable levels by heating (120 °C) in inert atmosphere [88,89]. Varying amounts of Zn and Br in commercial samples of have also been quantified [80]. Soxhlet extraction with methanol was sufficient to remove Zn and reduce Br to a value consistent with functional chain ends.

Combining RBS spectra with knowledge of film composition allows elemental profiling as a function of depth for complex, layered structures including transistors and solar cells. Attempts to manufacture a Ca/Ag cathode on a PHT:PCBM layer instead formed Ca–Ag alloy under processing conditions [156]. Phase separation of PHT-commodity polymer blends has been followed, monitoring sulfur distribution. When PHT was allowed to crystallize first, it was expelled to the surface of the film; when the matrix polymer crystallized first, a homogeneous distribution through the film was maintained [157]. Also of note is that the instability of the ITO interface with acidic PEDOT:PSS in diodes has been investigated by monitoring the distribution of migrating indium [158]. Etching of the substrate was observed, accelerated on heating, and was attributed to the inherent acidity of PSS.

1.8.4
X-Ray

Blending with PCBM disrupts polymer crystallization, giving amorphous spectra in pristine films with low polymer content [159]. Annealing promotes assembly and crystallization of the polymer. Recall that homogeneous thiophene films, as discussed above, tend to form nanocrystalline domains with isotropic orientation. Thin PHT:PCBM films (70–110 nm) on Si substrates are much more polarized, with only the 1 0 0 lamellar reflection seen perpendicular to the substrate surface [160]. Grazing incidence small-angle X-ray (GIXS, GIXRD) also confirms this orientation on PEDOT:PSS substrates, with π-stacking parallel to the surface [96,161]. Annealing spin-cast thin (65 nm) films intensifies this reflection but does not otherwise alter the spectrum [162]. Debye–Scherrer analysis determined that mean crystallite size was largest after short annealing (5 min) at 125 °C – suggesting a relation with T_{g2}. The final structure is solvent dependent. Pristine films cast from chlorobenzene form crystallites >50% larger than when cast from $CHCl_3$ [117]. However, GIXRD study of annealed films finds that the largest crystallites grow in $CHCl_3$ cast films and the overall crystallinity becomes nearly double than that of the other film (21%). It is noteworthy that crystallinity correlates with intensity of the 2.06 eV intermolecular exciton absorption. Absorption in the low energy region of the spectrum (2.0–2.5 eV) is negligibly perturbed by amorphous polythiophene; thus, the efficiency of photon absorption correlates with crystallinity.

Anisotropy dependence on solvent was examined, monitoring the 100 reflection intensity and incident beam angle. Films cast from $CHCl_3$ show bimodal distribution with maxima near 0° and at 90° from domains oriented both perpendicular and parallel to the substrate. Cast from trichlorobenzene, π-stacking is exclusively parallel to the substrate with large diffraction at 90° [163]. Varying film thickness revealed maximum anisotropy near the surface. Rocking curves showed consistent surface normal 100 orientation, but background attributed to disordered domains increased with film thickness [1].

1.8.5
Other Techniques (SIMS, UPS)

Infiltration of HT-PHT into nanocrystalline TiO_2 has been studied by secondary ion mass spectrometry (SIMS) depth profiling of carbon and metal ions [79]. Very little

(>0.5%) material was found intercalated in the inorganic network, even with high (~65%) porosity. Careful fractionation and use of an intermediate M_w fraction increased void filling to 22%.

1.9
Summary and Future Directions

Assembly and morphology of HT-regioregular polythiophenes have been examined to understand the advantages and limitations of the material. Compiled thermal data suggests inherent, distinct processes for the side chain and backbone. Properties of solar cells incorporating HT-PT were surveyed. Evolution of device parameters during annealing for high M_w (>10 000) polymers seems well correlated with corresponding thermal processes. Low M_w (<10 000) behaves differently, suggesting facile reorganization (shifting thermal process to lower temperature) or greater order (e.g., crystallinity). The onset of polymer assembly seems to be correlated with T_{m1} and enhanced molecular mobility. Between T_{m1} and T_{g2}, the polymer crystallizes without macroscopic PCBM crystal growth, implying that a large interface area is maintained. Above T_{g2}, PCBM migration coincides with larger crystal growth, while above T_{m2} extreme demixing is expected. Both these effects limit interface area. Studies generally find the maximal efficiency obtained after short annealing, precluding long anneals under the same conditions. The degree of polymer assembly, size, and distribution of crystallites (in both phases) can vary widely, making correlation with device properties difficult.

The relation between interface area and phase demixing remains unclear, morphological and electronic consequence for cell performance only slowly being resolved. Metrological studies will clarify morphology evolution on solvent evaporation, thermal processes governing solid-state reorganization, phase demixing, and interfacial and bulk structure. Polarization-dependent spectroscopies should help understand assembly and growth of anisotropic polymer domains and the electronic consequences of polarization. With a clear model of optimal structure, it will be possible to design photovoltaic devices with maximal performance for these materials. It will also become clear whether or not "healing" damaged devices is feasible, for example, by heating to a liquid crystalline state and cooling. Understanding limitations will speed development of the next-generation organic photovoltaics and narrow the performance gap with silicon-based devices.

Appendix 1.A. Survey of Photovoltaic Cells Incorporating Regioregular Polythiophenes (2001–2006)

Samples were selected only where HT-regioregularity was certain, and broad-spectrum photocurrent was measured. Omissions are possible. Listed by ascending power conversion efficiency.

Appendix 1.A. Survey of Photovoltaic Cells Incorporating Regioregular Polythiophenes (2001–2006) | 31

PAT/PCBM blends.

PCE, η (power[a] mW cm^{-2})	Area (cm^2)	Anode structure (quality or thickness)	Active layer (thickness)	Cathode structure (thickness)	Deposition (solvent)	Anneal temperature/ time	FF (%)	V_{oc} (V)	I_{sc}	J_{sc} (mA cm^{-2})	Year (reference)
0.2% (100)	—	ITO/PEDOT:PSS (—/100 nm)	1:4 PHT:PCBM (150 nm)	Al (100 nm)	Spin cast (CHCl$_3$/toluene)	—/—	0.30	0.48	—	1.28	2003 [176]
0.99% (100)	0.07	SWNT/PEDOT:PSS (282 Ω cm^{-2}/—)	1:1 PHT:PCBM (800 nm)	Ga:In (—)	Drop cast (Ph-Cl)	Annealed (to max η)	0.30	0.50	6.65	—	2005 [177]
1.12% (100)	0.2	Au/PEDOT:PSS (8 nm/70 nm)	PHT:PCBM (—)	Ti (200 nm)	Spin cast (—)	—/—	0.33	0.55	—	1.12	2004 [178]
1.4% (—)[b]	0.075	Au/PEDOT:PSS (50 nm/250 nm)	1:1.5 PHT:PCBM (150 nm)	Ti/Al (20 nm/80 nm)	Spin cast (Ph-Cl)	80 °C/~2 min	0.53	0.58	—	4.6	2005 [179]
1.4% (100)	0.045	ITO/PEDOT:PSS (25 Ω cm^{-2}/—)	1:1 PHT:PCBM (70 nm)	Al (1500 nm)	Spin cast (Ph-Cl)	140 °C/4 min	0.52	0.45	—	5.5	2005 [103]
1.54% (100)	0.5	ITO/PEDOT:PSS (—/100 nm)	1:3 PHT:PCBM (100–150 nm)	Al (80 nm)	Spin cast (Ph-Cl)	—	0.39	0.60	6.61	—	2005 [180]
2.03% (100)	0.1	ITO/ZnO nanorod (—)	PHT:PCBM (200 nm)	Ag (80 nm)	Spin cast (CHCl$_3$)	200 °C/1 min	0.43	0.475	—	10.0	2005 [181]
2.09% (97)	—	ITO/PEDOT:PSS (?/?)	PHT:PCBM (110 nm)	LiF/Al (1.5 nm/100 nm)	Spin cast (Ph-Cl)	120 °C/4 min	0.62	0.58	—	5.64	2006 [76]

(Continued)

PAT/PCBM blends. (Continued)

PCE, η (power[a], mW cm^{-2})	Area (cm^2)	Anode structure (quality or thickness)	Active layer (thickness)	Cathode structure (thickness)	Deposition (solvent)	Anneal temperature/ time	FF (%)	V_{oc} (V)	I_{sc}	J_{sc} (mA cm^{-2})	Year (reference)
2.14% (100)	0.01	ITO/PEDOT:PSS (25 Ω cm^{-2}/—)	1:1 PHT:1 PCBM (—)	Ga:In (—)	Spin cast (CHCl$_3$)	—	0.52	0.515	—	8.0	2006 [104]
2.15% (100)	0.09	ITO/PEDOT:PSS (8–12 Ω cm^{-1}/ 30–35 nm)	1:2 PCBM: PHT (80 nm)	LiF/Al (6 nm/—)	Spin cast (toluene)	96 °C/1 min	0.455	0.58	—	8.16	2004 [182]
2.39% (100)	—	ITO/PEDOT:PSS (20 Ω cm^{-2}/ 100 nm)	1:1 PHT:PCBM (—)	Al (100 nm)	Spin cast (CHCl$_3$)	130 °C/20 s	0.63	0.60	—	6.35	2004 [106]
2.7% (100)	—	ITO/PEDOT:PSS (—/—)	1:1 PCBM: PHT (70 nm)	LiF/Al (—/—)	Spin cast (DCB)	120 °C/60 min	0.61	0.615	—	7.2	2005 [113]
2.7% (100)	0.025–0.04	ITO/PEDOT:PSS (20 Ω cm^{-2}/ 30 nm)	1:2 PHT:PCBM (200 nm)	LiF/Al (0.6 nm/ 100 nm)	Spin cast (Ph-Cl)	100 °C/15 min	0.64	0.58	7.24	—	2006 [159]
2.8% (100)	0.04	ITO/PEDOT:PSS (?/60 nm)	1:3 PHT:PCBM (350 nm)	Ca/Al (100 nm)	Spin cast (—)	—	0.55	0.58	8.7	—	2002 [183]
2.8% (100)[b]	—	ITO/PEDOT:PSS (15 Ω cm^{-2}/—)	1:1 PHT:PCBM (—)	Sm/Al (8 nm/ 100 nm)	Spin cast (CHCl$_3$)	130 °C/4 min	0.65	0.60	—	7.4	2006 [114]
2.8% (100)	—	ITO/PEDOT:PSS (—/—)	1:1 PHT:PCBM (100 nm)	Al (100 nm)	Spin cast (Ph-Cl)	150 °C/10 min	0.55	0.61	—	8.35	2006 [184]
3% (—)	0.04	ITO/PEDOT:PSS (—/80 nm)	1:1 PHT:PCBM (100 nm)	Ca/Ag (>100 nm)	Spin cast (xylene)	—	0.50	0.55	10	—	2005 [185]

Appendix 1.A. Survey of Photovoltaic Cells Incorporating Regioregular Polythiophenes (2001–2006)

3% (100)	0.045	ITO/PEDOT:PSS (25 Ω cm^{-2}/40)	1:1 PHT:PCBM (—)	Al (1500)	Spin cast (DCB)	140°C/15 min	0.53	0.61	—	9.4	2005 [167]
3.1% (100)[b]	—	ITO/PEDOT:PSS (—)	1:2 PHT:PCBM (350 nm)	Al (—)	Spin cast (—)	75°C/4 min	0.37	0.54	—	15.2	2004 [186]
3.18% (100)	0.105	ITO/V$_2$O$_5$ (—/3 nm)	1:1 PHT:PCBM (—)	Ca/Al (25 nm/100 nm)	Spin cast (DCB)	110°C/10 min	0.591	0.59	8.83	—	2006 [187]
3.33% (100)	0.105	ITO/MoO$_3$ (—/5 nm)	1:1 PHT:PCBM (—)	Ca/Al (25 nm/100 nm)	Spin cast (DCB)	110°C/10 min	0.619	0.60	8.94	—	2006 [187]
3.4% (100)	—	ITO/PEDOT:PSS (60 Ω cm^{-2}/100 nm)	1:1 PHT:PCBM (100–200 nm)	Al (80 nm)	Spin cast (Ph-Cl)	100°C/10 min	0.49	0.63	10.9	—	2005 [175]
3.5% (80)[b]	0.05–0.08	ITO/PEDOT:PSS (15 Ω cm^{-2}/100 nm)	PHT:PCBM (100–1201 nm)	LiF/Al (0.6 nm/60 nm)	Spin cast (DCB)	75°C/4 min[c]	0.60	0.55	8.5	—	2003 [174]
3.6% (100)	0.025–0.04	ITO/PEDOT:PSS (20 Ω cm^{-2}/30 nm)	1:2 PHT:PCBM (200 nm)	LiF/Al (0.6 nm/100 nm)	Spin cast (Ph-Cl)	100°C/5 min	0.56	540	11.8	—	2006 [188]
3.6% (100)	0.28	ITO/PEDOT:PSS (—/50 nm)	1:1 PCBM:PHT (70–80 nm)	LiF/Al (0.8 nm/100 nm)	Spin cast (Ph-Cl)	100°C/30 min	0.63	0.613	—	9.44	2005 [124]
3.7% (100)	0.078	ITO/PEDOT:PSS (15 Ω/38 nm)	1:1 PCBM:PHT (225 nm)	LiF/Al (0.7 nm/100 nm)	Spin cast (DCB)	80°C/4 min	0.61	0.6	—	10.4	2006 [105]
3.7% (100)[d]	—	ITO/PEDOT:PSS (—/—)	1:1 PHT:PCBM (300 nm)	LiF/Al (—)	Spin cast (DCB)	110°C/4 min	0.60	0.59	—	10.5	2006 [189]
3.8% (—)	0.05	ITO	PHT:PCBM	Al	Spin cast	—	0.65	0.6	—	10	2006 [233]

(Continued)

PAT/PCBM blends. (Continued)

PCE, η (power[a], mW cm^{-2})	Area (cm^2)	Anode structure (quality or thickness)	Active layer (thickness)	Cathode structure (thickness)	Deposition (solvent)	Anneal temperature/ time	FF (%)	V_{oc} (V)	I_{sc}	J_{sc} (mA cm^{-2})	Year (reference)
4.0 (100)	—	ITO/PEDOT:PSS (—/30 nm)	1:1 PHT:PCBM (63 nm)	Ca/Al (25 nm/ 80 nm)	Spin cast (DCB)	110 °C/10 min	0.617	0.607	10.631	—	2005 [190]
4.1% (100)	0.075	ITO/PEDOT:PSS (15 Ω/50 nm)	1:1 PCBM: PHT (200 nm)	LiF/Al (0.7 nm/ 100 nm)	Spin cast (DCB)	80 °C/4 min	0.61	0.6	—	9.2	2005 [127]
4.37% (100)	0.11	ITO/PEDOT:PSS (—/25 nm)	1:1 PHT:PCBM (210–230 nm)	Ca/Al (25 nm/ 80 nm)	Spin cast (DCB)	110 °C/10 min	0.674	0.61	—	10.6	2005 [191]
4.9% (80)	0.19	ITO/PEDOT:PSS (10 Ω cm^{-2}/—)	0.8:1 PCBM: PHT (—)	LiF/Al (3–4 nm/ 100 nm)	Spin cast (Ph-Cl)	155 °C/5 min	0.54	0.60	—	11.1	2005 [192]
5.0% (80)	0.148	ITO/PEDOT:PSS (—/40 nm)	1:0.8 PHT: PCBM (—)	Al (100 nm)	Spin cast (Ph-Cl)	150 °C/30 min	0.68	0.63	9.5	—	2005 [44]

[a] AM 1.5 illumination unless otherwise noted.
[b] White light.
[c] Forward bias applied.
[d] Halogen lamp (uncorrected).

Appendix 1.A. Survey of Photovoltaic Cells Incorporating Regioregular Polythiophenes (2001–2006) | 35

PAT/C$_{60}$ or non-PCBM fullerene blends.

PCE, η (power[a], mW cm^{-2})	Area (cm^2)	Anode quality (thickness)	Active layer (thickness, nm)	Cathode (thickness)	Deposition (solvent)	Anneal temperature/ time	FF (%)	V$_{oc}$ (V)	I$_{sc}$ (mA)	J$_{sc}$ (mA cm^{-2})	Year (reference)
—(—)	0.01	ITO (10 Ω cm^{-1})	1:1 C$_{60}$: PTCBI[b]/PHT (100 nm/—)	Au (—)	Spin cast (CHCl$_3$)	200 °C/15 min	—	0.34	6.8	—	2006 [193]
0.82% (80)	—	ITO/PEDOT: PSS (—/—)	PHT/C$_{60}$ (120 nm/ 50 nm)	—(—)	Spin cast (Ph-Cl)	220 °C/5 min	0.47	0.42	—	4.2	2006 [194]
0.91% (100)	0.02	ITO (—)	C$_{60}$/PHT (100 nm/—)	Au (—)	Spin cast (CHCl$_3$)	—/—	0.43	0.41	5.2	—	2004 [195]
0.92% (—)	0.01	ITO (10 Ω cm^{-1})	1:1 C$_{60}$: H$_2$Pc[c]/PHT (200 nm/—)	Au (—)	Spin cast (CHCl$_3$)	150 °C/30 min	—	0.43	6.1	—	2006 [193]
1.03% (100)	0.01	ITO/ZnO (10 Ω cm^{-2}/50 nm)	C$_{60}$:PHT (—)	Au (—)	Spin cast (CHCl$_3$)	—/—	0.40	0.43	6.0	—	2005 [129,196]
1.15% (—)	0.01	ITO (—)	C$_{60}$/PHT (100 nm/—)	Au (—)	Spray (CHCl$_3$)	230 °C/—	0.45	0.33	0.043	—	2006 [197]
>2% (80)	0.05–0.08	ITO/PEDOT: PSS (15 Ω cm^{-2}/ 100 nm)	1:2 PHT: PCBG[d] (100–120 nm)	LiF/Al (0.6 nm/ 60 nm)	Spin cast (DCB)	140 °C/10 min	0.3	0.52	5.2	—	2005 [198]
2.22% (100)[e]	0.043	ITO/PEDOT: PSS (10 Ω cm^{-1}/—)	1:2 C$_{60}$:PHT (60 nm)	Al (—)	Spin cast (DCB)	100 °C/5 min	0.56	0.43	0.176	—	2006 [128]

(*Continued*)

PAT/C$_{60}$ or non-PCBM fullerene blends. *(Continued)*

PCE, η (power^a, mW cm^{-2})	Area (cm^2)	Anode quality (thickness)	Active layer (thickness, nm)	Cathode (thickness)	Deposition (solvent)	Anneal temperature/ time	FF (%)	V$_{oc}$ (V)	I$_{sc}$ (mA)	J$_{sc}$ (mA cm^{-2})	Year (reference)
2.3% (80)^e	—	ITO/PEDOT: PSS (15 Ω cm^{-2}/ 40 nm)	1:2 PHT: DPM-12^f (50–60 nm)	Al (—)	Spin cast (CHCl$_3$)	Annealed (to max η)	0.58	0.65	—	4.7	2005 [199]
2.50 (20)^e	0.0314	ITO/PEDOT: PSS (—/—)	3:2 PHT:C$_{60}$ (100 nm)	Al (10 nm)	Spin cast (CHCl$_3$)	55 °C/30 min	0.35	0.63	—	2.52	2002 [200]

^{a}AM 1.5 illumination unless otherwise noted.
^{b}PTCBI = perylene tetracarboxylic bis-benzimidazole.
^{c}H$_2$Pc = metal-free phthalocyanine.
^{d}PCBG = C$_{61}$-butyric acid glycidol ester.
^{e}White light.
^{f}DPM-12 = 1,1-bis(4,4′-dodecyloxyphenyl)-(5,6)C$_{61}$.

Appendix 1.A. Survey of Photovoltaic Cells Incorporating Regioregular Polythiophenes (2001–2006) | **37**

PAT/carbon nanotube blends.

PCE, η (power[a], mW cm^{-2})	Area (cm^2)	Anode quality (thickness)	Active layer (thickness, nm)	Cathode (thickness)	Polymer deposition (solvent)	Anneal temperature/time	FF	V$_{oc}$ (V)	I$_{sc}$ (mA)	J$_{sc}$ (mA cm^{-2})	Year (reference)
— (AM 0)	6.45	ITO (10 Ω cm^{-1})	POT/1% SWNT:POT (—)	Al (100 nm)	Spray (CHCl$_3$)	—/—	—	0.98	0.12	—	2005 [201]
0.04% (100)	—	ITO (—)	POT/POT: SWNT (—)	Al (—)	Spin cast (CHCl$_3$ or toluene)	—/—	0.40	0.75	0.12	—	2002 [202]
0.06 (100)	—	ITO (—)	1% SWNT: POT (—)	Al (—)	Spin cast (CHCl$_3$)	—/—	0.40	0.75	0.2	—	2003 [203]
0.22% (100)	0.05	ITO/PEDOT: PSS (10 Ω cm^{-1}/40 nm)	1% SWNT: POT (100 nm)	Al (—)	Spin cast (CHCl$_3$)	120 °C/5 min	0.60	0.75	0.5	—	2006 [204]
2.5% (100)	0.04	SWNT/PEDOT: PSS (200 Ω cm^{-2})	5:4 PHT: PCBM (—)	Al (100 nm)	Spin cast (Ph-Cl)	120 °C/10 min	0.52	0.605	—	7.8	2006 [135]

[a] AM 1.5 illumination unless otherwise noted.

PAT/miscellaneous organic blends.

PCE, η (power[a], mW cm^{-2})	Area (cm^2)	Anode quality (thickness)	Active layer (thickness, nm)	Cathode (thickness)	Polymer deposition (solvent)	Anneal temperature/time	FF	V_{oc} (V)	I_{sc} (mA)	J_{sc} (mA cm^{-2})	Year (reference)
0.13% (100)	0.045	ITO/PEDOT:PSS (25Ω cm^{-2}/350 nm)	3:2 PHT:F8BT (30–100 nm)	LiF/Al (1 nm/150 nm)	Spin cast (p-xylene)	—/—	0.4	0.4	0.63	—	2004 [205]
0.56% (100)	?	FTO/TiO$_2$ (100Ω cm^{-2}/50 nm)	PTPTB/PHT (4/50 nm)	Ag (70 nm)	Spin cast (THF)	—/—	0.63	0.67	—	1.33	2005 [206]
0.85% (100)	0.04	FTO/TiO$_2$ (10Ω cm^{-2}/350 nm)	1:0.3 PHT:NK2097 (60 nm)	Au/PEDOT:PSS/Au (2 nm/—/25 nm)	Spin cast (CHCl$_3$)	—/—	0.55	0.59	—	2.6	2005 [207]
1.00% (100)	0.04	ITO/In (8Ω cm^{-1}/5 nm)	PV/PDDT (34 nm/62 nm)	Au (25 nm)	Spin cast (CHCl$_3$)	None/—	0.56	0.41	—	4.4	2004 [208]
1.48% (20)	—	ITO/PEDOT:PSS (—/—)	PHT/chlorophyll (—)	Al (1000 nm)	Spin cast (—)	—/—	0.32	0.7	—	1.1	2005 [209]

[a] AM 1.5 illumination unless otherwise noted.

Appendix 1.A. Survey of Photovoltaic Cells Incorporating Regioregular Polythiophenes (2001–2006)

Hybrid organic/inorganic nanocomposites.

PCE, η (power[a], mW cm^{-2})	Area (cm^2)	Anode quality (thickness)	Active layer (thickness, nm)	Cathode (thickness)	Polymer deposition (solvent)	Anneal temperature/ time	FF	V_{oc} (V)	I_{sc} (mA)	J_{sc} (mA cm^{-2})	Year (reference)
— (150)[b]	—	FTO (—)	TiO$_2$/PBT/PGE[c] (—)	Pt (—)	Dip coat (CHCl$_3$)	—/—	0.68	0.51	3.9×10^{-4}	—	2001 [210]
— (AM 0)	—	ITO/PEDOT:PSS (10 Ω cm^{-2}/—)	5% CdSe-AET-SWNT:POT (—)	Al (—)	Spray (—)	—/—	—	0.75	—	1.6×10^{-4}	2005 [211]
— (30)[b]	—	ITO/PEDOT:PSS (—/60 nm)	CdTe nanorods + PHT (20–40 nm)	Al (150 nm)	Spin cast (CHCl$_3$/Py)	120 °C	0.16	0.74	—	3.5×10^{-4}	2004 [212]
0.02% (80)	—	ITO (13 Ω cm^{-1})	TiO$_2$/HMPER[d]/PHT (8000 nm/—/—)	LiF/Al (0.6 nm/60 nm)	Spin cast (Ph-Cl)	—/—	0.26	0.7	—	0.08	2005 [213]
0.04% (100)[b]	—	ITO/PEDOT:PSS (14 Ω cm^{-1}/30–35 nm)	1:6 PbSe dots/PHT (100–130 nm)	Al (1000 nm)	Spin cast (CHCl$_3$)	—/—	—	0.3–0.4	—	0.2	2005 [214]
0.06% (1000)[b]	—	FTO/TiO$_2$ (20 Ω cm^{-2}/70 nm)	TiO$_2$/POT (300 nm/30–60 nm)	Au (50 nm)	Spin cast (CHCl$_3$)	100 °C/30 min	0.35	0.72	0.25	—	2003 [215,216]
0.0125% (80)[b]	4	FTO (12.5–14.5 Ω cm^{-1})	TiO$_2$/PHT (800 nm/—)	Au (40 nm)	Drop cast (CHCl$_3$)	—/—	—	0.65	—	0.02	2006 [217]
0.15% (60)[b]	5	ITO (—)	TiO$_2$/Ru dye/POT (—)	Au (—)	Spin cast (toluene)	—/—	0.44	0.65	0.17	—	2001 [218]
0.17% (100)	—	ITO/PEDOT:PSS (14 Ω cm^{-1}/75 nm)	90% POT + TiO$_2$ (—)	LiF/Al (1 nm/100 nm)	Spin cast (THF)	—/—	0.41	450	—	0.7	2004 [219]
0.20% (100)	4.2	ITO/ZnO nanorod (—)	Z907/PHT (—/80 nm)	Au (—)	Spin cast (Ph-Cl)	—/—	—	0.25	—	2	2005 [220]

(*Continued*)

Hybrid organic/inorganic nanocomposites. *(Continued)*

PCE, η (power[a], mW cm^{-2})	Area (cm^2)	Anode quality (thickness)	Active layer (thickness, nm)	Cathode (thickness)	Polymer deposition (solvent)	Anneal temperature/time	FF	V_{oc} (V)	I_{sc} (mA)	J_{sc} (mA cm^{-2})	Year (reference)
0.26% (100)	25	FTO (10 Ω cm^{-1})	TiO$_2$/2.5:1 PHT:ZnTP (500 nm/100 nm)	Au (25 nm)	Spin cast (—)	—/—	0.48	0.50	—	1.11	2005 [221]
0.42% (100)	—	ITO/PEDOT:PSS (10 Ω cm^{-2}/30 nm)	2:3 PHT:TiO$_2$ particles (100 nm)	Al (—)	Spin cast (xylene)	110 °C/24 h	0.36	0.44	2.76	—	2004 [222–224]
0.85% (100)	4	FTO (10 Ω cm^{-1})	TiO$_2$/2.5:1 PHT:ZnTP (350 nm/60 nm)	Au/PEDOT/Au (2 nm/—/25 nm)	Spin cast (CHCl$_3$)	—/—	0.55	0.59	—	2.6	2005 [207]
0.9% (—)	10	ITO/PEDOT:PSS (—/—)	26% nc-ZnO + PHT (200 nm)	Al (100 nm)	Spin cast (CHCl$_3$)	80 °C/5 min	0.55	0.69	—	2.19	2006 [125]
1.9% (80)	19	ITO/PEDOT:PSS (10 Ω cm^{-2}/—)	4:1 POT:C$_{60}$ + 1.6% Ag (150 nm)	LiF/Al (3–4 nm/100 nm)	Spin cast (Ph-Cl)	—/—	0.43	0.56	—	6.3	2005 [225]
2.6% (92)	—	ITO (—)	1:9 PHT:CdSe nanorods (—)	Al (—)	Spin cast (TCB)	150 °C/30 min	0.50	0.62	—	8.79	2006 [226]

[a] AM 1.5 illumination unless otherwise noted.
[b] White light.
[c] PGE = polymeric gel electrolyte.
[d] HMPER = N,N'-bis-2-(1-hydroxy-4-methylpentyl)-3,4,9,10-perylene bis(dicarboxamide).

References

1 Kline, R.J., McGehee, M.D. and Toney, M.F. (2006) Highly oriented crystals at the buried interface in polythiophene thin-film transistors. *Nature Materials*, **5**, 222–228.
2 McCullough, R.D. (1998) The chemistry of conducting polythiophenes. *Advanced Materials*, **10** (2), 93–116.
3 McCullough, R.D. and Ewbank, P.C. (1998) Regioregular, head-to-tail coupled poly(3-alkylthiophene) and its derivatives, in *Handbook of Conducting Polymers* (eds T.A. Skotheim, R.L. Elsenbaumer and J.R. Reynolds) Marcel Dekker, New York, pp. 225–258.
4 McCullough, R.D. (1999) The chemistry of conducting polythiophenes: from synthesis to self-assembly to intelligent materials, in *Handbook of Oligo- and Polythiophenes* (ed. D. Fichou), Wiley-VCH Verlag GmbH, Weinheim, Germany, pp. 1–44.
5 Brabec, C.J., Sariciftci, N.S. and Hummelen, J.C. (2001) Plastic solar cells. *Advanced Functional Materials*, **11** (1), 15–26.
6 Gazotti, W.A., Nogueira, A.F., Girotto, E.M., Micaroni, L., Martini, M., das Neves, S. and De Paoli, M.A. (2001) Optical devices based on conductive polymers, in *Handbook of Advanced Electronic and Photonic Materials and Devices*, (Nalwa, H.S. ed.), Academic Press, NY, vol. **10**, pp. 53–98.
7 Gregg, B.A. (2003) Excitonic solar cells. *Journal of Physical Chemistry B*, **107**, 4688–4698.
8 Gledhill, S.E., Scott, B. and Gregg, B.A. (2005) Organic and nano-structured composite photovoltaics: an overview. *Journal of Materials Research*, **20** (12), 3167–3179.
9 Mozer, A.J. and Sariciftci, N.S. (2006) Conjugated polymer photovoltaic devices and materials. *Comptes Rendus Chimie*, **9**, 568–577.
10 Chen, T.-A., Wu, X. and Rieke, R.D. (1995) Regiocontrolled synthesis of poly (3-alkylthiophenes) mediated by Rieke zinc: their characterization and solid-state properties. *Journal of the American Chemical Society*, **117** (1), 233–244.
11 Curtis, M.D. (2001) Sequence length distributions (microstructure) of regioregular poly(3-alkylthiophene)s and related conjugated polymers and their use in simulating p–p* absorption peak profiles. *Macromolecules*, **34** (22), 7905–7910.
12 Zhang, R., Li, B., Iovu, M.C., Jeffries-El, M., Sauve, G., Cooper, J., Jia, S., Tristram-Nagle, S., Smilgies, D.M., Lambeth, D.N., McCullough, R.D. and Kowalewski, T. (2006) Nanostructure dependence of field-effect mobility in regioregular poly(3-hexylthiophene) thin film field effect transistors. *Journal of the American Chemical Society*, **128**, 3480–3481.
13 Sheina, E.E., Liu, J., Iovu, M.C., Laird, D.W. and McCullough, R.D. (2004) Chain growth mechanism for regioregular nickel-initiated cross-coupling polymerizations. *Macromolecules*, **37** (10), 3526–3528.
14 Yokoyama, A., Miyakoshi, R. and Yokozawa, T. (2004) Chain-growth polymerization for poly(3-hexylthiophene) with a defined molecular weight and a low polydispersity. *Macromolecules*, **37**, 1169–1171.
15 Iovu, M.C., Sheina, E.E., Gil, R.R. and McCullough, R.D. (2005) Experimental evidence for the quasi-"living" nature of the Grignard metathesis method for the synthesis of regioregular poly (3-alkylthiophenes). *Macromolecules*, **38** (21), 8649–8656.
16 Jeffries-El, M. and McCullough, R.D. Regioregular polythiophenes, in *Handbook of Conducting Polymers*, 3rd edn, (Skotheim, T.A., Reynolds, J.R. Eds.), CRC Press, Boca Raton, FL, vol. **1**, 9/1–9/49.
17 Jones, A.G., Balocco, C., King, R. and Song, A.M. (2006) Highly tunable, high-

throughput nanolithography based on strained regioregular conducting polymer films. *Applied Physics Letters*, **89** (1), 013119/1–013119/3.

18 Sirringhaus, H., Brown, P.J., Friend, R.H., Nielsen, M.M., Bechgaard, K., Langeveld-Voss, B.M.W., Spiering, A.J.H., Janssen, R.A.J., Meijer, E.W., Herwig, P. and de Leeuw, D.M. (1999) Two-dimensional charge transport in self-organized, high-mobility conjugated polymers. *Nature*, **401** (6754), 685–688.

19 Bäcklund, T.G., Sandberg, H.G.O., Österbacka, R., Stubb, H., Torkkeli, M. and Serima, R. (2005) A novel method to orient semiconducting polymer films. *Advanced Functional Materials*, **15** (7), 1095–1099.

20 Nagamatsu, S., Takashima, W., Kaneto, K., Yoshida, Y., Tanigaki, N., Yase, K. and Omote, K. (2003) Backbone arrangement in "friction-transferred" regioregular poly(3-alkylthiophene)s. *Macromolecules*, **36**, 5252–5257.

21 Kim, D.H., Jang, Y., Park, Y.D. and Cho, K. (2006) Controlled one-dimensional nanostructures in poly(3-hexylthiophene) thin film for high-performance organic field-effect transistors. *Journal of Physical Chemistry B*, **110**, 15763–15768.

22 Brinkmann, M. and Wittmann, J.-C. (2006) Orientation of regioregular poly(3-hexylthiophene) by directional solidification: a simple method to reveal the semicrystalline structure of a conjugated polymer. *Advanced Materials*, **18** (7), 860–863.

23 Kim, D.H., Han, J.T., Park, Y.D., Jang, Y., Cho, J.H., Hwang, M. and Cho, K. (2006) Single-crystal polythiophene microwires grown by self-assembly. *Advanced Materials*, **18**, 719–723.

24 Ma, Z., Geng, Y. and Yan, D. (2007) Extended-chain lamellar packing of poly(3-butylthiophene) in single crystals. *Polymer*, **48** (1), 31–34.

25 Yue, S., Berry, G.C. and McCullough, R.D. (1996) Intermolecular association and supramolecular organization in dilute solution. 1. Regioregular poly(3-dodecylthiophene). *Macromolecules*, **29**, 933–939.

26 Yamamoto, T., Komarudin, D., Maruyama, T., Arai, M., Lee, B.-L., Suganuma, H., Asakawa, N., Inoue, Y., Kubota, K., Sasaki, S., Fukuda, T. and Matsuda, H. (1998) Extensive studies on p-stacking of poly(3-alkylthiophene-2,5-diyl)s and poly(4-alkylthiazole-2,5-diyl)s by optical spectroscopy, NMR analysis, light scattering analysis, and X-ray crystallography. *Journal of the American Chemical Society*, **120** (9), 2047–2058.

27 Yamamoto, T., Oguro, D. and Kubota, K. (1996) Viscometric and light scattering analyses of $CHCl_3$ solutions of poly(3-alkylthiophene-2,5-diyl)s. *Macromolecules*, **29** (5), 1833–1835.

28 Heffner, G.W. and Pearson, D.S. (1991) Molecular characterization of poly(3-hexylthiophene). *Macromolecules*, **24**, 6295–6299.

29 Kiriy, N., Jaehne, E., Adler, H.-J., Schneider, M., Kiriy, A., Gorodyska, G., Minko, S., Jehnichen, D., Simon, P., Fokin, A.A. and Stamm, M. (2003) One-dimensional aggregation of regioregular polyalkylthiophenes. *Nano Letters*, **3** (6), 707–712.

30 Kiriy, N., Jaehne, E., Kiriy, A. and Adler, H.-J. (2004) Conformational transitions and aggregations of regioregular polyalkylthiophenes. *Macromolecular Symposia*, **210**, 359–367.

31 Fukunaga, T., Harada, K., Takashima, W. and Kaneto, K. (1997) Observation of molecular alignment of 3-n-octadecylthiophene by scanning tunneling microscope. *Japanese Journal of Applied Physics, Part 1: Regular Papers, Short Notes & Review Papers*, **36** (7A), 4466–4467.

32 Mena-Osteritz, E., Meyer, A., Langeveld-Voss, B.M.W., Janssen, R.A.J., Meijer, E.W. and Bauerle, P. (2000) Two-dimensional crystals of poly

(3-alkylthiophene)s: direct visualization of polymer folds in submolecular resolution. *Angewandte Chemie – International Edition*, **39** (15), 2680–2684.

33 Liu, J., Loewe, R.S. and McCullough, R.D. (1999) Employing MALDI-MS on poly(alkylthiophenes): analysis of molecular weights, molecular weight distributions, end-group structures, and end-group modifications. *Macromolecules*, **32**, 5777–5785.

34 Matthews, J.R., Goldoni, F., Schenning, A.P.H.J. and Meijer, E.W. (2005) Non-ionic polythiophenes: a non-aggregating folded structure in water. *Chemical Communications*, **44**, 5503–5505.

35 Chang, C.-C., Pai, C.-L., Chen, W.-C. and Jenekhe, S.A. (2005) Spin coating of conjugated polymers for electronic and optoelectronic applications. *Thin Solid Films*, **479** (1–2), 254–260.

36 Prosa, T.J., Winokur, M.J., Moulton, J., Smith, P. and Heeger, A.J. (1992) X-ray structural studies of poly(3-alkylthiophenes): an example of an inverse comb. *Macromolecules*, **25** (17), 4364–4372.

37 McCullough, R.D., Tristram-Nagle, S., Williams, S.P., Lowe, R.D. and Jayaraman, M. (1993) Self-orienting head-to-tail poly(3-alkylthiophenes): new insights on structure–property relationships in conducting polymers. *Journal of the American Chemical Society*, **115**, 4910–4911.

38 Kokubo, H., Kuroda, S.-I., Sasaki, S. and Yamamoto, T. (2001) Electrochemical deposition of regioregular head-to-tail poly(3-hexylthiophene-2,5-diyl) and characterization of the obtained film. *Japanese Journal of Applied Physics, Part 2: Letters*, **40** (3A), L228–L230.

39 Prosa, T.J., Winokur, M.J. and McCullough, R.D. (1996) Evidence of a novel side chain structure in regioregular poly(3-alkylthiophenes). *Macromolecules*, **29** (10), 3654–3656.

40 Causin, V., Marega, C., Marigo, A., Valentini, L. and Kenny, J.M. (2005) Crystallization and melting behavior of poly(3-butylthiophene), poly(3-octylthiophene), and poly(3-dodecylthiophene). *Macromolecules*, **38** (2), 409–415.

41 Winokur, M.J. and Chunwachirasiri, W. (2003) Nanoscale structure–property relationships in conjugated polymers: implications for present and future device applications. *Journal of Polymer Science, Part B: Polymer Physics*, **41** (21), 2630–2648.

42 Yazawa, K., Inoue, Y., Yamamoto, T. and Asakawa, N. (2006) Twist glass transition in regioregulated poly(3-alkylthiophene)s. Los Alamos National Laboratory, *Physical Review B*, **74** (9), 094204/1–094204/12.

43 Prosa, T.J., Moulton, J., Heeger, A.J. and Winokur, M.J. (1999) Diffraction line-shape analysis of poly(3-dodecylthiophene): a study of layer disorder through the liquid crystalline polymer transition. *Macromolecules*, **32** (12), 4000–4009.

44 Ma, W., Yang, C., Gong, X., Lee, K. and Heeger, A.J. (2005) Thermally stable efficient polymer solar cells with nanoscale control of the interpenetrating network morphology. *Advanced Functional Materials*, **15**, 1617–1622.

45 Chen, S.A. and Ni, J.M. (1991) Thermal analysis of ferric chloride-doped poly(3-butylthiophene) and poly(3-dodecylthiophene). *Polymer Bulletin*, **26** (6), 673–680.

46 Pal, S. and Nandi, A.K. (2003) Cocrystallization behavior of poly(3-alkylthiophenes): influence of alkyl chain length and head to tail regioregularity. *Macromolecules*, **36** (22), 8426–8432.

47 Pal, S. and Nandi, A.K. (2005) Cocrystallization mechanism of poly(3-alkyl thiophenes) with different alkyl chain length. *Polymer*, **46** (19), 8321–8330.

48 Pal, S. and Nandi, A.K. (2005) Thermodynamic behavior of poly(3-alkyl thiophene) blends: equilibrium cocrystal

49 Wentzel, P. and Du Pasquier, A. (2005) Plasticized conjugated polymers: a possible route to higher voltage solar cells. *Materials Research Society Symposium Proceedings*, **836**, 75–80.

50 Camaioni, N., Catellani, M., Luzzati, S. and Migliori, A. (2002) Morphological characterization of poly(3-octylthiophene):plasticizer:C_{60} blends. *Thin Solid Films*, **403–404**, 489–494.

51 Bolognesi, A., Botta, C. and Martinelli, M. (2001) Oriented poly(3-alkylthiophene) films: absorption, photoluminescence and electroluminescence behavior. *Synthetic Metals*, **121** (1–3), 1279–1280.

52 Zhokhavets, U., Goldhahn, R., Gobsch, G., Al-Ibrahim, M., Roth, H.-K., Sensfuss, S., Klemm, E. and Egbe, D.A.M. (2003) Anisotropic optical properties of conjugated polymer and polymer/fullerene films. *Thin Solid Films*, **444**, 215–220.

53 Zhokhavets, U., Goldhahn, R., Gobsch, G. and Schliefke, W. (2003) Dielectric function and one-dimensional description of the absorption of poly(3-octyl-thiophene). *Synthetic Metals*, **138**, 491–495.

54 Zhokhavets, U., Gobsch, G., Hoppe, H. and Sariciftci, N.S. (2004) A systematic study of the anisotropic optical properties of thin poly(3-octylthiophene)-films in dependence on growth parameters. *Thin Solid Films*, **451–452**, 69–73.

55 Yang, C.Y., Soci, C., Moses, D. and Heeger, A.J. (2005) Aligned rrP3HT film: structural order and transport properties. *Synthetic Metals*, **155** (3), 639–642.

56 Satol, M., Yamasaki, H., Aoki, S. and Yoshino, K. (1987) Anisotropy in electrical and thermal properties of drawn polythiophene film. *Polymer Communications*, **28** (5), 144–146.

57 Breiby, D.W., Sato, S., Samuelsen, E.J. and Mizoguchi, K. (2003) Electron spin resonance studies of anisotropy in semiconducting polymeric films. *Journal of Polymer Science, Part B: Polymer Physics*, **41** (23), 3011–3025.

58 Krebs, F.C., Hoffmann, S.V. and Jørgensen, M. (2003) Orientation effects in self-organized, highly conducting regioregular poly(3-hexylthiophene) determined by vacuum ultraviolet spectroscopy. *Synthetic Metals*, **138** (3), 471–474.

59 Onoda, M., Manda, Y. and Yoshino, K. (1990) Anisotropy of absorption and photoluminescence spectra of stretched poly(3-alkylthiophene). *Japanese Journal of Applied Physics, Part 1: Regular Papers, Short Notes & Review Papers*, **29** (8), 1490–1494.

60 Danno, T., Kuerti, J. and Kuzmany, H. (1991) Optical anisotropy and resonance Raman scattering of poly (alkylthiophenes). *Synthetic Metals*, **41** (3), 1251–1254.

61 Marumoto, K., Nagano, Y., Sakamoto, T., Ukai, S., Ito, H. and Kuroda, S. (2006) ESR studies of field-induced polarons in MIS diode structures with self-organized regioregular poly(3-hexylthiophene). *Colloids and Surfaces, A: Physicochemical and Engineering Aspects*, **284–285**, 617–622.

62 Heil, H., Finnberg, T., Schmechel, R. and von Seggern, H. (2002) Influence of mechanical rubbing of polyhexylthiophene layers on the field-effect mobility using differently treated isolator surfaces. *Materials Research Society Symposium Proceedings*, **734**, 225–230.

63 Heil, H., Finnberg, T., von Malm, N., Schmechel, R. and von Seggern, H. (2003) The influence of mechanical rubbing on the field-effect mobility in polyhexylthiophene. *Journal of Applied Physics*, **93**, 1636–1641.

64 Hao, X.T., Hosokai, T., Mitsuo, N., Kera, S., Mase, K., Okudaira, K.K. and Ueno, N. (2006) Electronic density tailing outside p-conjugated polymer surface. *Applied Physics Letters*, **89** (18), 182113/1–182113/3.

65 DeLongchamp, D.M., Vogel, B.M., Jung, Y., Gurau, M.C., Richter, C.A., Kirillov, O.A., Obrzut, J., Fischer, D.A., Sambasivan, S., Richter, L.J. and Lin, E.K. (2005) Variations in semiconducting polymer microstructure and hole mobility with spin-coating speed. *Chemistry of Materials*, **17** (23), 5610–5612.

66 Aasmundtveit, K.E., Samuelsen, E.J., Guldstein, M., Steinsland, C., Flornes, O., Fagermo, C., Seeberg, T.M., Pettersson, L.A.A., Inganäs, O., Feidenhans'l, R. and Ferrer, S. (2000) Structural anisotropy of poly(alkylthiophene) films. *Macromolecules*, **33** (8), 3120–3127.

67 Breiby, D.W., Samuelsen, E.J., Konovalov, O. and Struth, B. (2003) Drying behaviour of thick poly(octylthiophene) solutions. *Synthetic Metals*, **135–136**, 345–346.

68 Yamamoto, T. and Kokubo, H. (1999) Electrochemical deposition of films of p-doped regioregular poly(3-hexylthiophene-2,5-diyl). *Chemistry Letters* **28** (12), 1295–1296.

69 Yamamoto, T. and Kokubo, H. (2005) Heteroaromatic and aromatic conjugated polymers synthesized by organometallic coupling – preparation and selected electrochemical properties. *Electrochimica Acta*, **50** (7–8), 1453–1460.

70 Berson, S., Bettignies, R.D. and Guillerez, S. (2006) Poly(3-hexylthiophene) fibres for photovoltaic applications. PROC Poster.

71 Derue, G., Coppee, S., Gabriele, S., Surin, M., Geskin, V., Monteverde, F., Leclere, P., Lazzaroni, R. and Damman, P. (2005) Nanorubbing of polythiophene surfaces. *Journal of the American Chemical Society*, **127** (22), 8018–8019.

72 Bolognesi, A., Botta, C., Mercogliano, C., Marinelli, M., Porzio, W., Angiolini, L. and Salatelli, E. (2003) Oriented thin films from soluble polythiophenes. *Polymers for Advanced Technologies*, **14** (8), 537–543.

73 Yamamoto, T., Honda, Y., Sata, T. and Kokubo, H. (2004) Electrochemical behavior of poly(3-hexylthiophene). Controlling factors of electric current in electrochemical oxidation of poly(3-hexylthiophene)s in a solution. *Polymer*, **45** (6), 1735–1738.

74 Trznadel, M., Pron, A., Zagorska, M., Chrzaszcz, R. and Pielichowski, J. (1998) Effect of molecular weight on spectroscopic and spectroelectrochemical properties of regioregular poly(3-hexylthiophene). *Macromolecules*, **31**, 5051–5058.

75 Brown, P.J., Thomas, D.S., Kohler, A., Wilson, J.S., Kim, J.-S., Ramsdale, C.M., Sirringhaus, H. and Friend, R.H. (2003) Effect of interchain interactions on the absorption and emission of poly(3-hexyl-thiophene). *Physical Review B: Condensed Matter and Materials Physics*, **67** (6), 064203/1–064203/16.

76 Cugola, R., Giovanella, U., Di Gianvincenzo, P., Bertini, F., Catellani, M. and Luzzati, S. (2006) Thermal characterization and annealing effects of polythiophene/fullerene photoactive layers for solar cells. *Thin Solid Films*, **511–512**, 489–493.

77 Choulis, S.A., Kim, Y., Nelson, J., Bradley, D.D.C., Giles, M., Shkunov, M. and McCulloch, I. (2004) High ambipolar and balanced carrier mobility in regioregular poly(3-hexylthiophene). *Applied Physics Letters*, **85** (17), 3890–3892.

78 Naber, R.C.G., Mulder, M., de Boer, B., Blom, P.W.M. and de Leeuw, D.M. (2006) High charge density and mobility in poly (3-hexylthiophene) using a polarizable gate dielectric. *Organic Electronics*, **7**, 132–136.

79 Bartholomew, G.P. and Heeger, A.J. (2005) Infiltration of regioregular poly [2,2′-(3-hexylthiopene)] into random nanocrystalline TiO$_2$ networks. *Advanced Functional Materials*, **15** (4), 677–682.

80 Erwin, M.M., McBride, J., Kadavanich, A.V. and Rosenthal, S.J. (2002) Effects of impurities on the optical properties of poly-3-hexylthiophene thin films. *Thin Solid Films*, **409** (2), 198–205.

81 Kokubo, H., Yamamoto, T., Kondo, H., Akiyama, Y. and Fujimura, I. (2003)

Purification of head-to-tail-type regioregular poly(3-hexylthiophene), HT-P3HexTh, and investigation of the effects of polymer purity on the performance of organic field-effect transistors. *Japanese Journal of Applied Physics, Part 1: Regular Papers, Short Notes & Review Papers*, **42**, 6627–6628.

82 Nielsen, K.T., Bechgaard, K. and Krebs, F.C. (2005) Removal of palladium nanoparticles from polymer materials. *Macromolecules*, **38** (3), 658–659.

83 Nielsen, K., Bechgaard, K. and Krebs, F. (2006) Effective removal and quantitative analysis of Pd, Cu, Ni, and Pt catalysts from small-molecule products. *Synthesis*, **10**, 1639–1644.

84 Abdou, M.S.A., Orfino, F.P., Son, Y. and Holdcroft, S. (1997) Interaction of oxygen with conjugated polymers: charge transfer complex formation with poly (3-alkylthiophenes). *Journal of the American Chemical Society*, **119** (19), 4518–4524.

85 Rep, D.B.A., Huisman, B.H., Meijer, E.J., Prins, P. and Klapwijk, T.M. (2001) Charge-transport in partially-ordered regioregular poly(3-hexylthiophene) studied as a function of the charge-carrier density. *Materials Research Society Symposium Proceedings*, **660**, JJ7.9/1–JJ7.9/6.

86 Meijer, E.J., Mangnus, A.V.G., Huisman, B.H., t' Hooft, G.W., de Leeuw, D.M. and Klapwijk, T.M. (2004) Photoimpedance spectroscopy of poly(3-hexyl thiophene) metal–insulator–semiconductor diodes. *Synthetic Metals*, **142**, 53–56.

87 Abdou, M.S.A. and Holdcroft, S. (1993) Mechanisms of photodegradation of poly (3-alkylthiophenes) in solution. *Macromolecules*, **26** (11), 2954–2962.

88 Mattis, B.A., Chang, P.C. and Subramanian, V. (2003) Effect of thermal cycling on performance of poly(3-hexylthiophene) transistors. *Materials Research Society Symposium Proceedings*, **771**, 369–374.

89 Mattis, B.A., Chang, P.C. and Subramanian, V. (2006) Performance recovery and optimization of poly(3-hexylthiophene) transistors by thermal cycling. *Synthetic Metals*, **156** (18–20), 1241–1248.

90 Russell, D.M., Kugler, T., Newsome, C.J., Li, S.P., Ishida, M. and Shimoda, T. (2006) Dedoping of organic semiconductors. *Synthetic Metals*, **156**, 769–772.

91 Mas-Torrent, M., Den Boer, D., Durkut, M., Hadley, P. and Schenning, A.P.H.J. (2004) Field effect transistors based on poly(3-hexylthiophene) at different length scales. *Nanotechnology*, **15**, S265–S269.

92 Zen, A., Pflaum, J., Hirschmann, S., Zhuang, W., Jaiser, F., Asawapirom, U., Rabe, J.P., Scherf, U. and Neher, D. (2004) Effect of molecular weight and annealing of poly(3-hexylthiophene)s on the performance of organic field-effect transistors. *Advanced Functional Materials*, **14** (8), 757–764.

93 Brabec, C.J., Johannson, H., Padinger, F., Neugebauer, H., Hummelen, J.C. and Sariciftci, N.S. (2000) Photoinduced FT-IR spectroscopy and CW-photocurrent measurements of conjugated polymers and fullerenes blended into a conventional polymer matrix. *Solar Energy Materials & Solar Cells*, **61** (1), 19–33.

94 Sariciftci, N.S., Smilowitz, L., Heeger, A.J. and Wudl, F. (1992) Photoinduced electron transfer from a conducting polymer to buckminsterfullerene. *Science*, **258** (5087), 1474–1476.

95 Rispens, M.T., Meetsma, A., Rittberger, R., Brabec, C.J., Sariciftci, N.S. and Hummelen, J.C. (2003) Influence of the solvent on the crystal structure of PCBM and the efficiency of MDMO-PPV:PCBM 'plastic' solar cells. *Chemical Communications*, 2116–2118.

96 Erb, T., Zhokhavets, U., Gobsch, G., Raleva, S., Stuehn, B., Schilinsky, P., Waldauf, C. and Brabec, C.J. (2005) Correlation between structural and optical properties of composite polymer/ fullerene films for organic solar cells. *Advanced Functional Materials*, **15** (7), 1193–1196.

97 Chikamatsu, M., Nagamatsu, S., Yoshida, Y., Saito, K., Yase, K. and Kikuchi, K. (2005) Solution-processed n-type organic thin-film transistors with high field-effect mobility. *Applied Physics Letters*, **87** (20), 203504/1–203504/3.

98 Yang, X., Duren, J.K.J.v., Rispens, M.T., Hummelen, J.C., Janssen, R.A.J., Michels, M.A.J. and Loos, J. (2004) Crystalline organization of a methanofullerene as used for plastic solar-cell applications. *Advanced Materials*, **16** (9–10), 802–806.

99 Sensfuss, S., Konkin, A., Roth, H.K., Al-Ibrahim, M., Zhokhavets, U., Gobsch, G., Krinichnyi, V.I., Nazmutdinova, G.A. and Klemm, E. (2003) Optical and ESR studies on poly(3-alkylthiophene)/fullerene composites for solar cells. *Synthetic Metals*, **137**, 1433–1434.

100 Shrotriya, V., Ouyang, J., Tseng, R.J., Li, G. and Yang, Y. (2005) Absorption spectra modification in poly(3-hexylthiophene):methanofullerene blend thin films. *Chemical Physics Letters*, **411** (1–3), 138–143.

101 Ai, X., Beard, M.C., Knutsen, K.P., Shaheen, S.E., Rumbles, G. and Ellingson, R.J. (2006) Photoinduced charge carrier generation in a poly(3-hexylthiophene) and methanofullerene bulk heterojunction investigated by time-resolved terahertz spectroscopy. *Journal of Physical Chemistry B*, **110** (50) 25462–25471.

102 Huang, J., Li, G. and Yang, Y. (2005) Influence of composition and heat-treatment on the charge transport properties of poly(3-hexylthiophene) and [6,6]-phenyl C_{61}-butyric acid methyl ester blends. *Applied Physics Letters*, **87** (11), 112105/1–112105/3.

103 Kim, Y., Choulis, S.A., Nelson, J., Bradley, D.D.C., Cook, S. and Durrant, J.R. (2005) Composition and annealing effects in polythiophene/fullerene solar cells. *Journal of Materials Science*, **40** (6), 1371–1376.

104 Du Pasquier, A., Miller, S. and Chhowalla, M. (2006) On the use of Ga–In eutectic and halogen light source for testing P3HT–PCBM organic solar cells. *Solar Energy Materials & Solar Cells*, **90** (12), 1828–1839.

105 Moule, A.J., Bonekamp, J.B. and Meerholz, K. (2006) The effect of active layer thickness and composition on the performance of bulk-heterojunction solar cells. *Journal of Applied Physics*, **100** (9), 094503/1–094503/7.

106 Chirvase, D., Parisi, J., Hummelen, J.C. and Dyakonov, V. (2004) Influence of nanomorphology on the photovoltaic action of polymer–fullerene composites. *Nanotechnology*, **15** (9), 1317–1323.

107 Mihailetchi, V.D., van Duren, J.K.J., Blom, P.W.M., Hummelen, J.C., Janssen, R.A.J., Kroon, J.M., Rispens, M.T., Verhees, W.J.H. and Wienk, M.M. (2003) Electron transport in a methanofullerene. *Advanced Functional Materials*, **13** (1), 43–46.

108 Inoue, K., Ulbricht, R., Madakasira, P.C., Sampson, W.M., Lee, S., Gutierrez, J., Ferraris, J. and Zakhidov, A.A. (2005) Temperature and time dependence of heat treatment of RR-P3HT/PCBM solar cell. *Synthetic Metals*, **154**, 41–44.

109 Klimov, E., Li, W., Yang, X., Hoffmann, G.G. and Loos, J. (2006) Scanning near-field and confocal Raman microscopic investigation of P3HT–PCBM systems for solar cell applications. *Macromolecules*, **39** (13), 4493–4496.

110 Savenije, T.J., Kroeze, J.E., Yang, X. and Loos, J. (2006) The formation of crystalline P3HT fibrils upon annealing of a PCBM:P3HT bulk heterojunction. *Thin Solid Films*, **511–512**, 2–6.

111 Savenije, T.J., Kroeze, J.E., Yang, X. and Loos, J. (2005) The effect of thermal treatment on the morphology and charge carrier dynamics in a polythiophene–fullerene bulk heterojunction. *Advanced Functional Materials*, **15** (8), 1260–1266.

112 Swinnen, A., Haeldermans, I., vande Ven, M., D'Haen, J., Vanhoyland, G., Aresu, S.,

D'Olieslaeger, M. and Manca, J. (2006) Tuning the dimensions of C_{60}-based needlelike crystals in blended thin films. *Advanced Functional Materials*, **16**, 760–765.

113 Yang, X., Loos, J., Veenstra, S.C., Verhees, W.J.H., Wienk, M.M., Kroon, J.M., Michels, M.A.J. and Janssen, R.A.J. (2005) Nanoscale morphology of high-performance polymer solar cells. *Nano Letters*, **5**, 579–583.

114 Mihailetchi, V.D., Xie, H., de Boer, B., Koster, L.J.A. and Blom, P.W.M. (2006) Charge transport and photocurrent generation in poly(3-hexylthiophene): methanofullerene bulk-heterojunction solar cells. *Advanced Functional Materials*, **16** (5), 699–708.

115 Goris, L., Poruba, A., Purkrt, A., Vandewal, K., Swinnen, A., Haeldermans, I., Haenen, K., Manca, J.V. and Vanecek, M. (2006) Optical absorption by defect states in organic solar cells. *Journal of Non-Crystalline Solids*, **352** (9–20), 1656–1659.

116 Yang, C., Hu, J.G. and Heeger, A.J. (2006) Molecular structure and dynamics at the interfaces within bulk heterojunction materials for solar cells. *Journal of the American Chemical Society*, **128** (36), 12007–12013.

117 Zhokhavets, U., Erb, T., Gobsch, G., Al-Ibrahim, M. and Ambacher, O. (2006) Relation between absorption and crystallinity of poly(3-hexylthiophene)/fullerene films for plastic solar cells. *Chemical Physics Letters*, **418** (4–6), 347–350.

118 Brabec, C.J., Cravino, A., Meissner, D., Sariciftci, N.S., Fromherz, T., Rispens, M.T., Sanchez, L. and Hummelen, J.C. (2001) Origin of the open circuit voltage of plastic solar cells. *Advanced Functional Materials*, **11** (5), 374–380.

119 Camaioni, N., Ridolfi, G., Casalbore-Miceli, G., Possamai, G. and Maggini, M. (2002) The effect of mild thermal treatment on the performance of poly(3-alkylthiophene)/fullerene solar cells. *Advanced Materials*, **14** (23), 1735–1738.

120 Hiorns, R.C., Bettignies, R.d., Leroy, J., Bailly, S., Firon, M., Sentein, C., Khoukh, A., Preud'homme, H. and Dagron-Lartigau, C. (2006) High molecular weights, polydispersities, and annealing temperatures in the optimization of bulk-heterojunction photovoltaic cells based on poly(3-hexylthiophene) or poly(3-butylthiophene). *Advanced Functional Materials*, **16**, 2263–2273.

121 Hoppe, H., Sariciftci, N.S., Egbe, D.A.M., Mühlbacher, D. and Koppe, M. (2005) Plastic solar cells based on novel PPE–PPV-copolymers. *Molecular Crystals and Liquid Crystals*, **426**, 255–263.

122 Wegewijs, B., Grozema, F.C., Siebbeles, L.D.A., de Haas, M.P. and de Leeuw, D.M. (2001) Charge carrier dynamics in bulk poly(3-hexyl thiophene) as a function of temperature. *Synthetic Metals*, **119** (1–3), 431–432.

123 Moliton, A. and Nunzi, J.-M. (2006) How to model the behaviour of organic photovoltaic cells. *Polymer International*, **55**, 583–600.

124 de Bettignies, R., Leroy, J., Firon, M. and Sentein, C. (2005) Study of P3HT:PCBM bulk heterojunction solar cells: influence of components ratio and of the nature of electrodes on performances and lifetime. *Proceedings of SPIE – The International Society for Optical Engineering*, **5938**, 59380C/1–59380C/14 (Organic Photovoltaics VI).

125 Beek, W.J.E., Wienk, M.M. and Janssen, R.A.J. (2006) Hybrid solar cells from regioregular polythiophene and ZnO nanoparticles. *Advanced Functional Materials*, **16** (8), 1112–1116.

126 Koster, L.J.A., Mihailetchi, V.D. and Blom, P.W.M. (2006) Bimolecular recombination in polymer/fullerene bulk heterojunction solar cells. *Applied Physics Letters*, **88** (5), 052104/1–052104/3.

127 Moule, A.J., Bonekamp, J.B., Ruhl, A., Klesper, H. and Meerholz, K. (2005) The effect of active layer thickness on the

efficiency of polymer solar cells. *Proceedings of SPIE – The International Society for Optical Engineering*, **5938**, 593808/1–593808/7 (Organic Photovoltaics VI).

128 Umeda, T., Noda, H., Shibata, T., Fujii, A., Yoshino, K. and Ozaki, M. (2006) Dependences of characteristics of polymer solar cells based on bulk heterojunction of poly(3-hexylthiophene) and C_{60} on composite ratio and annealing temperature. *Japanese Journal of Applied Physics, Part 1: Regular Papers, Short Notes & Review Papers*, **45** (6A), 5241–5243.

129 Umeda, T., Hashimoto, Y., Mizukami, H., Fujii, A. and Yoshino, K. (2005) Fabrication of interpenetrating semilayered structure of conducting polymer and fullerene by solvent corrosion method and its photovoltaic properties. *Japanese Journal of Applied Physics, Part 1: Regular Papers, Short Notes & Review Papers*, **44**, 4155–4160.

130 Mizukami, H., Umeda, T., Noda, H., Shibata, T., Fujii, A. and Ozaki, M. (2006) Surface and interface morphology observation and photovoltaic properties of C_{60}/conducting polymer interpenetrating heterojunction devices. *Journal of Physics D: Applied Physics*, **39**, 1521–1524.

131 Kymakis, E. and Amaratunga, G.A.J. (2005) Solar cells based on composites of donor conjugated polymers and carbon nanotubes. *Optical Science and Engineering*, **99**, 351–365.

132 Kymakis, E. and Amaratunga, G.A.J. (2006) Electrical properties of single-wall carbon nanotube–polymer composite films. *Journal of Applied Physics*, **99**, 084302/1–084302/7.

133 Itoh, E., Suzuki, I. and Miyairi, K. (2005) Field emission from carbon-nanotube-dispersed conducting polymer thin film and its application to photovoltaic devices. *Japanese Journal of Applied Physics, Part 1: Regular Papers, Short Notes & Review Papers*, **44**, 636–640.

134 Waclawik, E.R., Bell, J.M., Goh, R.G.S., Musumeci, A. and Motta, N. (2006) Self-organization in composites of poly(3-hexylthiophene) and single-walled carbon nanotubes designed for use in photovoltaic applications. *Proceedings of SPIE – The International Society for Optical Engineering*, **6036**, 603607/1–603607/11.

135 Rowell, M.W., Topinka, M.A., McGehee, M.D., Prall, H.-J., Dennler, G., Sariciftci, N.S., Hu, L. and Gruner, G. (2006) Organic solar cells with carbon nanotube network electrodes. *Applied Physics Letters*, **88** (23), 233506/1–233506/3.

136 Newman, C.R., Frisbie, C.D., da Silva Filho, D.A., Bredas, J.-L., Ewbank, P.C. and Mann, K.R. (2004) Introduction to organic thin film transistors and design of n-channel organic semiconductors. *Chemistry of Materials*, **16** (23), 4436–4451.

137 Mena-Osteritz, E. (2002) Superstructures of self-organizing thiophenes. *Advanced Materials*, **14**, 609–616.

138 Grevin, B., Rannou, P., Payerne, R., Pron, A. and Travers, J.P. (2003) Multi-scale scanning tunneling microscopy imaging of self-organized regioregular poly(3-hexylthiophene) films. *Journal of Chemical Physics*, **118**, 7097–7102.

139 Tracz, A., Pakula, T. and Jeszka, J.K. (2005) Zone casting – a universal method of preparing oriented anisotropic layers of organic materials. *Materials Science*, **22** (4), 415–421.

140 Yabu, H. and Shimomura, M. (2005) Preparation of self-organized mesoscale polymer patterns on a solid substrate: continuous pattern formation from a receding meniscus. *Advanced Functional Materials*, **15** (4), 575–581.

141 Kang, E.T., Neoh, K.G. and Tan, K.L. (1993) X-ray photoelectron spectroscopic studies of electroactive polymers. *Advances in Polymer Science*, **106** (Polymer Characteristics), 135–190.

142 Losito, I., Sabbatini, L. and Gardella, J.A., Jr (2003) Electron, ion, and mass spectrometry, in *Comprehensive Desk*

Reference of Polymer Characterization and Analysis (Brady, R.F. Ed.), Oxford University Press, pp. 375–407.

143 Ponjee, M.W.G., Reijme, M.A., Langeveld-Voss, B.M.W., Gon, A.W.D.v.d. and Brongersma, H.H. (2002) Molecular surface structure of poly(3-hexylthiophene) studied by low energy ion scattering. *Surface Science*, **512**, 194–200.

144 Ponjee, M.W.G., Reijme, M.A., Denier van der Gon, A.W., Brongersma, H.H. and Langeveld-Voss, B.M.W. (2002) Intermolecular segregation of siloxane in P3HT: surface quantification and molecular surface-structure. *Polymer*, **43**, 77–85.

145 Kang, E.T., Neoh, K.G. and Tan, K.L. (1991) X-ray photoelectron spectroscopic studies of poly(2,2′-bithiophene) and its complexes. *Physical Review B: Condensed Matter and Materials Physics*, **44** (19), 10461–10469.

146 Naudin, E., Dabo, P., Guay, D. and Belanger, D. (2002) X-ray photoelectron spectroscopy studies of the electrochemically n-doped state of a conducting polymer. *Synthetic Metals*, **132** (1), 71–79.

147 Ahn, H., Oblas, D.W. and Whitten, J.E. (2004) Electron irradiation of poly(3-hexylthiophene) films. *Macromolecules*, **37** (9), 3381–3387.

148 Coakley, K.M., Liu, Y., McGehee, M.D., Frindell, K.L. and Stucky, G.D. (2003) Infiltrating semiconducting polymers into self-assembled mesoporous titania films for photovoltaic applications. *Advanced Functional Materials*, **13** (4), 301–306.

149 Schlebusch, C., Kessler, B., Cramm, S. and Eberhardt, W. (1996) Organic photoconductors and C_{60}. *Synthetic Metals*, **77** (1–3), 151–154.

150 Drees, M., Davis, R.M. and Heflin, J.R. (2005) Improved morphology of polymer–fullerene photovoltaic devices with thermally induced concentration gradients. *Journal of Applied Physics*, **97** (3), 036103/1–036103/3.

151 Rep, D.B.A., Morpurgo, A.F., Sloof, W.G. and Klapwijk, T.M. (2003) Mobile ionic impurities in organic semiconductors. *Journal of Applied Physics*, **93**, 2082–2090.

152 Lyon, J.E., Cascio, A.J., Beerbom, M.M., Schlaf, R., Zhu, Y. and Jenekhe, S.A. (2006) Photoemission study of the poly(3-hexylthiophene)/Au interface. *Applied Physics Letters*, **88**, 222109/1–222109/3.

153 Cascio, A.J., Lyon, J.E., Beerbom, M.M., Schlaf, R., Zhu, Y. and Jenekhe, S.A. (2006) Investigation of a polythiophene interface using photoemission spectroscopy in combination with electrospray thin-film deposition. *Applied Physics Letters*, **88** (6), 062104/1–222109/3.

154 Park, Y.D., Cho, J.H., Kim, D.H., Jang, Y., Lee, H.S., Ihm, K., Kang, T.-H. and Cho, K. (2006) Energy-level alignment at interfaces between gold and poly(3-hexylthiophene) films with two different molecular structures. *Electrochemical and Solid-State Letters*, **9** (11), G317–G319.

155 Blockhuys, F., Claes, M., Grieken, R.V. and Geise, H.J. (2000) Assessing the molecular weight of a conducting polymer by grazing emission XRF. *Analytical Chemistry*, **72**, 3366–3368.

156 de Bettignies, R., Leroy, J., Firon, M. and Sentein, C. (2006) Accelerated lifetime measurements of P3HT:PCBM solar cells. *Synthetic Metals*, **156** (7–8), 510–513.

157 Goffri, S., Mueller, C., Stingelin-Stutzmann, N., Breiby, D.W., Radano, C.P., Andreasen, J.W., Thompson, R., Janssen, R.A.J., Nielsen, M.M., Smith, P. and Sirringhaus, H. (2006) Multicomponent semiconducting polymer systems with low crystallization-induced percolation threshold. *Nature Materials*, **5** (12), 950–956.

158 Jong, M.P.d., IJzendoorn, L.J.v. and Voigt, M.J.A.d. (2000) Stability of the interface between indium-tin-oxide and poly(3,4-ethylenedioxythiophene)/poly.styrenesulfonate in polymer light-

emitting diodes. *Applied Physics Letters*, **77** (14), 2255–2257.

159 Vanlaeke, P., Swinnen, A., Haeldermans, I., Vanhoyland, G., Aernouts, T., Cheyns, D., Deibel, C., D'Haen, J., Heremans, P., Poortmans, J. and Manca, J.V. (2006) P3HT/PCBM bulk heterojunction solar cells: relation between morphology and electro-optical characteristics. *Solar Energy Materials & Solar Cells*, **90** (14), 2150–2158.

160 Erb, T., Raleva, S., Zhokhavets, U., Gobsch, G., Stuhn, B., Spode, M. and Ambacher, O. (2004) Structural and optical properties of both pure poly(3-octylthiophene) (P3OT) and P3OT/fullerene films. *Thin Solid Films*, **450** (1), 97–100.

161 Zhokhavets, U., Erb, T., Hoppe, H., Gobsch, G. and Serdar Sariciftci, N. (2006) Effect of annealing of poly(3-hexylthiophene)/fullerene bulk heterojunction composites on structural and optical properties. *Thin Solid Films*, **496**, 679–682.

162 Erb, T., Zhokhavets, U., Hoppe, H., Gobsch, G., Al-Ibrahim, M. and Ambacher, O. (2006) Absorption and crystallinity of poly(3-hexylthiophene)/fullerene blends in dependence on annealing temperature. *Thin Solid Films*, **511–512**, 483–485.

163 Chang, J.-F., Clark, J., Zhao, N., Sirringhaus, H., Breiby, D.W., Andreasen, J.W., Nielsen, M.M., Giles, M., Heeney, M. and McCulloch, I. (2006) Molecular-weight dependence of interchain polaron delocalization and exciton bandwidth in high-mobility conjugated polymers. *Physical Review B: Condensed Matter and Materials Physics*, **74** (11), 115318/1–115318/12.

164 Jen, K.Y., Miller, G.G. and Elsenbaumer, R.L. (1986) Highly conducting, soluble, and environmentally-stable poly(3-alkylthiophenes). *Journal of the Chemical Society, Chemical Communications*, **17**, 1346–1347.

165 Zhao, Y., Yuan, G., Roche, P. and Leclerc, M. (1995) A calorimetric study of the phase transitions in poly(3-hexylthiophene). *Polymer*, **36**, 2211–2214.

166 Pal, S., Roy, S. and Nandi, A.K. (2005) Temperature variation of DC conductivity of poly(3-alkyl thiophenes) and their cocrystals. *Journal of Physical Chemistry B*, **109**, 18332–18341.

167 Kim, Y., Choulis, S.A., Nelson, J., Bradley, D.D.C., Cook, S. and Durrant, J.R. (2005) Device annealing effect in organic solar cells with blends of regioregular poly(3-hexylthiophene) and soluble fullerene. *Applied Physics Letters*, **86** (6), 063502/1–063502/3.

168 Zen, A., Saphiannikova, M., Neher, D., Grenzer, J., Grigorian, S., Pietsch, U., Asawapirom, U., Janietz, S., Scherf, U., Lieberwirth, I. and Wegner, G. (2006) Effect of molecular weight on the structure and crystallinity of poly(3-hexylthiophene). *Macromolecules*, **39** (6), 2162–2171.

169 Chen, S.-A., Ni, J.-M. and Hua, M.-Y. (1997) Thermal undoping behavior of FeCl$_3$-doped poly(3-octylthiophene). *Journal of Polymer Research*, **4** (4), 261–265.

170 Yang, C., Orfino, F.P. and Holdcroft, S. (1996) A Phenomenological model for predicting thermochromism of regioregular and nonregioregular poly(3-alkylthiophenes). *Macromolecules*, **29** (20), 6510–6517.

171 Meille, S.V., Romita, V., Caronna, T., Lovinger, A.J., Catellani, M. and Belobrzeckaja, L. (1997) Influence of molecular weight and regioregularity on the polymorphic behavior of poly (3-decylthiophenes). *Macromolecules*, **30**, 7898–7905.

172 Payerne, R., Brun, M., Rannou, P., Baptist, R. and Grévin, B. (2004) STM studies of poly(3-alkylthiophene)s: model systems for plastic electronics. *Synthetic Metals*, **146** (3), 311–315.

173 Wang, Y., Archambault, N., Marold, A., Weng, L., Lucht, B.L. and Euler, W.B. (2004) Observation of two-step thermochromism in poly(3-docosylthiophene): DSC and reflection spectroscopy. *Macromolecules*, **37**, 5415–5422.

174 Padinger, F., Rittberger, R.S. and Sariciftci, N.S. (2003) Effects of postproduction treatment on plastic solar cells. *Advanced Functional Materials*, **13** (1), 85–88.

175 Al-Ibrahim, M., Ambacher, O., Sensfuss, S. and Gobsch, G. (2005) Effects of solvent and annealing on the improved performance of solar cells based on poly(3-hexylthiophene): fullerene. *Applied Physics Letters*, **86** (20), 201120/1–201120/3.

176 Chirvase, D., Chiguvare, Z., Knipper, M., Parisi, J., Dyakonov, V. and Hummelen, J.C. (2003) Temperature dependent characteristics of poly(3-hexylthiophene)–fullerene based heterojunction organic solar cells. *Journal of Applied Physics*, **93** (6), 3376–3383.

177 Du Pasquier, A., Unalan, H.E., Kanwal, A., Miller, S. and Chhowalla, M. (2005) Conducting and transparent single-wall carbon nanotube electrodes for polymer–fullerene solar cells. *Applied Physics Letters*, **87** (20), 203511/1–203511/3.

178 Al-Ibrahim, M., Sensfuss, S., Uziel, J., Ecke, G. and Ambacher, O. (2004) Comparison of normal and inverse poly(3-hexylthiophene)/fullerene solar cell architectures. *Solar Energy Materials & Solar Cells*, **85** (2), 277–283.

179 Glatthaar, M., Niggemann, M., Zimmermann, B., Lewer, P., Riede, M., Hinsch, A. and Luther, J. (2005) Organic solar cells using inverted layer sequence. *Thin Solid Films*, **491** (1–2), 298–300.

180 Al-Ibrahim, M., Roth, H.K., Zhokhavets, U., Gobsch, G. and Sensfuss, S. (2005) Flexible large area polymer solar cells based on poly(3-hexylthiophene)/fullerene. *Solar Energy Materials & Solar Cells*, **85** (1), 13–20.

181 Olson, D.C., Piris, J., Collins, R.T., Shaheen, S.E. and Ginley, D.S. (2005) Hybrid photovoltaic devices of polymer and ZnO nanofiber composites. *Thin Solid Films*, **496** (1), 26–29.

182 Inoue, K., Ulbricht, R.W., Madakasira, P.C., Sampson, W.M., Lee, S., Gutierrez, J.M., Ferraris, J.P. and Zakhidov, A.A. (2004) Optimization of postproduction heat treatment for plastic solar cells. *Proceedings of SPIE – The International Society for Optical Engineering*, **5520**, 256–262.

183 Schilinsky, P., Waldauf, C. and Brabec, C.J. (2002) Recombination and loss analysis in polythiophene based bulk heterojunction photodetectors. *Applied Physics Letters*, **81**, 3885–3887.

184 Kim, H., So, W.-W. and Moon, S.-J. (2006) Effect of thermal annealing on the performance of P3HT/PCBM polymer photovoltaic cells. *Journal of the Korean Physical Society*, **48**, 441–445.

185 Schilinsky, P., Asawapirom, U., Scherf, U., Biele, M. and Brabec, C.J. (2005) Influence of the molecular weight of poly(3-hexylthiophene) on the performance of bulk heterojunction solar cells. *Chemistry of Materials*, **17**, 2175–2180.

186 Riedel, I. and Dyakonov, V. (2004) Influence of electronic transport properties of polymer–fullerene blends on the performance of bulk heterojunction photovoltaic devices. *Physica Status Solidi a: Applied Research*, **201** (6), 1332–1341.

187 Shrotriya, V., Li, G., Yao, Y., Chu, C.-W. and Yang, Y. (2006) Transition metal oxides as the buffer layer for polymer photovoltaic cells. *Applied Physics Letters*, **88** (7), 073508/1–073508/3.

188 Vanlaeke, P., Vanhoyland, G., Aernouts, T., Cheyns, D., Deibel, C., Manca, J., Heremans, P. and Poortmans, J. (2006) Polythiophene based bulk heterojunction solar cells: morphology and its implications. *Thin Solid Films*, **511–512**, 358–361.

189 Mihailetchi, V.D., Xie, H., de Boer, B., Popescu, L.M., Hummelen, J.C., Blom, P.W.M. and Koster, L.J.A. (2006) Origin of the enhanced performance in poly(3-hexylthiophene):[6,6]-phenyl C$_{61}$-butyric acid methyl ester solar cells upon slow drying of the active layer. *Applied Physics Letters*, **89** (1), 012107/1–012107/3.

190 Li, G., Shrotriya, V., Yao, Y. and Yang, Y. (2005) Investigation of annealing effects and film thickness dependence of polymer solar cells based on poly(3-hexylthiophene). *Journal of Applied Physics*, **98** (4), 043704/1–043704/5.

191 Li, G., Shrotriya, V., Huang, J., Yao, Y., Moriarty, T., Emery, K. and Yang, Y. (2005) High-efficiency solution processable polymer photovoltaic cells by self-organization of polymer blends. *Nature Materials*, **4** (11), 864–868.

192 Reyes-Reyes, M., Kim, K. and Carroll, D.L. (2005) High-efficiency photovoltaic devices based on annealed poly(3-hexylthiophene) and 1-(3-methoxycarbonyl)-propyl-1-phenyl-(6,6) C$_{61}$ blends. *Applied Physics Letters*, **87**, 083506/1–083506/3.

193 Umeda, T., Hashimoto, Y., Mizukami, H., Noda, H., Fujii, A., Ozaki, M. and Yoshino, K. (2006) Improvement of sensitivity in long-wavelength range in organic thin-film solar cell with interpenetrating semilayered structure. *Japanese Journal of Applied Physics, Part 1: Regular Papers, Short Notes & Review Papers*, **45** (1B), 538–541.

194 Kim, K., Liu, J. and Carroll, D.L. (2006) Thermal diffusion processes in bulk heterojunction formation for poly-3-hexylthiophene/C$_{60}$ single heterojunction photovoltaics. *Applied Physics Letters*, **88** (18), 181911/1–181911/3.

195 Fujii, A., Shirakawa, T., Umeda, T., Mizukami, H., Hashimoto, Y. and Yoshino, K. (2004) Interpenetrating interface in organic photovoltaic cells with heterojunction of poly(3-hexylthiophene) and C$_{60}$. *Japanese Journal of Applied Physics, Part 1: Regular Papers, Short Notes & Review Papers*, **43** (8A), 5573–5576.

196 Umeda, T., Hashimoto, Y., Mizukami, H., Shirakawa, T., Fujii, A. and Yoshino, K. (2005) Improvement of characteristics on polymer photovoltaic cells composed of conducting polymer–fullerene systems. *Synthetic Metals*, **152**, 93–96.

197 Noda, H., Umeda, T., Mizukam, H., Fujii, A. and Ozaki, M. (2006) Fabrication of organic photovoltaic cells with interpenetrating heterojunction of conducting polymer and C$_{60}$ by spray method. *Japanese Journal of Applied Physics, Part 1: Regular Papers, Short Notes & Review Papers*, **45**, 2792–2793.

198 Drees, M., Hoppe, H., Winder, C., Neugebauer, H., Sariciftci, N.S., Schwinger, W., Schaeffler, F., Topf, C., Scharber, M.C., Zhu, Z. and Gaudiana, R. (2005) Stabilization of the nanomorphology of polymer–fullerene "bulk heterojunction" blends using a novel polymerizable fullerene derivative. *Journal of Materials Chemistry*, **15** (48), 5158–5163.

199 Riedel, I., von Hauff, E., Parisi, J., Martin, N., Giacalone, F. and Dyakonov, V. (2005) Diphenylmethanofullerenes: new and efficient acceptors in bulk-heterojunction solar cells. *Advanced Functional Materials*, **15** (12), 1979–1987.

200 Camaioni, N., Garlaschelli, L., Geri, A., Maggini, M., Possamai, G. and Ridolfi, G. (2002) Solar cells based on poly(3-alkyl) thiophenes and [60]fullerene: a comparative study. *Journal of Materials Chemistry*, **12** (7), 2065–2070.

201 Landi, B.J., Raffaelle, R.P., Castro, S.L. and Bailey, S.G. (2005) Single-wall carbon nanotube–polymer solar cells. *Progress in Photovoltaics*, **13** (2), 165–172.

202 Kymakis, E. and Amaratunga, G.A.J. (2002) Single-wall carbon nanotube/conjugated polymer photovoltaic devices. *Applied Physics Letters*, **80**, 112–114.

203 Kymakis, E., Alexandrou, I. and Amaratunga, G.A.J. (2003) High open-circuit voltage photovoltaic devices from carbon-nanotube–polymer composites. *Journal of Applied Physics*, **93**, 1764–1768.

204 Kymakis, E., Koudoumas, E., Franghiadakis, I. and Amaratunga, G.A.J. (2006) Post-fabrication annealing effects in polymer–nanotube photovoltaic cells. *Journal of Physics D: Applied Physics*, **39**, 1058–1062.

205 Kim, Y., Cook, S., Choulis, S.A., Nelson, J., Durrant, J.R. and Bradley, D.D.C. (2004) Organic photovoltaic devices based on blends of regioregular poly(3-hexylthiophene) and poly(9,9-dioctylfluorene-*co*-benzothiadiazole). *Chemistry of Materials*, **16** (23), 4812–4818.

206 Liu, Y., Summers, M.A., Edder, C., Frechet, J.M.J. and McGehee, M.D. (2005) Using resonance energy transfer to improve exciton harvesting in organic–inorganic hybrid photovoltaic cells. *Advanced Materials*, **17**, 2960–2964.

207 Takahashi, K., Nakanishi, T., Yamaguchi, T., Nakamura, J.-i. and Murata, K. (2005) Performance enhancement by blending merocyanine photosensitizer in TiO_2/polythiophene solid-state solar cells. *Chemistry Letters*, **34** (5), 714–715.

208 Nakamura, J.-i., Suzuki, S., Takahashi, K., Yokoe, C. and Murata, K. (2004) The photovoltaic mechanism of a polythiophene/perylene pigment two-layer solar cell. *Bulletin of the Chemical Society of Japan*, **77** (12), 2185–2188.

209 Yun, J.-J., Jung, H.-S., Kim, S.-H., Han, E.-M., Vaithianathan, V. and Jenekhe, S.A. (2005) Chlorophyll-layer-inserted poly(3-hexylthiophene) solar cell having a high light-to-current conversion efficiency up to 1.2%. *Applied Physics Letters*, **87**, 123102/1–123102/3.

210 Kaneko, M., Takayama, K., Pandey, S.S., Takashima, W., Endo, T., Rikukawa, M. and Kaneto, K. (2001) Photovoltaic cell using high mobility poly(alkylthiophene)s and TiO_2. *Synthetic Metals*, **121**, 1537–1538.

211 Landi, B.J., Castro, S.L., Ruf, H.J., Evans, C.M., Bailey, S.G. and Raffaelle, R.P. (2005) CdSe quantum dot-single wall carbon nanotube complexes for polymeric solar cells. *Solar Energy Materials & Solar Cells*, **87** (1–4), 733–746.

212 Kim, J.Y., Chung, I.J., Kim, Y.C. and Yu, J.-W. (2004) Nanocrystal-conjugated polymer photovoltaic devices. *Journal of the Korean Physical Society*, **45**, 231–234.

213 Zafer, C., Karapire, C., Serdar Sariciftci, N. and Icli, S. (2005) Characterization of N,N′-bis-2-(1-hydroxy-4-methylpentyl)-3,4,9,10-perylene bis(dicarboximide) sensitized nanocrystalline TiO_2 solar cells with polythiophene hole conductors. *Solar Energy Materials & Solar Cells*, **88**, 11–21.

214 Jiang, X., Lee, S.B., Altfeder, I.B., Zakhidov, A.A., Schaller, R.D., Pietryga, J.M. and Klimov, V.I. (2005) Nanocomposite solar cells based on conjugated polymer/PbSe quantum dot. *Proceedings of SPIE – The International Society for Optical Engineering*, **5938**, 59381F/1–59381F/9 (Organic Photovoltaics VI).

215 Huisman, C.L., Goossens, A. and Schoonman, J. (2003) Aerosol synthesis of anatase titanium dioxide nanoparticles for hybrid solar cells. *Chemistry of Materials*, **15**, 4617–4624.

216 Huisman, C.L., Goossens, A. and Schoonman, J. (2003) Preparation of a nanostructured composite of titanium dioxide and polythiophene: a new route towards 3D heterojunction solar cells. *Synthetic Metals*, **138**, 237–241.

217 Qiao, Q., Beck, J., Lumpkin, R., Pretko, J. and McLeskey, J.T. (2006) A comparison of fluorine tin oxide and indium tin oxide as the transparent electrode for P3OT/TiO_2 solar cells. *Solar Energy Materials & Solar Cells*, **90** (7–8), 1034–1040.

218 Gebeyehu, D., Brabec, C.J., Padinger, F., Fromherz, T., Spiekermann, S., Vlachopoulos, N., Kienberger, F., Schindler, H. and Sariciftci, N.S. (2001) Solid state dye-sensitized TiO_2 solar cells with poly(3-octylthiophene) as hole transport layer. *Synthetic Metals*, **121**, 1549–1550.

219 Slooff, L.H., Wienk, M.M. and Kroon, J.M. (2004) Hybrid TiO$_2$: polymer photovoltaic cells made from a titanium oxide precursor. *Thin Solid Films*, **451–452**, 634–638.

220 Peiró, A.M., Ravirajan, P., Govender, K., Boyle, D.S., O'Brien, P., Bradley, D.D.C., Nelson, J. and Durrant, J.R. (2005) The effect of zinc oxide nanostructure on the performance of hybrid polymer/zinc oxide solar cells. *Proceedings of SPIE – The International Society for Optical Engineering*, **5938**, 593819/1–593819/8 (Organic Photovoltaics VI).

221 Takahashi, K., Takano, Y., Yamaguchi, T., Nakamura, J.-I., Yokoe, C. and Murata, K. (2005) Porphyrin dye-sensitization of polythiophene in a conjugated polymer/TiO$_2$ p–n hetero-junction solar cell. *Synthetic Metals*, **155** (1), 51–55.

222 Kwong, C.Y., Djurisic, A.B., Chui, P.C., Cheng, K.W. and Chan, W.K. (2004) Influence of solvent on film morphology and device performance of poly(3-hexylthiophene):TiO$_2$ nanocomposite solar cells. *Chemical Physics Letters*, **384** (4–6), 372–375.

223 Kwong, C.Y., Choy, W.C.H., Djurisic, A.B., Chui, P.C., Cheng, K.W. and Chan, W.K. (2004) Poly(3-hexylthiophene):TiO$_2$ nanocomposites for solar cell applications. *Nanotechnology*, **15**, 1156–1161.

224 Kwong, C.Y., Djurisic, A.B., Chui, P.C. and Chan, W.K. (2004) Nanocomposite solar cells: influence of particle concentration, size, and shape on the device performance. *Proceedings of SPIE – The International Society for Optical Engineering*, **5520**, 176–183.

225 Kim, K. and Carroll, D.L. (2005) Roles of Au and Ag nanoparticles in efficiency enhancement of poly(3-octylthiophene)/C$_{60}$ bulk heterojunction photovoltaic devices. *Applied Physics Letters*, **87** (20), 203113/1–203113/3.

226 Sun, B. and Greenham, N.C. (2006) Improved efficiency of photovoltaics based on CdSe nanorods and poly(3-hexylthiophene) nanofibers. *Physical Chemistry Chemical Physics*, **8** (30), 3557–3560.

227 Michinobu, T., Nakanishi, T., Hill, J.P., Funahashi, M. and Ariga, K. (2006) Room Temperature Liquid Fullerenes: An Uncommon Morphology of C60 Derivatives. *Journal of the American Chemical Society*, **128** (132), 10384–10385.

228 Nakanishi, T., Schmitt, W., Michinobu, T., Kurth, D.G. and Ariga, K. (2005) Hierarchical supramolecular fullerene architectures with controlled dimensionality. (48). 5982–5984.

229 Nakanishi, T., Miyashita, N., Michinobu, T., Wakayama, Y., Tsuruoka, T., Ariga, K. and Kurth, D.G. (2006) Perfectly Straight Nanowires of Fullerenes Bearing Long Alkyl Chains on Graphite. *Journal of the American Chemical Society*, **128** (19), 6328–6329.

230 Martens, T., D'Haen, J., Munters, T., Beelen, Z., Goris, L., Manca, J., D'Olieslaeger, M., Vanderzande, D., De Schepper, L. and Andriessen, R. (2003) Disclosure of the nanostructure of MDMO-PPV:PCBM bulk hetero-junction organic solar cells by a combination of SPM and TEM. *Synthetic Metals* **138** (1–2), 243–247.

231 Hoppe, H., Glatzel, T., Niggemann, M., Schwinger, W., Schaeffler, F., Hinsch, A., Lux-Steiner, M.Ch. and Sariciftci, N.S. (2006)Efficiency limiting morphological factors of MDMO-PPV:PCBM plastic solar cells. *Thin Solid Films*, **511–512**, 587–592.

232 Mårdalen, J., Samuelsen, E.J., Konestabo, O.R., Hanfland, M. and Lorenzen, M. (1998)Conducting polymers under pressure: synchrotron x-ray determined structure and structure related properties of two forms of poly(octylthiophene). *Journal of Physics: Condensed Matter* **10** (32), 7145–7154.

233 Laird, D. Plextronics NREL Certified.

2
Fluorene-Containing Polymers for Solar Cell Applications

David Jones

2.1
Introduction

Energy generation by polymer-based photovoltaics has made significant gains in recent years [1] with the best bulk heterojunctions (BHJs) currently achieving power conversion efficiencies (PCEs) of up to 5.2% with poly-3-hexylthiophene: [6,6]-phenyl-C$_{61}$ butyric acid methyl ester (P3HT:PCBM) blends [2]. The concept of using donor–acceptor dyads in the formation of internal donor–acceptor heterojunctions is relatively new but significant advances have been made in understanding the origin of the photogenerated current and factors that influence the final device performance [3,4]. Devices may be formed from small molecules, polymers, or combinations of both as indicated with PCBM [5]. The best polymer/polymer bulk heterojunctions have power conversion efficiencies of up to 1.5% [6–8], which still lags well behind the efficiencies of substituted fullerene-containing bulk heterojunctions at 5.2%. Polymers, however, hold potential advantages over small molecules in device assembly, as they allow cheaper processing routes that avoid costly high-vacuum deposition, for example, inkjet printing.

The development of both solid-state dye-sensitized solar cells (DSSCs) and bulk heterojunctions is being pursued and similar problems are seen in both fields [9]. Significant development is required to improve the open-circuit voltage (V_{oc}) and short-circuit current (J_{sc}) with an improved fill factor (FF). Much progress has been made in the last few years. Recent reviews have examined the broad developments and highlighted the areas needing more attention; these can be listed as (i) low-bandgap materials to harvest more of the red and infrared portions of the solar spectrum; (ii) controlled morphology to generate optimal interfacial area for exciton dissociation while maintaining a bicontinuous structure allowing percolation of the separated charges to the collecting electrodes; (iii) new materials with higher charge mobilities; (iv) ability to tailor the absolute HOMO and LUMO levels to match the donor with acceptor to maximize device efficiencies; and (v) develop a better understanding of device architecture to optimize light harvesting, electron or hole blocking, and optical spacer layers and minimize device fabrication costs [9–11].

Two measures of device performance are the power conversion efficiency or energy conversion efficiency (ECE) and the external quantum efficiency (EQE) or incident photon to current efficiency (IPCE). The PCE or ECE measures the overall conversion into power of photon flux incident on a device, that is, the overall photon to current efficiency for all photons falling on a device (Equation 2.1). The EQE or IPCE is a measure of the photon to electron conversion efficiency at a particular irradiation wavelength (Equation 2.2). Device PCEs or EQEs, where available, will be quoted in this chapter [12].

$$\text{PCE or ECE} = \text{FF} \times J_{sc} \times \frac{V_{oc}}{P_{in}}, \text{ where fill factor (FF)} = \frac{J_m V_m}{J_{sc} V_{oc}}, \quad (2.1)$$

$$\text{IPCE or EQE} = 1240 \frac{J_{sc}}{\lambda_i P_{in}}. \quad (2.2)$$

This chapter examines the role of polyfluorene (Figure 2.1), and polymers containing a fluorene monomer unit, in the development of organic photovoltaics (OPVs) and their increasing importance in the development of new materials for improving device performance. Copolymers containing fluorene units have been used in bulk heterojunctions and dye-sensitized solar cells as both hole and electron transport materials. Blends incorporating fluorene polymers have been extensively used as models to study the effect of phase separation and thin-film morphology on device performance. Polymers or copolymers with high bandgaps have been omitted if they have not been examined in polymer solar cells, while potentially interesting low-bandgap polymers have been included where they may be of interest as red–IR absorbers.

Polyfluorenes are ubiquitous in the design of new materials for organic light-emitting diodes (OLEDs) [13] where the optical bandgap lies in the UV with electroluminescence in the blue. There has been significant investigation into the properties of fluorene polymers, especially to understand the decomposition mechanism that leads to green electroluminescence. It is, therefore, obvious that the optical bandgap for pure dioctyl-polyfluorene (PFO or F8), with an absorption onset at 2.95 eV (421 nm) and peak maximum at 3.25 eV (382 nm), lies in the UV [14] and is too high for efficient solar collection where the maximum solar flux is around 800 nm. However, the high nondispersive hole transport mobilities of polyfluorenes make them attractive as hole transport materials (HTMs or p-type materials) [15].

A simplified energy level diagram for a bulk heterojunction solar cell is shown in Figure 2.2. The maximum wavelength of light absorbed is the lower of the two bandgaps (HOMO–LUMO) for the donor or acceptor. To facilitate a comparison,

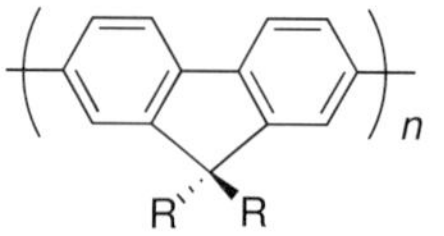

Figure 2.1 Generic structure of polydialkylfluorenes.

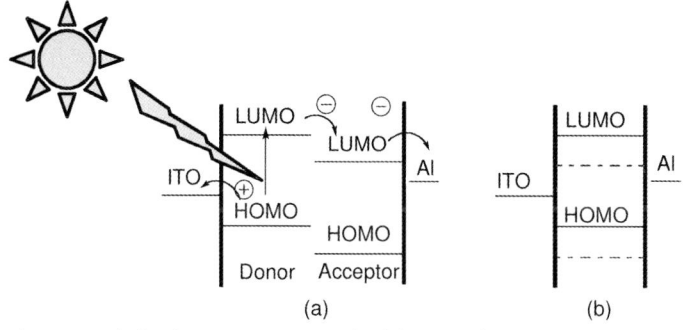

Figure 2.2 Idealized open-circuit energy level diagrams for (a) bilayer and (b) bulk heterojunction (acceptor, dashed lines) solar cells. Light absorption has been indicated to occur in the donor but may occur in the acceptor or both phases. Maximum V_{oc} is generally agreed to be the difference between the HOMO of the donor and the LUMO of the acceptor minus 0.3–0.5 eV required to drive charge separation at the donor–acceptor interface.

energy levels for the major materials discussed in this chapter have been collated in Table 2.1. It should be remembered that the numbers vary significantly between publications, especially when electronic bandgaps are used to estimate LUMO levels in the absence of electrochemical values (see Table 2.1) [16]. In a simple device, PFO would act as the donor or HTM, but the low HOMO level (approximately −5.8 eV) is lower than that for many electron transport materials (ETMs or n-type materials), which limits its use. Therefore, few solar cells with pure polyfluorene polymers have been examined directly.

What would be an ideal bulk heterojunction? An idealized single-layer device would absorb the majority of the available light in the solar spectrum down to around 950 nm or 1.3 eV. Assuming a driving force of 0.3–0.4 eV for charge separation, the maximum V_{oc} of a solar cell would then be 1.0 V. An upper limit for the donor material HOMO level of −5.2 eV (dashed line in Table 2.1) has been suggested to avoid air oxidation and facilitate material handling and device assembly [17]. Once a donor material with a 1.3 eV bandgap has been chosen, the LUMO of the acceptor is pinned to be 1.0 eV above the donor HOMO. Both the donor and the acceptor should have a high photoabsorption cross section and complementary absorption profiles to cover the entire solar spectrum. The material blends should phase separate on a 20–30 nm scale at RT to generate a bicontinuous interpenetrating network, delivering separated charges to the electrodes. Electrodes should provide ohmic contacts.

To this end, fluorene copolymers have been investigated extensively as potential materials for optimized polymer bulk heterojunctions.

2.1.1
Bulk Heterojunctions

Polymer/polymer bulk heterojunctions for organic photovoltaics were first described for CN-PPV and MEH-PPV (acceptor and donor, respectively) [4,18].

Table 2.1 HOMO–LUMO values reported in the literature for a range of fluorine-containing copolymers compared to relevant electrode materials, electroactive polymers, and molecular materials.

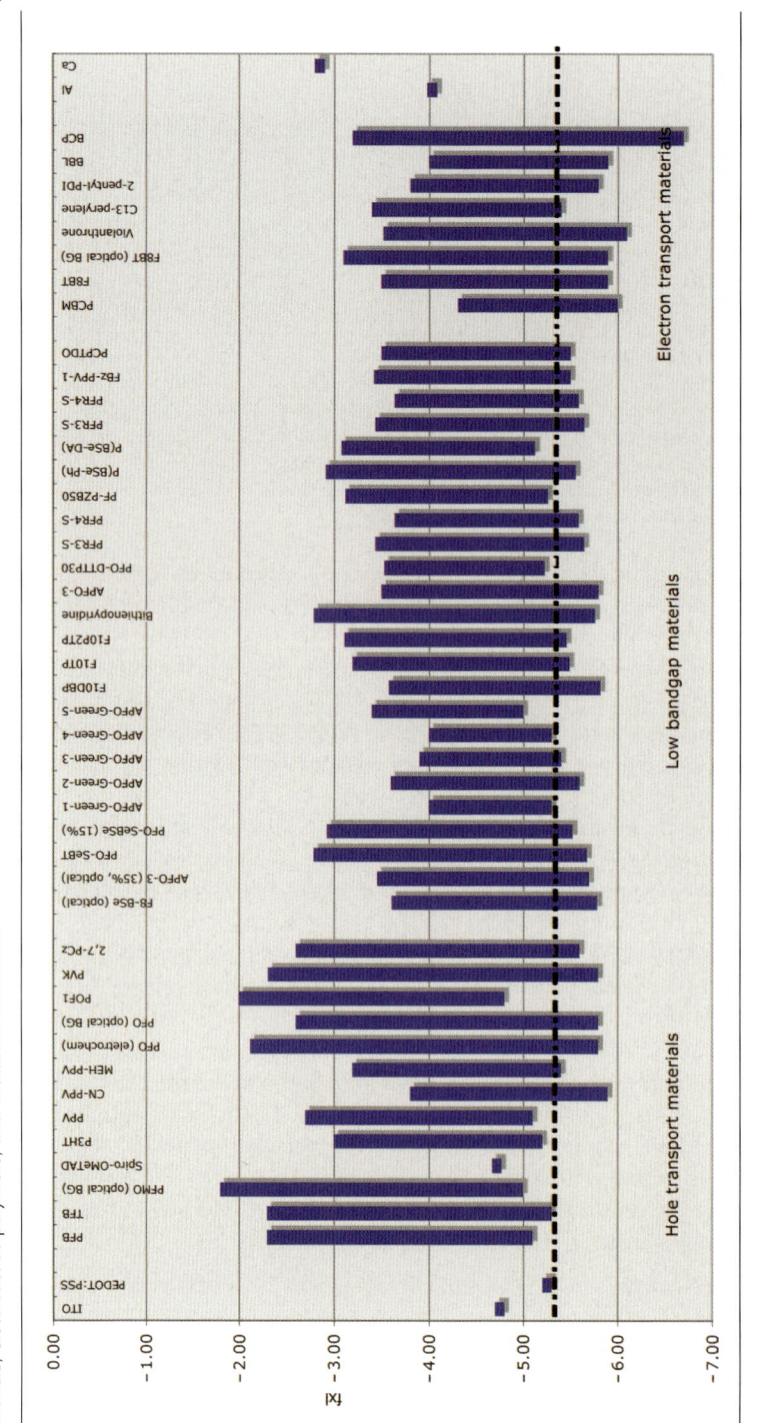

Abbreviations are as used in the main text.

Bulk heterojunctions were part of an attempt to increase the interfacial surface area between blend components and therefore enhance interfacial charge separation. Bulk heterojunction device efficiency has been found to be strongly influenced by thin-film morphology and, as theoretical studies have suggested, the best performance should be obtained when phase separation is of the order of the diffusion length of the exciton in the material and a bicontinuous network is formed allowing a continuous percolation of the separated species to the collecting electrodes [19].

This chapter covers polyfluorenes, copolymers containing fluorene monomers, and their use in photovoltaic devices. As significant practical and theoretical work has been completed on F8BT and PFB blends, these will be discussed in detail and reference made to other materials when reported.

2.2
Fluorene-Containing Materials

2.2.1
Polyfluorene-Containing Photovoltaics

For a polyfluorene to be used in an OLED as an HTM, the HOMO of the acceptor must be lower than that of the polyfluorene used (Table 2.1). Suitable electron transport materials (acceptors, n-type materials), where the HOMO is lower than that of polyfluorene, are poly(benzimidazobenzophenanthroline) ladder (BBL) (Figure 2.3), violanthrone (Figure 2.4), and PCBM. The photoabsorption cross section of PCBM is low and devices would be expected to be of very low efficiency while BBL, added as a gallium chloride adduct, is difficult to handle giving only bilayer devices [6].

Devices have been made by forming blends of PF2/6 (2-ethylhexyl-substituted polyfluorene) with violanthrone (Figure 2.4) with external quantum efficiencies of up to 2.5% measured at 610 nm or 3.0% measured at 410 nm for a 4:1 blend of violanthrone:PF2/6 after thermal annealing [20]. To broaden the spectral response of the device, a ternary blend was then examined with mixtures including a phenylene vinylene (PPV) derivative, Covion Super Yellow®. Device efficiencies under monochromatic irradiation depended upon post-film formation annealing; however, the improvement in EQEs after annealing was not as dramatic as for the binary blend reaching 1.6% under irradiation at 400 nm. Overall power conversion efficiencies for the devices were not reported.

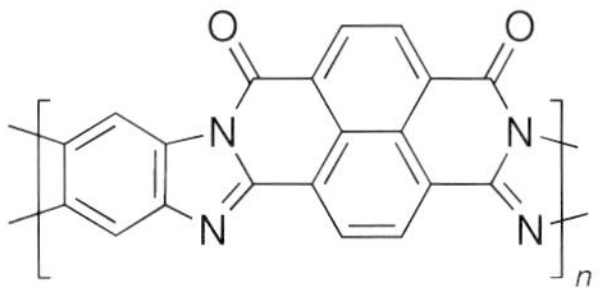

Figure 2.3 Chemical structure of BBL, an excellent electron transport material.

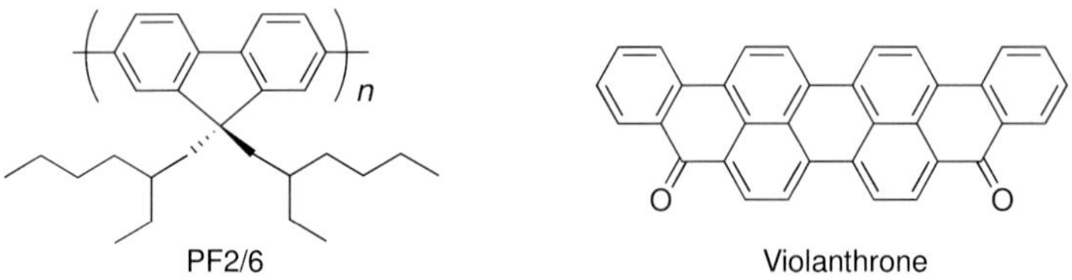

Figure 2.4 Structures of PF2/6 and violanthrone.

Photodiodes constructed from a PPV and a perylenemonodicarboxiimide (PI) substituted PFO (F8/FPI) have been examined as a method to enhance light harvesting by employing resonance energy transfer between the PFO and the PI red dye. EQEs approaching 7% in both blue and green regions of the spectrum were measured and were a significant enhancement compared to devices prepared without pendant PI groups [21]. Overall device efficiencies were not reported, but the poor conversions have been explained by radical anion trapping, as insufficient charge carriers were present. An examination of Table 2.1 indicates that the lowest energy point in the system would be the PI and insufficient concentrations of the dye would not lead to an efficient percolation network to carry electrons to the metal electrode.

When electron-deficient phenyloxadiazoles were included as fluorene side chains to enhance charge separation and electron transport, device efficiencies (PCEs) increased from $\eta_e = 0.00027\%$ for poly(9,9-dioctyl)fluorene (PFO) to $\eta_e = 0.0013\%$ for phenyloxadiazole side chain substituted PFO (Figure 2.6) [22].

The role of pure polyfluorenes in bulk heterojunctions is clearly limited; however, copolymers containing the fluorene unit play an essential role in the generation of hole or electron transport materials, especially in the synthesis of low-bandgap polymer sensitizers for bulk heterojunction solar cells [23].

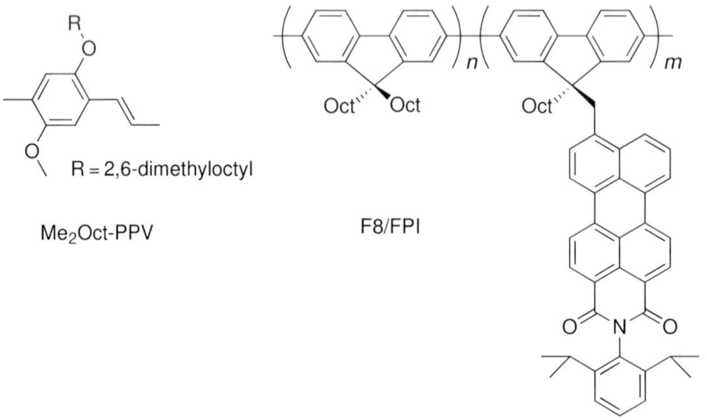

Figure 2.5 Perylenemonodicarboxiimide tethered from PFO used as an acceptor in Me$_2$OctPPV:F8/FPI blends.

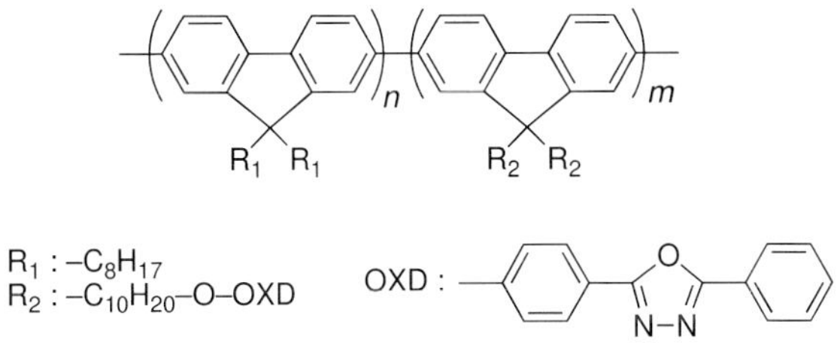

[Figure 2.6 structure]

R$_1$: –C$_8$H$_{17}$
R$_2$: –C$_{10}$H$_{20}$–O–OXD

OXD : [phenyloxadiazole structure with N–N]

Figure 2.6 Pendant phenyloxadiazole-substituted polyfluorenes used in OPVs. The oxadiazole unit was introduced to aid charge separation and electron transport ($n = m = 0.5$).

2.2.2
Polyfluorene Copolymers

2.2.2.1 Electron Transport Materials

There are a few reported electron transport materials generally used in bulk heterojunction photovoltaic devices [8], including the polymeric BBL discussed above and a few molecular materials such as PCBM or phthalocyanins. As the LUMO for the acceptor is pinned to the HOMO of the donor, this places a requirement on the electron transport material to have a relatively good electron affinity, that is, low-lying LUMO. The most frequently examined polymeric electron transport material in organic photovoltaics is the alternating copolymer of benzothiadiazole with dioctylfluorene, F8BT (Figure 2.7).

The introduction of electron-deficient comonomers, such as a benzothiadiazole (Figure 2.7), into fluorene copolymers increases the ability to reduce the polymer, making it a better electron acceptor. This reduces polymer bandgaps to give a better overlap of the solar spectrum. Electron and hole transport properties of F8BT have been measured and reported to be dispersive at $\mu_e = 2 \times 10^{-3}\,\mathrm{cm^2\,V^{-1}\,s^{-1}}$ and $\mu_h = 1 \times 10^{-3}\,\mathrm{cm^2\,V^{-1}\,s^{-1}}$ [24,25]. Theoretical modeling indicates that the HOMO level of the copolymer remains almost unaffected by the incorporation of the comonomer, while the LUMO energy reduces by around 1.1 eV [26]. The HOMO is predominantly located on the conjugated backbone while the LUMO is benzothiadiazole based with a strong heteroatom component (a key finding in later attempts to change polymer bandgaps). The first singlet excited state (with high oscillator strength) is therefore equivalent to a charge transfer to the BT unit, 2.26 eV (550 nm), while the second singlet excited state is equivalent to a π–π* transition on PFO, 3.39 eV (366 nm) [27]. The highly localized electron densities in the LUMO on BT

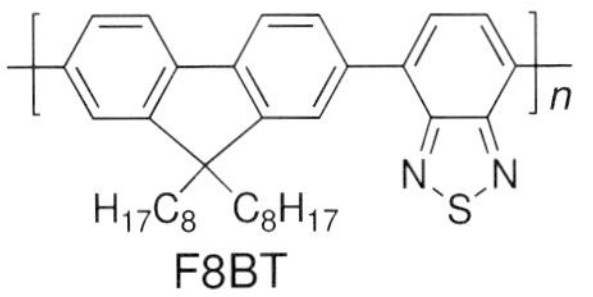

[F8BT structure with H$_{17}$C$_8$ C$_8$H$_{17}$ substituents and N-S-N benzothiadiazole unit]

F8BT

Figure 2.7 Structure of poly-2,7-(9,9'-dioctyl-9H-fluorene)-alt-4,7-benzo[c][1,2,5]-thiadiazole, F8BT.

units would be expected to decease the electron mobility in F8BT but mobilities of the order of $\approx 10^{-3}$ cm^2 V^{-1} s^{-1} have been recorded [25]. Electron mobility is dispersive and this figure can be regarded as that of the fastest electrons.

2.2.2.2 Hole Transport Materials

Poly(9,9-dioctyl)fluorene has a nondispersive hole transport mobility of 4×10^{-4} cm^2 V^{-1} s^{-1}, which makes it a very attractive material for photoactive devices [15]. Theoretical modeling has indicated that HOMO–LUMO levels can be significantly modified by inclusion of comonomers containing heteroatoms. The HOMO level of the resulting alternating copolymers remains relatively unchanged if the acceptor retains a benzene type conjugation, as in 5,8-disubstituted quinoxaline or 4,7-disubstituted benzo[c][1,2,5]thiadiazoles, while the LUMO can be varied. Inclusion of comonomers, such as thiophene [28] or 2,5-disubstituted-1,3,4-thiadiazoles, with heteroatoms in the linking ring can significantly affect both HOMO and LUMO levels [29].

One of the best molecular HTMs is spiro-OMeTAD [30,31], and polymeric versions have been synthesized by the inclusion of a triarylamine functionality into PFO. The inclusion of triarylamine comonomers raised the low level of HOMO of PFO to a level matching that of the ITO anode (Table 2.1).

Some typical triarylamine-containing copolymers used in bulk heterojunctions are shown in Figure 2.8. Hole transport in poly(2,7-dioctyl-9H-fluorenyl-alt-4,4'-(N-4-butylphenyl)-diphenylamine) TFB has been shown to be nondispersive with mobilities up to $\mu_h = 10^{-2}$ cm^2 V^{-1} s^{-1} [32]. The analogous poly(2,7-dioctyl-9H-fluorenyl-alt-4,4'-(N^1,N^4-bis(4-butylpheny))-diphenylbenzene-1,4-diamine) PFB has been shown to yield the best solar cell device efficiencies in polymer blends with F8BT [33]. Hole transport in PFB is nondispersive and mobilities have been measured at $\mu_h = 1.6 \times 10^{-2}$ cm^2 V^{-1} s^{-1} compared to PFO with a mobility of $\mu_h = 4.9 \times 10^{-2}$ cm^2 V^{-1} s^{-1} [34]. Redox properties of the polymer can be adjusted

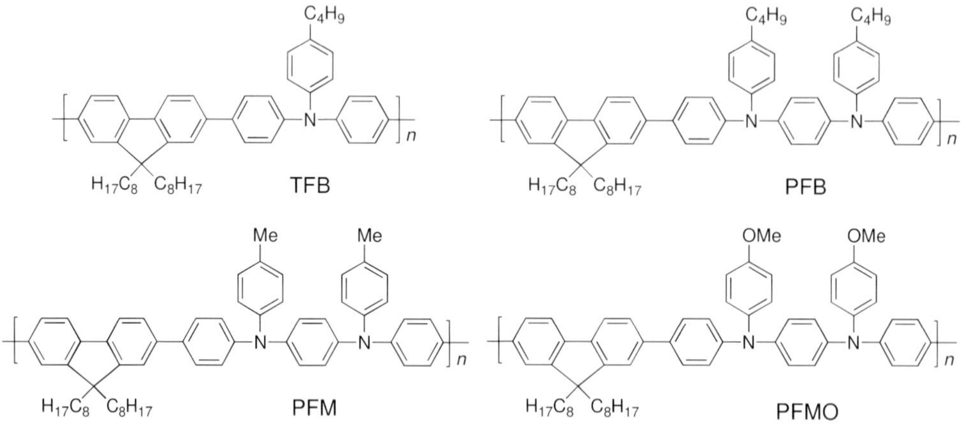

Figure 2.8 Hole transport copolymers of fluorene-containing triarylamine units. The abbreviations are those commonly used for these polymers.

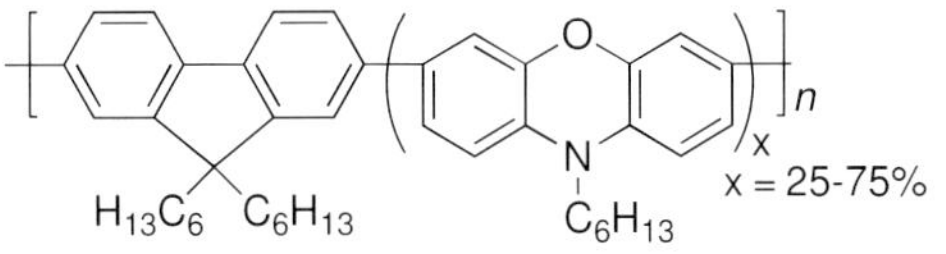

Figure 2.9 Phenoxazine-containing copolymers of fluorene, POF1.

by varying the arylamine substituents. The replacement of butyl in PFB with a methoxy substituent generates PFMO, which has redox properties similar to spiro-OMeTAD (Figure 2.12), the best molecular hole transport material used in solar cells. PFMO also has a nondispersive hole transport with mobilities reported at $\mu_h = 5 \times 10^{-4}\,cm^2\,V^{-1}\,s^{-1}$ [35].

Electrochemical studies of copolymers formed from dihexylfluorene and hexylphenoxazine show no reduction potential ranging from 0 to $-3.0\,eV$ versus SCE, indicating that they are inherent p-type materials (Figure 2.9) [36]. Field effect charge carrier mobilities (μ_h) of these polymers are of the order of $<6.0 \times 10^{-4}\,cm^2\,V^{-1}\,s^{-1}$; however, photovoltaic devices have not been reported with these materials.

Copolymerization of F8 with dithiophene (Figure 2.10) generated a polymer with hole transport mobilities of $\mu_h = 10^{-4}\,cm^2\,V^{-1}\,s^{-1}$, but transport was highly dispersive [37]. Therefore, polymers containing triphenylamine units have higher hole transport mobilities, which were nondispersive.

2.2.3
Devices

While examining organic solar cell devices in this chapter, we examine the nature of active materials but not electrodes, interfaces, or blocking layers even though these are an important part in device construction [38].

2.2.3.1 Bulk Heterojunctions

If we consider the polymer combinations PFB:F8BT and P3HT:F8BT, we can examine the possibility of forming photovoltaic devices. Using the information in Table 2.1, we generate open-circuit energy level diagrams for both combinations of materials (Figure 2.11) using either PFB or P3HT as the hole transport material. Both combinations indicate good material matches, with an estimated V_{oc} for a PFB:F8BT blend (Figure 2.11a) estimated by $|E_{LUMO}\text{-}ETM| - |E_{HOMO}\text{-}HTM|$ or $3.50 - 5.10\,V = -1.6\,eV$ where there is at least a 0.3 eV offset between the LUMOs of the HTM and the ETM to drive charge separation. The measured V_{oc} value for a bilayer

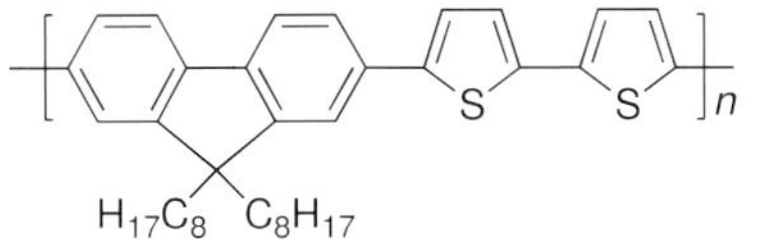

Figure 2.10 Dithiophene-containing copolymers of fluorene used as hole transport materials, F8DT.

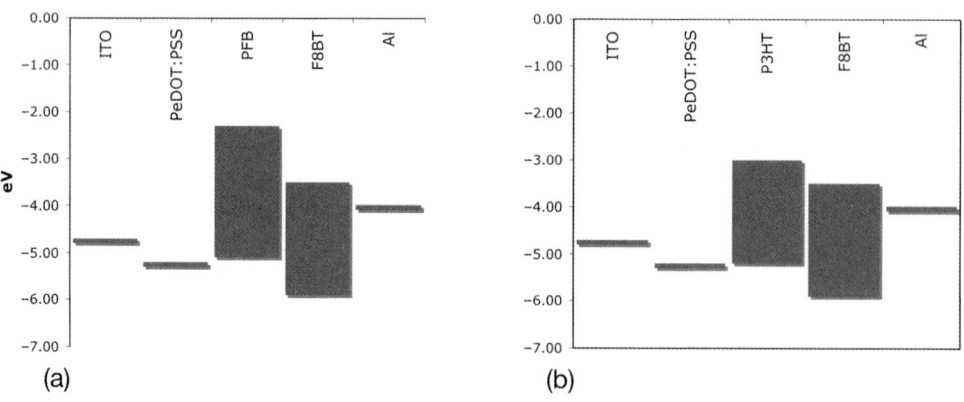

Figure 2.11 Energy level diagrams for combinations of materials with data taken from Table 2.1: (a) indicates the formation of a device incorporating PFB as a HTM, while (b) indicates the inclusion of poly-3-hexylthiophene (P3HT).

device is 1.5 V [39]. For device b (Figure 2.11) the expected V_{oc} is 1.8 V while the measured value for a blend is 1.0 V [24]. From a simple examination of the material energy levels, these devices may, under optimized conditions, be expected to give good overall efficiencies, as the HTMs and ETMs are well matched; however, maximum external quantum efficiencies for devices a and b are around 4% in blend films (a value up to 24% has been reported at 3.6 eV for an PFB/F8BT bilayer device) [40].

Measured EQEs in optimized P3HT/PCBM blend devices are generally above 50% over the range of 450–600 nm [41]. The poor device performance for blends containing F8BT as the ETM, even though the materials are well matched energetically and good fluorescence quenching is reported, indicates a problem with charge transport or device morphology with poor percolation networks unable to deliver the separated charges to the electrodes for collection.

For the P3HT:F8BT blend, the overall power conversion efficiency was reported at only 0.15% under optimized conditions, which in this case was reportedly attributable to the low electron transport mobility in the F8BT.

2.2.3.2 Dye-Sensitized Solar Cells

An alternative form of a modern photovoltaic device is the dye-sensitized solar cell, where a dye acts as a photosensitizer injecting photoexcited electrons into a mesoporous nanostructured titanium dioxide [30]. The dye is reduced by hole transfer to a liquid ionic electrolyte containing I^-/I_3^-, which is then reduced at the metal cathode. Device efficiencies for DSSCs have been reported to be up to 10.2%, but for a long-term use the liquid electrolyte needs to be hermetically sealed within the cell.

To overcome the problems of hermetically sealing the cell to contain the liquid electrolyte, there has been an intense search for "solid" hole transport

Figure 2.12 Molecular hole transport materials, spiro-OMeTAD and an analogue containing a pseudocrown ether substituted diarylamine substituted fluorene.

materials as a replacement [31,42]. Notable advances have been made with the use of the molecular HTM spiro-OMeTAD (Figure 2.12) in conjunction with amphiphilic dyes. Device efficiencies up to 4.0% have been reported. A number of additives, such as LiN(SiMe$_3$)$_2$ and *tert*-butylpyridine, were required in these "solid-state" devices to optimize device efficiency by modifying the electric fields and to control charge separation and recombination at the interface [43]. Introduction of a pseudocrown ether ion-solvating functionality, either on the ruthenium dye [44] or on the HTM (Figure 2.12) [45], has allowed the simplification of film formation by reducing the number of additives while still retaining an amorphous film [45,46].

The development of polymeric HTMs as replacements for spiro-OMeTAD (Figure 2.13), with ion-solvating functional groups, allowed for the formation of thin layers without the need for secondary additives such as pyridine and the formation of multiplayer devices with reported efficiencies up to 0.8% [46].

The reported HOMO level (−4.9 eV) for polymer **B** (R = −OMe, no lithium) was similar to that of spiro-OMeTAD (−4.8 eV). As the oxidation potential remains

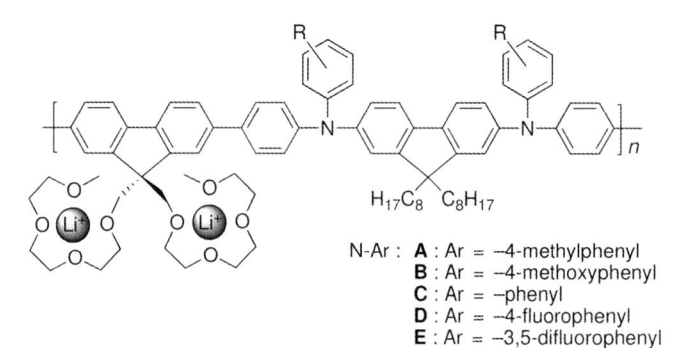

N-Ar : **A** : Ar = −4-methylphenyl
B : Ar = −4-methoxyphenyl
C : Ar = −phenyl
D : Ar = −4-fluorophenyl
E : Ar = −3,5-difluorophenyl

Figure 2.13 Polymeric triarylamine-substituted polyfluorenes with alternative pseudocrown ether substitution used as ion-solvating polymers for solid DSSCs.

relatively unchanged by the substitution of an alkyl group with a tetraethylene glycol group, the substitution of the MeO group on the aminophenyl group has allowed control of the polymer redox potential, leading to a redox cascade to be formed increasing the interfacial charge recombination times; however, device efficiencies have not been reported [47].

2.3
Bulk Heterojunction Device Performance

Device performance in bulk heterojunctions is related to a number of physical processes:

- photoabsorption efficiency to generate excitons;
- migration rates of the excitons to an interface where charge separation can occur;
- efficiency of charge separation;
- mobility of the charges through the n- and p-type materials;
- formation of a bicontinuous interpenetrating network to deliver the charges to the electrodes.

Photoluminescence quenching can give a quick indication of the first three properties where complete quenching is desired and tends to indicate that excited states form and charge separate before decaying nonradiatively. Where photoluminescence quenching is seen, the poor device performance is then related to device morphology, with poor collection of the separated charges or poor charge transport through the materials. Spectroscopic studies can provide a deeper understanding of the absorption and charge separation processes, but understanding and controlling morphology development is more difficult.

2.3.1
Morphology

Understanding and controlling device morphology is central to maximizing device efficiencies, and polymer architecture has been controlled in other systems by deposition techniques and post-film formation annealing [48]. It has been shown that charge generation is good in bulk heterojunction blend films formed from F8BT and PFB, but poor charge transport through films (related to morphology) leads to poor device performance [49]. Theoretical modeling for polymer blends has indicated potential directions to idealized morphologies. A maximum is found if device efficiency is plotted against interfacial area. Well-formed thin films must have phase separation that allows good percolation networks to enable charge transport through the thin film and sufficient interfacial surface to allow good charge separation [50]. Phase separation for F8BT:PFB blend films of the order of 20 nm has been suggested as delivering maximized device efficiencies.

Owing to the relative energy levels of the HOMO and LUMO in PFB and F8BT blends, they are ideally set up to form solar cells with charge separation at the

2.3 Bulk Heterojunction Device Performance

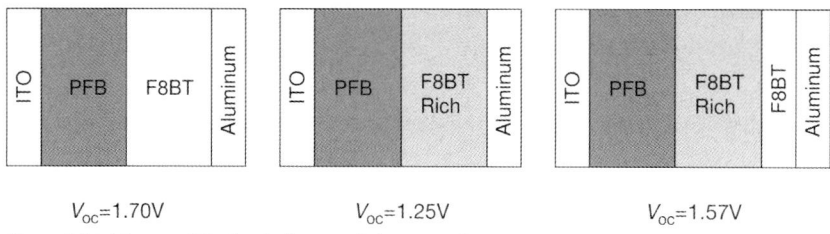

V_{oc}=1.70V V_{oc}=1.25V V_{oc}=1.57V

Figure 2.14 Measured V_{oc} levels for varied device configurations indicating the need for a hole blocking layer between the PFB-rich layer and the aluminum electrode.

interface [4,39], and EQEs of up to 22% have been recorded in blends at 335 nm. They have been extensively studied in bulk heterojunctions using a variety of techniques and are ideally suited to illustrate the principles of blend composition, morphology, and available experimental techniques used to investigate device performance. Phase separation depends on the conditions of film formation, and will not be covered here in detail, but is affected by molecular weights, viscosities, component relative solubilities, solvents, and spin speed or relative rate of solvent evaporation [51]. Equally, device V_{oc} depends on the device structure. The reported V_{oc} value of simple bilayer device, for example, ITO|PFB|F8BT|Al, of 1.7 V is close to the expected maximum of |LUMO-acceptor| − |HOMO-donor|, or $3.5 - 5.3 = -1.8$ eV (Figure 2.14). For blends where there are direct paths from both phases to the electrodes, there are short circuits reducing the V_{oc}, for example, an Al|F8BT:PFB|PFB|ITO device with a reported V_{oc} of 1.25 V. Introduction of a blocking layer, which effectively reduces short circuits, improves V_{oc}, for example, an ITO|PFB|F8BT:PFB|F8BT|Al device where the reported $V_{oc} = 1.57$ V [52].

If device performance is determined by thin-film morphology, a detailed understanding of the film is required. Detailed XPS studies of spin-cast TFB:F8BT films (TFB is similar to PFB) have indicated a complex phase morphology with a TFB wetting layer on the ITO glass electrode and TFB and F8BT regions forming micron-sized domains separate from the wetting layer (Figure 2.15) [53]. The F8BT domains are partially capped by TFB-rich domains [54]. Many different techniques have been applied in an attempt to better understand film morphology. A brief survey of these methods is detailed below.

2.3.1.1 Techniques for Probing Thin-Film Morphology

Friend and coworkers have extensively examined phase separation in F8BT:TFB or F8BT:PFB (Figure 2.16) films used in OLED or photovoltaic devices [33]. Phase

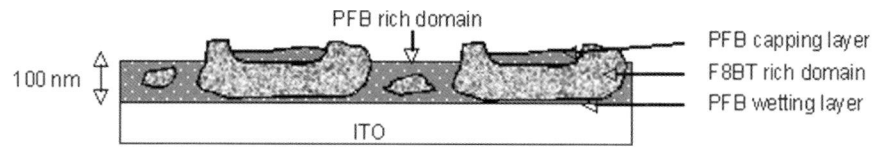

Figure 2.15 Proposed model of the PFB:F8BT blend thin film.

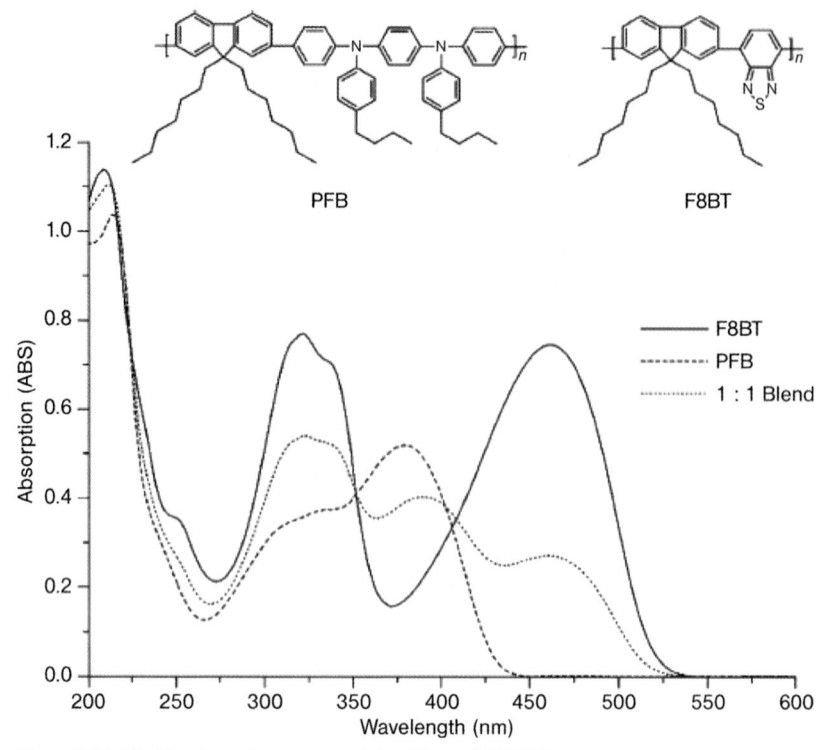

Figure 2.16 UV–Vis absorption spectra of thin films of F8BT:PFB (1:1 blend, solid line), PFB (dotted line), and F8BT (dashed line). Structures of PFB and F8BT are shown (from Ref. [60]).

separation during thin-film formation is ultimately driven by the different solubilities of the individual polymers in the solvent used. Therefore, film formation can strongly depend on the volatility of the solvent or substrate temperature and hence the rate of film formation. In fact, bilayer devices have been formed from blend mixtures with large solubility differences and where solvent evaporation is slow allowing almost complete phase segregation [40]. Blends of F8BT:TFB formed by spin casting from xylene solution, where the solubility of the F8BT is significantly lower than TFB, have been examined by AFM [53,55]. Morphology mapping indicated lateral phase separation on a 0.1–5.0 µm scale, depending on blend composition, vertical phase separation of up to 50 nm for thin films up to 100 nm depth. For 50:50 blends, lateral phase separation is 2.0–5.0 mm with vertical depth profiles varying by up to 50 nm. Photoluminescence images of the thin-film 50:50 blends under blue excitation show luminescence only from the F8BT phase, which is mapped to the high regions identified on the AFM image. In this case, electroluminescence images indicate the most intense emission from the phase boundaries.

Raman spectroscopy offers the possibility of phase composition mapping [56] owing to large Raman scattering cross section for phonon modes in individual conjugated polymers that are coupled to the π–π^* transitions and the fact that the probe can be focused down to around 0.5 µm for $\lambda = 633$ nm. Phase composition of 50 : 50 F8BT:PFB blends mapped by Raman spectroscopy indicates F8BT-rich (high) phases with around 80% F8BT, while the PFB-rich (low) phases are 50% PFB. That is, the PFB-rich phases have more F8BT than the F8BT have PFB. These figures represent spatial averages across large spot sizes and through thin films.

2.3.1.2 Relating Film Morphology to Device Performance

The ability to correlate morphology and phase separation to device performance remains a challenge and new techniques need to be developed to probe blend film structure/performance down to the nanometer range. There has, however, been an increased use of techniques that can map charge generation, fluorescence or photovoltaic performance against blend film structure. But most of these techniques suffer from resolution problems below the 100 nm level. Almost all of the techniques described below will need dramatic improvement to be able to map new structures on a 20 nm scale, where better device performance is accompanied by generation of interpenetrating percolation networks as predicted by theory.

Mapping the photovoltaic performance onto film morphology and relating this to blend composition, as determined by Raman spectroscopy, has indicated a drop in EQE with increased phase separation and the EQEs can also be related to the interfacial surface area (mm2.mm^{-2}), with higher EQEs resulting from higher interfacial surface area [50,57].

Thin-film morphology changes with blend composition have been mapped onto device performance using fluorescence quenching. In these studies, charge transport was the limiting process to improved device performance [57]. Blend ratios from 1 : 500 to 500 : 1 of PFB to F8BT were examined with the best EQEs reported for a blend ratio of 1 : 5 PFB to F8BT, while the highest photoluminescence quenching yield was reported at a ratio of 5 : 1 PFB to F8BT. Phase separation occurs on the 0.1–1.0 µm range. As the best device EQEs occur where photoluminescence quenching is low, charge transport efficiency was suggested as a limiting factor in these devices as a result of poor percolation network formation.

It is important to note that the nanoscale morphology of the F8BT- or TFB-rich phase in these blends remains unknown and, therefore, it is difficult to fully relate device performance with the observed large-scale morphology.

Coupling AFM with other detection systems has led to a series of powerful mapping techniques adding significant information to our understanding of thin-film formation and, therefore, photovoltaic performance. Sub-100 nm lateral resolution can be obtained in scanning Kelvin probe microscopy (SKPM). SKPM uses a noncontact AFM tip with a conductive coating to measure the local surface potential. Surface potential mapping of an F8BT:PFB thin-film blend on ITO is shown in Figure 2.17 [58], with films formed by direct spin coating (Figure 2.17a–c) and in a saturated solvent atmosphere (Figure 2.17d–f), that is, slower evaporation and therefore larger domain sizes. Figure 2.17a and d represents AFM surface topography images indicating the general micron scale features of PFB:F8BT thin films. Figure 2.17b and e indicates surface potential mapping for a dark film, while Figure 2.17c and f indicates surface potential mapping under film illumination at

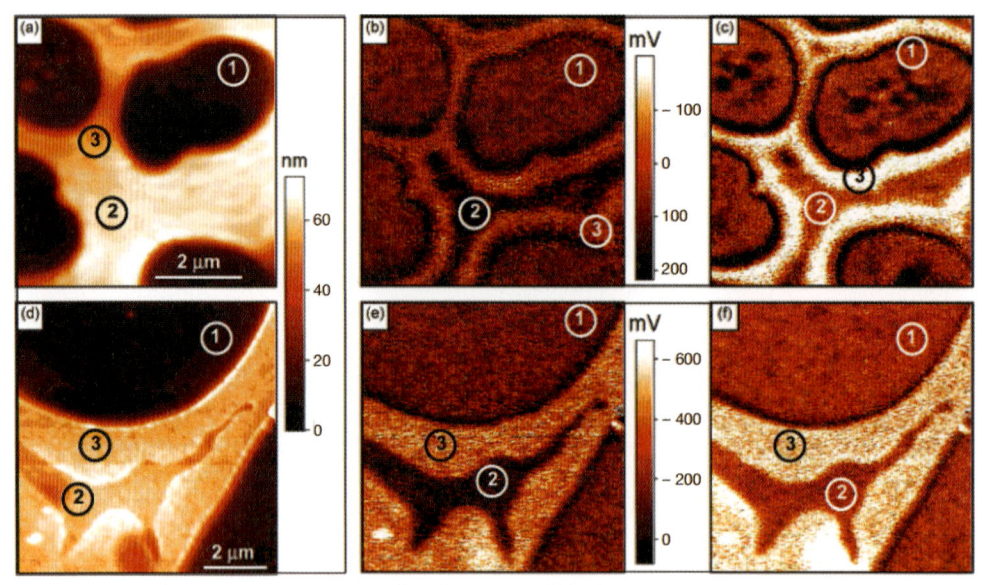

Figure 2.17 SKPM images of thin-film PFB:F8BT blends on an ITO anode prepared by spin coating (a–c) and in a saturated solvent atmosphere (d–f): topography (a, d) and surface potential measured in the dark (b, e) and under illumination (c, f). Domains marked 1, 2, and 3 are, respectively, PFB-rich, bulk, and interfacial F8BT-rich regions (from Ref. [58]).

473 nm. There is expected to be no residual charge in the dark mapped film and the surface potential should be constant, the variation observed indicates some residual charge failing to recombine. Surface potential mapping under illumination indicates the three main domains with the interfacial F8BT-rich region possessing the higher negative charge. The potential map has been correlated with the film morphology indicated in Figure 2.17 and device efficiency losses are explained by the presence of the capping layer reducing effective contact of the electron transport material with the cathode.

In a similar technique, the surface electrostatic charge buildup has been mapped by electrostatic force microscopy (EFM) [59]. This allows charging rates to be related to film morphology and measured EQEs. The mapping indicated that most charge generation is within the main phases rather than the boundaries and correlates with results from near-field scanning optical microscopy (NSOM) results discussed below.

The previous techniques were able to map a region or indicate domains within bulk heterojunction thin films, but they were unable to directly map morphology with current generation under illumination. NSOM measurements are possible and directly map morphological and optical properties [51,60,61]. Using this technique, the sample is scanned by a tapered optical fiber (tip 50–500 nm) and shear force or AFM feedback modes are used to keep the tip within 10 nm of the surface. When the photocurrent generated by the optical scanning is measured, the technique is called near-field scanning photocurrent microscopy (NSPM) and has been used to map

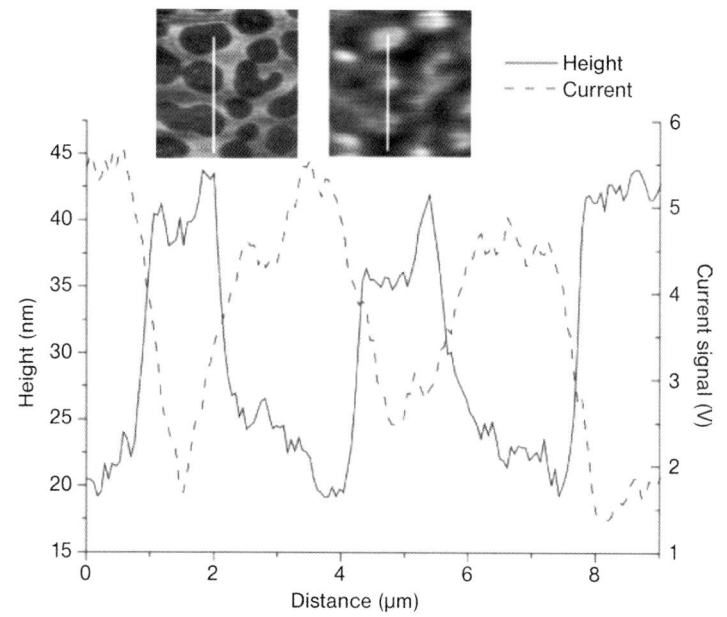

Figure 2.18 Cross-sectional traces of an AFM image (left inset; black low, white high) and near-field scanning photocurrent map (right inset; black low, white high) indicating a negative correlation of height with photogenerated current for a TFB:F8BT blend film (from Ref. [60]).

morphology against photocurrent generation in PFB:F8BT (1:1) thin films, with illumination at 409 nm via a transparent cathode [60,62]. Typical results generated from such a study are depicted in Figure 2.18 and show a direct correlation of topography with generated photocurrent, the higher domains generate lower currents while the lower domain generate higher current values. As the domains have been identified in earlier studies, it is possible to relate photocurrent to composition. For PFB:F8BT thin films, the higher photocurrents are associated with PFB-rich regions, accounting for up to 80% of the observed photocurrents (Figure 2.18). The differences observed in the two major domains have been ascribed to the microstructure of the domains where phase separation on the 10–20 nm scale, or a scale comparable with the exciton diffusion length, dominates. There is more F8BT in the PFB-rich regions than there is PFB in the F8BT-rich regions; this leads to a better charge separation in the PFB-rich regions and, hence, a higher photocurrent in this better mixed domain.

Vertical phase morphology in TFB:F8BT blend films for OLED devices has also been probed by examining thin films with UV–Vis spectroscopy after successive oxygen plasma etching to reveal successive layers [54]. It has been found that TFB-rich domains are effectively homogeneous throughout their depth, containing about 20–30% of F8BT. The F8BT domains are, as indicated above, more complex with a TFB wetting layer on the PEDOT:PSS and often a capping layer of TFB above

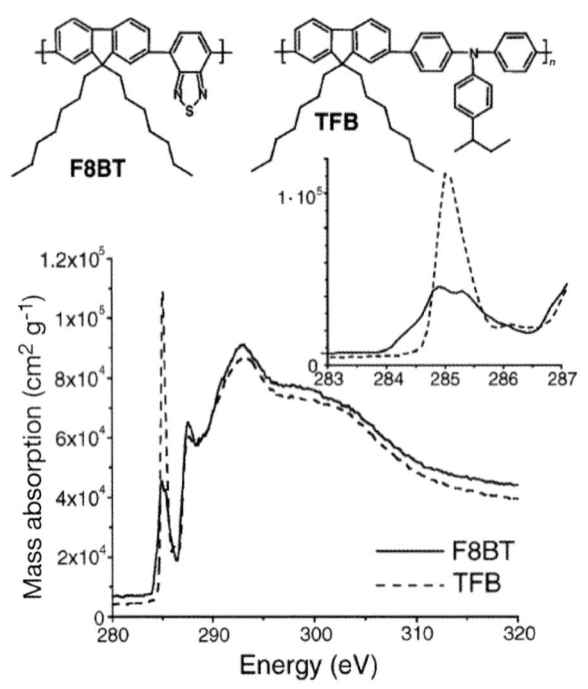

Figure 2.19 NEXAFS spectra of TFB and F8BT. The inset NEXAFS spectra show an enlargement of the region where the π^* transitions are found (from Ref. [63]).

the predominantly F8BT (60–70%) domain. The model of the thin-film structure is shown in Figure 2.15.

Scanning transmission X-ray microscopy (STXM) coupled with near-edge absorption fine structure (NEXAFS) has been used to map phase domains and estimate the composition and film thickness of F8BT/TFB blends [63]. This powerful technique allows film scanning and composition mapping to a resolution of around 50 nm. Comparison of AFM and STXM images shows similar phase domains of 1–5 µm, while the knowledge gained from NEXAFS measurements on pristine films of the component materials allows compositional mapping from analysis of the differential absorption of characteristic π^* transitions (Figure 2.19). Measurements indicate a phase interface of around 200 nm (independent of film thickness) and a TFB concentration of around 80% in TFB-rich domains, while the composition of F8BT domains is more complicated with around 60% F8BT in the center of F8BT-rich domains with concentrations of up to 90% in the phase interface (Figure 2.20).

Secondary ion mass spectroscopy (SIMS) has been used to profile the layer structure of blend films indicating vertical phase separation in APFO-3:PCBM films (see Figure 2.22) and specific variations with substrate, silica, gold, or PEDOT:PSS on ITO [64].

The vertical analysis of phase-separated regions in blend films remains problematic with results depending on averaged compositions obtained by examining

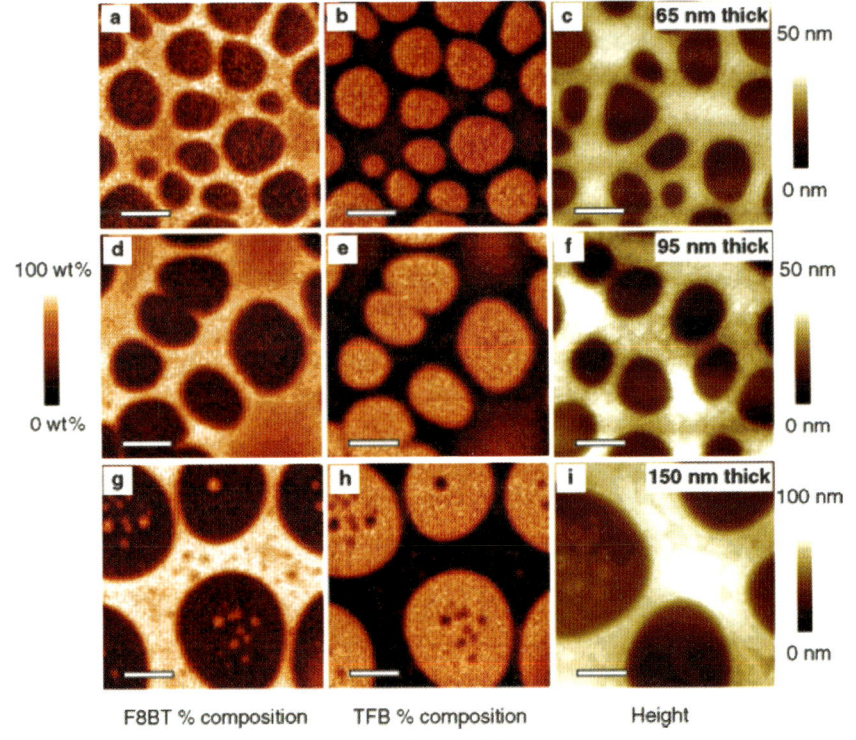

Figure 2.20 5 nm × 5 nm STXM compositional maps (left and center columns) and AFM images (right column) of F8BT:TFB blend films. The left column corresponds to the F8BT wt% composition maps, while the center column corresponds to the TFB wt% composition maps: (a), (b), and (c) are of a 65 nm thick film; (d), (e), and (f) are of a 95 nm thick film; (g), (h), and (i) are of a 150 nm thick film. The scale bar is 1 nm (from Ref. [63]).

the whole blend as in NEXAFS, or slow removal of successive layers and analysis of the exposed surface by plasma etching [54] or SIMS. Although vertical separation has been indicated to be important in OLEDs [65], mapping the variations is difficult. Environmental scanning electron microscopy (ESEM) has been shown to be of some use in differentiating phases in thin-film cross sections, but more work is required [66]. Both environmental scanning microscopy in transmission mode [67] and quantitative secondary electron imaging [68] have been able to differentiate PFB/F8BT phases.

2.3.1.3 Inkjet Printing
Phase separation of polymer blends can also be effected by the method of deposition. EQEs of F8BT:PFB thin films deposited on ITO via inkjet printing at 40 °C are consistently higher than spin-coated films increasing from 4 to 9% when irradiated at 400 nm [69]. This has been explained by the control of phase separation with the

features of greater than 1 μm in spin-cast films down to <300 nm with inkjet-printed films. EQEs of spin-cast F8BT:TFB films are negligible but measurements on inkjet-printed films are significant (7.5% at 375 nm) indicating the ability to fundamentally improve material performance by morphology control. Inkjet printing increases the surface roughness of deposited films owing to the "coffee stain effect," whereby the droplets on drying leave a crater shaped drop.

2.3.1.4 Microemulsions of Blends

An alternative route to intimate mixing and nanoscale phase separation is to introduce the material as blends formed as aqueous dispersions of particles from microemulsions [70]. The intimate mixing found in the individual 49–53 nm diameter carried through to the annealed film and EQEs up to 1.5% were comparable with previous thin-film studies. The approach may be important where multilayer devices are needed and where layers with similar solubilities need to be sequentially formed. The use of an aqueous microemulsion of the upper layer material may allow bilayer formation where this would not be possible otherwise.

In a further step, the aqueous dispersions of the polymer nanoparticles have been electroplated from solution to form a thin film that, surprisingly, was insoluble in chloroform [71]. Device performance was, however, not improved, with EQEs of only 0.35%.

2.4
Low-Bandgap Materials

The most successful combination of materials recorded to date for organic solar cells is the combination of P3HT and [60]PCBM, with device efficiencies over 5% achieved with post-film formation annealing [2,72]. The P3HT acts as the hole transport material, while the PCBM acts as the electron transport material owing to the high electron affinity of the C_{60}. In the search for low-bandgap materials to optimize the solar collection efficiency of PCBM-containing blends, two parameters have been suggested: LUMO level of the HTM at 0.3–0.5 eV above the LUMO level of PCBM (−4.3 eV) and a bandgap of 1.5–1.8 eV [11].

These parameters have been defined for materials to be used in conjunction with PCBM and has led to the development of a number of polymers with many being copolymers with fluorene units. The polymers act as hole transport materials in place of the pure P3HT. It is interesting to note that many of these new low-bandgap polymers have energy levels that make them attractive as electron transport materials, but few mobility measurements have been reported on these new copolymers.

2.4.1
New Low-Bandgap Materials

Alternating polyfluorene copolymers (APFOs) have been extensively studied for the generation of low-bandgap polymers, especially for red OLEDS [73], where HOMO

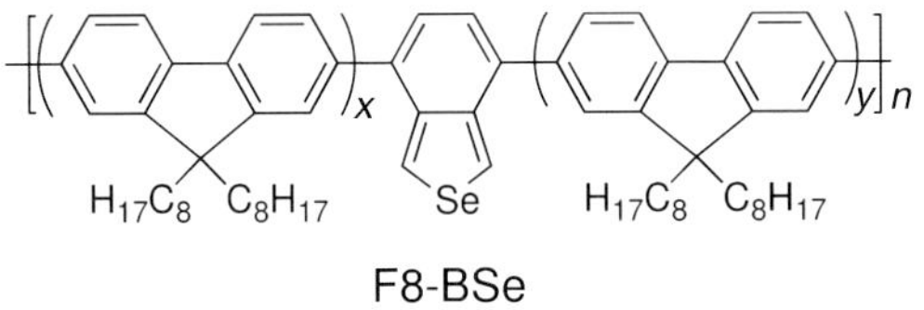

2.4 Low-Bandgap Materials

F8-BSe

Figure 2.21 Selenium analogue of F8BT, F8BSe. Polymers with increasing levels of BSe up to a 50:50 mix ($x + y = 1$) have been synthesized.

and LUMO levels can be systematically altered by manipulating an electron-rich comonomer [29].

Substitution of sulfur with selenium in F8BT, to generate the selenium analogue F8BSe, allows a direct comparison on polymer properties with heteroatom substitution. It has been suggested that the heavier heteroatom would allow lowering of the LUMO energy level [74]. The selenium analogue of F8BT, F8BSe ($x + y = 1$) (Figure 2.21), has been synthesized and used in polymeric OLEDs. The UV–Vis absorption onset for F8BSe is seen at ≈560 nm compared to 525 nm for F8BT (Figure 2.16), indicating the expected lowering of the LUMO on selenium substitution.

The inclusion of two thiophene or two selenophene units beside the BT or BSe in alternating copolymers has been examined.

The inclusion of the 4,7-di(thiophen-2-yl)benzo[c][1,2,5]thiadiazole unit (DBT) into PFO copolymer, PFO-DBT(X) ($X = 1-35\%$), has generated a series of polymers with DBT content up to 35% (Figure 2.22) [75]. Copolymers with low concentrations of DBT have been used for polymeric OLEDs. Inclusion of the thiophene units changes the UV–Vis absorption onset to ≈615 nm (2.01 eV) and results in a HOMO level of −5.4 eV. Photovoltaic device efficiencies for blend films incorporating PFO-DBT35 with PCBM 1:2 are up to ECE $\eta_e = 1.95\%$ [76]. For these systems, the J_{sc} increases linearly with illumination intensities up to 5 suns (500 mW cm^{-2}) opening up the possibility of solar cell use in conjunction with a solar heliostat.

The 50:50 alternating copolymer PFO-DBT50 ($x + y = 1$), also reported as APFO-3, has been extensively studied in blends with PCBM, delivering reported PCEs of up to 2.4% [77,78]. HOMO–LUMO levels in APFO-3 have been reported

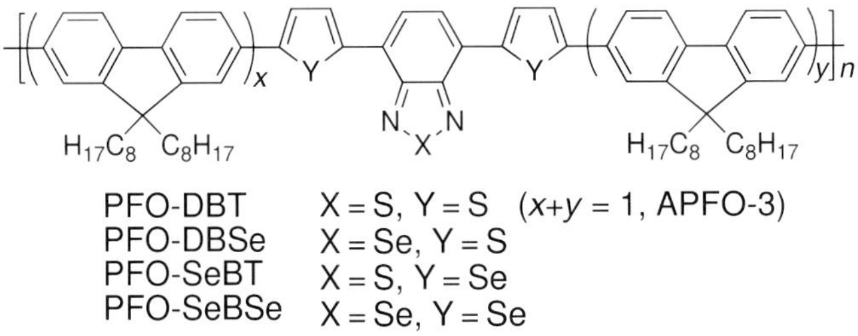

PFO-DBT   X = S, Y = S (x+y = 1, APFO-3)
PFO-DBSe   X = Se, Y = S
PFO-SeBT   X = S, Y = Se
PFO-SeBSe X = Se, Y = Se

Figure 2.22 Low-bandgap polymers based on F8BT, modified by selenium substitution of the sulfur or inclusion of two thiophene or selenophene units.

to be −5.8 and −3.5 eV, respectively, with an absorption onset at 650 nm [79], while hole transport mobilities have been reported to be $\eta_h = 8.8 \times 10^{-4}\,cm^2\,V^{-1}\,s^{-1}$ [80]. The type of photoexcitation in the blend has been examined, using continuous wave photoabsorption and transient absorption spectroscopic techniques, and indicate polaron formation showing dispersive recombination with lifetimes of about a millisecond and a second assigned as a coulombically bound intrachain polaron pair showing nondispersive recombination with a lifetime of about 20 ms [81,82]. The temperature dependence of APFO-3:PCBM devices has indicated a more robust device structure than PPV:PCBM blends with peak J_{sc} at 120 °C. J_{sc}, FF, and PCE all increase with temperature, while V_{oc} decreases [83]. To allow optical modeling of APFO-3:PCBM blend devices and thereby appropriate film thickness of the components to be deposited, spherical ellipsometry [84] has been used to determine the dielectric function of APFO-3 for wavelengths from the IR to UV [85]. An averaged value of $n \approx 2$ was determined for the wavelength range of interest. Modeling has allowed device behavior to be examined, allowing prediction of optimum material thin-film thickness and suggested reasons, and therefore solutions, to poor device performance. One suggested limiting factor in APFO-3:PCBM blends is low hole mobilities in the blend film.

The effect of fluorene alkyl group substitution has been examined by replacing the octyl groups in APFO-3 with dihexyl, didodecyl, or hexyl/ethylhexyl [86]. Only didodecyl substitution altered polymer absorption, causing a slight redshift to λ_{max} 570 nm from 550 nm. The overall device performance was lower than APFO-3, PCEs of 1.4 and 2.1%, owing to a lower J_{sc}, 2.4 mA cm^{-2}, compared to 3.55 mA cm^{-2} for devices formed from APFO-3.

Polymers with increasing ratios of selenophene–BT–selenophene (PFO-SeBT(X), X = 1, 5, 10, 15, 30, or 50%) or selenophene–BSe–selenophene (PFO-SeBSe(X), X = 1, 5, 10, or 30%) (Figure 2.22) have been examined as HTMs in blends with PCBM. Measured bandgap for PFO-SeBT1 was reported at 2.89 eV (optical 1.89 eV) with a UV–Vis absorption onset at ≈700 nm, while for the fully selenated polymer, PFO-SeBSe15, the bandgap was reported at 2.6 eV (optical 1.78 eV) with a UV–Vis absorption onset at ≈710 nm. The best photovoltaic device made from either of these materials in blends was with PFO-SeBT30 in a 1 : 4 blend with PCBM. Device ECE reached 1% under AM1.5 illumination (78.2 mW cm^{-2}) with $V_{oc} = 1.0$ V.

HOMO–LUMO levels for the thiophene–BSe–thiophene containing polymers have not been reported, but the UV–Vis absorption onset is at ≈710 nm and photovoltaic devices made with PFO-DBSe35 in blends with PCBM 1 : 3 achieved energy conversion efficiencies of up to 0.91% under AM1.5 100 mW cm^{-2} [87]. As with devices made with APFO-3, the ECE remains relatively constant, even up to 500 mW cm^{-2}.

Pyrrole-containing polymers have been examined as low-bandgap analogues of APFO-3 (Figure 2.23) [77]. The polymer, **P1**, had an absorption maximum of 504 nm higher than the thiophene analogues at 552 nm.

Inclusion of quinoxaline, phenazine, or thienopyrazine units (Figure 2.24), generates polymers with bandgaps in the range of 2.24–2.35 eV (or 1.76–2.37 eV optical) [88]. The variation in HOMO level between a phenylene type linker and

P1: R = hexyl, R' = 2-ethylhexyl

Figure 2.23 Low-bandgap polymers containing N-substituted pyrrole.

a thiophene type linker is seen with an increase in the HOMO level from −5.82 to −5.46 eV for F10DBP and F10TP, respectively. A bandgap of 2.24 eV is large compared to the target level of 1.5–1.8 eV and further modification is required.

Inclusion of a further two thiophene units on either side of the thienopyrazine has been examined by other researchers to generate a statistical copolymer, DFO-DTTP30 (30 mol% dithenylthiazopyrazine unit, R = Me) (Figure 2.26) [89]. The bandgap has been measured at 1.69 eV and the change is mainly the result of increasing the HOMO level to −5.22 from −5.46 eV in F10TP. Photovoltaic devices made with blends of DFO-DTTP30 and PCBM reached a maximum η_e of 0.83%.

The search for an "ideal" [11] polymer to match the electronic properties of PCBM has lead to the isolation of a series of dithienyl-pyrazino[4,7,c]benzothiadiazole-containing polymers, APFO-Green-1–3, where the polymers now absorb down

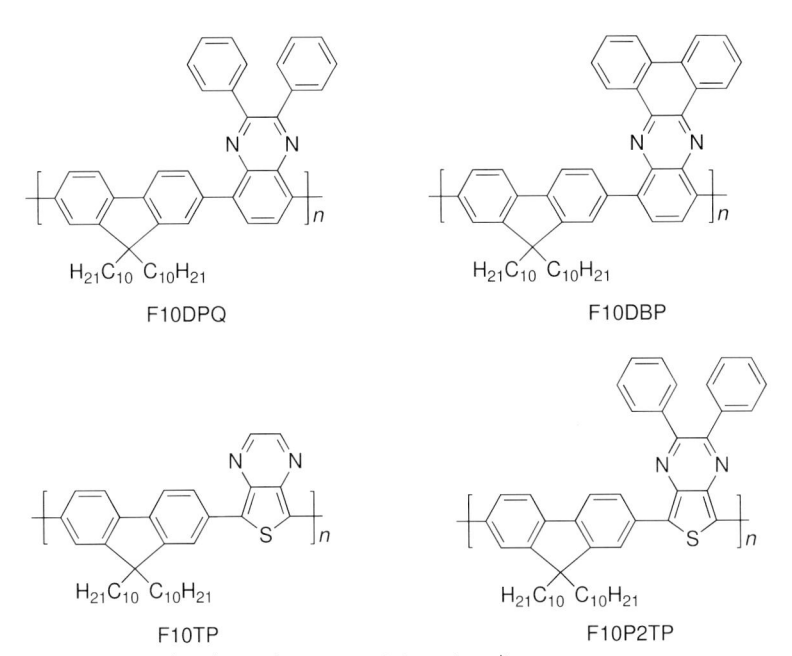

Figure 2.24 Low-bandgap polymers containing quinoxaline, phenazine, and thienopyrazine functionalities; copoly (2,3-diphenylquinoxaline-*alt*-didecylfluorene) F10DPQ, copoly (dibenzo[a,c]phenazine-*alt*-9,9-didecylfluorene) F10DBP, copoly (thieno[3,4-b]pyrazine-*alt*-9,9-didecylfluorene) F10TP, and its 2,3-diphenyl analogue F10P2TP.

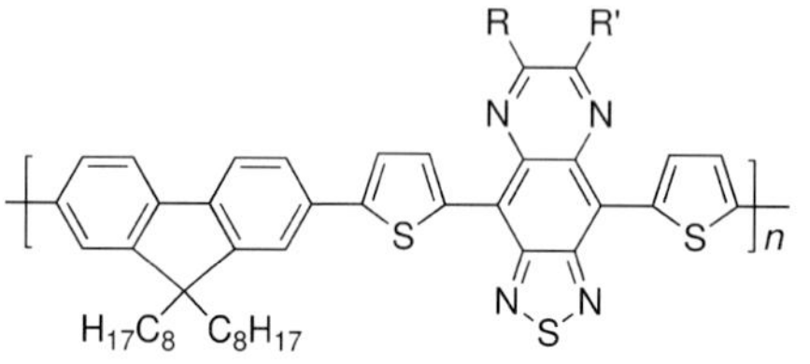

APFO-Green–1 : R = R' = phenyl
APFO-Green–3 : R = 2-ethylheptyl
R' = hexyl
APFO-Green–4 : R = R' = 4-(2-ethylhexyloxy)phenyl

Figure 2.25 Low-bandgap polymers designed as hole transport materials for PCBM–polymer blends.

to >1 μm (Figure 2.25). However, the lower optical bandgap (1.3 eV) and LUMO (4.0 eV) for APFO-Green-1 resulted in the requirement of an alternative electron acceptor to PCBM, as the relative difference in LUMO levels was found to be insufficient to drive charge separation, PCBM being replaced by a C_{60} analogue, BTPF [90], or a C_{70} analogue [79,91]. Power conversion efficiencies for blend devices (η_e) up to 0.3% were reported with highest EQEs of 25% for devices reported at 400 nm. One reason for the poor device performance was that the low absorption of APFO-Green-1 in the 500–700 nm region of the spectrum. The low-bandgap polymers, 1.21 eV estimated from an absorption onset at 1020 nm, have high hole mobilities, measured after annealing at 195 °C for polymers of molecular weight $M_n = 3600$, at up to 0.03 cm^2 V^{-1} s^{-1} [92].

APFO-Green dyes 3 and 4 have bandgaps of 1.5 and 1.3 eV, respectively, and perform well with the C_{70} analogue of BTPF [90], BTPF70 [93], with best device efficiencies up to 0.6%. EQEs of 5% at 900 nm indicate significant absorption of these dyes in the near-IR [79,93–96].

Inclusion of thiophene units on either side of a thieno[3,4-b]pyrazine unit (cf. F10TP in Figure 2.24) resulted in the formation of a new series of alternating copolymers with interesting optoelectronic properties (Figure 2.26) [96]. When DFO-DTTP30 was tested in a photovoltaic device, a PCE of 0.8% was recorded in a 1 : 1

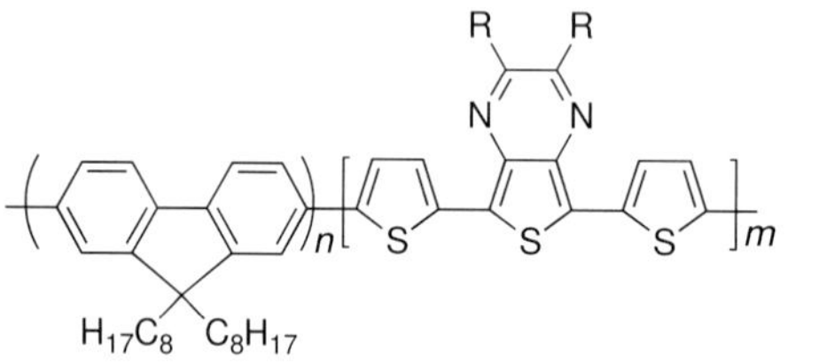

DFO-DTTP30 : n = 0.7, m = 0.3, R = Me
APFO-Green-2 : n = m = 1, R = Ph
APFO-Green-5 : n = m = 1, R = 4-(2-ethylhexyloxy)phenyl)

Figure 2.26 Low-bandgap polymers incorporating a dithiophenylthienopyrazine unit DFO-DTTP30 (30 mol% dithienylthienopyrazine unit), APFO-Green-2, and APFO-Green-5.

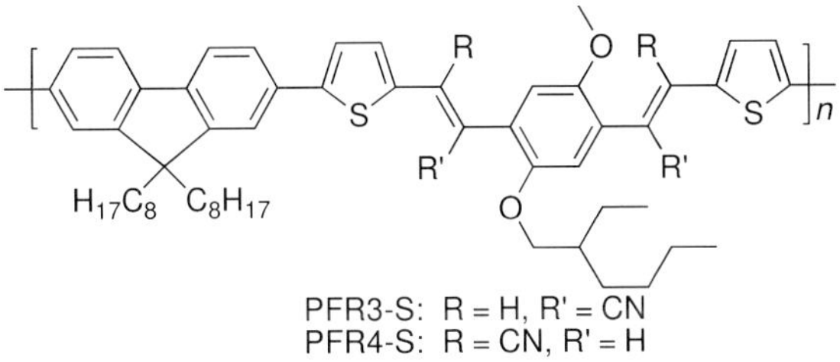

2.4 Low-Bandgap Materials | 81

Figure 2.27 Formation of low-bandgap polymers by the incorporation of a cyano-substituted bisthienylphenyl-enedivinylene unit into an alternating block copolymer with F8.

blend with PCBM. The absorption onset was at ≈700 nm with λ_{max} at 581 nm. The alternating copolymers ($m = n = 1$) have been labeled APFO-Green-2 (R = Ph) and APFO-Green-5 (R = 4-(2-ethylhexyloxy)phenyl) [97]. APFO-Green-2 has an optical bandgap onset of 850 nm when mixed with PCBM but has a low device efficiency, max PCE of 0.96%, owing to poor EQEs for wavelengths >750 nm. APFO-Green-5 (R = R′ = 4-(2-ethylhexyloxy)phenyl) [95] has a better solubility and absorption than APFO-Green-2 and when mixed with PCBM, with maximum device efficiencies up to 2.20%. FFs decrease with increasing film thickness indicating charge transport limitations. These results are notable, however, as they represent the best device efficiencies to date for low-bandgap polymers in blends with PCBM.

Functional solar cells have been synthesized by combining MEH-PPV as the donor with cyano-MEH-PPV as the acceptor, utilizing the high charge mobilities of PPV type materials and extensive knowledge of the effects of the cyano substitution on lowering the LUMO with little effect on the HOMO, thereby lowering the optical bandgap. Copolymers containing fluorene, thiophene, and cyano-PPV monomers, PFR3-S and PFR4-S (Figure 2.27) [98], have been tested in bulk heterojunctions with C_{60} (1 : 1) and initial results indicate device efficiencies of up to $\eta_e = 0.98$ and 1.02% for PFR3-S and PFR4-S, respectively [99]. In these devices, PFR3-S or PFR4-S was acting as an HTM, while they have also been reported as an ETM in a BHJ with MDMO-PPV [7]. Blend devices of PFR3-S with MDMO-PPV (1 : 1 blend) were shown to have PCEs of up to 1.5%. EQEs were as high as 50% at 540 nm with an absorption onset at around 600 nm.

PF-PZB50 ($x = 50\%$) has been used as an electron donor material in blends for photovoltaics with PCBM (1 : 3), achieving a maximum efficiency of $\eta_e = 0.53\%$. With an optical bandgap of 2.14 eV and LUMO level of −3.21 eV, this material lies significantly above the ideal bandgap for a device material [100] (Figure 2.28).

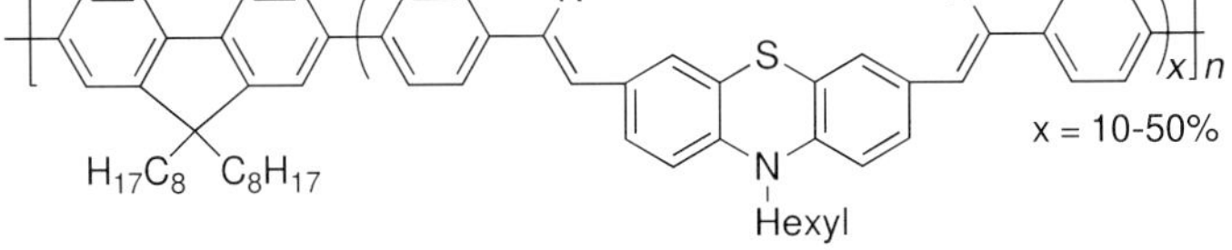

Figure 2.28 Low-bandgap HTM for photovoltaics including a divinyl-phenothiazine unit, PF-PZBx.

Figure 2.29 New benzylidene-fluorene based low-bandgap polymers with useful low-lying LOMO levels.

A number of materials have been synthesized but not yet examined in bulk heterojunction solar cells. The low-bandgap polymers formed by copolymerizing a phenylene divinylene unit with a benzylidene-fluorene generates potentially useful polymers with bandgaps of 2.29–2.08 eV and LUMO level at −3.30 and −3.42 eV (Figure 2.29) [101]. FBz-PPV-1 has an absorption onset of ≈650 nm. The low-lying HOMO–LUMO levels for these polymers allow them to be potentially matched to a number of ETMs in place of F8BT. As with many of the above low-bandgap polymers, there is little information on the hole and electron transport mobilities for these new polymers.

2.4.2
Alternative Structures

2.4.2.1 Carbazoles

The substitution of the substituted methylene in fluorenes with substituted amines generates a range of carbazole analogues to the fluorenes. Polycarbazoles or carbazole-containing molecules are intrinsic hole transport materials that have been examined in OLED devices owing to their oxidative stability (Figure 2.30) [102,103]. Polyvinyl-carbazol (PVK) or directly linked poly-3,6-carbazols (PCz) have high bandgaps, similar to PFO (Table 2.1), and therefore by themselves are not ideally suited for use in solar cells; however, some notable exceptions have recently been reported.

PVK has been used as an HTM in DSSCs resulting in overall power conversion efficiencies of up to 2.4% [104]. These results are notable when compared to be the best optimized devices with spiro-OMeTAD at 3.2% [31,42].

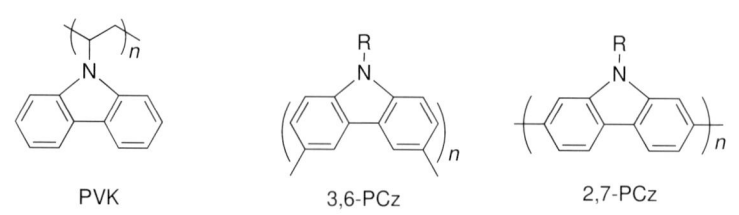

Figure 2.30 Polycarbazoles: nonconjugatively linked PVK (poly-N-vinylcarbazole) and conjugatively linked 3,6- and 2,7-polycarbazoles.

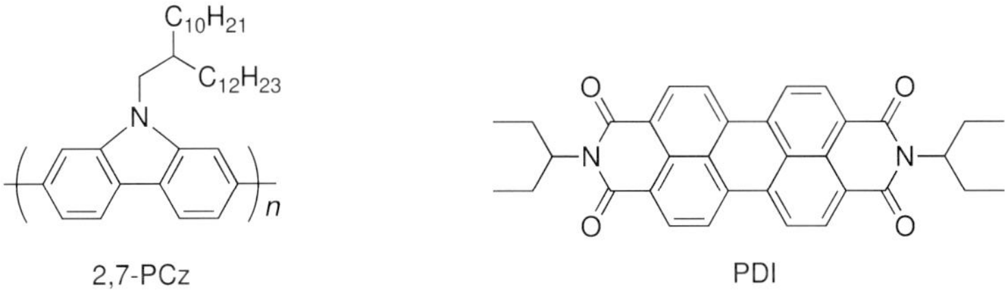

Figure 2.31 Efficient donor–acceptor pair for photovoltaic devices based on 2,7-polycarbazoles (2,7-PCz) with di-2-pentyl-substituted perylenetetracarbodiimides (2-pentyl-PDI).

Carbazole-containing bulk heterojunctions were showed performance until recently when electron transport materials were coupled with electron transport materials, which are highly absorbing in the visible [105]. Bulk heterojunctions formed from blends of 2,7-PCz (R = 2-decyltetradecyl to give the polymer solubility) with 2-pentyl-PDI (Figure 2.31) show the absorption onset at ≈600 nm and EQE values up to 15% between 480 and 580 nm [106]. Overall power conversion efficiencies vary with thin-film morphology reaching 0.63% under AM 1.5 global illumination. Poor thin-film morphology resulting in poor charge separation was thought responsible for the low PCEs.

Copolymers of 2,7-carbazolenevinylene with thiophene-containing units have been examined as a route to low-bandgap materials for blending with PCBM (Figure 2.32) [107]. Increasing the number of thiophene units in the comonomer (not all shown) changed the copolymer absorption onset from 500 to ≈725 nm for PCPTDO. The polymer HOMO levels showed no significant variation with HOMO (PCV) = −5.6 eV and HOMO(PCPTDO) = −5.5 eV, while the LUMO levels dropped

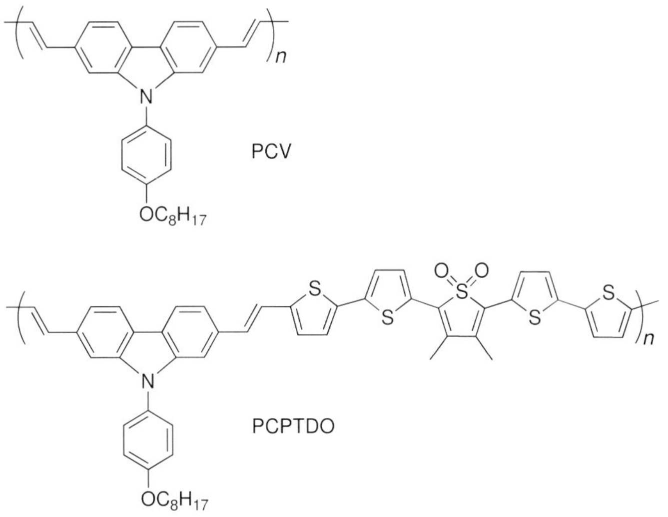

Figure 2.32 Polymers containing the 2,7-carbazolenevinylene monomer as low-bandgap materials.

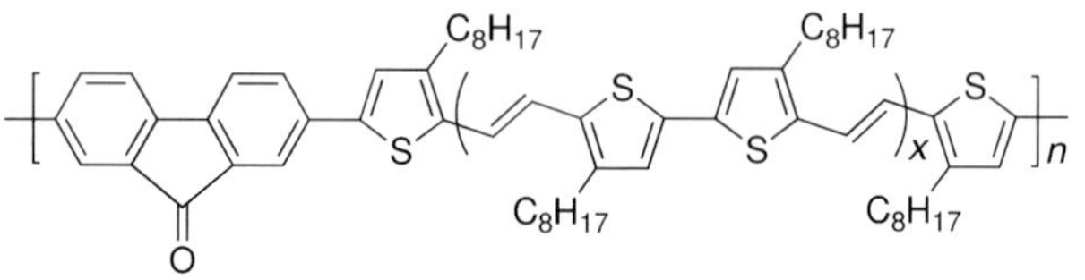

Figure 2.33 Fluorenone-containing copolymer with conjugated thiophene-containing units.

from −2.8 to −3.5 eV. Power conversion efficiencies for 1:4 blends of PVC or PCPTDO with PCBM were 0.4 or 0.8%, respectively.

2.4.2.2 Fluorenones
Fluorenone substitution for fluorene (Figure 2.33) has been shown to be effective in active solar cell formation with cell efficiencies reaching 1.1% [108]. Optical bandgaps of the copolymers with thiophene units were down to 1.52 eV (PTVF). EQE values were up to 20% at 400 nm.

2.5
Future Directions

A number of reports on the theoretical limits on device performance for bulk heterojunctions have recently appeared. Design parameters have been suggested to direct the development of new materials, allowing devices to potentially reach theoretical limits for thin-film solar cells [109]. It has been indicated that the new materials need to be polymers of high intrinsic charge mobility and HOMO and LUMO levels of the donor need to be pinned relative to the LUMO of the acceptor. That is, the LUMO of the donor needs to be ≤0.3 eV above the LUMO of the acceptor to allow the required driving force for charge separation, while an optical bandgap of 1.3–1.5 eV would allow a closer match of the absorption profile with the solar spectrum [11,110].

Calculations on a number of systems indicate that the HOMO lies extensively on the conjugated polymer, while the LUMO can be very localized on electron-rich comonomers, the nature of the substituents or heteroatoms can have a significant impact on the absolute position of the LUMO, and the substitution of heteroatoms from the second or third row can significantly lower LUMO levels [103].

2.5.1
Controlled Morphology

Common to many reports and used to explain poor device performance is the lack of the ability to control morphologies, thereby hindering good charge separation or collection. Morphology has been controlled by varying basic parameters used in spin coating, film casting or annealing. Spin coating from different solvents gave different morphologies; for example, thin film of MDMO-PPV:PCBM (1:4) spin cast from toluene has a variation in surface height of up to 12 nm, while the equivalent film spin cast from chlorobenzene shows surface height variations of <3 nm [111]. Recently, similar results have been demonstrated for APFO-3:PCBM (1:3) films, where spin coating from a chloroform:chlorobenzene (80:1) led to smoother films and better device efficiency compared to chloroform alone [82]. The use of toluene or xylene mixtures led to decreased device performance. Improved performance was associated with finer morphology with increased EQEs, while larger domain sizes (greater surface roughness) and lower performance were observed for toluene and xylene solvent mixtures. Although V_{oc} was not significantly altered in the devices, a significant variation in J_{sc} was recorded, from 3.2 to 4.6 mA cm^{-2}, leading to a slight increase in PCE, from 2.1 to 2.3%.

Post-film deposition annealing has also been shown to directly affect device performance for some material composition, for example, P3HT:PCBM, while evaporation rates have been shown to affect device performance in similar cells where it has been suggested that slow evaporation allows time for self-organization, increased charge mobilities, and better efficiencies [112]. Ultimately, however, the speed of film formation required for reel-to-reel processing dictates that extended annealing times are simply not suitable and contact times of only seconds are optimal. Controlled morphologies must be self-generating and much work needs to be done.

The morphological control therefore requires a better understanding of phase separation in the materials used or an access to new polymer architectures, which may be available by the use of conjugated block copolymers, rod–coil block copolymers, or the use of supramolecular self-assembly.

Conjugated triblock copolymers have not yet been reported for fluorene-containing polymer blocks as either donor or acceptor material. Conjugated triblock copolymers have recently been reported for P3HT:cyano-PPV:P3HT systems (donor:acceptor:donor) [113].

Quadruple hydrogen bonded self-assembled polymers using ureidopyrimidinones capped fluorene oligomers have been demonstrated, giving polymers of high molecular weights. (UPy-F6$_n$-UPy)$_x$, $n = 1, 3, 5,$ or 7 (Figure 2.34). Energy transfer experiments after endcapping the polymers with acceptors has indicated good transfer, Up-PV = 5%. However, devices have not yet been examined [114].

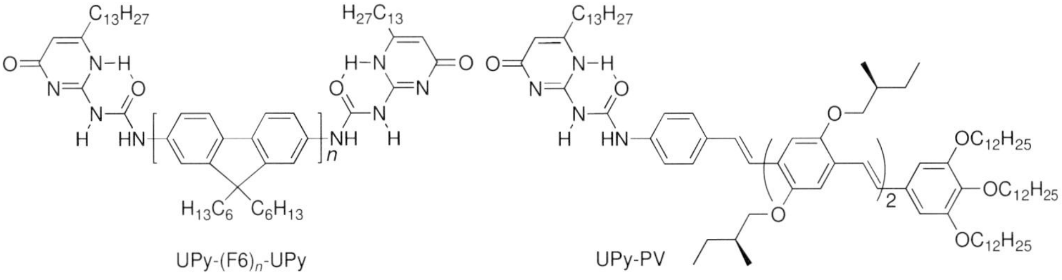

Figure 2.34 Supramolecular assembly of donor–acceptor blocks as a route to controlled architecture in organic photovoltaics.

2.6
Conclusions

The performance of polyfluorene containing polymers in photovoltaics has not yet led to the development of high-efficiency devices. Poor blend morphology has hindered the formation of the fine scale structure required to effectively allow charge separation, polaron formation, and charge collection. This has been demonstrated for F8BT:PFB blends, with the highest charge generation occurring in areas where photoluminescence had not been effectively quenched [57]. In fact, fluorescence quenching is relatively poor in recorded blends using F8BT as an ETM and is reflected in the generally low EQE values for polyfluorene-containing blends [24,57]. Therefore, to improve efficiencies in devices containing polyfluorene-containing blends, the following need to be achieved:

- Better methods of controlling blend morphologies leading to the formation of bicontinuous interpenetrating networks to allow charge collection.
- Techniques to allow block copolymer formation.
- Analytical techniques to allow morphologies to be probed down to the 20–30 nm level.
- Low-bandgap electron transport materials with high absorption coefficients.
- Electron and hole transport mobilities on reported low-bandgap materials to examine their potential to be used as electron transport materials.

Polymers containing fluorene units remain versatile building blocks in the toolbox for the development of new materials for polymer solar cells.

References

1 See *MRS Bulletin*, 2005, **30**, special issue dedicated to solar cells; Spanggaard, H. and Krebs, F.C. (2004) *Solar Energy Materials and Solar Cells*, **83**, 125146;
2 Shaheen, S.E., Ginley, D.S. and Jabbour, G.E. (2005) *MRS Bulletin*, **30**, 10–19; Peumans, P., Yakimov, A. and Forrest, S.R. (2003) *Journal of Applied Physiology*, **93**, 3693–3723; Nunzi, J.-M. (2002) *Comptes Rendus Physique*, **3**, 523–542.
3 Reyes-Reyes, M., Kim, K., Dewald, J., Lopez-Sandoval, R., Avadhanula, A., Curran, S. and Carroll, D.L. (2005) *Organic Letters*, **7**, 5749–5752.
4 Yu, G., Gao, J., Hummelen, J.C., Wudl, F. and Heeger, A.J. (1995) *Science*, **270**, 1789–1791.
5 Halls, J.J.M., Walsh, C.A., Greenham, N.C., Marseglia, E.A., Friend, R.H., Moratti, S.C. and Holmes, A.B. (1995) *Nature*, **376**, 498–500.
6 Rand, B.P., Xue, J.G., Yang, F. and Forrest, S.R. (2005) *Applied Physics Letters*, **87**, 233508.
7 Alam, M.M. and Jenekhe, S.A. (2004) *Chemistry of Materials*, **16**, 4647–4656.
8 Koetse, M.M., Sweelssen, J., Hoekerd, K.T., Schoo, H.F.M., Veenstra, S.C., Kroon, J.M., Yang, X.N. and Loos, J. (2006) *Applied Physics Letters*, **88**, 083504.
9 Kulkarni, A.P., Tonzola, C.J., Babel, A. and Jenekhe, S.A. (2004) *Chemistry of Materials*, **16**, 4556–4573.

10 Nelson, J. (2002) *Current Opinion in Solid State & Materials Science*, **6**, 87–95.
11 Janssen, R.A.J., Hummelen, J.C. and Sariciftci, N.S. (2005) *MRS Bulletin*, **30**, 33–36.
12 Scharber, M.C., Wuhlbacher, D., Koppe, M., Denk, P., Waldauf, C., Heeger, A.J. and Brabec, C.L. (2006) *Advanced Materials*, **18**, 789–794.
13 Wallace, G.G., Dastoor, P.C., Officer, D.L. and Too, C.O. (2000) *Chemical Innovation*, **30**, 14–22.
14 Leclerc, M. (2001) *Journal of Polymer Science, Part A: Polymer Chemistry*, **39**, 2867–2873.
15 Janietz, S., Bradley, D.D.C., Grell, M., Giebeler, C., Inbasekaran, M. and Woo, E.P. (1998) *Applied Physics Letters*, **73**, 2453–2455.
16 Redecker, M., Bradley, D.D.C., Inbasekaran, M. and Woo, E.P. (1998) *Applied Physics Letters*, **73**, 1565–1567.
17 Bredas, J.L., Silbey, R., Boudreaux, D.S. and Chance, R.R. (1983) *Journal of the American Chemical Society*, **105**, 6555–6559.
18 Thompson, B.C., Kim, Y.G. and Reynolds, J.R. (2005) *Macromolecules*, **38**, 5359–5362.
19 Yu, G. and Heeger, A.J. (1995) *Journal of Applied Physiology*, **78**, 4510–4515.
20 Watkins, P.K., Walker, A.B. and Verschoor, G.L.B. (2005) *Nano Letters*, **5**, 1814–1818.
21 Cabanillas-Gonzalez, J., Virgili, T., Lanzani, G., Yeates, S., Ariu, M., Nelson, J. and Bradley, D.D.C. (2005) *Physical Review B: Condensed Matter*, **71**, 014211; Cabanillas-Gonzalez, J., Nelson, J., Bradley, D.D.C., Ariu, M., Lidzey, D.G. and Yeates, S. (2003) *Synthetic Metals*, **137**, 1471–1472; Cabanillas-Gonzalez, J., Yeates, S. and Bradley, D.D.C. (2003) *Synthetic Metals*, **139**, 637–641.
22 Russell, D.M., Arias, A.C., Friend, R.H., Silva, C., Ego, C., Grimsdale, A.C. and Mullen, K. (2002) *Applied Physics Letters*, **80**, 2204–2206.
23 Huang, S.P., Liao, J.L., Tseng, H.E., Jen, T.H., Liou, J.Y. and Chen, S.A. (2006) *Synthetic Metals*, **156**, 949–953.
24 Chen, P., Yang, G.Z., Liu, T.X., Li, T.C., Wang, M. and Huang, W. (2006) *Polymer International*, **55**, 473–490.
25 Kim, Y., Cook, S., Choulis, S.A., Nelson, J., Durrant, J.R. and Bradley, D.D.C. (2004) *Chemistry of Materials*, **16**, 4812–4818.
26 Campbell, A.J., Bradley, D.D.C. and Antoniadis, H. (2001) *Applied Physics Letters*, **79**, 2133–2135.
27 Cornil, J., Gueli, I., Dkhissi, A., Sancho-Garcia, J.C., Hennebicq, E., Calbert, J.P., Lemaur, V., Beljonne, D. and Bredas, J.L. (2003) *Journal of Chemical Physics*, **118**, 6615–6623.
28 Jespersen, K.G., Beenken, W.J.D., Zaushitsyn, Y., Yartsev, A., Andersson, M., Pullerits, T. and Sundstrom, V. (2004) *Journal of Chemical Physics*, **121**, 12613–12617.
29 Onoda, M., Tada, K., Zakhidov, A.A. and Yoshino, K. (1998) *Thin Solid Films*, **331**, 76–81.
30 Wu, W.C. and Chen, W.C. (2006) *Journal of Polymer Research*, **13**, 441–449.
31 Bach, U., Lupo, D., Comte, P., Moser, J.E., Weissortel, F., Salbeck, J., Spreitzer, H. and Gratzel, M. (1998) *Nature*, **395**, 583–585.
32 Grätzel, M. (2005) *MRS Bulletin*, **30**, 23–27; Redecker, M., Bradley, D.D.C., Inbasekaran, M., Wu, W.W. and Woo, E.P. (1999) *Advanced Materials*, **11**, 241–246; Fong, H.H., Papadimitratos, A. and Malliaras, G.G. (2006) *Applied Physics Letters*, **89**, 172116.
33 Morteani, A.C., Dhoot, A.S., Kim, J.S., Silva, C., Greenham, N.C., Murphy, C., Moons, E., Cina, S., Burroughes, J.H. and Friend, R.H. (2003) *Advanced Materials*, **15**, 1708–1712.
34 Poplavskyy, D., Nelson, J. and Bradley, D.D.C. (2004) *Macromolecular Symposia*, **212**, 415–420.
35 Campbell, A.J., Bradley, D.D.C. and Antoniadis, H. (2001) *Journal of Applied Physiology*, **89**, 3343–3351.
36 Zhu, Y., Babel, A. and Jenekhe, S.A. (2005) *Macromolecules*, **38**, 7983–7991.

37 Ravirajan, P., Haque, S.A., Durrant, J.R., Bradley, D.D.C. and Nelson, J. (2005) *Advanced Functional Materials*, **15**, 609–618.

38 Zhang, F.L., Gadisa, A., Inganas, O., Svensson, M. and Andersson, M.R. (2004) *Applied Physics Letters*, **84**, 3906–3908.

39 Ramsdale, C.M., Barker, J.A., Arias, A.C., MacKenzie, J.D., Friend, R.H. and Greenham, N.C. (2002) *Journal of Applied Physiology*, **92**, 4266–4270.

40 Arias, A.C., Corcoran, N., Banach, M., Friend, R.H., MacKenzie, J.D. and Huck, W.T.S. (2002) *Applied Physics Letters*, **80**, 1695–1697.

41 Kim, Y., Cook, S., Tuladhar, S.M., Choulis, S.A., Nelson, J., Durrant, J.R., Bradley, D.D.C., Giles, M., McCulloch, I., Ha, C.-S. and Ree, M. (2006) *Nature Materials*, **5**, 197–203.

42 Kroeze, J.E., Hirata, N., Schmidt-Mende, L., Orizu, C., Ogier, S.D., Carr, K., Gratzel, M. and Durrant, J.R. (2006) *Advanced Functional Materials*, **16**, 1832–1838.

43 Schmidt-Mende, L., Zakeeruddin, S.M. and Gratzel, M. (2005) *Applied Physics Letters*, **86**, 013504; Kruger, J., Plass, R., Gratzel, M. and Matthieu, H.-J. (2002) *Applied Physics Letters*, **81**, 367–369.

44 Kuang, D.B., Klein, C., Ito, S., Moser, J.E., Humphry-Baker, R., Zakeeruddin, S.M. and Gratzel, M. (2007) *Advanced Functional Materials*, **17**, 154–160; Kuang, D.B., Klein, C., Snaith, H.J., Moser, J.E., Humphry-Baker, R., Comte, P., Zakeeruddin, S.M. and Gratzel, M. (2006) *Nano Letters*, **6**, 769–773.

45 Park, T., Haque Saif, A., Potter Robert, J., Holmes Andrew, B. and Durrant James, R., (2003) *Chemical Communications*, 2878–2879.

46 Haque, S.A., Park, T., Xu, C., Koops, S., Schulte, N., Potter, R.J., Holmes, A.B. and Durrant, J.R. (2004) *Advanced Functional Materials*, **14**, 435–440.

47 Hirata, N., Kroeze, J.E., Park, T., Jones, D., Haque, S.A. and Holmes, A.B. (2006) *Chemical Communications*, 535–537.

48 Moons, E. (2002) *Journal of Physics: Condensed Matter*, **14**, 12235–12260; Hoppe, H. and Sariciftci, N.S. (2006) *Journal of Materials Chemistry*, **16**, 45–61; Van Hutten, P.F. and Hadziioannou, G. (2001) *Monatshefte für Chemie*, **132**, 129–139.

49 Snaith, H.J. and Friend, R.H. (2004) *Thin Solid Films*, **451–52**, 567–571.

50 Shikler, R., Chiesa, M. and Friend, R.H. (2006) *Macromolecules*, **39**, 5393–5399.

51 Arias, A.C., MacKenzie, J.D., Stevenson, R., Halls, J.J.M., Inbasekaran, M., Woo, E.P., Richards, D. and Friend, R.H. (2001) *Macromolecules*, **34**, 6005–6013.

52 Snaith, H.J., Greenham, N.C. and Friend, R.H. (2004) *Advanced Materials*, **16**, 1640–1645.

53 Kim, J.S., Ho, P.K.H., Murphy, C.E. and Friend, R.H. (2004) *Macromolecules*, **37**, 2861–2871.

54 Xia, Y.J. and Friend, R.H. (2006) *Advanced Materials*, **18**, 1371–1376.

55 Bjorstrom, C.M., Magnusson, K.O. and Moons, E. (2005) *Synthetic Metals*, **152**, 109–112.

56 Stevenson, R., Arias, A.C., Ramsdale, C., MacKenzie, J.D. and Richards, D. (2001) *Applied Physics Letters*, **79**, 2178–2180.

57 Snaith, H.J., Arias, A.C., Morteani, A.C., Silva, C. and Friend, R.H. (2002) *Nano Letters*, **2**, 1353–1357.

58 Chiesa, M., Burgi, L., Kim, J.S., Shikler, R., Friend, R.H. and Sirringhaus, H. (2005) *Nano Letters*, **5**, 559–563.

59 Coffey, D.C. and Ginger, D.S. (2006) *Nature Materials*, **5**, 735–740.

60 McNeill, C.R., Frohne, H., Holdsworth, J.L. and Dastoor, P.C. (2004) *Nano Letters*, **4**, 2503–2507.

61 Stevenson, R., Milner, R.G., Richards, D., Arias, A.C., Mackenzie, J.D., Halls, J.J.M., Friend, R.H., Kang, D.J. and Blamire, M. (2001) *Journal of Microscopy*, **202**, 433–438; Milner, R.G., Arias, A.C., Stevenson, R., Mackenzie, J.D., Richards, D., Friend, R.H., Kang, D.J. and Blamire, M. (2002) *Materials Science & Technology*, **18**,

759–762; Chappell, J., Lidzey, D.G., Jukes, P.C., Higgins, A.M., Thompson, R.L., O'Connor, S., Grizzi, I., Fletcher, R., O'Brien, J., Geoghegan, M. and Jones, R.A.L. (2003) *Nature Materials*, 616–621; McNeill, C.R., Fell, C.J.R., Holdsworth, J.L. and Dastoor, P.C. (2005) *Synthetic Metals*, **153**, 85–88.

62 McNeill, C.R., Frohne, H., Holdsworth, J.L. and Dastoor, P.C. (2004) *Synthetic Metals*, **147**, 101–104.

63 McNeill, C.R., Watts, B., Thomsen, L., Belcher, W.J., Greenham, N.C. and Dastoor, P.C. (2006) *Nano Letters*, **6**, 1202–1206.

64 Bjorstrom, C.M., Nilsson, S., Bernasik, A., Budkowski, A., Andersson, M., Magnusson, K.O. and Moons, E. (2007) *Applied Surface Science*, **253**, 3906–3912.

65 Corcoran, N., Arias, A.C., Kim, J.S., MacKenzie, J.D. and Friend, R.H. (2003) *Applied Physics Letters*, **82**, 299–301.

66 Ramsdale, C.M., Bache, I.C., MacKenzie, J.D., Thomas, D.S., Arias, A.C., Donald, A.M., Friend, R.H. and Greenham, N.C. (2002) *Physica E: Low-Dimensional Systems & Nanostructures*, **14**, 268–271.

67 Dobberstein, H., Homsy, O.A. and Thiel, B.L. (2005) *Microscopy and Microanalysis*, **11**, 32–33.

68 Williams, S.J., Morrison, D.E., Thiel, B.L. and Donald, A.M. (2005) *Scanning*, **27**, 190–198.

69 Xia, Y.J. and Friend, R.H. (2005) *Macromolecules*, **38**, 6466–6471.

70 Kietzke, T., Neher, D., Kumke, M., Montenegro, R., Landfester, K. and Scherf, U. (2004) *Macromolecules*, **37**, 4882–4890; Kietzke, T., Neher, D., Landfester, K., Montenegro, R., Guntner, R. and Scherf, U. (2003) *Nature Materials*, **2**, 408–412.

71 Snaith, H.J. and Friend, R.H. (2004) *Synthetic Metals*, **147**, 105–109.

72 Kim, J.Y., Kim, S.H., Lee, H.H., Lee, K., Ma, W.L., Gong, X. and Heeger, A.J. (2006) *Advanced Materials*, **18**, 572–576; Reyes-Reyes, M., Kim, K. and Carroll, D.L. (2005) *Applied Physics Letters*, **87**, 083506.

73 Beaupre, S. and Leclerc, M. (2002) *Advanced Functional Materials*, **12**, 192–196; Charas, A., Morgado, J., Martinho, J.M.G., Alcacer, L. and Cacialli, F. (2002) *Synthetic Metals*, **127**, 251–254; Charas, A., Barbagallo, N., Morgado, J. and Alcacer, L. (2001) *Synthetic Metals*, **122**, 23–25.

74 Yang, R.Q., Tian, R.Y., Hou, Q., Yang, W. and Cao, Y. (2003) *Macromolecules*, **36**, 7453–7460.

75 Hou, Q., Xu, Y.S., Yang, W., Yuan, M., Peng, J.B. and Cao, Y. (2002) *Journal of Materials Chemistry*, **12**, 2887–2892.

76 Zhou, Q.M., Hou, Q., Zheng, L.P., Deng, X.Y., Yu, G. and Cao, Y. (2004) *Applied Physics Letters*, **84**, 1653–1655.

77 Svensson, M., Zhang, F., Inganas, O. and Andersson, M.R. (2003) *Synthetic Metals*, **135**, 137–138.

78 Jonsson, S.K.M., Carlegrim, E., Zhang, F., Salaneck, W.R. and Fahlman, M. (2005) *Japanese Journal of Applied Physics, Part 1: Regular Papers, Short Notes & Review Papers*, **44**, 3695–3701.

79 Andersson, L.M. and Inganas, O. (2006) *Applied Physics Letters*, **88**, 082103.

80 Gadisa, A., Zhang, F.L., Sharma, D., Svensson, M., Andersson, M.R. and Inganas, O. (2007) *Thin Solid Films*, **515**, 3126–3131.

81 Aarnio, H., Westerling, M., Osterbacka, R., Svensson, M., Andersson, M.R. and Stubb, H. (2006) *Chemical Physics*, **321**, 127–132; Aarnio, H., Westerling, M., Osterbacka, R., Svensson, M., Andersson, M.R., Pascher, T., Pan, J.X., Sundstrom, V. and Stubb, H. (2005) *Synthetic Metals*, **155**, 299–302.

82 Zhang, F.L., Jespersen, K.G., Bjorstrom, C., Svensson, M., Andersson, M.R., Sundstrom, V., Magnusson, K., Moons, E., Yartsev, A. and Inganas, O. (2006) *Advanced Functional Materials*, **16**, 667–674.

83 Zhang, F., Lacic, S., Svensson, M., Andersson, M.R. and Inganas, O. (2006) *Solar Energy Materials and Solar Cells*, **90**, 1607–1614.

84 Campoy-Quiles, M., Etchegoin, P.G. and Bradley, D.D.C. (2005) *Synthetic Metals*, **155**, 279–282.

85 Persson, N.K., Schubert, M. and Inganas, O. (2004) *Solar Energy Materials and Solar Cells*, **83**, 169–186; Persson, N.K., Arwin, H. and Inganas, O. (2005) *Journal of Applied Physiology*, **97**, 034503.

86 Inganas, O., Svensson, M., Zhang, F., Gadisa, A., Persson, N.K., Wang, X. and Andersson, M.R. (2004) *Applied Physics A: Materials Science & Processing*, **79**, 31–35.

87 Iuo, J., Hou, Q., Chen, J.W. and Cao, Y. (2006) *Synthetic Metals*, **156**, 470–475.

88 Ashraf, R.S., Hoppe, H., Shahid, M., Gobsch, G., Sensfuss, S. and Klemm, E. (2006) *Journal of Polymer Science, Part A: Polymer Chemistry*, **44**, 6952–6961; Ashraf, R.S., Shahid, M., Klemm, E., Al-Ibrahim, M. and Sensfuss, S. (2006) *Macromolecular Rapid Communications*, **27**, 1454–1459.

89 Xia, Y.J., Luo, J., Deng, X.Y., Li, X.Z., Li, D.Y., Zhu, X.H., Yang, W. and Cao, Y. (2006) *Macromolecular Chemistry and Physics*, **207**, 511–520.

90 Wang, X., Perzon, E., Delgado, J.L., De la Cruz, P., Zhang, F., Langa, F., Andersson, M. and Inganas, O. (2004) *Applied Physics Letters*, **85**, 5081–5083.

91 Perzon, E., Wang, X.J., Zhang, F.L., Mammo, W., Delgado, J.L., de la Cruz, P., Inganas, O., Langa, F. and Andersson, M.R. (2005) *Synthetic Metals*, **154**, 53–56; Perzon, E., Wang, X.J., Admassie, S., Inganas, O. and Andersson, M.R. (2006) *Polymer*, **47**, 4261–4268.

92 Chen, M.X., Crispin, X., Perzon, E., Andersson, M.R., Pullerits, T., Andersson, M., Inganas, O. and Berggren, M. (2005) *Applied Physics Letters*, **87**, 252105.

93 Wang, X.J., Perzon, E., Oswald, F., Langa, F., Admassie, S., Andersson, M.R. and Inganas, O. (2005) *Advanced Functional Materials*, **15**, 1665–1670.

94 Wang, X.J., Perzon, E., Mammo, W., Oswald, F., Admassie, S., Persson, N.K., Langa, F., Andersson, M.R. and Inganas, O. (2006) *Thin Solid Films*, **511**, 576–580.

95 Zhang, F.L., Mammo, W., Andersson, L.M., Admassie, S., Andersson, M.R., Inganas, L., Admassie, S., Andersson, M.R. and Ingands, O. (2006) *Advanced Materials*, **18**, 2169–2173.

96 Persson, N.K., Sun, M., Kjellberg, P., Pullerits, T. and Inganas, O. (2005) *Journal of Chemical Physics*, **123**, 204718.

97 Zhang, F.L., Perzon, E., Wang, X.J., Mammo, W., Andersson, M.R. and Inganas, O. (2005) *Advanced Functional Materials*, **15**, 745–750.

98 Cho, N.S., Hwang, D.H., Jung, B.J., Lim, E., Lee, J. and Shim, H.K. (2004) *Macromolecules*, **37**, 5265–5273.

99 Lee, S.K., Cho, N.S., Kwak, J.H., Lim, K.S., Shim, H.K., Hwang, D.H. and Brabec, C.J. (2006) *Thin Solid Films*, **511**, 157–162.

100 Cho, N.S., Park, J.H., Lee, S.K., Lee, J., Shim, H.K., Park, M.J., Hwang, D.H. and Jung, B.J. (2006) *Macromolecules*, **39**, 177–183.

101 Yu, Z.Q., Tan, S.T., Zou, Y.P., Fan, B.H., Yuan, Z.L. and Li, Y.F. (2006) *Journal of Applied Polymer Science*, **102**, 3955–3962.

102 Grazulevicius, J.V., Strohriegl, P., Pielichowski, J. and Pielichowski, K. (2003) *Progress in Polymer Science*, **28**, 1297–1353; Strohriegl, P. and Grazulevicius, J.V. (2002) *Advanced Materials*, **14**, 1439–1452; Jenekhe, S.A., Lu, L.D. and Alam, M.M. (2001) *Macromolecules*, **34**, 7315–7324.

103 Yasuda, T., Imase, T. and Yamamoto, T. (2005) *Macromolecules*, **38**, 7378–7385.

104 Ikeda, N. and Miyasaka, T. (2005) *Chemical Communications*, 1886–1888.

105 Bernede, J.C., Derouiche, H. and Djara, V. (2005) *Solar Energy Materials and Solar Cells*, **87**, 261–270; Shi, M.M., Chen, H.Z., Sun, J.Z., Ye, J. and Wang, M. (2003) *Chemical Physics Letters*, **381**, 666–671; Jin, H., Hou, Y.B., Meng, X.G. and Teng, F. (2006) *Chemical Physics*, **330**, 501–505; Jin, H., Hou, Y.B., Meng, X.G. and Teng, F. (2006) *Chinese Journal of Polymer Science*, **24**, 553–558.

106 Li, J.L., Dierschke, F., Wu, J.S., Grimsdale, A.C. and Mullen, K. (2006) *Journal of Materials Chemistry*, **16**, 96–100.

107 Leclerc, N., Michaud, A., Sirois, K., Morin, J.F. and Leclerc, M. (2006) *Advanced Functional Materials*, **16**, 1694–1704.

108 Demadrille, R., Firon, M., Leroy, J., Rannou, P. and Pron, A. (2005) *Advanced Functional Materials*, **15**, 1547–1552.

109 Forrest, S.R. (2005) *MRS Bulletin*, **30**, 28–32.

110 Gregg, B.A. (2003) *Journal of Physical Chemistry B*, **107**, 4688–4698.

111 Shaheen, S.E., Brabec, C.J., Sariciftci, N.S., Padinger, F., Fromherz, T. and Hummelen, J.C. (2001) *Applied Physics Letters*, **78**, 841–843.

112 Shrotriya, V., Yao, Y., Li, G. and Yang, Y. (2006) *Applied Physics Letters*, **89**, 063505.

113 Tu, G.L., Li, H.B., Forster, M., Heiderhoff, R., Balk, L.J. and Scherf, U. (2006) *Macromolecules*, **39**, 4327–4331.

114 Dudek, S.P., Pouderoijen, M., Abbel, R., Schenning, A. and Meijer, E.W. (2005) *Journal of the American Chemical Society*, **127**, 11763–11768.

3
Carbazole-Based Conjugated Polymers as Donor Material for Photovoltaic Devices

Wojciech Pisula, Ashok K. Mishra, Jiaoli Li, Martin Baumgarten, and Klaus Müllen

3.1
Introduction

In a world of limited energy resources, organic photovoltaics is raising increasing attention. Although organic light-emitting diodes (LEDs) have established their place in the market with a multibillion euro business, the organic solar cell research and device development still have to be dramatically improved to compete with inorganic, mainly silicon-based, solar cells. While the efficiency of inorganic solar cells has been increased to 25–40% and new techniques as thin layer production (∼20%) decreasing the cost of production and widening the applicability will likely lead to even further applications and broaden their use, organic solar cells so far operate at a maximum of 4–5% efficiency at this moment and therefore can hardly compete in the market. Many research efforts are spent to improve the efficiency and better understand the basic processes and barriers faced so far. While there is high demand, the inorganic semiconductors are characterized by high production costs due to high temperatures and vacuum steps necessary for their processing. In comparison, solution processable *organic semiconducting materials* represent a promising alternative and a rather new approach. They provide many desirable properties such as low price, lightweight, flexibility, and semitransparency. These features have captured the interest of scientists and engineers both in academy and industry. Apart from possible economic advantages, organic materials possess low specific weight and are mechanically flexible – properties that are desirable for a solar cell to guarantee easy integration of the modules on any surface, from glass windows to textiles and clothes. Several geometries for organic photovoltaic devices have been investigated to date. On the whole, the solar cell setup consists of equivalent layer techniques as used for LEDs; it just operates in the reverse direction, that is, electricity is not used to create the light, but the light is used to create electricity (Figure 3.1).

The first generation of organic photovoltaic devices was based on a single active layer sandwiched between two metal electrodes with different work functions. The efficiencies were generally poor, in the range of 10^{-3} to 10^{-2}% [1,2]. In 1986, Tang

Organic Photovoltaics: Materials, Device Physics, and Manufacturing Technologies.
Edited by Christoph Brabec, Vladimir Dyakonov, and Ullrich Scherf
Copyright © 2008 WILEY-VCH Verlag GmbH & Co. KGaA, Weinheim
ISBN: 978-3-527-31675-5

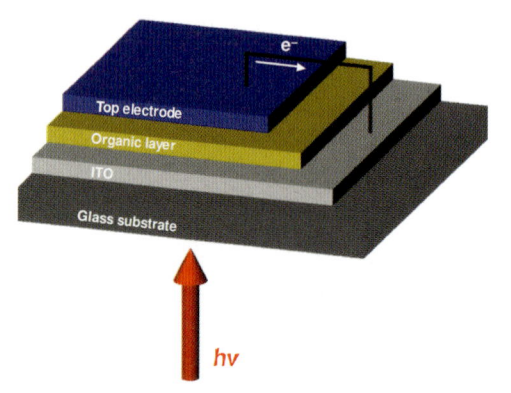

Figure 3.1 Schematic illustration of a photovoltaic cell.

first introduced the concept of heterojunction with a bilayer solar cell and achieved an efficiency of 1% [3]. In this case, two organic layers (a phthalocyanine derivative as p-type semiconductor and a perylene derivative as n-type semiconductor) were sandwiched between two electrodes. This result was for many years the record efficiency for an organic solar cell. The next encouraging breakthrough was achieved with Sariciftci's report in 1992 on the photophysics of mixtures (bulk heterojunction) of conjugated polymers with fullerene [4]. It was clearly shown that the forward electron transfer from excited conjugated polymers to fullerene was faster than 50 fs, while the back electron transfer rate is on the order of 1 µs at room temperature [5]. Later, work based on this bulk heterojunction concept developed greatly the efficiency values of organic solar cells [6–13], approaching 5% for devices containing regioregular polythiophene and fullerene derivatives [14,15].

Construction of an organic PV device requires the use of two types of materials, that is, a donor (p-type) material and an acceptor material (n-type). Many conjugated polymers in their undoped, semiconducting state are electron donors upon photoexcitation (electrons promoted to the antibonding π^* band). A substituted fullerene, such as [6,6]-phenyl-C_{61}-butyricacid-methylester (PCBM), is most commonly used as an acceptor [13,16,17]. Most of the work on such devices have used soluble polythiophenes [11,18] or derivatives of poly(phenylene vinylene) (PPV) [19,20] as the donor material.

The first heterojunction photovoltaic devices based on these conjugated polymers adopted a bilayer structure (Figure 3.2a). The efficiency of this kind of devices was low because of the small charge-generating interface. The superior solubility of functionalized fullerenes compared to C_{60} made it possible to produce high fullerene

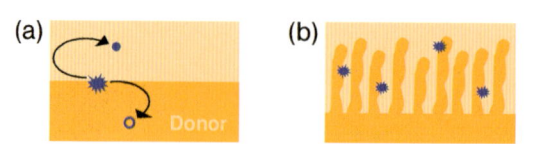

Figure 3.2 Charge generation at D/A interface of (a) bilayer and (b) bulk heterojunction.

content composite films. With the introduction of the concept of "bulk heterojunction" (Figure 3.2b) into photovoltaic devices, significant improvements in the efficiency have been achieved using phase-separated composite materials, processed through control of the morphology of the phase separation into an interpenetrating network (bulk heterojunction). The power efficiency of solar cells made from MEH-PPV/fullerene composites has increased dramatically [21]. In parallel, the groups of Heeger in Santa Barbara [22] and Friend in Cambridge [23] developed an approach using acceptor-type conjugated polymers in an interpenetrating polymer–polymer composite with MEH-PPV, yielding polymeric PV devices with efficiencies comparable to fullerene mixed devices.

PPV and polythiophene derivatives are in many cases considered as donors with fullerene derivatives acting as acceptors, while other new combinations of donors and acceptors obtain much less attention. However, poly(N-vinylcarbazole) is the most widely investigated and applied semiconductor in the carbazole family due to its photoconductive properties and ability to form charge transfer complexes arising from the electron-donating character of the carbazole moiety [24]. One could expect an improvement in the mobility of the charge carriers if carbazole groups are covalently incorporated in the main chain to form conjugated backbones. There are two ways to connect carbazole units, the first is to link the repeat units at the 3- and 6-positions to yield poly(3,6-carbazole)s, and the second to connect via the 2- and 7-positions to provide poly(2,7-carbazole)s. Oligomers of 3,6-carbazole were first synthesized in 1967 by Ambrose [25]. Siove et al. in 2004 succeeded in synthesizing a high molecular weight polymer by oxidative polymerization [26]. A field effect transistor (FET) based on poly(N-butyl-3,6-carbazole) showed a good charge carrier mobility of 10^{-3} cm^2 V^{-1} s^{-1} [27]. Poly(3,6-carbazole)s have also been used as blue emitters in LED, showing a much higher stability than the more widely studied polyfluorene [26,28].

Compared to poly(3,6-carbazole)s, poly(2,7-carbazole)s have been less investigated. Nevertheless, this isomer could be very interesting for electroactive and photoactive devices, since carbazole units linked at the 2- and 7-positions should lead to materials having a higher conjugation length when compared to the 3,6- isomer [29–31]. Both the 2- and 7- positions are *meta* to the *ortho–para* directing amino group of the carbazole unit, however, rendering the synthesis of 2,7-linked carbazole-based materials not as straightforward as that of 3,6-carbazole-based materials. The 2,7-homopolymers of carbazole previously reported had low molecular weights due to their poor solubility, limiting their use in thin-film devices. Our group has developed an efficient way to synthesize 2,7-dibromocarbazole derivatives in only three steps from commercially available reagents with an overall yield of approximately 50% [32]. This new approach represents a real improvement, compared to the previously published synthesis of 2,7-dibromocarbazole [33], and made it possible to use soluble high molecular weight poly(2,7-carbazole)s in photoelectronic devices. LEDs based on poly(2,7-carbazole)s have proved to be among the best blue light emitters, with luminance values as high as 800 cd m^{-2} at about 10 V [34]. However, the use of polycarbazoles as donor materials in solar cells has remained till now an unexplored area of research, which is described in this chapter.

The focus of this chapter is to study the application of 2,7-linked carbazole-based conjugated polymers as donor materials in photovoltaic devices. In this section, the historical synthetic approaches for these kind of conjugated polymers are depicted starting with the transition from 3,6- to 2,7-poly(carbazole)s. It has been observed that poly(2,7-carbazole)s are more promising materials since they have an extended conjugation length, which is beneficial for charge migration along the chains. Another important issue is the introduction of suitable substituents, as for instance long branched alkyl chain, which increases the processibility of the material. On the contrary, the solubility control by the attachment of longer side chains has unwanted effects since it reduces the amount of light-absorbing chromophores in the active layer, increases the insulating effects, and hampers the facile photoinduced electron transfer from donor to acceptor material as well as charge transport. Therefore, a photovoltaic device with pronounced efficiency is a great challenge for the macromolecular design. One way to optimize the molecular structure is the use of ladder-type polymers that can absorb in the visible region where the solar spectrum is most prominent due to their more extended conjugation. A series of ladder-type carbazole-based polymers have been tested in solar cells to provide yet more information on the correlation between device performance, molecular structures, and supramolecular order.

3.2
Synthesis of Carbazole-Based Polymers

Carbazole is one of the widely used building blocks in the field of conjugated polymers for electronic application for several reasons: (i) carbazole is a cheap starting material; (ii) it is a fully aromatic system very stable to environmental and chemical degradation; (iii) the nitrogen group can be easily substituted to provide solubility to the polymer; and (iv) the carbazole can be easily substituted at 3,6-positions via electrophilic substitution reactions.

Historically, polyvinylcarbzole is the one of the important polymers because of its photoconductive properties even applied in Xerox process or copying machines and the ability to donate electrons; however, there the carbazole is not in the main chain but just a pendant group [24].

The main chain chemistry of carbazole-based materials started in 1977 with the discovery of metal-like conductivity in doped polyacetylene [35]. The first easy way to make polymers out of 9-substituted carbazole was the electropolymerization of carbazole itself, which resulted only in oligomers, mainly dimers linked through 3,6-position [36–39]. Similar results were obtained by oxidation of carbazole with chemical agents [40–42]. Later, it was found that the formation of mainly dimers by these methods is due to the stability of biscarbazylium radical cations or dications [43,44] and that the oxidative route is not suitable for getting higher molecular weight polycarbazoles. In a different approach, the reductive polymerization of 3,6-dihalocarbazole was also carried out using Grignard reactions, electrochemical or chemical methods like palladium- or nickel-catalyzed reaction (Scheme 3.1). These methods resulted in moderate molecular weight polymers (about $M_w = 10^4 \, \text{g mol}^{-1}$).

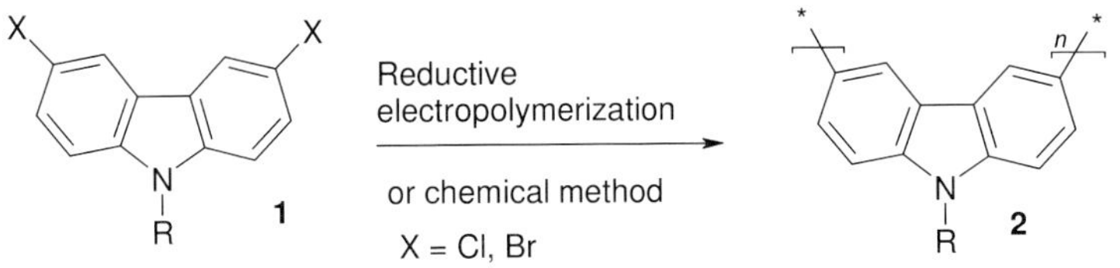

3.2 Synthesis of Carbazole-Based Polymers

Scheme 3.1 Synthesis of 3,6-polycarbazole **2** via dihalocarbazoles **1**.

Only recently, Zhang *et al.* reported a high molecular weight 3,6-polycarbazole by using standard Yamamoto coupling with reverse addition of reagents [45].

Another development in polycarbazole chemistry was initiated by Leclerc *et al.* who for the first time introduced the 2,7-polycarbazole [46]. Since all the phenylene units are linked in *para* fashion, the 2,7-polycarbazole has more extended conjugation length as compared to the 3,6-polycarbazole. However, the synthesis of 2,7-dihalo-carbazole is not straightforward since it starts with the preparation of 2-nitrobiphenyl unit carrying functional groups at the 4,4′-positions **5**. Then 2-nitrobiphenyl can be converted into 2,7-functionalized carbazole by Cadogan ring closure reaction to **6** (Scheme 3.2) [46].

One of the disadvantages of chloro substituents compared to bromo or iodo functionality is its difficult conversion into boronic acid derivatives. The boronic acid derivative is very useful in making alternating copolymers by Suzuki-type polymerization. Only random polymers can be made using nickel-catalyzed Yamamoto polymerization. However, the alternating copolymer is useful in many electronic devices because of better packing over random polymers. Later, the synthetic route for iodo as well as triflate derivatives was also reported by different groups [46,47]. In all the cases, multistep synthesis was involved that limits the application of these methods. We reported a very efficient method to synthesize 2,7-dibromocarbazole (**9**) in two steps (Scheme 3.3), starting directly from the dibromobiphenyl **7**, which is easily converted into the 2-nitrobiphenyl **8** in high yield [32].

Scheme 3.4 depicts some alternating copolymers synthesized by the Suzuki type of copolymerization. The Suzuki polymerization is indeed a great tool to synthesize a wide variety of alternating copolymers.

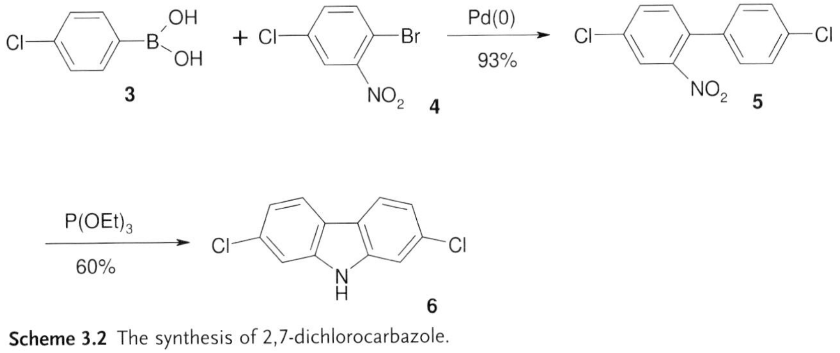

Scheme 3.2 The synthesis of 2,7-dichlorocarbazole.

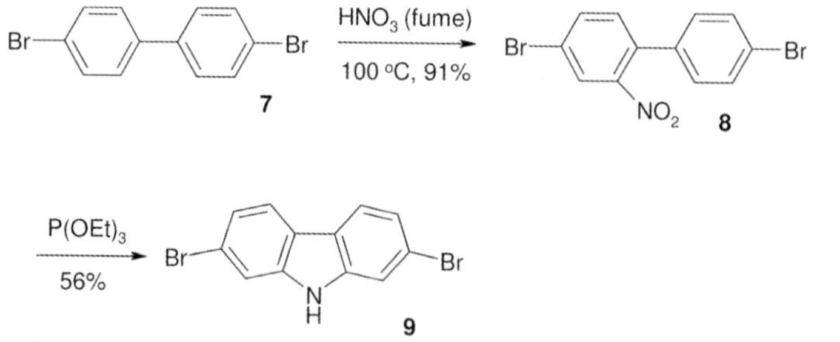

Scheme 3.3 Synthetic route for 2,7-dibromocarbazole.

Many attempts have been made to access soluble high molecular weight polycarbazoles to be useful in electronic applications. The 2,7-dichlorocarbazole (**10**) was alkylated using NaH as a base and alkyl bromide as alkylating agent and then polymerized by Yamamoto polycondensation reaction to give 2,7-polycarbazole (**12**) (Scheme 3.5).

However, the resulting polymers only reached number average molecular weight of 5000–11 000. Later it was found that the low molecular weight formed during polymerization reaction is mainly due to the poor solubility of the polymer chains. We synthesized a high molecular weight soluble polycarbazole **14** by changing the ethyl hexyl alkyl chain to 2-decyltetradecyl substituents (Scheme 3.6) [48]. The M_n value for this polymer was 3.9×10^4 with polydispersity index of 1.9. The better solubility of the polymer backbone was found to be the reason for the increase in molecular weight.

The alkyl-substituted polycarbazole polymers are not very stable under UV light and form radical cations that are very well stabilized by the nitrogen. Thus, it seems advantageous to replace the alkyl substituent by aryl substituents, which was reported

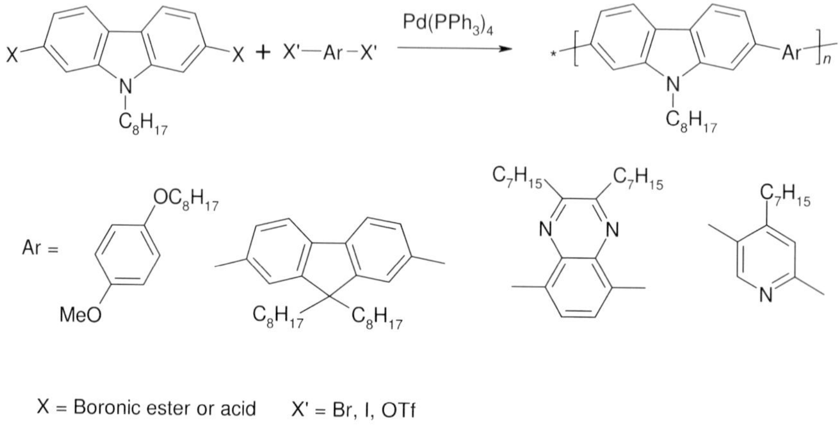

X = Boronic ester or acid   X' = Br, I, OTf

Scheme 3.4 The synthesis of alternating copolymers by Suzuki polymerization.

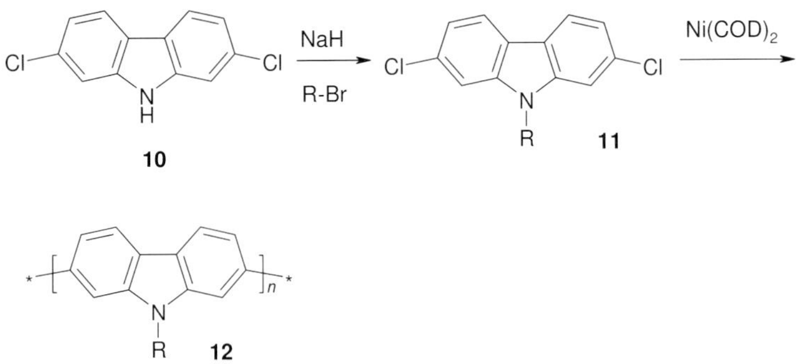

Scheme 3.5 Synthesis of 2,7-polycarbazole.

for the synthesis of soluble arylated poly(2,7-carbazole) **16** (Scheme 3.7). In this case, a high number average molecular weight of 9.1×10^4 was achieved.

The optical and electronic properties of conjugated polymers depend on the conformation of the main chain. Torsion of the monomeric units along the conjugated main chain leads to reduced conjugation and this is accompanied by drastic changes in the optical properties, along with a reduction of the photoluminescence quantum yield. One solution to this problem is the ladderization of the polymer backbone by appropriate carbon bridges [49]. However, this often leads to poor solubility of the polymers since planar conjugated π-systems show high tendency toward stacking. The other alternative is to introduce solubilizing side chains on the polymer backbone so as to impart solubility to the material. Ideally, it should be possible to design structures tuned for a particular absorption or emission wavelength without compromise concerning their solubility. The planar and rigid

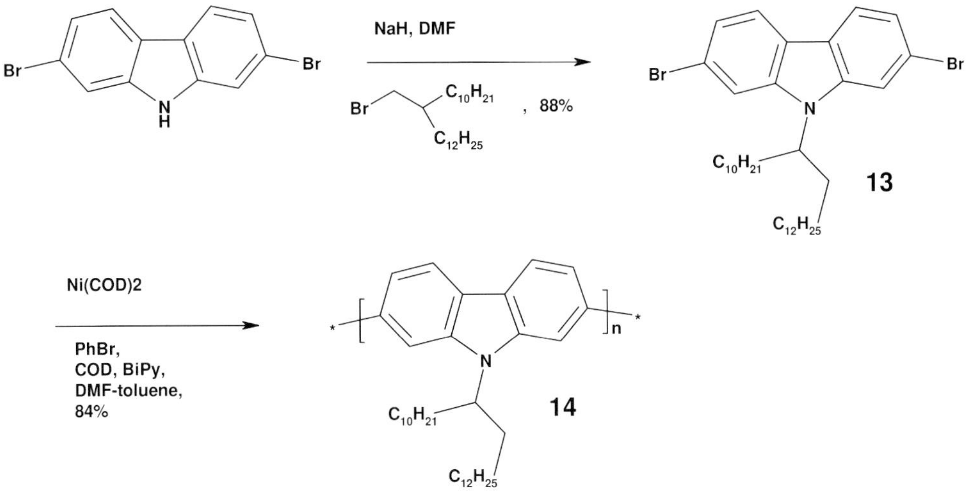

Scheme 3.6 Synthesis of alkyl-substituted soluble poly(2,7-carbazole).

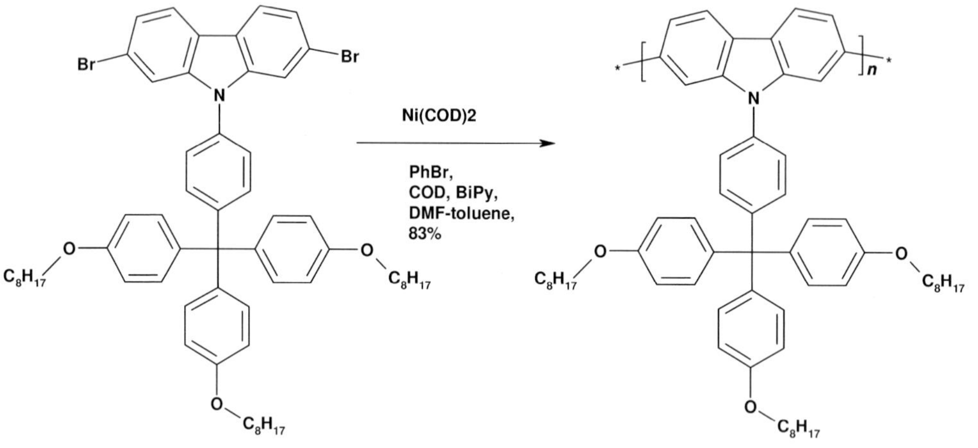

Scheme 3.7 Synthesis of aryl-substituted soluble poly(2,7-carbazole).

structure of ladderized LPPP-type polymers are excellent for optoelectronics applications but they suffer from disadvantages such as poor charge injection and susceptibility of the bridgehead 9-position toward oxidation to generate ketones, which lead to undesirable changes in their optical properties [50]. Materials that avoid both of these problems are nitrogen-bridged ladder-type polymer since the nitrogen bridge is not oxidizable to the corresponding ketone and improves the hole-accepting and transporting properties by raising the highest occupied molecular orbital (HOMO). The first synthesis of the nitrogen-bridged analogue of LPPP was reported by Scherf and coworkers who prepared the precursor alternating copolymer **19** by Suzuki polycondensation reaction of carbazole diboronic ester **17** and diketobenzene **18** and then converted it into the ladder polymer **21** by the polymer analogous addition of methyllithium followed by ring closure with boron trifluoride (Scheme 3.8) [51].

The ring closure was assisted by the electron-donating properties of the carbazole nitrogen that enhanced the susceptibility of the carbazole 3- and 6-positions toward electrophilic attack. The number average and weight average molecular weight of polymer **21** were found to be 35 300 and 69 100 g mol^{-1}, respectively, which corresponds to an average degree of polymerization of 45 units by GPC (gel permeation chromatography) analysis against polystyrene as a standard. Absorption maximum and emission maximum of polymer **21** in chloroform solution was found to be at 460 and 468 nm, respectively, with a very small Stokes shift of 8 nm, which is characteristic for ladder-type polymers.

In a similar attempt, we synthesized a carbazole-based ladder-type polymer with higher carbazole content by taking the advantage that the 3,6-positions of the carbazole can be easily functionalized by electrophilic substitution reaction (Scheme 3.9) [50]. Then, the precursor polymer **23** was made by the Suzuki polycondensation of diacyl carbazole **22** with carbazole diboronic ester **17**. Finally,

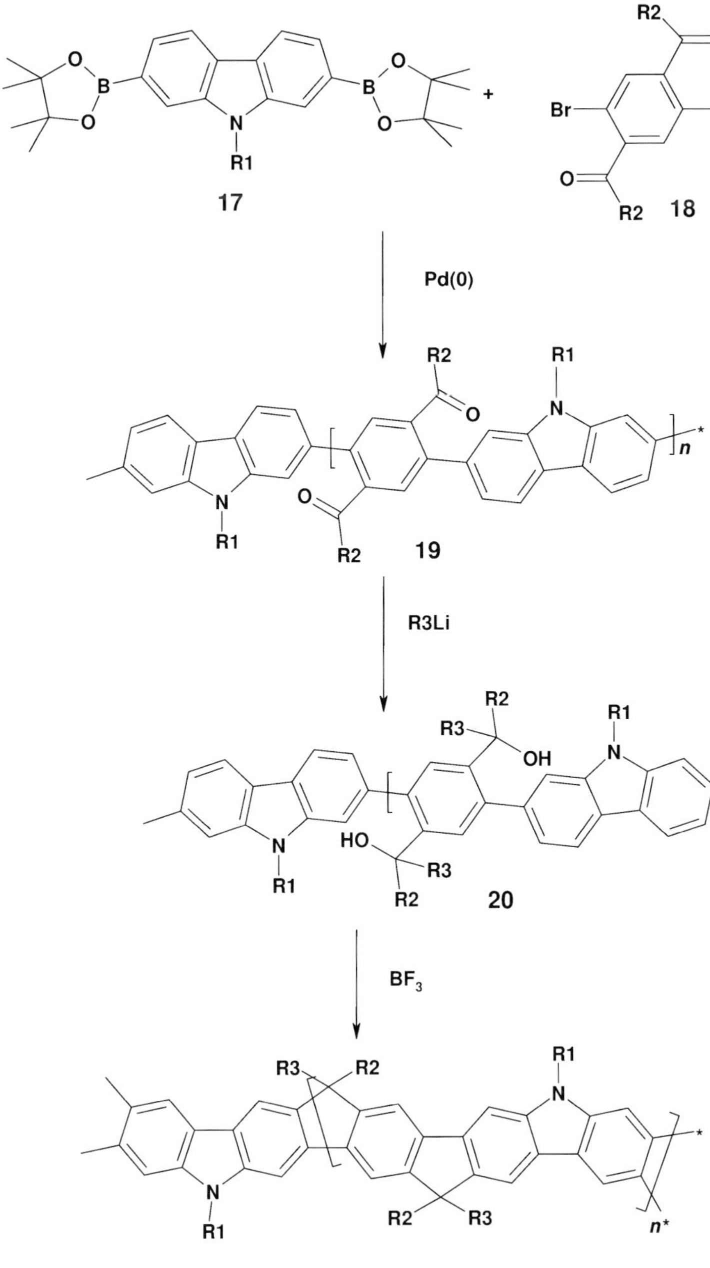

Scheme 3.8 Synthesis of carbazole-based ladder-type polymer **21** reported by Scherf and coworkers [51].

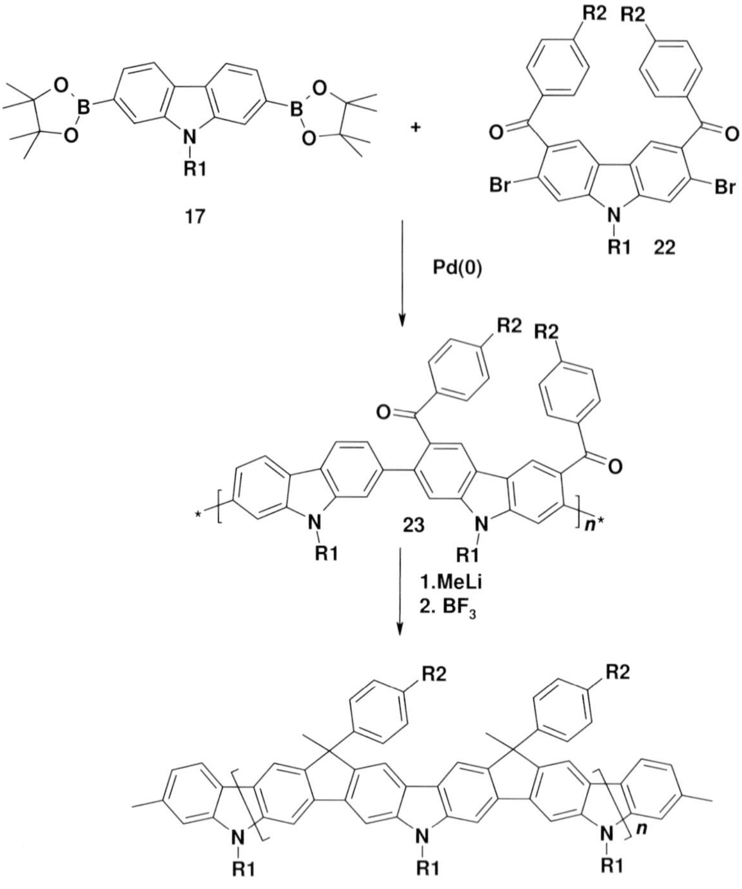

Scheme 3.9 Synthesis of carbazole-based ladder-type polymer **24**.

addition of methyllithium and ring closure by boron triflouride resulted in ladder polymer **24**. This polymer was anticipated to be better hole transporter than ladder-type polymer **21** made by Scherf and coworkers because of the higher density of nitrogen in the polymer backbone as well as due to their asymmetry (all nitrogen are at same side of polymer), potentially useful for selective binding with an electrode or a nanoparticle or for layer-by-layer self-assembly. The GPC measurements of polymer **24** displayed M_n and M_w of 13 500 and 23 000 g mol^{-1}, respectively, with a polydispersity index of 1.7. The optical properties were measured in chloroform solvent and found to be highly dependent on the concentration of the polymer in solution. This concentration-dependent optical behavior was a clear indication of the aggregation of the polymer chains in solution. At high concentrations of the polymer (0.1 mg ml^{-1}), emission spectra were featureless in shape, and when the concentration was reduced to 1%, the broad featureless emission disappeared and a new blue-

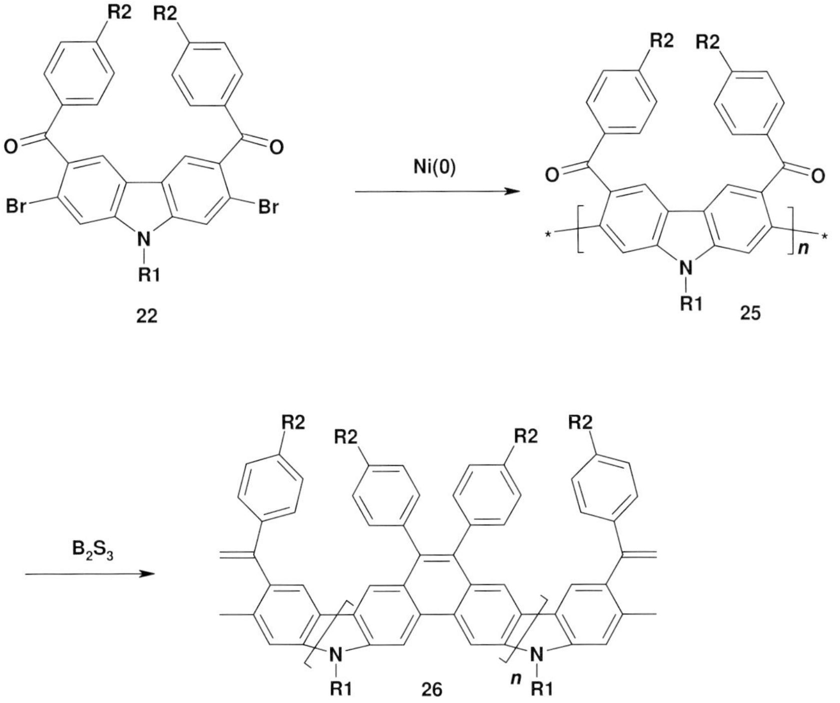

Scheme 3.10 Synthesis of carbazole-based ladder-type polymer 26.

shifted sharp emission maximum at 473 nm appeared. This demonstrated that the polymer chains were interacting in solution and forming aggregates. The bulky aryl substituents were present only on one side of the polymer chain, which proved insufficient to prevent aggregation. Polymer 24 therefore shows green fluorescence in solution.

A similar polymer was also synthesized where the five-membered ring was replaced by a six-membered ring in the ladder-type polymer 26 [50]. In this case, the precursor monomer was diacyl carbazole 22 that was polymerized by Yamamoto polycondensation reaction to give precursor polymer 25 and then ring closure with B_2S_3 treatment gave the desired polymer 26 (Scheme 3.10). Polymer 26 was synthesized with M_n and M_w values of 11 000 and 61 000 g mol^{-1}, respectively, against polystyrene standard. In contrast to polymer 24, polymer 26 displayed a well-resolved photoluminescence spectrum in solution, not showing concentration dependence. The absence of aggregation induced broadening that was attributed to the bulky substituents and the helical structure of the polymer backbone unlike polymer 24.

Due to more extended conjugation, polymer 26 demonstrated a bathochromic shift of the emission maximum to 467 nm with secondary bands at 499 and 534 nm, so that the fluorescence was yellowish-green (see Figure 3.3).

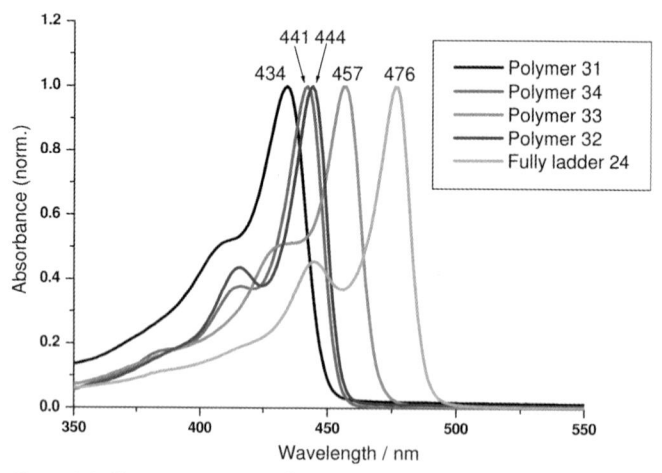

3 Carbazole-Based Conjugated Polymers as Donor Material for Photovoltaic Devices

Figure 3.3 Absorption spectra of nitrogen-bridged polymers in THF solution.

As mentioned above, the fully ladder-type nitrogen-bridged polymers have the following disadvantages:

1. The solubility is so low that it is difficult to obtain good films from solution processing.
2. Aggregate formation leads to low-energy emission in the solid state.
3. The emission maximum is bathochromically shifted (around 470 nm) and hence not suitable for use as a blue emitter.

To overcome these problems, a series of semiladder-type polymers were synthesized by our group. The semiladder-type polymer has several advantages over fully ladder-type polymers:

(i) Emission maximum of bridged poly(phenylene)s has been shown to be tunable systematically over the blue region of the spectrum, from polyfluorene (421 nm) to poly(ladder-type pentaphenylene) (445 nm).
(ii) A single bond introduces some flexibility between two monomers and reduces the planarity, which enhances solubility as well as reduces the aggregation.
(iii) The monomers are highly pure and well characterized prior to polymerization such that the probability of chemical defects by incomplete reactions is less likely.
(iv) There is improved thermal stability of polymers as well as better electrical transport properties compared to polymers with a metalinkage or a nonconjugated group in the backbone.

For the synthesis of higher analogues of carbazole-based monomers, it is very important to select appropriate precursor molecules that can be easily end functionalized; otherwise it becomes difficult to achieve the desired substitution pattern as shown by the directionality of different monomers toward an electrophile (Scheme 3.11).

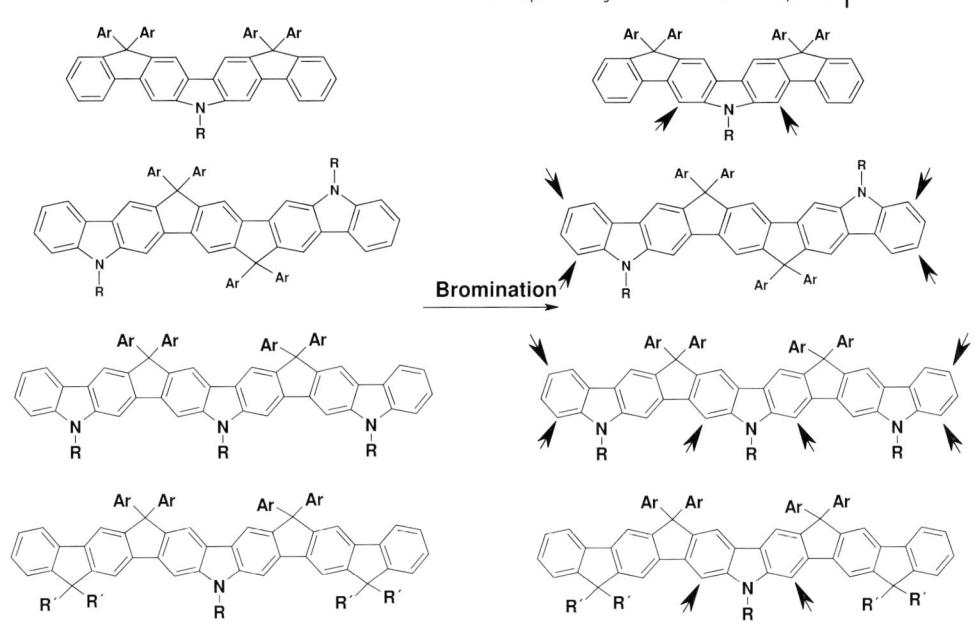

Scheme 3.11 Preferred sites for electrophilic substitution reaction of carbazole-based monomers.

Therefore, a new synthetic methodology was developed [52,53], where the precursor molecule was first end functionalized with halogens preferably with chlorine atom because of two main reasons: (i) chlorine is less susceptible for exchange with phenyllithium than bromine during the conversion of the ester or keto group into alcohol as well as it is stable toward Suzuki coupling and hence avoids polymerization; (ii) chlorine is preferred over bromine in Yamamoto polymerization because carbon–chlorine bond is stronger than C–Br, and so may be less susceptible to dehalogenation at the relatively long reaction times required for Yamamoto polycondensations at 70 °C. The decrease in dehalogenated side product would cause an increase in the molecular weight of the polymer.

Nitrogen-bridged semiladder-type polymers have been synthesized starting from poly(ladder-type tetraphenylene), poly(ladder-type pentaphenylene), and poly(ladder-type hexaphenylene) according to Scheme 3.12 [52,53], which depicts the general synthesis of the nitrogen-bridged (ladder-type tetraphenylene) **27** and nitrogen-bridged (ladder-type hexaphenylene) monomers (**28** and **29**), starting from common precursor **22**.

Similarly, the nitrogen-bridged ladder-type pentaphenylene **30** was made starting from 2,5-dibromo-terephthalic acid dimethyl ester and 2-chloro-9-(2-ethylhexyl)-7-(4,4,5,5-tetramethyl-[1,3,2]dioxaborolan-2-yl)carbazole as shown in Scheme 3.13.

Finally the nitrogen-bridged ladder-type polymers were synthesized from the corresponding "macromonomers" as shown in Scheme 3.14. In all cases, high number average molecular weight polymers were obtained ranging from 23 000 g mol^{-1} for polymers **32** to 46 000 g mol^{-1} for polymer **31** with PPP as GPC standard.

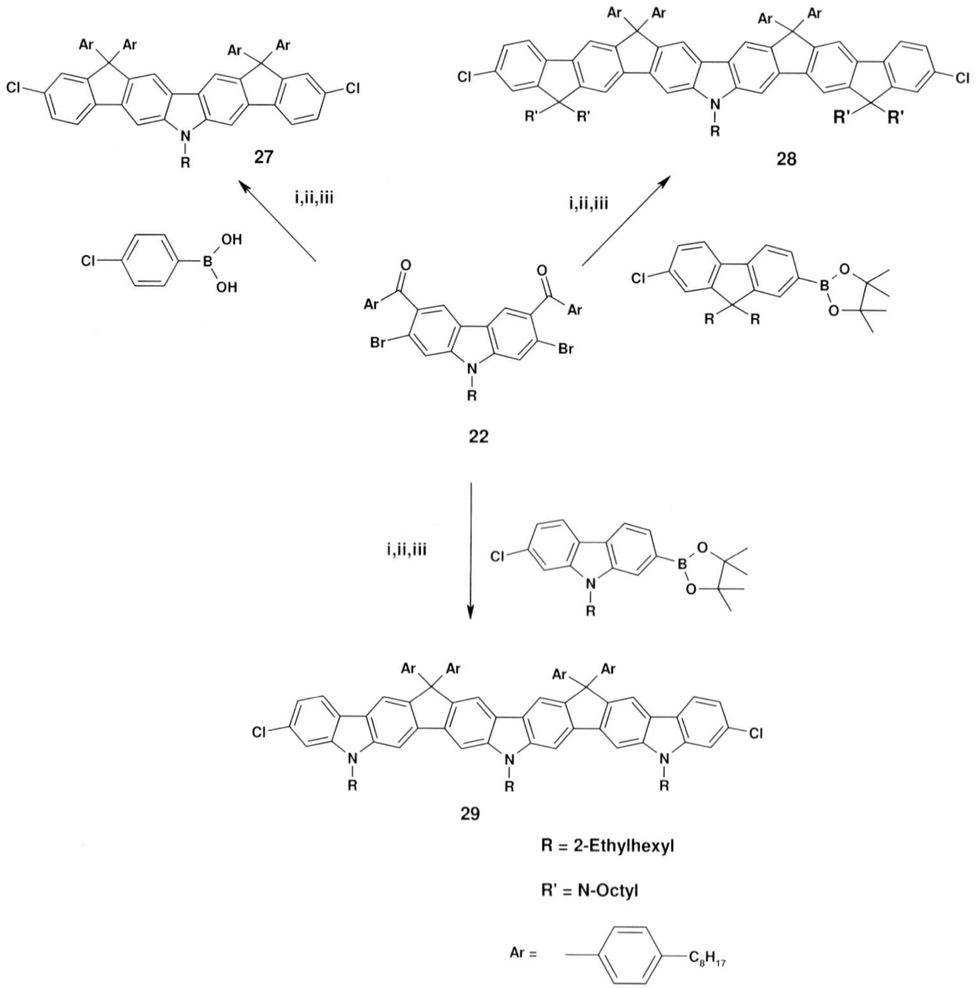

Scheme 3.12 The synthetic route toward nitrogen-bridged ladder-type tetraphenylene and nitrogen-bridged ladder-type hexaphenylene. (i) Pd (PPh$_3$)$_4$, THF/H$_2$O (ii) ArLi, and (iii) BF$_3$·OEt$_3$.

The above-presented new synthetic routes lead to semiladder polymers with nitrogen atoms at the bridges being considered to improve hole-transporting properties and to extend the conjugation length by additional benzene rings or nitrogen atoms inside the monomer backbone. The introduction of these additional units within the backbone has effects on the optical properties that are essential for the application of these materials in optoelectronic devices. Therefore, absorption or photoluminescence are discussed in terms of the macromolecular structure.

Figures 3.3 and 3.4 display the absorption spectra and photoluminescence spectra of polymers **31**, **32**, **33**, **34** and fully ladder-type polymer **24** [52,53]. The absorption

Scheme 3.13 Synthetic route for nitrogen-bridged ladder-type pentaphenylene monomer.

maxima varied from 434 nm for nitrogen-bridged polymer **31** to 476 nm for fully ladder-type polymer **24** with other polymers falling in between. In a similar way, the emission maxima varied from 446 nm for polymer **31** to 481 nm for fully ladder polymer **24**.

Therefore, it is demonstrated that it is possible to tune the emission color from a pure blue emitter to blue-green and finally green by increasing the conjugation length. The conjugation within the polymer can be tuned by increasing the benzene

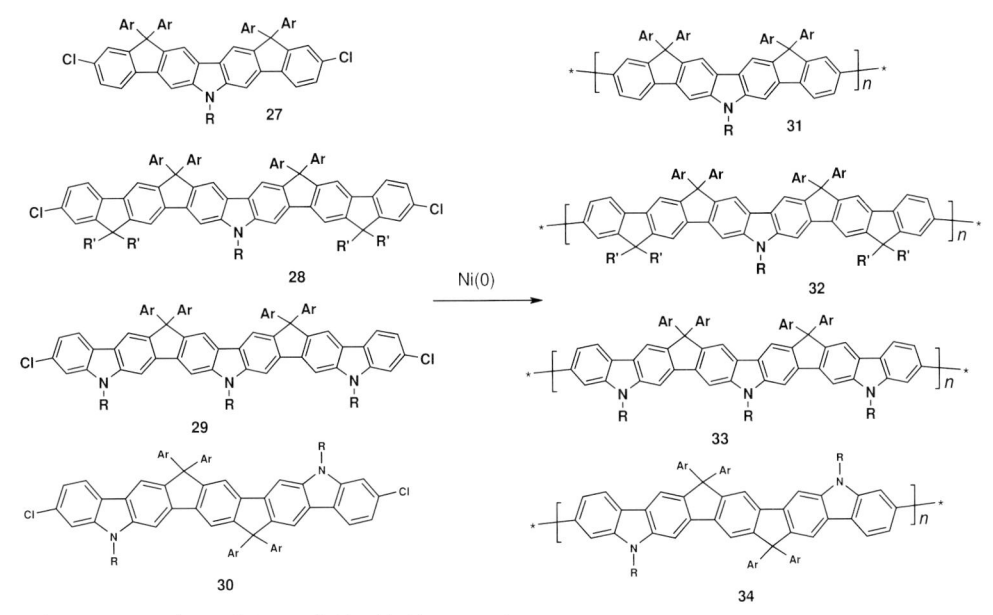

Scheme 3.14 Synthesis of nitrogen-bridged ladder-type polymers.

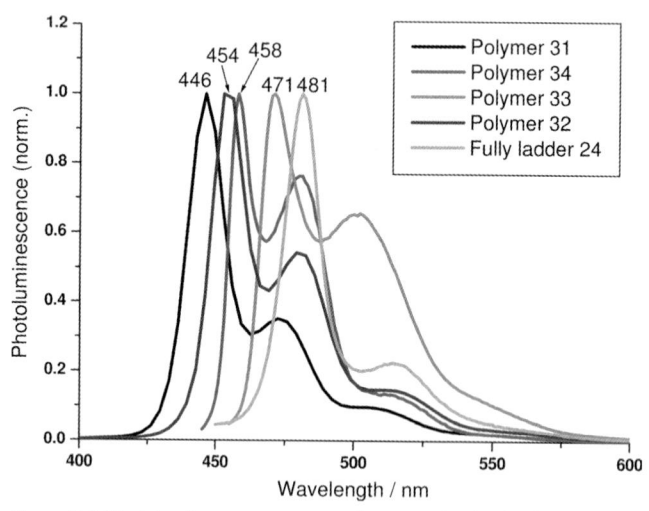

Figure 3.4 Photoluminescence spectra of nitrogen-bridged polymers in THF solution.

unit in the ladderized monomer as shown in the previous section or by introducing heteroatoms such as nitrogen. For example, polymer **31** emitted at 446 nm and has one nitrogen and four benzene ring in each repeat unit, whereas polymer **32** emitted at 454 nm with the same number of nitrogen atoms per monomer but six benzene units in the repeat unit. This increase of 8 nm resulted from the two extra benzene rings. In the same way, the emission maxima depend upon the number of nitrogens in the repeat unit such as polymer **33** and polymer **32** where both have the same number of benzene units but the former has three nitrogen atoms and the latter has one nitrogen atom in each repeat unit. Hence, polymer **33** demonstrated a red shift in both the absorption and emission maxima by 13 and 17 nm, respectively. Polymer **34**, which has two nitrogens in each repeat unit with five benzene unit, showed a blue shift of 3 nm compared to polymer **32** that has one nitrogen and six benzene rings in the repeat unit. However, the photoluminescence spectrum showed the reverse trend where polymer **34** displayed emission maxima at 458 nm compared to polymer **32** that demonstrated primary emission at 454 nm.

This suggests that the presence of nitrogen bridges has more impact on the emission maxima than on the absorption maxima. Fully ladder-type polymer **24** showed surprisingly less intense secondary band emission (Figure 3.4) compared to all other nitrogen-bridged ladder-type polymers.

So far we have discussed the hole transporting (p-type) conjugated polymers. As already outlined above, one of the most important applications of these polymers is solar cells. They are often based on polymer donors and molecular acceptors in blends, so-called bulk heterojunctions. A general drawback of the bulk heterojunction design is the fact that the transport and collection of charges in a disordered nanoscale blend can be hindered by phase boundaries and discontinuities. One way to overcome this drawback is to covalently link the donor and acceptor in a single

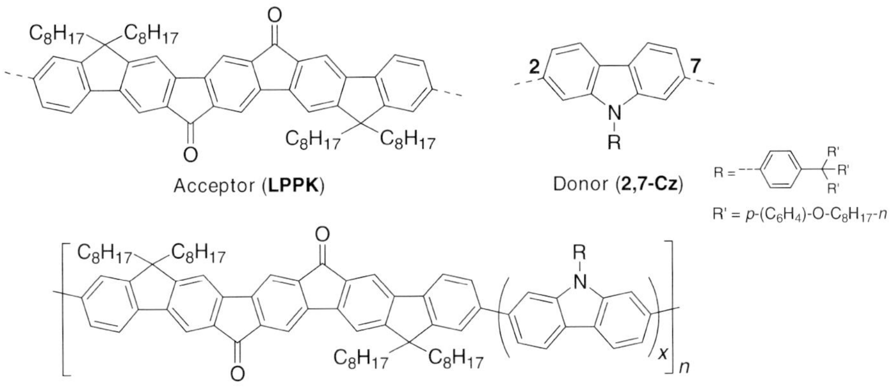

3.2 Synthesis of Carbazole-Based Polymers

Scheme 3.15 Structures of 2,7-poly(carbazole) and **LPPK** alternating copolymers (**35** and **36**).

polymer chain. This method has great advantages because the intrinsic tendency of each segment in copolymers to aggregate in an individual phase provides a means to create a well-ordered nanoscale morphology (e.g., spheres, cylinders, lamellar), governed by the relative volume fractions. Moreover, the built-in intramolecular charge transfer can facilitate manipulation of the electronic structure and lead to low-bandgap semiconducting polymers. In particular, through design and synthesis of new polymer structures, donor–acceptor (D–A) conjugated polymers can be extended to systems with efficient photoinduced charge transfer, charge separation, and charge transport processes for photovoltaic devices.

Recently, we reported the synthesis of two copolymers **35** and **36** consisting of a different number of carbazole units as donor and ladder-type pentaphenylene with two ketone groups at the bridges (**LPPK**) as an excellent electron-accepting (n-type) material (Scheme 3.15) [54]. Figure 3.5 shows the UV–Vis absorption and photoluminescence spectra of **36** in a dilute chloroform solution and a drop-casted film on quartz glass at room temperature, together with the absorption spectra in different solvents. Compound **36** showed two absorption maxima at around 345 and 408 nm in solution, which correspond to the π–π* transitions of the copolymer's backbone. A weak absorption band at 560 nm was apparent corresponding to the n-π* transition in the carbonyl of **LPPK**.

The effect of the solvent polarity on the absorption of **36** is displayed in Figure 3.5b. By changing the solvent from toluene to chloroform, the peak at around 550 nm shifted to 560 nm. Comparing UV–Vis spectra of **36** in thin film with that of the dilute solution, it can be clearly seen that the absorption spectra of thin films were apparently broader, the position of the main absorption peaks were red shifted, and the absorption intensity at 550 nm was stronger, which might be due to the stronger π-stacking of the **LPPK** units in the main chain. It is postulated that this feature is associated with the formation of close-packed self-organized layers by strong interchain interaction in this π-conjugated system in the solid state. There was no

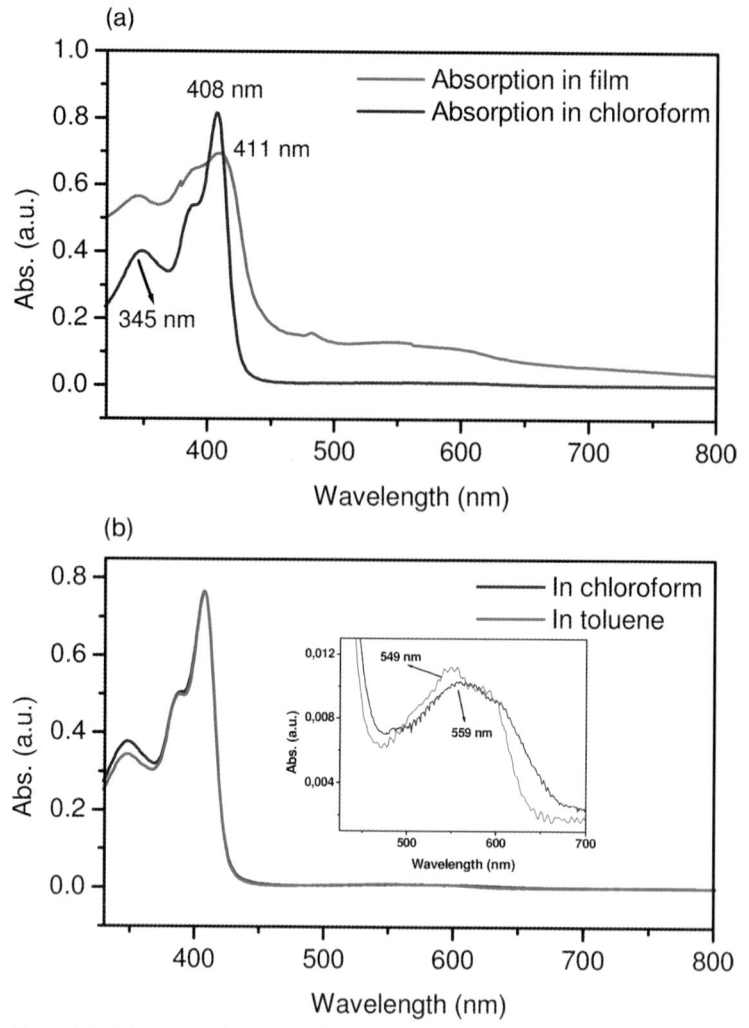

Figure 3.5 (a) UV–Vis absorption of **36** in a dilute chloroform solution and drop-casted film on quartz glass; (b) absorption spectra of **36** in different solvents.

emission from the **2,7-Cz** subunits, implying complete charge transfer from **2,7-Cz** donor to **LPPK** acceptor moieties, while the luminescence quenching of acceptor moiety in **36** compared to pure acceptor **LPPK** monomer indicated the charge transfer involved: When copolymer **36** had the same concentration as the pure monomer **LPPK**, stronger emission quenching of **36** was found, which supports the existence of a charge separated state upon photoexcitation in the D–A polymer.

Modifications of the carbazole polymer structure open a great opportunity to control the absorption and photoluminescence of the material. The optical properties are directly related to the chemical structure and the type of introduced units.

However, the change in the chemical design of the polymer has further consequences on structural and physical properties, as, for example, on their supramolecular organization. In terms of the application of conjugated polymers for application in solar cells, besides the optical properties that are relevant for light harvesting (see Section 3.4), the interaction between the macromolecules and their self-assembly are important for the exciton diffusion and charge carrier mobility. Therefore, the study of the supramolecular structure of conjugated polymers in relation to their design became an important issue in the development of organic semiconductors for solar cells, which is described in more detail in the next section.

3.3
Supramolecular Order of Carbazole-Based Polymers

It has been shown in the literature that the performance of devices based on conjugated polymers as organic semiconductors is strongly dependent not only on the macroscopic order and morphology but also on the packing of individual conjugated backbones that takes place via π-stacking interaction [55–57]. In many cases, the self-assembly and the degree of order in the bulk and thin films were related to the structure of the polymer backbone [58,59]. However, for carbazole-based polymers little is known about the supramolecular organization. X-ray diffractometry is a powerful tool to gain a deeper insight into the self-assembly and is usually applied for powders and thin films. An additional option is the investigation of the superstructure of mechanically aligned filaments by using a two-dimensional X-ray scattering detector [60,61]. This analytical method opens the opportunity to distinguish not only between reflections at different scattering angles but also between scattering intensities in different planes, indicating the relative arrangement of the single building blocks with the respect to the supramolecular assembly [62,63]. The experimental setup is illustrated schematically in Figure 3.6, where the filament, which is prepared via extrusion, is positioned vertically toward the detector. To analyze the obtained scattering pattern, it is helpful to make a distinction between the meridional and equatorial plane of the two-dimensional pattern. In general, conjugated polymer chains are aligned along the mechanical shearing direction and thus along the filament axis, defined as a "c" axis, due to their large molecular aspect ratio, whereby "a" and "b" are randomly distributed perpendicular to "c" (Figure 3.6 inset). Despite many conjugated polymers, especially polymers possessing a rigid backbone as, for example, ladder-type polymers, do not reveal any thermal phase transition, these materials show little softening at higher temperatures without changing the organization but allowing the mechanical orientation using a homebuilt mini-extruder.

Figures 3.6 and 3.7 show five typical two-dimensional wide-angle X-ray scattering (2D-WAXS) patterns for carbazole-based polymers with different macromolecular design. For instance, the pattern in Figure 3.6 has been obtained for a 2,7-polycarbazole decorated with 2-decyl-tetradecyl (**14**) alkyl chains. The high intensity and distinct reflections point toward a pronounced organization of **14** in the extruded

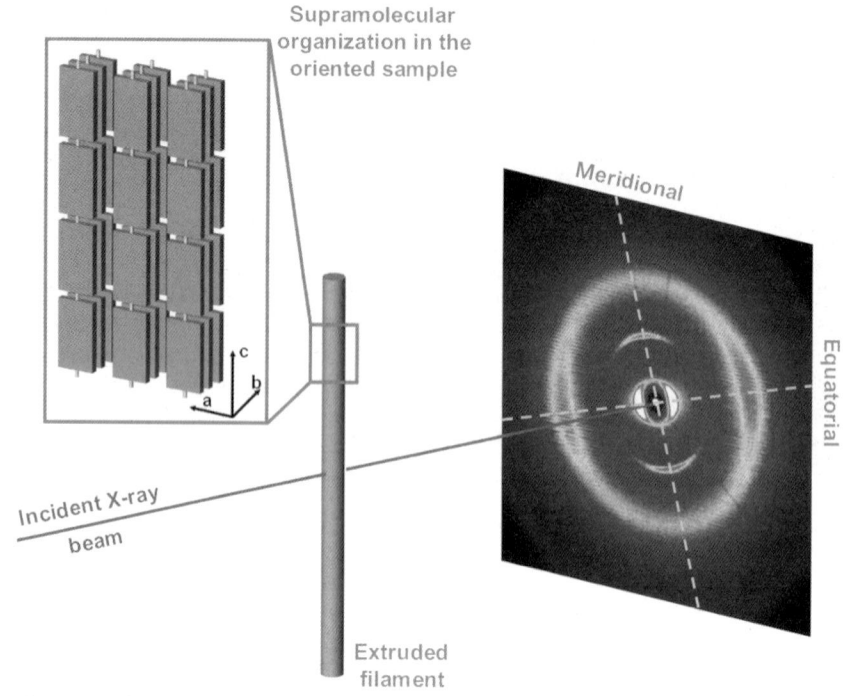

Figure 3.6 Schematic illustration of the setup of two-dimensional X-ray scattering experiments on the basis of the pattern of **14**.

sample. The sharp equatorial small-angle reflections are characteristic for the alignment of the macromolecules along the shearing direction and correspond in this example to a distance of 2.3 nm. This value is related to the chain-to-chain distance of the conjugated polymer chains that is affected primarily by the length of the attached side chains. For comparison, a 2,7-polycarbazole with shorter 2-ethylhexyl chains gave a distance of only 1.8 nm. Furthermore, the equatorial wide-angle reflections at a correlation distance of 0.46 nm were attributed to the π-stacking spacing between the polymer units, which is a typical value for this type of conjugated polymers. The meridional reflections correspond to the repeating distance of 0.81 nm between the carbazole units along the polymer backbone and are in agreement with the theoretical value.

As an additional example, the 2D-WAXS pattern of a more rigid ladder-type polymer **33** carrying aryl groups is presented in Figure 3.7a, revealing a similar X-ray pattern as observed for **14** but with somewhat broader reflections, which corresponded to the stacking distance and the sequence repetition. The packing parameters between the rigid blocks changed due to the difference in polymer architecture and a different alkyl density. From the equatorial small-angle reflection, it was possible to derive a chain-to-chain distance of 1.51 nm for **33**. An identical stacking distance of 0.48 nm was indicated by the wide-angle reflections as observed for **14**. The correlation between single units of 2.3 nm could not be derived in a

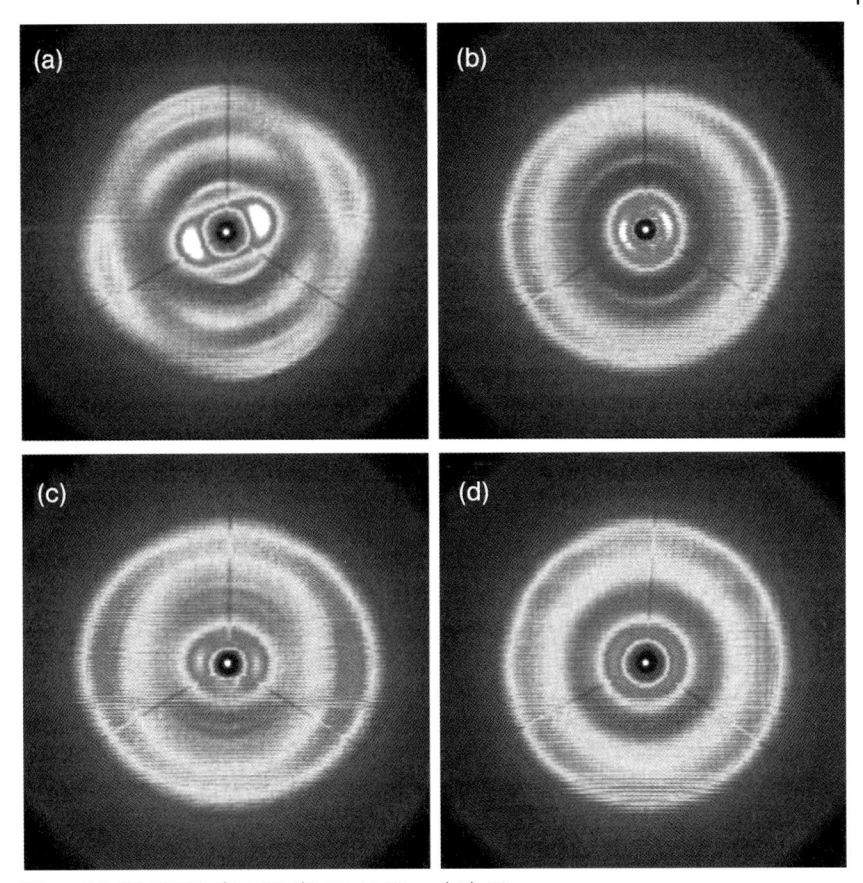

Figure 3.7 2D-WAXS of (a) **33**, (b) **16**, (c) **36**, and (d) **32**.

straightforward way, since these reflection positions were superimposed by the equatorial scattering intensity corresponding to 1.51 nm. However, multiple higher order reflections, which correlated with the distance between monomer units, appeared in the pattern. For instance, the second higher order reflection in the meridional was related to 1.15 nm, which was the half value of 2.3 nm. Reflections up to the sixth order were obvious, confirming a pronounced order of the polymer chains. The organization of **33** is illustrated schematically in Figure 3.8. All essential periodicities are pointed out in the drawing. Due to simplifications, the alkyl side chains are shown as stiff. But in reality, the alkyl substituents are flexible and rather disordered, filling the periphery of the rigid conjugated macromolecules stacked on top of each other. The aryl groups not only play an important role in the solubility of the compound but also influence the packing. Since these groups possess a rather high steric demand due to their out-of-plane position toward the polymer backbone of **33**, one can assume a flipping of the conjugated chain during self-assembly, leading to the most favorable arrangement as illustrated in the inset of Figure 3.8. Therefore, the

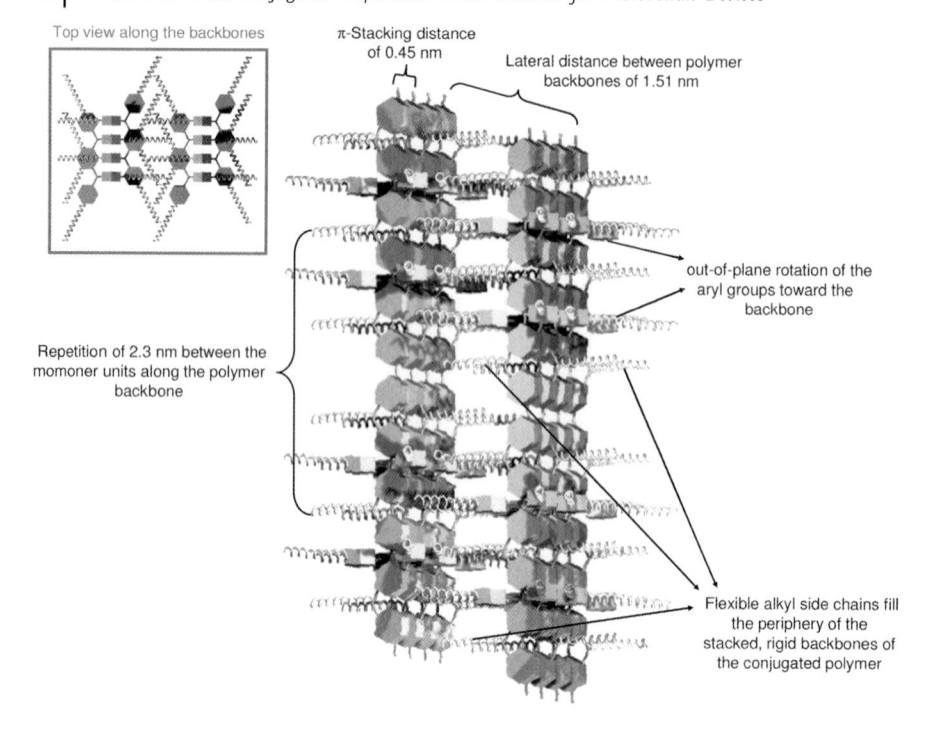

114 *3 Carbazole-Based Conjugated Polymers as Donor Material for Photovoltaic Devices*

Top view along the backbones

π-Stacking distance of 0.45 nm

Lateral distance between polymer backbones of 1.51 nm

Repetition of 2.3 nm between the momoner units along the polymer backbone

out-of-plane rotation of the aryl groups toward the backbone

Flexible alkyl side chains fill the periphery of the stacked, rigid backbones of the conjugated polymer

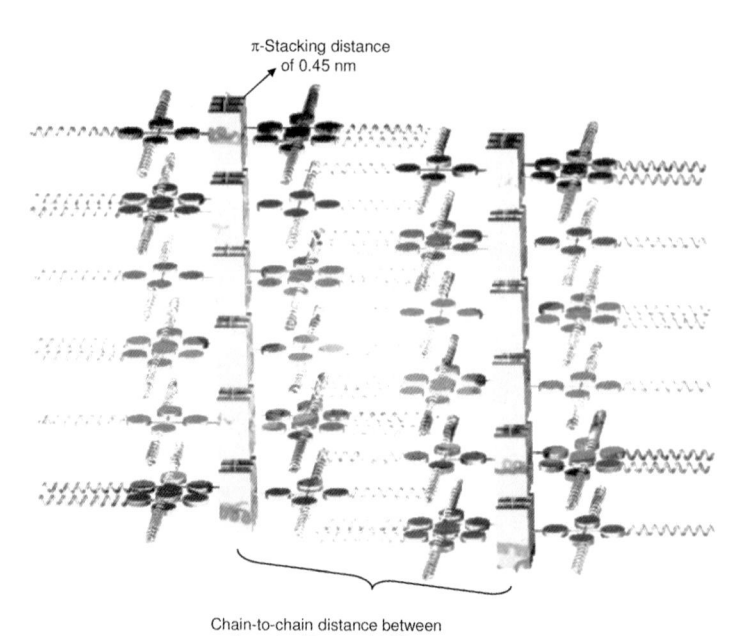

π-Stacking distance of 0.45 nm

Chain-to-chain distance between polymer backbones of 1.92 nm

Figure 3.8 Schematic illustration of the packing of **33** (upper) and **16** (lower). The alkyl side chains are simplified.

high steric requirements of the aryl groups do not necessarily decrease the π-orbital overlap, which is important for an unhindered interchain charge carrier transport.

An identical effect has been observed for **16** and **36** carrying even bulkier side arms (Figure 3.7b and c). The X-ray patterns indicate in both cases a well-organized conjugated polymer by the distinct and strong reflections in the equatorial and meridional plane. A chain-to-chain distance between the conjugated polymer chains of 2.36 nm for **16** and 1.92 nm for **36** and a π-stacking period of 0.45 nm for the two systems were determined. In fact, the appearance of a strong scattering intensity typical for the π-stacking interactions between backbones is quite surprising. One might assume initially that the pendant 4-[tris-(4-octyloxyphenyl)methyl]phenyl substituents hindered a stacking. However, the packing during self-assembly takes place due to a most probable flipping of the polymers as it is the case for **33**. In contrast to **33**, the backbones of **16** and **36** are not that rigid and planar, allowing a spontaneous twisting during stacking and thus diminishing in this way the sterically demanding influence of the 4-[tris-(4-octyloxyphenyl)methyl]phenyl (see schematic illustration in Figure 3.8). The relatively pronounced meridional intensity scattering correlated with 0.80 nm, which was in accordance to the monomer of **16** repeating distance along the conjugated backbone, confirmed a well-organized superstructure.

The three described cases showed that conjugated backbones with sterically demanding substituents such as aryl groups are able to pack efficiently. Since these groups are attached to only one side of the polymer backbone, the packing is achieved by flipping. Figure 3.7d exhibits an example for **32** with poor organization of the conjugated chains represented by the isotropic reflections. Therefore, it can be concluded that aryl groups can actually disturb the π-stacking of macromolecules **32**, if these substituents are attached on both sides of the backbone.

Moreover, it is possible to understand the relationship between polymer architecture and macroscopic order by comparing the 2D-WAXS results in more detail for different chemical structures of the carbazole-based polymers presented in Scheme 3.14. The most significant difference in the long-range organization was observed between **33** and **32**, whereby **33** revealed the higher degree of order and **32** showed only isotropic reflections in the X-ray pattern. This was even more interesting, since both polymers possess an almost identical polymer design. Compound **33** carries only carbazole units within the polymer backbone, which are substituted by short, branched alkyl side chains. On the contrary, **32** consists of only one carbazole and two fluorene units per monomer block. Additionally, each of the two fluorene units is substituted by two *n*-octyl side chains that increase the steric requirements at the polymer backbone. It has been shown for small molecular weight species that longer alkyl side chains, which were branched close to the rigid conjugated building block, decreased the π-interaction between the single molecules and thus increased the macroscopic disorder. Therefore, the two octyl side chains at the two fluorene units decrease the interaction between the rigid polymer blocks of **32** and lead to higher disorder compared to compound **33**. Furthermore, the interaction between carbazole units was considered to be stronger than between fluorenes due to the heteroatoms. Since **14** and **33** consist only of carbazoles, these compounds possess the most pronounced macroscopic organization among the presented polymers. On the contrary, the slight decrease in the order for **31** could be explained by the highest aryl fraction in relation to the molecular weight of each monomer unit and thus a larger "density" of aryl groups along the conjugated backbone. Since the aryl groups rotate out of plane toward the polymer, they are sterically more demanding and decrease the degree of π-interaction. However, the X-ray results indicated a well-defined structure of **31** in the extruded filaments, since the polymer backbone is substituted by less flexible side chains than the other investigated compounds **14**, **32**, and **33**. It is important to note that the π-stacking distance was almost identical for polymers **14**, **31**, and **33** and was thus not affected by the polymer architecture.

This section described the role of the macromolecular chemical design on the packing of the carbazole-based polymer chains and their self-organization. The successful implementation of the material in a device represents the final proof of principle. Therefore, the above-presented concept of the chemical design of the polymer structure and the related properties such as optical behavior and supramolecular order are reflected on the device performance. However, the operation of a solar cell is much more complex, since additional parameters have an additional impact on the efficiency. In the next section, the fabrication and performance of solar cells, based on conjugated polymers consisting of carbazole units, is described starting with the application of a soluble polycarbazole in a heterojunction morphology.

3.4
Photovoltaic Devices

3.4.1
Polycarbazole

Although soluble high molecular weight poly(2,7-carbazole)s have proved to be among the best blue light emitters in LEDs with luminance values as high as 800 cd m^{-2} at about 10 V [64], the use of these conjugated polymers as donor materials in solar cells has remained till now an unexplored area of research, which is discussed in this section.

One important issue for the fabrication of photovoltaic is the solubility of the materials. Therefore, 2-decyltetradecyl alkyl chains were introduced to improve the processibility of poly(2,7-carbazole). It has been shown previously that this is an efficient way to increase the solubility of perylene-based dyes [65] and two-dimensional polycyclic aromatic hydrocarbons [66]. The resulting **14** is highly soluble in common organic solvents such as chloroform or toluene and shows excellent film-forming properties. The potential applicability of **14** in solar cells was confirmed by the determination of the frontier molecular orbital energy levels by CV measurements. A thin film on an ITO substrate was tested in a three-electrode system. The onset oxidation potential versus ferrocene was 0.8 V, corresponding to a HOMO energy level of −5.6 eV. The bandgap calculated from the absorption spectrum of the thin film is 3.0 eV, so the LUMO is at −2.6 eV. Therefore, the HOMO energy level of the carbazole-based polymers is lower than that of polythiophene (e.g., −5.2 eV for

P3HT [67]) and PPV-based materials (e.g., −5.3 eV for MDMO-PPV [68]) being less easily oxidized.

Dyes based on perylene tetracarboxydiimide (PDI) have been used as acceptors in solar cells since the pioneering work of Tang [3]. The perylene-based materials have a low bandgap (~2 eV) and hence show perfect absorption compensation for **14**. On the contrary, PCBM is the most successful acceptor material in organic photovoltaics, especially when paired with conjugated polymers [5,14,15]. Therefore, both PDI and PCBM were chosen as potential acceptor materials in combination with **14**. These two acceptors were compared with regard to light absorption, charge transfer, matching of energy levels, and device performance.

Since the bandgap of **14** is relatively large (~3 eV), a material with a strong absorption in the visible range of 2–3 eV is required as an acceptor to achieve a device with high efficiency. PDI reveals an absorption maximum at 524 nm in CHCl$_3$ and at 537 nm in a thin film (Figure 3.9a). PCBM shows an absorption peak at 332 nm with an extended tail to 600 nm. The positions of the absorption maxima of the corresponding pure materials have a significant effect on the absorption of the blends. The blend of **14** and PDI as thin film achieves a broad absorption from 300 to 600 nm, covering the strongest emission of the solar spectrum (Figure 3.9b), whereas the **14**/PCBM film shows strong absorption only below 400 nm, attaining a much smaller overlap with the solar spectrum. Therefore, PDI compensates **14** more efficiently than PCBM does in terms of light absorption.

The fluorescence quenching of the donor material by adding acceptor material was used to verify the charge transfer from the donor to the acceptor. Fluorescence measurements on thin-film blends of **14** with PDI or PCBM at a ratio of 1 : 1 were taken. The peak of **14** at 429 nm is almost fully quenched by adding the same weight of PDI or PCBM. The HOMO and LUMO levels of PDI were estimated to be at −5.8 and −3.8 eV, respectively. The corresponding levels for PCBM have been reported to be at −6.0 eV and approximately −4.1 to −4.3 eV [69,70]. Therefore, both materials can potentially be used as acceptor materials with **14** as the donor. Photovoltaic devices

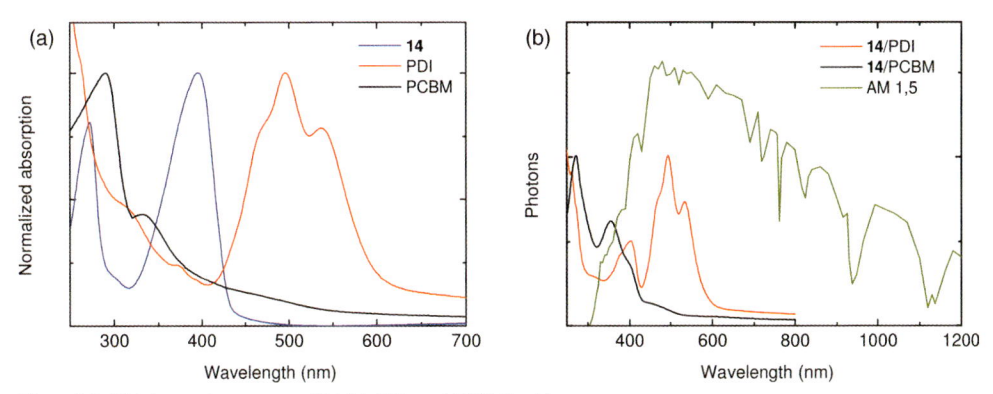

Figure 3.9 UV absorption spectra of (a) **14**, PDI, and PCBM in thin films and (b) blend thin films at a D/A ratio of 1 : 4; the AM 1.5 G solar spectrum is shown for comparison.

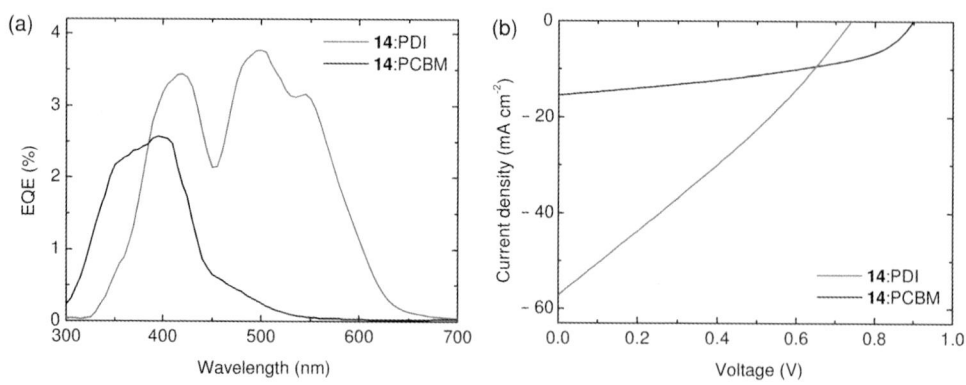

Figure 3.10 Photovoltaic performance of devices ITO/14:PDI (1:1)/Ag (red curve) and ITO/14:PCBM (1:1)/Ag (black curve): (a) EQE–λ curve, and (b) I–V curve under solar light AM 1.5 G with light intensity of 100 W m^{-2}.

were fabricated based on blends of **14**/PDI and **14**/PCBM at a D/A ratio of 1 : 1. The EQE–λ spectrum of the device using PCBM exhibits only one peak near 400 nm with a value of 2.5% (Figure 3.10). In contrast, the device using PDI reveals a broad EQE–λ spectrum with two strong peaks showing EQE values of 3.4% at 420 nm and 3.8% at 500 nm, which corresponds to the absorption of **14** and PDI, respectively (Figure 3.10). The broader photoresponse leads to a higher short-circuit current (I_{sc}) (57 µA cm^{-2}) for devices with PDI in comparison to solar cells based on PCBM (15 µA cm^{-2}). This results in a much higher overall efficiency for the device using PDI.

The higher photocurrent of the PDI-based device arises from more photons being absorbed by the active layer, as shown in the EQE–λ spectrum. The relative area covered by the spectrum represents the relative number of photons that contribute to the photocurrent. For the PDI-based device, this area is larger than that of the device with PCBM, thus leading to a higher I_{sc}. Therefore, the device using **14** as the donor and PDI as the acceptor was further optimized by varying the D/A ratio. The ratio of donor/acceptor not only determines the interface area at which the excitons are separated into electrons and holes, but also effects the transport pathways for the charge carriers. Therefore, the D/A ratio is a very important parameter that can profoundly influence the device behavior. In general, it is necessary to optimize the D/A ratio to achieve the best performance. In the presented case, blends of **14**/PDI with varying weight ratios from 1 : 1 to 1 : 5 were tested (Figure 3.11).

It is obvious that the PDI content strongly affects the photocurrent (Figure 3.11). I_{sc} increases from 57 µA cm^{-2} at a 1 : 1 ratio to 264 µA cm^{-2} at 1 : 4, that is, about four times higher. The V_{oc} remains 0.7 V for different ratios, except for D/A = 1 : 5. The fill factor (FF) increases from 1 : 1 to 1 : 2, and then remains the same for ratios of 1 : 3 and 1 : 4. The best performance is achieved for the blend containing 20 wt% of **14** and 80 wt% of PDI (1 : 4), corresponding to an overall efficiency of 0.67%, and a maximum EQE of 15.7% at 495 nm. When the content of PDI increases further, the film quality becomes poor since PDI does not possess good film-forming ability

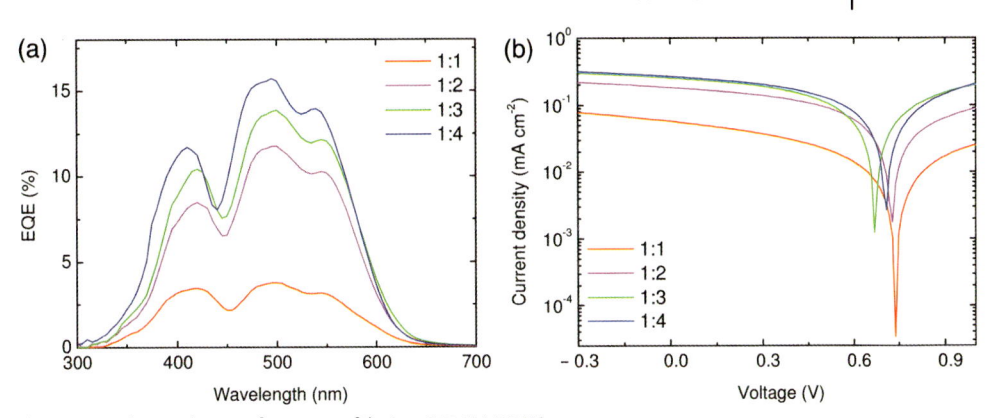

Figure 3.11 Photovoltaic performance of devices ITO/**14**:PDI/Ag at different D/A ratios: (a) EQE–λ curve, and (b) I–V curve with logarithmic ordinate under solar light AM1.5 G with light intensity of 100 W m^{-2}.

and tends to crystallize, thus leading to poor performance with a low V_{oc} of 0.32 V for the device containing 83% PDI (D/A = 1 : 5).

Since the relation between morphology and device performance is crucial during device fabrication, scanning electron microscopy (SEM) was applied as a powerful tool to investigate the internal film morphology in great detail in both top view and the section view modes (Figure 3.12). Pure films of **14** are homogeneous without any apparent texture. In comparison, the PDI film revealed significantly higher roughness with foliage-shaped crystals all over the surface and some cylindrical crystals protruding out of the surface. After **14** and PDI were blended, the films showed quite diverse morphologies varying with D/A ratios (Figure 3.12).

The section view images show the tendency of the morphology to change from "bilayer" to "homogeneous" as the PDI content is increased (Figure 3.12). At a ratio of 1 : 1 between **14** and PDI, some discontinuous foliage-like crystals float on the featureless layer. In view of the volume ratio, there must be PDI present in this featureless layer, indicating the existence of a homogeneous blend of **14** and PDI. The discontinuous foliage-like crystals of PDI develop into a continuous layer when the content of PDI increases to 66 wt% (D/A = 1 : 2), thus forming a clear bilayer structure composed of an upper layer characterized by closely packed foliage-like crystals, with a cross-sectional area of ∼0.5–1 µm^2, and a bottom layer of a homogeneous blend of **14** and PDI. When more PDI is contained within the blend, the size of the foliage crystals increase up to ∼5–10 µm^2 at 75 wt% (1 : 3) and to ∼10–20 µm^2 at 80 wt% (1 : 4). In the corresponding section images from 66 to 80 wt% PDI, the upper layer gradually becomes blurred and merges into the bottom layer. The bilayer structure develops into one "homogeneous" layer for the film containing 83 wt% (1 : 5) PDI, where the surface is composed of both crystalline domains (cylindrical and foliage-like crystals) and featureless domains. The cylindrical crystals on the surface that are similar to those in a pure PDI film can be observed in the bottom layer for all

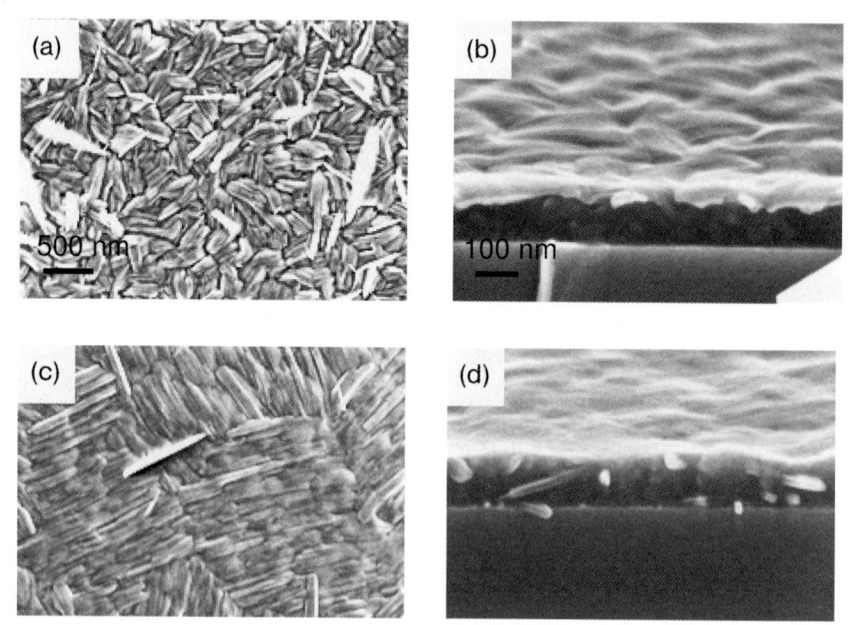

Figure 3.12 SEM pictures for top view (a and c) and cross-sectional view (b and d) of **14**:PDI blends = 1 : 2 (a and b) and 1 : 4 (c and d) thin films spin coated from chloroform solution.

the D/A ratios, and are especially notable when the content of PDI is higher than 75%. In conclusion, the blend films contain two kinds of morphologies, a featureless region consisting of a miscible blend of **14** and PDI, and a crystalline PDI-rich domain.

The formation of the bilayer structure can be explained by the different solubilities of **14** and PDI. When the films are prepared by spin coating from chloroform solution onto silicon wafer, during quick solvent evaporation **14** separates out from the solution first and accumulates on the substrate because of its lower solubility than PDI and higher density. However, the solubility is not the only driving force for the separation of the two compounds. Since the deposition is a very rapid process, dynamics of molecular diffusion in the solution should also be considered. Just before the solidification of **14**, the PDI molecules that are miscible with **14** and have not separated in this short time interval start to precipitate together with **14** to form the bottom layer on the substrate. Afterward, PDI crystals develop on the surface from the remaining PDI-rich solution. This is confirmed by the SEM images, revealing a top layer similar to a pure PDI film. Correlating the morphology with the device performance, one can conclude that with a decrease in the phase separation scale, the device performance enhances significantly. The phase separation directly influences charge separation and charge transport. The smaller the degree of phase separation, the larger is the D/A interface area and the greater is the charge separation. The efficiency of charge transport is determined by the percolation pathways formed by the donor and the acceptor in the blend. In conclusion, a homogeneous film rather than a bilayer structure is preferred in the **14**/PDI system to maximize efficiency.

The above section described a well-soluble poly(2,7-carbazole) (**14**) being incorporated as a donor material in solar cells for the first time using PDI as an electron acceptor. The good fit in orbital energy levels and absorption spectra led to high efficiency (overall efficiency of 0.7% under solar light) in the non-fullerene-containing system. This result indicates that conjugated polymers with high bandgap, which is in contrast to the low-bandgap polymers whose bandgap is normally less than 2 eV, can also be applied as materials to build efficient solar cell if appropriate electron acceptors are chosen. The device was further optimized by thermal treatment at 120 °C. A significant improvement in I_{sc} (and η) was observed for devices containing less than 66 wt% PDI, which was correlated with the improvement in crystallization of PDI. For devices having more than 75 wt% PDI, the thermal treatment did not show any apparent effect on the performance of the device. This is in accordance with the observed morphologies of the film before and after annealing, where the development of crystals was suppressed by the more miscible blends. The results indicate that charge separation is decisive in these systems and a good charge transport property is expected for **14**, making this kind of materials promising as donors for solar cells.

3.4.2
Ladder-Type Polymers Based on 2,7-Carbazole

Poly(2,7-carbazole) was shown to be a novel promising donor material in solar cells [13], but it absorbs mainly in the violet or ultraviolet region, which limits its application as a good light-harvesting component in photovoltaic cells. The incorporation of a long, branched alkyl chain at the polycarbazole backbone on one hand increases its solubility, but on the other hand reduces the amount of light-absorbing chromophore in the active layer. The high density of alkyl chains also enlarges the insulating effects [71–73], which prevents the intermolecular contact of semiconducting materials and hampers the facile photoinduced electron transfer from donor to acceptor material as well as the charge transport. It has been shown that the introduction of a ladder-type structure is an efficient way to red-shift the absorption spectra of carbazole-based polymers. Due to their more extended conjugation, ladder-type polycarbazoles absorb in the visible region, where the solar spectrum is most prominent [50,51]. On the basis of this, the carbazole-based ladder-type polymers **31–33** were designed and synthesized (Scheme 3.14). All three polymers **31–33** are highly soluble in common organic solvents such as $CHCl_3$ and THF with good film-forming properties. The photoluminescence of the polymers was effectively quenched by adding PCBM or PDI, which implied efficient charge transfer from the polymers **31–33** to PCBM and PDI. Cyclic voltammetry experiments indicated the redox behavior of polymers **31–33** in thin films revealing two oxidation peaks. The corresponding HOMO levels of **31–33** are summarized in Table 3.1. Polymer **33**, which has the largest density of electron-rich nitrogen on the backbone, shows the highest HOMO level and hence is the strongest electron donor among the three compounds, whereas **31** with the shortest conjugation unit possesses the lowest

Table 3.1 Electrochemical properties and energy levels of polymers **31–33**.

Polymer	E^{ox}_{ons} (V)	HOMO (eV)	LUMO (eV)	E_g (optical, eV)
31	0.73	−5.54	−2.82	2.72
32	0.66	−5.46	−2.76	2.70
33	0.60	−5.40	−2.78	2.62

HOMO level. The bandgaps were calculated from the UV spectra allowing the estimation of the LUMO levels (both are listed also in Table 3.1 for **31–33**). These values lead to a larger potential difference in HOMO (D)–LUMO (A) than for P3HT. Since the difference between the HOMO of the donor and the LUMO of the acceptor determines the open-circuit voltage, a higher V_{oc} is expected for devices based on the ladder-type polymers **31–33** than for those based on P3HT and PPV.

The UV absorption of the polymers **31–33** shows a 40 nm bathochromic shift compared to **14**, suggesting that PCBM is a promising acceptor material in solar cells using these polymers, as the overlap with the solar spectrum is larger. The orbital energy values allow the paring of these polymers with both PDI and PCBM. Therefore, photovoltaic devices using PCBM as the acceptor material and polymers **31–33** as donor materials were investigated. The polymer **31** was adopted as the model compound to optimize the D/A ratio in a ladder-type polymer/PCBM system. Four different ratios varying from 1 : 2 to 1 : 8 (D/A = 1 : 2, 1 : 4, 1 : 6, 1 : 8) were studied. A thickness of 100 nm was used for the active layer. The highest efficiency was achieved for the device containing 20 wt% polymer and 80 wt% PCBM, corresponding to an overall efficiency of 0.72% under solar light and an EQE_{max} of 12% at 450 nm. The V_{oc} was 0.93 V, greater than the V_{oc} of 0.6 V obtained from a P3HT/PCBM device, which was the most successful organic solar cell reported so far [74]. An increase of ∼0.1 V in V_{oc} was observed for the device using a 1 : 4 blend compared to the device using a 1 : 2 blend, and no further improvement was recorded on increasing the PCBM content. The optimum D/A ratio was thus determined to be 1 : 4 for the polymer/PCBM device. Deviation from this ratio decreased the performance of the device, especially when the content of PCBM was lower than 66 wt% or higher than 88 wt%. On the basis of these results, devices using **32** and **33** as the donor material and PCBM as the acceptor compound were fabricated at the same D/A ratio of 1 : 4, which is assumed to also be optimal also for these polymers (Figure 3.13).

The EQE curves (Figure 3.13a) show one main peak near the absorption maximum of the donor and a shoulder peak with an extended tail to 600 nm, arising from the absorption of PCBM. The device with **31** shows the highest EQE, while the solar cell with **33** has the broadest spectrum, which result in comparable photocurrent and overall efficiency, as concluded in Table 3.2. For **32**, a lower efficiency and EQE were observed, whereas it exhibits the highest V_{oc} among the three polymers.

To understand the variation in the device performance, one has to consider the chemical structures of the three polymers. The highest efficiency is found for **33** and the lowest efficiency for **32**, though both possess a similar poly(ladder-type hexaphenylene)s design. However, **33** has three nitrogen bridges as in carbazole and two carbon

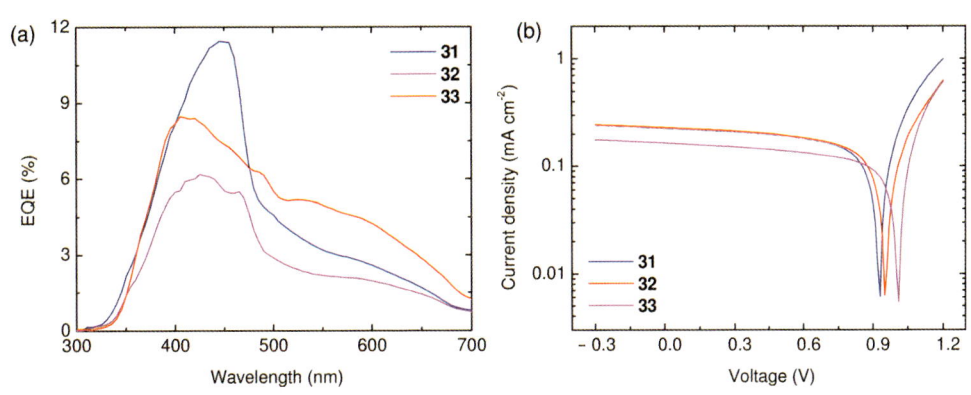

Figure 3.13 Photovoltaic performance of devices ITO/polymer 31–33:PCBM (1:4)/Ag, (a) EQE spectra and (b) semilogarithmic plot of the I/V characteristics. The I–V curves were tested under solar light AM1.5 G with light intensity of 150–160 W m^{-2}.

bridges as in fluorene per repeat unit, whereas **32** has only one nitrogen bridge and four carbon bridges. It has been shown that the nitrogen bridge is more efficient than carbon bridge in facilitating charge migration along the chain. Therefore, enhanced charge transport is expected for **33** than **32**, indicated by higher photocurrents I_{sc} observed for the device containing **33** (0.23 mA cm^{-2}) than **32** (0.16 mA cm^{-2}). The WAXS experiments revealed that polymer **33** also possesses the most pronounced macroscopic organization. Although the packing behavior will be somewhat different in a thin film than in an extruded fiber, the tendency toward self-organization depends mainly upon the molecular structure. So, the small decrease in efficiency for the device based on polymer **31** can reasonably be attributed to its reduced macroscopic order in comparison to **33**, which will decrease the exciton diffusion length and interchain charge transport. The chain-to-chain distance between conjugated backbones is a further important aspect that can be related to the device performance. This lateral distance depends only on the length of the substituted alkyl side chains, which fill the periphery between the conjugated backbones. Figure 3.14a shows the equatorial X-ray scattering intensity distribution from the obtained 2D patterns (Figure 3.7) for the investigated polymers. It is obvious that the distance between the macromolecules changes for different polymers, while the π-stacking remains identical for all compounds. Therefore, the longer the disordered, insulating side chains are, the larger is the chain-to-chain distance, the more hindered the interchain charge carrier transport,

Table 3.2 Photovoltaic parameters derived from Figure 3.13.

Polymer/PCBM (1:4)	I_{sc} (mA cm^{-2})	V_{oc} (V)	FF	Efficiency (%)	EQE$_{max}$ (%)
31	0.23	0.93	0.55	0.72	11
32	0.16	1.01	0.55	0.56	6
33	0.23	0.95	0.54	0.74	9

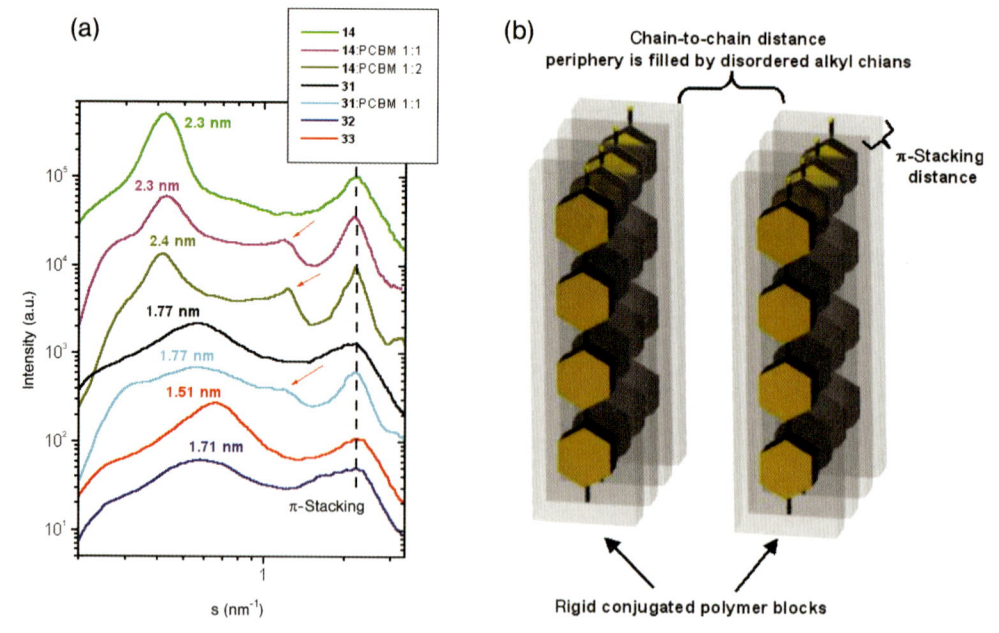

Figure 3.14 The chain-to-chain and π-stacking distance between conjugated polymers **14** and **31–33**, which is correlated with the photovoltaic performance, and schematic illustration of the periodicities observed in the plot.

and the shorter the exciton diffusion length between conjugated units. To verify if this distance changes in the presence of PCBM, extruded samples of polymer:PCBM blends were prepared and investigated. Characteristic equatorial plots of three different blends are added to Figure 3.14a for comparison with the pure polymers. In all cases, independent of the blend ratio, an identical chain-to-chain distance was observed as determined for the pure polymers. Since the chain-to-chain distance did not change, one can assume that the PCBM molecules do not intercalate between the rigid polymer backbones. This in turn can be attributed to the different solubility of the polymers and PCBM leading to a phase separation taking place during the evaporation of the solvent. Additionally, the long alkyl substituents dilute the concentration of the chromophore, decreasing in this way the light absorption and also the semiconducting fraction. In our study, this relation was quite obvious, since **14** showed the highest long-range order and the largest lateral distance between chains and a blue shift in absorption maximum and hence the lowest photovoltaic performance. By contrast, compounds **32** and **33** revealed a significantly smaller chain-to-chain distance and consequently a higher performance.

The typical V_{oc} of these ladder-type polymer/PCBM devices are in the range of 0.9–1.0 V, which is higher than that reported for P3HT/PCBM (0.6 V) [74] cell or PPV/PCBM cell (0.8 V) [75]. This is in accordance with the HOMO of P3HT (−5.2 eV) or PPV (−5.1 eV), being higher than the HOMO of polymers **31–33** (−5.4 to −5.5 eV).

However, the photocurrent of the device based on ladder-type polymers is lower than that of P3HT-based devices due to the better matching of light absorption of P3HT with solar spectrum and its stronger ability for self-organization in thin films, which prolongs the diffusion length of excitons and induces a higher charge carrier mobility. When comparing the difference in molecular structure between the polymers **31–33** and P3HT, it could be noted that apart from the different conjugation units, the substituents for polymers **31–33** are more bulky and larger than the hexyl substituent on P3HT. The large aryl/alkyl chains prohibit the efficient interchain transport of charges, implying the importance of the right balance between solubilizing substituents and required processability to lower the number of insulating side group and side chains. In comparison with the **14**/PCBM device, which showed an overall efficiency of 0.07% and an EQE_{max} of 4%, the efficiency values of **31–33** based devices are improved by approximately a factor of 10. The above results clearly demonstrate that with minor alternations of the monomer structure, it is possible to substantially improve the efficiency of polymer-based solar cells. In the present study, it was shown that tuning of the absorption to absorb more of the solar energy without altering the HOMO level results in a considerably enhanced efficiency.

This section presented the successful application of new ladder-type carbazole-based polymers as donor materials in photovoltaic devices and indicated ways to improve the device performance by molecular design, that is, maintaining the HOMO level while bathochromically shifting the absorption by adopting a more rigid ladder-type structure. A high ratio of nitrogen bridges with small alkyl substituents was a desirable feature in terms of both adjusting the absorption and maintaining a low lateral interchain separation, which was necessary for obtaining high current and efficiency values.

3.5
Conclusions

This chapter describes the successful application of conjugated polymers based on 2,7-carbazole as donor materials in organic bulk heterojunction solar cells. It has been shown that these donor materials perform efficiently in photovoltaic devices in combination with suitable acceptors. For a series of 2,7-carbazole-based polymers as donor compounds with slightly different design, a structure–performance relation was established, which is not only instructive for these materials but also serves as a guideline for improved macromolecular architectures. For the first time, a well-soluble poly(2,7-carbazole) was incorporated as a donor material in solar cells using perylentetracarboxydiimide (PDI) as an electron acceptor. The good fit in orbital energy levels and absorption spectra led to high efficiency. This result indicates that conjugated polymers with high bandgap can also be applied as materials to build efficient solar cells if appropriate electron acceptors are chosen. To enhance the light absorption ability, new ladder-type polymers based on pentaphenylene and hexaphenylene with one and three nitrogen bridges per repeat unit have been synthesized and applied in solar cell devices. The study suggests that the more nitrogen bridges are included in the conjugated backbone, the more red-shifted absorbance and emission is observed together with the better packing in the solid state, leading to improved device efficiency. This leads to the main conclusion that device performance can be enhanced by molecular design, for instance, maintaining the HOMO level while bathochromically shifting the absorption by adopting a more rigid ladder-type structure. Also, a high ratio of nitrogen bridges with small alkyl substituents was a desirable feature in terms of both adjusting the absorption and maintaining a low lateral interchain separation, which were essential for obtaining high current and efficiency values.

Since the efficiency of organic bulk heterojunction solar cells is not high in comparison to the already commercialized inorganic solar cells, there are further strategies necessary to improve the devices. In this chapter, the implementation of carbazole-based conjugated polymers in photovoltaics was presented as an alternative. On the basis of these results new macromolecular structures will be developed to match the requirements for the better performance of solar cells, as for instance extended light absorption and enhanced self-organization, in the expectation to further increase the efficiency of organic photovoltaics to reach the market goals of around 10%.

References

1 Wöhrle, D. and Meissner, D. (1991) *Advanced Materials*, **3**, 129.
2 Chamberlain, G.A. (1983) Organic solar cells: a review. *Solar Cells*, **8**, 47.
3 Tang, C.W. (1986) *Applied Physics Letters*, **48**, 183.
4 Sariciftci, N.S., Smilowitz, L., Heeger, A.J. and Wudl, F. (1992) *Science*, **258**, 1474.
5 Brabec, C.J., Zerza, G., Cerullo, G., De Silvestri, S., Luzzati, S., Hummelen, J.C. and Sariciftci, N.S. (2001) *Chemical Physics Letters*, **340**, 232.
6 Shaheen, S.E., Brabec, C.J., Sariciftci, N.S., Padinger, F., Fromherz, T. and Hummelen, J.C. (2001) *Applied Physics Letters*, **78**, 841.
7 Kroon, J.M., Wienk, M.M., Verhees, W.J.H. and Hummelen, J.C. (2002) *Thin Solid Films*, **403–404**, 223.
8 Munters, T., Martens, T., Goris, L., Vrindts, V., Manca, J., Lutsen, L., Ceunick, W.D., Vanderzande, D., Schepper, L.D., Gelan, J., Sariciftci, N.S. and Brabec, C.J. (2002) *Thin Solid Films*, **403–404**, 247.
9 Aernouts, T., Geens, W., Portmans, J., Heremans, P., Borghs, S. and Mertens, R. (2002) *Thin Solid Films*, **403**, 297.
10 Schilinsky, P., Waldauf, C. and Brabec, C.J. (2002) *Applied Physics Letters*, **81**, 3885.
11 Padinger, F., Rittberger, R.S. and Sariciftci, N.S. (2003) *Advanced Functional Materials*, **13**, 85.
12 Svensson, M., Zhang, F., Veenstra, S.C., Verhees, W.J.H., Hummelen, J.C., Kroon, J.M., Inganäs, O. and Andersson, M.R. (2003) *Advanced Materials*, **15**, 988.
13 Wienk, M.M., Kroon, J.M., Verhees, W.J.H., Knol, J., Hummelen, J.C., van Hall, P.A. and Janssen, R.A.J. (2003) *Angewandte Chemie–International Edition*, **42**, 3371.
14 Kim, Y., Cook, S., Tuladhar, S.M., Choulis, S.A., Nelson, J., Durrant, J.R., Bradley, D.D.C., Giles, M., Mcculloch, I., Ha, C. and Ree, M. (2006) *Nature Materials*, **5**, 197.
15 Ma, W., Yang, C., Gong, X., Lee, K. and Heeger, A.J. (2005) *Advanced Functional Materials*, **15**, 1617.
16 Benanti, T.L. and Venkataraman, D. (2006) *Photosynthesis Research*, **87**, 73.

17 Li, G., Shrotriya, V., Yao, Y. and Yang, Y. (2005) *Journal of Applied Physiology*, **98**, 43704.

18 Chirvase, D., Parisi, J., Hummelen, J.C. and Dyakonov, V. (2004) *Nanotechnology*, **15**, 1317.

19 Brabec, C.J., Padinger, F., Sariciftci, N.S. and Hummelen, J.C. (1999) *Journal of Applied Physiology*, **85**, 6866.

20 Shaheen, S.E., Radspinner, R., Peyghambarian, N. and Jabbour, G.E. (2001) *Applied Physics Letters*, **79**, 2996.

21 Yu, G., Gao, J., Hummelen, J.C., Wudl, F. and Heeger, A.J. (1995) *Science*, **270**, 1789.

22 Yu, G. and Heeger, A.J. (1995) *Journal of Applied Physiology*, **78**, 4510.

23 Halls, J.J.M., Walsh, C.A., Greenham, N.C., Marseglia, E.A., Friend, R.H., Moratti, S.C. and Holmes, A.B. (1995) *Nature*, **376**, 498.

24 Grazulevicius, J.V., Strohriegl, P., Pielichowski, J. and Pielichowski, K. (2003) *Progress in Polymer Science*, **28**, 1297.

25 Ambrose, J.F. and Nelson, R.F. (1967) *Journal of the Electrochemical Society*, **115**, 1159.

26 Siove, A. and Adès, D. (2004) *Polymer*, **45**, 4045.

27 Brihaye, O., Legrand, C., Chapoton, A., Chevrot, C. and Siove, A. (1993) *Synthetic Metals*, **55**, 5075.

28 Romero, D.B., Schaer, M., Leclerc, M., Adès, D., Siove, A. and Zuppiroli, L. (1996) *Synthetic Metals*, **80**, 271.

29 Belletête, M., Bédard, M., Leclerc, M. and Durocher, M. (2004) *Journal of Molecular Structure*, **679**, 9.

30 Belletête, M., Bédard, M., Bouchard, J., Leclerc, M. and Durocher, M. (2004) *Canadian Journal of Chemistry*, **82**, 1280.

31 Brière, J.F. and Côté M. (2004) *Journal of Physical Chemistry. B*, **108**, 3123.

32 Dierschke, F., Grimsdale, A.C. and Müllen, K. (2003) *Synthesis*, **16**, 2470.

33 Yamato, T., Hideshima, C., Suehiro, K., Tashiro, M., Prakash, G.K.S. and Olah, G.A. (1991) *Journal of Organic Chemistry*, **56**, 6248.

34 Morin, J.F., Beaupré, S., Leclerc, M., Lévesque, I. and D'Iorio, M. (2002) *Applied Physics Letters*, **80**, 341.

35 Shirakawa, H., Louis, E.J., Macdiarmid, A.G., Chiang, C.K. and Heeger, A.J. (1977) *Journal of the Chemical Society. Chemical Communications*, **16**, 578.

36 Desbene-Monvernay, A., Lacaze, P.C. and Dubois, J.E. (1981) *Journal of Electroanalytical Chemistry*, **129**, 229.

37 Mengoli, G., Musiani, M.M., Schreck, B. and Zeccin, S. (1988) *Journal of Electroanalytical Chemistry*, **246**, 73.

38 Cattarin, S., Mengoli, G., Musiani, M.M. and Schreck, B. (1988) *Journal of Electroanalytical Chemistry*, **246**, 87.

39 Marrec, P., Dano, C., Gueguen-Simonet, N. and Simonet, J. (1997) *Synthetic Metals*, **89**, 171.

40 Tucker, S.H. (1927) *Journal of the American Chemical Society*, **1**, 1388.

41 Bersford, P., Iles, D.H., Kricka, L.J. and Ledwith, A. (1974) *Journal of the Chemical Society, Perkin Transactions*, **1**, 276.

42 Siove, A., David, R., Ades, D., Roux, C. and Leclerc, M. (1995) *Journal De Chimie Physique*, **92**, 787.

43 Ambrose, J.F. and Nelson, R.F. (1967) *Journal of the Electrochemical Society*, **115**, 1159.

44 Ambrose, J.F., Carpenter, L.L. and Nelson, R.F. (1975) *Journal of the Electrochemical Society*, **122**, 876.

45 Zhang, Z.B., Fujiki, M., Tang, H.-Z., Motonaga, M. and Torimistu, K. (2002) *Macromolecules*, **35**, 1988.

46 Morin, J.-F. and Leclerc, M. (2001) *Macromolecules*, **34**, 4680.

47 Zotti, G., Schiavon, G., Zeccin, S., Morin, J.-F. and Leclerc, M. (2002) *Macromolecules*, **35**, 2122.

48 Li, J.L., Dierschke, F., Wu, J.S., Grimsdale, A.C. and Müllen, K. (2006) *Journal of Materials Chemistry*, **16**, 96.

49 Scherf, U. and Müllen, K. (1991) *Makromolecular Chemistry, Rapid Communications*, **12**, 489.

50 Dierschke, F., Grimsdale, A.C. and Müllen, K. (2004) *Macromolecular Chemistry and Physics*, **205**, 1147.

51 Patil, S., Scherf, U. and Kadashchuk, A. (2003) *Advanced Functional Materials*, **13**, 609.

52 Mishra, A.K., Graf, M., Grasse, F., Jacob, J., List, E.J.W. and Müllen, K. (2006) *Chemistry of Materials*, **18**, 2879.

53 Mishra, A.K., Li, J., Pisula, W., Dierschke, F., Jacob, J., Grimsdale, A.C. and Müllen, K., submitted.

54 Zhang, M., Yang, C., Mishra, A.K., Pisula, W., Zhou, G., Schmaltz, B., Baumgarten, M. and Müllen, K. Chemical Communications, online.

55 Ong, B.S., Wu, Y., Liu, P. and Gardner, S. (2004) *Journal of the American Chemical Society*, **126**, 3378.

56 McCulloch, I., Heeney, M., Bailey, C., Genevicius, K., MacDonald, I., Shkunov, M., Sparrowe, D., Tierney, S., Wagner, R., Zhang, W., Chabinyc, M.L., Kline, R.J., McGehee, M.D. and Toney, M.F. (2006) *Nature Materials*, **5**, 328.

57 Zhang, M., Tsao, H.N., Pisula, W., Yang, C., Mishra, A.K. and Müllen, K. (2007) *Journal of the American Chemical Society*, **129**, 3472.

58 Chabinyc, M.L., Toney, M.F., Kline, R.J., McCulloch, I. and Heeney, M. (2007) *Journal of the American Chemical Society*, **129**, 3226.

59 Chang, J.-F., Clark, J., Zhao, N., Sirringhaus, H., Breiby, D.W., Andreasen, J.W., Nielsen, M.M., Giles, M., Heeney, M. and McCulloch, I. (2006) *Physical Review B*, **74**, 115318.

60 Pisula, W., Tomovic, Z., Simpson, C.D., Kastler, M., Pakula, T. and Müllen, K. (2005) *Chemistry of Materials*, **17**, 4296.

61 Grenier, C.R.G., Pisula, W., Joncheray, T.J., Müllen, K. and Reynolds, J.R. (2007) *Angewandte Chemie-International Edition*, **46**, 714.

62 Percec, V., Glodde, M., Peterca, M., Rapp, A., Schnell, I., Spiess, H.W., Bera, T.K., Miura, Y., Balagurusamy, V.S.K., Aqad, E. and Heiney, P.A. (2006) *Chemistry – A European Journal*, **12**, 6298.

63 Pisula, W., Dierschke, F. and Müllen, K. (2006) *Journal of Materials Chemistry*, **16**, 4058.

64 Morin, J.F., Beaupré, S., Leclerc, M., Lévesque, I. and D'Iorio, M. (2002) *Applied Physics Letters*, **80**, 341.

65 (a) Langhals, H., Demming, S. and Potrawa, T., (1991) *Journal Fur Praktische Chemie*, **333**, 733–748. (b) Langhals, H., Ishmael, R. and Yuruk, O., (2000) *Tetrahedron*, **56**, 5435–5441.

66 Pisula, W., Kastler, M., Wasserfallen, D., Pakula, T. and Müllen, K. (2004) *Journal of the American Chemical Society*, **126**, 8074.

67 Al-Ibrahim, M., Roth, H.K., Zhokhavets, U., Gobsch, G. and Sensfuss, S. (2005) *Solar Energy Mater & Solar Cells*, **85**, 13.

68 Al-Ibrahim, M., Konkin, A., Roth, H.K., Egbe, D.A.M., Klemm, E., Zhokhavets, U., Gobsch, G. and Sensfuss, S. (2005) *Thin Solid Films*, **474**, 201.

69 Scharber, M.C., Mühlbacher, D., Koppe, M., Denk, P., Waldauf, C., Heeger, A.J. and Brabec, C.J. (2006) *Advanced Materials*, **18**, 789.

70 Wienk, M.M., Struijk, M.P. and Janssen, R.A.J. (2006) *Chemical Physics Letters*, **422**, 488.

71 Egbe, D.A.M., Nguyen, L.H., Hoppe, H., Mühlbacher, D. and Sariciftci, N.S. (2005) *Macromolecular Rapid Communications*, **26**, 1389.

72 Chen, Z.K., Huang, W., Wang, L.H., Kang, E.T., Chen, B.J., Lee, C.S. and Lee, S.T. (2000) *Macromolecules*, **33**, 9015.

73 Yamamoto, T. and Lee, B.L. (2002) *Macromolecules*, **35**, 2993.

74 van Duren, J.K.J., Yang, X., Loos, J., Bulle-Lieuwma, C.W.T., Sieval, A.B., Hummelen, J.C. and Janssen, R.A.J. (2004) *Advanced Functional Materials*, **14**, 425.

75 Schilinsky, P., Asawapirom, U., Scherf, U., Biele, M. and Brabec, C.J. (2005) *Chemistry of Materials*, **17**, 2175.

4
New Construction of Low-Bandgap Conducting Polymers
Zhengguo Zhu, David Waller, and Christoph J. Brabec

4.1
Introduction

Conducting polymers [1] are organic polymeric materials that possess electronic properties similar to those of metals or inorganic semiconductors. They surpass their inorganic counterparts in solubility and processability, which makes possible the fabrication of exceptionally thin devices, promising revolutionary advancements in many technologies such as flexible and large-area displays [1–3], field effect transistors [4–6], and photovoltaics [7–9]. Polyacetylene, the first conducting polymer, was discovered in the 1970s [10,11]. After that, many new conducting polymers were soon synthesized and the scope of this class of material was quickly expanded to include the poly(arylenene), poly(arylene vinylene), and poly(arylene ethynylenes), where the arylenes are pyrrole, benzene, or thiophene. Incorporation of arylenes into the polymer chains affords more stable polymers and also provides anchoring points for side chains, offering additional control of polymer properties such as solubility. As conducting polymers continued to evolve, more sophisticated aromatic units such as fluorenes were incorporated into the main chains and various polymer and copolymers were developed (Figure 4.1(3)) [12]. One interesting feature of fluorene is that its 9-carbon can be readily functionalized with alkyl groups to increase the solubility of the polymer without causing additional twist in the main chain. For the past 10 years, both main chains and side chains are being constantly diversified giving rise to numerous structures and rich combinations of properties, providing an increasingly clear picture of structure–property relationship and allowing for more efficient design and optimization of structures to address various application needs. One of the most important characteristics of a conducting polymer is its bandgap, which is defined as the energy difference between the HOMO and LUMO of a polymer. A small bandgap is desired for increasing the conductivity and enhancing the nonlinear optical properties of the polymers [13,14]. Also, a small bandgap allows the material to absorb light up to a longer wavelength, which is highly desired for photovoltaic applications, where a good coverage of the solar spectrum is required for

Organic Photovoltaics: Materials, Device Physics, and Manufacturing Technologies.
Edited by Christoph Brabec, Vladimir Dyakonov, and Ullrich Scherf
Copyright © 2008 WILEY-VCH Verlag GmbH & Co. KGaA, Weinheim
ISBN: 978-3-527-31675-5

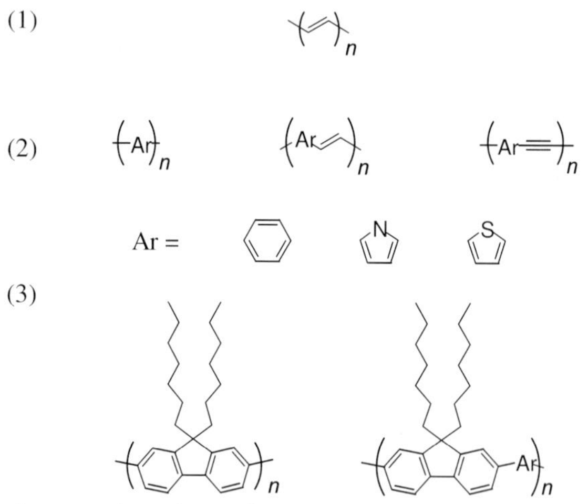

Figure 4.1 The evolution of conduction polymers.

high power conversion efficiency [15,16]. Furthermore, the fabrication of OLEDs that operate at the IR region also requires low-bandgap materials [13].

Since the methods of creating aryl–aryl, aryl–vinyl–aryl, and aryl–ethynyl–aryl linkage were well established through the synthesis of many poly(arylenene), poly(arylene vinylene), and poly(arylene ethynylenes), most of the recent designs of new low-bandgap polymers have been the result of derivatization of these classes of polymers. A general strategy has involved the synthesis of low-bandgap segments and their incorporation into the well-known conducting polymer frameworks such as polyfluorenes or polythiophenes. A representative list of such building blocks is summarized in Figure 4.2. One important class of such low-bandgap fragments consists of electron-deficient heterocycles that have a strong tendency to take a quinoidlike form, such as isothianaphthene [17,18], benzothiadiazole (BT, BBT) [19–21], quinoxaline (QU, PQ) [22–25], and pyrazine (TP, PQ) [26,27], covalently attached to two electron-rich moieties (e.g., thiophenes) at both sides. Arylene vinylene units containing donor–acceptor functional groups also afford low-bandgap building blocks. Acenes such as pentacenes, which already have a low bandgap, were also functionalized and incorporated into conducting polymers that are of low bandgap. New electron-donating monomers with forced planarity promoting electron delocalization were also prepared allowing for synthesis of either homopolymers or polymers containing alternating donor–accepter units across the main chain, resulting in small bandgaps.

4.2
Low-Bandgap Polymers Containing 4,7-Di-2-Thienyl-2,1,3-Benzothiadiazole Moieties

4,7-Di-2-thienyl-2,1,3-benzothiadiazole (**1**, Figure 4.3) can be readily synthesized by Suzuki coupling of 2-thiopheneboronic acid with 4,7-dibromo-1,2,3-benzothiadiazole [28,29] or Stille coupling of 2-(tributylstannyl)thiophene with 4,7-dibromo-1,2,3-

4.2 Low-Bandgap Polymers Containing 4,7-Di-2-Thienyl-2,1,3-Benzothiadiazole Moieties

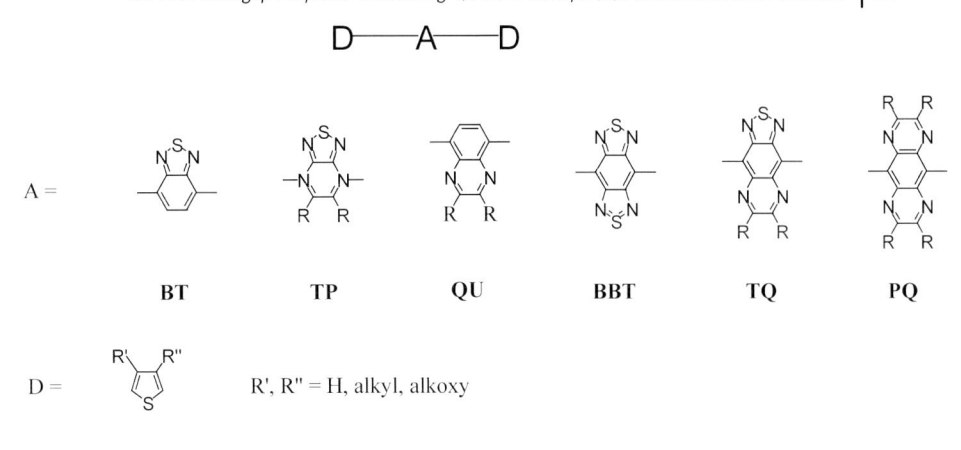

Figure 4.2 Representative building blocks for low-bandgap polymers.

benzothiadiazole [22]. This molecular material itself has an absorption maximum of 445 nm in chloroform. When it was electrically polymerized, the resulting polymer was found to possess a bandgap of 1.1 eV, which is significantly narrower than that of polythiophene (2.1 eV) [19], but due to the lack of side chains, the polymer is not soluble. Similarly, 4,7-bis(thieno[3,4-b]-1,4-dioxin-5-yl)-2,1,3-benzothiadiazole (**2**) and 4,7-bis[3,4-bis(hexyloxy)-2-thienyl]-2,1,3-benzothiadiazole (**3**) were prepared [21]. The absorption maximum of the prepared 4,7-bis(thieno[3,4-b]-1,4-dioxin-5-yl)-2,1,3-benzothiadiazole and 4,7-bis[3,4-bis(hexyloxy)-2-thienyl]-2,1,3-benzothiadiazole-2 were found to be 481 and 456 nm in dichloromethane, respectively.

Soluble polymers containing 4,7-di-2-thienyl-2,1,3-benzothiadiazole moieties were prepared by incorporating 3-alkylthiophenes or 3-alkoxythiophenes. 4,7-Di-2-thienyl-

Figure 4.3 4,7-Di-2-thienyl-2,1,3-benzothiadiazoles.

Table 4.1 Low-bandgap polymers containing 4,7-?di-2-thienyl-2,1,3-benzothiadiazole moieties.

Structure	Bandgap (eV)
PB4BT R = n-octyl,	1.96
PB34BT R = n-octyl,	NA
PB3BT R = n-octyl, 3,7,11-trimethyldodecyl	2.01
4	2.10
5	2.10
6	1.65
P6, R = n-octyl,	1.65
PDDBT, R = n-decanyl	1.38

Table 4.1 (Continued)

Structure	Bandgap (eV)
P1, R = n-octyl, R′ = n-dodecyl	1.85
PTPTB, R = H, R′ = n-dodecyl	1.6

2,1,3-benzothiadiazole polymers containing alkyl side chains such as poly[4,7-bis(4-octyl-2-thienyl)-2,1,3-benzothiadiazole] (PB4TB, Table 4.1), poly[4,7-bis(3-octyl-2-thienyl)-2,1,3-benzothiadiazole] (PB3TB, Table 4.1), and poly[4-(3-octyl-2-thienyl)-7-(4-octyl-2-thienyl)-2,1,3-benzothiadiazole] (PB34TB, Table 4.1) were synthesized by oxidative coupling using iron(III) chloride. These polymers are soluble; the optical bandgap of PB4TB and PB3TB was found to be 1.96 and 2.01 eV, respectively, in solid state, which is significantly larger than that of electropolymerized 4,7-di-2-thienyl-2,1,3-benzothiadiazole [30]. This increase in bandgap was caused by the steric hindrance of the alkyl substitutes, forcing the benzothiadiazole and thiophene rings in the main chain to be further away from coplanarity. Recently,\ the synthesis and characterization of polymers 4–6 (Table 4.1) were reported. These polymers possess a bandgap of 2.10, 2.10, and 1.65 eV, respectively, as determined by a combination of UV–Vis spectroscopy and ultraviolet photoelectron spectroscopy (UPS) [31]. The much lower bandgap of **6** compared to others indicates that the coplanarity of the two thiophenes adjacent to the benzothiadiazole with the acceptor is much more important in lowering the bandgap, and polymers with side chains further away from the acceptor possess much smaller bandgap. Soluble copolymers of 4,7-bis(5-bromo-4-octyl-2-thienyl)-2,1,3-benzothiadiazole with 1-dodecyl-2,5-bis(trimethylstannyl)-1H-pyrrole (P1) and 4,7-bis(5-bromo-4-octyl-2-thienyl)-2,1,3-benzothiadiazole with thiophene (P6) were also synthesized, the bandgap of which are 1.89 and 1.65 eV, respectively, in solid state. This decrease in bandgap relative to PB4TB and PB3TB is due to the incorporation of a stronger electron donor for the former and the ease of steric hindrance for the latter [32]. Recently, an analogue of P6, poly[4,7-bis(4-decanyl-2-thienyl)-2′,1′,3′-benzothiadiazole-thiophene-2,5-] (PDDBT), was synthesized by a Stille coupling reaction. This polymer was found to have a bandgap of 1.38 eV. This further decrease is likely due to an increase in solubility as a result of longer side chains and an increase in molecular weight. Bulk heterojunction photovoltaic cells from PDDBT and [6,6]-phenyl-C_{61}-butyric acid methyl ester (PCBM) showed power conversion efficiencies up to 0.13% under AM 1.5 conditions and a photocurrent response up to 880 nm [33]. Removal of the alkyl groups from the thiophenes in P1 results in a polymer containing alternating electron-rich N-dodecyl-2,5-bis(2′-thienyl)pyrrole

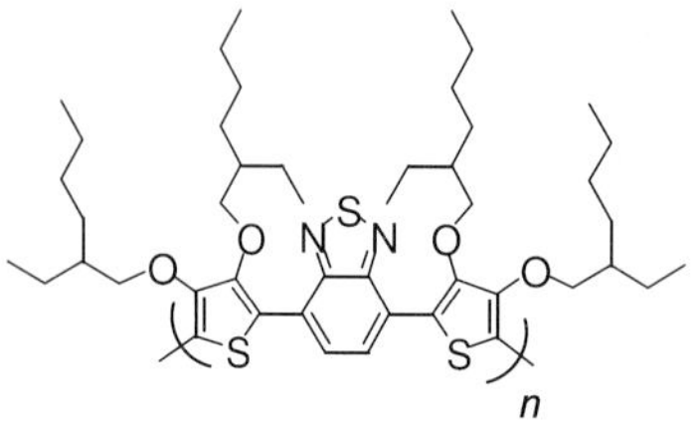

PBEHTB **7** **poly(heptyl4-PTBT)**, R = *n*-heptyl

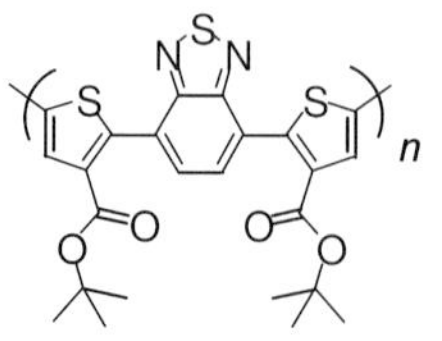

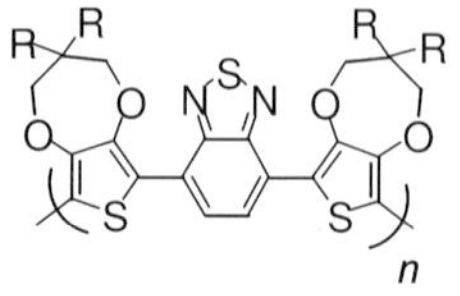

Figure 4.4 4,7-Di-2-thienyl-2,1,3-benzothiadiazole containing additional functional groups.

(TPT) and electron-deficient 2,1,3-benzothiadiazole (B) units (PTPTB, Figure 4.2). Due to the decrease in steric hindrance and thus increased conjugation, PTPTB has a bandgap of 1.6 eV [34]. However, the solubility of PTPTB is quite low. Bulk heterojunction photovoltaic cells prepared from PTPTB and [6,6]-phenyl-C$_{61}$-butyric acid methyl ester (PCBM) gave power conversion efficiencies up to 1% under AM 1.5 illumination [16,35].

A homopolymer of 4,7-bis[3,4-bis[(2-ethylhexyl)oxy]-2-thienyl]-2,1,3-benzothiadiazole (PBEHTB, Figure 4.4) was synthesized by reductive condensation of 4,7-bis(5-bromo-3,4-bis[2-ethylhexyloxy]-2-thienyl)-2,1,3-benzothiadiazole by reductive polycondensation using Ni(COD)$_2$ [36,37]. The polymer absorbs light up to 800 nm, and solar cells prepared using this material blended with PCBM have photoresponse up to 1 μm with an open-circuit voltage of 0.77 V, a short-circuit current of 2.8 mA cm^{-2}, and a power conversion efficiency of 0.9%. A series of soluble polymers were prepared from 2,2'-(2,1,3-benzothiadiazole-4,7-diyl)bis[5-bromo-3-thiophenecarboxylic acid, bis(1,1-dimethylpentyl) ester] (7, Figure 4.5). The solubility of these polymers drastically changes after thermal treatment due to loss of the alkyl side groups and the formation of carboxylic functional group [38].

A low-bandgap, conjugated polymer, poly[4,7-bis(3',3'-diheptyl-3,4-propylenedioxythienyl)-2,1,3-benzothiadiazole] (poly(heptyl4-PTBT), Figure 4.5), consisting of alternating electron-rich diheptyl-substituted propylene dioxythiophene and electron-deficient 2,1,3-benzothiadiazole units, was found to have an optical bandgap of 1.55 eV at solid state. A bulk heterojunction solar cell with poly(heptyl4-PTBT) as electron donor and PCBM as electron acceptor showed open-circuit voltage of 0.37 eV, short circuit of 3.15 mA cm^{-2}, and power conversion efficiency of 0.35%

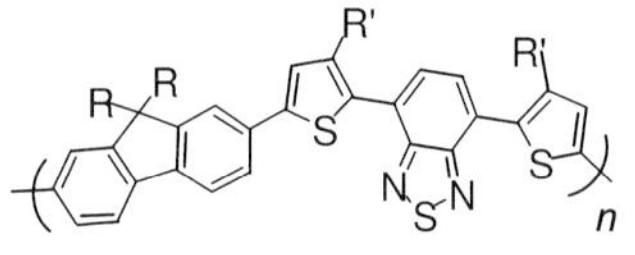

APFO-3, R = n-octyl, R' = H
PF-co-DTB, R = *n*-octyl, R' = *n*-decyloxy

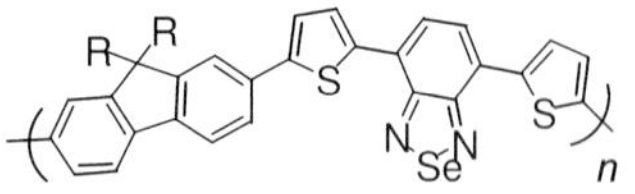

PFO-DBSe, R = n-octyl, R' = H

Figure 4.5 Copolymers of fluorene and 4,7-di-2-thienyl-2,1,3-benzothiadiazole and their analogues.

under AM 1.5G conditions. A solid-state, dye-sensitized solar cell with a SnO$_2$:F/TiO$_2$/N$_3$ dye/poly(heptyl4-PTBT)/Pt device configuration was fabricated using poly(heptyl4-PTBT) as hole transport material. This device exhibited power conversion efficiency of 3.1% [39].

Polymers containing alternating 2,7-(9,9-dialkyl-fluorene) and 4,7-di-2-thienyl-2′,1′,3′-benzothiadiazole (APFO-3 and PF-co-DTB, Figure 4.5) were synthesized by Suzuki coupling protocols [40–43]. This class of polymers has an absorption maximum around 545 nm. Solar cells prepared from a mixture of APFO-3 and PCBM has an open-circuit voltage as high as 1.03 V and a power conversion efficiency as high as 2.84% [43,44]. Poly{(9,9-dioctylfluorene)-2,7-diyl-alt-[4,7-bis(3-decyloxythien-2-yl)-2,1,3-benzothiadiazole]-5′,5″-diyl} (PF-co-DTB) was synthesized in a similar manner, and this polymer was found to have a bandgap of 1.78 eV [45]. This narrowing of bandgap relative to that of APFO-3 can be attributed the alkoxy side chains that may enhance the electron-donating strength of the thiophenes. Bulk heterojunction solar cells prepared from a blend of PF-co-DTB and PCBM give power conversion efficiency of 1.6%, with an open-circuit voltage of 0.76 V. When PF-co-DTB is blended with (6,6)-phenyl-C$_{71}$-butyric acid Me ester (C$_{70}$-PCBM), the resulting solar cells with 2.4% power conversion efficiency are realized [46]. 4,7-Di-2-thienyl-2,1,3-benzoselenadiazole (DBSe) was synthesized from the corresponding 4,7-di-2-thienyl-2,1,3-benzothiadiazole. A copolymer of 9,9-dioctylfluorene and 4,7-di-2-thienyl-2,1,3-benzoselenadiazole (DBSe) was prepared (PFO–DBSe, Figure 4.6). PFO–DBSe absorbs and emits at significantly longer wavelength than the corresponding 2,1,3-benzothiodiazole polymers (i.e., APFO-3). Light-emitting diodes prepared using PFO–DBSe have electroluminescence maximum at 727 nm. Bulk heterojunction photovoltaic cells prepared with the copolymer as the electron donor and PCBM acceptor gave energy conversion efficiency of 0.91% under AM 1.5 conditions [47].

4,7-Di-2-thienyl-2′,1′,3′-benzothiadiazoles were recently incorporated into a platinum metallopolyyne, where this low-bandgap segment was linked with bis(tributylphosphine)platinums through ethynylene linkages (8, Figure 4.6). The resulting polymer has a bandgap of 1.85 eV. Solar cells prepared from a blend of this metallopolyyne and a fullerene derivative showed a power conversion efficiency of 4.1% [48].

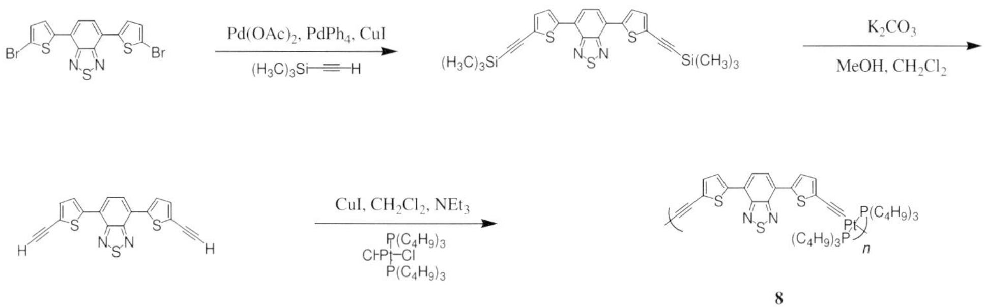

Figure 4.6 The synthesis of a platinum metallopolyyne polymer containing 4,7-di-2-thienyl-2′,1′,3′-benzothiadiazoles.

4.3
Low-Bandgap Polymers Containing 4,8-Di-2-Thienyl-Benzo[1,2-c:4,5-c′]bis[1,2,5]thiadiazole Segments

4,8-Di-2-thienyl-benzo[1,2-c:4,5-c′]bis[1,2,5]thiadiazole was synthesized by coupling 4,7-dibromo-5,6-dinitro-2,3,1-benzothiadiazole with tributyl(2-thienyl)tin, followed by the reduction of the nitro groups into amine, and then followed by the treatment of the resulting diamine with N-thionylaniline (Figure 4.7) [19,20,22]. This material itself already has an absorption maximum of 702 nm. Crystal structure of 4,8-di-2-thienyl-benzo[1,2-c:4,5-c′]bis[1,2,5]thiadiazole showed that the thiophene and bis[1,2,5]thiadiazole are coplanar, indicating strong electron delocalization. Its electropolymerization on indium-tin oxide (ITO) electrodes produces an insoluble polymer with an optical bandgap smaller than 0.5 eV. A soluble low-bandgap copolymer (**9**, Figure 4.7) was prepared by coupling 4,7-bis(5-bromothiophen-2-yl)-benzo[1,2-c;4,5-c′]bis-[1,2,5]thiadiazole (synthesized by the bromination of 4,8-dithien-2-yl-benzo[1,2-c;4,5-c′]bis[1,2,5]thiadiazole) with 3-(3,7,11-trimethyldodecyl)-2,5-bis(trimethylstannyl)thiophene [31]. This polymer was found to have a bandgap of 0.65 eV. Likewise, polymer **10** (Figure 4.7) was prepared in a similar manner, which has a bandgap of 0.67 eV.

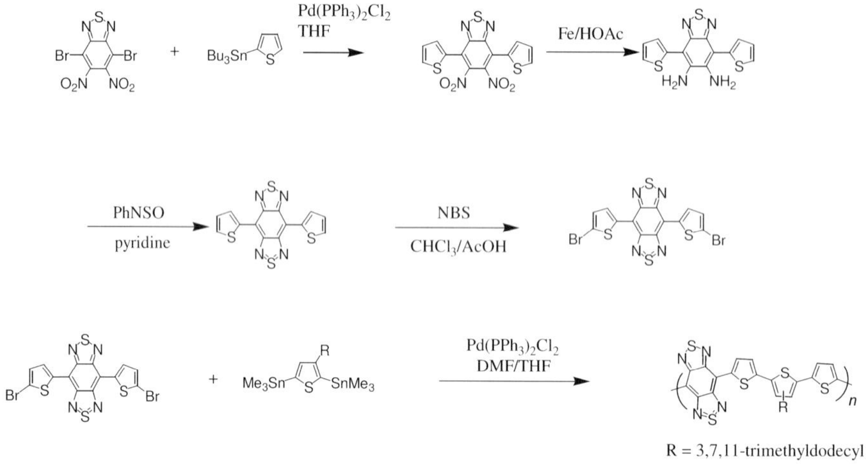

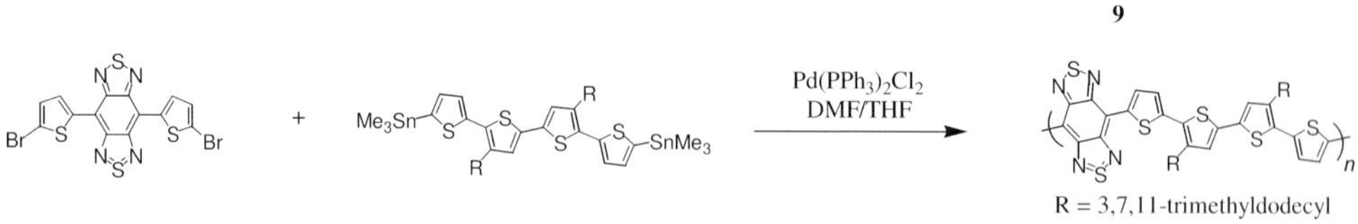

Figure 4.7 Polymers containing 4,8-di-2-thienyl-benzo[1,2-c:4,5-c′]bis[1,2,5]thiadiazole.

4.4
Low-Bandgap Polymers Containing 4,9-Di-2-Thienyl[1,2,5]thiadiazolo[3,4-g]quinoxalines

4,9-Di-2-thienyl[1,2,5]thiadiazolo[3,4-g]quinoxalines were prepared by the condensation of 4,7-di-2-thienyl-2,1,3-benzothiadiazole-5,6-diamine with 1,2-diketones (Figure 4.8) [22]. Treatment of the resulting bis[1,2,5]thiadiazole with NBS affords the corresponding dibromide. A series of low-bandgap polymers were prepared by coupling the dibromide with 2,2′-(9,9-dioctyl-9H-fluorene-2,7-diyl)bis[4,4,5,5-tetramethyl-1,3,2-dioxaborolane] (Figure 4.9) [49,50]. These polymers possess a bandgap in the range of 1.2–1.5. For example, APFO-Green-1 has an electrical bandgap of 1.3 eV, a HOMO of −5.3 eV, and a LUMO of −4.0 eV. Due to its low LUMO level, which is very close to that of PCBM, a PCBM derivative with stronger electron affinity (and thus lower LUMO level, such as BRPF60, Figure 4.1) is required for constructing working solar cells. Power conversion efficiency of 0.3% was obtained with blends of APFO-Green-1:BTPF60 and was improved to 0.7% when a mixture of APFO-Green-1:BTPF70 was used to increase the light absorption in the visible region. Cells prepared from APFO-Green-3 and APFO-Green-4 give power conversion efficiency of 0.59 and 0.37%, respectively. APFO-Green-1 was used as an active material in thin-film transistors, and field effect mobility of 3×10^{-3} cm^2 V^{-1} s^{-1} and current on/off ratio of 10^4 were realized. Light-emitting diodes were also fabricated with this material, which has electroluminescence spectrum peaking at about 1 μm [51]. APFO-3 and APFO-Green-1 were spin coated on a layer of rubbed conducting

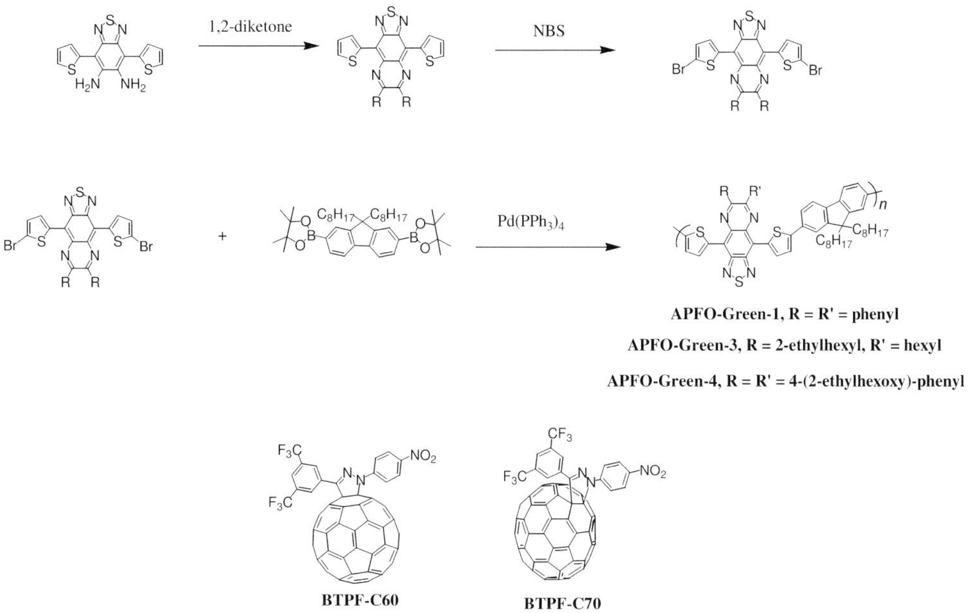

Figure 4.8 Polymers containing 4,9-di-2-thienyl[1,2,5]thiadiazolo[3,4-g]quinoxalines moieties.

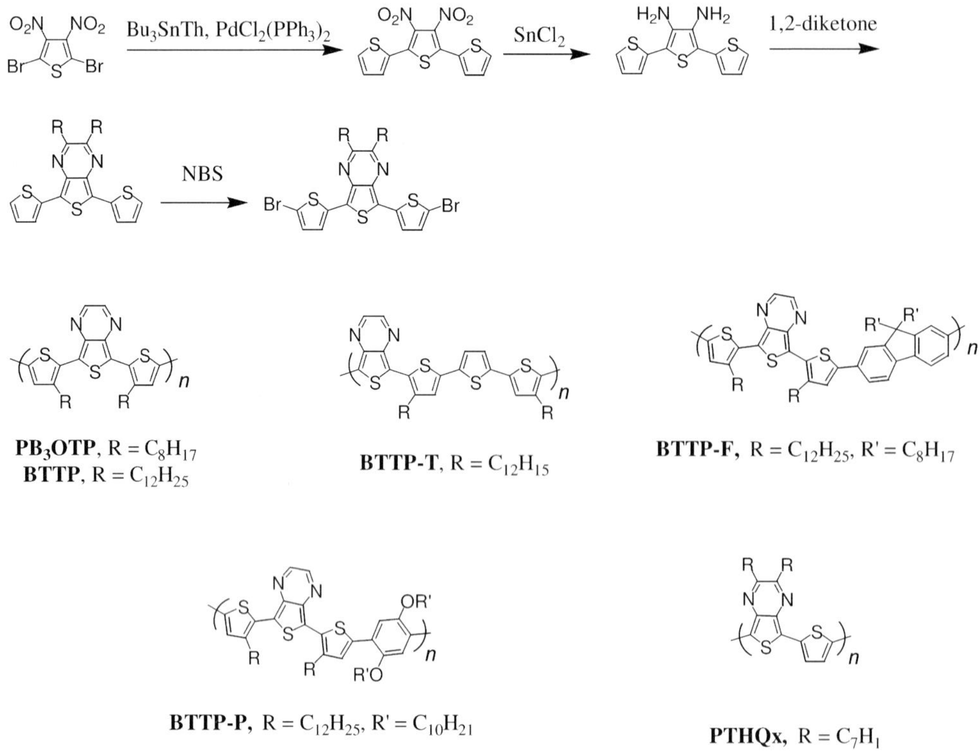

Figure 4.9 Polymers containing thieno[3,4-b]pyrazines moieties.

polymer poly(3,4-ethylene dioxythiophene)-poly(styrene sulfonate) and built into OLED devices, and polarized electroluminescence was observed with an emission maximum of 705 nm (red) and 950 nm (near IR), respectively [52].

4.5
Low-Bandgap Polymers Containing Thieno[3,4-b]pyrazines

Thieno[3,4-b]pyrazine is another successful acceptor used in the preparation of low-bandgap polymers. 3′,4′-Dinitro-2,2′:5′,2″-terthiophene was prepared by Stille coupling of 2,5-dibromo-3,4-dinitrothiophene with tributyl(thienyl-2-yl)stannane and reduced to the corresponding diamine, 3′,4′-diamino-2,2′:5′,2″-terthiophene, with SnCl$_2$. The condensation of the diamine with 1,2-diketal affords 5,7-di-thiophen-2-yl-thieno[3,4-b]pyrazine with a variety of substituents at the 2,3-positions of the pyrazine. This material can then be oxidative polymerized to prepare low-bandgap polymers. A polymer consisting of thiophene-thieno[3,4-b]pyrazine-thiophene repeating units (Figure 4.9, PB$_3$OTP) was found to have an optical bandgap of 1.3 eV. Bulk heterojunction solar cell made of a mixture of this material and [6,6]-phenyl-C$_{61}$-

butyric acid methyl ester (C$_{60}$-PCBM) had a maximum IPCE of 6% at about 660 nm polymer [53]. An alternative route to low-bandgap polymers containing thiophene-thieno[3,4-b]pyrazine-thiophene involves the Stille or Suzuki coupling of the 5,7-bis(5-bromo-2-thienyl)-thieno[3,4-b]pyrazine. Recently, a series of conjugated thieno[3,4-b]pyrazine-based donor–acceptor copolymers were synthesized by Stille and Suzuki couplings (Figure 4.9). The copolymers poly(5,7-bis(3-dodecylthiophen-2-yl)thieno[3,4-b]pyrazine) (BTTP), poly(5,7-bis(3-dodecylthiophen-2-yl)thieno[3,4-b]pyrazine-*alt*-2,5-thiophene) (BTTP-T), poly(5,7-bis(3-dodecylthiophen-2-yl)thieno[3,4-b]pyrazine-*alt*-9,9-dioctyl-2,7-fluorene) (BTTP-F), and poly(5,7-bis(3-dodecylthiophen-2-yl)thieno[3,4-b]pyrazine-*alt*-1,4-bis(decyloxy)phenylene) (BTTP-P) were found to have absorption maxima at 667–810 nm and small optical bandgaps of 1.1–1.6 eV. They possess HOMO levels of 4.6–5.04 eV. The field effect mobilities of holes of BTTP-T and BTTP-F were found to be 4.2×10^{-4} and 1.6×10^{-3} cm^2 V^{-1} s^{-1}, respectively [54]. By coupling 5,8-dibromo-2,3-diheptyl-quinoxaline with 2,5-thiophenediylbis[trimethylstannane], a donor–acceptor conjugated copolymer, poly[(thiophene-2,5-diyl)-*alt*-(2,3-diheptylquinoxaline-5,8-diyl)] (PTHQx), was obtained. This material was found to have a hole mobility of 3.6×10^{-3} cm^2 V^{-1} s^{-1}, and an on/off current ratio of 6×10^5 was observed in p-channel OFETs made from spin-coated PTHQx thin films [55].

Derivatized thieno[3,4-b]pyrazines were synthesized to increase stability and solubility. An alternating polyfluorene containing 2,3-diphenyl-5,7-di-thiophen-2-yl-thieno[3,4-b]pyrazine and 9,9-dioctyl-9*H*-fluorene (APFO-Green-2, Figure 4.10) [56,57] was found to have a bandgap of 1.55. Cells fabricated from APFO-Green-2 and PCBM have photoresponse at a long wavelength of 850 nm and external quantum efficiency as high as 10% at 650 nm. A short-circuit current of 3.0 mA cm^{-2}, an open-circuit voltage of 0.78 V, and a power conversion efficiency of 0.9% were obtained. One of the drawbacks of this polymer was that it had poor solubility. By attaching 2-ethyl-hexyloxy groups to the phenyls on 5,7-di-thiophen-2-yl-thieno[3,4-b]pyrazine, a polymer (APFO-Green-5) [58] with much better solubility and high molecular weight was obtained. APFO-Green-5 has a higher hole mobility than APFO-Green-2 (about 8×10^{-4} cm^2 V^{-1} s^{-1} versus 8×10^{-6} cm^2 V^{-1} s^{-1} with APFO-Green-2), and a power conversion efficiency as high as 2.2% was obtained. By attaching multiple solublizing side chains to thienyl or phenyl groups on 2,3-diphenyl-5,7-dithiophen-2-yl-thieno[3,4-b]pyrazine, soluble low-bandgap polymers without fluorene units (PBEHTT and PTBEHT, Figure 4.10) have also been recently synthesized. Solar cells that have photoresponse up to 1 mm and power conversion efficiency up to 1.1% have been constructed with these materials [36,59]. PBEHTT was recently used in the fabrication of tandem cells, which has open-circuit voltage as high as 1.4 and has broad absorption over the solar spectrum [60].

Recently, a series of polymers containing 5,7-di-(thiophen-2-yl)-thieno[3,4-b]-pyrazines with substitutions at the pyrazine and 3-alkylthiophenes were prepared (RISO-GREEN-1–3, Figure 4.10). The optical bandgap of these polymers was estimated to be 1.3 eV [61]. The condensation of 3′,4′-diamino-[2,2′,5′,2″]terthiophene with phenanthrene-9,10-quinone affords 11-thia-9,13-diaza-cyclopenta[b]triphenylene. This compound when brominated and coupled with 3-(3,7,11-

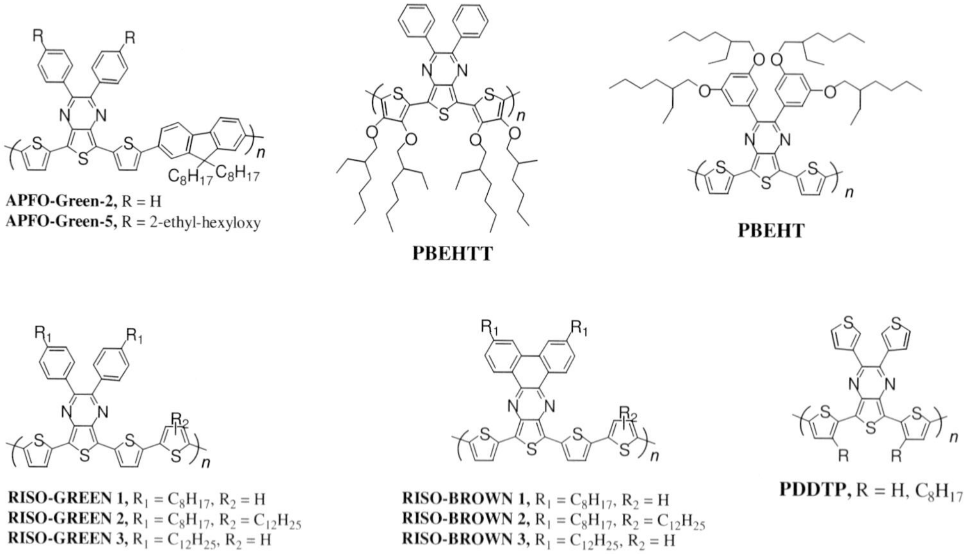

APFO-Green-2, R = H
APFO-Green-5, R = 2-ethyl-hexyloxy

PBEHTT

PBEHT

RISO-GREEN 1, $R_1 = C_8H_{17}$, $R_2 = H$
RISO-GREEN 2, $R_1 = C_8H_{17}$, $R_2 = C_{12}H_{25}$
RISO-GREEN 3, $R_1 = C_{12}H_{25}$, $R_2 = H$

RISO-BROWN 1, $R_1 = C_8H_{17}$, $R_2 = H$
RISO-BROWN 2, $R_1 = C_8H_{17}$, $R_2 = C_{12}H_{25}$
RISO-BROWN 3, $R_1 = C_{12}H_{25}$, $R_2 = H$

PDDTP, R = H, C_8H_{17}

Figure 4.10 Low-bandgap polymers containing 2,3-diphenyl-thieno[3,4-b]pyrazine.

trimethyl-dodecyl)-2,5-bis-trimethylstannyl-thiophene gives RISO-BROWN 1–3 (Figure 4.10) that possess similarly good absorption coverage over solar spectrum. Solar cells prepared from RISO-GREEN-1 were found to have photoresponse up to 900 nm.

5,7-Di(thien-2-yl)thieno[3,4-b]pyrazine di(thien-2-yl) substituents at the 2,3-positions of thieno[3,4-b]pyrazine have also been synthesized and electropolymerized [62]. The resulting poly[2,3-di(thien-3-yl)-5,7-di(thien-2-yl)thieno-[3,4-b]pyriazine] reflects green light in the neutral state and is transmitive in the oxidized state. Dialkyl-substituted 2,3-di(thien-3-yl)-5,7-di(thien-2-yl)thieno[3,4-b]pyrazine, 5,7-bis (3-octyl-thien-2-yl)-2,3-di(thien-3-yl)thieno[3,4-b]pyrazine was synthesized and polymerized by both electrochemical and chemical methods, providing soluble polymers (PDDTP) [63,64]. Such polymers absorb both blue and red light and reflect green color in neutral form. In its oxidized form, they are transmissive in the visible region but have strong absorption in the near-IR. A bandgap of about 1.3 eV was determined by electrochemical and spectroelectrochemical methods. These polymers were used in the fabrication of full-color electrochromic devices [65].

4.6
Arylene Vinylene Based Low-Bandgap Polymers

It is well known that poly(thienylenevinylene) has a low bandgap of 1.7–1.8 eV [13]. New low-bandgap poly(arylene vinylene)s often contain donor–acceptor–donor seg-

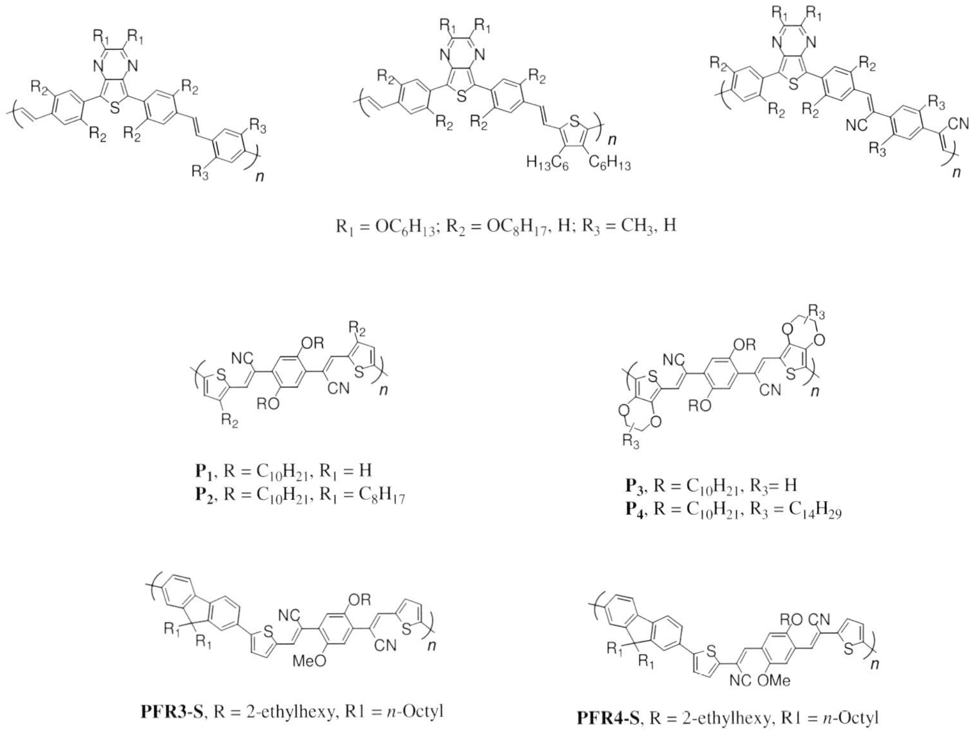

4.6 Arylene Vinylene Based Low-Bandgap Polymers | 141

$R_1 = OC_6H_{13}$; $R_2 = OC_8H_{17}$, H; $R_3 = CH_3$, H

P₁, R = C₁₀H₂₁, R₁ = H
P₂, R = C₁₀H₂₁, R₁ = C₈H₁₇

P₃, R = C₁₀H₂₁, R₃ = H
P₄, R = C₁₀H₂₁, R₃ = C₁₄H₂₉

PFR3-S, R = 2-ethylhexy, R1 = n-Octyl

PFR4-S, R = 2-ethylhexy, R1 = n-Octyl

Figure 4.11 Polymers containing low-bandgap arylene vinylene segments.

ments, where the acceptor is frequently heteroaromatic that favors a quinoidal structure. A series of thieno[3,4-*b*]pyrazine-containing poly(heteroarylenevinylene) synthesized by Knoevenagel polycondensation are shown in Figure 4.11. The bandgap of the resulting polymers ranges from 1.56 to 2.08 eV, with the ones containing thienylenevinylenes having the lower value [66]. Donor–acceptor-type monomer bis(1-cyano-2-thienylvinylene)phenylenes were oxidatively polymerized using FeCl₃. The resulting polymers (P1–P4, Figure 4.11) were soluble, having a bandgap of 1.82–2.10 eV (P2). The high bandgap of P2 is most likely caused by the steric hindrance of R1 (*n*-octyl). However, due to low charge carrier mobilities, these materials only give low power conversion efficiency when built into a solar cell [67]. Two bis(2-thienylvinyl)benzene copolymers with fluorene derivatives, poly{9,9-dioctylfluorene-2,7-diyl-*alt*-2,5-bis(2-thienyl-1-cyanovinyl)-1-(2′-ethylhexyloxy)-4-methoxy-benzene-5′,5′-diyl} (PFR3-S) and poly{9,9-dioctylfluorene-2,7-diyl-*alt*-2,5-bis(2-thienyl-2-cyanovinyl)-1-(2′-ethylhexyloxy)-4-methoxy-benzene-5″,5′-diyl} (PFR4-S), were synthesized and were found to have absorption maximum at 460 and 537 nm, respectively, at solid state [68]. Solar cells prepared with blends of these materials with PCBM have open-circuit voltage of 0.85 and 0.81 V and power conversion efficiency of 0.98 and 1.02% for PFR3-S and PFR4-S, respectively [69].

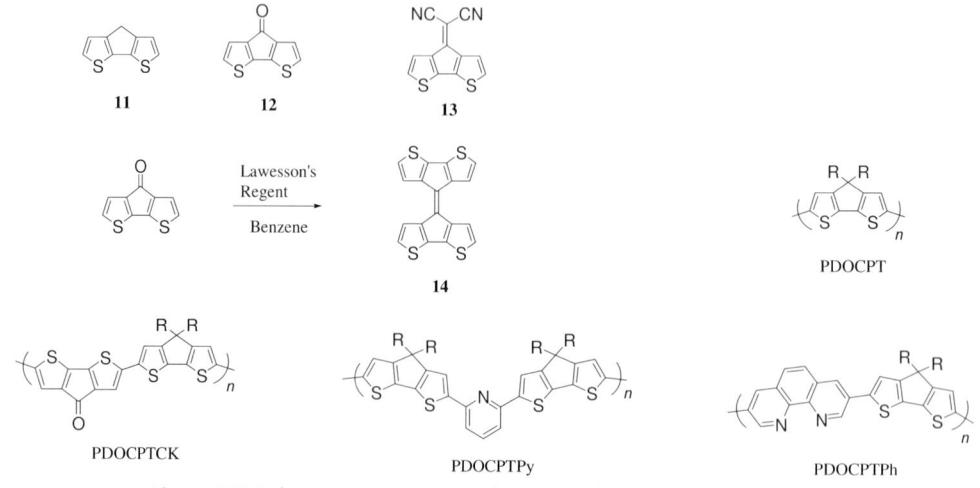

4.7
Low-Bandgap Polymers Containing 4H-Cyclopenta[2,1-b;3,4-b′]dithiophene or Its Analogues

4H-Cyclopenta[2,1-b;3,4-b′]dithiophene (**11**, Figure 4.12) can be considered to be a 2,2′-bithiophene with the 3,3′-positions of the thiophenes tied to a bridging carbon. The two thienyl groups in 4H-cyclopenta[2,1-b;3,4-b′]dithiophene are forced to be planar by the additional 4-carbon that connects to the 3,3′-positions of the thiophenes. Initially, 4H-cyclopenta[2,1-b:3,4-b′]dithiophene was electropolymerized. Due to the forced coplanarity of the two thienyl subunits, 4H-cyclopenta[2,1-b:3,4-b′]dithiophene was found to have a low oxidation potential and the resulting polymers were found to be highly conductive [70–73]. Electrooxidation of cyclopenta[2,1-b:3,4-b′]dithiophen-4-one (**12**) produces the corresponding polymer that has a bandgap of <1.5 eV [74]. Likewise, 4-dicyanomethylene-4H-cyclopenta[2,1-b:3,4-b′]dithiophene (**13**) was polymerized, resulting in a polymer with a bandgap of 0.8 eV [75]. Δ4,4′-Dicyclopenta[2,1-b:3,4-b′]dithiophene, which contains two cyclopentadithiophenes connected by a double bond at the 4-positions, was prepared by refluxing cyclopenta[2,1-b:3,4-b′]dithiophen-4-one in benzene in the presence of Lawesson's reagent. Electrochemical oxidation of this material results in a polymer that has an onset of absorption >1600 nm and a very low bandgap <0.8 eV [76].

The 4-carbon of the 4H-cyclopenta[2,1-b:3,4-b′]dithiophene can be readily functionalized by alkyl groups to increase solubility without causing additional twisting of the repeating units in the resulting polymers. Homopolymers of 4,4-dialkyl-4H-cyclopenta[2,1-b;3,4-b′]dithiophene already possess much longer wavelength absorption than the regular polythiophenes [77–79]. In solution, these soluble homopolycyclopentadithiophenes containing alkyl side chains have absorption maxima around 550–610 nm. At solid state, poly(4,4-dialkylcyclopentadithiophene (PDOCPT, Figure 4.9) has an optical bandgap of 1.8 eV. Alternating polymers of 4,4-dialkylcy-

Figure 4.12 Polymers containing 4H-cyclopenta[2,1-b;3,4-b′]dithiophene.

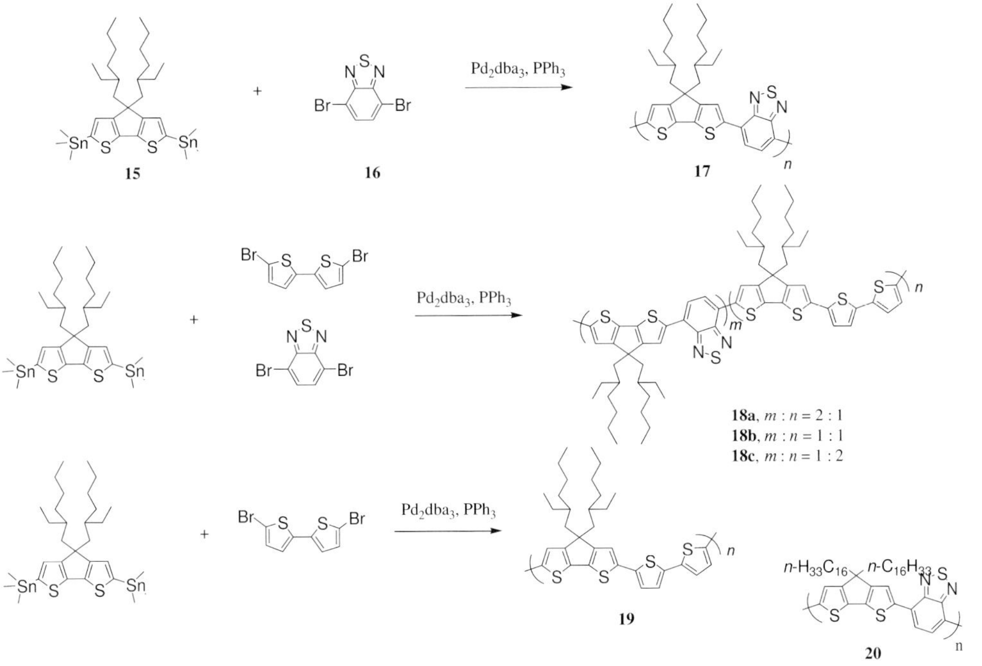

4.7 Low-Bandgap Polymers Containing 4H-Cyclopenta[2,1-b;3,4-b′]dithiophene or Its Analogues

Figure 4.13 Polymers containing 4H-cyclopenta[2,1-b;3,4-b′] dithiophene alternating with 2,1,3-benzothiadiazole or 5,5′-[2,2′] bithiophene.

clopentadithiophene with pyridine, phenanthroline, and cyclopenta[2,1-b:3,4-b′] dithiophen-4-one were also prepared. Poly(4,4-dialkylcyclopentadithiophene-co-pyridine) (PDOCPTPy), poly(4,4-dialkylcyclopentadithiophene-co-phenanthroline) (PDOCPTPh), and poly(cyclopenta[2,1-b:3,4-b′]dithiophen-4-one) (PDOCPTCK) have a bandgap of 2.5, 2.5, and 2.0 eV, respectively [80].

By coupling 4H-cyclopenta[2,1-b;3,4-b′]dithiophene donor with 2,1,3-benzothiadiazole acceptor, poly[2,6-(4,4-bis-(2-ethylhexyl)-4H-cyclopenta[2,1-b;3,4-b′]dithiophene)-alt-4,7-(2,1,3-benzothiadiazole)] (PCPDTBT, **17**, Figure 4.13), a polymer containing alternating donor and acceptor units was obtained [81,82]. This polymer has an optical bandgap of 1.4 eV, a HOMO of −5.3 eV, and a LUMO of −3.57 eV as determined by cyclic voltammetry. The field effect mobility of the polymer was found to be as high as 3×10^{-3} cm^2 s^{-1} V^{-1} [82]. Cells prepared with this polymer gave power conversion efficiency of 3.5%, which is significantly higher than those of other known low-bandgap polymers. This efficiency was further improved to 5.5% using a processing condition involving the use of alkane dithiol additives to induce desirable morphology [83]. Tandem solar cells prepared with this material have reached a power conversion efficiency of 6% [84]. Random copolymers containing three different repeating units, (4,4-bis-(2-ethylhexyl)-4H-cyclopenta[2,1-b;3,4-b′]dithiophene), 4,7-(2,1,3)-benzothiadiazole, and 5,5′-[2,2′]bithiophene (**18a–c**, Figure 4.13), were also synthesized, and it was found that their absorption characteristics can be tuned by

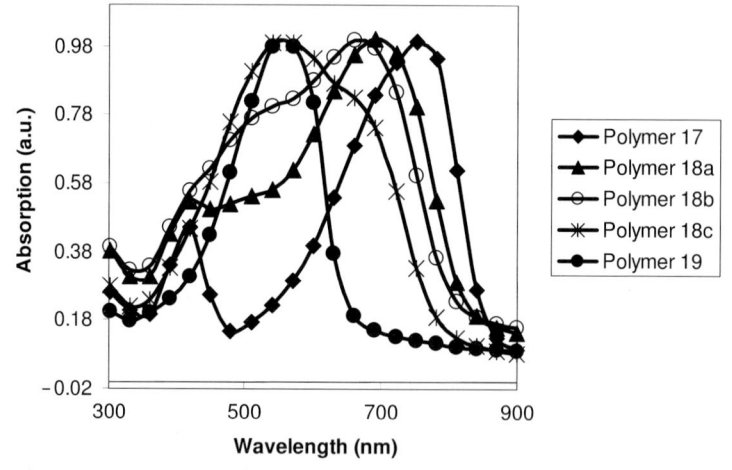

Figure 4.14 Comparison of absorption characteristics of polymers **17**, **18a–c** and polymer **19**.

adjusting the ratio of the two electron-donating units, (4,4-bis-(2-ethylhexyl)-4H-cyclopenta[2,1-b;3,4-b']dithiophene) and 5,5'-[2,2']bithiophene. By adjusting this ratio, very nice absorption coverage over the visible spectrum can be realized (Figure 4.14). The more 2,1,3-benzothiadiazole is incorporated in the polymer, the longer is the absorption wavelength. Electrochemical characterization indicates sufficiently deep HOMO/LUMO levels that enable a high photovoltaic device open-circuit voltage when fullerene derivatives are used as electron transporters. Field effect transistors made of these materials show hole mobility in the range of 5×10^{-4}–3×10^{-3} cm^2 V^{-1} s^{-1}, which promises good device fill factor. An analogue of PCPDTBT, with 4,4-dihexadecyl-substituent at the 4H-cyclopenta[2,1-b:3,4-b'] dithiophenes (**20**, Figure 4.13), was also synthesized and was found to have mobility as high as 0.17 cm^2 V^{-1} s^{-1}. Field effect transistors built with this material have on/off ratio of 10^5 [85]. This high mobility is likely caused by the increase in the order due to the replacement of branched side chains with straight side chains.

Another way of attaching solubilizing alkyl groups to 4H-cyclopenta[2,1-b;3,4-b'] dithiophene is through a methylene group that is linked to the 4-carbon with an alkene linkage. For example, 4-n-dodecylidene-4H-cyclopenta(2,1-b;3,4-b')dithiophene was prepared by Wittig condensation between dodecylidenetriphenylphosphine ylide and cyclopenta(2,1-b;3,4-b')dithiophene-4-one (Figure 4.15) [86]. Deprotonation of 4H-cyclopenta[2,1-b;3,4-b']dithiophene followed by treatment with CS$_2$, and then with methyl iodide, affords [bis(methylthio)methylene]-4H-cyclopenta [2,1-b:3,4-b']dithiophene (**23**). 4-Dialkylmethylene-4H-cyclopenta[2,1-b;3,4-b']dithiophene such as **24** was prepared by the reaction of alkyl Grignard reagent with [bis(methylthio)methylene]-4H-cyclopenta[2,1-b:3,4-b']dithiophene **23** [87]. Oxidative polymerization of alkylidene-4H-cyclopenta[2,1-b;3,4-b']dithiophene **21** and **24** gives the corresponding polymers **22** and **25**. These polymers have an absorption maximum of 599 and 621 nm, respectively, in solution, significantly longer than the poly (dialky-4H-cyclopenta[2,1-b;3,4-b']dithiophene) such as **26**. From solution to solid

4.7 Low-Bandgap Polymers Containing 4H-Cyclopenta[2,1-b;3,4-b']dithiophene or Its Analogues | 145

	22	25	26
Absorption maximum solution (nm)	599	621	566
Absorption maximum film (nm)	628	699	567

Figure 4.15 The synthesis of 4H-cyclopenta[2,1-b;3,4-b'] dithiophene polymers carrying "in-plane" alkylidene chains.

state, there is much more significant red shift for the polymers containing alkylidene side chains. These red shifts can be attributed to improved solid-state π-stacking due to the "in-plane" placement of the alkylidene groups.

Bithiophenes with the 3,3'-positions tied to a heteroatom (e.g., N or S) were also prepared (DTP and DTT, Figure 4.16). Oxidation of N-alkyl-dithieno[3,2-b:2',3'-d] pyrroles by either chemical or electrochemical approach affords poly(dithieno[3,2-b:2',3'-d]pyrrole)s, which fluoresce red and have a bandgap of approximately 1.7 eV [88]. Poly(dithieno[3,2-b:2',3'-d]thiophene) (poly-DTT) was found to have a

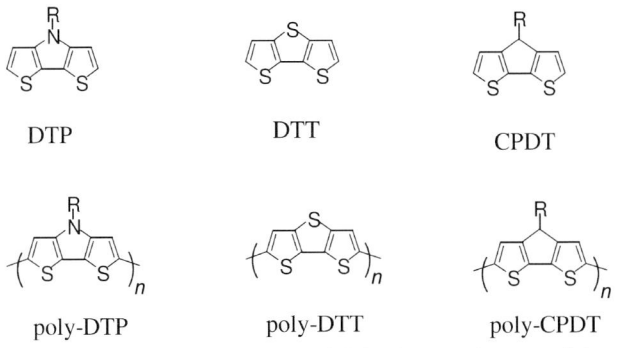

Figure 4.16 4H-Cyclopenta[2,1-b;3,4-b']dithiophene analogues and their polymers.

bandgap of 1.9 eV [89], while the bandgap of poly(cyclopenta-[2,1-*b*:3,4-*b*′]dithiophene) (poly-CPDT) is 1.7 eV [77].

4.8
Low-Bandgap Polymers Based on Other Types of Building Blocks

Several low-bandgap polymers containing thiophene-*S*,*S*-dioxide have been prepared. For example, 5,5′-(1,1-dioxido-2,5-thiophenediyl)bis[2,3-dihydro-thieno[3,4-*b*]-1,4-dioxin] (EDT-OEDT, Figure 4.17) was electrically polymerized, and the resulting polymer has a bandgap of 1.95 eV, while the homopolymer of 2,2′,2″,3,3′,3″-hexahydro-5,5′:7′,5′-terthieno[3,4-*b*]-1,4-dioxin 6′,6′-dioxide (EDT-EDO-EDT) has a bandgap of 1.7 [90]. A polymer prepared from the oxidative polymerization of 3″,4″-dihexyl-2,2′:5′,2″:5″,2−:5 2(-quinquethiophene 1″,1″-dioxide has an absorption maximum of 585 nm in chloroform [91], significantly longer than that of P3HT. These results indicate the potential of thienyl *S*,*S*-dioxide to serve as electron-withdrawing building blocks for low-bandgap polymers.

Acenes such as pentacenes are readily available low-bandgap molecular materials, but they are not easily functionalized due to low solubility and limited stability. Recently, a dibromopentacene monomer containing bulky solubilizing groups (TIPSEBr2P, Figure 4.18) was synthesized. The polymerization of TIPSEBr2P with 1,4-diethynyl-2,5-dialkoxybenzene afforded polymers with bandgap <1.7 eV [92]. In comparison, the bandgap of the triisopropylsilylethyny-substituted pentacene (TIPSEP) is 1.86 eV.

Poly(3,4-ethylenedioxythiphene) (PEDOT) has been one of the most well-known and frequently used conducting polymers. It has a bandgap ranging from 1.6 to 1.7 [93]. Its building block, 3,4-ethylenedioxythiophene, has been used as a donor in the preparation of donor–acceptor-type polymers, as has been shown in previous sections. Besides this, several PEDOT analogues have been investigated or developed. For example, both poly(3,4-ethylenedithiathiophene) (PEDTT, Figure 4.19) [94,95] and poly(3,4-ethylenediselena)thiophene (PEDST) [96] were synthesized. PEDTT and PEDST were found to have optical bandgap of 2.15 and 1.79 eV, respectively. Moreover, a series of poly(3,4-propylenedioxythiophene) (PProDOT) derivatives were also synthesized to achieve soluble PEDOT analogues [97–100]. Thieno[3,4-*b*]thiophene

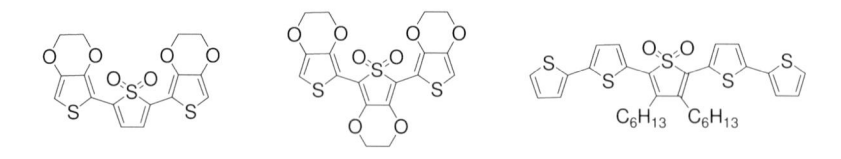

EDT-O-EDT   EDT-EDO-EDT

Figure 4.17 Thienyl *S*,*S*-dioxide containing monomers for low-bandgap polymers.

Figure 4.18 Synthesis of low-bandgap polymers based on pentacenes.

(T34bT) was oxidatively polymerized to give a polymer with a bandgap of 1.0–1.1 eV [101,102]. Electropolymerization of thieno[3,4-*b*]furan (T34bF) affords a polymer with similar bandgap (1.0 eV) [103].

In summary, low-bandgap conducting polymers have significant implications in optoelectronics. New and advanced applications of these materials often require simultaneous optimization of multiple characteristics such as bandgap, mobility, solubility, and electron affinity. New low-bandgap conducting polymers are being designed and synthesized to address such needs. More rationally designed and assembled complex repeating units are being employed to allow fine-tuning of manifold polymer properties. Generally, repeating units that promote a planar main chain structure increase the effective π-conjugation length and are thus desirable for lowering of bandgap. Incorporating alternating donor and acceptor repeating units effectively reduces the HOMO/LUMO energy difference and shifts the absorption spectrum to longer wavelength.

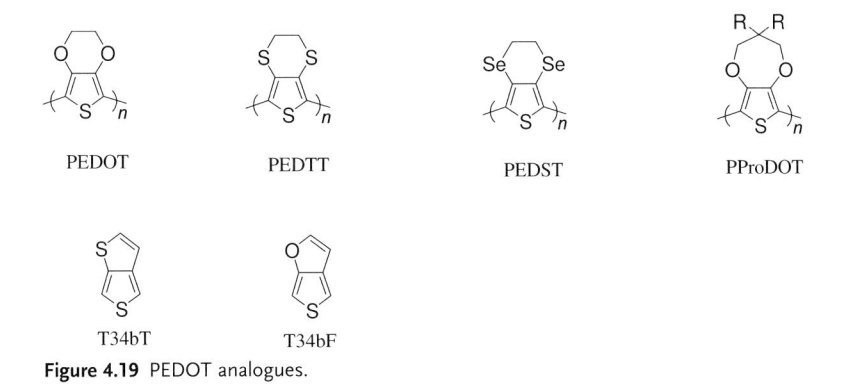

Figure 4.19 PEDOT analogues.

References

1 Heeger, A.J. (2001) *Angewandte Chemie – International Edition*, **40**, 2591–2611.
2 Friend, R.H., Gymer, R.W., Holmes, A.B., Burroughes, J.H., Marks, R.N., Taliani, C., Bradley, D.D.C., Dos Santos, D.A., Bredas, J.L., Logdlund, M. and Salaneck, W.R. (1999) *Nature*, **397**, 121–128.
3 Burroughes, J.H., Bradley, D.D.C., Brown, A.R., Marks, R.N., Mackay, K., Friend, R.H., Burns, P.L. and Holmes, A.B. (1990) *Nature*, **347**, 539–541.
4 Dimitrakopoulos, C.D. and Malenfant, P.R.L. (2002) *Advanced Materials*, **14**, 99–117.
5 Katz, H.E. and Bao, Z. (2000) *Journal of Physical Chemistry B*, **104**, 671–678.
6 Horowitz, G. (2004) *Journal of Materials Research*, **19**, 1946–1962.
7 Yu, G., Gao, J., Hummelen, J.C., Wudl, F. and Heeger, A.J. (1995) *Science*, **270**, 1789–1791.
8 Halls, J.J.M., Walsh, C.A., Greenham, N.C., Marseglia, E.A., Friend, R.H., Moratti, S.C. and Holmes, A.B. (1995) *Nature*, **376**, 498–500.
9 Sirringhaus, H., Brown, P.J., Friend, R.H., Nielsen, M.M., Bechgaard, K., Langeveld-Voss, B.M.W., Spiering, A.J.H., Janssen, R.A.J., Meijer, E.W., Herwig, P. and de Leeuw, D.M. (1999) *Nature*, **401**, 685–688.
10 Ito, T., Shirakawa, H. and Ikeda, S. (1974) *Journal of Polymer Science, Polymer Chemistry Edition*, **12**, 11–20.
11 Chiang, C.K., Fincher, C.R., Jr, Park, Y.W., Heeger, A.J., Shirakawa, H., Louis, E.J., Gau, S.C. and MacDiarmid, A.G. (1977) *Physical Review Letters*, **39**, 1098–1101.
12 Scherf, U. and List, E.J.W. (2002) *Advanced Materials*, **14**, 477–487.
13 Roncali, J. (1997) *Chemical Reviews*, **97**, 173–205.
14 van Mullekom, H.A.M., Vekemans, J.A.J.M., Havinga, E.E. and Meijer, E.W. (2001) *Materials Science & Engineering R: Reports*, **R32**, 1–40.
15 Brabec, C.J., Shaheen, S.E., Fromherz, T., Padinger, F., Hummelen, J.C., Dhanabalan, A., Janssen, R.A.J. and Sariciftci, N.S. (2001) *Synthetic Metals*, **121**, 1517–1520.
16 Dhanabalan, A., Van Duren, J.K.J., Van Hal, P.A., Van Dongen, J.L.J. and Janssen, R.A.J. (2001) *Advanced Functional Materials*, **11**, 255–262.
17 Lorcy, D. and Cava, M.P. (1992) *Advanced Materials*, **4**, 562–564.
18 Musmanni, S. and Ferraris, J.P. (1993) *Journal of the Chemical Society, Chemical Communications*, 172–174.
19 Karikomi, M., Kitamura, C., Tanaka, S. and Yamashita, Y. (1995) *Journal of the American Chemical Society*, **117**, 6791–6792.
20 Kitamura, C., Tanaka, S. and Yamashita, Y. (1996) *Chemistry Letters*, 63–64.
21 Raimundo, J.-M., Blanchard, P., Brisset, H., Akoudad, S. and Roncali, J.(2000) *Chemical Communications*, 939–940.
22 Kitamura, C., Tanaka, S. and Yamashita, Y. (1996) *Chemistry of Materials*, **8**, 570–578.
23 Yamamoto, T., Zhou, Z.-H., Kanbara, T., Shimura, M., Kizu, K., Maruyama, T., Nakamura, Y., Fukuda, T., Lee, B.-L. *et al.* (1996) *Journal of the American Chemical Society*, **118**, 10389–10399.
24 Yamamoto, T., Kanbara, T., Ooba, N. and Tomaru, S. (1994) *Chemistry Letters*, 1709–1712.
25 Yamamoto, T., Sugiyama, K., Kushida, T., Inoue, T. and Kanbara, T. (1996) *Journal of the American Chemical Society*, **118**, 3930–3937.
26 Kitamura, C., Tanaka, S. and Yamashita, Y. (1994) *Journal of the Chemical Society, Chemical Communications*, 1585–1586.
27 Tanaka, S. and Yamashita, Y.(1995) *Synthetic Metals*, **69**, 599–600.
28 Zhang, X., Yamaguchi, R., Moriyama, K., Kadowaki, M., Kobayashi, T., Ishi-i, T., Thiemann, T. and Mataka, S. (2006) *Journal of Materials Chemistry*, **16**, 736–740.

29 Kato, S.-I., Matsumoto, T., Ishi-i, T., Thiemann, T., Shigeiwa, M., Gorohmaru, H., Maeda, S., Yamashita, Y. and Mataka, S. (2004) *Chemical Communications*, 2342–2343.

30 Jayakannan, M., Van Hal, P.A. and Janssen, R.A.J. (2001) *Journal of Polymer Science, Part A: Polymer Chemistry*, **40**, 251–261.

31 Bundgaard, E. and Krebs, F.C. (2006) *Macromolecules*, **39**, 2823–2831.

32 Jayakannan, M., Van Hal, P.A. and Janssen, R.A.J. (2002) *Journal of Polymer Science, Part A: Polymer Chemistry*, **40**, 2360–2372.

33 Xia, Y., Deng, X., Wang, L., Li, X., Zhu, X. and Cao, Y. (2006) *Macromolecular Rapid Communications*, **27**, 1260–1264.

34 van Duren, J.K.J., Dhanabalan, A., van Hal, P.A. and Janssen, R.A.J. (2001) *Synthetic Metals*, **121**, 1587–1588.

35 Brabec, C.J., Winder, C., Sariciftci, N.S., Hummelen, J.C., Dhanabalan, A., Van Hal, P.A. and Janssen, R.A.J. (2002) *Advanced Functional Materials*, **12**, 709–712.

36 Janssen, R.A.J., Wienk, M.M., Turblez, M.G.R., Struijk, M.P. and Fonrodona, M. (2006) *PMSE Preprints*, **95**, 94–96.

37 Wienk, M.M., Struijk, M.P. and Janssen, R.A.J. (2006) *Chemical Physics Letters*, **422**, 488–491.

38 Edder, C., Armstrong, P.B., Prado, K.B. and Frechet, J.M.J. (2006) *Chemical Communications*, 1965–1967.

39 Shin, W.S., Kim, S.C., Lee, S.-J., Jeon, H.-S., Kim, M.-K., Naidu, B.V.K., Jin, S.-H., Lee, J.-K., Lee, J.W. and Gal, Y.-S. (2007) *Journal of Polymer Science, Part A: Polymer Chemistry*, **45**, 1394–1402.

40 Cao, Y., Hou, Q., Niu, Y.-H., Yang, R.-Q., Xu, Y.-S., Luo, J. and Yang, W. (2002) *Huanan Ligong Daxue Xuebao, Ziran Kexueban*, **30**, 1–10.

41 Svensson, M., Zhang, F., Inganas, O. and Andersson, M.R. (2003) *Synthetic Metals*, **135–136**, 137–138.

42 Admassie, S., Inganaes, O., Mammo, W., Perzon, E. and Andersson, M.R. (2006) *Synthetic Metals*, **156**, 614–623.

43 Inganaes, O., Svensson, M., Zhang, F., Gadisa, A., Persson, N.K., Wang, X. and Andersson, M.R. (2004) *Applied Physics A: Materials Science & Processing*, **79**, 31–35.

44 Zhang, F., Jespersen, K.G., Bjoerstroem, C., Svensson, M., Andersson, M.R., Sundstroem, V., Magnusson, K., Moons, E., Yartsev, A. and Inganaes, O. (2006) *Advanced Functional Materials*, **16**, 667–674.

45 Shi, C., Yao, Y., Yang, Y. and Pei, Q. (2006) *Journal of the American Chemical Society*, **128**, 8980–8986.

46 Yao, Y., Shi, C., Li, G., Shrotriya, V., Pei, Q. and Yang, Y. (2006) *Applied Physics Letters*, **89**, 153507/153501–153507/153503.

47 Luo, J., Hou, Q., Chen, J. and Cao, Y. (2006) *Synthetic Metals*, **156**, 470–475.

48 Wong, W.-Y., Wang, X.-Z., He, Z., Djurisic, A.B., Yip, C.-T., Cheung, K.-Y., Wang, H., Mak, C.S.K. and Chan, W.-K. (2007) *Nature Materials*, **6**, 521–527.

49 Perzon, E., Wang, X., Admassie, S., Inganaes, O. and Andersson, M.R. (2006) *Polymer*, **47**, 4261–4268.

50 Wang, X., Perzon, E., Mammo, W., Oswald, F., Admassie, S., Persson, N.-K., Langa, F., Andersson, M.R. and Inganaes, O. (2006) *Thin Solid Films*, **511–512**, 576–580.

51 Chen, M.X., Perzon, E., Robisson, N., Joensson, S.K.M., Andersson, M.R., Fahlman, M. and Berggren, M. (2004) *Synthetic Metals*, **146**, 233–236.

52 Gadisa, A., Perzon, E., Andersson, M.R. and Inganas, O. (2007) *Applied Physics Letters*, **90**, 113510/113511–113510/113513.

53 Campos, L.M., Tontcheva, A., Guenes, S., Sonmez, G., Neugebauer, H., Sariciftci, N.S. and Wudl, F. (2005) *Chemistry of Materials*, **17**, 4031–4033.

54 Zhu, Y., Champion, R.D. and Jenkhe, S.A. (2006) *Macromolecules*, **39**, 8712–8719.

55 Champion, R.D., Cheng, K.-F., Pai, C.-L., Chen, W.-C. and Jenekhe, S.A. (2005) *Macromolecular Rapid Communications*, **26**, 1835–1840.

56 Perzon, E., Wang, X., Zhang, F., Mammo, W., Delgado, J.L., de la Cruz, P., Inganaes, O., Langa, F. and Andersson, M.R. (2005) *Synthetic Metals*, **154**, 53–56.

57 Zhang, F., Perzon, E., Wang, X., Mammo, W., Andersson, M.R. and Inganaes, O. (2005) *Advanced Functional Materials*, **15**, 745–750.

58 Zhang, F., Mammo, W., Andersson, L.M., Admassie, S., Andersson, M.R. and Inganaes, O. (2006) *Advanced Materials*, **18**, 2169–2173.

59 Wienk, M.M., Turbiez, M.G.R., Struijk, M.P., Fonrodona, M. and Janssen, R.A.J. (2006) *Applied Physics Letters*, **88**, 153511/153511–153511/153513.

60 Hadipour, A., de Boer, B., Wildeman, J., Kooistra, F.B., Hummelen, J.C., Turbiez, M.G.R., Wienk, M.M., Janssen, R.A.J. and Blom, P.W.M. (2006) *Advanced Functional Materials*, **16**, 1897–1903.

61 Petersen, M.H., Hagemann, O., Nielsen, K.T., Jorgensen, M. and Krebs, F.C. (2007) *Solar Energy Materials & Solar Cells*, **91**, 996–1009.

62 Sonmez, G. and Wudl, F. (2005) *Journal of Materials Chemistry*, **15**, 20–22.

63 Sonmez, G., Sonmez, H.B., Shen, C.K.F., Jost, R.W., Rubin, Y. and Wudl, F. (2005) *Macromolecules*, **38**, 669–675.

64 Sonmez, G., Sonmez, H.B., Shen, C.K.F. and Wudl, F. (2004) *Advanced Materials*, **16**, 1905–1908.

65 Sonmez, G., Shen, C.K.F., Rubin, Y. and Wudl, F. (2004) *Angewandte Chemie – International Edition*, **43**, 1498–1502.

66 Shahid, M., Ashraf, R.S., Klemm, E. and Sensfuss, S. (2006) *Macromolecules*, **39**, 7844–7853.

67 Colladet, K., Fourier, S., Cleij, T.J., Lutsen, L., Gelan, J., Vanderzande, D., Nguyen, L.H., Neugebauer, H., Sariciftci, S., Aguirre, A., Janssen, G. and Goovaerts, E. (2007) *Macromolecules*, **40**, 65–72.

68 Cho, N.S., Hwang, D.-H., Jung, B.-J., Lim, E., Lee, J. and Shim, H.-K. (2004) *Macromolecules*, **37**, 5265–5273.

69 Lee, S.K., Cho, N.S., Kwak, J.H., Lim, K.S., Shim, H.-K., Hwang, D.-H. and Brabec, C.J. (2006) *Thin Solid Films*, **511–512**, 157–162.

70 Cunningham, D.D., Galal, A., Pham Chiem, V., Lewis, E.T., Burkhardt, A., Laguren-Davidson, L., Nkansah, A., Ataman, O.Y., Zimmer, H. and Mark, H.B., Jr (1988) *Journal of the Electrochemical Society*, **135**, 2750–2754.

71 Berlin, A., Brenna, E., Pagani, G.A., Sannicolo, F., Zotti, G. and Schiavon, G. (1992) *Synthetic Metals*, **51**, 287–297.

72 Kalaji, M., Murphy, P.J. and Williams, G.O. (1999) *Synthetic Metals*, **101**, 123.

73 Mills, C.A., Taylor, D.M., Murphy, P.J., Dalton, C., Jones, G.W., Hall, L.M. and Hughes, A.V. (1999) *Synthetic Metals*, **102**, 1000–1001.

74 Lambert, T.L. and Ferraris, J.P. (1991) *Journal of the Chemical Society, Chemical Communications*, 752–754.

75 Ferraris, J.P. and Lambert, T.L. (1991) *Journal of the Chemical Society, Chemical Communications*, 1268–1270.

76 Loganathan, K., Cammisa, E.G., Myron, B.D. and Pickup, P.G. (2003) *Chemistry of Materials*, **15**, 1918–1923.

77 Zotti, G., Schiavon, G., Berlin, A., Fontana, G. and Pagani, G. (1994) *Macromolecules*, **27**, 1938–1942.

78 Asawapirom, U. and Scherf, U. (2001) *Macromolecular Rapid Communications*, **22**, 746–749.

79 Coppo, P., Cupertino, D.C., Yeates, S.G. and Turner, M.L. (2003) *Macromolecules*, **36**, 2705–2711.

80 Wu, C.-G., Hsieh, C.-W., Chen, D.-C., Chang, S.-J. and Chen, K.-Y. (2005) *Synthetic Metals*, **155**, 618–622.

81 Zhu, Z., Waller, D., Gaudiana, R., Morana, M., Muehlbacher, D., Scharber, M. and Brabec, C. (2007) *Macromolecules*, **40**, 1981–1986.

82 Muehlbacher, D., Scharber, M., Morana, M., Zhu, Z., Waller, D., Gaudiana, R. and Brabec, C. (2006) *Advanced Materials*, **18**, 2884–2889.

83 Peet, J., Kim, J.Y., Coates, N.E., Ma, W.L., Moses, D., Heeger, A.J. and Bazan, G.C. (2007) *Nature Materials*, **6**, 497–500.

84 Kim, J.Y., Lee, K., Coates, N.E., Moses, D., Nguyen, T.-Q., Dante, M. and Heeger, A.J. (2007) *Science*, **317**, 222–225.

85 Zhang, M., Tsao, H.N., Pisula, W., Yang, C., Mishra, A.K. and Muellen, K. (2007) *Journal of the American Chemical Society*, **129**, 3472–3473.

86 Coppo, P., Adams, H., Cupertino, D.C., Yeates, S.G. and Turner, M.L. (2003) *Chemical Communications*, 2548–2549.

87 Buennagel, T.W., Galbrecht, F. and Scherf, U. (2006) *Macromolecules*, **39**, 8870–8872.

88 Ogawa, K. and Rasmussen, S.C. (2006) *Macromolecules*, **39**, 1771–1778.

89 Jow, T.R., Jen, K.Y., Elsenbaumer, R.L., Shacklette, L.W., Angelopoulos, M. and Cava, M.P. (1986) *Synthetic Metals*, **14**, 53–60.

90 Berlin, A., Zotti, G., Zecchin, S., Schiavon, G., Cocchi, M., Virgili, D. and Sabatini, C. (2003) *Journal of Materials Chemistry*, **13**, 27–33.

91 Barbarella, G., Favaretto, L., Sotgiu, G., Zambianchi, M., Arbizzani, C., Bongini, A. and Mastragostino, M. (1999) *Chemistry of Materials*, **11**, 2533–2541.

92 Okamoto, T. and Bao, Z. (2007) *Journal of the American Chemical Society*, **129**, 10308–10309.

93 Kiebooms, R.H.L., Goto, H. and Akagi, K. (2001) *Macromolecules*, **34**, 7989–7998.

94 Wang, C., Schindler, J.L., Kannewurf, C.R. and Kanatzidis, M.G. (1995) *Chemistry of Materials*, **7**, 58–68.

95 Spencer, H.J., Skabara, P.J., Giles, M., McCulloch, I., Coles, S.J. and Hursthouse, M.B. (2005) *Journal of Materials Chemistry*, **15**, 4783–4792.

96 Pang, H., Skabara, P.J., Gordeyev, S., McDouall, J.J.W., Coles, S.J. and Hursthouse, M.B. (2007) *Chemistry of Materials*, **19**, 301–307.

97 Aubert, P.-H., Argun, A.A., Cirpan, A., Tanner, D.B. and Reynolds, J.R. (2004) *Chemistry of Materials*, **16**, 2386–2393.

98 Cirpan, A., Argun, A.A., Grenier, C.R.G., Reeves, B.D. and Reynolds, J.R. (2003) *Journal of Materials Chemistry*, **13**, 2422–2428.

99 Gaupp, C.L., Welsh, D.M., Rauh, R.D. and Reynolds, J.R. (2002) *Chemistry of Materials*, **14**, 3964–3970.

100 Chwendeman, I., Hickman, R., Soenmez, G., Schottland, P., Zong, K., Welsh, D.M. and Reynolds, J.R. (2002) *Chemistry of Materials*, **14**, 3118–3122.

101 Lee, B., Seshadri, V. and Sotzing, G.A. (2005) *Langmuir*, **21**, 10797–10802.

102 Lee, K. and Sotzing, G.A. (2001) *Macromolecules*, **34**, 5746–5747.

103 Kumar, A., Buyukmumcu, Z. and Sotzing, G.A. (2006) *Macromolecules*, **39**, 2723–2725.

B
Acceptors

5
Fullerene-Based Acceptor Materials
David F. Kronholm and Jan C. Hummelen

5.1
Introduction and Overview

Fullerene derivatives have been adopted widely as n-type semiconductors in solution-processed bulk heterojunction (BHJ) organic photovoltaic (OPV) devices since the first use of a fullerene derivative in this application was published in 1995 [1]. The derivative used in this work was phenyl-C_{61}-butyric acid methyl ester ([60]PCBM; IUPAC: methyl 5-(3H-cyclopropa[1,9](C_{60}-I_h)[5,6]fullerenyl)-5-phenylpentanoate; see structure **1** in Figure 5.1) and it remains the most commonly used fullerene derivative, as well as the most commonly used n-type, in solution-processed OPV. While various performance improvements have been obtained for OPV devices by modifying the p-type material and processing conditions leading to the present promising power conversion efficiencies (η) of about 5% [2], few substitutions of [60]PCBM as the n-type material have shown improvements. Figure 5.2 shows the number of publications per year with the keyword PCBM (as [60]PCBM is commonly known), which gives an indication of how the usage of [60]PCBM in OPV has steadily increased as improved power conversion efficiencies have been obtained for devices incorporating it.

Fullerenes have several intrinsic properties that lead to good performance in BHJ devices and PCBM appears to essentially preserve these properties. An array of PCBM homologues exploring different fullerenes (C_{70}, C_{84}), modifications in addend structure, cage-opened ketolactams, and modified addends and other derivative types have been synthesized and tested in BHJ OPV, with only a few cases showing incremental improvement in certain device architectures.

The short history of fullerene derivatives in state-of-the-art performing devices can be summarized as follows: all world record OPV devices incorporated [60]PCBM, until a brief period in 2003 where [70]PCBM (**2** in Figure 5.1) was used in an $\eta = 3.0\%$ device [3], following which the advent of poly-3-hexylthiophene (P3HT) led to record devices with $\eta \sim 4.0\%$, incorporating again [60]PCBM [4]. Recently, tandem cells with $\eta > 5.0\%$ incorporating both [60]PCBM and [70]PCBM have been reported [5] and devices based on low-bandgap polymers with $\eta > 5.0\%$ incorporating [60]PCBM [6]

Organic Photovoltaics: Materials, Device Physics, and Manufacturing Technologies.
Edited by Christoph Brabec, Vladimir Dyakonov, and Ullrich Scherf
Copyright © 2008 WILEY-VCH Verlag GmbH & Co. KGaA, Weinheim
ISBN: 978-3-527-31675-5

(1) [60]PCBM   **(2)** [70]PCBM   **(3)** [84]PCBM

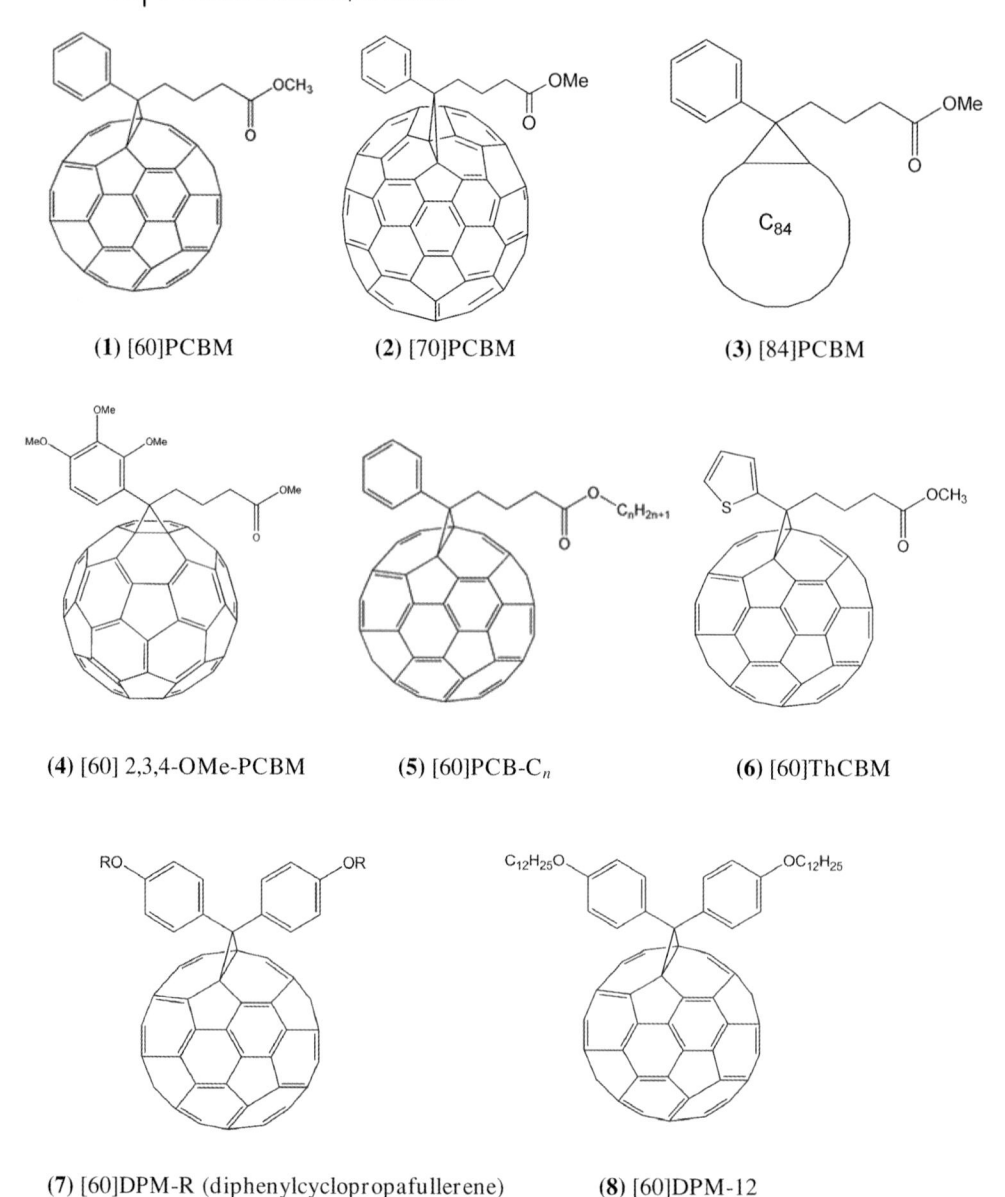

(4) [60] 2,3,4-OMe-PCBM   **(5)** [60]PCB-C$_n$ **(6)** [60]ThCBM

(7) [60]DPM-R (diphenylcyclopropafullerene)   **(8)** [60]DPM-12

Figure 5.1 Fullerene derivatives referred to in this chapter.

and [70]PCBM[2] have also been reported. To our knowledge, no record solution-processed OPV device has yet been published, incorporating an n-type material other than [60]PCBM or [70]PCBM.

We focus here on what has been learned of the best-performing n-type materials, which thus far have been PCBM analogues, and through the attempts at improving

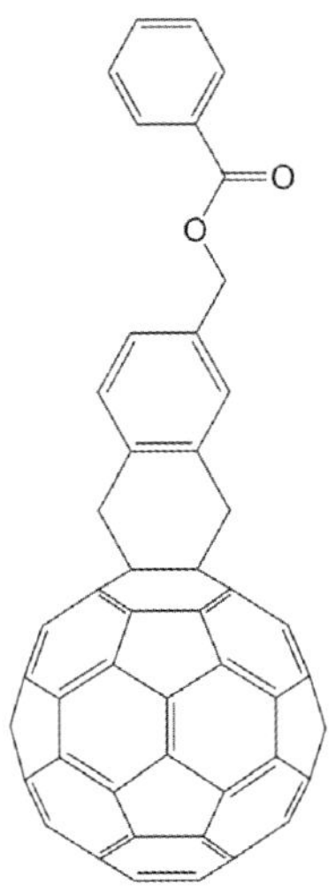

5.1 Introduction and Overview | 157

(9) Cyclohexafullerene

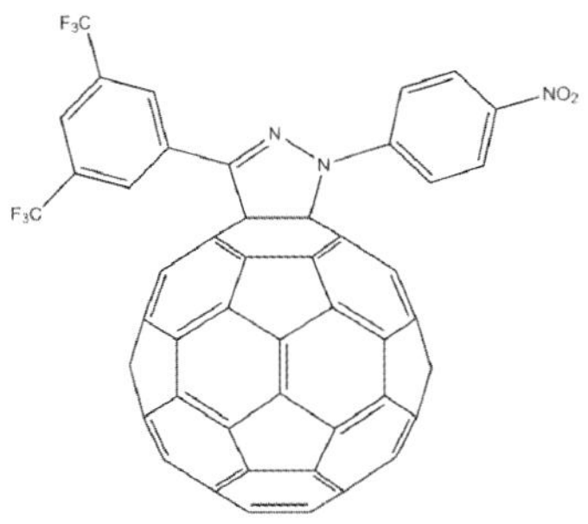

(10) Pyrazolinofullerene ([60]PBTF)

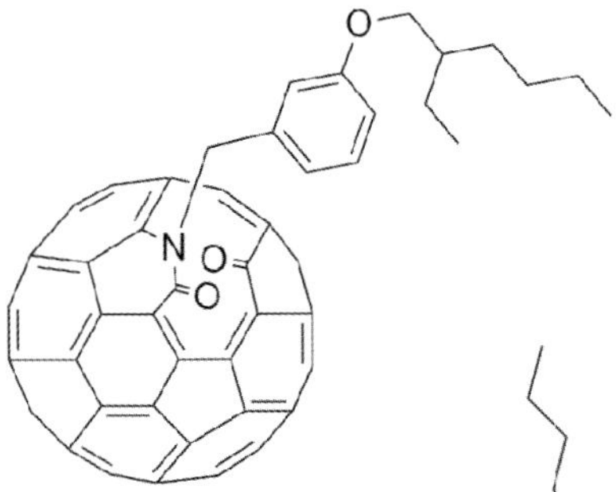

(11) Ketolactam

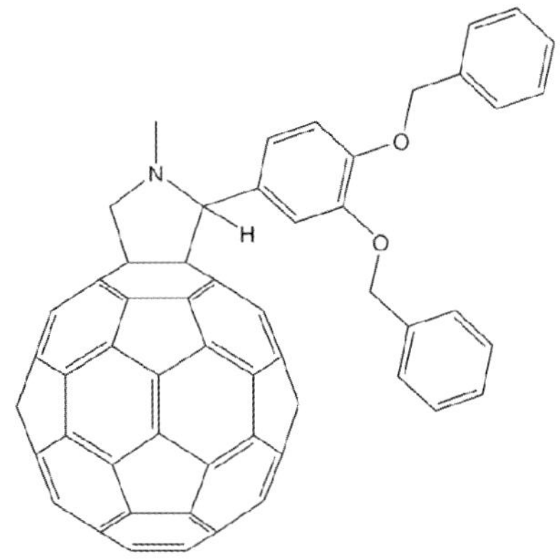

(12) Fulleropyrrolidine

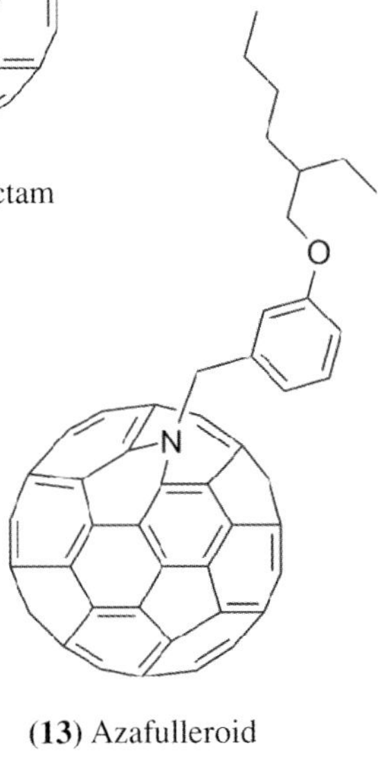

(13) Azafulleroid

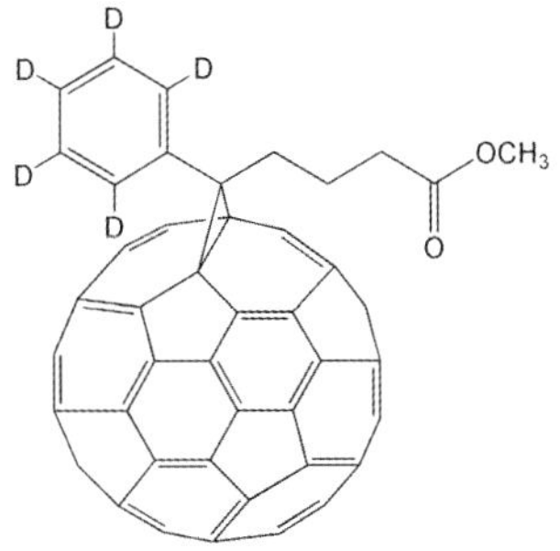

(14) d$_5$-[60]PCBM (deuterated PCBM)

Figure 5.1 (*Continued*)

Keyword "PCBM" in title or abstract

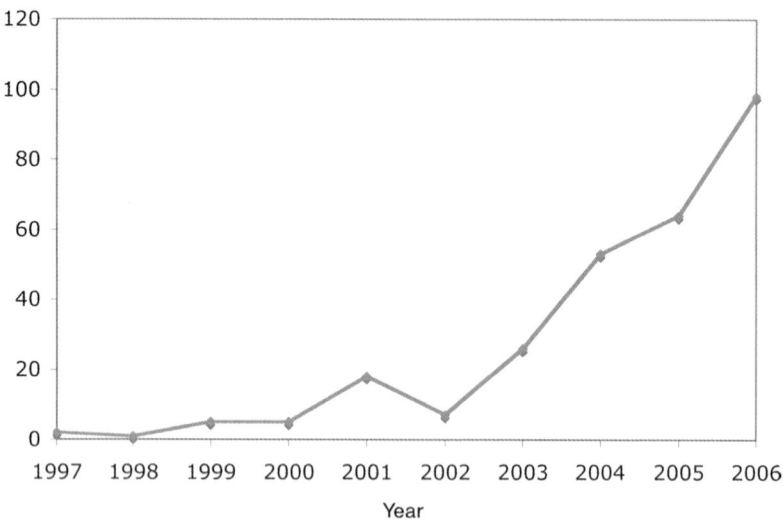

Figure 5.2 Number of occurrences of PCBM in title or abstract; from Web of Science (http://scientific.thomson.com/products/wos/).

the fullerene n-type material in solution-processed BHJ OPV devices. We point out the key findings in the major axes of development rather than list in detail all of the various derivatives that have been designed and tested. Since the start of the field on 1995, quite a number of reviews have appeared on BHJ photovoltaics research: Brabec et al. [7,8]; Nelson [9]; Spanggaard and Krebs [10]; Sariciftci [11]; Benanti and Venkataraman [12]; Hoppe and Sariciftci [13]; Mozer and Sariciftci [14]; and Günes et al. [15]. In addition, one review by Rispens and Hummelen focused specifically on the role of fullerenes and their derivatives in OPV [16].

It is relevant first to consider the native fullerenes, as they are the electron-accepting moiety, and which electronic and physical properties mainly govern the properties of the fullerene derivative. It will be seen that fullerenes demonstrate several desirable properties for use in bulk heterojunction OPV, though improvement by derivatization is key to their use in solution-processed OPV.

5.2
Fullerenes as n-Type Semiconductors

5.2.1
Electron Accepting and Transport

- Fullerenes exhibit ultrafast (subpicosecond) forward photoinduced electron transfer in combination with various conjugated polymers [17]. The process between poly(2-methoxy-5-(3′,7′-dimethyl-octyloxy))-p-phenylene vinylene(MDMO-PPV) and [60]PCBM (instead of C_{60}, to obtain an intimate mixture in which all photoluminescence is quenched optimally) has been studied in detail using femtosecond laser spectroscopy [18]. In that case, a 30 fs forward process, nine orders of magnitude faster than the back transfer was found. In hindsight, it is interesting to note that in that work the MDMO-PPV:[60]PCBM mixtures were still processed from toluene. Later on, it was found that even more optimal morphology, at least in terms of OPV performance, could be obtained upon spin casting the blend from chlorobenzene, which increased η by approximately a factor of 3 (see below) [19].

- Isotropic (in the case of C_{60} derivatives) or relatively isotropic (in the case of C_{70} and C_{84} derivatives) electron accepting owing to the 3D symmetry of the fullerene acceptor; this has positive implications on the orientation factor in exciton diffusion, charge transfer, and charge carrier mobility. It is conceivable that the isotropic nature of the fullerenes can also play a positive role in breaking up the initial intimate D^+–A^- ion pair as well as in electron transport across grain boundaries.

- Good electron mobilities (up to $6\,cm^2\,V^{-1}\,s^{-1}$) [20].

- The exciton diffusion length in pure C_{60} has been estimated to be on the order of 40 nm [21].

The first three properties above are related, in part at least, to the low internal reorganization energy (λ_i) upon uptake of an electron [22]. C_{60} hardly changes shape, which may help to minimize the external ("solvent") reorganization energy (λ_s) as well. The sum (λ) of λ_i and λ_s plays a crucial role in the energetics (i.e., the activation barrier) of charge transfer [23,24].

The isotropic nature also extends to the macroscopic level, as one expects crystallites to show rather isotropic electron mobility. This in turn has strong positive implications for the bulk heterojunction, where electron accepting and transport is demanded in all three dimensions. The kinetics of precipitation also benefits from the degree of symmetry and this feature of relatively fast precipitation kinetics arising from the little need for orientation for crystal packing is desirable for a robust and uniform film formation. This can be seen in the well-known propensity of fullerenes to cluster in solution [25]; the strength of the intermolecular attraction is on the order of a hydrogen bond, but it is of a completely different nature in terms of dimensionality/directionality.

5.2.2
Other Electronic Properties

- As a starting point, fullerenes have adequate lowest unoccupied molecular orbital (LUMO) levels (i.e., are strong enough acceptors) with respect to most conjugated polymer (donor) LUMO levels for efficient photoinduced electron transfer. See Table 5.1 for first reduction potential values for a few PCBM homologues. For efficient OPV, the donor–acceptor LUMO–LUMO offset is to be optimized for each individual donor–acceptor pair to avoid unnecessary loss of V_{oc} while maintaining enough driving force for the photoinduced electron transfer. This point will be

5 Fullerene-Based Acceptor Materials

Table 5.1 First reduction potentials of several PCBM homologues.

	$E_{1/2}$, first reduction potential (V)
[60]PCBM	−1.078 [74] (versus Fc/Fc$^+$)
[70]PCBM	−1.089 [74] (versus Fc/Fc$^+$)
[84]PCBM	−0.730 [74] (versus Fc/Fc$^+$)
[60]ThCBM	−1.08 [66] (versus Fc/Fc$^+$)

Values taken from references in this work.

addressed further below. Furthermore, the highest occupied molecular orbital (HOMO) energy levels of C_{60} and C_{70} are low enough to avoid possible energy transfer (instead of charge transfer) between the donor and acceptor. The HOMO (donor)− LUMO(acceptor) energy difference determines the upper limit for open-circuit voltage (V_{oc}) of the device (ohmic electrodes taken).

- Fullerenes have a relatively high dielectric constant ε ($\varepsilon_{C_{60}} = 4.4$) [26], much higher than the commonly known π-conjugated polymers. A high dielectric constant allows the intimate D^+–$A^−$ ion pair, formed upon exciton dissociation at the donor–acceptor interface, to stabilize. The high dielectric constant of the fullerenes stems from the high polarizability, as a result of the large delocalized π electron system. The high fullerene ε lifts the average dielectric constant of the blend. The importance of the dielectric constant can be seen through the comparison, at maximum power point, of MDMO-PPV:[60]PCBM cells in which the donor:acceptor ratio is steadily increased from 1:1 to 1:4. The cells made with higher fullerene content perform better in terms of charge generation efficiency as the result of higher ε of [60]PCBM, in combination with the improved hole mobility of the polymer [27].

Owing mainly to the unique symmetry and electronics of a fullerene, the combination of the above properties is particularly well-suited to the solution-processed bulk heterojunction application, except for the fact that the self-organization tendency of the pristine fullerenes is too high, leading to the low solubility in most solvents and undesired strong phase separation (i.e., the formation crystalline island, instead of bicontinuous interpenetrating networks) in the blend.

Less than optimal properties in OPV application include:

- Insolubility in solvents other than aromatics and low solubility in these.

- Relatively low-optical absorption in the solar spectral range (see Figure 5.3a for absorption spectrum of C_{60} and [60]PCBM).

- Until now, for optimal performance, the electron acceptor strength of fullerenes is too high (or in other words, the LUMO is too low) in combination with most donors that have been tested. Moreover, starting after C_{70}, with increasing size, the fullerenes become even better electron acceptors. Tuning the LUMO energy level in a substantial way in fullerene derivatives has been found to be quite challenging (see below).

5.2 Fullerenes as n-Type Semiconductors

(a) Absorption spectra of C$_{60}$ and [60]PCBM

(b) Absorption spectra of C$_{60}$ and PCBM homologues

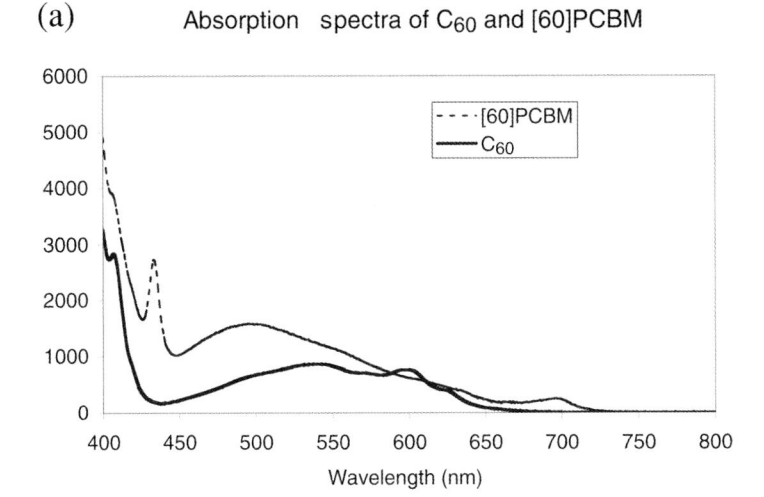

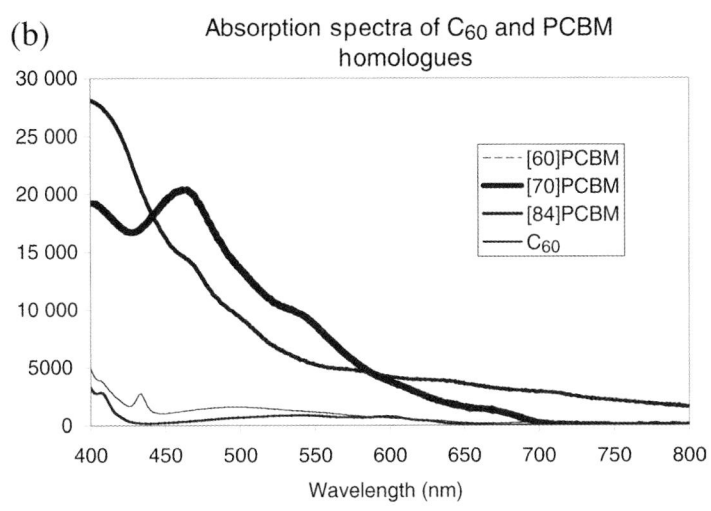

Figure 5.3 Absorption spectra of C$_{60}$ and PCBM homologues. Data from Ref. [74] of this work, except C$_{60}$ (from Solenne BV).

- For fullerenes higher than C$_{60}$, isomer problems quickly arise, as even though C$_{70}$ is present as only one isomer, derivatives are formed as isomeric mixtures. Fullerenes higher than C$_{70}$ come as a mixture of constitutional isomers (and quite a few are even chiral), leading to an even more complex isomeric mixture of products upon derivatization. For example, 10 [28–31] of the 24 theoretically possible isomers of C$_{84}$ [32] have been isolated up to now. The ability to purify at large scale single fullerenes, even as mixtures of isomers, for fullerenes higher than C$_{70}$ (i.e., C$_{76}$, C$_{78}$, and C$_{84}$) has also remained, even after more than a decade of work, very challenging.

The first three properties not surprisingly represent the axes around which derivative development has centered. The last decade has taught us that finding improvements in these properties is challenging, mainly because it appears that a deviation too strongly beyond the native fullerene leads to diminished performance as one of the desirable properties of the native fullerene is affected. In other words, except for solubility, the native fullerene appears as something of a local (at least) optimum in n-type semiconductors for BHJ devices. PCBM, being a relatively compact derivative, while providing the needed increase in solubility, appears to preserve the other important properties adequately.

5.3
[60]PCBM

Since experience at this point shows that [60]PCBM has a broad acceptability in performance, it is perhaps instructive to consider its properties in relation to C_{60} to lay the groundwork for comparing other derivatives. Below we compare some of the key properties of PCBM with C_{60} to illustrate the extent to which it preserves or deviates from the properties of C_{60}.

[60]PCBM is a single isomer, synthesized by the addition of diazoalkane [33]. The solubility of [60]PCBM is higher than that of C_{60} [34] in most aromatic solvents (see Table 5.2 for approximate solubilities of several PCBM homologues and published solubility data for C_{60} and C_{70} in several solvents), allowing for adequate (i.e., leading to the desired functional morphology) solution processing, depending on the choice of solvent. This has been demonstrated with a variety of donor materials. A significant advance in η was obtained (from ~1 to 2.5%) by using chlorobenzene in place of toluene [19], which altered the morphology of the MDMO-PPV:[60]PCBM blend.

The electron mobility of [60]PCBM in blends with P3HT and MDMO-PPV (as determined from space charge limited current measurements in MIM-type devices)

Table 5.2 Practical dissolution guidelines for several PCBM homologues[a], C_{60}, and C_{70} (mg ml^{-1}).

Solvent	[60]PCBM	[70]PCBM	[60]ThCBM	C_{60}	C_{70}
Toluene	10	20	5	3[b]	1[c]
p-Xylene	5	10	5	6[d]	4[c]
o-Xylene	15	30	10	9[d]	11[e]
Chlorobenzene	25	40	10	7[b]	—
Chloroform	25	30	20	0.2[b]	—
ODCB	30	70	20	27[b]	36[c]

[a]Determined by HPLC analysis of liquid phase after stirring with excess solid for 3 days at 25 °C. These values do not necessarily reflect thermodynamic solubilities, but are rather concentration values of what is easily dissolved with stirring at room temperature. Data from Solenne BV.
[b]Ref. [34].
[c]Sivaraman, N. (1994) *Fullerene Science and Technology*, 2, 233.
[d]Scrivens, W.A. et al. (1993) *Journal of Chemical Society: Chemical Communications*, 1207.
[e]Zhou, X. et al. (1994) *Carbon*, 32, 935.

is typically on the order of 10^{-4}–10^{-3} cm^2 V^{-1} s^{-1} [35] (though thermal annealing or slow film formation is necessary to achieve these values with P3HT). The field effect mobility is often used as an indicator of relative film electron mobilities. The recently established values of 0.2 and 0.1 cm^2 V^{-1} s^{-1} for the field effect mobility of pure [60]PCBM and pure [70]PCBM films, respectively [36,37], is in fact quite high for an organic n-type material and is more than adequate, as the hole mobility of the p-types tested to date is typically on the order of 10^{-5}–10^{-2} cm^2 V^{-1} s^{-1} and is thus current-limiting in most OPV devices. A C$_{60}$ field effect mobility as high as 6 cm^2 V^{-1} s^{-1} has been measured for pure epitaxially grown films of sublimed C$_{60}$ [20]. Derivatization can be seen to effect a reduction in electron mobility, in general, as would be expected on the basis of considerations of the crystal properties.

A key physical property of [60]PCBM with implications for the bulk electronic and physical properties, in particular electron mobility, is the crystal structure. Single-crystal structures of [60]PCBM crystallized from chlorobenzene and 1,2-dichlorobenzene (ODCB) were analyzed and it was found, surprisingly, that ball-to-ball distances of C$_{60}$ and [60]PCBM are quite similar (~10 Å) for crystals grown in chlorobenzene [38] and that in fact the shortest distance is shorter for [60]PCBM than C$_{60}$ (9.85 compared to 10 Å). Solvates were formed from both solvents; C$_{60}$ and C$_{70}$ are known to readily form solvates as well [39]. ODCB gave a ball-to-ball distance significantly larger in one dimension (~13 Å), owing to the solvate layer. In OPV devices based on MDMO-PPV, the V_{oc} was also nearly identical for the ODCB and chlorobenzene films, with about a 20% lower short-circuit current for the ODCB-cast film. Taken together, it appears clear that crystal-packing considerations are important and this work gives one indication of the sensitivity of mobility to increase in ball-to-ball distance of 10–13 Å (for one dimension).

The dielectric constant of C$_{60}$ is not altered drastically upon derivatization to [60]PCBM: ε_r (C$_{60}$) = 4.4 [26]; ε_r ([60]PCBM) = 3.9 [40]. The relatively high dielectric constant of [60]PCBM has a positive effect on charge carrier formation in blends with donor polymers. This effect was discussed for blends in general [41] and for MDMO-PPV:[60]PCBM blends in particular [27].

To the best of our knowledge, the exciton diffusion length has thus far been determined neither for [60]PCBM nor for any other fullerene derivative. In fact, this is highly remarkable, because this parameter is of key importance in relation to the optimal blend morphology. While most OPV researchers seem to almost neglect the contribution of the fullerene acceptor to the PV action of the device, it is obvious that both electron transfer (from the excited donor) and hole transfer (from the excited acceptor) take place in a BHJ device. The ratio between both generation mechanisms is directly related to the combination of the two absorption profiles of the components. In MDMO-PPV:[60]PCBM devices, the contribution of the acceptor in the light absorption and subsequent charge generation is quite important. In accord, a 20% improvement in cell performance (η) was obtained upon shifting from [60]PCBM to the much stronger absorbing (in the 450–700 nm region) [70]PCBM (see Figure 5.3b for absorption spectra of [60]PCBM and [70]PCBM). As is discussed below, [70]PCBM is present as multiple isomers and a stronger fullerene solvent was needed to obtain the right morphology for MDMO-PPV:[70]PCBM. This can be rationalized in terms of precipitation kinetics, but it could just as well be understood in terms of the need for smaller [70]PCBM domains owing to a (most likely) shorter exciton diffusion length in [70]PCBM compared to the more isotropic single isomer [60]PCBM. Such data are lacking to date, however.

There is a considerable confusion in the literature on the precise energy of the LUMO of [60]PCBM. Values ranging anywhere from 4.5 down to 3.7, relative to the vacuum level, have been used, also for C_{60}. Very few groups have actually tried to measure the value and a value is most often taken from the literature. Because [60]PCBM could not be evaporated under high vacuum in our hands, UPS and inverse PES data on [60]PCBM are still lacking. Early investigations on pure C_{60} under ultrahigh vacuum conditions yielded the nowadays generally accepted values for the HOMO and LUMO of 6.1 and 3.8 eV, respectively, thus leading to an electronic bandgap of 2.3 eV [42]. The optical bandgap of C_{60} (in *solution* in toluene) is 1.9 eV, that is, the exciton binding energy is on the order of 0.4 eV. Because in cyclic voltammetry and DPV measurements (i.e., in solution) [60]PCBM consistently shows a ~90 mV more negative first reduction potential than C_{60}, we put the value for the LUMO level energy of [60]PCBM at 3.7 eV. The optical bandgap of [60]PCBM in toluene solution can be estimated to be ~1.73 eV by taking the onset of absorption at the 697 nm peak (using the tangent at inversion point at $\lambda = 704$ nm). This peak is missing in the solution spectrum of C_{60}. The tail of the lowest energy absorption of C_{60} in toluene does not significantly extend above 700 nm, hence the optical bandgap of C_{60} in toluene solution is *larger* than that of [60]PCBM. With these values at hand, and taking an exciton binding energy of 0.4 eV, we estimate the HOMO of [60]PCBM to be at ~5.9 eV. In the literature, however, simply the HOMO value of C_{60} (i.e., 6.1 eV) is used usually for [60]PCBM as well.

The linear optical absorption spectrum of [60]PCBM is quite similar to that of C_{60}. See Figure 5.3a for a comparison. The typical C_{60} absorption at 550 nm is shifted 50 nm to higher energy, while the lowest energy transition shows a small redshift, as mentioned above. Attempts at improving optical absorption by altering the addend moiety have been unsuccessful, probably owing to the tendency of such derivatives to deviate too strongly from the desirable properties of the native fullerene, notably the crystal structure, which affects electron mobility. Putting an antennalike group on the fullerene in such way as to help the absorption *and* the OPV performance, in combination with a (separate) donor material is likely to be *very* difficult, because (1) fullerene-based electron transport in the fullerene phase is likely to become (sterically) hindered by the antenna + linkage to the ball; (2) tendencies for energy transfer, electron transfer, triplet formation, and so on of the kind that these processes *hinder* the desired photoinduced electron transfer between the donor material and the fullerene have to be avoided completely; and (3) the antenna should not act as a trap for either holes or electrons.

The comparison of [60]PCBM to C_{60} leads to a few observations: the addend provides adequate solubility for solution processing, provided a solvent giving desired supersaturation levels for precipitation in film formation is chosen. Other important properties are either unaffected or affected little enough or in a positive way for adequate performance: LUMO is slightly increased (i.e., in combination with most presently known donors, improved with respect to the V_{oc} of the OPV device) and the electron mobility is somewhat reduced by the chemical derivatization, though [60]PCBM appears to preserve adequately the crystal packing of C_{60} and give sufficient electron mobility relative to most donor materials. The optical absorption is only mildly reduced upon derivatization and the kinetics of forward and reverse electron transfer seems unaffected. The dielectric constant is lowered from 4.4 to 3.9.

The properties that would appear the most promising as axes of improvement would be optical absorption and LUMO level, as the other important properties appear to be adequate for [60]PCBM compared to the native fullerene. However, these properties also require some significant changes in the properties of the native fullerene, as they derive directly from the properties of the native fullerene, and therefore modification of the addend moiety is unlikely to achieve the desired results.

5.4
Variations in Fullerene Derivative and Effect on OPV Device

The key parameters determining device performance as affected by variations of the fullerene derivative n-type are morphology, electron mobility, and LUMO. We summarize below a few of the considerations in how the derivative type affects these device properties and what has been observed in device performance when the derivative type is varied.

5.4.1
Morphology Considerations – Solubility and Miscibility of the Fullerene Derivative

The concept of the bulk heterojunction was introduced to eliminate the problem of inherently short exciton diffusion lengths in molecular donor and acceptor materials [43,44].

In the present theoretical concept of the bulk heterojunction (which probably has not been achieved to date in its true ideal form), there is an upper limit of domain size governed by exciton diffusion length limitation, in each domain, and a lower limit of domain size governed by charge recombination, mostly of the intimate ion pair, with concomitant percolation of both domains for efficient charge collection. Hoppe et al. [45] provide a good review of the work on varying morphology, based on the study of MDMO-PPV:[60]PCBM cells. However, observation shows some discrepancies, as for the two most well-studied systems, P3HT domain sizes in optimized OPV devices seem to be larger than the polymer exciton diffusion length, and in MDMO-PPV systems, the "polymer" domains – in reality consisting of an approximately 1:1 very intimately mixed MDMO-PPV:[60]PCBM phase – would be expected to show a high level of charge recombination. Morphology is determined by the phase diagram of the multicomponent system, which in turn is largely determined by the nature of the addend, in terms of how it acts as solubilizing agent in the solvent and its affinity with the donor material.

5.4.2
Solubility and Supersaturation in the Donor/Acceptor Blend

Most devices require 50% or more of the n-type material in the blend, requiring significant loading in the solvent used for film formation. Instead of absolute solubility as a determining factor in precipitation thermodynamics and kinetics, the more accurate physical parameter is the level of supersaturation in the system as the solvent evaporates. Supersaturation can be controlled by varying the solubilizing addend on the fullerene, the solvent, or both. In other words, poorer solvents (or less soluble derivatives) typically lead to faster precipitation and rougher morphologies (or larger n-type domains), better solvents (or more soluble derivatives) give less driving force for precipitation and lead to smoother morphologies (or smaller n-type domains).

Precipitation kinetics, not just the supersaturation thermodynamics, however, is also likely to play a role. An observation consistent with the influence of precipitation kinetics can be found in the work of Wienk et al. [3], in which [70]PCBM, although showing higher solubilities than [60]PCBM as a single component in various solvents (see Table 5.2 for a comparison), was apparently found to precipitate faster than [60]PCBM when forming films combined with MDMO-PPV. It was observed that larger domain sizes were obtained when [70]PCBM was used in place of [60]PCBM, using chlorobenzene as the solvent, and ODCB (a better solvent) gave a smoother morphology and better performance. However, as pointed out above, there is likely an interplay as well with a lower exciton diffusion length (relative to [60]PCBM), which could lead to a different optimal n-type domain size. It is interesting to recall at this point the results on single-crystal structure of [60]PCBM described above. It is not clear in the case of [70]PCBM whether the ball-to-ball distance is unaffected by crystal solvates formed with ODCB, or whether it is affected similarly, and other factors outweigh this effect. It suggests, however, that generalizing over different fullerene derivatives the propensity of particular solvents to influence mobility through solvate effects is not possible because of the large parameter space affecting final device performance.

Improvement of solubility was accomplished by changing the methyl ester in [60]PCBM for longer chain alkyl moieties. More soluble homologues of [60]PCBM were tested by substitution at the ester moiety with alkyl chains of varying length of $n = 2, 3$, and 8 carbon atoms [46] and later $n = 4, 8, 12$, and 16 carbon atoms [47]. See PCB-C_n (5) in Figure 5.1. Improvement of solubility was accomplished but this resulted in OPV device improvement only when poorer (and subsequently found to be less optimal) fullerene solvents were used for spin coating. More soluble derivatives may give rise to more intimate morphology, with risk of high recombination rates. Moreover, there was likely a decrease in electron mobility, brought on by increased ball-to-ball distances in the fullerene domain. The effect of increasing addend size on electron mobility of pure fullerene derivative films can be seen in the application of the derivative DPM-12 [48]; (8) in Figure 5.1. This molecule showed a pure-film electron mobility a factor of 40 less than [60]PCBM that was thought to be because of smaller domain sizes, which is likely a result of greatly increased solubility.

To the best of our knowledge, the electron mobility for cyclohexafullerenes (9) of the group of Fréchet [49], pyrazolinofullerenes (10) by Langa et al. [50–53], the substituted diphenylcyclopropafullerenes (7) by the group of Martin [48,54], or the fulleropyrrolidine (12) of Maggini [55] has not been determined so far (all structures shown in Figure 5.1).

In a direct comparison between [60]PCBM and a series of cyclohexafullerenes in optimized blend compositions with regioregular P3HT, the blend with the benzoate (9) of C_{60} hydroxydihydronaphthalene yielded, within experimental error, the same result (η) as the blend with [60]PCBM [49]. Replacing [60]PCBM with DPM-12 in blends with either MDMO-PPV or P3HT resulted in slightly higher V_{oc} values. This is somewhat surprising because the first reduction potentials of the acceptors are all the same within experimental error. The power conversion efficiency of the devices with DPM-12 is slightly lower, however, because of the lower electron mobility in the DPM-12 phase compared to [60]PCBM [48]. The pyrazolinofullerenes such as PBTF (10) that were tested in BHJ devices in combination with APFO-Green-1 gave a lower V_{oc}, as expected from the increased electron affinity of PBTF (relative to that of [60]PCBM). But overall better performing devices were obtained with PBTF in this study (i.e., $\eta = 0.3$ versus 0.17%), presumably because (a) the increased driving force necessary in combination with this relatively weak donor and (b) the obtained morphology under the circumstances used were more optimal for BHJ activity [50]. Substitution of [60]PBTF by its [70]PBTF analogue later gave a further increase in η to 0.7% [51,56]. In a comparison between [60]PCBM on the one side and [60]PBTF and [70]PBTF on the other, in combination with two APFO polymers (APFO-Green-1 and APFO 3), it was found that while [60]PCBM increased the hole mobility in those polymers, the PBTF acceptors had the opposite effect [52,53]. The use of fulleropyrrolidine (12) in Figure 5.1, in combination with a number of conjugated polymers, among which rr-P3HT was included, led to quite inefficient devices ($\eta < 0.6\%$). However, because this acceptor was not compared directly in this study with [60]PCBM, there might be other hidden variables that led to these results [55].

In our experience, it is vitally important to consider the solubility of the fullerene derivative as a single component in different solvents, but the only generality possible is that it can be used somewhat as a guideline to predict precipitation rates during film formation, *for a given derivative*. That is, once it is known that the morphology is hindered by too large an n-type domain size, a stronger fullerene solvent can be chosen, but between derivatives it is almost impossible to generalize more broadly from single component solubilities when the different parameters of precipitation kinetics, solvate formation, and the behavior of the multiphase p-type/n-type/solvent systems are considered. A caveat to this is that the range of solubility provided by [60]PCBM appears to be adequate with the p-type materials studied thus far, including the recent low-bandgap materials [6], especially when solubilizing additives are considered [2,57]. On the basis of the above, it appears that finding optimal morphology for a given new derivative must still be an almost entirely experimental process, where solvents spanning the range of fullerene solubility (toluene, o-xylene, xylenes, chlorobenzene, ODCB, 1-chloronaphthalene) may need to be tested, possibly combined with different evaporation rates. Phase diagram work on a given p-type/

n-type/solvent system, done little to date, would also be very informative to serve as a guideline to the precise control of morphology. Taking all of the above into account suggests that the most likely course for efficient progress is to alter morphology not through alteration of the addend moiety of the fullerene derivative but through other means, such as additives and control of the precipitation process by physical means such as solvent type and evaporation rates [57].

It can be expected that an optimum solubility exists for a given p-type/solvent, based on the premise that an optimum morphology exists. However, to date it appears that the basic PCBM moiety provides enough solubility in combination with typical donor materials so that with adequate solvent choice and control of solvent evaporation during film formation, the desired morphology can be formed. Efforts at increasing solubility gave no improvement, once optimal solvents were found. However, the trend that arises from the work to date is that larger addend sizes that provide increased solubility (relative to PCBM) in aromatics seem to result in an unnecessary increase in solubility, which can lead to too small a domain size and recombination losses, and impacts electron mobility likely as a result of increased ball-to-ball distances in the crystal structure.

5.4.3
Miscibility

Miscibility presents an even more complex situation, since based on observations in the two most well-known systems to date, the miscibility of the p-type and n-type materials is quite different but still offers relatively good performance, for very different reasons. In one case, the miscibility of PCBM with MDMO-PPV provides a 400-fold increase in hole mobility compared to that of the pristine polymer [58,59]. In contrast, pristine rr-P3HT has a hole mobility close to that of [60]PCBM. When a P3HT:[60]PCBM mixed layer is spun from a low-boiling solvent, the initial hole mobility is quite low, but simple thermal annealing results in full restoration of it, as a result of regained interchain interaction in the polymer phase [60]. Using a slowly evaporating solvent in the spin-cast process gives rise to polymer domains with good interchain interaction without the need for further treatment, owing to the longer period given for optimal phase separation and self-organization of the polymer chains. The hole mobility in slowly formed P3HT:PCBM films is even higher than in thermally annealed films [57,61–63]. These two donor:acceptor systems show very different miscibility behavior of the donor and acceptor. MDMO-PPV: [60] PCBM films exhibit a ~1:1 mixed phase at the intersection of the p-type and n-type domain [64,65]. P3HT:[60]PCBM films show little evidence of a mixed phase; all microscopic investigations point to much more phase segregated morphology, with strong indications for "whisker" formation, typical for pure P3HT. Therefore, it is now impossible to know a priori how exactly the miscibility of the two components should be altered to provide the best performance with a given donor: acceptor tandem.

Miscibility relies on the molecular affinity between p-type and n-type and so alterations in molecular structure may hold promise, even though phase diagrams for the various systems could point to miscible regions, which could be used to guide the precipitation in a controlled manner to minimize demixing.

On the basis of the observation that polythiophenes show a more pronounced tendency to phase-separate with fullerene derivatives upon precipitation of the BHJ film, ThCBM, (6) in Figure 5.1, was developed to address the issue of miscibility with polythiophene donor materials [66]. ThCBM has been found to maintain the LUMO level of PCBM (see Table 5.1) and show a slight improvement in a P3HT cell compared to [60]PCBM, although its morphology has not yet been fully characterized [67]. The solubility of ThCBM is somewhat lower than that of [60]PCBM (see Table 5.2), but electron mobility is identical.

5.4.4
Morphology Fixation and Insoluble Fullerene Layers

Cross-linkable fullerene derivatives have been prepared for two different applications related to OPV. First, a thermally cross-linkable fullerene derivative (a bis-diacetylene substituted [60]fulleropyrrolidine) was made and used for making a (luminescence quenching) layer with a sharp interface in combination with a conjugated polymer spun on top of it. This clean two-layer system (without mutual diffusion) was used to adequately determine exciton diffusion length in the polymer [68,69]. Another cross-linkable fullerene derivative ([60]PCBG, in which G is the glycidyl ester instead of the methyl ester of [60]PCBM), was used in an effort to fix the donor:acceptor BHJ blend morphology in a chemical or thermal posttreatment step. When devices made from P3HT:[60]PCBG were heated for 10 min at 140 °C, unexpected serious degradation in device performance of the P3HT:[60]PCBG cell was observed, however. Hence, the results are contrary to the normally improved efficiency upon thermal annealing in the corresponding P3HT:[60]PCBM cells [70].

5.4.5
Optical Absorption of the Fullerene Derivative

Improving the light absorption is one strategy in increasing η and as the n-type material contributes to the optical absorption in most BHJ systems, the low absorption of C_{60} is a drawback. Various attempts have been made to improve the absorption in the visible by altering the addend moiety [71,72]. These strategies have thus far yielded no improvement in device performance probably for the reasons described above: increased addend size and crystal-packing effects on mobility, difficulties in keeping recombination to a minimum, and competitive electron accepting of the addend moiety with the C_{60}. Baffreau et al. [71] did, however, manage to minimize electron accepting of a perylene dicarboximide antennae moiety through stronger electron donating groups, though P3HT OPV devices gave low short-circuit currents, leading to $\eta = 0.27\%$. The most effective development with respect to increasing optical absorption has been the n-types based on higher fullerenes (C_{70}, C_{84}), as they have a significantly stronger absorption in the visible region.

5.4.6
More Strongly Absorbing Fullerene Derivatives: [70]PCBM and [84]PCBM

C_{70} has a significantly higher absorption coefficient at longer wavelengths than C_{60} and C_{84} even higher. See Figure 5.3b. The general trend is that as molecular weight of the fullerene increases, absorption at longer wavelengths increases. PCBM homologues based on C_{70} and C_{84} have been synthesized and tested in OPV. As C_{76} and C_{78} are challenging to purify as pure components, no fullerene derivatives based on them have yet been synthesized and tested. The key finding in the tests on higher fullerene derivatives was that [70]PCBM does indeed give improvement in systems where the fullerene derivative contributes to the absorption of the device film. MDMO-PPV:[70]PCBM devices gave a 20% improvement in η compared to MDMO-PPV:[60]PCBM devices [3].

[70]PCBM is present as one major and two minor isomers and is used in OPV in this mixed isomer form. [70]PCBM has shown to be fairly robust in terms of performance with different p-types and processing conditions: a substantial improvement with MDMO-PPV; slight improvement with P3HT [73] and promise in conjunction with P3HT as a subcell in a tandem architecture [5]; and improvement owing to the enhanced optical absorption compared to [60]PCBM with newer low-bandgap materials when morphology is controlled with an alkanedithiol additive [2]. This effect is consistent with the system, first studied with [70]PCBM, MDMO-PPV described above, making either a stronger solvent or a solubilizing agent necessary for optimal morphology when substituting [70]PCBM for [60]PCBM in devices, owing perhaps to a faster precipitation kinetics for [70]PCBM compared to [60]PCBM, even though, as described above, exciton diffusion lengths could also play a role, among other considerations.

[84]PCBM, (3) in Figure 5.1, comes as a mixture of mainly three isomers and has panchromatic absorption (extending into the near infrared (NIR); see Figure 5.3b), poorer solubility than [60]PCBM, and gave poor performance in MDMO-PPV devices, with an η of 0.25% [74], owing to a significantly lower LUMO and also apparently a lack of enhancement (as observed with [60]PCBM) of the hole mobility of the MDMO-PPV. This last finding was based on the observation that though the LUMO is 350 mV lower than [60]PCBM, the V_{oc} was 500 mV lower. Interestingly, however, the lower LUMO may lead to the observed improved stability in air [75] and depending on the HOMO level of the donor, [84]PCBM may still have promise in certain systems.

The state of the art at this time appears that the increase in optical absorption holds promise for the use of higher fullerene derivatives, though with present p-type materials, the LUMO levels of the fullerenes higher in molecular weight than C_{70} appear to limit potential with current donor materials. [70]PCBM appears to be applicable to a range of p-type and processing conditions, when adequate adjustment of morphology is established relative to optimal conditions for [60]PCBM.

5.4.7
LUMO Variation

As the V_{oc} of OPV devices has been shown to correlate directly with the energy difference between HOMO of the donor and LUMO of the acceptor, an optimum

LUMO of the fullerene derivative exists for an optimum V_{oc}. In 2001, the hypothesis that the V_{oc} of a BHJ cell is related to the energy difference between the donor HOMO and the acceptor LUMO was proven correct upon the preparation and investigation of a series of four C_{60} related structures [76,77]. In order of the increasing acceptor strength (less negative first reduction potential), these were [60]PCBM, an azafulleroid, (13) in Figure 5.1, C_{60}, and a ketolactam, (11) in Figure 5.1. These compounds were provided with solubilizing groups to have adequate solubility to be tested in combination with MDMO-PPV, using one standard solvent for spin coating. As long as electrode materials were used that gave rise to ohmic contacts, the V_{oc} scaled in a 1:1 relationship with the change in HOMO(D) − LUMO(A) energy difference. Ketolactams are, till now, fullerene-type materials in which the electron affinity is most increased (i.e., by 70 mV compared to the pristine fullerene; 160 mV relative to the corresponding PCBM). Later on, Kooistra et al. confirmed that *increasing* the LUMO of the acceptor also results in an increase in V_{oc} and vice versa [78]. In this work, LUMO was increased by adding electron donating groups (methoxy- and methylthio-) to the phenyl moiety of PCBM. This strategy yielded modest increases in LUMO (up to 44 mV), with the larger number of methoxy addends to the phenyl giving the most increase, irrespective of their position on the phenyl ring. Most derivatives tested, however, showed poor solubility and the resulting morphology and device performance appeared rather inadequate. The processing conditions were not optimized, however, because the study was focused on the V_{oc} trends. See Table 5.3 for a comparison of first reduction potentials and resulting V_{oc} in BHJ OPV devices using MDMO-PPV as the donor for the work described above on stronger acceptors than [60]PCBM (series 1) and representative compounds from the study of Kooistra et al. on weaker acceptors (series 2); Figure 5.4 shows graphically the correlation between first reduction potential and V_{oc} values from Kooistra et al. The dependence of V_{oc} on the HOMO(D) − LUMO(A) energy difference from the standpoint of variation in HOMO(D) was also confirmed [6,79].

Though work on the improvement in LUMO is presently at an early stage, the above work suggests that it is possible to increase the LUMO energy level and that this does have the desired effect on V_{oc}. With the new low-bandgap polymers that exhibit a high current, increases in V_{oc} thus have the potential to make a substantial

Table 5.3 Variation of V_{oc} with LUMO of acceptor.

	$E_{1/2}$, first reduction potential (V)	V_{oc} (V) w/MDMO-PPV
[60]PCBM (1) (series 1)	−0.69 [76] (versus NHE)	0.74 [76]
Azafulleroid (13) (series 1)	−0.67 [76] (versus NHE)	0.63 [76]
Ketolactam (11) (series 1)	−0.53 [76] (versus NHE)	0.53 [76]
2,3,4-OMe-PCBM (4) (series 2)	−1.118 [76] (versus Fc/Fc$^+$)	0.94 [78]
2-OMe-PCBM (series 2)	−1.104 [78] (versus Fc/Fc$^+$)	0.87 [78]
[60]PCBM (1) (series 2)	−1.084 [78] (versus Fc/Fc$^+$)	0.84 [78]

Data taken from references in this work. Series 1 refers to devices described in Ref. [76]; series 2 refers to devices described in Ref. [78].

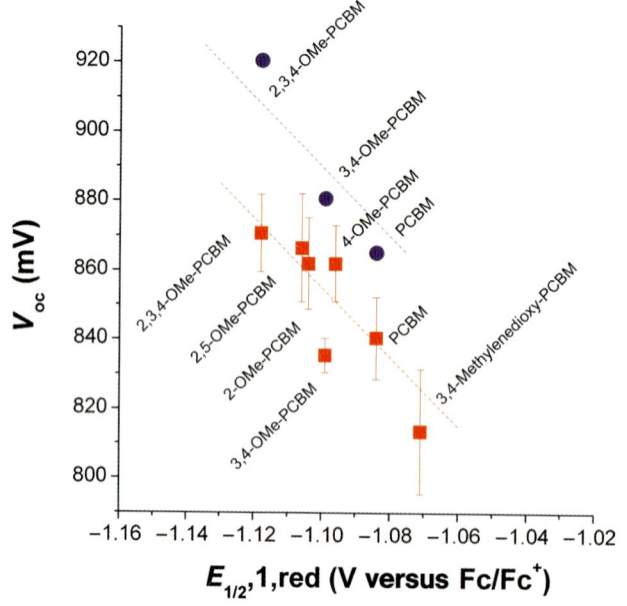

Figure 5.4 Variation of V_{oc} with LUMO of the fullerene acceptor. Circles correspond to MDMO-PPV:fullerene derivative devices fabricated with ODCB; squares correspond to MDMO-PPV: fullerene derivative devices fabricated with chlorobenzene. Graph reprinted with permission (from Ref. [78]).

step-change in device efficiencies. The challenge lies in maintaining the solubility, charge carrier mobility, and other properties relative to the PCBMs.

5.4.8
Deuterated PCBM

A deuterated version of [60]PCBM, (**14**) in Figure 5.1, was prepared to allow SIMS investigation of the MDMO-PPV:[60]PCBM blend [80]. First, it was assured that replacing [60]PCBM by [60]PCBM-d_5 has no noticeable effect. Second, the SIMS investigation revealed that the blend morphology is isotropic from top to bottom in the thin film.

5.5
Practical Considerations and Potential in Commercial Devices

5.5.1
Powder Morphology and Dissolution

A key concern is adequate dissolution of the fullerene derivative in the blend solvent. Particle characteristics play a key role in the speed of dissolution and care must be taken during solution processing to ensure fullerene derivative is adequately dissolved. Mechanical mixing helps ensure fast and complete dissolution relative to other methods, and the solution should always be filtered with a submicron filter to ensure a minimal amount of particulate matter, which could seed nucleation. Visually, a solution may appear clear, but still contain submicron particles owing to the strong tendency of fullerenes to cluster in solution. For each method of film formation (spin coating, doctor blading, screen printing, ink-jet printing, etc.), the formulation of the solution containing the n-type and p-type components (either with or without other additives for morphology control, viscosity, speed of film formation, etc.) has to be optimized for each n-type/p-type tandem. We feel the approach most likely to minimize effort and maximize chances of success should take into consideration solvent choice, solubilizing additive, or possibly extension of a proven derivative with alkyl moieties, for example, at the ester moiety of a PCBM as in PCB-C_n, (5) in Figure 5.1 (though with a note of caution that increasing carbon chain length is likely to affect morphology and/or electron mobilities).

5.5.2
Stability of the Fullerene Derivative and the Device Film

As fullerenes are exceptional radical scavengers, it is not surprising that autoxidation and photooxidation takes place. However, the rate of oxidation is surprisingly small (compared to, for example, tocopherol), considering that fullerenes react with radicals with very high rate constants. This is likely owing to the stability of oxide reaction products and most fullerene derivatives are relatively air stable as long as they are kept sealed in opaque containers. We have observed that [60]PCBM and [70]PCBM remain usable in OPV for a year or more when kept sealed in an opaque container, without the necessity to purge with an inert gas. For longer storage, purging the container with N_2 or Ar may be desired.

It has also been noted that this radical scavenging ability may, in fact, act as a preservative to help preserve the donor:acceptor blend film from autoxidation. MDMO-PPV degraded a factor of 10^3 slower when PCBM was added [81], exhibiting only about 10% degradation after 2000 h of illumination, and the authors concluded that the degradation was too slow to be the primary reason for degradation in device performance, at least in the system studied. The mechanism whereby fullerenes generate singlet oxygen has long been known and much has been made of the fact that the quantum yield of C_{60}, C_{70}, and many derivatives is approximately unity. However, Hamano *et al.* [82] showed that the quantum yield of singlet oxygen generation diminishes upon two epoxidations (which are among the products of fullerene singlet oxygen photooxygenation) to 0.1. The *net* generation of singlet O_2 is also quite small, and in fact smaller than very low photosensitizing agents such as tocopherol [83], owing probably to this and other quenching effects. Therefore, evidence suggests that fullerenes in fact may exert a net protective effect on the p-type because of their strong radical scavenging ability and it should be considered an open question as to how fullerene oxidation products affect device performance. The potential (photo)chemical threads for polymer:fullerene blends have been investigated [84–86]. In the meantime, it has become clear that whole BHJ-based devices can show degradation in many other ways than just inside the active layer. In many other studies, the degradation of performance of whole BHJ-based devices was investigated. Since a great number of variables play a role here, this field of research is considered outside the scope of this overview.

5.5.3
Impurities

Impurities that arise during synthesis must be eliminated to varying degrees for a good device performance. Main impurities include unreacted fullerenes, oxides, and multiadduct derivatives. The standard grade of PCBM used most commonly until recently was >99.5%, but >99% (both are molar percentages, by HPLC area) has also been used with success in some systems. In general, the lower purity grades should be compared with the higher purity grades to ensure best performance.

5.5.4
Commercial-Scale Application

Since the fullerene derivative is used in proportions of at least 1 : 1 (with P3HT) and 1 : 2 to 1 : 3 for certain low-bandgap polymers [2], the cost is an important factor at least in the economics of the active layer. Owing mainly to the fact that film thicknesses are typically on the order of 300 nm, little enough of the fullerene derivative is needed per meter square so that no constraint exists for commercialization even based on current volume prices, which are on order of €100 g^{-1} for [60]PCBM. Fullerenes are also now readily available at sufficiently low prices to allow for the preparation of commercial-scale quantities of fullerene derivatives. The main challenges of scale-up of the addition chemistry used for PCBM synthesis have been addressed. One ton per year production, to support order of magnitude 100 MW OPV production at cost targets required, is fully achievable at this point in time. There are no inherent obstacles, with respect to at least the most commonly used PCBM-type materials, for further upscaling to support full-size mass production of OPV devices. The toxicity profile of fullerenes, important from the viewpoint of a quantitative structure activity relationship (QSAR) analysis used in predicting potential toxicity – for example, in the environment – also appears to raise few concerns, based on LD$_{50}$ values in rodents, and *in vitro* genotoxicity and mutagenicity of mixed C$_{60}$ and C$_{70}$ [87]. Earlier suggestions that fullerenes may exhibit some cytotoxicity made on the basis of results with "nano-C$_{60}$," as 30–100 nm particles of C$_{60}$ prepared by solvent precipitation were termed [88] (which importantly should not be considered as necessarily relevant to the molecularly dissolved form), now have been explained as the result of tetrahydrofuran impurities emanating from the preparation of the C$_{60}$ nanoparticles [89]. A recent overview of fullerene and fullerene derivative toxicity can be found in Kolosnjaj *et al.* [90].

References

1. Hummelen, J.C., Yu, G., Gao, J., Wudl, F. and Heeger, A.J. (1995) *Science*, **270**, 1789–1791.
2. Peet, J., Kim, J.Y., Coates, N.E., Ma, W.L., Moses, D., Heeger, A.J. and Bazan, G.C. (2007) *Nature Materials*, **6**, 497–500.
3. Wienk, M.M., Kroon, J.M., Verhees, W.J.H., Knol, J., Hummelen, J.C., van Hal, P.A. and Janssen, A.J. (2003) *Angewandte Chemie – International Edition*, **42**, 3371.
4. Li, G., Shrotriya, V., Huang, J., Yao, Y., Moriarty, T., Emery, K. and Yang, Y. (2005) *Nature Materials*, **4**, 864–868.
5. Kim, Y., Lee, K., Coates, N.E., Moses, D., Nguyen, T., Dante, M. and Heeger, A.J. (2007) *Science*, **317** (5835), 222–225.
6. Muhlbacher, D., Scharber, M., Morana, M., Zhu, Z., Waller, D., Gaudiana, R. and Brabec, C. (2006) *Advanced Materials*, **18**, 2884–2889.
7. Brabec, C.J., Sariciftci, N.S. and Hummelen, J.C. (2001) *Advanced Functional Materials*, **11**, 15–26.
8. Brabec, C.J. and Sariciftci, N.S. (2001) *Monatshefte für Chemie*, **132**, 421.
9. Nelson, J. (2002) *Current Opinion in Solid State and Materials Science*, **6**, 87–95.
10. Spanggaard, H. and Krebs, F.C. (2004) *Solar Energy Materials and Solar Cells*, **83**, 125.
11. Sariciftci, N.S. (2004) *Materials Today* (September), 36.
12. Benanti, T.L. and Venkataraman, D. (2006) *Photosynthesis Research*, **87**, 73.
13. Hoppe, H. and Sariciftci, N.S. (2006) *Journal of Materials Chemistry*, **16**, 45.
14. Mozer, A.J. and Sariciftci, N.S. (2006) *Comptes Rendus Chimie*, **9**, 568.
15. Günes, S., Neugebauer, H. and Sariciftci, N.S. (2007) *Chemical Reviews*, **107**, 1324.
16. Rispens, M.T. and Hummelen, J.C. (2002) *Fullerenes: From Synthesis to Optoelectronic Properties* (eds D.M. Guldi and N. Martin), Kluwer, The Netherlands, pp. 387–435.
17. Kraabel, B., McBranch, D., Sariciftci, N.S., Moses, D. and Heeger, A.J. (1994) *Physical Review B: Condensed Matter*, **50**, 18543–18552.
18. Brabec, C.J., Zerda, G., Cerullo, G., De Silvestri, S., Luzzati, S., Hummelen, J.C. and Sariciftci, N.S. (2001) *Chemical Physics Letters*, **340**, 232–236.
19. Shaheen, S.E., Brabec, C.J., Padinger, F., Fromherz, T., Hummelen, J.C. and Sarciftci, N.S. (2001) *Applied Physics Letters*, **78**, 841.
20. Schwödiauer, R., Andreev, A.Y., Singh, B., Marjanovic, N., Matt, G., Günes, S., Sariciftci, N.S., Ramil, A.M., Sitter, H. and Bauer, S. (2005) *Materials Research Society Symposium Proceedings*, **871E**, I 4.9.1.
21. Peumans, P., Yakimov, A. and Forrest, S.R. (2003) *Journal of Applied Physics*, **93**, 3693.
22. Guldi, D.M. (2003) *Spectrum*, **16**, 8.
23. Marcus, R.A. (1965) *Journal of Chemical Physics*, **43**, 679–701.
24. Marcus, R.A. (1993) *Angewandte Chemie – International Edition in English*, **32**, 1111–1121.
25. Aksenova, V.L., Avdeeva, M.V., Tropina, T.V., Priezzheva, V.B. and Schmelzera, J.W.P. (2006) *Journal of Molecular Liquids*, **127** (1–3), 142–144.
26. Hebard, A.F., Haddon, R.C., Fleming, R.M. and Kortan, A.R. (1991) *Applied Physics Letters*, **59**, 2109.
27. Mihailetchi, V.D., Koster, L.J.A., Blom, P.W.M., Melzer, C., de Boer, B., van Duren, K.J. and Janssen, R.A.J. (2005) *Advanced Functional Materials*, **15**, 795.
28. Dennis, T.J.S., Kai, T., Tomiyama, T. and Shinohara, H. (1998) *Chemical Communications*, 619.
29. Dennis, T.J.S., Kai, T., Asato, K., Tomiyama, T., Shinohara, H., Yoshida, T., Kobayashi, Y., Ishiwatari, H., Miyake, Y., Kikuchi, K. and Achiba, Y. (1999) *Journal of Physical Chemistry A*, **103**, 8747.
30. Tagmatarchis, N., Avent, A.G., Prassides, K., Dennis, T.J.S. and Shinohara, H. (1999) *Chemical Communications*, 1023.

31 Tagmatarchis, N., Okada, K., Yoshida, T., Kobayashi, Y. and Shinohara, H. (2001) *Chemical Communications*, 1366.

32 Fowler, P.W. and Manolopoulos, D.E. (1995) *An Atlas of Fullerenes*. International Series of Monographs on Chemistry, Clarendon Press, Oxford.

33 Hummelen, J.C., Knight, B.W., LePeq, F., Wudl, F.J., Yao, J. and Wilkens, C.L. (1995) *Organic Chemistry*, **60** (3), 532–538.

34 Ruoff, R.S., Tse, D.S., Malhotra, R. and Lorents, D.C. (1993) *Journal of Physical Chemistry*, **97**, 3379–3383.

35 Mihailetchi, V.D., Xie, H., de Boer, B., Koster, J.A. and Blom, P.W.M. (2006) *Advanced Functional Materials*, **16**, 699–708.

36 Singh, T.B., Marjanovic, N., Stadler, P., Auinger, M., Matt, G.J., Gunes, S. and Sariciftci, N.S. (2005) *Journal of Applied Physics*, **97**, 083714.

37 Wöbkenberg, P., Bradley, D.D.C., Kronholm, D., Kooistra, F.B., Hummelen, J.C., Cölle, M., de Leeuw, D.M. and Anthopoulos, T.D. (2007) High mobility n-channel organic transistors based on soluble C_{60} and C_{70} fullerene derivatives. Poster Presentation, ECME.

38 Rispens, M.T., Meetsma, A., Rittberger, R., Brabec, C.J., Sariciftci, N.S. and Hummelen, J.C. (2003) *Chemical Communications*, 2116–2118.

39 Talyzin, A. and Jansson, U. (2000) *Journal of Physical Chemistry B*, **104** (21), 5064–5071.

40 Mihailetchi, V.D., van Duren, J.K.J., Blom, P.W.M., Hummelen, J.C., Janssen, R.A.J., Kroon, J.M., Rispens, M.T., Verhees, W.J.H. and Wienk, M.M. (2003) *Advanced Functional Materials*, **13**, 43.

41 Mihailetchi, V.D., Koster, L.J.A., Hummelen, J.C. and Blom, P.W.M. (2004) *Physical Review Letters*, **93**, 216601.

42 Lof, R.W., van Veenendaal, M.A., Koopmans, B., Jonkman, H.T. and Sawatzky, G.A. (1992) *Physical Review Letters*, **68**, 3924.

43 Yu, G., Gao, J., Hummelen, J.C., Wudl, F. and Heeger, A.J. (1995) *Science*, **270**, 1789.

44 Halls, J.J.M., Walsh, C.A., Greenham, N.C., Marseglia, E.A., Friend, R.H., Moratti, S.C. and Holmes, A.B. (1995) *Nature*, **376**, 498.

45 Hoppe, H., Niggemann, M., Winder, C., Kraut, J., Hiesgen, R., Hinsch, A., Meissner, D. and Sariciftci, S. (2004) *Advanced Functional Materials*, **14** (10), 1005–1011.

46 Sanchez, L., Knol, J. and Hummelen, J.C. (2001) Presented at Solar '01 & ENPHO '01, April 3–8, Cairo, Egypt.

47 Zheng, L., Zhou, Q., Deng, X., Yuan, M., Yu, G. and Cao, Y. (2004) *Journal of Physical Chemistry B*, **108** (32), 11921–11926.

48 Riedel, I., von Hauff, E., Parisi, J., Martín, N., Giacalone, F. and Dyakonov, V. (2005) *Advanced Functional Materials*, **15**, 1979.

49 Backer, S.A., Sivula, K., Kavulak, D.F. and Frechet, J.M.J. (2007) *Chemistry of Materials*, **19**, 2927–2929.

50 Wang, X., Perzon, E., Delgado, J.L., de la Cruz, P., Zhang, F., Langa, F., Andersson, M.R. and Inganäs, O. (2004) *Applied Physics Letters*, **85**, 5081.

51 Wang, X., Perzon, E., Oswald, F., Langa, F., Admassie, S., Andersson, M.R. and Inganäs, O. (2005) *Advanced Functional Materials*, **15**, 1665.

52 Gadisa, A., Wang, X., Admassie, S., Perzon, E., Oswald, F., Langa, F., Andersson, M.R. and Inganäs, O. (2006) *Organic Electronics*, **7**, 195.

53 Andersson, L.M. and Inganäs, O. (2006) *Applied Physics Letters*, **88**, 082103.

54 Segura, J.L., Giacalone, F., Gomez, R., Martin, N., Guldi, D.M., Luo, C., Swartz, A., Riedel, I., Chirvase, D., Parisi, J., Dyakonov, V., Sariciftci, N.S. and Padinger, F. (2005) *Materials Science and Engineering C*, **25**, 835–842.

55 Camaioni, N., Ridolfi, G., Casalbore-Miceli, G., Possamai, G. and Maggini, M. (2002) *Advanced Materials*, **14**, 1735.

56 Wang, X., Perzon, E., Mammo, W., Oswald, F., Admassie, S., Persson, N.-K., Langa, F.S., Andersson, M.R. and Inganäs, O. (2006) *Thin Solid Films*, **511–512**, 576–580.

References

57 Huang, J., Li, G. and Yang, Y. (2005) *Applied Physics Letters*, **87**, 112105.
58 Melzer, C., Koop, E.J., Mihailetchi, V.D. and Blom, P.W.M. (2004) *Advanced Functional Materials*, **14**, 865.
59 Dennler, G., Mozer, A.J., Juška, G., Pivrikas, A., Österbacka, R., Fuchsbauer, A. and Sariciftci, N.S. (2006) *Organic Electronics*, **7**, 229.
60 Mihailetchi, V.D., Xie, H., de Boer, B., Koster, L.J.A. and Blom, P.W.M. (2006) *Advanced Functional Materials*, **16**, 699.
61 Mihailetchi, V.D., Xie, H., de Boer, B., Popescu, L.M., Hummelen, J.C., Blom, P.W.M. and Koster, L.J.A. (2006) *Applied Physics Letters*, **89**, 012107.
62 Li, G., Shrotriya, V., Huang, J., Yao, Y., Moriarty, T., Emery, K. and Yang, Y. (2005) *Nature Materials*, **4**, 864.
63 Li, G., Yao, Y., Yang, H., Shrotriya, V., Yang, G. and Yang, Y. (2007) *Advanced Functional Materials*, **17**, 1636.
64 Hoppe, H., Niggeman, N., Winder, C., Kraut, J., Hiesgen, R., Hinsch, A., Meissner, D. and Sariciftci, N.S. (2004) *Advanced Functional Materials*, **14**, 1005.
65 McNeill, C.R., Watts, B., Thomsen, L., Belcher, W.J., Kilcoyne, A.L.D., Greenham, N.C. and Dastoor, P.C. (2006) *Small*, **2**, 1432.
66 Popescu, L.M., van't Hof, P., Sieval, A.B., Jonkman, H.T. and Hummelen, J.C. (2006) *Applied Physics Letters*, **89**, 213507.
67 Popescu, L.M. *et al.*, unpublished data.
68 Markov, D.E., Amsterdam, E., Blom, P.W.M., Sieval, A.B. and Hummelen, J.C. (2005) *Journal of Physical Chemistry A*, **109**, 5266–5274.
69 Markov, D.E., Hummelen, J.C., Blom, P.W.M. and Sieval, A.B. (2005) *Physical Review B: Condensed Matter*, **72**, 045216.
70 Drees, M., Hoppe, H., Winder, C., Neugebauer, H., Sariciftci, N.S., Schwinger, W., Schäffler, F., Topf, C., Scharber, M.C., Zhud, Z. and Gaudiana, R. (2005) *Journal of Materials Chemistry*, **15**, 5158.
71 Baffreau, J., Leroy-Lhez, S., Hudhomme, P., Groeneveld, M.M., van Stokkum, I.H.M. and Williams, R.M. (2006) *Journal of Physical Chemistry A*, **110**, 49, 13123–13125.
72 Canteenwala, T., Padmawar, P.A. and Chiang, L.Y. (2005) *Journal of the American Chemical Society*, **127**, 26–27.
73 Popescu, L.M. *et al.*, unpublished data.
74 Kooistra, F.B., Mihailetchi, V.D., Popescu, L.M., Kronholm, D., Blom, P.W.M. and Hummelen, J.C. (2006) *Chemistry of Materials*, **18** (13), 3068–3073.
75 Anthopoulos, T.D., Kooistra, F.B., Wondergem, H.J., Kronholm, D., Hummelen, J.C. and de Leeuw, D.M. (2006) *Advanced Materials*, **18**, 1679–1684.
76 Brabec, C.J., Cravino, A., Meissner, D., Sariciftci, N.S., Fromherz, T., Rispens, M.T., Sanchez, L. and Hummelen, J.C. (2001) *Advanced Functional Materials*, **11**, 374.
77 Brabec, C.J., Cravino, A., Meissner, D., Sariciftci, N.S., Rispens, M.T., Sanchez, L., Hummelen, J.C. and Fromherz, T. (2002) *Thin Solid Films*, **403–404**, 368–372.
78 Kooistra, F.B., Knol, J., Kastenberg, F., Popescu, L.M., Verhees, W.J.H., Kroon, J.M. and Hummelen, J.C. (2007) *Organic Letters*, **9** (4), 551–554.
79 Gadisa, A., Svensson, M., Andersson, M.R. and Inganäs, O. (2004) *Applied Physics Letters*, **84**, 1609.
80 Bulle-Lieuwmaa, C.W.T., van Duren, J.K.J., Yang, X., Loos, J., Sieval, A.B., Hummelen, J.C. and Janssen, R.A.J. (2004) *Applied Surface Science*, **231–232**, 274–277.
81 Chambon, S., Rivaton, A., Gardette, J.L. and Firon, M. (2007) *Solar Energy Materials and Solar Cells*, **91**, 394–398.
82 Hamano, T., Okuda, K., Mashino, T., Hirobe, M., Arakane, K., Ryu, A., Mashiko, S. and Nagano, T. (1997) *Chemical Communications*, 21.
83 Sieval, A.B. *et al.*, unpublished data.
84 Neugebauer, H., Brabec, C.J., Hummelen, J.C., Janssen, R.A.J. and Sariciftci, N.S. (1999) *Synthetic Metals*, **102**, 1002–1003.

85 Neugebauer, H., Brabec, C.J., Hummelen, J.C. and Sariciftci, N.S. (1999) *Solar Energy Materials and Solar Cells*, **61**, 35–42.

86 Hummelen, J.C., Knol, J. and Sanchez, L. (2001) *Proceedings of SPIE – International Society for Optical Engineering*, **4108**, 76–84.

87 Mori, T., Takada, H., Ito, S., Matsubayashi, K., Miwa, N. and Sawaguchi, T. (2006) *Toxicology*, **225**, 48–54.

88 Oberdörster, E. (2004) *Environmental Health Perspectives*, **112** (10), 1058–1062.

89 Isakovica, A., Markovicb, Z., Nikolicb, N., Todorovic-Markovicb, B., Vranjes-Djuricb, S., Harhajic, L., Raicevicc, N., Romcevicd, N., Vasiljevic-Radovice, D., Dramicaninb, M. and Trajkovic, V. (2006) *Biomaterials*, **27**, 5049–5058.

90 Kolosnjaj, J., Szwarc, H. and Moussa, F. (2007) Toxicity studies of fullerenes and derivatives, in *Bioapplications of Nanomaterials* (ed. W.C.W. Chan), Landes Bioscience.

6
Hybrid Polymer/Nanocrystal Photovoltaic Devices
Neil C. Greenham

6.1
Introduction

The challenge of producing cheap, efficient photovoltaic devices places many demands on the active materials. In devices based on donor–acceptor composites, the nanostructure plays a vital role in determining the efficiency. A large area of donor–acceptor interface is needed to dissociate photogenerated excitons, but an efficient charge extraction requires a morphology that allows efficient transport of charge carriers to the electrodes. Much effort has been devoted to optimizing nanostructures in molecular and polymeric films by controlling the processing conditions during the film deposition, relying on the kinetics of solvent removal, phase separation, or annealing to achieve the desired morphology. Using nanoparticles of inorganic semiconductors in a photovoltaic device allows an alternative approach in which the nanoparticles provide stable preformed structures on length scales of 2–100 nm that can then be incorporated with an organic component to give a photovoltaic film with desirable exciton-dissociation and charge-transport properties.

Semiconductor nanoparticles can now be formed in a huge range of shapes, sizes, and materials [1,2]. Here, we will focus on inorganic semiconductors nanoparticles formed by chemical methods, providing particles that are solution-processable and therefore suitable for processing over large areas together with materials such as conjugated polymers. Solubility is typically achieved by the presence of organic ligands on the semiconductor surface. One of the main advantages of semiconductor nanoparticles for optoelectronic applications is that their bandgap is size-dependent through the quantum confinement effect. In the simplest model for a cubic particle of dimension $2r$, the bandgap scales with size as

$$E_g = E_0 + \frac{h^2 \pi^2}{8r^2}\left(\frac{1}{m_e^*} + \frac{1}{m_h^*}\right),$$

Organic Photovoltaics: Materials, Device Physics, and Manufacturing Technologies.
Edited by Christoph Brabec, Vladimir Dyakonov, and Ullrich Scherf
Copyright © 2008 WILEY-VCH Verlag GmbH & Co. KGaA, Weinheim
ISBN: 978-3-527-31675-5

where E_0 is the bulk semiconductor bandgap and m_e^* and m_h^* are the electron and hole effective masses, respectively. For spherical particles, the same scaling of the quantum confinement with r^{-2} is observed. More detailed theories take into account the leakage of electron and hole wave functions into the finite barriers at the nanocrystal surface [3], the Coulomb interaction between electron and hole [4], the presence of multiple bands [5,6], the effect of the crystal field [7] and slight nonsphericity [8], and the effect of exchange interactions [9]. A huge amount of spectroscopy has been performed to study these intriguing "artificial atom" systems. For the design of photovoltaic materials, however, it is sufficient to appreciate that the bandgap can be significantly raised from its bulk value in quantum-confined systems of dimensions below about 5 nm and that these effects are particularly strong in semiconductors with low effective masses. It is important to note that quantum confinement controls the bandgap by changing both the electron affinity and the ionization potential of the nanoparticle [10]. This allows the energies of charge carriers in a photovoltaic device to be tuned, in addition to the absorption energy.

The structure of a generic polymer/nanocrystal photovoltaic device is shown in Figure 6.1. To form a device, nanoparticles are first mixed with a conjugated polymer in solution. A mixed polymer/nanocrystal film of thickness around 100–200 nm is then deposited from solution onto a transparent substrate coated with a transparent hole-collecting electrode. For prototype devices, spin coating is typically used, but other solution deposition methods, including printing techniques, may also be used. A top electron-accepting electrode is then deposited by vacuum evaporation.

The device operation starts by absorption of a photon in the active layer. The conjugated polymer typically dominates the absorption, as conjugated polymers have absorption coefficients usually larger than those of inorganic semiconductors, even when the inorganic semiconductor is in nanocrystalline form. Excitons generated in the polymer diffuse to the interface with a nanoparticle, where they undergo charge transfer. Most inorganic semiconductors have significantly higher electron affinities than conjugated polymers, and hence act as electron acceptors, leaving a hole on the

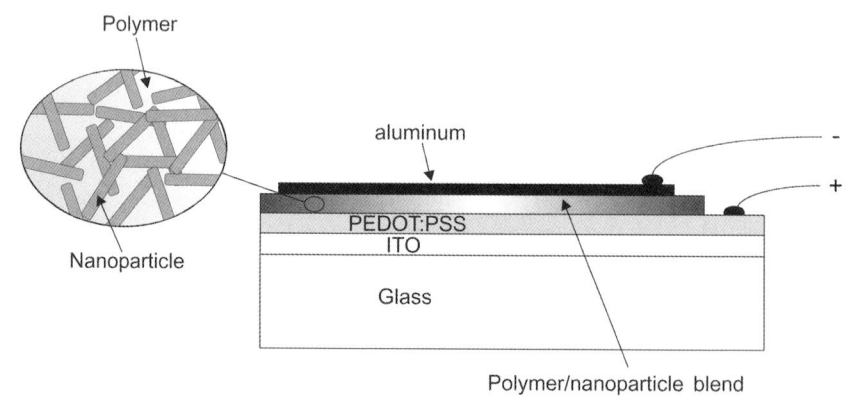

Figure 6.1 Schematic structure of a polymer/nanoparticle photovoltaic device.

polymer and an electron on the nanocrystal. Absorption in the nanocrystals is also possible, followed by hole transfer to the polymer. Electrons and holes must then be further separated and transported out of the device, either by diffusion or under the influence of the internal electric field. Hopping of electrons between nanoparticles is, therefore, key to device operation. As with all nanostructured photovoltaics, the challenge is to achieve morphologies that allow an efficient dissociation of excitons, whilst also providing pathways for charge carriers to reach the electrodes without recombination. As we will see, nanoparticles offer some unique advantages in addressing this challenge.

Many of the physical processes involved are similar to those taking place in dye-sensitized solar cells, where an absorbing dye is adsorbed onto the surface of a sintered network of nanocrystalline electron acceptor [11]. In those devices, the electron acceptor is deposited first onto the substrate (typically followed by a high-temperature annealing step), after which the deposition of the absorbing and hole-transporting components takes place. Here, though, we will concentrate on devices in which the entire active layer is deposited from solution in one step.

In this chapter, we will examine the various nanocrystal materials systems that have been studied in the context of photovoltaic devices with conjugated polymers, giving a brief review of synthesis methods and showing how the control of shape and electronic structure influences the device performance. We will then go on to examine in more detail the various physical processes occurring in polymer/nanocrystal photovoltaics, discussing what factors limit the performance of current devices. The structures of some of the conjugated polymers used in these devices are shown in Table 6.1.

6.2
Classes of Polymer/Nanocrystal Device

6.2.1
Devices Based on CdSe Nanoparticles

CdSe nanoparticles were the first to be used with conjugated polymers in photovoltaic devices [12] and devices using CdSe nanoparticles remain the most extensively studied. CdSe nanoparticles have a number of advantages: they absorb at a useful energy for solar energy conversion, they are a good electron acceptor for conjugated polymers, and shape control is relatively easy to achieve. The bulk bandgap of CdSe is 1.74 eV, which can be increased to 2.5 eV in spherical nanoparticles as small as 2 nm in diameter. Absorption spectra of spherical CdSe nanoparticles of various sizes are shown in Figure 6.2.

6.2.1.1 Synthesis of CdSe Nanoparticles
The classic synthesis for highly crystalline CdSe nanoparticles with narrow size distributions is the method of Murray et al. [13]. The ligand used is tri-n-octylphosphine

Table 6.1 Chemical structures of common conjugated polymers.

Structure	Abbreviation	Name
	MEH-PPV	Poly(2-methoxy-5-(2'-ethyl)-hexyloxy-p-phenylene vinylene)
	P3HT	Poly(3-hexylthiophene)
	OC$_1$C$_{10}$-PPV	Poly(2-methoxy-5-(3',7'-dimethyl-octyloxy)-p-phenylene vinylene)
	APFO-3	Poly(2,7-(9,9-dioctyl-fluorene)-alt-5,5-(4',7'-di-2-thienyl-2',1',3'-benzothiadiazole))
	MEH-CN-PPV	Poly(2-methoxy-5-(2'-ethylhexyloxy)-α,α'-di-cyano-p-xylylidene-alt-2-methoxy-5-(2'-ethylhexyloxy)-p-xylylidene)

Table 6.1 (Continued)

Structure	Abbreviation	Name
(chemical structure with O–C$_6$H$_{13}$, CN, NC groups)	DHeO-CN-PPV	Poly(2,5-dihexyloxy-α,α'-dicyano-p-xylylidene-alt-2,5-dihexyloxy-p-xylylidene)

oxide (TOPO), which is heated to around 360 °C in a three-necked flask purged with argon. Selenium is mixed with trioctylphosphine or tributylphosphine [14] to form a trialkylphosphine selenide complex. This is then mixed with dimethyl cadmium and the mixture is then injected rapidly into the hot TOPO. An immediate nucleation of nanoparticles takes place and the temperature of the reaction mixture then drops, suppressing further nucleation. Provided the reactant concentration does not drop too low, a narrow size distribution can be achieved since the smaller particles grow faster than the larger ones [15]. Further growth can be achieved by adding additional reactant solution, giving good control over particle growth for diameters between 2 and 5 nm. The particles can be precipitated by adding methanol and recovered by centrifugation.

A more recent synthesis method developed by Peng and Peng has been widely adopted as it removes the need to use the highly toxic and pyrophoric dimethyl cadmium [16]. In this method, cadmium oxide is heated together with TOPO and

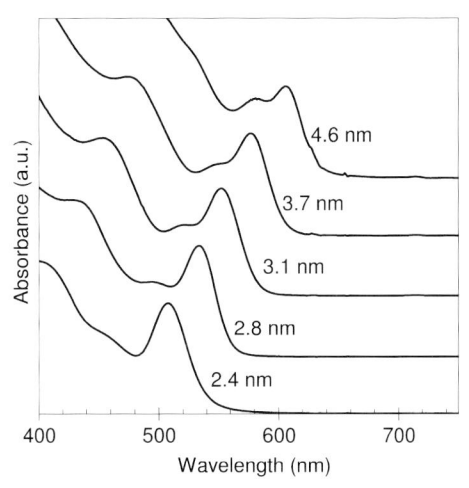

Figure 6.2 Absorption spectra of spherical CdSe nanocrystals of the approximate diameters shown.

a phosphonic acid ligand, such as tetradecylphosphonic acid (TDPA), to form a Cd–TDPA complex.

$$\text{CdO} + 2\left(\begin{array}{c}\text{O}\\\|\\\text{HO}-\text{P}-\text{C}_{18}\text{H}_{37}\\|\\\text{OH}\end{array}\right) \rightleftharpoons \text{Cd}^{2+}\left(\begin{array}{c}\text{O}\\\|\\\text{O}^{-}-\text{P}-\text{C}_{18}\text{H}_{37}\\|\\\text{OH}\end{array}\right)_2 + \text{H}_2\text{O}$$

Nanocrystal nucleation then proceeds by reheating the solution to around 300 °C followed by injection of a selenium–trioctylphosphine complex.

$$\text{Cd}^{2+}\left(\begin{array}{c}\text{O}\\\|\\\text{O}^{-}-\text{P}-\text{C}_{18}\text{H}_{37}\\|\\\text{OH}\end{array}\right)_2 + \begin{array}{c}\text{Se}\\\|\\\text{C}_8\text{H}_{17}-\text{P}-\text{C}_8\text{H}_{17}\\|\\\text{C}_8\text{H}_{17}\end{array} \rightleftharpoons$$

$$\text{CdSe} + \begin{array}{c}\text{O}\\\|\\\text{C}_8\text{H}_{17}-\text{P}-\text{C}_8\text{H}_{17}\\|\\\text{C}_8\text{H}_{17}\end{array} + \left(\begin{array}{ccc}\text{O} & \text{O} & \text{O}\\\| & \| & \|\\\text{C}_{18}\text{H}_{37}-\text{P}-\text{O}-\text{P}-\text{O}-\text{P}-\text{C}_{18}\text{H}_{37}\\| & | & |\\\text{OH} & \text{C}_{18}\text{H}_{37} & \text{OH}\end{array}\right)_n$$

In both synthetic routes, the shape of the nanoparticles can be controlled by carefully choosing the reaction conditions, reactant concentrations, and the ligands used [17]. The growth of nonspherical particles relies on the anisotropic crystal structure of the wurtzite form of CdSe, which has a unique c axis. Preferential growth along this axis leads to the formation of "quantum rods," with diameters of typically 4–6 nm and lengths that can exceed 100 nm. Anisotropic shapes are favored by a rapid growth reaction, leading to a shape that is determined by the kinetics of growth along the different crystallographic directions. Control of the growth rate is achieved by adding alkylphosphonic acids, which bind more strongly to cadmium than TOPO does, to the reaction mixture. High concentrations of the reactant monomers are also important to achieve rod growth [17,18]. Shape control can also be achieved, by varying the monomer concentration, in the CdO route described above [16].

Further three-dimensional control of nanoparticle shape in CdSe can lead to branched structures, including tetrapods where four limbs extend tetrahedrally from a central core [19]. These structures rely on the presence of two crystal polytypes for CdSe that are relatively close in energy. In addition to the hexagonal wurtzite structure mentioned above, CdSe can exist in a cubic zinc blende structure. Tetrapods are formed by the nucleation of tetrahedrally faceted zinc blende cores, followed by the

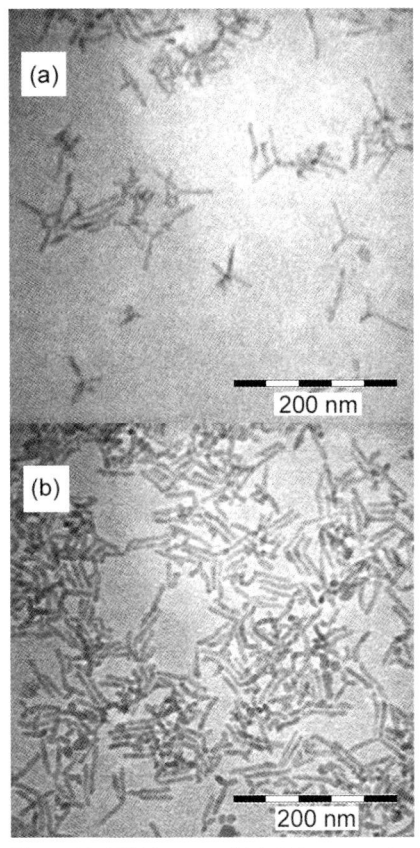

Figure 6.3 TEM images of (a) CdSe tetrapods and (b) CdSe nanorods [20].

growth of hexagonal wurtzite "legs." The growth of tetrapods seems to be favored by higher reactant concentrations than those used for rods [20], but in practice for CdSe it is difficult to control the fraction of rods and tetrapods formed in the synthesis. In particular, it should be noted that for CdSe all samples containing tetrapods also contain a significant fraction of other branched structures and rods [19]. TEM images of CdSe rods and tetrapods are shown in Figure 6.3. With CdTe (see Section 6.2.3), however, shape control is much easier and this has been attributed to the slightly larger energy difference between the crystal polytypes in CdTe than in CdSe, making selective growth and nucleation easier to achieve [21].

For photovoltaic applications, it is usually necessary to remove bulky ligands from the nanocrystal surface. This can be achieved by refluxing the nanocrystals in pyridine, followed by precipitation with hexanes, to give particles with pyridine as a weakly bound ligand [12]. The particles can then be re-dissolved in a solvent such as chloroform and mixed with the polymer solution for spin coating. It is believed that during the spin coating and device preparation processes, the remaining pyridine is removed from the surface of the nanoparticles, leaving the nanoparticle surface directly in contact with the polymer [22].

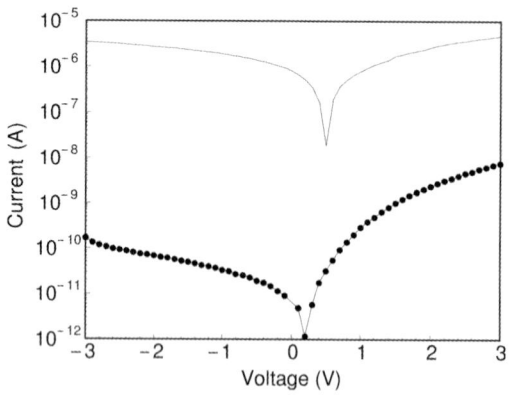

Figure 6.4 Current–voltage characteristics for an MEH-PPV/CdSe photovoltaic device, under illumination (solid line) and in the dark (circles). The illumination was at 514 nm with an intensity of 5 W m^{-2}. The active area of the device was 7.3 mm^2 and the CdSe concentration was 90 wt% [12].

6.2.1.2 Devices Using CdSe Nanoparticles

The first polymer/nanoparticle devices were made by mixing CdSe nanoparticles of approximately 5 nm diameter with the conjugated polymer poly(2-methoxy-5-(2'-ethyl)-hexyloxy-p-phenylene vinylene) (MEH-PPV) [12,23]. The best efficiencies were achieved using a high concentration of nanoparticles (90% by weight), corresponding to a volume fraction of approximately 50%. An efficient exciton dissociation at the polymer/nanocrystal interface was observed and quantum efficiencies of up to 10% were obtained. Current–voltage characteristics for these devices are shown in Figure 6.4. Power conversion efficiencies were low (<0.1%), which was attributed to the difficulty of transporting electrons out of the device by hopping from nanocrystal to nanocrystal. TEM images of the blend films used (Figure 6.5) show

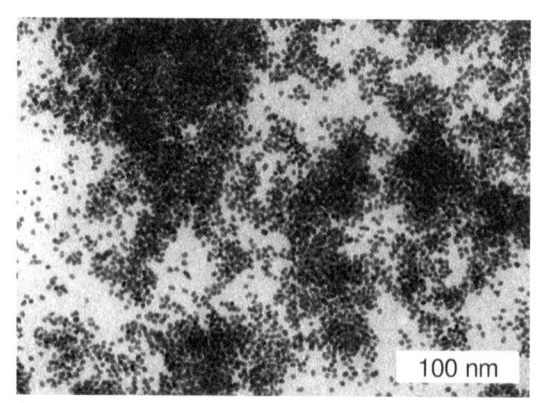

Figure 6.5 TEM image of an MEH-PPV film containing 65 wt% of pyridine-treated CdSe nanoparticles [12].

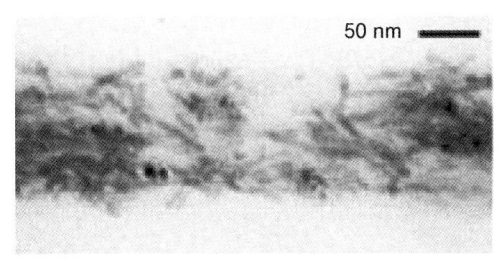

Figure 6.6 TEM cross-section of a P3HT/CdSe nanorod film containing 40 wt% nanorods. Reproduced with permission from Ref. [25]. Copyright 2002 American Association for the Advancement of Science.

phase separation between the polymer and the nanocrystal components, providing pathways for electron and hole transport through the film.

A significant advance was reported by Huynh *et al.*, who used CdSe nanorods mixed with regioregular poly(3-hexylthiophene) (P3HT) [24,25]. This polymer is known to exhibit good hole-transport properties [26] and the use of rods allowed improved transport of electrons out of the device. Figure 6.6 shows a TEM cross section of the device, which contains 60 nm long nanorods. Quantum efficiencies of 55% were achieved, together with power conversion efficiencies of 1.7%. Current–voltage curves for these devices under AM 1.5 G illumination are shown in Figure 6.7. Sun and Greenham later reported that the power efficiency of this type of device could be increased to 2.6% by changing the solvent to 1,2,4-trichlorobenzene, which evaporates slowly during the spin-coating process (Figure 6.8) [27]. This allows a favorable ordering to occur in the P3HT component of the blend, which improves the hole mobility [28].

One potential problem associated with the use of nanorods is that the rods tend to lie in the plane of the film, which is not the direction in which the charges are to be transported. This can be overcome by using samples containing tetrapod-shaped nanoparticles, which by virtue of their shape are unable to lie flat in the plane of

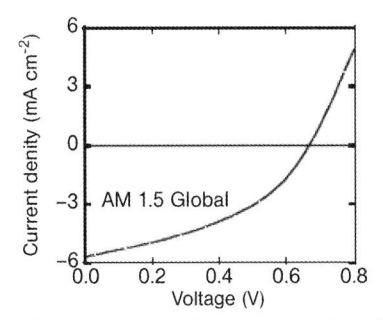

Figure 6.7 Current density versus voltage for a P3HT/CdSe nanorod device under AM 1.5 G conditions. Reproduced with permission from Ref. [25]. Copyright 2002 American Association for the Advancement of Science.

6 Hybrid Polymer/Nanocrystal Photovoltaic Devices

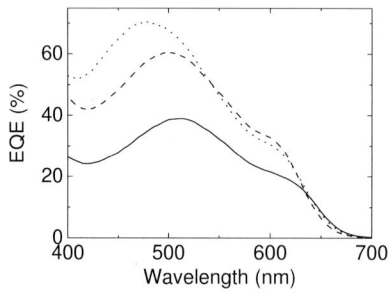

Figure 6.8 Action spectra for photovoltaic devices fabricated from a blend of 90 wt% CdSe nanorods and P3HT using different solvents: chloroform (solid line), thiophene (dashed line), and 1,2,4-trichlorobenzene (dotted line) [27].

the film. Using poly(2-methoxy-5-(3′,7′-dimethyl-octyloxy)-p-phenylene vinylene) (OC$_1$C$_{10}$-PPV) as the hole transporter, Sun and Greenham showed that by changing the electron transporter from CdSe nanorods to CdSe tetrapods, quantum efficiencies could be increased from 23 to 45% [20]. By changing to the high boiling point solvent 1,2,4-trichlorobenzene, they were able to further improve the power efficiency, achieving values of 2.8% in the best devices (Figure 6.9) [29]. In these devices, it appears that the tetrapods are preferentially segregated toward the top surface of the film, which is beneficial for efficient electron collection and high open-circuit voltage (see Section 6.3.5).

Gur et al. investigated the performance of devices using P3HT with hyperbranched CdSe nanoparticles [30], which exhibit a dendritic structure with many branch points (Figure 6.10) [31]. They compared the performance as a function of CdSe loading between nanorods and hyperbranched particles. They found that hyperbranched particles gave better performance than rods, reaching optimized efficiencies of 2.2%. In particular, they noted that whilst nanorod devices required high CdSe fractions to

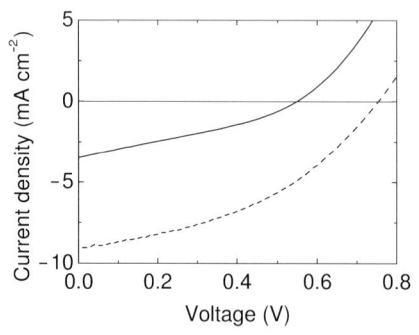

Figure 6.9 Current density versus voltage for photovoltaic devices fabricated with CdSe and OC$_1$C$_{10}$-PPV using different solvents: chloroform (solid line) and 1,2,4-trichlorobenzene (dashed line). The devices were illuminated under a simulated AM 1.5 G spectrum at 88.9 mW cm^{-2} [29].

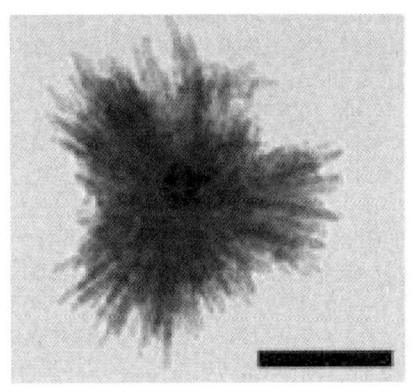

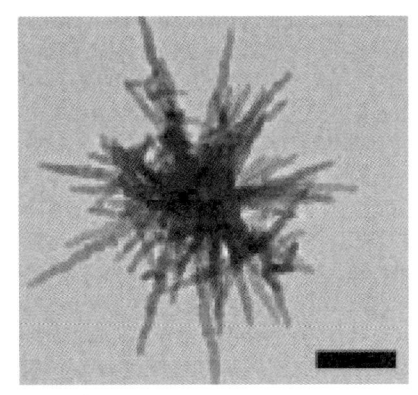

Figure 6.10 Transmission electron micrographs showing the structure of CdSe (left) and CdTe (right) hyperbranched nanocrystals. Scale bar, 100 nm. Reproduced with permission from Ref. [30]. Copyright 2007 American Chemical Society.

achieve percolation for electron transport, devices using hyperbranched particles gave good performance at lower CdSe fractions, indicating that percolation between hyperbranched particles is much easier to achieve.

Another class of conjugated polymer attractive for photovoltaics is the polyfluorenes. These materials have been extensively used in polymer LEDs [32] and exhibit good stability combined with excellent processing properties. The fluorene unit can be copolymerized with a range of different electron-accepting or electron-donating moieties to give donor or acceptor polymers. One red-absorbing polyfluorene of particular interest for photovoltaics is poly(2,7-(9,9-dioctylfluorene)-*alt*-5,5-(4′,7′-di-2-thienyl-2′,1′,3′-benzothiadiazole)) (APFO-3). This polymer has been used as an electron donor in conjunction with fullerene derivatives as the electron acceptor [33]. Recently, it has been shown that this polymer can be used in conjunction with CdSe tetrapods, giving AM 1.5 solar power conversion efficiencies of 2.4% [34]. The internal quantum efficiency is over 40% in the spectral range from 510 to 590 nm with a maximum value of 44% at 565 nm. Once again, the choice of solvent is important in determining the power efficiency, with *p*-xylene giving better performance than chloroform. The improved efficiency was tentatively assigned to the low solubility of APFO-3 in *p*-xylene, leading to a vertically segregated structure where a pure or nearly pure polymer layer is formed on the bottom electrode.

6.2.2
Devices Based on Metal Oxide Nanoparticles

Sintered films of titanium dioxide nanoparticles form the basis for dye-sensitized solar cells, where they are known to act as electron acceptors from organic dyes [11]. It is, therefore, natural to explore whether TiO_2 and other metal oxide nanoparticles can be used in place of CdSe in solution-processed composite devices. Using metal oxides offers advantages as they are less toxic than typical II–VI semiconductors and are relatively easy to synthesize in large quantities. However, they have large bandgaps and therefore do not contribute usefully to the absorption of light in a photovoltaic device. They also present challenges for processing since they tend to have a polar surface, so they are difficult to combine with most conjugated polymers in solution at high weight fractions. Owing to the different crystal structure mechanisms for shape control are different from those in II–VI particles, but much progress has been made recently in achieving particle shapes optimized for photovoltaics.

For the specific case of TiO_2, although there has been extensive work on devices in which polymers are infiltrated into nanostructured TiO_2 films formed on substrates, for purely solution-processed devices it is difficult to form stable solutions of TiO_2 nanoparticles, particularly with solvents that are compatible with polymers [35]. Sonication of TiO_2 with heated P3HT solutions in xylene has helped obtain power conversion efficiencies of 0.4% [36,37] and similar efficiencies have now been achieved using MEH-PPV [38]. Owing to the processing difficulties for TiO_2 nanoparticles, we will concentrate on devices using ZnO nanoparticles from now onward.

6.2.2.1 Synthesis of ZnO Nanoparticles

Zinc oxide nanoparticles for photovoltaics are synthesized by the method of Pacholski et al., which involves the hydrolysis and condensation of zinc acetate dihydrate by potassium hydroxide [39]. At low concentrations of reactants, roughly spherical nanocrystals of approximately 5 nm diameter are formed. By increasing the reactant concentration, it is possible to obtain rods, which are believed to form by oriented attachment of small particles. The particles can be stabilized in solution by adding a less polar solvent such as chloroform or dichlorobenzene and this solution can then be mixed with polymer solution for device fabrication [40].

6.2.2.2 Devices Based on ZnO Nanoparticles

Devices based on ZnO nanoparticles have been developed largely by Beek, Janssen, and coworkers. The first devices were based on blends of OC_1C_{10}-PPV with 5 nm diameter ZnO nanoparticles, as shown in Figure 6.11, and the efficiency was found to be optimized at about 67 wt% ZnO [40]. Estimated power efficiencies were in the range of 1.4–1.6%, with a V_{oc} of just over 800 mV, and fill factors between 0.5 and 0.6 (Figure 6.12). Using higher ZnO fractions up to 75 wt% did not improve the power efficiency since the increased J_{sc} was accompanied by a decrease in V_{oc}. Above 75%, the film quality was poor and both J_{sc} and V_{oc} dropped. The relationship between morphology and device performance for this type of device was subsequently investigated in more detail [41]. The effect of using ZnO rods rather than spherical particles was investigated, but this did not lead to an improvement in device efficiency.

Other polymers have been used with ZnO nanoparticles; for example, P3HT [42] and APFO-3 [43]. With P3HT, efficiencies were optimized at 26 vol% of ZnO, with annealing at 80 °C. Estimated power efficiencies of 0.92% were obtained, with the efficiency being limited by coarse mixing, giving a rough film containing regions of polymer that do not contribute to charge generation. Using APFO-3 with 67 wt% ZnO gives a power efficiency of 0.45%, limited by the relatively low V_{oc} (510 mV) and

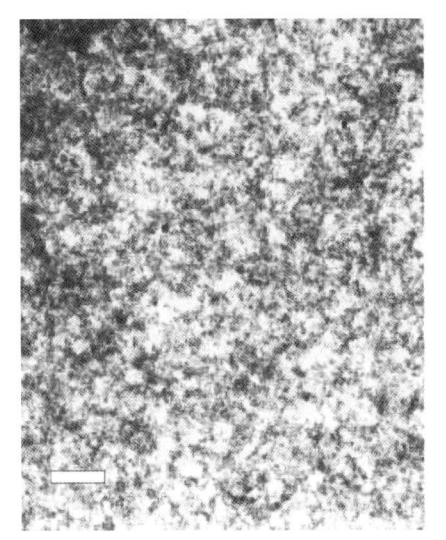

Figure 6.11 Transmission electron micrograph of an OC$_1$C$_{10}$-PPV:ZnO nanoparticle blend (75 wt% ZnO fraction). Scale bar, 200 nm. Dark regions indicate ZnO-rich domains and lighter regions indicate polymer-rich domains. Reproduced with permission from Ref. [40]. Copyright 2004 Wiley-VCH.

fill factor (0.36). The low efficiency was attributed to nanoparticle aggregation, producing an unfavorable morphology that allowed the passage of a high dark current through the ZnO, thus reducing the open-circuit voltage.

In summary, whilst zinc oxide nanoparticles are an attractive material for many reasons, the difficulty in processing them together with polymers to yield well-

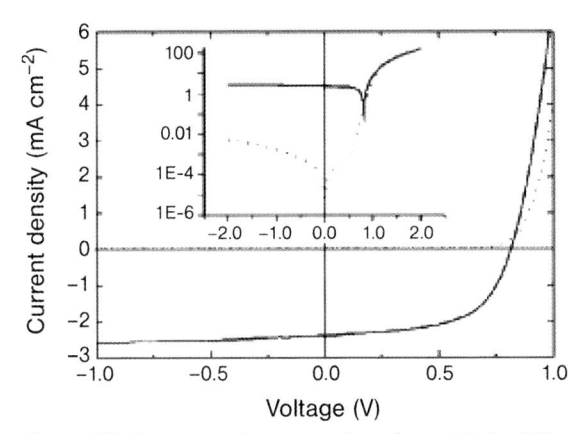

Figure 6.12 Current density versus voltage for an OC$_1$C$_{10}$-PPV:ZnO nanoparticle blend device (67 wt% ZnO fraction), in the dark (dotted line) and under white light illumination at approximately 0.71 sun equivalent intensity. Reproduced with permission from Ref. [40]. Copyright 2004 Wiley-VCH.

defined morphologies has made it difficult to achieve high efficiencies. Future directions may include surface modification of the particles to optimize morphologies with high fractions of ZnO.

6.2.3
Devices Based on Low-Bandgap Nanoparticles

The advantage of using low-bandgap nanoparticles in combination with polymers is that absorbing long wavelengths in the nanoparticles may allow a larger fraction of the incident solar spectrum to be absorbed. The obvious choice of semiconductor to obtain a lower bandgap than CdSe is CdTe. CdTe nanoparticles have a bandgap that can be tuned in the range from 1.7 to 2.0 eV [13]. The same synthetic routes as for CdSe can be used [16] and shape control is easier than with CdSe. Early attempts to use CdTe nanorods with MEH-PPV gave disappointing power efficiencies of 0.05% [44]. Strong PL quenching was observed and this was attributed to charge transfer on the basis of an electron affinity of 3.7 eV for the nanoparticles, estimated from cyclic voltammetry, which should be sufficient to give electron transfer from the polymer to the CdTe. However, there is a considerable variation in the electron affinity measured for CdTe nanoparticles by different methods [45–47] and it is therefore difficult to prove conclusively whether morphology or charge transfer is the problem in these devices. Spectroscopic measurements on a composite of a substituted polythiophene with CdTe nanoparticles have shown no evidence of long-lived charge-separated states, but this was tentatively attributed to a rapid charge-pair recombination process [48]. The same paper raised the possibility that the CdTe nanoparticles might have an electron affinity so low that it is energetically favorable for the hole, rather than the electron, to transfer from polymer to nanoparticle.

Since shape control is easy to achieve in CdTe, it is natural to investigate photovoltaics based on CdTe tetrapods. Zhou et al. performed a systematic study of photovoltaic devices using MEH-PPV blended with ternary $CdSe_xTe_{1-x}$ tetrapods [49]. They observed a steady decrease in power efficiency from 1.1% with CdSe to 0.003% with CdTe and attributed the low efficiency observed with CdTe tetrapods to a type-I band alignment, leading to an energy transfer from the polymer to CdTe rather than charge transfer. Gur et al. investigated devices made by first assembling a layer of CdTe tetrapods on ITO, followed by the deposition of P3HT layers of various thicknesses [50]. Different spectral responses and current–voltage behaviors were observed depending on whether the P3HT completely covered the polymer, but device efficiencies remained low.

The CdTe/polymer system has thus not yet been found to be promising for photovoltaics, although the system is close enough to the borderline for energy/electron transfer so that a careful tuning of polymer and/or nanocrystal energy levels may lead to improved efficiencies in the future.

Further lowering of the nanoparticle bandgap is possible using semiconductors such as PbS. Zhang et al. have reported photovoltaics using annealed blends of MEH-PPV with PbS nanoparticles [51]. The first exciton absorption of the nanoparticles was at 0.95 eV. Unfortunately, the power conversion efficiency was very low – about

0.001%. Watt *et al.* have developed a novel surfactant-free synthetic route where PbS nanoparticles are synthesized *in situ* within an MEH-PPV film [52]. Devices based on this composite showed encouraging low-intensity AM 1.5 power conversion efficiencies of 0.7% [53]. However, no evidence of a contribution to the photocurrent from absorption in PbS was presented. The exact role played by PbS in these devices is not clear, but time-of-flight measurements show that the presence of PbS nanoparticles significantly increases the electron mobility [54]. In general, when using nanoparticles of low bandgap, it is difficult to achieve a type-II heterojunction with a polymer and energy transfer from polymer to nanoparticle is the most likely outcome.

In a simple photovoltaic device, lowering the bandgap below a certain limit will cause a decrease in efficiency, since the maximum open-circuit voltage is limited by the bandgap, and the energy of absorbed high-energy photons will be wasted as the electron and hole relax to the band edges. An interesting approach, which might avoid this limitation, is based on the phenomenon of "multiple exciton generation," in which a single high-energy absorbed photon can lead to multiple electron–hole pairs. If these carriers can be extracted efficiently, then by choosing a suitable bandgap, a larger fraction of the available energy can be extracted.

In bulk semiconductors, the process of multiple exciton generation is inefficient; however, in nanoparticles, the momentum conservation laws are relaxed and more efficient generation of multiple carrier pairs might be expected [55]. This phenomenon has been studied in detail using nanocrystals including PbSe, CdSe, and PbS [56–58]. The presence of excited states within the particles is detected in a pump-probe transient absorption measurement, as illustrated in Figure 6.13. A single electron–hole pair within a nanoparticle will decay relatively slowly (on a timescale of tens of nanoseconds), whereas multiple electron–hole pair states will relax much more quickly (subnanosecond) by an Auger recombination process. A fast component in the decay, therefore, indicates the presence of multiple electron–hole pairs. At low enough excitation intensities, the probability of these multiple electron–hole pairs being generated by sequential excitation by different photons can be made negligible. Under these conditions, the pump photon energy is then increased and when sufficient energy is available so that one photon can generate more than one exciton, a fast decay component is observed, confirming the presence of multiple exciton generation [56].

Klimov has performed a detailed balance calculation of the effect of multiple exciton generation on the efficiency of a polymer/nanoparticle photovoltaic device, assuming that the additional carriers can be extracted efficiently [59]. For PbSe nanoparticles, where the threshold for multiple exciton generation is at $3E_g$, the maximum power conversion efficiency is 36% with a bandgap of $E_g = 0.35$ eV, compared to 31% in the absence of multiple exciton generation. If the threshold for multiple exciton generation is lowered to its theoretical minimum of $2E_g$, then the maximum efficiency rises to 42% with a bandgap of 0.45 eV. In practice, it is not yet clear whether the additional carriers can be extracted quickly enough to compete with the rapid Auger recombination processes occurring within the nanocrystals (see Section 6.3.2). Photodetectors based on PbSe/MEH-PPV blends have been investigated and quantum efficiencies in excess of 100% have been seen at high reverse biases [60]. Although it was suggested

194 | *6 Hybrid Polymer/Nanocrystal Photovoltaic Devices*

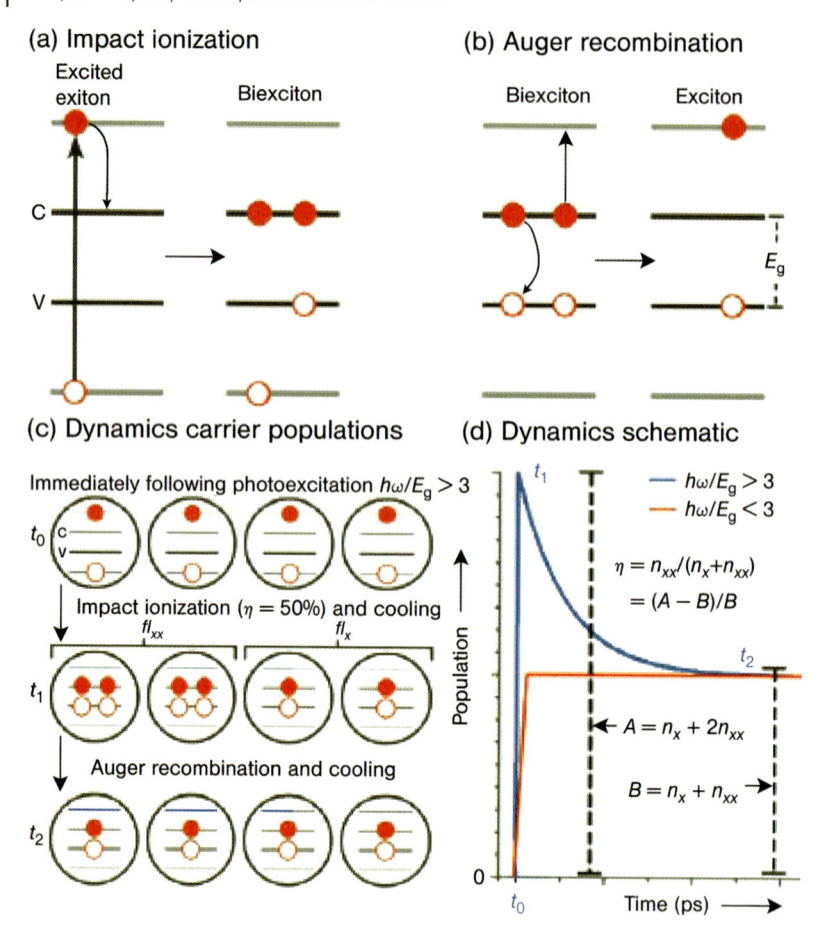

Figure 6.13 Dynamics of carrier generation and relaxation in nanoparticles. (a) Multiple exciton generation. (b) Auger recombination processes. The diagrams show electrons (filled circles), holes (empty circles), conduction band (labeled C), and valence band (labeled V). (c) Immediately following high photon energy excitation ($\hbar\omega/E_G > 3$ for this system), highly excited excitons form in some nanoparticles. A fraction of these (n_{xx}) undergoes multiple exciton generation to create biexcitons, while others simply cool to the band edge, remaining as single excitons (n_x). Biexcitons then undergo Auger recombination to produce a single exciton at long times. (d) Representative time-resolved data in which carrier populations are monitored in the pump intensity regime of $N_{e-h} < 1$ for low (bottom line) and high (top line) pump photon energy (here either below or above $3E_g$). The fast relaxation component in the top trace is because of Auger recombination of biexcitons that have been generated via multiple exciton generation. Its amplitude provides a direct measure of the efficiency of multiple carrier generation $\eta = n_{xx}/(n_x + n_{xx}) = (A - B)/B$. Reproduced with permission from Ref. [56]. Copyright 2004 American Institute of Physics.

that this may be caused by multiple carrier generation, a more likely explanation is persistent photoconductivity owing to trapping effects [61].

6.2.4
Polymer Brush Devices

The ideal structure for a nanostructured photovoltaic device would have connected pathways to the electrodes. It is clear from the devices discussed above that depositing the polymer and nanoparticles in a single spin-coating process does not always lead to the optimum nanostructure. In the ideal structure, electrons and holes generated anywhere within the device would be able to find a pathway to the respective electrode without getting stuck at dead-ends within the network. One way to approach the optimum nanostructure is to use a polymer brush as the hole-transporting component in the device. In a polymer brush, polymer chains are grown directly from a substrate and if the density of attachment sites on the substrate is high enough, then the chains show some degree of alignment perpendicular to the substrate. Growing high-quality conjugated polymer brushes is challenging, but hole-transporting brushes can be grown by attaching hole-transporting side groups to a conventional polymer. This strategy was demonstrated by Snaith *et al.*, who used polyacrylate brushes with triphenylamine side groups, grown from ITO, as shown in Figure 6.14 [62]. These brushes demonstrated improved hole transport properties compared to spin-coated films of the same polymer [63]. By soaking the brushes in a solution containing CdSe nanocrystals, it was possible to infiltrate nanocrystals into the brush film, followed by evaporation of a top electrode to form a photovoltaic device. The efficiency of these devices was relatively low, owing to the low fraction of incident light absorbed in the device. Since triphenylamine has a high bandgap, the polymer does not contribute usefully to the absorption in the visible range and the nanoparticles must, therefore, provide the absorption. In principle, the absorption could be improved using thicker devices, but it is currently difficult to grow brushes thicker than about 50 nm. From the measured absorption of the device, it is possible to estimate an internal quantum efficiency (electrons collected per photon absorbed) and values as high as 50% are found. This indicates that the internal structure of the films gives very good exciton dissociation and charge collection. This is backed up by atomic force microscopy of the films, which suggests the formation of nanocrystal "channels" in a polymer brush network, as shown in Figure 6.14b and c.

6.2.5
All-Nanoparticle Devices

The presence of polymer as the hole transporter in a composite photovoltaic device is convenient for film formation, but is not necessarily required to achieve a functioning device. Gur *et al.* have demonstrated solution-processed photovoltaics, based on bilayers of CdTe and CdSe nanorods [64]. Both layers are deposited by spin coating, with a brief annealing step after the deposition of the CdTe layer to allow the CdSe layer to be deposited on top. As discussed in Section 6.2.3, CdTe has a lower electron

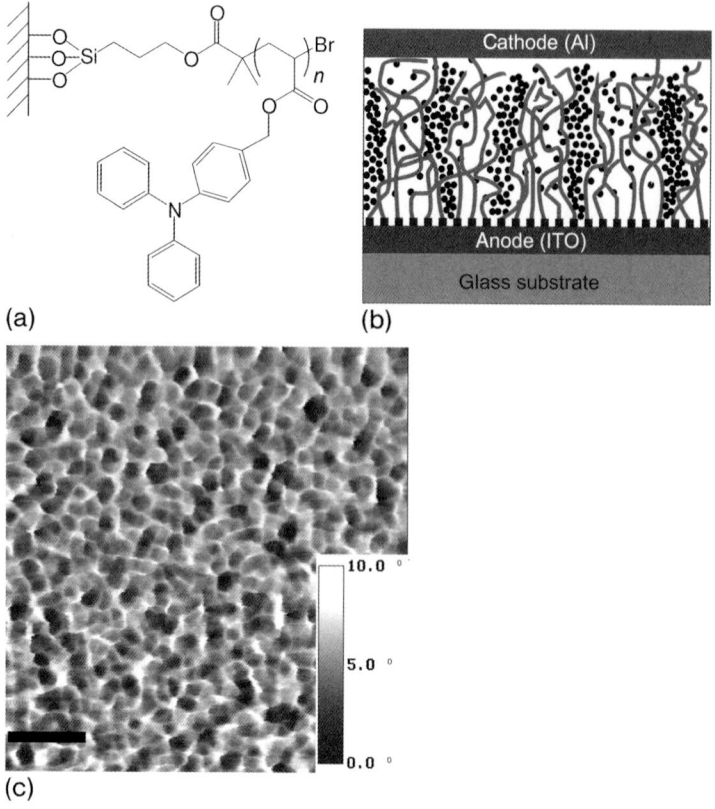

Figure 6.14 (a) Chemical structure of a hole-transporting polymer brush. (b) Schematic structure of a polymer brush/CdSe nanocrystal photovoltaic device [62]. (c) Atomic force micrograph of the surface of a polymer brush film infiltrated with CdSe nanoparticles [62]. Scale bar, 100 nm.

affinity than CdSe and it is clear from this work that the interface between CdSe and CdTe nanorods forms a type-II heterojunction where electron transfer from CdTe to CdSe occurs. Efficiencies of 2.1% are achieved, which can be enhanced to 2.9% by further heating to sinter the nanoparticles. These devices show encouraging stability under illumination at open-circuit conditions in air.

6.3
Physical Processes in Polymer/Nanoparticle Devices

Efficiency improvements are clearly still needed to take polymer/nanocrystal devices toward applications. Efficiencies above 5% will require careful optimization of each of the processes involved in device operation, so that carrier and energy losses are minimized. It is therefore important to understand the physics of each of these processes in some detail. Our current understanding of the detailed physics of absorption, exciton transport, charge transfer, charge separation, and recombination in these devices is summarized in the following sections.

6.3.1
Absorption and Exciton Transport

Any nanostructured photovoltaic device needs to absorb a large fraction of the incident photons and to transport them to sites where they can be dissociated. The optical design of thin-film photovoltaics is not straightforward, since the reflective top electrode tends to force a node in the optical absorption profile at that interface. Improvements have been demonstrated using spacer layers to move the active material into a region of high optical field [65], but these are not specific to polymer/nanoparticle devices and hence will not be discussed further here. It is clear that efficient absorption can be achieved using thick films, but this can lead to transport problems; high oscillator strengths for absorption in the active layer are thus desirable. In many polymer composite photovoltaic devices, including polymer/ phenyl C_{61}-butyric acid methyl ester (PCBM) or polymer/metal oxide nanoparticle devices, the majority of the absorption happens in the hole-accepting polymer. However, using relatively low-bandgap nanoparticles such as CdSe allows the electron-accepting component of the device to contribute to the absorption and this contribution (which is size-tunable) is particularly noticeable where the nanoparticle absorption extends below the polymer absorption edge.

Exciton transport can occur in both polymer and nanocrystal components of the device. Models for exciton transport in conjugated polymer films have been extensively discussed in the literature and will not be reviewed here. The process of exciton diffusion in films of nanoparticles has been studied by Kagan *et al.* [66,67]. Förster transfer provides a good model for the energy transfer process and is easier to apply in nanoparticle films than in polymer films, since it is straightforward to identify the emitting and absorbing spectroscopic units. Since the Förster process depends on the overlap between the emission spectrum of the energy donor and the absorption spectrum of the energy acceptor, the energy transfer dynamics are highly sensitive to the amount of energetic disorder (inhomogeneous broadening) in the sample. Owing to the relatively large size of the nanoparticles and the strong $1/R^6$ distance-dependence of the energy transfer rate, nearest-neighbor energy transfer is found to dominate. Arranging an energy gradient in a layered nanocrystal film can be used to give directional energy transfer over larger distances [68,69]. In principle, this is an attractive way to transport energy to a charge dissociation site in a photovoltaic device; however, it has not yet been possible to implement this idea in a geometry that also allows charge separation and efficient charge carrier collection.

The obvious route for charge generation is absorption in the donor–(acceptor) material, followed by exciton diffusion to an interface and then electron–(hole) transfer to the acceptor–(donor). In the case of polymer/CdSe devices, where the polymer has a higher bandgap than the nanocrystals, another possibility exists: absorption in the polymer, followed by energy transfer to the nanocrystal, followed by

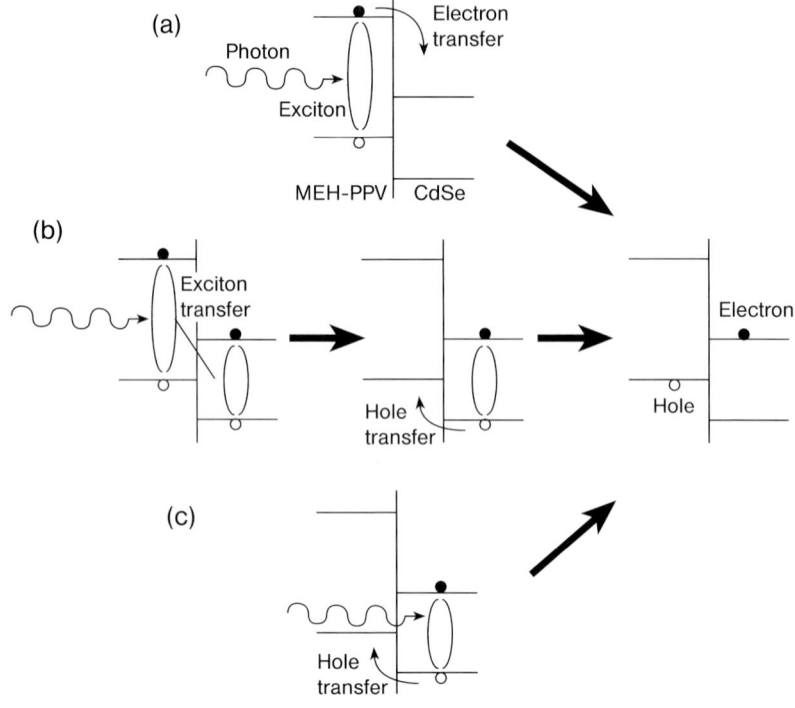

Figure 6.15 Routes for exciton and charge transfer in MEH-PPV/CdSe blends. (a) Absorption in the polymer, followed by electron transfer onto the nanocrystal. (b) Absorption in the polymer, followed by exciton transfer onto the nanocrystal, followed by hole transfer onto the polymer. (c) Absorption in the nanocrystal, followed by hole transfer onto the polymer. Reproduced with permission from Ref. [12]. Copyright 1996 American Institute of Physics.

hole transfer back to the polymer, as shown in Figure 6.15 [12]. As has been pointed out by Liu et al., this process can certainly occur in organic dye/C_{60} derivative systems, giving more efficient energy transport to the interface than would happen by simple exciton transport alone [70]. The extent to which this process actually happens in polymer/nanoparticle blends is still under investigation.

6.3.2
Charge Transfer

The energetics required for charge transfer at the polymer/nanoparticle interface is well established and is similar to that for other donor–acceptor systems. Charge transfer is favored by a type-II heterojunction, where the electron affinity of the acceptor is larger than that of the donor and the ionization potential of the donor is less than that of the acceptor, as shown in Figure 6.15. For the charge-separated state to be lower in energy than the exciton on the donor or the acceptor, the offsets in both electron affinity and ionization potential must be larger than the exciton binding energy of the polymer minus the coulombic binding energy of the charge-separated state [71]. Nanoparticles provide an interesting opportunity to study this process, since the electron affinity of the nanoparticles can be tuned by the changing nanocrystal size. Ginger and Greenham studied the photoluminescence quenching in various conjugated polymers as a function of the concentration of CdSe nanoparticles in the film, as shown in Figure 6.16 [72]. Various sizes of spherical nanocrystals were used in the hope of tuning the LUMO energy of the nanoparticle through the position at which charge transfer became unfavorable. In the dialkoxy-PPV derivative MEH-PPV, which was used for the first polymer/nanoparticle photovoltaics [12], PL quenching was observed with all nanocrystal sizes. Charge transfer (as opposed to energy transfer) was confirmed by photoinduced absorption measurements showing the presence of long-lived charges. Substituting cyano groups at the vinylene position is known to increase the electron affinity of the polymer by approximately 0.5 eV, making electron transfer to nanocrystals less favorable. However, in the blends of nanocrystals with the cyano-substituted analogue of MEH-PPV, MEH-CN-PPV (see Table 6.1), PL quenching and charge generation were still observed. Interestingly, using the closely related polymer DHeO-CN-PPV, no PL quenching was observed, despite the fact that this polymer should have very similar energy levels to those of MEH-CN-PPV. It is argued that the symmetric dihexyloxy side chains in this polymer provide an insulating barrier that separates the conjugated backbone of the polymer from the nanocrystal surface, thus reducing the charge transfer rate to a level that does not compete efficiently with exciton decay. Thus, we see that on energetic grounds, CdSe nanocrystals have a sufficiently high electron affinity to act as electron acceptors from a wide range of conjugated polymers, although a physical barrier between the polymer and the nanocrystal can in some cases interfere with the kinetics of the charge transfer, preventing efficient charge generation.

The dynamics of charge transfer at the polymer/nanoparticle interface has proved difficult to study directly. Experimentally, the technique of choice is pump-probe spectroscopy, in which excitons are created by a short laser pulse and the formation of charged species is monitored by probing their absorptions (in the near-IR) as a function of time after excitation. For organic dyes adsorbed on the surface of TiO_2, the charge transfer process has been found to be extremely rapid (<50 fs), indeed sufficiently fast to compete with electronic relaxation and intersystem crossing processes in the dye [73]. In polymer/nanocrystal blends, the interpretation of transient absorption spectra is more complex, since the observed dynamics represent a combination of exciton diffusion and charge transfer kinetics. For ZnO nanoparticle/polymer blends, absorption owing to charges is seen to appear on timescales of less than 10 ps (Figure 6.17) [74]. However, for CdSe/polymer blends, it has proved difficult to detect the spectral signature of charges on ps timescales [75], despite the fact that photoinduced absorptions because of long-lived separated charges are easily detectable on μs–ms timescales [72].

Time-resolved luminescence measurements can also provide information about the charge transfer process, since charge transfer competes with the usual radiative and nonradiative decay processes. Time-correlated single-photon counting, streak camera measurements, and luminescence upconversion are all useful techniques for

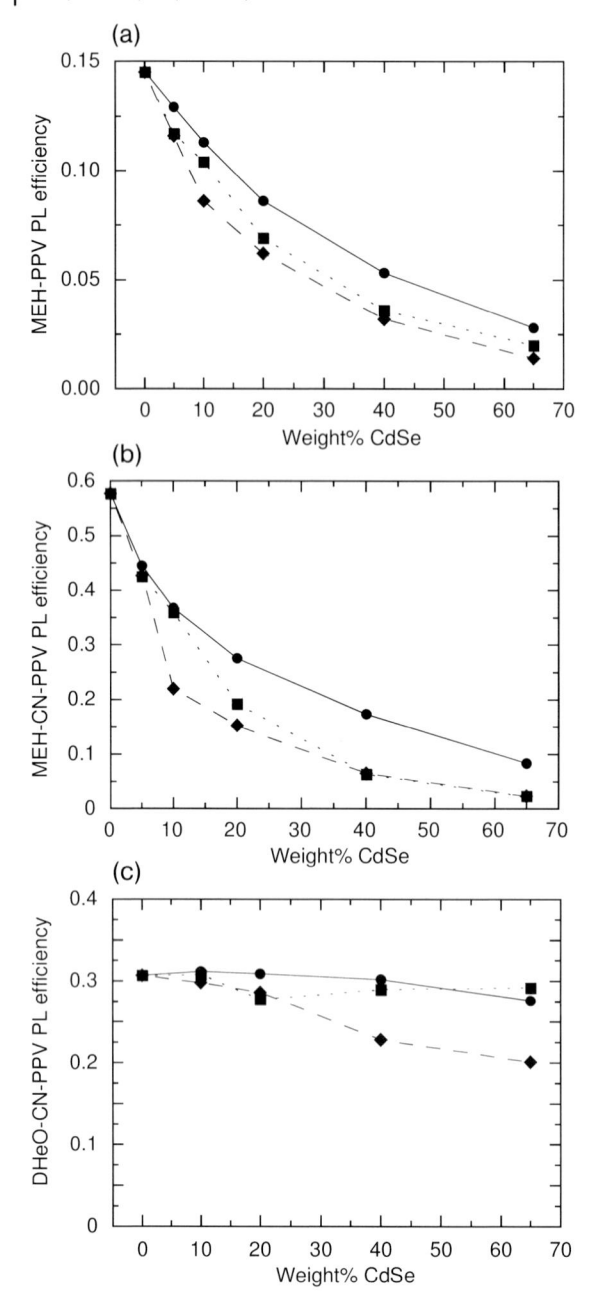

Figure 6.16 Photoluminescence efficiencies of blends of (a) MEH-PPV, (b) MEH-CN-PPV, and (c) DHeO-CN-PPV with CdSe nanocrystals of diameter 2.5 nm (squares), 3.3 nm (circles), and 4.0 nm (diamonds) [72].

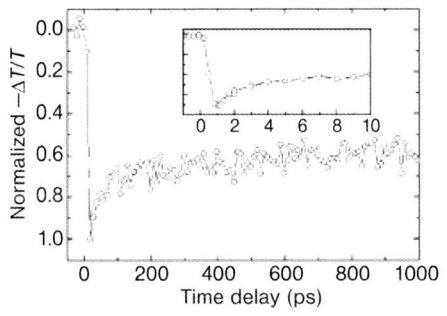

Figure 6.17 Transient absorption at 0.56 eV (because of charged states) in an OC_1C_{10}-PPV:ZnO blend, after excitation at 2.43 eV with 200 fs pulses. Reproduced with permission from Ref. [40]. Copyright 2004 Wiley-VCH.

measuring luminescence decay, in order of increasing experimental complexity but improving time resolution. Charge transfer will manifest itself as a decrease in the photoluminescence lifetime of the polymer. The exact form of the luminescence decay is complex, since excitons are generated at different distances from the polymer/nanocrystal interfaces. For example, if there is a population of excitons generated on polymer segments immediately adjacent to an interface, this will cause a reduction in the initial magnitude of the transient signal if the charge transfer occurs on timescales that are shorter than the time resolution of the experiment. In many cases the luminescence lifetime of a polymer is sensitive to the polymer morphology owing to interchain effects. Adding nanoparticles to the polymer can change these interactions even in the absence of charge transfer, which complicates the analysis of the luminescence decay dynamics. Furthermore, luminescence decay does not distinguish between quenching by energy transfer or charge transfer to a nonluminescent material.

Time-correlated single-photon counting has been used to study charge transfer between OC_1C_{10}-PPV and CdSe tetrapods (Figure 6.18) [29]. In films where the

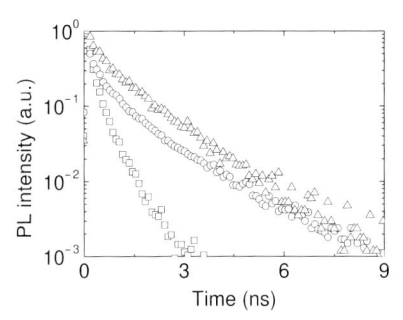

Figure 6.18 Photoluminescence decay at 580 nm in blends of CdSe tetrapods and OC_1C_{10}-PPV (4 : 6 w:w) spin coated from 1,2,4-trichlorobenzene (circles) or chloroform (squares). Data for a pristine OC_1C_{10}-PPV film are also shown (triangles) [29].

polymer and the nanoparticles are intimately mixed, the luminescence lifetime is clearly reduced. In films spin coated from 1,2,4-trichlorobenzene, however, an initial rapid component of luminescence decay was found, followed by a slower component with similar lifetime to that of the pure polymer film. This was consistent with the proposed structure involving vertical phase separation of the nanoparticles to the top of the film, leaving some regions of the polymer, which are sufficiently pure so that excitons do not encounter an interface within their lifetime.

6.3.3
Charge Separation and Recombination

Once charge transfer has occurred at a polymer/nanoparticle interface, the final state produced leaves an electron on the nanocrystal and a hole on the polymer. However, the electron and hole are not "free carriers" since they are still coulombically bound with a binding energy that exceeds kT at room temperature. They can be further separated by thermal (diffusive) motion or by the influence of an internal field in the device. The alternative fate to forming free carriers is that the electron and hole may recombine at the interface, known as geminate recombination as the electron recombines with the same hole with which it was originally generated.

It is now established that in organic photovoltaics the dominant effect determining the quantum efficiency and current–voltage curve is the efficiency of geminate pair separation [76,77]. Devices including inorganic semiconductor nanoparticles have a potential advantage over all-organic devices, as the dielectric constants of typical inorganic semiconductors are much larger than those of organics. Since the coulombic binding energy is reduced, this can greatly increase the fraction of charge pairs, which escape to form free charges. The most widely used theory for charge carrier separation is the one proposed by Braun, which treats the donor–acceptor material as an effective medium and accounts for the drift–diffusion motion in the applied and local field, assuming a finite recombination lifetime for charge carriers at short distances [78]. In real donor–acceptor materials, the process of charge separation is complicated by the existence of a distinct microstructure distinguishing the donor and acceptor materials, and Monte Carlo modeling has shown that the feature size is important in determining the charge carrier separation efficiency [79,80]. The delineation between donor and acceptor materials is particularly obvious in polymer/nanoparticle blends where the two components can be imaged directly. Modeling of the charge separation in these systems has not been fully explored, particularly in the case of the two materials that have very different dielectric constants.

The rate of recombination of charge pairs at the polymer/nanocrystal interface is critical in determining the charge collection efficiency. Indeed, if this rate were zero, then the efficiency of charge collection would be unity even if the charge separation process were very slow. Very little is known about the molecular-scale structure at the polymer/nanocrystal interface and hence it is difficult to produce a theoretical approach to treat recombination at that interface even in the ideal case. The situation is further complicated by the likely existence of surface states and/or trap states in the nanocrystal. Experimentally, the recombination process can be followed in a photoinduced absorption experiment by measuring the decay of the infrared absorption features associated with charges. In a quasi-steady state photoinduced absorption experiment, dynamical information is obtained from the dependence of the signal on the modulation frequency of the excitation. Ginger and Greenham showed that the recombination in MEH-PPV/CdSe blends did not follow a simple monoexponential behavior and that to obtain a good fit to the frequency-dependence of the photoinduced absorption signal required the presence of decay components with lifetimes ranging from tens of microseconds to several milliseconds (Figure 6.19) [72].

One attractive route to slow down recombination is to arrange for the electron and hole formed after charge separation to be spatially separated. This can be achieved by introducing a ligand on the nanocrystal that has an electron affinity intermediate between the electron affinity of the nanocrystal and that of the polymer. Electron

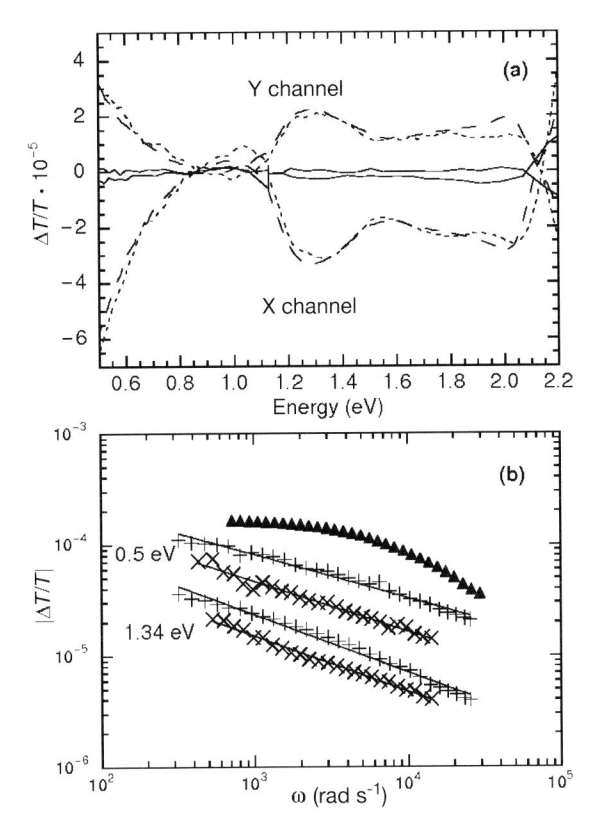

Figure 6.19 (a) Room-temperature photoinduced absorption spectra of MEH-PPV (solid line), a blend of MEH-PPV containing 40 wt% weight of 4.0 nm CdSe nanocrystals (long dashes), and a blend of MEH-PPV and 40 wt% of 2.5 nm CdSe nanocrystals (short dashes). (b) Frequency dependence of the 1.34 eV photoinduced absorption signal in pristine MEH-PPV at 10 K (triangles), along with that of the 0.5 and 1.34 eV features from the blend with 4.0 nm nanocrystals (×) and the blend with 2.5 nm nanocrystals (+). The straight lines are power-law fits to the data, with exponents from −0.4 to −0.5 [72].

transfer can then take place in a stepwise fashion, producing a final state where the electron on the nanocrystal and the hole on the polymer have negligible wavefunction overlap and therefore can only recombine by thermal excitation of one or other carrier onto the intermediate ligand. Unfortunately, though, this strategy has a number of practical disadvantages. First, the driving force for both polymer-to-ligand and ligand-to-nanocrystal electron transfer must be sufficient to give rapid transfer, and this leads to an irreversible loss of energy, lowering the amount of energy available in the collected charges. Second, the nanocrystal must be well covered by ligand to prevent direct polymer–nanocrystal contact, but this then makes it difficult to transfer electrons from nanoparticle to nanoparticle, since the ligand acts as a spatial and energetic barrier to electron transport. Although there have been some reports of electroactive ligands attached to semiconductor nanoparticles [81,82], these do not have the required energy level alignment to achieve the two-step electron transfer described above.

6.3.4
Charge Transport

Electron transport takes place within the nanocrystalline component of a hybrid photovoltaic device. Valuable information can be obtained by examining the electron transport properties of films of nanoparticles. Ginger and Greenham studied transport in sandwich-structure devices in which the active layer comprised a film of spherical CdSe nanocrystals (Figure 6.20) [83]. It was found that space-charge-

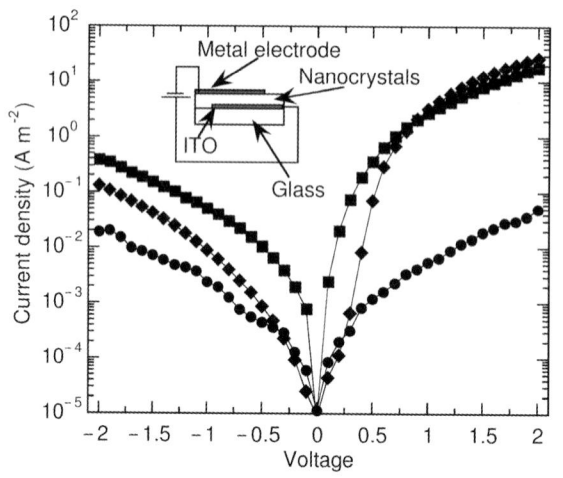

Figure 6.20 Current density versus voltage for three different ITO/CdSe/metal devices. The device structure is shown inset. The CdSe layer for each device is composed of a 190 nm thick film of 3.4 nm diameter nanocrystals separated by TOPO surfactant. Diamonds indicate the current–voltage curve for an Al contact device, squares for a Ca contact device, and circles for an Au contact device [83].

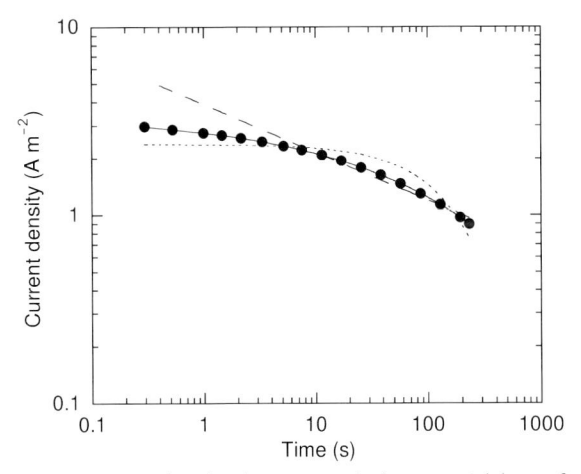

Figure 6.21 Log–log plot showing stretched exponential decay of current (circles) with time over three orders of magnitude for an ITO/CdSe/Al device made from 3.3 nm diameter nanocrystals. The solid line is a fit to a stretched exponential decay. Also shown for comparison are least-square fits to the data for exponential (dotted line) and power-law (dashes) decays [83].

limited currents could be achieved using either calcium or aluminum as the electron-injecting electrode, giving initial mobilities on the order of 10^{-5} cm^2 V^{-1} s^{-1}. The currents at constant voltage decayed in a stretched exponential manner with time (Figure 6.21), which was attributed to the gradual filling of deep-trap states with a density comparable with the space-charge density. This caused the effective mobility (taking into account both mobile charges and immobile deep-trapped charges) to decrease with time. Illumination with the above bandgap light reset the current close to its initial value. The mechanism for this "persistent photoconductivity" is that photogenerated holes end up close to trapped electrons (with which they eventually recombine), thus canceling out the local space charge and allowing more electrons to be injected and transported through the device. The dominance of trapping effects made it difficult to identify clear trends in transport properties with the size or surface treatment of the ligands. Similarly low mobilities with dominant trapping effects were observed in films of CdSe nanocrystals with ZnS shells [84].

A number of authors have studied in-plane transport in CdSe nanocrystal films. Morgan *et al.* found a power-law decay of the current with time and interpreted the results in terms of long-range Coulomb interactions between the charge carriers [85]. The conductivity measured in this type of device was found to increase with annealing [86].

The measurements described above are conducted at relatively low carrier densities, where trapping effects might be expected to dominate. Valuable additional information about electron transport mechanisms can be gained from temperature- and field-dependent measurements at higher carrier concentrations.

This has been achieved by Guyot-Sionnest and coworkers, who used electrochemical techniques to control the carrier concentration in CdSe nanocrystal films, in which the particles are cross-lined by 1,4-phenylenediamine [87]. At low temperatures (<120 K) and low fields, their results fit well into the theory of Efros and Shklovskii [88], which considers variable range hopping in the presence of Coulomb interactions. At higher temperatures, the nearest neighbor jumps start to dominate and the conductivity shows an Arrhenius-like thermal activation. The complex interplay between hopping, coulombic effects, and energetic disorder in these systems has been modeled by Chandler *et al.* using dynamical Monte Carlo techniques [89].

Despite the relatively high conductivities seen in doped nanoparticle films, the fact remains that the transport through the nanocrystalline component of photovoltaic device exhibits relatively low mobilities, likely caused by deep-trap states and further influenced by the complex morphology of the nanocrystal network. Even in rod and tetrapod structures, in which individual nanoparticles might be expected to span a significant fraction of the thickness of the device, observed dark currents are orders of magnitude lower than would be expected on the basis of the intrinsic transport properties of bulk CdSe [50,90]. Fortunately, it is not necessary to have particularly high mobilities to achieve efficient charge extraction in a thin nanostructured photovoltaic device [91].

6.3.5
Electrical Characteristics and Morphology

The ideal photovoltaic device would have the structure shown in Figure 6.22, with an interdigitated donor–acceptor layer designed to achieve efficient electron and hole extraction, surrounded by pure layers designed to suppress dark currents and hence optimize the open-circuit voltage. The internal asymmetry of this type of device can give further improvements in open-circuit voltage owing to diffusion currents caused by the spatially nonuniform charge generation profile [76,92]. We have seen in Section 6.2 various strategies to achieve the desired morphology, typically by controlling the processing conditions. The nanocrystals tend to segregate toward the topside of the film during spin coating and this is desirable for efficient device operation. Indeed, when spin coating mixtures of nanoparticles with small organic molecules, Coe-Sullivan *et al.* found that it is possible to obtain a dense monolayer of nanoparticles at the top surface of the film [93]. The morphologies obtained in

Figure 6.22 Ideal interdigitated structure for a hybrid photovoltaic device.

polymer/nanocrystal photovoltaic blends have been studied in some detail [94], but the physical processes leading to the formation of phase-separated morphologies in these blends are still not well understood. Factors that need to be considered include differential solubilities of polymers and nanocrystals, nanocrystal aggregation in concentrated solution, differential evaporation in solvent mixtures, entropic effects, spinodal decomposition, convective instabilities during spin coating, and surface energies at the substrate and air interfaces. The formation of controlled nanostructures is a rich area for future study.

The electrical characteristics of P3HT/CdSe nanorod devices have been studied in detail by Huynh et al. [90]. They analyzed their data in terms of a Schottky model incorporating series and shunt resistances and investigated the role of space charge effects at high intensities. They noted that the photocurrent is voltage-dependent, limiting the validity of the models applied. Whilst this type of model can provide a useful parameterization of device performance that can guide choice of materials and electrodes, detailed understanding of the electrical properties is likely to require more detailed numerical modeling taking into account the field-dependence of the photocurrent and the fact that charge generation takes place throughout the device. Numerical modeling based on a continuum approach has been developed for all-organic photovoltaic devices [77,95] and has recently been applied to polymer/ZnO nanoparticle devices [96]. To model the true effect of the nanostructure on charge separation and the effects of percolation on both dark current and photocurrent, microscopic modeling using Monte Carlo approaches is likely to be necessary [79,80].

6.4 Conclusions

We have seen that charge separation occurs at the polymer/nanocrystal interface in a wide range of materials systems. The challenges for efficient photovoltaic operation include achieving the desired nanostructure to allow efficient charge extraction, and minimizing energy losses within the device. Polymer/nanoparticle photovoltaics offer unique opportunities to address these challenges since the nanoparticle shape can be designed before film deposition to optimize the nanostructure, and the electronic energy levels can be varied widely by using size tuning with different materials. The field is not well developed compared to many other organic photovoltaic systems and there are good prospects for increasing power conversion efficiencies beyond their current level of about 3%.

Acknowledgments

I am grateful to my current and former graduate students, postdocs, and collaborators for their contributions to this work, as well as to the other groups around the world who have allowed their figures to be reproduced in this chapter.

References

1 Weller, H. (1993) *Advanced Materials*, **5**, 88.
2 Yin, Y. and Alivisatos, A.P. (2005) *Nature*, **437**, 664.
3 Rodina, A.V., Efros, A.L. and Alekseev, A.Y. (2003) *Physical Review B: Condensed Matter*, **67**, 155312.
4 Brus, L.E. (1984) *Journal of Chemical Physics*, **80**, 4403.
5 Ekimov, A.I., Hache, F., Schanneklein, M.C., Ricard, D., Flytzanis, C., Kudryavtsev, I.A., Yazeva, T.V., Rodina, A.V. and Efros, A.L. (1993) *Journal of the Optical Society of America B: Optical Physics*, **10**, 100.
6 Efros, A.L., Rosen, M., Kuno, M., Nirmal, M., Norris, D.J. and Bawendi, M. (1996) *Physical Review B: Condensed Matter*, **54**, 4843.
7 Efros, A.L. (1992) *Physical Review B: Condensed Matter*, **46**, 7448.
8 Efros, A.L. and Rodina, A.V. (1993) *Physical Review B: Condensed Matter*, **47**, 10005.
9 Franceschetti, A. and Zunger, A. (1997) *Physical Review Letters*, **78**, 915.
10 Brus, L.E. (1983) *Journal of Chemical Physics*, **79**, 5566.
11 O'Regan, B. and Grätzel, M. (1991) *Nature*, **353**, 737.
12 Greenham, N.C., Peng, X. and Alivisatos, A.P. (1996) *Physical Review B: Condensed Matter*, **54**, 17628.
13 Murray, C.B., Norris, D.J. and Bawendi, M.G. (1993) *Journal of the American Chemical Society*, **15**, 8706.
14 Bowen Katari, J.E., Colvin, V.L. and Alivisatos, A.P. (1994). *Journal of Physical Chemistry*, **98**, 4109.
15 Peng, X., Wickham, J. and Alivisatos, A.P. (1998) *Journal of the American Chemical Society*, **120**, 5343.
16 Peng, Z.A. and Peng, X.G. (2001) *Journal of the American Chemical Society*, **123**, 183.
17 Peng, X.G., Manna, L., Yang, W.D., Wickham, J., Scher, E., Kadavanich, A. and Alivisatos, A.P. (2000) *Nature*, **404**, 59.
18 Peng, Z.A. and Peng, X.G. (2001) *Journal of the American Chemical Society*, **123**, 1389.
19 Manna, L., Scher, E.C. and Alivisatos, A.P. (2000) *Journal of the American Chemical Society*, **122**, 12700.
20 Sun, B., Marx, E. and Greenham, N.C. (2003) *Nano Letters*, **3**, 961.
21 Manna, L., Milliron, D.J., Meisel, A., Scher, E.C. and Alivisatos, A.P. (2003) *Nature Materials*, **2**, 382.
22 Erwin, M.M., Kadavanich, A.V., McBride, J., Kippeny, T., Pennycook, S. and Rosenthal, S.J. (2001) *European Physical Journal D*, **16**, 275.
23 Bozano, L., Carter, S.A., Scott, J.C., Malliaras, G.G. and Brock, P.J. (1999) *Applied Physics Letters*, **74**, 1132.
24 Huynh, W.U., Peng, X.G. and Alivisatos, A.P. (1999) *Advanced Materials*, **11**, 923.
25 Huynh, W.U., Dittmer, J.J. and Alivisatos, A.P. (2002) *Science*, **295**, 2425.
26 Sirringhaus, H., Brown, P.J., Friend, R.H., Nielsen, M.M., Bechgaard, K., Langeveldvoss, B.M.W., Spiering, A.J.H., Janssen, R.A.J., Meijer, E.W., Herwig, P. and de Leeuw, D.M. (1999) *Nature*, **401**, 685.
27 Sun, B.Q. and Greenham, N.C. (2006) *Physical Chemistry Chemical Physics*, **8**, 3557.
28 Chang, J.F., Sun, B.Q., Breiby, D.W., Nielsen, M.M., Solling, T.I., Giles, M., McCulloch, I. and Sirringhaus, H. (2004) *Chemistry of Materials*, **16**, 4772.
29 Sun, B., Snaith, H.J., Dhoot, A.S., Westenhoff, S. and Greenham, N.C. (2005) *Journal of Applied Physics*, **97**, 014914.
30 Gur, I., Fromer, N.A., Chen, C.P., Kanaras, A.G. and Alivisatos, A.P. (2007) *Nano Letters*, **7**, 409.
31 Kanaras, A.G., Sonnichsen, C., Liu, H.T. and Alivisatos, A.P. (2005) *Nano Letters*, **5**, 2164.
32 Morteani, A.C., Dhoot, A.S., Kim, J.S., Silva, C., Greenham, N.C., Friend, R.H., Murphy, C., Moons, E., Ciná, S. and Burroughes, J.H. (2003) *Advanced Materials*, **15**, 1708.

33 Svensson, M., Zhang, F.L., Veenstra, S.C., Verhees, W.J.H., Hummelen, J.C., Kroon, J.M., Inganas, O. and Andersson, M.R. (2003) *Advanced Materials*, **15**, 988.

34 Wang, P., Abrusci, A., Wong, H.M.P., Svensson, M., Andersson, M.R. and Greenham, N.C. (2006) *Nano Letters*, **6**, 1789.

35 Salafsky, J.S. (1999) *Physical Review B: Condensed Matter*, **59**, 10885.

36 Kwong, C.Y., Choy, W.C.H., Djurisic, A.B., Chui, P.C., Cheng, K.W. and Chan, W.K. (2004) *Nanotechnology*, **15**, 1156.

37 Kwong, C.Y., Djurisic, A.B., Chui, P.C., Cheng, K.W. and Chan, W.K. (2004) *Chemical Physics Letters*, **384**, 372.

38 Zeng, T.W., Lin, Y.Y., Lo, H.H., Chen, C.W., Chen, C.H., Liou, S.C., Huang, H.Y. and Su, W.F. (2006) *Nanotech*, **17**, 5387.

39 Pacholski, C., Kornowski, A. and Weller, H. (2002) *Angewandte Chemie – International Edition*, **41**, 1188.

40 Beek, W.J.E., Wienk, M.M. and Janssen, R.A.J. (2004) *Advanced Materials*, **16**, 1009.

41 Beek, W.J.E., Wienk, M.M., Kemerink, M., Yang, X.N. and Janssen, R.A.J. (2005) *Journal of Physical Chemistry B*, **109**, 9505.

42 Beek, W.J.E., Wienk, M.M. and Janssen, R.A.J. (2006) *Advanced Functional Materials*, **16**, 1112.

43 Wong, H.M.P., Wang, P., Abrusci, A., Svensson, M., Andersson, M.R. and Greenham, N.C. (2007) *Journal of Physical Chemistry C*, **111**, 5244.

44 Kumar, S. and Nann, T. (2004) *Journal of Materials Research*, **19**, 1990.

45 Rajh, T., Micic, O.I. and Nozik, A.J. (1993) *Journal of Physical Chemistry*, **97**, 11999.

46 Bae, Y., Myung, N. and Bard, A.J. (2004) *Nano Letters*, **4**, 1153.

47 Poznyak, S.K., Osipovich, N.P., Shavel, A., Talapin, D.V., Gao, M.Y., Eychmuller, A. and Gaponik, N. (2005) *Journal of Physical Chemistry B*, **109**, 1094.

48 van Beek, R., Zoombelt, A.P., Jenneskens, L.W., van Walree, C.A., Donega, C.D., Veldman, D. and Janssen, R.A.J. (2006) *Chemistry – A European Journal*, **12**, 8075.

49 Zhou, Y., Li, Y.C., Zhong, H.Z., Hou, J.H., Ding, Y.Q., Yang, C.H. and Li, Y.F. (2006) *Nanotech*, **17**, 4041.

50 Gur, I., Fromer, N.A. and Alivisatos, A.P. (2006) *Journal of Physical Chemistry B*, **110**, 25543.

51 Zhang, S., Cyr, P.W., McDonald, S.A., Konstantatos, G. and Sargent, E.H. (2005) *Applied Physics Letters*, **87**, 233101.

52 Watt, A., Thomsen, E., Meredith, P. and Rubinsztein-Dunlop, H. (2004) *Chemical Communications*, 2334.

53 Watt, A.A.R., Blake, D., Warner, J.H., Thomsen, E.A., Tavenner, E.L., Rubinsztein-Dunlop, H. and Meredith, P. (2005) *Journal of Physics D*, **38**, 2006.

54 Watt, A., Eichmann, T., Rubinsztein-Dunlop, H. and Meredith, P. (2005) *Applied Physics Letters*, **87**, 253109.

55 Nozik, A.J. (2002) *Physica E*, **14**, 115.

56 Schaller, R.D. and Klimov, V.I. (2004) *Physical Review Letters*, **92**, 186601.

57 Schaller, R.D., Petruska, M.A. and Klimov, V.I. (2005) *Applied Physics Letters*, **87**, 253102.

58 Ellingson, R.J., Beard, M.C., Johnson, J.C., Yu, P.R., Micic, O.I., Nozik, A.J., Shabaev, A. and Efros, A.L. (2005) *Nano Letters*, **5**, 865.

59 Klimov, V.I. (2006) *Applied Physics Letters*, **89**, 123118.

60 Qi, D.F., Fischbein, M., Drndic, M. and Selmic, S. (2005) *Applied Physics Letters*, **86**, 093103.

61 Campbell, I.H. and Crone, B.K. (2007) *Journal of Applied Physics*, **101**, 024502.

62 Snaith, H.J., Whiting, G.L., Sun, B.Q., Greenham, N.C., Huck, W.T.S. and Friend, R.H. (2005) *Nano Letters*, **5**, 1653.

63 Whiting, G.L., Snaith, H.J., Khodabakhsh, S., Andreasen, J.W., Breiby, D., Nielsen, M.M., Greenham, N.C., Friend, P.H. and Huck, W.T.S. (2006) *Nano Letters*, **6**, 573.

64 Gur, I., Fromer, N.A., Geier, M.L. and Alivisatos, A.P. (2005) *Science*, **310**, 462.

65 Kim, J.Y., Kim, S.H., Lee, H.H., Lee, K., Ma, W.L., Gong, X. and Heeger, A.J. (2006) *Advanced Materials*, **18**, 572.

66 Kagan, C.R., Murray, C.B., Nirmal, M. and Bawendi, M.G. (1996) *Physical Review Letters*, **76**, 1517.

67 Kagan, C.R., Murray, C.B. and Bawendi, M.G. (1996) *Physical Review B: Condensed Matter*, **54**, 8633.

68 Achermann, M., Petruska, M.A., Crooker, S.A. and Klimov, V.I. (2003) *Journal of Physical Chemistry B*, **107**, 13782.

69 Klar, T.A., Franzl, T., Rogach, A.L. and Feldmann, J. (2005) *Advanced Materials*, **17**, 769.

70 Liu, Y.X., Summers, M.A., Scully, S.R. and McGehee, M.D. (2006) *Journal of Applied Physics*, **99**, 093521.

71 Halls, J.J.M., Cornil, J., dos Santos, D.A., Silbey, R., Hwang, D.H., Holmes, A.B., Bredas, J.L. and Friend, R.H. (1999) *Physical Review B: Condensed Matter*, **60**, 5721.

72 Ginger, D.S. and Greenham, N.C. (1999) *Physical Review B: Condensed Matter*, **59**, 10622.

73 Watson, D. and Meyer, G. (2005) *Annual Review of Physical Chemistry*, **56**, 119.

74 Beek, W.J.E., Wienk, M.M. and Janssen, R.A.J. (2005) *Journal of Materials Chemistry*, **15**, 2985.

75 Sun, B., Westenhoff, S., Dhoot, A.S., Silva, C. and Greenham, N.C. (2004) *Proceedings of SPIE*, **5513**, 76.

76 Barker, J.A., Ramsdale, C.M. and Greenham, N.C. (2002) *Physical Review B: Condensed Matter*, **67**, 075205.

77 Mihailetchi, V.D., Koster, L.J.A., Hummelen, J.C. and Blom, P.W.M. (2004) *Physical Review Letters*, **93**, 216601.

78 Braun, C.L. (1984) *Journal of Chemical Physics*, **80**, 4157.

79 Watkins, P.K., Walker, A.B. and Verschoor, G.L.B. (2005) *Nano Letters*, **5**, 1814.

80 Marsh, R.A., Groves, C. and Greenham, N.C. (2007) *Journal of Applied Physics*, **101**, 083509.

81 Milliron, D.J., Alivisatos, A.P., Pitois, C., Edder, C. and Frechet, J.M.J. (2003) *Advanced Materials*, **15**, 58.

82 Odoi, M.Y., Hammer, N.I., Sill, K., Emrick, T. and Barnes, M.D. (2006) *Journal of the American Chemical Society*, **128**, 3506.

83 Ginger, D.S. and Greenham, N.C. (2000) *Journal of Applied Physics*, **97**, 1361.

84 Hikmet, R.A.M., Talapin, D.V. and Weller, H. (2003) *Journal of Applied Physics*, **93**, 3509.

85 Morgan, N.Y., Leatherdale, C.A., Drndic, M., Jarosz, M.V., Kastner, M.A. and Bawendi, M. (2002) *Physical Review B: Condensed Matter*, **66**, 075339.

86 Drndic, M., Jarosz, M.V., Morgan, N.Y., Kastner, M.A. and Bawendi, M.G. (2002) *Journal of Applied Physics*, **92**, 7498.

87 Yu, D., Wang, C.J., Wehrenberg, B.L. and Guyot-Sionnest, P. (2004) *Physical Review Letters*, **92**, 216802.

88 Efros, A.L. and Shklovskii, B.I. (1975) *Journal of Physics C*, **8**, P L49.

89 Chandler, R.E., Houtepen, A.J., Nelson, J. and Vanmaekelbergh, D. (2007) *Physical Review B: Condensed Matter*, **75**, 085325.

90 Huynh, W.U., Dittmer, J.J., Teclemariam, N., Milliron, D.J., Alivisatos, A.P. and Barnham, K.W.J. (2003) *Physical Review B: Condensed Matter*, **67**, 115326.

91 Mandoc, M.M., Koster, L.J.A. and Blom, P.W.M. (2007) *Applied Physics Letters*, **90**, 133504.

92 Ramsdale, C.M., Barker, J.A., Arias, A.C., MacKenzie, J.D., Friend, R.H. and Greenham, N.C. (2002) *Journal of Applied Physics*, **92**, 4266.

93 Coe-Sullivan, S., Steckel, J.S., Woo, W.K., Bawendi, M.G. and Bulovic, V. (2005) *Advanced Functional Materials*, **15**, 1117.

94 Huynh, W.U., Dittmer, J.J., Libby, W.C., Whiting, G.L. and Alivisatos, A.P. (2003) *Advanced Functional Materials*, **13**, 73.

95 Koster, L.J.A., Mihailetchi, V.D., Xie, H. and Blom, P.W.M. (2005) *Applied Physics Letters*, **87**, 203502.

96 Koster, L.J.A., van Strien, W.J., Beek, W.J.E. and Blom, P.W.M. (2007) *Advanced Functional Materials*, **17**, 1297.

C
Transport Layers

7
PEDOT-Type Materials in Organic Solar Cells
Andreas Elschner and Stephan Kirchmeyer

7.1
Introduction

Poly(3,4-alkylenedioxythiophenes) – PEDOT-type polymers – belong to the group of polymers containing 3,4-dialkoxythiophene structures. Mono- and dialkoxy-substituted thiophene derivatives were developed by Leclerc [1], and industrial scientists at Hoechst AG [2–4]. However, most polymers of mono- and dialkoxythiophenes exhibited low conductivity in the oxidized doped state and therefore could not be used for technical purposes. A breakthrough in the area of intrinsically conducting polymers (ICPs) was the polymerization of the bicyclic 3,4-ethylenedioxythiophene (EDOT) and its derivatives – electrochemically polymerized by Heinze *et al.* and chemically polymerized by Jonas *et al.* of the Bayer AG Corporate Research Laboratories [5]. Contrary to the polymers of monocyclic mono- and dialkoxythiophenes, poly-3,4-ethylenedioxythiophene (PEDOT) and most of its derivatives have a very stable and highly conductive oxidized "doped" state. The low HOMO–LUMO bandgap of conductive PEDOT [6] allowed the formation of a tremendously stable and highly conductive ICP. Technical use and commercialization consequently followed soon and today ICPs based on PEDOT [7] are commercially available [8].

The most prominent member of the group of poly(3,4-alkylenedialkoxythiophenes), also known under the trade name of Baytron, today plays a dominant role in antistatic and conductive coatings, electronic components, and displays. In particular, its widespread applications have been developed using the conducting properties of both the PEDOT complex with polystyrene sulfonic acid [9] (PEDOT:PSS, Baytron P) and the *in situ* polymerized layers of the EDOT (Baytron M) monomer (*in situ* PEDOT). Antistatic coating applications for PEDOT:PSS include, for example, photographic films, electronic packaging, CRT screens, and LCD polarizer films. Conductive films of PEDOT:PSS are found in inorganic electroluminescent devices and organic field effect transistors. Additionally, PEDOT:PSS layers function as the material of choice for hole injection in organic light-emitting diodes (OLEDs) [10,11] and organic solar cells (OSCs). *In situ* PEDOT is also well established in industry and is used as a polymeric cathode material for solid

Organic Photovoltaics: Materials, Device Physics, and Manufacturing Technologies.
Edited by Christoph Brabec, Vladimir Dyakonov, and Ullrich Scherf
Copyright © 2008 WILEY-VCH Verlag GmbH & Co. KGaA, Weinheim
ISBN: 978-3-527-31675-5

7.2
Chemical Structure and Impact on Electronic Properties

7.2.1
Chemical Structure of PEDOT-Type Materials

Poly(3,4-dialkoxythiophenes) contain monomer units with a 3,4-dioxthiophene structure. The thiophene ring has a planar conformation owing to its aromatic nature. Alkoxy groups of 3,4-substituted thiophenes are flexible and allow rotational motion. The bridging atoms of the 3,4-alkylenedioxy structure build a second ring fused with the thiophene unit with more or less fixed geometry. This ring will take the conformation with minimum steric and electronic energy, which will depend on ring size and substitution pattern. A few monomeric structures have been studied in detail, including structures obtained from X-ray diffraction of EDOT [16] (determined from the EDOT/cyclodextrine complex) and its tetradecyl derivative [17].

The mesomeric (+M) effect of oxygen atoms adjacent to the thiophene ring stabilizes positive charges generated during polymerization [16] and explains why EDOT readily polymerizes to yield the charged conductive PEDOT. The charge distribution of positively charged EDOT-oligomers calculated by *ab initio* methods [18] indicates that the degree of charge localization along the chain strongly depends on the presence of a counterion.

The electronic overlap between thiophene π-orbitals and oxygen σ-orbitals may help to stabilize positive charges depending on the ring geometry. In an unfavorable conformation, the participation of the oxygen in the overall charge stabilization will be low.

Compared to EDOT, the five-membered ring of methylene dioxythiophene (MDOT [19]) is almost planar (see Figure 7.1). It has not been possible to link MDOT by oxidative polymerization. Polymerized MDOT (PMDOT) made by reductive synthesis [20] is stable in the undoped state and unstable in the oxidatively doped state.

The double bonds in vinylene dioxythiophene (VDOT [21]) or benzo-ethylene-dioxythiophene (Benzo-EDOT [22–24]) apparently forces the ring into an electronically less favorable conformation. As a consequence, poly-VDOT is not accessible by chemical oxidative polymerization. Benzo-EDOT yields polymers that are easily reduced to a neutral state.

Poly(3,4-alkylenedioxythiophenes) with larger ring sizes have successfully been synthesized [25] easily from the neutral state. In addition to the ring size, steric effects that hinder an efficient intermolecular π–π packing may also destabilize the oxidized state in the solid [26].

In summary, poly(alkylenedioxythiophenes) with other than six-membered ring in general exhibit low conductivities apparently because of an unstable oxidized state.

| MDOT | EDOT | VDOT | Benzo-EDOT |

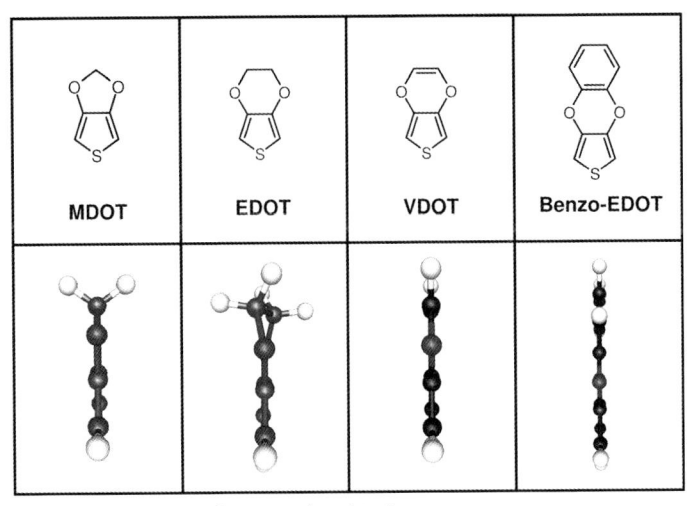

Figure 7.1 Structures of two-ring dioxythiophenes.

Most correlations of chemical structures to final polymer properties are still empirical and not fully mechanistically understood. However, it is obvious that PEDOT and a few substituted PEDOT derivatives exhibit extraordinary properties in the class of poly(alkylenedioxythiophenes).

7.2.2
Polymerization

In general, PEDOT and PEDOT-type polymers may be polymerized from the monomer by oxidative polymerization in the presence of a monomeric or polymeric charge balancing counterions. The synthetic procedure as well as the counterion will affect the resulting polymer morphology, its crystallinity, doping level, conductivity, and molecular weight.

The oxidative polymerization of EDOT in the presence of monomeric counterions leads to *in situ* PEDOT, a brittle and highly crystalline polymer with limited potential for further processing after polymerization. A widely used form of PEDOT is made by aqueous oxidative polymerization of EDOT in the presence of a polymeric counterion, usually polystyrene sulfonic acid (PSS or PSSA). PSS is a commercially available water-soluble polymer and can thus serve as a good dispersant for aqueous PEDOT. Polymerization with the oxidant sodium peroxodisulfate yields a PEDOT:PSS complex [9] in its conductive cationic form as an aqueous dispersion. This dispersion can easily be formulated with binders and additives and processed by various coating and printing techniques (Figure 7.2).

Although the overall reaction of EDOT to PEDOT should roughly parallel the oxidative polymerization mechanism discussed for alkylthiophenes [27] in general, a detailed look reveals that the total reaction path is rather complex [28,29].

PEDOT and PEDOT-like polymers may also be polymerized using transition metal-catalyzed coupling of activated organometallic derivatives [30]. This synthetic

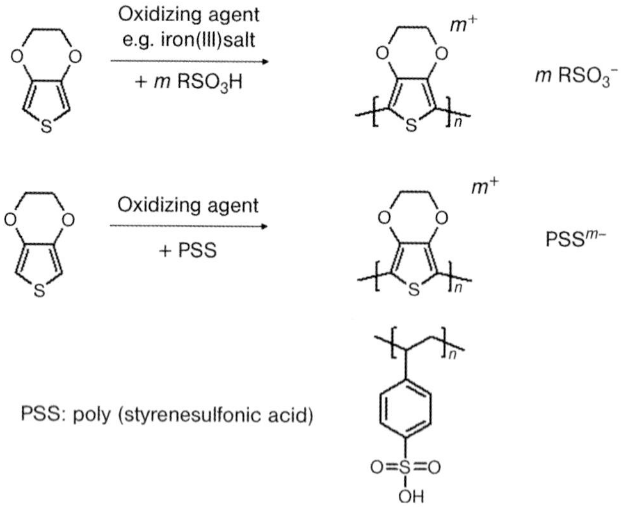

Figure 7.2 Polymerization of EDOT to in situ PEDOT and PEDOT:PSS.

method is of interest especially for monomers, which cannot be polymerized oxidatively. However, this method is not significant for PEDOT itself.

The molecular weight of PEDOT chains in in situ PEDOT and PEDOT:PSS has been discussed intensively. It appears that PEDOT segments formed during polymerization are most likely oligomeric rather than polymeric. It has not been possible to directly observe high molecular weight PEDOT polymers and the analysis of various PEDOT-containing polymers via MALDI-TOF mass spectroscopy strongly supports this assumption [31,32]. Several measurements with PEDOT:PSS or with substituted PEDOT derivatives, including neutral PEDOT molecules, indicate that the molecular weights of the individual PEDOT molecules do not exceed 1000–2500 Da, or about 6 to 18 repeating units.

7.2.3
Morphology: π–π Stacking and Crystallization

X-ray diffraction has been successfully used to determine the structure of in situ PEDOT films. Specifically, PEDOT polymeric salts of perchlorate [33], p-toluenesulfonate [34] and hexafluorophosphate [35] have been investigated. These studies reveal that PEDOT chains are π-stacked with a characteristic repeat distance of 0.34 nm. The counterions of PEDOT salts are incorporated between the π-stacks leading to a structure resembling the one shown in Figure 7.3, in which the distance between the stacked layers is determined by the size of the counterion.

Films of PEDOT:PSS are considered to be amorphous in contrast to in situ PEDOT. X-ray diffraction patterns do not reveal any significant structures here. Therefore, long-range order π–π stacking of PEDOT-segments can be excluded. Because of the planar structure of PEDOT, the face-to-face next neighbor alignment between the thiophene units appears to be a reasonable configuration.

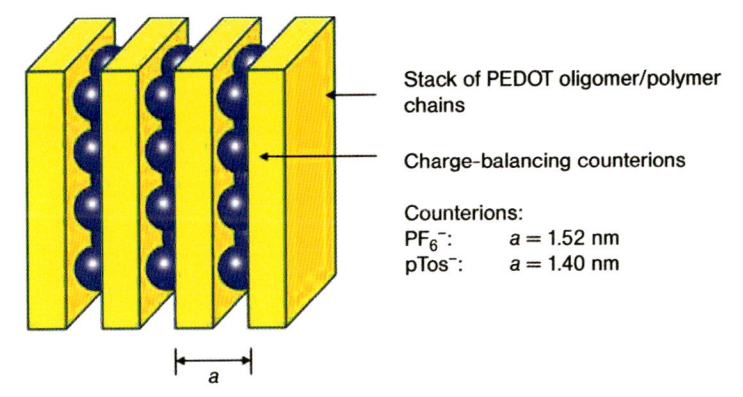

Figure 7.3 Molecular stacking in *in situ* PEDOT films.

7.2.4
Redox States of PEDOT

From oxidative polymerization PEDOT is generally obtained in the oxidized, doped form. Detailed studies of the reaction kinetics reveal a complex reaction mechanism [28] in which a dimerization of the formed EDOT radical cation is followed by the formation of higher oligomers and final oxidation to the doped polymer. This polymerization mechanism was established valid for *in situ* PEDOT [28] as well as polymerization of aqueous PEDOT microdispersions [29].

The nature of the oxidized state has been discussed intensively and appears to be an equilibrium of two distinct oxidation states. Both states, a paramagnetic polaronic state and a highly conductive, diamagnetic bipolaronic state, need to be stabilized by charge balancing counterions [36].

The oxidation state of PEDOT and its derivatives can be modified chemically and electrochemically after synthesis. Controlled chemical and electrochemical oxidation will result in a slightly higher positive charge, strong electrochemical oxidation, and chemical oxidation; using oxidizing agents such as hydrogen peroxide or hypochloride will result in chemical overoxidation by side reactions and therefore will destroy the polymer [37,38].

PEDOT salts like *p*-toluene sulfonate and tetrachloro ferrate, as well as PEDOT:PSS, can be reduced [30a] using hydrazine, hydroxylamine, and other reducing agents. However, residual charge moieties – easily detectable by IR spectroscopy – stay in the reduced PEDOT and cannot be removed completely. During the electrochemical reduction, the transparent pale blue PEDOT cations are reduced to a deep blue form [39]. Residual charges also remain during electrochemical reduction [40,41]. The neutral completely undoped PEDOT can either be made by organometallic synthesis means [42,43] or by oxidative polymerization of EDOT using iron-III-chloride with tightly controlled stoichiometry [44]. While oxidized and partly oxidized (doped) PEDOT is a conductor with a strong polar character, which is not soluble in moderately polar solvents, neutral PEDOT is nonconducting and soluble in organic solvents such as chloroform, dichloromethane, or tetrahydrofuran.

The modification of the redox state of PEDOT will also alter its morphology [35,45,46] and absorption spectrum. Hence its absorption can be tuned electrochemically. PEDOT as well as other PEDOT-type polymers therefore exhibit electrochromic properties, which can be utilized in appropriate devices [39,47,48]. Conductivity will change simultaneously and may be used to switch the area between source and drain in a transistorlike structure [49,50].

It has been proposed to modify the redox state of PEDOT to tune the hole-injection properties of PEDOT in polymeric LED and OSC [51–54]. However, in most cases the effort associated with tuning the oxidation state results in additional processing costs and therefore PEDOT is preferred in the pristine state.

7.3
PEDOT-Type Materials in Organic Solar Cells

Within the last decade, PEDOT has become an established material in OSCs. Most publications discussing this material relate on waterborne PEDOT:PSS implemented as a buffer layer in between the anode typically formed by a transparent conductive oxide (TCO) and the photoactive charge generating layer as depicted in Figure 7.4.

Investigations on high-conductive PEDOT:PSS as TCO substitution are still limited but progressing [55,56]. This new concept on alternative anodes offers the perspective to reduce the overall solar cell costs [57]. *In situ* PEDOT, see Section 2.2, has significant applications in the field of so-called "polymer" tantalum and aluminum capacitors [12] and printed wiring boards (in the so-called DMS-E process) [13–15]; however, these layers have not attracted considerable interest in OSCs probably because of processing difficulties. Table 7.1 depicts the most relevant thin-film properties of various PEDOT types being made commercially available by H.C. Starck.

7.3.1
Preparation of PEDOT Layers

Layers of *in situ* PEDOT are deposited by mixing EDOT (Baytron M), an oxidant, that is, Fe(III)tosylate (Baytron CB 40), and imidazole first [34,58]. The butanolic solution is spin-coated onto a substrate where EDOT polymerizes to PEDOT while drying. The remaining oxidant is removed by purging the films in deionized water followed by a second drying sequence. Several successive processing steps are needed, some of them being time critical, to obtain finished films limiting the technical relevance for *in situ* PEDOT.

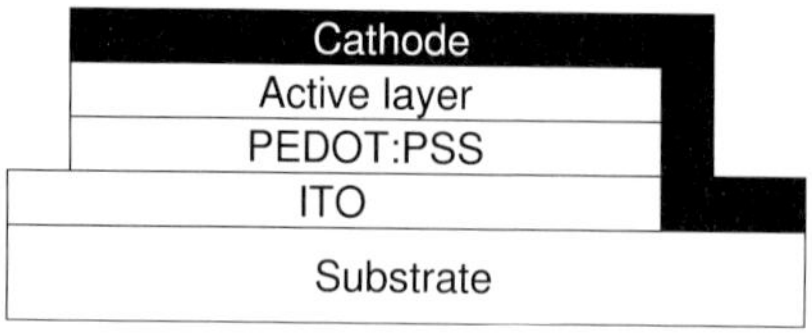

Figure 7.4 Cross section of a typical organic solar cell including a PEDOT:PSS buffer layer.

7.3 PEDOT-Type Materials in Organic Solar Cells

Table 7.1 Properties of commercially available PEDOT:PSS types relevant for OLEDs and OSCs [58].

Baytron P types	Composition PEDOT:PSS (by weight)	Conductivity σ (S cm^{-1})	Absorption constant k_1 550 nm	Work function, Φ (eV)[a]	Application
In situ PEDOT	Pure PEDOT	300–500	0.20	4.2	Electrode
PH 500	1:2.5	10–50	0.040	4.7–5.2	Buffer
PH 500 + 5 wt% DMSO	1:2.5	400–600	0.040	4.7–5.2	Electrode
Al 4083	1:6	10^{-3}	0.020	4.8–5.2	Buffer
CH 8000	1:20	10^{-5}	0.006	5.0–5.2	Buffer

[a]Reference for work function data (see Section 7.5.2).

Baytron P Al4083 and Baytron PH 500 are both aqueous dispersions of PEDOT:PSS ready to use. To obtain uniform films, the solution should be filtered carefully to remove dissolved particles, that is, dust and aggregated PEDOT:PSS. For small quantities a syringe equipped with a PVDF filter of pore size 0.45 μm or less is sufficient for most applications. The solution should be deposited on a thoroughly cleaned substrate by using common cleaning techniques such as plasma cleaning, UV ozonizing, or wet cleaning. Clean, fat-free surfaces are imperative for a proper PEDOT:PSS deposition. PEDOT:PSS dispersions can be deposited by multiple techniques including spin coating, doctor blading, dip coating, and ink jetting. Spin coating is regarded to be the most reliable technique to obtain uniform thin layers. The dependence of film layer thickness on spin speed is illustrated in Figure 7.5.

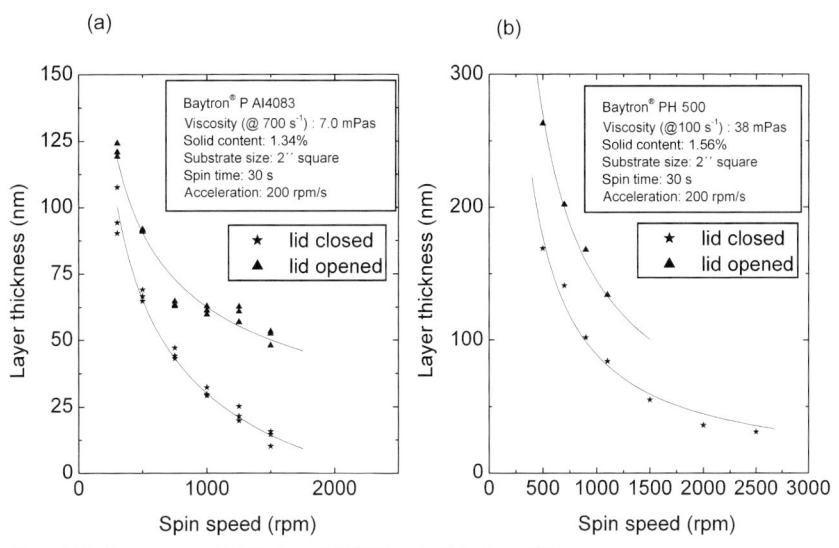

Figure 7.5 Spin curves of (a) Baytron P Al4083 and (b) Baytron PH 500 taken on a Carl Suss RC8 spin-coated and equipped with a 3″-lid in opened and closed modes. The insets disclose parameter settings.

The successive baking step depends on the PEDOT:PSS type being used. Baytron P Al4083 should immediately be baked for 5–15 min at 200 °C on a hot plate or in an oven after spin coating, whereas Baytron PH 500 should be first prebaked at a lower temperature of about 130 °C before increasing drying temperature to 200 °C when being reformulated with 5 wt% dimethylsulfoxide (DMSO) to enhance conductivity (see Section 7.4.2). PEDOT:PSS films are hygroscopic and will uptake moisture when handled in ambient after baking. Therefore, it might be beneficial to dry and further process the finished films in an inert atmosphere to achieve reproducible conditions in preparation.

7.4
High-Conductive PEDOT:PSS as TCO-Substitution in OSCs

At least one of the two electrodes embodied in thin-film photovoltaic (PV) cells has to be transparent. Thin layers made of transparent conducting oxides have traditionally been employed for PV cells to provide high conductivity at highly visible transparency, limiting series resistance and parasitic light absorbance. TCO layers are fabricated in vacuum by sputter deposition in the presence of reactive gases. Most prominent TCO materials employed in photovoltaics [57] are indium-tin oxide (ITO), zinc oxide doped with Al (AZO), and tin oxides doped with F [59] or Sb (ATO). Especially in OSCs, ITO has become the general material of choice. ITO is easily accessible as it is produced in large quantities for display industry. ITO has a smooth surface in contrast to other TCOs, making it favorable as a substrate for organic thin-film devices. Because of the high raw material costs, that is, the prices of indium have crossed the line of $1000 a kg in 2005, and high processing efforts needed to deposit ITO, ITO layers will constitute a significant proportion of the final costs in OSC production [60]. Therefore, alternative materials are being discussed such as low-cost TCOs, conductive polymers, and carbon single-wall nanotubes [61]. The last two materials exhibit the advantage to be depositable from solution and do not require any vacuum equipment.

First attempts to replace ITO by conductive polymers made of PEDOT:PSS in current-driven organic devices such as OLEDs [56,62] or OSCs [11,63,64] have been reported. Although the conductivity and the transparency of PEDOT:PSS layers are significantly lower compared to high-quality TCOs, the principal proof of employing conductive polymers in these applications has been demonstrated. A major concern is the anticipated voltage drop across the conductive transparent layer limiting device efficiency, especially at high current densities. In passive matrix OLEDs, current densities easily exceed 100 mA cm^{-2} in the pulsed mode predominantly at high frame refreshing rates. In contrast, photocurrents in OLED lamps or OSCs are significantly lower. Typically, maximal photocurrents in OSCs will not exceed 10–20 mA cm^{-2}. To reduce the voltage drop across the transparent conductor, a concept has been proposed to overcome this obstacle [11,65]. By depositing a mesh of metallic bus bars, the current will be laterally and uniformly distributed by reducing the overall serial resistance. This concept has been realized already in

7.4.1
Conductivity of PEDOT:PSS

The nature of charge transport in organic solids has been investigated intensively within the past decades [66] and should only be briefly outlined here. Like in metals and inorganic crystalline semiconductors, the model of extended states can be applied to crystalline organic semiconductors only [67]. This model cannot be transferred to amorphous polymers, however. Owing to the lack of crystalline order within this class of materials, the prerequisite for Bloch states as the eigenfunctions in Schrödingers one-electron equation is not fulfilled. Other models developed to explain charge transport in amorphous semiconductors or metallic particles dispersed in insulating materials have been adapted to polymers. These are based on charge-carrier hopping between localized states and have become in polymers the common accepted vehicle to explain transport phenomenon. But a satisfying general description of charge transport in conducting polymers, especially those with high densities of free positively charged carriers (holes) as PEDOT:PSS, is still missing.

First attempts to explain the charge transport phenomenon in solid PEDOT:PSS films have been reported by Alishin *et al.* [68]. The authors studied the conductivity and magnetoresistance of PEDOT:PSS as a function of temperature and found that both parameters increase with temperature. The temperature dependence of conductivity has been taken from commercial PEDOT:PSS types covering a wide range of conductivities (Figure 7.6) [69]. Again an increase in conductivity with temperature has been observed. An indication for metallic transport that predicts

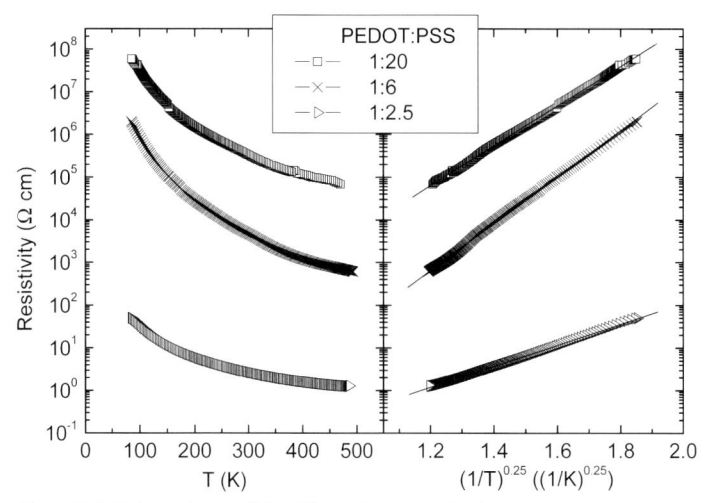

Figure 7.6 T-dependence of thin films of various PEDOT:PSS types of different PEDOT:PSS ratios [69].

a decrease of conductivity with increasing temperature cannot be found, not even for the high-conductive types. In contrast to earlier findings where the curves had been fitted by the equation for variable range hopping $\sigma(T) = \sigma(0) \exp(-(T/T_0)^{-\alpha})$ with $\alpha = 0.5$ [68], the curves in Figure 7.6 match slightly better for an exponent of $\alpha = 0.25$. We consider it to be rather speculative, however, to draw conclusions on the transport mechanism just by interpreting the slope of plot $\log(\alpha(T))$ over $\log(T)$ in a limited temperature range.

To get a better insight of the nature of conductivity in PEDOT:PSS films, other properties, especially the composition of the formulation and the resulting film morphology, have to be taken into account.

Commercial PEDOT:PSS types cover a wide range in conductivity reaching from 10^{-5} S cm^{-1} (Baytron P CH 8000) up to 500 S cm^{-1} (Baytron PH 500 formulated with 5 wt% DMSO). Three different approaches have been invoked to modify the conductivity by changing

- the PEDOT:PSS ratio between 1 : 2.5 and 1 : 20 by weight,
- the morphology of the polymer in its aqueous state,
- the morphology in its solid state.

In commercial PEDOT:PSS grades, the maximum PEDOT:PSS ratio is about 1 : 2.5 by weight. Various water soluble, electrically inert polymers can be easily added to reduce conductivity, such as polyvinylalcohol, polyvinylpyrilidone, polysiloxanes, polyacrylic acid, and more straightforward polystyrene sulfonic acid. The choice of the blend partner will be determined by the prospected application because the addition of an electrical inert polymer will not only reduce the conductivity but might also change mechanical thin-film properties such as uniformity, mechanical stability, adhesion, aging, wettability, and other electrical properties, such work function or dipole formation at the interface with adjacent layers as discussed in Section 7.5.

7.4.2
Morphology Impact on Conductivity

PEDOT:PSS dispersions are composed by hydrated (swollen) gel particles being formed by high molecular weight PSS interlinked by PEDOT chains via Coulomb interaction [8]. The conductivity of dried PEDOT:PSS films depends critically on the gel particle size distribution (PSD). By applying shear stress to the solution the thin-film conductivity increases (Figure 7.7a). This mechanical process will reduce average gel particle size [70] by tearing apart loosely joined gel particles as monitored by ultracentrifugation technique [71]. In Figure 7.7b, three PSDs originating from the same base dispersion are shown. These three formulations differ only in the amount of mechanical shear stress being applied. The maximum width of the PSDs decreases continuously when untreated material (X0) has been subjected to two (X2) or five (X5) shear stress cycles. The gel particle sizes depicted in PSDs reflect the aqueous swollen state. After dehydration, particles will shrink to about 5–10% of their original volume. According to Figure 7.7 a correlation between mean particle size and thin-film

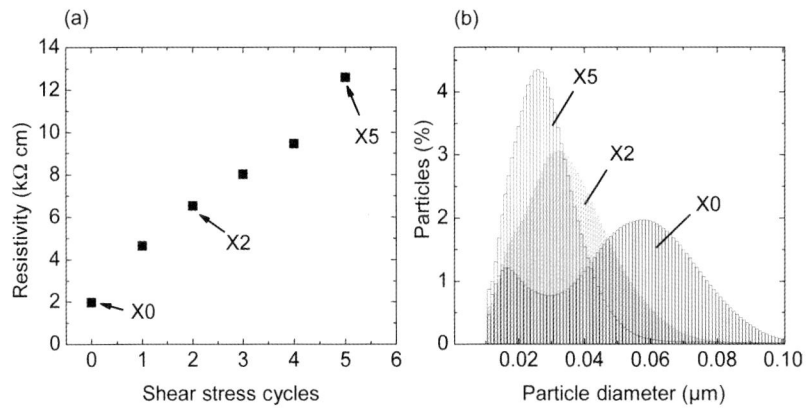

Figure 7.7 (a) Dependence of thin-film resistivity of PEDOT:PSS dispersions with a PEDOT:PSS ratio of 1:6 on the number of shear stress cycles being applied. Samples X0, X2, and X5 have been subjected to 0, 2, and 5 shear stress cycles, respectively. (b) Particle size distribution of dispersions X0, X2, and X5 determined by analytical ultracentrifugation (adapted from Ref. [71]).

conductivity is obvious: the larger the average particle size the higher the conductivity of dried films.

To our understanding, applying shear stress to the dispersion will only alter gel particle distribution and hence thin-film morphology. We have no hints that shear stress modifies the doping level of PEDOT as the absorption spectra of samples X0–X5 in Figure 7.7 remain unchanged. Neutralized PEDOT and oxidized (doped) PEDOT exhibit significant different absorption bands in the visible spectral region, as discussed in Section 7.4.3.

The drying process of aqueous PEDOT:PSS gel particles has to be discussed in more detail to obtain a better picture of the film morphology. PEDOT:PSS gel particles consist of high molecular weight PSS physically cross-linked by PEDOT oligomers [8]. The latter are considered to be mainly located in the center of the gel particle owing to their water insolubility. The outer part of the gel particle is PSS-rich because of the hydrophilic nature of PSS. When these particles merge to form a continuous film by water evaporation, the original dispersion of PEDOT-rich and -poor regimes will remain. This model gives rise to the assumption that the free-charge carrier mobility is hindered by energy barriers between regimes of higher conductivity. The energy barriers are considered to be formed by the PSS-rich contact surfaces of adjacent gel particles being statistically distributed in height and width as predicted for amorphous organic solids [72]. In the case of small particle distributions, free charge carriers have to overcome many energy barriers, whereas in the case of large particles the number of hurdles to overcome is reduced as schematically depicted in Figure 7.8.

A similar model has been proposed to explain the conductivity in polyaniline. Here, the conductivity is limited by energy barriers formed at interfaces between nanometallic primary particles [73].

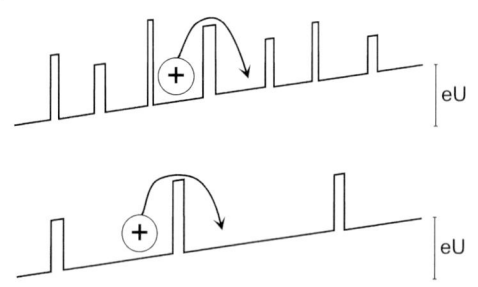

Figure 7.8 The mobility of holes in PEDOT:PSS is dominantly determined by energy barriers between adjacent particles. The larger the particles the less energy barriers to overcome.

This model on the morphology of PEDOT:PSS films and its impact on its properties have been generated by several experiments reported in the meantime:

(a) Greczynski et al. have shown by XPS that the surface of PEDOT:PSS films is PSS-rich and the outer sphere of gel particles consists dominantly of electrical insulating PPS [74].

(b) Higgins et al. [75] have investigated deuterated PEDOT:PSS films of different weight composition by small-angle neutron spectroscopy. As depicted in Figure 7.9, the volume fraction of PSS at the film surface is enriched relative to the bulk. For highly PSS-diluted compositions, this effect is more pronounced.

(c) Nardes et al. have investigated the anisotropic conductivity of PEDOT:PSS films. AFM images taken on thin-film cross sections unravelled structures similar to pancakes being stacked on top of each other. The conductivity was found to be low perpendicular and high parallel to the surface. The analysis combining morphology and conductivity is in accordance with Figure 7.8.

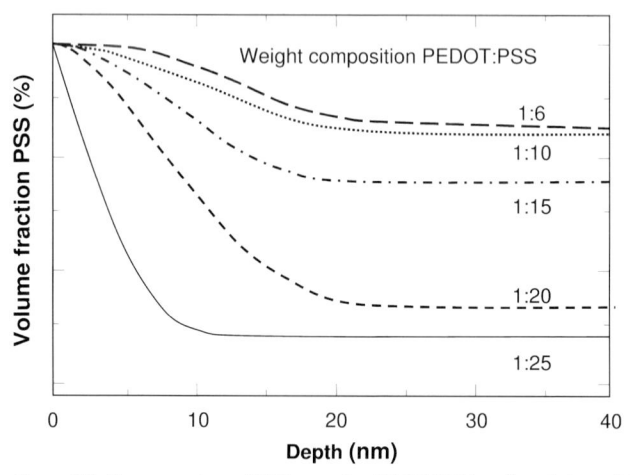

Figure 7.9 Concentration of PSS near the PEDOT:PSS surface (adapted from Ref. [75]).

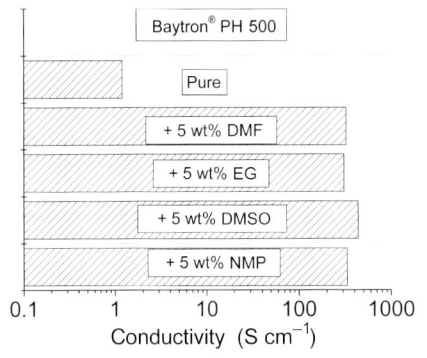

Figure 7.10 Conductivity increase of PEDOT:PSS films by the addition of high-boiling solvents.

(d) The conductivity of PEDOT:PSS films increases with baking temperature [77,78]. The degree of phase separation will be reduced owing to enhanced interpenetration of gel particles at elevated temperatures.

(e) The addition of water miscible, high-boiling solvents such as dimethylsulfoxide, ethyleneglycol (EG), N-methylpyrolidone (NMP), dimethylformamide (DMF), and others will increase conductivity by more than two orders of magnitude [79]. Typical data for conductivity changes by adding high-boiling solvents to Baytron PH 500 are shown in Figure 7.10.

In the meanwhile, the origin of this so-called "solvent effect" has been discussed by several groups [80–84]. The conductivity σ is given by $\sigma = n \times \mu \times e$, with n the density of free charge carriers being equivalent to the doping concentration of PEDOT/PSS, μ the hole mobility, and e the elementary charge. It is unreasonable to assume the free charge carrier density to increase by more than two orders of magnitude upon the addition of nonoxidizing solvents. The conductivity increase has been addressed to an increase of hole mobility owing to a screening effect by the remaining solvents [80], or morphological changes [82–84]. We favor the latter interpretation claiming the formation of a 3D-network.

When aqueous PEDOT:PSS including high-boiling solvents dries, water as the dominant solvent will evaporate first. The PEDOT:PSS gel particles will be finally dispersed solely in the high-boiling solvent. As the discussed high-boiling solvents are less polar than water, the anticipated phase separation between PEDOT-rich and PSS-rich regions will disappear or will be at least less pronounced. The resulting films will be more uniform and consequently charge transport hindering energy barriers between adjacent particles will not form. This morphological change will increase hole mobility.

(f) A 30 nm thick PEDOT:PSS film (Baytron P) has been floated on top of a copper grid and has been analyzed by Lang et al. with high angular annular dark field detector scanning transmission electron microscopy (HAADF-STEM) [85]. Figure 7.11 depicts aggregated PEDOT:PSS particles of circular shape being in close contact. Surprisingly, the joints between these particles generate higher

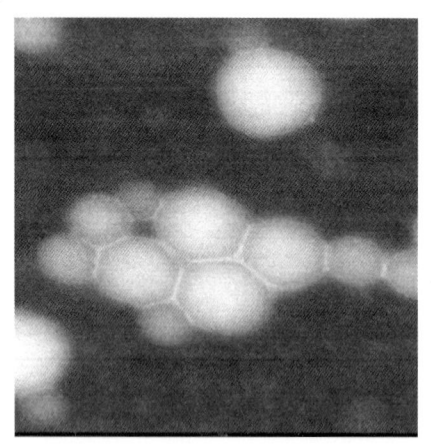

Figure 7.11 HAADF-STEM image of a 30 nm thick PEDOT:PSS film (Baytron P), exhibiting aggregated large PEDOT:PSS gel particles (reproduced with permission of Ref. [85]). The image size is 1 µm².

scattering intensities giving rise to the assumption that the layer thickness is enlarged or the material density is higher here.

7.4.3
Optical Properties of PEDOT:PSS

The optical properties of polythiophenes have been discussed in detail elsewhere [86–88]. Almost featureless absorption bands have been found in undoped polythiophene films indicating a HOMO–LUMO transition at approximately 2.0 eV, almost independent of the substitution pattern. In the case of electrical doping, new batochromic absorption bands become visible [86].

For thin films of PEDOT:PSS, the optical absorption has been studied as a function of doping level in electrochemical cells [39]. Figure 7.12i depicts the absorption spectra taken at different applied voltages oxidizing or reducing PEDOT:PSS. The main absorption peak at 2.2 eV can be assigned to partly neutralized PEDOT corresponding to the LUMO–HOMO transition schematically shown in Figure 7.12ii. By increasing the electrode potential this peak disappears and the deep blue color of the film vanishes to almost full transparency. The absorption in the IR increases accordingly as new electronic states are generated within the bandgap owing to cationic PEDOT forming polarons or even bipolarons at high oxidation levels (see Figure 7.12iii) [86].

PEDOT:PSS films are suitable for light out-coupling owing to the low absorption in the visible spectral range. The transmission spectra of 80 nm thick layers made of PH500 and Al4083 are shown in Figure 7.13a. In Figure 7.13b, the index of refraction and the absorption constant for the two PEDOT:PSS types are depicted, respectively. The data have been obtained for the incident light beam perpendicular to the polymer surface [56] not taking into account the optical anisotropy [89]. Note that the optical

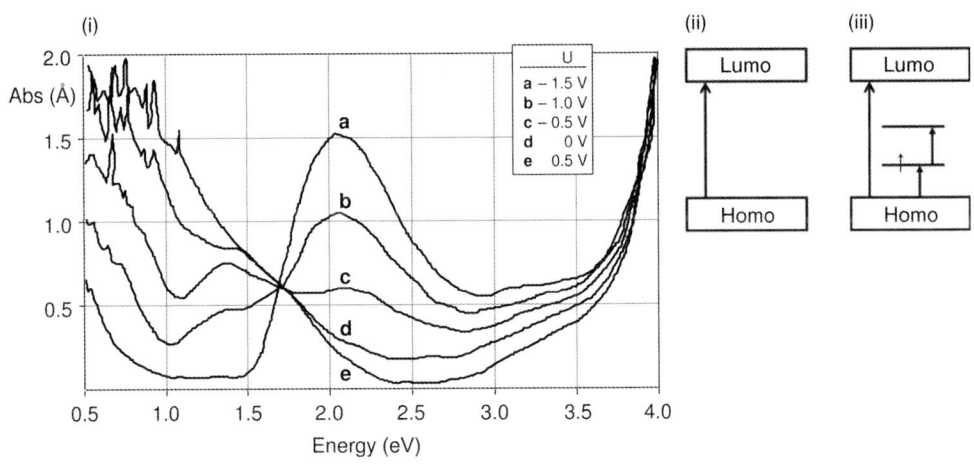

7.4 High-Conductive PEDOT:PSS as TCO-Substitution in OSCs

Figure 7.12 (i) Optical absorption spectra of a PEDOT electrochemical cell for different applied voltages: (a) − 1.5 V, (b) − 1.0 V, (c) − 0.5 V, (d) 0 V, and (e) + 0.5 V (adapted from Ref. [39]); (ii) schematic representation of energy levels and optical transitions of polythiophenes in the neutral; and (iii) the p-doped state [88].

constants of PH500 are not altered whether high-boiling solvents used to increase conductivity are added or not.

To compare high-conductive PEDOT:PSS layers with established TCOs, the internal transmission T/T_0 and the sheet resistance R_s have to be discussed in

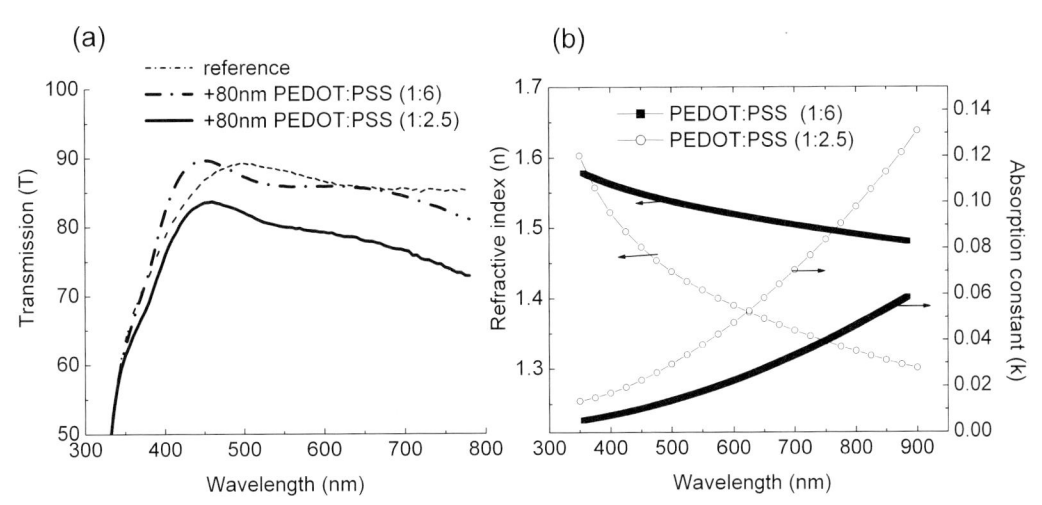

Figure 7.13 (a) Transmission (T) of 80 nm thick layers of Baytron P Al4083 and PH 500 on ITO-coated glass substrates (reference) in the VIS spectral range. The samples were mounted in front of a photointegrating sphere. (b) Spectral dependence of refractive index n and absorption constant k [58].

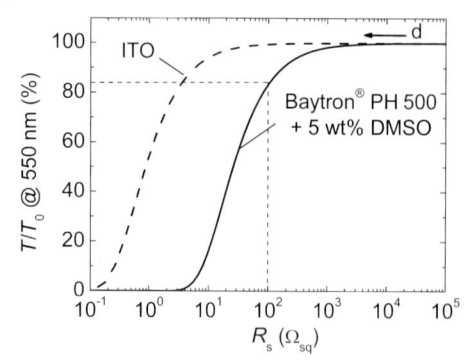

Figure 7.14 Calculated curves of internal transmission T/T_0 over sheet resistance R_s for ITO and high-conductive PEDOT:PSS.

parallel. T/T_0 and R_s depend on layer thickness d both according to

$$R_s = \frac{1}{\sigma \cdot d},$$

$$\frac{T}{T_0} = \exp\left(-\frac{4\pi \cdot k}{\lambda} d\right)$$

with conductivity σ, absorption constant k, and wavelength λ.

The dependence of T/T_0 on R_s has been calculated for increasing d by using absorption constants $k = 0.016$ and $k = 0.038$ and conductivity $\sigma = 6000$ and $\sigma = 500\,\text{S cm}^{-1}$ for ITO and high-conductive PEDOT:PSS, respectively (see Figure 7.14).

It is obvious that the conductive polymer is absorbing more light compared to ITO at the same level of R_s. Nevertheless, an internal transmission of 83% at a sheet resistance of $100\,\Omega/\text{square}$ is considered to be sufficient for transparent anodes especially in combination with metallic bus bars.

7.4.4
Long-Term Stability

Like all hydrocarbon materials, PEDOT-type polymers are subject to degradation when left in ambient conditions [90]. The light stability of PEDOT and its derivatives has been studied in detail [91]. The overall decay mechanism seems to be oxidation by oxygen [92–94] enhanced by light. Attack of the sulfur atom of the thiophene ring will yield nonconducting sulfoxide and sulfone structures, while the attack of a carbon atom next to the thiophene sulfur will yield a hydroxyl group that rearranges (Figure 7.15) [95]. In any case, the aging effect might also affect the performance of devices. Owing to the large electric field in very thin layers even at low voltages, the PSS in PEDOT:PSS has been suspected to be sensitive to decomposition by electrons [96].

To maintain the conductivity in PEDOT:PSS films over time, UV light exposure or elevated temperatures above $70\,°C$ in combination with oxygen have to be avoided.

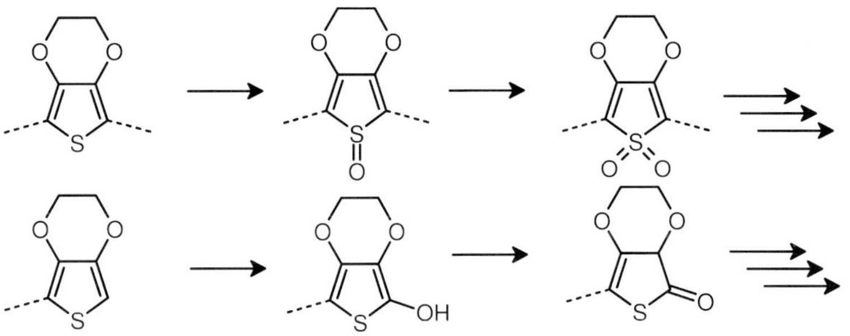

Figure 7.15 Possible degradation reactions of PEDOT.

It remains an open question whether the organic materials employed in OSCs will meet the requirements for lifetime stability. However, in practical applications like in capacitors or as antistatic layers especially PEDOT has proven to fulfill its function over years.

7.5
PEDOT-Type Materials as Hole-extracting Layers in OSCs

PEDOT and PEDOT-like polymers, in most cases PEDOT:PSS, are widely used in organic photovoltaic cells as buffer layers between the anode (usually ITO) and the photoactive layer [97] as depicted in Figure 7.4.

The role of PEDOT buffer layers has been studied in detail in various device structures. This includes polymeric photovoltaic cells [98–100] also in combination with titanium dioxide [101], dye-sensitized photovoltaic cells [102,103] also in combination with zinc compounds [104], silicon hybrid solar cells [105], and hybrid organic nanocrystal solar cells [106].

7.5.1
PEDOT:PSS as Buffer Layer in Solar Cells

The advantage of incorporating PEDOT:PSS buffer layers in OSCs stems from several properties that are closely related to its beneficial use in OLEDs [10,11,107].

First, PEDOT:PSS planarizes the ITO surface by smoothening surface imperfections and reducing the root-mean-square roughness, that is, from 10 nm for bare ITO to 3 nm for a 50 nm layer of PEDOT:PSS on ITO [108]. In particular, local spikes on the ITO surface of several tens of nanometers in size originating from the ITO deposition and causing electrical shorts in thin-film devices [109] will be partly covered by the polymer. Therefore, a PEDOT:PSS buffer layer will typically increase the yield of functional devices as the probability for electrical shortages within active layer is reduced [110,111].

Second, PEDOT:PSS buffer layers are considered to be advantageous as device performance will become independent of ITO precleaning steps and ITO work function. This is because PEDOT:PSS has a high density of free charge carriers allowing the Fermi levels of ITO and PEDOT:PSS to equilibrate as perfect metals would do. The potential drop at the interface is, therefore, determined by the work function difference of ITO and PEDOT:PSS as confirmed by ultraviolet photoelectron spectroscopy [51,112]. UV/Ozone or plasma treatment of ITO surfaces is commonly used to improve surface wetting. Oxygen plasma treatment has been shown to improve OLED performance despite an intermediate PEDOT:PSS buffer layer [113]. It is assumed that ITO treatment might have a beneficial influence on PEDOT:PSS film formation and morphology in that specific device configuration.

Third, the voltage drop across the buffer layer can be neglected. In the case of Baytron P Al4083, the material often used in OSCs, the conductivity of $1\,\mathrm{mS\,cm^{-1}}$ leads to a voltage drop of 0.1 mV for a 100 nm thick layer at a current density of $10\,\mathrm{mA\,cm^{-2}}$.

Fourth, the work function of PEDOT/PSS is on the order of 5.0–5.2 eV [112,114] leading to a built-in potential of 0.8–1.0 V when combined with Al cathodes ($\Phi =$ 4.2 eV) according to the simplifying metal–insulator–metal (MIM) model [115,116]. By introducing a 320 Å thick PEDOT:PSS layer in between ITO and copper phthalocyanine (CuPc), photocurrent characteristics are shifted by 0.5 V in double-heterostructure PV cells according to the difference in work function [111]. As the interaction at the interface of PEDOT:PSS with the active semiconducting layer is of crucial importance, this topic will be discussed in more detail below.

Although there are several advantages of introducing a PEDOT:PSS layer in OSCs, various obstacles related to this buffer layer have also been reported. These focus on the acidity of the aqueous dispersions and the water uptake of solid films.

The acidity of PEDOT:PSS in the pH range of 1–2 is suspected to dissolve indium ions from the ITO layer, which migrate from the anode into the buffer layer and even contaminate the photoactive layer [117,118]. Although a generally accepted evidence is still missing that traces of indium ions in PEDOT:PSS harm OSCs, several attempts have been made to protect ITO surfaces using self-assembly monolayers of alkylsilanes [119,120], α-quaterthiophene-2-phosphonate [121], or amorphous carbon layers [122]. Heil *et al.* claimed that a thin Au or pentacene layer placed in between ITO and PEDOT:PSS reduces light-induced degradation as observed in OLEDs [123]. None of these ideas have, however, been implemented in regular device processing to our best knowledge.

Conversely, the acidity of PEDOT:PSS is considered advantageous as it enables good contact formation at the interface by etching off all contaminations from the TCO surface during deposition.

Layers of PEDOT:PSS when thoroughly being dried will uptake water again if they are left unprotected in air. The release of water in a functional device is considered to be a severe source of degradation as water might corrode the metallic contacts or might oxidize the adjacent semiconductive organic layer. Therefore, PEDOT:PSS films are commonly baked in inert atmosphere at elevated temperatures to remove the water. The long-term stability of OLEDs with incorporated PEDOT:PSS layers indicate that cathode corrosion is negligible as long as the device is hermetically sealed. To avoid expensive encapsulation in OSCs, corrosive stable cathodes, that is,

Al top contacts are favored. Much progress has been made in designing semiconducting materials for OSCs that are less sensitive to oxidation. This will increase device stability no matter if traces of oxygen or water attacking the active layer penetrates from outside or are released from the PEDOT:PSS layer itself.

Various modifications of PEDOT:PSS have been discussed as alternative buffer layers in OSCs. Zhang *et al.* have compared various PEDOT:PSS types as thin layers in between ITO and an active layer comprised of a polyfluorene copolymer blended in proportion (1 : 4) with [6,6]-phenyl-C_{61}-butyric acid methylester (PCBM) in OSCs [99]. By employing PEDOT:PSS in a ratio of 1 : 2.5 and 1 : 6 and by adding conductivity enhancing sorbitol, the authors suggested the segregation of PSS to occur at the top of a PEDOT:PSS film. These surface modifications are considered responsible for the observed differences in open-circuit voltage (V_{oc}).

Single-wall carbon nanotube (SWCNT) films have been invoked to replace ITO as printable and flexible anodes in OSCs. Again, the introduction of a PEDOT:PSS will significantly improve device performance, whether as a buffer layer [124,125] or as a blend together with CNTs [126].

Williams *et al.* have investigated spin-casted PEDOT:PSS as p-layer in an organic–inorganic p-i-n stack on ITO in combination with amorphous silicon and microcrystalline silicon. A power efficiency (η) of 2.1% and V_{oc} of 0.883 V have been achieved, whereas without PEDOT:PSS only 0.21% and 0.176 V, respectively, have been achieved [127].

Improved carrier collection properties in OSCs were reported by Peumans *et al.* [111] by treating the PEDOT:PSS surface with a mild Ar or O_2 plasma. The impact of gas ions on the polymer surface leads to layer thinning and increased layer microroughness [128] and will most probably modify the chemical composition of PEDOT:PSS, too.

Frohne *et al.* have electrochemically modified the work function Φ of PEDOT:PSS anodes and claimed that the overall OSC performance has been improved by matching Φ with the oxidation potential of the semiconductor [54].

PEDOT:PSS, CuPc, and thin evaporated Au films were compared in parallel as buffer layers in bulk-heterojunction OSCs by Yoo *et al.* [100]. A significant increase for V_{oc} and η was found for the organic layers in contrast to bare ITO or Au.

Inorganic buffer layers of transition metal oxides have been discussed as an alternative to PEDOT:PSS, such as vanadium (III) oxide (V_2O_3) and molybdenum oxide (MoO_3) vacuum deposited on ITO in OSCs [129].

7.5.2
Electronic Effects at the PEDOT:PSS–Semiconductor Interface

The question which factor determines the V_{oc} in organic solar cells is a matter of controversy [130,131] as discussed by Kawano *et al.* [132]. The present understanding on V_{oc} is that if both electrodes establish ohmic contacts with the active layer, V_{oc} is mostly governed by the difference between the lowest unoccupied molecular orbital (LUMO) level of the acceptor and the highest occupied molecular orbital (HOMO) level of the donor in the photoactive layer. If contact is nonohmic, V_{oc} is determined by the difference of work functions of the two electrodes. The answer as to whether the contact between PEDOT:PSS and the active layer is ohmic or not is therefore of significant interest.

For the sake of simplicity, the energy alignment between a metal and an organic semiconductor has been traditionally discussed within the Schottky–Mott framework. This model assumes vacuum level alignment at the interface and band bending with Fermi level alignment. The energy barrier for injecting and extracting holes from a metal into the organic layer and vice versa, respectively, is defined within this model by the difference between the metal work function and the ionization potential of the organic layer. By employing experimental techniques such as UPS or Kelvin probe to investigate energy level alignment of metal/organic interfaces, it has been shown that the Schottky–Mott model can seldom be applied [133].

The work function of PEDOT:PSS has been determined first by UPS [114] and Kelvin probe [112] to be 5.0 ± 0.1 and 5.2 ± 0.1 eV, respectively. In the meantime, several laboratories have measured the work function of PEDOT:PSS, reporting values ranging from 4.7 [134] to 5.6 eV [135]. This confusing spread of reported data is attributed to different preparation conditions and different PEDOT:PSS types investigated. As PEDOT:PSS film deposition is normally done in ambient air, surface contaminations modifying the work function are unavoidable. Especially traces of water in the film depending on postbaking conditions will alter the work function significantly [135]. Additionally, different types of commercial and self-made PEDOT:PSS have been compared without noticing that the ratio of PEDOT:PSS might be different (see Table 7.1). Koch et al. have determined the work function of various PEDOT:PSS types in a comparative experiment, finding slightly higher values for the PSS-rich types [136].

The interface between the quasimetal PEDOT:PSS and organic semiconductors has been investigated in numerous experiments. As illustrated in Figure 7.16, the energy barrier for hole injection (ΔE) is not simply determined by the difference between the PEDOT:PSS work function (Φ) and the ionization potential (I_p) of the

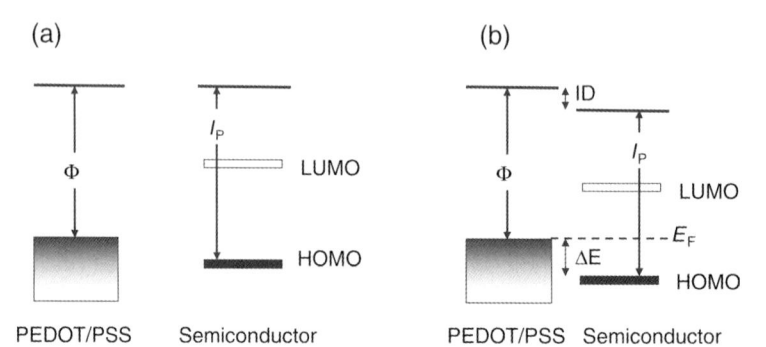

Figure 7.16 Energy level alignment at the PEDOT:PSS–semiconductor interface. The two layers are separated (a) and in contact (b). The formation of an interface dipole (ID) might significantly determine the energy barrier for hole injection ΔE.

semiconductor as predicted by the Schottky–Mott model, but that dipole layer formation at the interface will lead to a vacuum level shift ID [136–139].

ΔE will be calculated according to

$$\Delta E = \Phi - I_p - \text{ID}.$$

The dipole layer formation at the PEDOT:PSS–semiconductor interface is believed to be triggered by sulfate moieties within the PSS. Anionic semiconductor species generated within the interface formation are being counterbalanced by SO_3^- to provide a stable charge-transfer-like configuration [136]. The strength of the dipole determining ID scales linearly with Φ. Therefore, ΔE is effectively independent of Φ. This has been addressed as Fermi level pinning [138,140]. Tengstedt *et al.* have investigated P3HT, a commonly used hole-transporting material in OSCs and other organic semiconductors on various anodes, and found that the Fermi level pinning would occur if Φ exceeds a threshold defined by I_p and the semiconductor's polaronic relaxation energy [138].

However, it has to be kept in mind that energy barriers determined by UPS and Kelvin probe mimic only steady-state conditions. The conditions might change in real devices significantly when free charge carriers trapped at the interface will alter the energy level alignment by generating locally high electric fields. This additional effect on interface alignment is of explicit importance in bipolar OLED devices [139,141,142].

Reports on absolute values of energy barriers for hole extraction at PEDOT:PSS–semiconductor interfaces for heterojunction OSCs have not been reported to our best knowledge. In contrast to OLEDs or OFETs, the contact resistance is not anticipated to be a general problem in OSCs as the flow of holes is counterdirected in photovoltaic devices. No energy barrier will hinder holes from being extracted by a PEDOT:PSS anode owing to the fact that the I_p of all relevant hole conductors are generally in the range of 5.0–6.0 eV [133] and the electrical dipole at the interface is directed presumably toward the anode.

7.6
Conclusions

The conducting polymer PEDOT:PSS has become a well recognized, mature product within the last decade. The dispersion can be easily processed out of an environment-friendly solvent making it especially attractive for use in printed applications. PEDOT:PSS has been implemented to fulfill two functions in organic solar cells: First, as a buffer and hole-extraction layer mounted between the TCO anode and the active layer and second, as a conductive transparent layer to form the anode by itself. Numerous scientific groups have investigated the synthesis of the monomer, the polymerization reaction, and the solid-state properties of PEDOT:PSS in the meantime. These activities have provided the solid fundamentals necessary to understand and further improve organic electronic device performance. These activities are ongoing as there is still a lack of a detailed understanding of many aspects.

References

1 Daoust, G. and Leclerc, M. (1991) Structure-property relationships in alkoxy-substituted polythiophenes. *Macromolecules*, 24, 455–459.
2 Feldhues, M., Mecklenburg, T., Wegener, P. and Kämpf, G. (1986) EP 257 573 (Hoechst AG).
3 Kämpf, G. and Feldhues, M. (1987) EP 292 905 (Hoechst AG).
4 Feldhues, M., Kämpf, G., Litterer, H., Mecklenburg, T. and Wegener, P. (1989) Polyalkoxythiophenes soluble electrically conducting polymers. *Synthetic Metals*, 28, 487–489.
5 (a) Jonas, F., Heywang, G., Schmidtberg, W., Heinze, J. and Dietrich, M. (1988) EP 339 340 (Bayer AG); (b) Heywang, G. and Jonas, F. (1992) Poly(alkylenedioxythiophene)s – new, very stable conducting polymers. *Advanced Materials*, 4, 116–118.
6 Pei, Q., Zuccarello, G., Ahlskog, M. and Inganäs, O. (1994) Electrochromic and highly stable poly(3,4-ethylenedioxythiophene) switches between opaque blue-black and transparent sky blue. *Polymer*, 35, 1347–1351.
7 Groenendaal, B.L., Jonas, F., Freitag, D., Pielartzik, H. and Reynolds, J.R. (2000) Poly(3,4-ethylenedioxythiophene) and its derivatives: past, present, and future. *Advanced Materials*, 12, 481–494.
8 Kirchmeyer, S. and Reuter, K. (2005) Scientific importance, properties and growing applications of poly(3,4-ethylenedioxythiophene). *Journal of Materials Chemistry*, 15, 2077–2088.
9 Jonas, F. and Krafft, W. (1990) EP 440 957 (Bayer AG).
10 Jonas, F., Elschner, A., Wehrmann, R. and Quintens, D. (1996) EP 909464 (Bayer AG).
11 Carter, S.A., Angelopoulos, M., Karg, S., Brock, P.J. and Scott, J.C. (1997) Polymeric anodes for improved polymer light-emitting diode performance. *Applied Physics Letters*, 70, 2067–2069.
12 Jonas, F., Heywang, G. and Schmidtberg, W. (1988) EP 340 512 (Bayer AG).
13 Hupe, J., Wolf, G.D. and Jonas, F. (1995) A known principle with a novel basis. Through-hole contacting of printed circuit boards using conductive polymers. *Galvanotechnik*, 86, 3404–3411.
14 Wolf, G.-D., Jonas, F. and Schomaecker, R. (1994) EP 707 440 (Bayer AG).
15 Kirchmeyer, S. and Jonas, F. (1999) WO 2000 045625 (Bayer AG).
16 Storsberg, J. (2001) Dissertation, Mainz.
17 Storsberg, J. and Ritter, H. (1999) private communication.
18 Dkhissi, A., Beljonne, D., Lazzaroni, R., Louwet, F., Groenendaal, L. and Bredas, J.L. (2003) Density functional theory and Hartree–Fock studies of the geometric and electronic structure of neutral and doped ethylenedioxythiophene (EDOT) oligomers. *International Journal of Quantum Chemistry*, 91, 517–523.
19 Ahonen, H.J., Kankare, J., Lukkari, J. and Pasanen, P. (1997) Electrochemical synthesis and spectroscopic study of poly(3,4-methylenedioxythiophene). *Synthetic Metals*, 84, 215–216.
20 Brassat, L., Kirchmeyer, S. and Reuter, K. (2004) EP 1598358 (H.C. Starck GmbH & Co. KG)
21 Leriche, P., Blanchard, P., Frere, P., Levillain, E., Mabon, G. and Roncali, J. (2006) 3,4-Vinylenedioxythiophene (VDOT): a new building block for thiophene-based π-conjugated systems. *Chemical Communications*, 275–277.
22 Kirchmeyer, S., Klausener, A., Rauchschwalbe, G. and Reuter, K. (2001) EP 1275649 (Bayer AG).
23 Roquet, S., Leriche, P., Perepichka, I., Jousselme, B., Levillain, E., Frere, P. and Roncali, J. (2004) 3,4-Phenylenedioxythiophene (PheDOT): a novel platform for the synthesis of planar substituted π-donor conjugated systems. *Journal of Materials Chemistry*, 14, 1396–1400.

24 Perepichka, I.F., Roquet, S., Leriche, P., Raimundo, J.-M., Frere, P. and Roncali, J. (2006) Electronic properties and reactivity of short-chain oligomers of 3,4-phenylenedioxythiophene (PheDOT). *Chemistry – A European Journal*, **12**, 2960–2966.

25 Welsh, D.M., Kumar, A., Meijer, E.W. and Reynolds, J.R. (1999) Enhanced contrast ratios and rapid switching in electrochromics based on poly(3,4-propylenedioxythiophene) derivatives. *Advanced Materials*, **11**, 1379–1382.

26 Nielsen, C.B. and Bjornholm, T. (2005) Structure-property relations of regiosymmetrical 3,4-dioxy-functionalized polythiophenes. *Macromolecules*, **38**, 10379–10387.

27 Fichou, D. (1999) *Handbook of Oligo- and Polythiophenes*, Wiley-VCH Verlag GmbH, Weinheim, Germany.

28 (a) Tracht, U. (2002) personal communication; (b) Kirchmeyer, S. (2002) 5th International Symposium on Functional π-Electron Systems, Ulm, Germany.

29 Groenendaal, L., Louwet, F., Adriaensens, P., Carleer, R., Vanderzande, D. and Gelan, J. (2002) Mechanical and structural aspects of poly(3,4- ethylenedioxythio- phene)/polystyrenesulfonic acid. *Polymeric Materials Science and Engineering*, **86**, 52.

30 (a) Yamamoto, T. and Abla, M. (1999) Synthesis of non-doped poly(3,4-ethylenedioxythiophene) and its spectroscopic data. *Synthetic Metals*, **100**, 237; (b) Yamamoto, T., Abla, M., Shimizu, T., Komarudin, D., Lee, B.-L. and Kurokawa, E. (1999) Temperature dependent electrical conductivity of p-doped poly(3,4-ethylenedioxythiophene) and poly(3-alkylthiophene)s. *Polymer Bulletin*, **42**, 321.

31 Jonas, F., Groenendaal, L. and Pausch, J. (1999) unpublished results.

32 Reuter, K., Karbach, A., Ritter, H. and Wrubbel, N. (2003) EP 1 440 974 A2 (Bayer AG).

33 Granström, M. and Inganäs, O. (1995) Electrically conductive polymer fibres with mesoscopic diameters: 1. Studies of structure and electrical properties. *Polymer*, **36**, 2867.

34 Aasmundtveit, K.E., Samuelsen, E.J., Petterson, L.A.A., Inganäs, O., Johansson, T. and Feidenhans`l, R. (1999) Structure of thin films of poly(3,4-ethylenedioxythiophene). *Synthetic Metals*, **101**, 561.

35 Niu, L., Kvarnström, C., Fröberg, K. and Ivaska, A. (2001) Electrochemically controlled surface morphology and crystallinity in poly(3,4-ethylenedioxythiophene) films. *Synthetic Metals*, **122**, 425.

36 (a) Ahonen, H.J., Lukkari, J. and Kankare, J. (2000) n- and p-Doped poly (3,4-ethylenedioxythiophene): two electronically conducting states of the polymer. *Macromolecules*, **33**, 6787–6793; (b) Son, Y., Lim, C.-B., Choi, J.-S. and Lee, Y. (2000) A chemical process initiated by an electrochemical process of electrochromic conducting polymer PEDOT. *Molecular Crystals and Liquid Crystals Science and Technology Section A-Molecular Crystals and Liquid Crystals*, **349**, 347–350; (c) De Kok, M.M., Buechel, M., Vulto, S.I.E., Van De Weijer, P., Meulenkamp, E.A., De Winter, S.H.P., Mank, A.J.G., Vorstenbosch, H.J.M., Weijtens, C.H.L. and Van Elsbergen, V. (2004) Modification of PEDOT:PSS as hole injection layer in polymer LEDs. *Physica Status Solidi a: Applied Research*, **201**, 1342–1359; (d) Chiu, W.W., Travas-Sejdic, J., Cooney, R.P. and Bowmaker, G.A. (2005) Spectroscopic and conductivity studies of doping in chemically synthesized poly (3,4-ethylenedioxythiophene). *Synthetic Metals*, **155**, 80–88.

37 Zykwinska, A., Domagala, W., Czardybon, A., Pilawa, B. and Lapkowski, M. (2006) In-situ ESR spectroelectro- chemical studies of overoxidation behaviour of poly(3,4-butylenedioxythio- phene). *Electrochimica Acta*, **51**, 2135–2144.

38 Du, X. and Wang, Z. (2003) Effects of polymerization potential on the

properties of electrosynthesized PEDOT films. *Electrochimica Acta*, **48**, 1713–1717.

39 Gustafsson, J.C., Liedberg, B. and Inganäs, O. (1994) In situ spectroscopic investigations of electrochromism and ion transport in a poly (3,4-ethylenedioxythiophene) electrode in a solid state electrochemical cell. *Solid State Ionics*, **69**, 145–152.

40 Garreau, S., Louam, G., Lefrant, S., Buisson, J.P. and Froyer, G. (1999) Optical study and vibrational analysis of the poly (3,4-ethylenedioxythiophene) (PEDT). *Synthetic Metals*, **101**, 312.

41 Johansson, T., Petterson, L.A.A. and Inganäs, O. (2002) Conductivity of de-doped poly(3,4-ethylenedioxythiophene). *Synthetic Metals*, **129**, 269.

42 Tran-Van, F., Garreau, S., Louarn, G., Froyer, G. and Chevrot, C. (2001) A fully undoped oligo(3,4-ethylenedioxythiophene): spectroscopic properties. *Synthetic Metals*, **119**, 381.

43 Tran-Van, F., Garreau, S., Louarn, G., Froyer, G. and Chevrot, C. (2001) Fully undoped and soluble oligo(3,4-ethylenedioxythiophene)s: spectroscopic study and electrochemical characterization. *Journal of Materials Chemistry*, **11**, 1378–1382.

44 Reuter, K. and Kirchmeyer, S. (2001) EP 1 327 645 (Bayer AG).

45 Randriamahazaka, H., Noel, V. and Chevrot, C. (2002) Fractal dimension of the active zone for a p-doped poly(3,4-ethylenedioxythiophene) modified electrode towards a ferrocene probe. *Journal of Electroanalytical Chemistry*, **52**, 107–116.

46 Randriamahazaka, H., Noel, V. and Chevrot, C. (2003) Anomalous diffusion on the active zone of p-doped poly(3,4-ethylenedioxythiophene) modified electrodes. *Journal of Electroanalytical Chemistry*, **556**, 35–42.

47 Pielartzik, H., Heuer, H.-W., Wehrmann, R. and Bieringer, T. (1999) Intrinsically conductive, charge-injecting, electrochromic, and photochemically addressable plastics. *Kunststoffe*, **89**, 135–138 (in German, English translations, see Pielartzik, H., Heuer, H.-W., Wehrmann, R., Bieringer, T., *Engineering Plastics*, 1999, **89**, 135–138).

48 Heuer, H.-W., Wehrmann, R. and Kirchmeyer, S. (2002) Electrochromic window based on conducting poly(3,4-ethylenedioxythiophene)-poly(styrene sulfonate). *Advanced Functional Materials*, **12**, 89–94.

49 Nilsson, D., Chen, M., Kugler, T., Remonen, T., Armgarth, M., Remonen, T., Armgarth, M. and Berggren, M. (2002) Bi-stable and dynamic current modulation in electrochemical organic transistors. *Advanced Materials*, **14**, 51–54.

50 Anderson, P., Nielsson, D., Svensson, P.-O., Chen, M., Malmström, A., Remomen, T., Kugler, T. and Berggren, M. (2002) Active matrix displays based on all-organic electrochemical smart pixels printed on paper. *Advanced Materials*, **14**, 1460–1464.

51 Zhang, F., Petr, A., Peisert, H., Knupfer, M. and Dunsch, L. (2004) Electrochemical variation of the energy level of poly(3,4-ethylenedioxythiophene): poly (styrenesulfonate). *Journal of Physical Chemistry B*, **108** (45), 17301–17305.

52 Frohne, H., Müller, D.C. and Meerholz, K. (2002) Continuously variable hole injection in organic light emitting diodes. *Chemical Physics*, **3**, 707–711.

53 Yang, X., Müller, D.C., Neher, D. and Meerholz, K. (2006) Highly efficient polymeric electrophosphorescent diodes. *Advanced Materials*, **18**, 948–954.

54 Frohne, H., Shaheen, S.E., Brabec, C.J., Müller, D.C., Sariciftci, N.S. and Meerholz, K. (2002) Influence of the anodic work function on the performance of organic solar cells. *ChemPhysChem*, **3**, 795–799.

55 Zhang, F., Johansson, M., Andersson, M.R., Hummelen, J.C. and Inganäs, O. (2002) Polymer photovoltaic cells with conducting polymer anodes. *Advanced Materials*, **14**, 662–665.

56 Fehse, K., Walser, K., Leo, K., Lövenich, W. and Elschner, A. (2007) Highly conductive polymer anodes as replacement of inorganic materials for high efficiency organic light-emitting diodes. *Advanced Materials*, **19**, 441–444.

57 Genley, D.S., and Bright, C. (2000) Transparent conducting oxides. *MRS Bulletin*, **25**, 15–21.

58 Starck, H.C., http://www.baytron.com.

59 Yang, F., and Forrest, S.R. (2006) Organic solar cells using transparent SnO_2-F anodes. *Advanced Materials*, **18**, 2018–2022.

60 Shaheen, S.E., Ginley, D.S. and Jabbour, G.E. (2005) Organic-based photovoltaics: toward low-cost power generation. *MRS Bulletin*, **30**, 10–19.

61 Peltola, J., Weeks, C., Levitsky, I.A., Britz, D.A., Glatkowski, P., Trottier, M. and Huang, T. (2007) Carbon nanotube transparent electrodes for flexible displays. *Journal of the Society for Information Display*, **23/2**, 20–23.

62 Kim, W.H., Mäkinen, A.J., Nikolov, N., Shashidhar, R., Kim, H. and Kafafi, Z.H. (2002) Molecular organic light-emitting diodes using highly conducting polymers as anodes. *Applied Physics Letters*, **80**, 3844–3846.

63 Glatthaar, M., Niggemann, M., Zimmermann, B., Lewer, P., Riehle, M., Hinsch, A. and Luther, J. (2005) Organic solar cells using inverted sequence. *Thin Solid Films*, **491**, 298–300.

64 Aernouts, T., Vanlaeke, P., Geens, W., Poortmans, J., Heremans, P., Borghs, S., Mertens, R., Andriessen, R. and Leenders, L. (2004) Printable anodes for flexible organic solar cell modules. *Thin Solid Films*, **451–452**, 22–25.

65 Neyts, K., Marescaux, M., Nieto, A.U., Elschner, A., Lövenich, W., Fehse, K., Huang, Q., Walzer, K. and Leo, K. (2006) Inhomogeneous luminance in organic light emitting diodes related to electrode resistivity. *Journal of Applied Physics*, **100**, 114513/1–114513/4.

66 Conwell, E.M. (1997) *Handbook of Organic Conductive Molecules and Polymers*, vol. 4 (ed. H.S. Nalva), John Wiley & Sons, New York, pp. 1–45.

67 Pope, M., and Swenberg, C.E. (1997) *Electronic Processes in Organic Crystals and Polymers*, Oxford University Press, Chapter XIV.

68 Aleshin, A.N., Williams, S.R. and Heeger, A.J. (1998) Transport properties of poly (3,4-ethylenedioxythiophene)/poly (styrenesulfonate). *Synthetic Metals*, **94**, 173–177.

69 Elschner, A. (2001) unpublished results.

70 Elschner, A., Jonas, F., Kirchmeyer, S. and Wussow, K. High-resistivity PEDT/PSS for reduced crosstalk in passive matrix OELs. Proceedings IDW 2001, OEL3-3, Nagano, Japan.

71 Müller, H.G. (2004) Determination of very broad particle size distributions via interference optics in the analytical ultracentrifuge. *Progress in Colloid and Polymer Science*, **127**, 9–13.

72 Bässler, H. (1993) Charge transport in disordered organic photoconductors a Monte Carlo simulation study. *Physica Status Solidi a: Applied Research*, **175**, 15–56.

73 Pelster, R., Nimtz, G. and Wessling, B. (1994) Fully protonated polyaniline: Hopping transport on a mesoscopic scale. *Physical Review B: Condensed Matter*, **49**, 12718–12723.

74 (a) Greczynski, G., Kugler, T. and Salaneck, W.R. (1999) Characterization of the PEDOT-PSS system by means of X-ray and ultraviolet photoelectron spectroscopy. *Thin Solid Films*, **354**, 129–135; (b) Jönsson, S.K.M., Birgerson, J., Crispin, X., Greczynski, G., Osikowicz, W., Denier van der Gon, A.W., Salaneck, W.R. and Fahlman, M. (2003) The effects of solvents on the morphology and sheet resistance in poly(3,4-ethylenedioxythiophene)- polystyrenesulfonic acid (PEDOT-PSS) films. *Synthetic Metals*, **139**, 1–10.

75 Jukes, P.C., Martin, S., Higgins, A.M., Geoghegan, M., Jones, R.A.L., Langridge, R.S., Wehrum, A. and Kirchmeyer, S.

(2004) Controlling the surface composition of poly(3,4-ethylene dioxythiophene)-poly(styrene sulfonate) blends by heat treatment. *Advanced Materials*, **16**, 807–811.

76 Nardes, A.M., Kemerink, M., Janssen, R.A.J., Bastiaansen, J.A.M., Kiggen, N.M.M., Langeveld, B.M.W., van Breemen, A.L.J.M. and de Kok, M.M. (2007) Microscopic understanding of the anisotropic conductivity of PEDOT:PSS thin films. *Advanced Materials*, in print.

77 Elschner, A. (2001) presentations OLEDs, www.baytron.com.

78 Huang, J., Miller, P.F., de Mello, J.C., de Mello, A.J. and Bradley, D.D.C. (2003) Influence of thermal treatment on the conductivity and morphology of PEDOT/PSS films. *Synthetic Metals*, **139**, 569–572.

79 Jonas, F., Karbach, A., Muys, B., van Thillo, E., Wehrmann, R., Elschner, A. and Dujardin, R. (1995) Patent EP 686662 (Bayer AG).

80 Kim, J.Y., Jung, J.H., Lee, D.E. and Joo, J. (2002) Enhancement of electrical conductivity of poly(3,4-ethylenedioxythiophene)/poly(4-styrenesulfonate) by change of solvents. *Synthetic Metals*, **126**, 311–316.

81 Ouyang, J., Xu, Q., Chu, C.-W., Yang, Y., Li, G. and Shinar, J. (2004) On the mechanism of conductivity enhancement in poly(3,4-ethylenedioxythiophene):poly(styrene sulfonate) film through solvent treatment. *Polymer*, **45**, 8443–8450.

82 Snaith, H.J., Kenrick, H., Chiesa, M. and Friend, R.H. (2005) Morphological and electronic consequences of modifications to the polymer anode PEDOT/PSS. *Polymer*, **46**, 2573–2578.

83 Huang, J., Miller, P.F., Wilson, J.S., de Mello, A.J., de Mello, J.C. and Bradley, D.D.C. (2005) Investigation of the effects of doping and post-deposition treatments on the conductivity, morphology, and work function of poly(3,4-ethylenedioxythiophene)/poly(styrene sulfonate) films. *Advanced Functional Materials*, **15**, 290–296.

84 Crispin, X., Jakobsson, F.L.E., Crispin, A., Grim, P.C.M., Andersson, P., Volodin, A., van Haesendonck, C., Van der Auweraer, M., Salaneck, W.R. and Berggren, M. (2006) The origin of the high conductivity of poly(3,4-ethylenedioxythiophene)-poly(styrenesulfonate) (PEDOT-PSS) plastic electrodes. *Chemistry of Materials*, **18**, 4354–4360.

85 Lang, U. and Müller, E. (2007) private communication.

86 Ziegler, C. (1997) *Handbook of Organic Conductive Molecules and Polymers*, vol. 3 (ed. H.S. Nalva), John Wiley & Sons, New York, pp. 678–737.

87 Taliani C. and Gebauer, W. (1999) *Handbook of Oligo- and Polythiophenes* (ed. D. Fichou), Wiley-VCH Verlag GmbH, Weinheim, Germany, pp. 361–404.

88 Jiang, X.M., Österbacka, R., Korovyanko, O., An, C.P., Horovitz, B., Janssen, R.A.J. and Vardeny, Z.V. (2002) Spectroscopic studies of photoexcitations in regioregular and regiorandom polythiophene films. *Advanced Functional Materials*, **12**, 587–597.

89 Pettersson, L.A.A., Ghosh, S. and Inganäs, O. (2002) Optical anisotropy in thin films of poly(3,4-ethylenedioxythiophene)-poly(4-styrenesulfonate). *Organic Electronics*, **3**, 143–148.

90 Kolesov, I.S., and Münstedt, H. (2003) Stability of antielectrostatic coatings on basis of polyethylenedioxythiophene. *Materialwissenschaft und Werkstofftechnik*, **34**, 542–548.

91 Jeuris, K., Groenendaal, L., Verheyen, H., Louwet, F. and De Schryver, F.C. (2003) Light stability of 3,4-ethylenedioxythiophene-based derivatives. *Synthetic Metals*, **132**, 289–295.

92 Norrman, K., and Krebs, F.C. (2006) Lifetimes of organic photovoltaics: using TOF-SIMS and O-18(2) isotopic labelling to characterise chemical degradation mechanisms. *Solar Energy Materials and Solar Cells*, **90**, 213–227.

93 Vazquez, M., Bobacka, J., Ivaska, A. and Lewenstam, A. (2002) Influence of oxygen and carbon dioxide on the electrochemical stability of poly(3,4-ethylenedioxythiophene) used as ion-to-electron transducer in all-solid-state ion-selective electrodes. *Sensors and Actuators B: Chemical*, **82**, 7–13.

94 Kawano, K., Pacios, R., Poplavskyy, D., Nelson, J., Bradley, D.D.C. and Durrant, J.R. (2006) Degradation of organic solar cells due to air exposure. *Solar Energy Materials and Solar Cells*, **90**, 3520–3530.

95 Kang, H.S., Kang, H.-S., Lee, J.K., Lee, J.W., Joo, J., Ko, J.M., Kim, M.S. and Lee, J.Y. (2005) Humidity-dependent characteristics of thin film poly(3,4-ethylenedioxythiophene) field-effect transistor. *Synthetic Metals*, **155**, 176–179.

96 van der Gon, A.W.D., Birgerson, J., Fahlman, M. and Salaneck, W.R. (2002) Modification of PEDOT-PSS by low-energy electrons. *Organic Electronics*, **3**, 111–118.

97 Hoppe, H. and Sariciftci, N.S. (2004) Organic solar cells: An overview. *Journal of Materials Research*, **19**, 1924–1945.

98 Aernouts, T., Geens, W., Poortmans, J., Heremans, P., Borghs, S. and Mertens, R. (2002) Extraction of bulk and contact components of the series resistance in organic bulk donor–acceptor-heterojunctions. *Thin Solid Films*, **403**, 297–301.

99 Zhang, F.L., Gadisa, A., Inganaes, O., Svensson, M. and Andersson, M.R. (2004) Influence of buffer layers on the performance of polymer solar cells. *Applied Physics Letters*, **84**, 3906–3908.

100 Yoo, I., Lee, M., Lee, C., Kim, D.W., Moon, I.S. and Hwang, D.H. (2005) The effect of a buffer layer on the photovoltaic properties of solar cells with P3OT:fullerene composites. *Synthetic Metals*, **153**, 97–100.

101 Song, M.Y., Kim, K.J. and Kim, D.Y. (2005) Enhancement of photovoltaic characteristics using a PEDOT interlayer in TiO2/MEHPPV heterojunction devices. *Solar Energy Materials and Solar Cells*, **85**, 31–39.

102 Saito, Y., Azechi, T., Kitamura, T., Hasegawa, Y., Wada, Y. and Yanagida, S. (2004) Photo-sensitizing ruthenium complexes for solid state dye solar cells in combination with conducting polymers as hole conductors. *Coordination Chemistry Reviews*, **248**, 1469–1478.

103 Fukuri, N., Saito, Y., Kubo, W., Senadeera, G.K.R., Kitamura, T., Wada, Y. and Yanagida, S. (2004) Performance improvement of solid-state dye-sensitized solar cells fabricated using poly(3,4-ethylenedioxythiophene) and amphiphilic sensitizing dye. *Journal of the Electrochemical Society*, **15**, 1745–1748.

104 Liao, J.-Y. and Ho, K.-C. (2005) A photovoltaic cell incorporating a dye-sensitized ZnS/ZnO composite thin film and a hole-injecting PEDOT layer. *Solar Energy Materials and Solar Cells*, **86**, 229–241.

105 Williams, E.L., Wang, Q., Shaheen, S.E., Ginley, D.S. and Jabbour, G.E. (2004) Conducting polymer and thin-film silicon hybrid solar cells. *Polymeric Materials Science and Engineering*, **91**, 802–804.

106 Huynh, W.U., Dittmer, J.J. and Alivisatos, A.P. (2002) Hybrid nanorod-polymer solar cells. *Science*, **295**, 2425–2427.

107 Berntsen, A., Croonen, Y., Liedenbaum, C., Schoo, H., Visser, R.-J., Vleggaar, J. and Van de Weijer, P. (1998) Stability of polymer LEDs. *Optical Materials*, **9**, 125–133.

108 Jonda, C., Mayer, A.B.R., Stolz, U., Elschner, A. and Karbach, A. (2000) Surface roughness effects and their influence on the degradation of organic light emitting devices. *Journal of Materials Science*, **35**, 5645–5651.

109 Pichler, K. (1997) Conjugated polymer electroluminescence: technical aspects from basic devices to commercial products. *Philosophical Transactions of the Royal Society of London. Series A:*

Mathematical and Physical Sciences, **335**, 829–842.
110 Elschner, A., Bruder, F., Heuer, H.W., Jonas, F., Karbach, A., Kirchmeyer, S. and Thurm, S. (2000) PEDT/PSS for efficient hole-injection in hybrid organic light-emitting diodes. *Synthetic Metals*, **111**, 139–143.
111 Peumans, P. and Forrest, S.R. (2001) Very-high-efficiency double-heterostructure copper phthalocyanine/C$_{60}$ photovoltaic cells. *Applied Physics Letters*, **79**, 126–128; Erratum, *Applied Physics Letters*, 2002, **80**, 338.
112 Brown, T.M., Kim, J.S., Friend, R.H., Cacialli, F., Daik, R. and Feast, W.J. (1999) Built-in field electroabsorption spectroscopy of polymer light-emitting diodes incorporating a doped poly(3,4-ethylene dioxythiophene) hole injection layer. *Applied Physics Letters*, **75**, 1679–1681.
113 Kim, J.S., Friend, R.H. and Cacially, F. (1999) Improved operational stability of polyfluorene-based organic light-emitting diodes with plasma-treated indium-tin-oxide anodes. *Applied Physics Letters*, **74**, 3084–3086.
114 Kugler, Th., Salaneck, W.R., Rost, H. and Holmes, A.B. (1999) Polymer band alignment at the interface with indium tin oxide: consequences for light emitting devices. *Chemical Physics Letters*, **310**, 391–396.
115 Parker, I.D. (1994) Carrier injection and device characteristics in polymer light emitting diodes. *Journal of Applied Physics*, **75**, 1656.
116 Malliaras, G.G., Salem, J.R., Brock, P.J. and Scott, J.C. (1998) Photovoltaic measurement of the built-in potential in organic light emitting diodes and photodiodes. *Journal of Applied Physics*, **84**, 1583.
117 de Jong, M.P., van IJzendoorn, L.J. and de Voigt, M.J.A. (2000) Stability of the interface between indium-tin-oxide and poly(3,4- ethylenedioxythiophene)/poly (styrenesulfonate) in polymer light-emitting diodes. *Applied Physics Letters*, **77**, 2255–2257.
118 van Duren, J.K.J., Loos, J., Morrissey, F., Leewis, C.M., Kivits, K.P.H., van IJzendoorn, L.J., Rispens, M.T., Hummelen, J.C. and Janssen, R.A.J. (2002) In-situ compositional and structural analysis of plastic solar cells. *Advanced Functional Materials*, **12**, 665–669.
119 Malinsky, J.E., Jabbour, G.E., Shaheen, S.E., Anderson, J.D., Richter, A.G., Marks, T.J., Armstrong, N.R., Kippelen, B., Dutta, P. and Peyghambarian, N. (1999) Self-assembly processes for organic LED electrode passivation and charge injection balance. *Advanced Materials*, **11**, 227–231.
120 Wong, K.W., Yip, H.L., Luo, Y., Wong, K.Y., Lau, W.M., Low, K.H., Chow, H.F., Gao, Z.Q., Yeung, W.L. and Chang, C.C. (2002) Blocking reactions between indium-tin oxide and poly (3,4-ethylene dioxythiophene): poly(styrene sulphonate) with a self-assembly monolayer. *Applied Physics Letters*, **80**, 2788–2790.
121 Hanson, E.L., Guo, J., Koch, N., Schwartz, J. and Bernasek, S.L. (2005) Advanced surface modification of indium tin oxide for improved charge injection in organic devices. *Journal of the American Chemical Society*, **127**, 10058–10062.
122 Chen, B.J., Sun, X.W., Tay, B.K., Ke, L. and Chua, S.J. (2005) Improvement of efficiency and stability of polymer light-emitting devices by modifying indium tin oxide anode surface with ultrathin tetrahedral amorphous carbon film. *Applied Physics Letters*, **86**, 063506/1–063506/3.
123 Heil, H., Andress, G., Schmechel, R., von Seggern, H., Steiger, J., Bonrad, K., and Sprengard, R. (2005) Sunlight stability of organic light-emitting diodes. *Journal of Applied Physics*, **97**, 124501/1–124501/4.
124 van de Lagemaat, J., Barnes, T.M., Rumbles, G., Shaheen, S.E., Coutts, T.J., Weeks, C., Levitsky, I., Peltola, J. and

Glatkowski, P. (2006) Organic solar cells with carbon nanotubes replacing In2O3:Sn as the transparent electrode. *Applied Physics Letters*, **88**, 233503/1–233503/3.

125 Rowell, M.W., Topinka, M.A., McGehee, M.D., Prall, H.-J., Dennler, G., Sariciftci, N.S., Hu, L. and Gruner, G. (2006) Organic solar cells with carbon nanotube network electrodes. *Applied Physics Letters*, **88**, 233506/1–233506/3.

126 Reale, A., Brown, T.M., Di Carlo, A., Giannini, F., Brunetti, F., Leonardi, E., Lucci, M., Terranova, M.L., Orlanducci, S., Tamburri, E., Toschi, F. and Sessa, V. (2006) Nanocomposites for organic and hybrid organic-inorganic solar cells. *Proceedings of SPIE*, **6334**, 63340Y/1–63340Y/8.

127 Williams, E.L., Jabbour, G.E., Wang, Q., Shaheen, S.E., Ginley, D.S. and Schiff, E.A. (2005) Conducting polymer and hydrogenated amorphous silicon hybrid solar cells. *Applied Physics Letters*, **87**, 223504/1–223504/3.

128 Zhou, Y., Yuan, Y., Lian, J., Zhang, J., Pang, H., Cao, L. and Zhou, X. (2006) Mild oxygen plasma treated PEDOT:PSS as anode buffer layer for vacuum deposited organic light-emitting diodes. *Chemical Physics Letters*, **427**, 394–398.

129 Shortriya, V., Li, G., Yao, Y., Chu, C.-W. and Yang, Y. (2006) Transition metal oxides as the buffer layer for polymer photovoltaic cells. *Applied Physics Letters*, **88**, 073508/1–073508/3.

130 Brabec, C.J., Cravino, A., Meissner, D., Sariciftci, N.S., Fromherz, T., Rispens, M.T., Sanchez, L. and Hummelen, J.C. (2001) Origin of the open circuit voltage of plastic solar cells. *Advanced Functional Materials*, **11**, 374–380.

131 Mihailetchi, V.D., Blom, P.W.M., Hummelen, J.C. and Rispens, M.T. (2003) Cathode dependence of the open-circuit voltage of polymer:fullerene bulk heterojunction solar cells. *Journal of Applied Physics*, **94**, 6849–6854.

132 Kawano, K., Ito, N., Nishimori, T. and Sakai, J. (2006) Open circuit voltage of stacked bulk heterojunction organic solar cells. *Applied Physics Letters*, **88**, 073514/1–073514/3.

133 Ishii, H., Sugiyama, K., Ito, E. and Seki, K. (1999) Energy level alignment and interfacial electronic structures at organic/metal and organic/organic interfaces. *Advanced Materials*, **11**, 605–625.

134 Greczynski, G., Kugler, Th. and Salaneck, W.R. (2000) Energy level alignment in organic-based three-layer structures studied by photoelectron spectroscopy. *Journal of Applied Physics*, **88**, 7187–7191.

135 Koch, N., Vollmer, A. and Elschner, A. (2007) The influence of water on the work function of conducting poly(3,4-ethylenedioxythiophene)/poly(styrenesulfonate). *Applied Physics Letters*, **90**, 043512/1–043512/3.

136 Koch, N., Elschner, A., Rabe, J.P. and Johnson, R.L. (2005) Work function independent hole-injection barriers between pentacene and conducting polymers. *Advanced Materials*, **17**, 330–335.

137 Mäkinen, A.J., Hill, I.G., Shashidhar, R., Nikolov, N. and Kafafi, Z.H. (2001) Hole injection barriers at polymer anode/small molecule interfaces. *Applied Physics Letters*, **79**, 557–559.

138 Tengstedt, C., Osikowicz, W., Salaneck, W.R., Parker, I.D., Hsu, C.-H. and Fahlman, M. (2006) Fermi-level pinning at conjugated polymer interfaces. *Applied Physics Letters*, **88**, 053502/1–053502/3.

139 Koch, N., Elschner, A. and Johnson, R.L. (2006) Green polyfluorene-conducting polymer interfaces: energy-level alignment and device performance. *Journal of Applied Physics*, **100**, 024512/1–024512/5.

140 Koch, N., and Vollmer, A. (2006) Electrode-molecular semiconductor contacts. Work-function-dependent hole injection barriers versus Fermi - level pinning. *Applied Physics Letters*, **89**, 162107/1–162107/3.

141 Brewer, P.J., Lane, P.A., Huang, J., de Mello, A.J., Bradley, D.D.C. and de Mello, J.C. (2005) Role of electron injection in polyfluorene-based light emitting diodes containing PEDOT:PSS. *Physical Review B: Condensed Matter and Materials Physics*, **71**, 205209/1– 205209/6.

142 Seeley, A.J.A.B., Friend, R.H., Kim, J.-S. and Burroughes, J.H. (2004) Trap-assisted hole injection and quantum efficiency enhancement in poly (9,9′ dioctylfluorene-alt-benzothiadiazole) polymer light-emitting diodes. *Journal of Applied Physics*, **96**, 7643–7649.

8
The Dispersion Approach for Buffer Layers and for the Active Light Absorption Layer

Bjoern Zeysing and Bernhard Weßling

8.1
Introduction

The access to energy is one of the most important cornerstones for most modern societies. Especially in the light of limited resources of fossil fuels, the fast growing energy consumption of the world's population, and the climate change, the generation of energy from renewable sources is becoming more and more important. Consequently, the research for the generation of energy from renewable sources such as biomaterial, wind, water, and sunlight has gained increasing interest in the last few years. This fact can be seen when looking not only at the large number of scientific publications on these fields but also at the number of filed patents and the growing activities of start-up companies as well as established companies that are also investing in renewable energy research.

Photovoltaic devices can play a major role in the future of energy generation because, among other advantages, the generation of electric energy from sunlight does not involve production of greenhouse gases. The production of polymer-based photovoltaic devices has very attractive features such as the use of roll-to-roll processes generating cheap, flexible, and large-area devices.

8.2
Photovoltaic Devices

Photovoltaic devices transform – based on the photoelectric effect – sunlight into (electrical) energy. In these devices, photogeneration of charge carriers takes place in a light-absorbing material followed by a charge separation. Finally, the charges are transported to a conductive contact. The generated voltage can be used by connecting a consumer between the (metal) contacts.

The first solar cell was made of gold and selenium in 1883. About 60 years later, the silicon-based solar cell was invented, which still is the most commonly used solar cell today. The typical setup of a silicon-based solar cell is shown in Figure 8.1.

Organic Photovoltaics: Materials, Device Physics, and Manufacturing Technologies.
Edited by Christoph Brabec, Vladimir Dyakonov, and Ullrich Scherf
Copyright © 2008 WILEY-VCH Verlag GmbH & Co. KGaA, Weinheim
ISBN: 978-3-527-31675-5

8 The Dispersion Approach for Buffer Layers and for the Active Light Absorption Layer

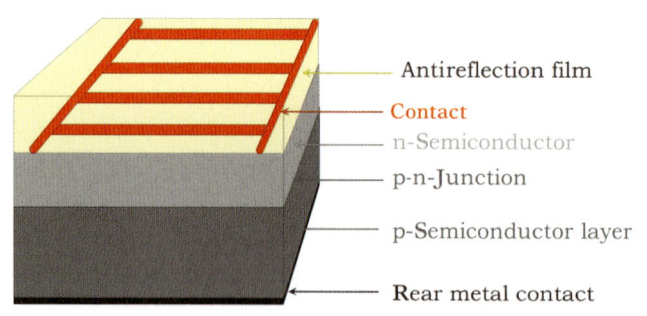

Figure 8.1 Typical setup of an inorganic solar cell.

The overall efficiency of the device is strongly dependent on the quality of the used silicon. The performance of monocrystalline silicon-based devices is two to three times higher than that of devices made of amorphous silicon [1].

Another setup for an inorganic photovoltaic device is the dye-sensitized solar cell (DSSC), which was first reported in 1991 by Grätzel and coworkers [2]. The advantages of the DSSC, which is based on a nanoporous TiO_2 electrode, are high efficiency and low production cost.

Upon excitation of a dye molecule by absorption of visible light, an electron is injected into the conduction band of TiO_2. The excited state of the dye molecule is being reduced by an electrolyte, while the electrolyte is in contact with an electrode. Although the efficiency of DSSCs using liquid electrolytes is quite high (about 10%), these devices exhibit serious sealing and stability problems. In recent years, it has been shown by different research groups that it is possible to circumvent these problems by using solid hole conductors like conductive polymers [3–5]. Polythiophene (PTh), polypyrrole (PPy), polyphenylenevinylene, polyaniline (PAni), and their derivatives have been used in photovoltaic devices as conductive polymers. The requirements for polymers in this application are a high conductivity and a good penetration into the pores of the nanoporous TiO_2.

In 1985, Tang reported one of the first organic solar cells. This two-layer cell was fabricated from copper phthalocyanine and a perylene tetracarboxylic derivative [6]. The power conversion efficiency of this cell was about 1%. In the past 20 years, research has been focused on enhancing the power conversion efficiencies. The power conversion efficiency is dependent on different factors: the open-circuit voltage, the short-circuit density, and the fill factor. By improving any of these factors, the overall performance of photovoltaic devices can be improved.

Recently, a completely new approach to building organic solar cells has been published [7]. Krebs *et al.* described the synthesis and performance of a single molecule that integrates all functionalities needed for a solar cell. The drawbacks are both the great synthetic effort (some 60 steps are needed for the synthesis) and the poor efficiency of devices derived from this molecule (about 8×10^{-5}%).

By using new materials and new device setups, the efficiency of the latest organic photovoltaic devices was improved up to 5% [8]. Still, the efficiency of inorganic devices – mostly based on silicon – is much higher. Till the end of 2006, the best inorganic devices had efficiencies of about 25% [9]. Recently, Spectrolab published a terrestrial solar cell with 40.7% efficiency [10]. These results were verified by the US Department of Energy's National Renewable Energy Laboratory (NREL). Due to its higher efficiency, about 95% of today's solar cell production is based on silicon devices [1].

Nevertheless, organic photovoltaic devices will have certain advantages over their inorganic counterparts. One of the major drawbacks for the use of solar energy on a larger scale is its relatively high price. To produce the energy more cost efficiently, either the efficiency has to be improved drastically or the price of the devices has to drop. The use of polymers has the potential to be far less expensive than the classic silicon-based solar cells, from both raw material and processing reasons. Other advantages of organic photovoltaics are the lightweight of the devices and the possibility to fabricate large-area and flexible devices relatively easily. Additionally, one can fine-tune some characteristics of the polymers by chemical modification.

In photovoltaic devices, polymers can be used for basically all possible applications. They can be used as hole conduction material, as antireflection layer, as contact layer, or to smooth the surface of one layer and therefore enhance the contact between two layers.

Before discussing the so far realized photovoltaic devices using polymers, a short introduction on the topic of conductive polymers, focusing on polyaniline, and on the topic of dispersion will be given.

8.3
Conductive Polymers

The emergence of conductive polymers as a research topic was in 1977, when Shirakawa, MacDiarmid, Heeger, and coworkers discovered that the conductivity of polyacetylene could be enhanced by several orders of magnitude upon reaction with iodine [11]. In the following years, a great variety of other conductive polymers were synthesized and their conductivity could also be enhanced upon reaction with different species (mostly acids). Most of the conductive polymers that are under investigation in R&D today are PPy, PTh, PAni, and their derivatives. Only one special derivative of PTh and PAni is – although still in small scale – in commercial use.

One of the features of conductive polymers is their insolubility in all solvents. This was proven on the basis of thermodynamics for polyaniline by one of the authors in numerous articles [12–14]. The reason for this is the polycation structure of the conductive polymers. Therefore, conductive polymers can only be processed as dispersions.[1] Neutral conjugated polymers can be processed either from solution (true or oversaturated solutions) or from dispersion.

---

1) It must be well understood that here we are using the term "conductive polymer" in its precise wording and understanding, that is, the polymers are conductive, which means that they are "doped," and they have reacted with the oxidative agent (e.g., iodine) or the acid used for rendering the neutral conjugated polymer conductive. The neutral form, the nonconductive (sometimes semiconductive) conjugated polymer, may be soluble in some solvents.

The differentiation between dispersions and solutions is not a purely academic one – it affects the fundamental properties of the solution/dispersion and all processing and application aspects. The key difference between solutions and dispersions lies in the thermodynamics: solutions are equilibrium systems, while dispersions are nonequilibrium systems. Regardless of whether we like this fact or not, we have to deal with nonequilibrium systems and try to work with the special properties and requirements of dispersions. At Ormecon, the entire research in the field of conductive polymers is based on the dispersion theory published by one of the authors years ago [15].

One of the major characteristics of nonequilibrium systems is their nonlinear behavior, that is, the sudden and often unpredictable and nonlinear change in one certain property (e.g., viscosity or conductivity) upon the linear change of one single parameter in small steps (like the concentration of one component). This behavior is most challenging for formulating and manufacturing dispersions – not only conductive polymer dispersions.

On the contrary, for the manufacturing of organic photovoltaic cells – as well as for other applications – there is an advantage that is mostly overlooked by most scientists and technicians when dealing with dispersions:

- When using true solutions incorporating the key additive (like a neutral conjugated polymer), in the beginning, the process liquid may be a true solution in the stage of thermodynamic equilibrium and the conjugated polymer fully unfolded; however, when drying, the polymer needs to refold, and monomer unit/solvent molecule interactions (which are the basis for the system being a solution) have to be replaced by monomer unit/monomer unit interactions, until all solvent molecules that had solvated the polymer backbone are driven off – a very complicated and poorly reproducible process;

- Drying a polymer solution involves two-phase transitions – from solution to dispersion and from liquid to solid! Both phase transitions are not thermodynamic equilibrium but nonequilibrium processes. Therefore, the fact that solutions are thermodynamic equilibrium systems and dispersions are not is overcompensated by the fact that for the formation of the layers, the solutions need to undergo two nonequilibrium processes, the phase transitions. In addition, in each drying step, the three-dimensional form of the polymer particle needs to be formed, starting from a totally random uncoiled polymer chain that was in solution – another nonequilibrium process involving poor reproducibility of the semiconductive properties.

- However, when starting from a dispersion, all the conductive or semiconductive properties of the particles have already been formed in the previous (synthetic) step where this property building can be nicely controlled – no need to worry about the formation of particles and monomer unit/monomer unit interactions any more!

The conductivity of conductive polymers is some orders of magnitude lower than for metals. Although there are some examples in literature for polymers with conductvities up to $1000\,S\,cm^{-1}$ and higher [16–20], these polymers are either extremely air and/or moisture sensitive (and therefore not suitable for most industrial needs), not processable by solution or dispersion, or cannot be synthesized reproducibly. Commercially available conductive polymers have a conductivity up to 500 S cm^{-1}.

Two different types of polymer dispersions reach this conductivity also on a commercial scale: PAni dispersions from Ormecon and polyethylenedioxythiophene (PEDOT) dispersions from HC Starck (after formulation). Although the conductivity of these polymers is lower than the conductivity of metals, the unique property combination, low weight, good processability, high transparency, and the possibility to create flexible devices make these conductive polymers a very interesting and, for industrial application, promising alternative.

8.3.1
Polyaniline

Polyaniline is one of the best investigated conductive polymers due to the availability and low price of the monomer, the relatively straightforward polymerization in one single step with a high yield, as well as the high conductivity, high stability, and rich redox chemistry of the polymer. When mentioning polyaniline most people mean the emeraldine salt, which is only one of the six possible forms of the polyaniline as is shown in Figure 8.2.

Since the leucoemeraldine forms are not stable in air and the pernigraniline as well as the emeraldine base has a very low conductivity, the (green) emeraldine salt is by far the most frequently used form of polyaniline.

Polyaniline can conveniently be synthesized either electrochemically or chemically. The chemical synthesis occurs by mixing an oxidizing agent like a persulfate with aniline and a dopant in an aqueous solution. However, the precipitated polymer has to be dispersed after the synthesis. Different approaches for performing the nontrivial dispersion process have been published [13,14,21].

In the dispersion polymerization a steric stabilizer is added to the reaction mixture, preventing the polyaniline from precipitating [22]. The drawback of this method is the insulating behavior of the steric stabilizers, lowering the conductivity of the polyaniline. Additionally, it is hardly possible to purify the polyaniline by removing side products and oligomers.

The most common process for preparing polyaniline dispersions is to polymerize aniline in diluted protonic acid and purify the precipitated polyaniline after synthesis. In the next step the dopant of the polyaniline is removed by stirring the emeraldine salt in a strongly basic aqueous solution. The resulting emeraldine base is purified and then dispersed in *m*-cresol and redoped with camphor sulfonic acid [16,17,23].

Another route used by Ormecon begins with the synthesis and purification of the polyaniline powder (emeraldine salt). In a series of dispersion steps, the primary particles are separated and the polymer can be dispersed in various different solvents, ranging from water over polar and weakly polar organic solvents to unpolar solvents. The processing of polyaniline powder to the ready-to-use dispersion is shown in Figure 8.3.

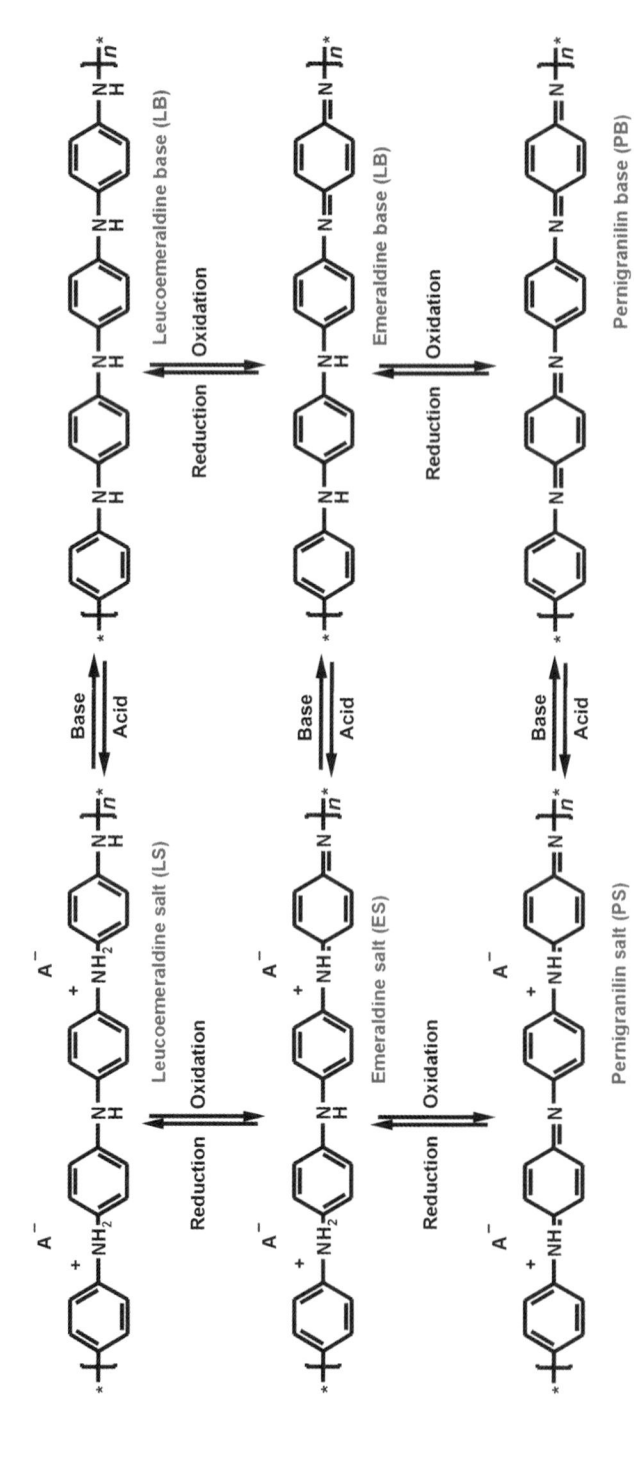

248 | *8 The Dispersion Approach for Buffer Layers and for the Active Light Absorption Layer*

Figure 8.2 Different states of polyaniline.

1. Synthesis
- agglomerated particles in μm range
- conductivity about 5 S cm⁻¹

2. Predispersion
- separation of larger agglomerates
- Conductivity about 50 S cm⁻¹

3. Dispersion
- separation of primary particles (10 nm)
- Conductivity up to 500 S cm⁻¹

4. Formulation
- Adjustment of certain parameters
- Conductivity up to 500 S cm⁻¹

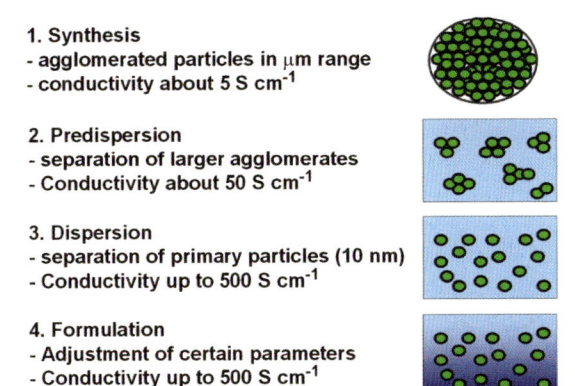

● Primary particle

Figure 8.3 Processing of conductive polymers by dispersion.

During the special Ormecon dispersion process, the polyaniline undergoes an insulator-to-metal transition, becoming an organic metal [24]. Thermopower measurements also show that dispersed conductive polymers have essentially the same thermopower behavior as metals [25]. After the dispersion process the charge transport in polyaniline takes place by a hopping mechanism between metallic regions, which are surrounded by amorphous polymer [26]. These metallic regions were shown by microwave absorption studies to have a diameter of about 8 nm. The 8 nm sized core is surrounded by a nonmetallic, amorphous shell of about 1 nm in thickness.

Additionally, the conductivity of the polyaniline is enhanced by the dispersion steps by roughly two orders of magnitude. Directly after the synthesis, the conductivity of the polyaniline powder is about $5\,\mathrm{S\,cm^{-1}}$. In a series of dispersion steps, about 10 nm large, globular primary particles (shown in Figure 8.4) are separated, while the conductivity is enhanced up to $500\,\mathrm{S\,cm^{-1}}$. In a final formulation step, certain parameters such as surface tension and wetting properties of the dispersion can be adjusted, ideally without changing the conductivity.

The intrinsic conductivity of PAni is still a subject of research interest. A recent publication in cooperation with one of us indicated that the metallic core could have a conductivity of several $10\,000\,\mathrm{S\,cm^{-1}}$ (which could indicate that the surrounding shell may even be insulating) [27].

It is very important to keep in mind that the (macroscopic) conductivity of the polyaniline is enhanced by about two orders of magnitude without changing the chemical composition. The only changes affecting the conductivity are structural changes in the polyaniline particles. We will discuss the importance of the structure/morphology of the polymer in another paragraph in more detail.

With this process polyaniline can be produced with a great variety of different dopants (basically all known acids) and – depending on the process – be dispersed in many different solvents. The combination of dopants, solvents, and process conditions chosen for a specific product define the properties of the polymer

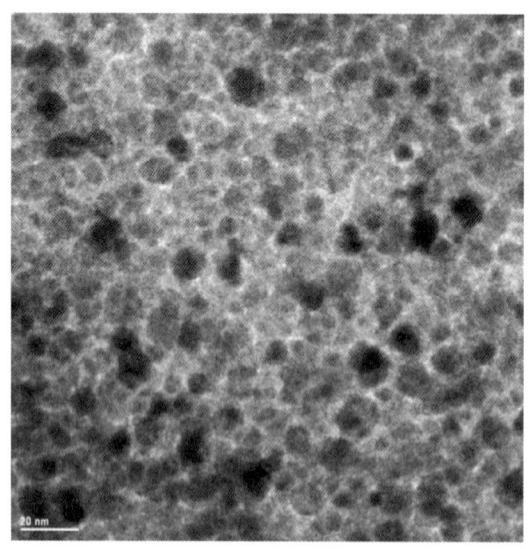

Figure 8.4 TEM of a polyaniline film.[2]

dispersion. One very important property for charge transfer processes with other materials is the work function. By varying the above-mentioned parameters, we can create polyaniline dispersions leading to layers with a work function in the range between 4.5 and 5.6 eV.

8.4
Polymers in Photovoltaic Devices

Polymers are used for different purposes and to different degrees in photovoltaic devices. One approach is to substitute one or more inorganic materials in a given photovoltaic device by a polymer, creating polymer/inorganic devices. Most of the research in the field of polymer photovoltaic devices is based on this approach. The advantage of the polymer/inorganic devices is to combine the high efficiencies of the inorganic device with low-cost processability of the organic moieties. The other approach is to build new devices only using organic (polymeric) materials. In literature, the separation between all organic and organic/inorganic hybrid devices is oftentimes not very clear. In most cases photovoltaic devices using at least one organic material are called "plastic solar cell."

Especially in the light of the quickly evolving the field of the ITO (indium-tin oxide) replacement (by conductive polymers), it seems to be possible to replace the metal

---

2) A free-standing polyaniline film was made from commercially available polyaniline dispersion from Ormecon. The film was cut by ultra-microtome and the resulting very thin films were investigated by TEM.

or ITO electrodes in the near future with polymer-based alternatives, creating "all-organic devices."

8.4.1
ITO Replacement

The replacement of ITO by conductive polymers is being pursued for different applications (such as transparent conductive electrode, touch panel, hole injection layer, among others) for a couple of reasons. First, the price of indium (and therefore the price of ITO more or less in parallel) has risen in the past 5 years by a factor of 10, mainly due to the increasing demand for flat panel displays [28].

Second, completely flexible films generated by conductive polymers are better suited for some applications (like touch panels) than the brittle ITO layers. It was shown that the resistance of ITO on polyethylene terephthalate (PET) substrates is strongly dependent on the bending of the substrate, after a certain radius of curvation of the ITO covered PET foil is exceeded [29]. The resistance of a PEDOT-covered foil was completely unaffected by bending in the same setup.

The third reason for ITO replacement is the work function of ITO. While the work function is in the range of −4.5–4.7 eV, depending on the preparation method and surface treatment of the ITO, a higher work function value would be desirable for efficient hole injection in organic hole transport materials (e.g., in OLEDs) [30].

The fourth disadvantage is the relatively rough surface of ITO, which causes contact problems in some device setups. Both the increase of the work function and the smoothing of the surfaces can be achieved by spin coating a layer of conductive polymer on top of the ITO.

While conductive polymers have the potential to outperform ITO in regard to the above-mentioned disadvantages of ITO, there are still some problems to solve. On the one hand, conductive polymers are also not cheap. The processing of conductive polymer dispersions can, however, be much more cost efficient than the sputtering of ITO, and therefore the whole process could be cheaper using conductive polymers.

On the other hand, the largest shortcoming of conductive polymers has not been completely overcome: the conductivity is about 10 times lower than the conductivity of ITO, and a degradation of the conductivity takes place over time. While the conductivity of ITO is about 6000 S cm^{-1}, the best commercially available conductive polymer dispersion has a value of 500 S cm^{-1}. Since the polymer dispersion has to be formulated to suit the needs of each given application, the value of 500 S cm^{-1} is usually reduced during the formulation step. However, for some applications a conductivity well below 1000 S cm^{-1} would be sufficient to replace the ITO, as will be discussed in the next paragraph.

For applications like the transparent conductive electrode, the color of the conductive polymers is a major issue. Neither of the two commercially available polymers is colorless: PEDOT is blue, while PAni is green. To make a transparent conductive electrode based on one of the polymers, the film thickness of the polymer layer has to be quite low (below 100 nm) to be sufficiently transparent. Together with the requirement of the surface resistivity, the formulated polymer dispersion needs to allow for a conductivity of 200–300 S cm^{-1}.

First products based on PEDOT [31] or PAni [32] already fulfill the requirements and are being introduced to the market for this application.

In December 2003, Fujitsu announced the development of a touch panel based on conductive polymers, which offers according to their press release a higher durability and lower production cost than the ITO-based devices [33].

In other fields conductive polymers already have replaced or are in the process of replacing ITO. Fehse and coworkers showed that the replacement of ITO in OLEDs by a conductive polymer (PEDOT/PSSH) leads to higher efficiencies for green and blue OLEDs, while the efficiency for red OLEDs was comparable to the ITO-based device [30]. For PAni, similar results have been found by Fehse and coworkers [34].

Although conductive polymers have the potential for replacing ITO in many applications, a lot of research has to be done to overcome some still existing shortcomings of the polymers such as degradation and relatively low conductivity.

8.4.2
Polymer Photovoltaic Devices

The main challenge of using polymers in photovoltaic devices is to make use of their before-mentioned advantages while trying to work around their disadvantages such as relatively low conductivity and hole mobility. Another disadvantage of polymers is their very short exciton diffusion length [35]. Therefore, the thickness of the charge transport and the sensitizing layer has to be very small.

Several different approaches to plastic solar cells have been tried so far: polymer/polymer, polymer/fullerene, polymer/perylene, and polymer/quantum dot [36–39]. A typical setup of a plastic solar cell is depicted in Figure 8.5. The device is oftentimes based on a heterojunction between an electron-donating material and an electron-accepting material. The blending of both materials into one phase-segregated mixture is a good option for shortening the travel distance for excitons [40].

Additionally, it is important for the device performance that the donor and acceptor have to be chosen in a way that the process of the photoinduced charge separation is much faster than the recombination process. An often used system uses poly(3-hexylthiophene) as donor and a fullerene derivative (PCBM) as acceptor

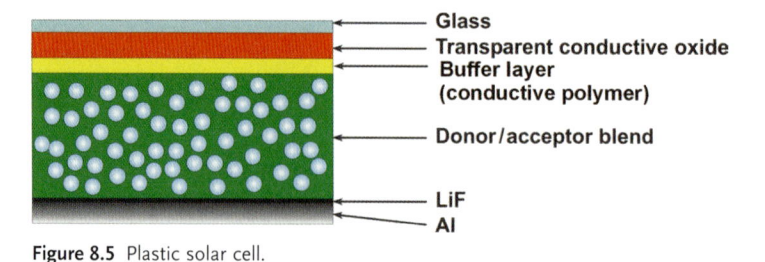

Figure 8.5 Plastic solar cell.

for photovoltaic devices. It was published that the electron transfer from the excited polymer to the fullerene derivative takes less than 50 fs, while the back electron transfer takes about 1 µs at room temperature [40,41].

A good example for the attempt to combine easy processability with an as high as possible performance is the photovoltaic device fabricated by Man et al. via layer-by-layer deposition of sulfonated polyaniline (SPAN) and a ruthenium complex containing poly(p-phenylenevinylene) (PPV) [42]. It was shown by Chan and coworkers that the combination of a ruthenium complex and PPV could work as a photosensitizer and charge transport material [43]. The sulfonated polyaniline was chosen because of its good dispersion (although the authors believed it being a "solution") properties and moderate conductivity.

The fabrication method for the photovoltaic device was purely based on dipping processes, promising a highly convenient and very well adaptable industrial application process. Only by dipping an ITO substrate into different solutions/dispersions in a defined sequence, a precise number of layers could be deposited on the substrate. These multilayer films ensured a small thickness of charge transport and sensitizing layer and therefore a short travel distance for the excitons. The films were sandwiched between an indium-tin oxide and an aluminum electrode to form photovoltaic cells. However, the power conversion efficiency of the devices was only in the range of 10^{-3}%. Although the processing is very straightforward and inexpensive, the efficiency has to be improved to be of practical use for industrial application.

Zhu and coworkers fabricated a photovoltaic device based on a DSSC, where polyaniline acts as a hole conductor [44]. The setup is shown in Figure 8.6.

The authors investigated the performance of the device using polyaniline dispersions with different conductivities, ranging from 10^{-7} to 300 S cm^{-1} as hole conductors. For the experiments they used emeraldine base in N-methylpyrrolidone (NMP) with a conductivity of $10^{-7} \text{ S cm}^{-1}$ (1), 4-dodecylbenzenesulfonic acid doped polyaniline in chloroform with a conductivity of 3.5 S cm^{-1} (2), 10-camphorsulfonic acid doped polyaniline in m-cresol, having a conductivity of 297 S cm^{-1} (3), and mixtures of (1) and (2), as well as mixtures of (2) and (3).

To rank the performance of the different polymers, the current–voltage characteristics and the incident photon to current conversion of the devices were measured. The performance of the devices was improved by using polyaniline dispersions with higher conductivity from 10^{-7} to 3.5 S cm^{-1} and dropped again by using polyaniline with conductivity over 3.5 S cm^{-1}.

SEM investigations of the surface morphology of polyaniline-coated TiO_2 films showed that the most homogeneous surface with the smallest clusters was made by the 4-dodecylbenzene sulfonic acid doped polyaniline with a conductivity

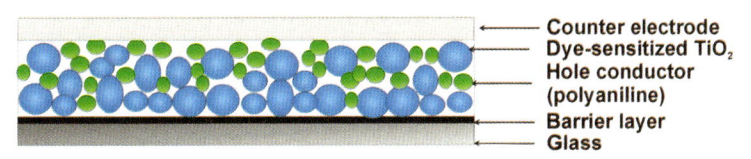

Figure 8.6 Setup of DSSC with conductive polymer.

of 3.5 S cm^{-1}. These results show clearly that the morphology of the polymer particles is more important for the device performance than the conductivity alone.

However, one should keep in mind that the authors compare three different polymers in three different solvents (even if all polymers are polyanilines with "only" different dopants) and disregard any solvent-related effects. Therefore, the results have to be carefully evaluated.

The same authors reported in a different publication about DSSCs based on titanium dioxide with polyaniline doped with 4-dodecylbenzene sulfonic acid as hole conductor blended with LiI and 4-*tert*-butylpyridine (*t*BP) [45]. The blending of the polymer with LiI and *t*BP significantly improved the photovoltaic behavior of the devices, clearly indicating an interaction of LiI and *t*BP with the polyaniline. The energy conversion efficiency improved up to 6.6 times – compared with the devices without LiI and *t*BP – to a value of 1.15%.

The authors interpret in these results that the lithium ions are responsible for an improved interfacial electron transfer, while the *t*BP supresses the dark current at the semiconductor/electrolyte junction.

Murakoshi *et al.* investigated DSSC in a comparable setup, using photodeposited polypyrrole instead of polyaniline [46]. It was shown that polypyrrole also works as a hole transport material. The energy conversion efficiency of the resulting photovoltaic device was 0.1%. Since the experimental details do not exactly match the ones from Zhu and coworkers, it is hard, in general, to compare the performance of polyaniline and polypyrrole based on these experimental results.

As previously mentioned, another use for conductive polymers is as a contact layer. Smith *et al.* used two different conductive polymers in their organic photovoltaic device for different applications [47]. A PEDOT/PSSH (polyethylenedioxythiophene/polystyrenesulfonic acid) layer was used to provide a more uniform contact between ITO and the active layer and also to improve hole injection. The active layer consisted of a conjugated, regioregular polymer (poly(3-dodecyl-2,5-thienylene vinylene)) acting as an electron donor and a C_{60} derivative acting as an electron acceptor. Both materials were dissolved in chlorobenzene and spin coated as a mixture on top of the PEDOT/PSSH layer.

The performance of the devices varied with changing ratios between the regioregular polymer and the C_{60} derivative. The reason for the performance difference was found in AFM investigations to be a rough layer surface generated by an island formation of the different materials. Both pure materials formed smooth layers. All blends formed rough films, where the roughness depended on the ratio of the two materials. The authors assume that the lack of control over the morphology is the main problem leading to the observed low power conversion efficiency of only 0.24%. The lack of control over morphology is a consequence of the inappropriate strategy of trying to work with solutions while disregarding that a solution has to undergo several nonequilibrium process steps when drying to form a dry solid layer. A dispersion – for which the morphology can be controlled by formulation and manufacturing procedures – would be the tool to choose to overcome these problems.

The assumption of Smith and coworkers that the lack of control over the morphology is the main reason for the poor performance of their devices may be completely right, as experiments from Carroll and coworkers demonstrate. They used a similar system, only changing the polythienylene vinylene for a regioregular poly(3-hexylthiophene) (P3HT) [48]. The experiments were aimed at improving the control over the morphology of the fullerene derivative mesophase by annealing of the loaded devices. This should enhance the device performance by creating more order in the nanophased material. It could be shown by HRTEM that the fullerene derivative formed crystalline structures after annealing. According to the authors, the ordered structures lead to an improved matching of electron mobility and hole mobility, enhancing the overall device performance.

The fact is that the annealing of the loaded devices led to an enhancement of the efficiency by up to 120%. The overall efficiency of the system was improved up to 5.2%.

Using the same materials, Schilinksy *et al.* showed that the molecular weight of the P3HT has a large influence on the overall device performance [49]. Reducing M_n of P3HT below 10 000 resulted in a decreased device performance. It was shown by absorption and photocurrent measurements that P3HT with low M_n formed less ordered (lamellar) structures. The presence of the fullerene derivative did not hinder the formation of these structures, which seem to be very important for the device performance.

Several companies and institutes working with polymer organic solar cells tested the performance of various buffer layers. Most of these tests were made using PEDOT–PSSH. A couple of investigations were made using other conductive polymers; among them are PAni dispersions provided by Ormecon.

The majority of the published investigations, however, were in the first test stage and only very few were made in depth. Although it is very hard to make a general conclusion based on so few and in part not comparable data, in some cases (depending on a lot of parameters such as the work function provided, the dispersion, and so on) the layers deposited from Ormecon polyaniline performed better than the simple PEDOT–PSSH dispersions [50,51]. More details are not available due to know-how protection restrictions.

8.5
The Dispersion Approach as a Productive Tool for Photoactive Layer Deposition

The importance of the morphology of the polymer on the performance was also shown by a basic study that was initiated and contributed by one of us. The main goal of the study was the investigation of the field effect mobility of a certain polymer dependent on the solvent used, resulting either in a solution or in a dispersion of the polymer in the solvent [52]. The polymer poly(2-methoxy-5-(3′,7′-dimethyloctyloxy)-1,4-phenylenevinylene) (MDMO-PPV) was used as the organic semiconducting layer in a field effect transistor (FET) setup (Figure 8.7).

A highly p-doped Si substrate was used as a gate electrode. About 200 nm insulating oxide was thermally grown on one side of the substrate, while the other side was covered with an Al layer as gate connection. Source and drain electrodes

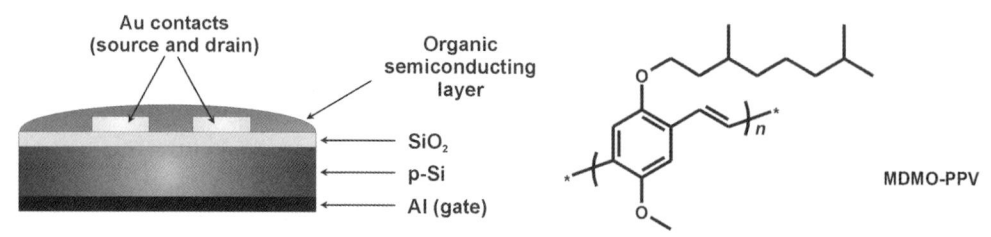

Figure 8.7 FET setup and chemical structure of polymeric organic semiconductor.

were formed by interdigitating fingers of TiW/Au. A layer of the organic semiconductor – either in toluene or chlorobenzene – was spin coated on the substrate.

Toluene and chlorobenzene were chosen as solvents to investigate the differences in the polymer properties using on the one hand toluene as a "good" solvent for MDMO-PPV and on the other hand chlorobenzene as a "bad" solvent for MDMO-PPV, most probably resulting in different interchain interactions and therefore different morphologies of the polymer. In an earlier publication, it was shown by investigation of photoluminescence spectra of spin-cast films that the degree of interchain interactions in MDMO-PPV is dependent on the used solvent [53].

Light-scattering measurements of MDMO-PPV in toluene and chlorobenzene were taken in three different concentrations to compare the aggregation of the polymer in both solvents. The measured parameter was the particle size.

One difference between solutions and dispersions of polymers is the behavior of the particle size upon dilution of the solution/dispersion. The dilution of solutions leads to an increase in particle size, while the particle size is not affected by dilution in the case of dispersions.

In toluene, the particle size of MDMO-PPV (number distribution) changes upon dilution from 35 nm over 6 nm to more than 6 μm at the lowest concentration. While at the highest concentration two fractions can be distinguished; one of the fractions decreases after the first and disappears after the second dilution, clearly indicating a solution process. The particle size of MDMO-PPV (number distribution) in chlorobenzene changes from 66 to 8 nm with decreasing concentration. This is an indication that MDMO-PPV forms colloidal systems at least over a wide concentration range.

The field effect hole mobilities – calculated from the saturation regime of the source–drain current – of the MDMO-PPV spin cast from toluene (5×10^{-6} cm^2 V^{-1} s^{-1}) were much lower than the mobility gained by spin coating MDMO-PPV from chlorobenzene (3×10^{-5} cm^2 V^{-1} s^{-1}). On the basis of the results of Nguyen and coworkers, who investigated the morphology and interchain interactions of a conjugated polymer poly(2-methoxy-5-(2'-ethyl-hexyloxy)-1,4-phenylene vinylene) (MEH-PPV) in different solvents before and after spin coating, we can assume that the morphology of and molecular arrangement within the particles in the dispersion is not changed dramatically during the spin coating process [54]. Therefore, we can assume that the degree of interchain interaction of MDMO-PPV, which was spin cast from chlorobenzene (as a dispersion), is higher than the MDMO-PPV spin cast from toluene (as a solution).

These experiments show that the performance of a certain polymer layer in an FET device is dependent on whether it was deposited from solution or dispersion. More generally, some important properties of the polymer like charge transfer are dependent on the morphology of the polymer, which is crucially dependent on the processing history. There is no doubt that the morphology can be better controlled in dispersions and for deposits made from dispersions. As these also showed the higher mobility in the experiments described above, it is promising to investigate this dispersion approach more than that has been previously done, and shift the focus from trying to achieve solutions to manufacturing controlled dispersions.

The results of Shi *et al.* who found a strong dependence of the quantum efficiency of polymer LEDs on the morphology of the used polymer – in this case MEH-PPV – also strongly point in the same direction [55].

In many other reports the importance of a control of the polymer morphology is also stressed, however, without pointing at the need to use dispersions rather than solutions for the control of morphology. Romano and coworkers investigated the charge carrier mobility in PPVs and concluded that the relatively low mobility of $0.1\,\mathrm{cm}^2\,\mathrm{V}^{-1}\,\mathrm{s}^{-1}$ could be easily enhanced by at least a factor of 10 by a better control over the chemical purity and especially the structure of the polymer [56].

8.6
Discussion of the Influence of Polymer Morphology on Device Performance

As described in the previous paragraphs, many experimental evidences show that important properties of (conductive) polymers, such as UV–Vis spectra, conductivity, and so on, strongly depend on the morphology of the polymers. Since these properties are very relevant for the performance of assembled devices such as polymer LEDs or organic photovoltaic devices, it is a key point to understand the relationship between polymer morphology and physical properties of the polymer. Only a deep understanding of this relationship enables us to tailor the properties of conductive polymers to suit the needs for high-performance organic photovoltaic devices.

It is important to focus both on the synthesis of the polymer and on the dispersion process that defines the polymer morphology. The combination of both processes enables us to test the full potential of each given polymer. In this light, large parts of the literature tend to deal only with the synthesis of a polymer–dopant system and define the properties of this system by measuring one dispersion in one given solvent. The results of these experiments do *not* represent the properties of the polymer–dopant system but only a small facet of it.

8.7
Summary

In the last few years, a great variety of different polymers have been tested in photovoltaic devices, among them polyaniline. Although polyaniline-based devices did not generally outperform devices based on other polymers, polyaniline has the advantage of being useful for different purposes. Due to its good dispersability, it can penetrate porous structures and be used as hole-conducting material and solid electrolyte in DSSCs. For similar reasons dispersions of polyaniline form relatively smooth films and can act as a contact layer to enhance contact between other two layers. The decent conductivity of polyaniline also allows for its use as hole-conducting material. Finally, the work function of polyaniline dispersions can be adjusted over a wide range (-4.3 to -5.6 eV), giving the opportunity to influence the charge transfer processes between polyaniline and other materials.

Although the "classic" silicon-based solar cells by far outperform the polymer solar cells in terms of power conversion efficiency, polymers in photovoltaic devices have certain advantages such as low costs, good processability, and allowance for the production of large-area and flexible devices. In the near future, the combination of inorganic and polymeric materials in one device could prove to be a real competitor to the "classic" solar cell. The all-organic solar cell is the choice device for special applications, where flexible and/or large-area devices are necessary. Besides that, the efficiencies achieved today are too low to be a good alternative to the other devices unless there are major advantages, especially in the mobility values of the used polymer.

To enhance the performance of the polymer photovoltaic devices, the energy conversion efficiency has to be improved drastically. As was shown in previous chapters, the control over the morphology of the polymer is probably one of the key issues. It is at least as important (probably more) as the choice of the polymer itself and requires a good understanding of the material properties and the dispersion process.

References

1 http://www.solarserver.de.
2 O'Reagan, B. and Grätzel, M. (1991) *Nature*, **353**, 737.
3 Kajihara, K., Tanaka, K., Hirao, K. and Soga, N. (1997) *Japanese Journal of Applied Physics*, **56**, 5537.
4 van Hal, P.A., Christiaans, M.P.T., Wienk, M.M., Kroon, J.M. and Janssen, R.A.J. (1999) *Journal of Physical Chemistry B*, **103**, 4352.
5 Murakoshi, K., Kogure, R., Wada, Y. and Yanagida, S. (1997) *Chemistry Letters*, **26** (5), 471.
6 Tang, C.W. (1986) *Applied Physics Letters*, **48**, 183.
7 Hagemann, O., Jørgensen, M. and Krebs, F.C. (2006) *Journal of Organic Chemistry*, **71**, 5546.
8 Ma, W.L., Yang, Y.C., Gong, X., Lee, K. and Heeger, A.J. (2005) *Advanced Functional Materials*, **15**, 1617.
9 Yakuphanoglu, F., Aydin, M.E. and Kilicoglu, T. (2006) *Journal of Physical Chemistry B*, **110**, 9782.
10 http://www.spectrolab.com.
11 Shirakawa, H., Louis, E., MacDiarmid, A., Chiang, C. and Heeger, A. (1977) *Chemical Communications*, 578.
12 Wessling, B. (1999) *Handbook of Nanostructured Materials and Nanotechnology*, vol. 5 (ed. H.S. Nalwa), Academic Press, p. 525.
13 Wessling, B. (1998) *Handbook of Conducting Polymers* (eds T. Skotheim, R.L. Elsenbaumer and J.R. Reynolds), Dekker, New York.

14 Wessling, B. (1997) *Handbook of Organic Conductive Molecules and Polymers*, vol. 3 (ed. H.S. Nalwa), John Wiley & Sons, Inc., Chichester, p. 497.
15 Wessling, B. (1991) *Synthetic Metals*, 45, 119.
16 Lee, K., Cho, S., Park, S.H., Heeger, A.J., Lee, C.-W. and Lee, S.-H. (2006) *Nature*, 44, 65.
17 Lee, S.-H., Lee, D.-H., Lee, K. and Lee, C.-W. (2005) *Advanced Functional Materials*, 15, 1495.
18 Winther-Jensen, B. and West, K. (2004) *Macromolecules*, 37, 4538.
19 Akagi, K., Katayama, S., Ito, M., Shirakawa, H. and Araya, K. (1989) *Synthetic Metals*, 28, P D51.
20 Coustel, N., Foxonet, N., Ribet, J.L., Bernier, P. and Fischer, J.E. (1991) *Macromolecules*, 24, 5867.
21 Wessling, B. (1984) US-PS. 4935164, EPC-0168620, Patent No. 85107027.6, Zipperling Kessler and Co.
22 Stejskal, J. (2001) *Journal of Polymeric Materials*, 18, 255.
23 Beadle, P.M., Nicolau, Y.F., Banka, E., Rannou, P. and Djurado, D. (1998) *Synthetic Metals*, 95, 29.
24 Wessling, B., Srinivasan, D., Rangarajan, G., Mietzner, T. and Lennartz, W. (2000) *European Physical Journal*, E2, 207.
25 Subramaniam, C.K., Kaiser, A.B., Gilberd, P.W. and Wessling, B. (1993) *Journal of Polymer Science Part B: Polymer Physics*, 31, 1425.
26 Pelster, R., Nimtz, G. and Wessling, B. (1994) *Physical Review B: Condensed Matter*, 49, 12718.
27 Krinichnyi, V.I., Tokarev, S.V., Roth, H.-K., Schrödner, M. and Wessling, B. (2006) *Synthetic Metals*, 156, 1368.
28 Walzer, K., Maennig, B., Pfeiffer, M. and Leo, K. (2007) *Chemical Reviews*, 107, 1233.
29 Paetzold, R., Heuser, K., Henseler, D., Roeger, S., Wittmann, G. and Winnaker, A. (2003) *Applied Physics Letters*, 82, 3342.
30 Fehse, K., Walzer, K., Leo, K., Lövenich, W. and Elschner, A. (2007) *Advanced Materials*, 19, 441.
31 www.dupont.com.
32 www.nissanchem.co.jp.
33 http://www.fujitsu.com.
34 Fehse, K., personal communication.
35 Man, K.Y.K., Wong, H.L., Chan, W.K., Kwong, C.Y. and Djurisic, A.B. (2004) *Chemistry of Materials*, 16, 3, 365.
36 Huynh, W.U., Dittmer, J.J. and Alivisatos, A.P. (2002) *Science*, 295, 2425.
37 Brabec, C.J., Sariciftci, N.S. and Hummelen, J.C. (2001) *Advanced Functional Materials*, 15, 11.
38 Breeze, A.J., Salomon, A., Ginley, D.S., Gregg, B.A., Tillmann, H. and Hörhold, H.-H. (2002) *Applied Physics Letters*, 81, 3085.
39 Yakimov, A. and Forrest, S.R. (2002) *Applied Physics Letters*, 80, 1667.
40 Shaheen, S., Brown, K., Miedaner, A., Curtis, C., Parilla, P., Gregg, B. and Ginley, D. (2003) PROC Conference Paper for National Center for Photovoltaics and Solar Program Review Meeting.
41 Brabec, C.J., Zerza, G., Cerullo, G., De Silvestri, S., Luzzati, S., Hummelen, J.C. and Sariciftci, S. (2001) *Chemical Physics Letters*, 340, 232.
42 Man, K.Y.K., Wong, H.L., Chan, W.K., Djurisic, A.B., Beach, E. and Rozeveld, S. (2006) *Langmuir*, 22, 3368.
43 Chan, W.K., Gong, X. and Ng, W.Y. (1997) *Applied Physics Letters*, 71, 2919.
44 Tan, S., Zhai, J., Xue, B., Wan, M., Meng, Q., Jiang, L. and Zhu, D. (2004) *Langmuir*, 20, 2937.
45 Tan, S., Zhai, J., Wan, M., Meng, Q., Li, Y., Jiang, L. and Zhu, D. (2004) *Journal of Physical Chemistry B*, 108, 18693.
46 Murakoshi, K., Kogure, R., Wada, Y. and Yanagida, S. (1997) *Chemistry Letters*, 26 (5), 471.
47 Smith, A.P., Smith, R.R., Taylor, B.E. and Durstock, M.F. (2004) *Chemistry of Materials*, 16, 4687.
48 Reyes-Reyes, M., Kim, K., Dewald, J., Lopez-Sandoval, R., Avadhanula, A., Curran, S. and Carrol, D.L. (2005) *Organic Letters*, 7 (26), 5749.
49 Schilinsky, P., Asawapirom, U., Scherf, U., Biele, M. and Brabec, C.J. (2005) *Chemistry of Materials*, 17, 2175.

50 Sariciftci, N.S., personal communication.
51 Brabec, C.J., personal communication.
52 Geens, W., Shaheen, S.E., Wessling, B., Brabec, C.J., Poortmans, J. and Sariciftci, N.S. (2002) *Organic Electronics*, **3**, 105.
53 Nguyen, T., Kwong, R.C., Thompso, M.E. and Schwartz, B.J. (2000) *Applied Physics Letters*, **76**, 2454.
54 Nguyen, T.-Q., Doan, V. and Schwartz, B.J. (1999) *Journal of Chemical Physics*, **110**, 4068.
55 Shi, Y., Liu, J. and Yang, Y. (2000) *Journal of Applied Physics*, **87** (9), 4254.
56 Hoofman, R.J.O.M., de Haas, M.P., Siebbeles, L.D.A. and Warman, J.M. (1998) *Nature*, **392**, 54.

II
Device Physics

A
Overview of the State-of-the-Art

9
Titanium Oxide Films as Multifunctional Components in Bulk Heterojunction "Plastic" Solar Cells

Kwanghee Lee, Jin Young Kim, and Alan J. Heeger

9.1
Introduction

Bulk heterojunction "plastic" solar cells are based on phase-separated blends of polymer semiconductors and fullerene derivatives [1–6]. Because of the self-assembly on the nanometer length scale, excitons formed after absorption of solar irradiation diffuse to a heterojunction prior to their recombination and are dissociated at the polymer/fullerene interface. Ultrafast charge transfer from semiconducting polymers to fullerenes guarantees that the quantum efficiency (QE) for charge transfer (CT) at the interface approaches unity [7–10], with electrons on the fullerene network and holes on the polymer network. After breaking the symmetry by using different metals for the two electrodes, electrons migrate toward the lower work function metal and holes migrate toward the higher work function metal. Carrier recombination prior to reaching the electrodes and low mobility limit both the device fill factor (FF) and the overall photon harvesting by reducing the optimum active layer thickness [11,12]. The carrier lifetime is largely controlled by the phase morphology between the donor and acceptor materials [6,10,13].

In this short review, we summarize the recent progress enabled by including one or more layers of titanium suboxide, TiO_x, into the architecture of bulk heterojunction solar cells. The use of TiO_x provides several important opportunities; specifically, the TiO_x layers enable higher performance from single cells, longer lifetime as a result of reduced sensitivity to oxygen and water vapor, and the fabrication of bulk heterojunction cells in the "tandem cell" architecture, a multilayer structure that is equivalent to two photovoltaic cells in series.

9.2
Sol–Gel Processed Titanium Oxide as an Optical Spacer in Polymer Solar Cells [14]

In the standard bulk heterojunction solar cell structure, the intensity of the light is zero at the metallic (e.g., Al) back electrode because of optical interference between

Organic Photovoltaics: Materials, Device Physics, and Manufacturing Technologies.
Edited by Christoph Brabec, Vladimir Dyakonov, and Ullrich Scherf
Copyright © 2008 WILEY-VCH Verlag GmbH & Co. KGaA, Weinheim
ISBN: 978-3-527-31675-5

the incident light (from the ITO side) and the back-reflected light [15–17]. Thus, as sketched in Figure 9.1a, a relatively large fraction of the active charge-separating layer is in a dead zone in which the photogeneration of carriers is significantly reduced. Moreover, this effect causes a larger fraction of electron–hole pairs to be produced near the ITO/PEDOT:PSS (poly-3,4-ethylenedioxythiophene:polystyrene sulfonic acid) electrode, a distribution known to reduce the photovoltaic conversion efficiency [18,19]. This optical interference effect is especially important for thin-film structures where layer thicknesses are comparable with the absorption depth and the wavelength of the incident light, as is the case with photovoltaic cells fabricated from semiconducting polymers.

To overcome these problems, one might simply increase the thickness of the active layer to absorb more light. Because of the low mobility of the charge carriers in the polymer/C_{60} composites, however, the increased internal resistance of thicker films will inevitably lead to a reduced fill factor.

An alternative approach is to change the device architecture with the goal of spatially redistributing the light intensity inside the device by introducing an optical spacer between the active layer and the Al electrode as sketched in Figure 9.1a [17]. Although this revised architecture would appear to solve the problem, the prerequisites for an ideal optical spacer limit the choice of materials; for instance, the layer must be a good acceptor and an electron transport material with a conduction band edge lower in energy than that of the lowest unoccupied molecular orbital (LUMO) of C_{60}, the conduction band edge must be above (or close to) the Fermi energy of the collecting metal electrode, and the material must be transparent to light with wavelengths within the solar spectrum.

Titanium dioxide (TiO_2) is a promising candidate as an electron acceptor and transport material as confirmed by its use in dye-sensitized Grätzel cells [20,21], hybrid polymer/TiO_2 cells [22–24], and multilayer Cu-phthalocyanine/dye/TiO_2 cells [17,25]. Typically, however, crystalline TiO_2 is used either in the anatase phase or the rutile phase, both of which require treatment at temperatures ($T > 450\,°C$) that are inconsistent with the device architecture shown in Figure 9.1; the polymer/C_{60} composite cannot survive such high temperatures.

As demonstrated in Ref. [14], we have used a solution-based sol–gel process to fabricate a titanium oxide (TiO_x) layer on top of the polymer–fullerene active layer (Figure 9.1b). By introducing the TiO_x optical spacer, polymer photovoltaic cells with power conversion efficiencies (PCEs) can be increased by approximately 50% compared to similar devices fabricated without the optical spacer.

Dense TiO_x films were prepared using a TiO_x precursor solution, as described in detail elsewhere [14]. The precursor solution was spin cast in air on top of the polymer–fullerene composite layer. The sample was then heated under vacuum at 80–90 °C for 10 min during which the precursor converted to the TiO_x layer via hydrolysis. The resulting TiO_x films were transparent and smooth with surface features smaller than few nanometer [14]. Because the titanium oxide layer was treated at temperatures below 100 °C, the film was amorphous as confirmed by X-ray diffraction (XRD). The typical XRD peaks of the anatase crystalline form appear only after sintering the spin-cast films at 500 °C for 2 h. Analysis by X-ray photoelectron

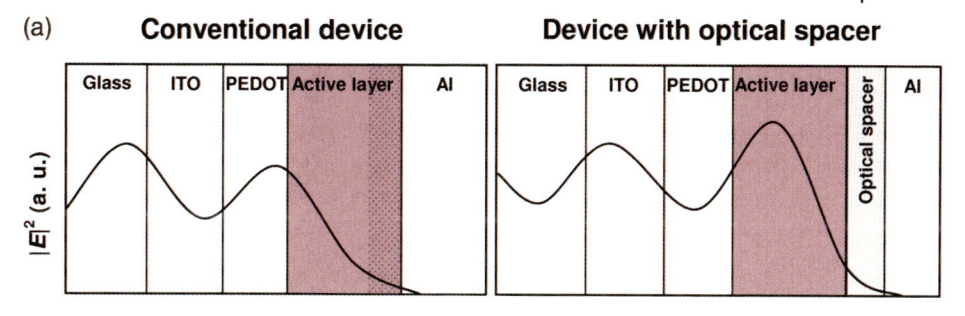

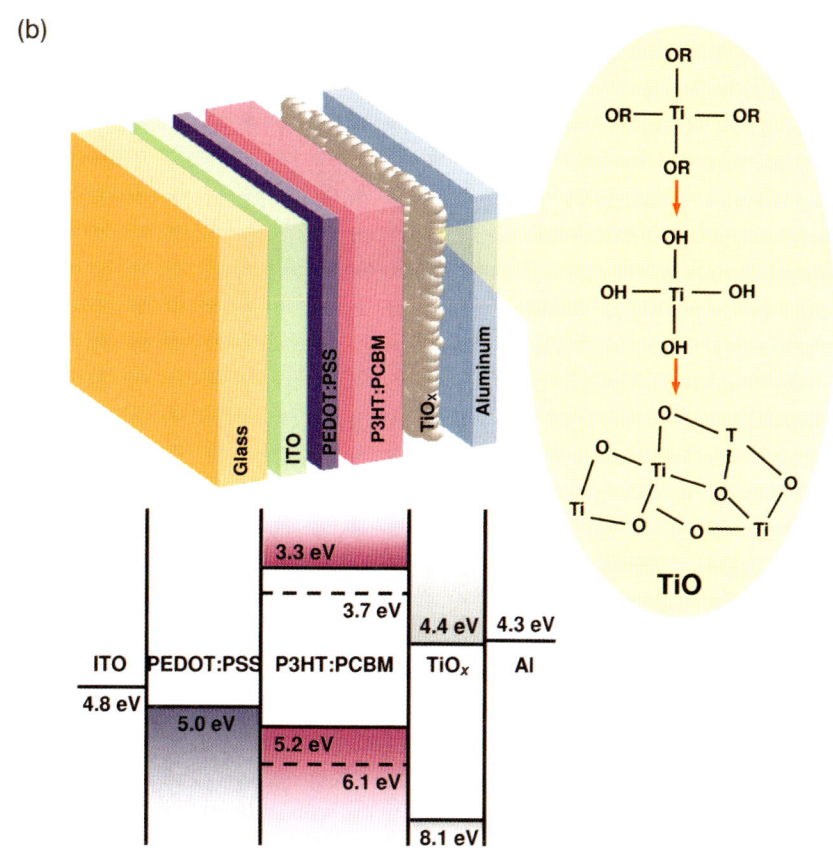

9.2 Sol–Gel Processed Titanium Oxide as an Optical Spacer in Polymer Solar Cells

Figure 9.1 (a) Distribution of the squared optical electric field strength $|E|^2$ inside the devices with a structure of ITO/PEDOT/active layer/Al (left) and ITO/PEDOT/active layer/optical spacer/Al. The dot region in the left figure denotes the dead zone as explained in the text. (b) Schematic illustration of the thin-film solar cell. The device consists of a P3HT:PCBM active layer sandwiched between an Al electrode and a transparent ITO electrode coated with PEDOT:PSS. The TiO$_x$ optical spacer layer is inserted between the active layer and an Al electrode. Also shown is a brief flow chart of the steps involved in the preparation of TiO$_x$ layer. The energy level of the single components of photovoltaic cell is also shown at the bottom. The TiO$_x$ optical spacer layer exhibits excellent band matching for cascading charge transfer.

spectroscopy (XPS) reveals an oxygen deficiency at the surface of the thin-film samples with Ti:O ratio as 42.1 : 56.4 (% ratio); hence we designate the composition as TiO$_x$.

In spite of the amorphous nature of the TiO$_x$ layer, the physical properties are excellent. The absorption spectrum of the film shows a well-defined absorption edge at $E_g \approx 3.7$ eV. Using optical absorption and cyclic voltammetry (CV) data, the energies of the bottom of the conduction band (LUMO) and the top of the valence band (highest occupied molecular orbital, HOMO) of the TiO$_x$ material were determined; see Figure 9.1b. This energy level diagram demonstrates that the TiO$_x$ layer satisfies the electronic structure requirements of the optical spacer.

Utilizing this TiO$_x$ layer as the optical spacer, donor–acceptor composite photovoltaic cells were fabricated using the phase-separated "bulk heterojunction" material comprising poly(3-hexylthiophene) (P3HT) as the electron donor and the fullerene derivative, [6,6]-phenyl-C$_{61}$ butyric acid methyl ester (PCBM) as the acceptor. The device structure is shown in Figure 9.1b.

Figure 9.2a compares the incident photon to current collection efficiency spectrum (IPCE) of devices fabricated with and without the TiO$_x$ optical spacer. The IPCE is defined as the number of photogenerated charge carriers contributing to the photocurrent per incident photon. The conventional device (without the TiO$_x$ layer) shows the typical spectral response of the P3HT:PCBM composites with a maximum IPCE of ~60% at 500 nm, consistent with previous studies [26–29]. For the device with the TiO$_x$ optical spacer, the results demonstrate substantial enhancement in the IPCE over the entire excitation spectral range; the maximum reaches almost 90% at 500 nm, corresponding to a 50% increase in IPCE. Since surface reflection from the glass is approximately 10%, the optical spacer enhances the efficiency to >90% for light incident with wavelengths in the range of 450–550 nm.

This enhancement is the result of increased absorption in the bulk heterojunction layer as a result of the TiO$_x$ optical spacer; the increased photogeneration of charge carriers results from the spatial redistribution of the light intensity. To clarify the role of the TiO$_x$ layer, the reflectance spectrum was measured from a "device" with glass/P3HT:PCBM/TiO$_x$/Al geometry using a glass/P3HT:PCBM/Al "device" as the reference (the P3HT:PCBM composite film thickness was approximately 100 nm in both). The ITO/PEDOT layers were omitted to avoid any complication arising from the conducting layers. Since the two "devices" are identical except for TiO$_x$ optical spacer layer, comparison of the reflectance yields information on the additional absorption, $\Delta\alpha(\omega)$, in the P3HT:PCBM composite film as a result of the spatial redistribution of the light intensity by the TiO$_x$ layer [30]

$$\Delta\alpha(\omega) \approx -\left(\frac{1}{2}d\right)\ln\left[\frac{I'(\omega)}{I(\omega)}\right], \tag{9.1}$$

where $I'(\omega)$ is the intensity of the reflected light from the device with the optical spacer and $I(\omega)$ is the intensity of the reflected light from an identical device with the optical spacer (d is the thickness of the P3HT:PCBM composite layer). The data demonstrate a clear increase in absorption over the entire spectrum. Moreover, since the spectral features of the P3HT:PCBM absorption are evident in both spectra, the increased absorption arises from a better match of the spatial distribution of the light intensity

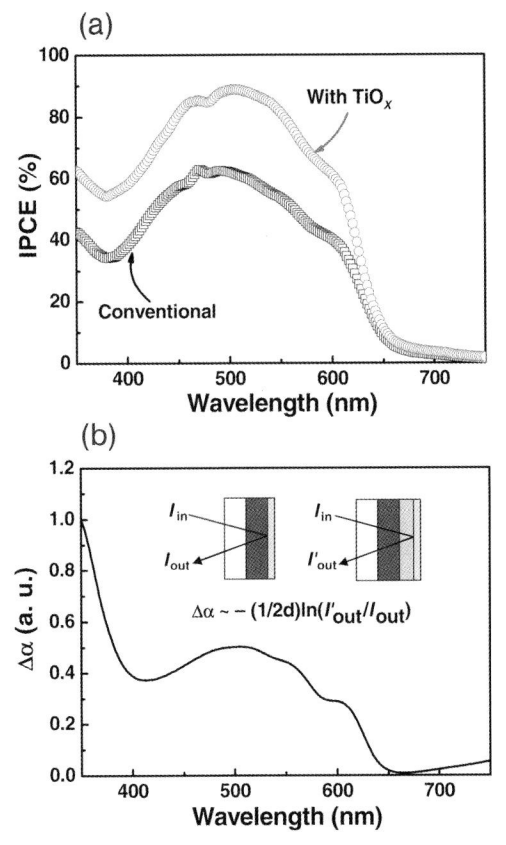

Figure 9.2 (a) The external quantum efficiency (EQE; same as the incident monochromatic photon to current collection efficiency, IPCE) spectra are compared for the two devices with and without TiO$_x$ optical spacer layer. (b) The change in the absorption spectrum resulting from the addition of the optical spacer is shown as the red circles; $\Delta\alpha(\omega) \approx -(1/2d)\ln[I'(\omega)/I(\omega)]$, where $I'(\omega)$ is the intensity of the reflected light from the device with the optical spacer and $I(\omega)$ is the intensity of the reflected light from an identical device with the optical spacer (d is the thickness of the P3HT:PCBM composite layer). The black squares represent $\alpha(\omega)$ as obtained using an aluminum mirror as the reference. The inset is a schematic description of the optical beam path in the samples.

to the position of the P3HT:PCBM composite film. These data demonstrated that the higher absorption is caused by the TiO$_x$ layer that functions as an optical spacer (as sketched in Figure 9.1a). As a result, the TiO$_x$ optical spacer increases the number of carriers per incident photon collected at the electrodes.

As shown in Figure 9.3, the enhancement in the device efficiency that results from the optical spacer can be directly observed in the current density versus voltage (J–V) characteristics under monochromatic illumination (Figure 9.3a) and under AM 1.5 simulated solar radiation (Figure 9.3b). The conventional device (without the TiO$_x$ layer) shows typical photovoltaic response under 25 mW cm^{-2} at 532 nm; the short-circuit current (I_{sc}) = 8.4 mA cm^{-2}, open-circuit voltage (V_{oc}) = 0.6 V, and the FF

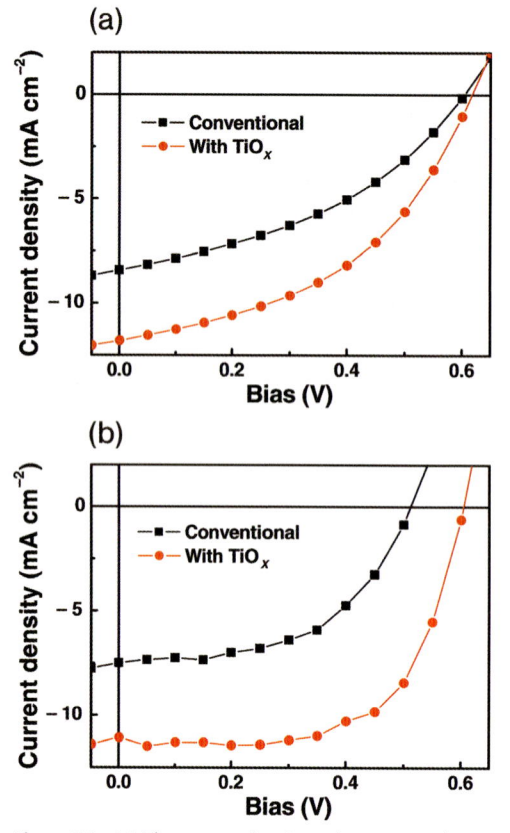

Figure 9.3 (a) The current density–voltage (J–V) characteristics of polymer solar cells with and without TiO$_x$ optical spacer illuminated with 25 mW cm^{-2} at 532 nm. (b) The current density–voltage (J–V) characteristics of polymer solar cells with and without TiO$_x$ optical spacer under AM 1.5 illumination from a calibrated solar simulator with an intensity of 90 mW cm^{-2}.

0.40. These values correspond to a power conversion efficiency of $\eta_e = 8.1\%$. For the device with the TiO$_x$ layer, the data demonstrate substantially improved device performance; I_{sc} increases 11.8 mA cm^{-2}, FF increases slightly to 0.45, while V_{oc} remains at 0.6 V. The corresponding power conversion efficiency (η_e) = 12.6%, which corresponds to ∼50% increase in the device efficiency, consistent with the IPCE measurements.

The corresponding data obtained under AM 1.5 illumination from a calibrated solar simulator with irradiation intensity of 90 mW cm^{-2} are shown in Figure 9.3b. The device without the TiO$_x$ layer again shows typical photovoltaic response with device performance comparable with the one reported in previous studies. For the device with the TiO$_x$ layer, the data again demonstrate substantially improved device performance; the corresponding power conversion efficiency increases to ∼5%.

9.3
Air-Stable Bulk Heterojunction Polymer Solar Cells

The degradation of devices fabricated from semiconducing polymers (e.g., polymer LEDs and bulk heterojunction solar cells) can be eliminated, or at least reduced to acceptable levels, by sealing the components inside an impermeable package using glass and/or metal [31–33]. Attempts to create flexible packaging using hybrid multilayer barriers comprised of inorganic oxide layers separated by polymer layers with total thickness of 5–7 µm have been reported with promising results [34–36]. Although such encapsulation methods can prevent (or at least reduce) oxygen and moisture permeation, they complicate the fabrication process and also result in increased thickness and loss of flexibility. To achieve the goal of printed "plastic" electronics, either the development of improved barrier materials for packaging or the development of devices with reduced sensitivity (or both) are required. Thus, the creation of new methods for enhancing device lifetime is an important goal that must be accomplished without interfering with the principal "flexible device" concepts; simple fabrication by solution processing, flexibility, and thin form factor.

Using the solution-based sol–gel process, described above as a collector and optical spacer for bulk heterojunction solar cells [14], the device lifetime, unpackaged and in air, is significantly enhanced. The device architecture is again that shown in Figure 9.1b.

To explore the barrier and oxygen scavenging effects of the TiO_x layer, the photoluminescence (PL) stability of polyfluorene (PF) was investigated with and without a protecting layer of TiO_x. Four kinds of films were studied with the following structures: glass/PF, glass/TiO_x/PF, glass/PF/TiO_x, and glass/TiO_x/PF/TiO_x; all were prepared by spin casting. The films were then heated for 15 h at 150 °C in air. The PF type materials typically degrade with the appearance of a long-wavelength emission (around 500–600 nm) after heating in air [37]. This longer wavelength emission arises from oxygen present in the luminescent polymer and the formation of keto defects.

The PL spectra of these films are shown in Figure 9.4a. The initial PL spectra of all the films are typical of PF without any peak in the region of 500–600 nm. The initial PL color was pure blue. When the films were heated for 15 h at 150 °C in air, the PF film without the TiO_x layer developed a pronounced peak in the PL emission spectrum in the 500–600 nm region; the emission color changes from blue to green. For the PF films covered by the TiO_x layer (glass/PF/TiO_x and glass/TiO_x/PF/TiO_x), the PL peak in the 500–600 nm spectral range was significantly reduced (almost completely eliminated); the emission color remained blue. Note that the TiO_x layer provides some benefit even when it is beneath the PF (i.e., glass/TiO_x/PF); the green emission peak is smaller than that emitted from the glass/PF film, demonstrating that the TiO_x layers play an additional oxygen-scavenging role as well as providing an oxygen barrier.

A direct evidence of the oxygen barrier and scavenging effects comes from XPS measurements comparing the oxygen concentration inside the PF in the glass/polymer and glass/polymer/TiO_x samples. After heating the samples for 48 h at

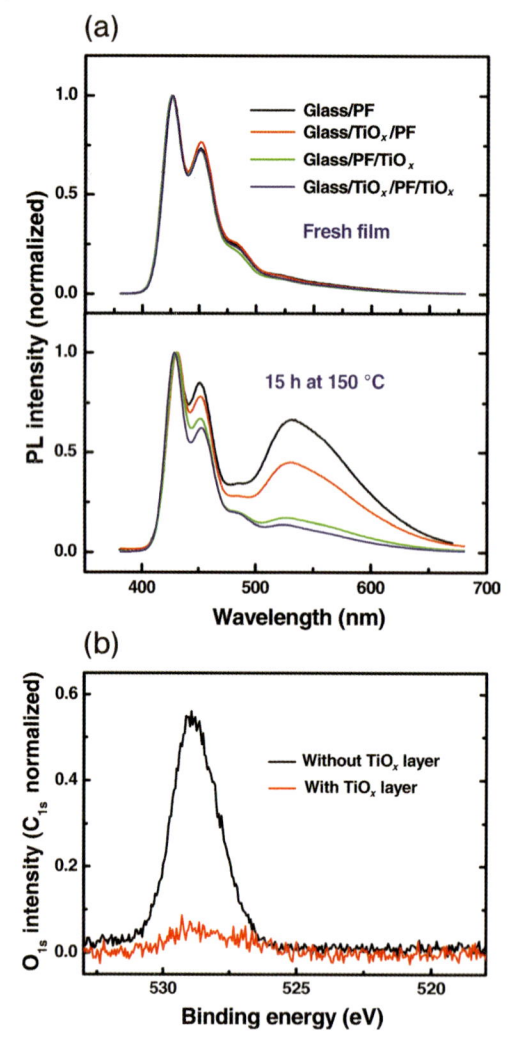

Figure 9.4 Spectroscopic evidence of the barrier and oxygen scavenging effects of the TiO$_x$ layer. (a) Comparison of the photoluminescence spectra of PF films in combination with the TiO$_x$ layers before and after annealing for 15 h in air. Four kinds of films are studied with the following structures: glass/PF (red line), glass/TiO$_x$/PF (green line), glass/PF/TiO$_x$ (blue line), glass/TiO$_x$/PF/TiO$_x$ (black line). (b) The relative ratios of O$_{1s}$/C$_{1s}$ in the polymers in the glass/polymer (black line) and glass/polymer/TiO$_x$ (red line) samples are compared after being normalized by C$_{1s}$ peak at 218 eV as measured by XPS. After annealing the samples for 48 h at 150 °C in air, the TiO$_x$ layers were removed by using the XPS depth profiling technique to directly compare the oxygen concentration inside the PF in the samples.

150 °C in air, the TiO$_x$ layer was removed by using the XPS depth profiling technique. The measured polymer layers of both samples were etched to a depth of around 10 nm to remove any surface oxygen. Figure 9.4b shows the relative ratio of O$_{1s}$/C$_{1s}$ inside the polymers without and with the TiO$_x$ layer. The polymer without the TiO$_x$

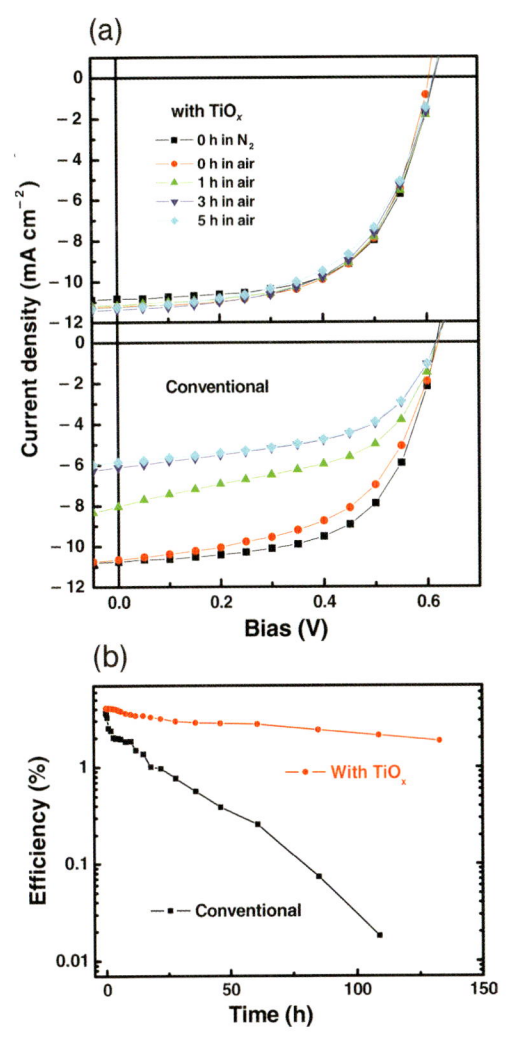

Figure 9.5 Device performance as a function of storage time for polymer solar cells. (a) The current density–voltage (J–V) characteristics of polymer solar cells with and without the TiO$_x$ layer measured after various storage periods in the air under AM 1.5 illumination from a calibrated solar simulator with an intensity of 100 mW cm^{-2}. (b) Comparison of the power conversion efficiencies as a function of storage time for polymer solar cells with and without the TiO$_x$ layer. Note that the characteristics of the devices were monitored with increasing storage time for the same devices.

layer shows a high intensity peak with an asymmetric feature, whereas this signal is hardly detectable in the polymer covered with TiO$_x$.

The TiO$_x$ layer improves the lifetime of polymer-based solar cells. Figure 9.5 compares the current density versus voltage (J–V) characteristics and the efficiency versus storage time of a photovoltaic cell with and without the TiO$_x$ layer (no external packaging or encapsulation were used). The conventional device shows typical photovoltaic response with device performance comparable with that reported in previous study [13]; the $I_{sc} = 10.7$ mA cm^{-2}, $V_{oc} = 0.62$, and the FF = 0.60. These values correspond to a power conversion efficiency ($\eta_e = I_{sc}V_{oc}FF/P_{inc}$, where P_{inc} is the intensity of incident light) of $\eta_e = 4.0\%$. When these conventional devices were stored in the ambient air, a dramatic decrease in I_{sc} was observed as the storage time increased. Note, however, that the V_{oc} remained almost constant at 0.62 V, indicating that the devices still functioned properly without catastrophic failure.

For the device with the TiO$_x$ layer, the initial performance was comparable with those of the conventional devices without the TiO$_x$ layer; $I_{sc} = 10.8$ mA cm^{-2}, V_{oc} 0.62 V, FF = 0.61, yielding $\eta_e = 4.1\%$. Note, however, that the conventional devices were fabricated by using postproduction heat treatment at 150 °C to improve the efficiency [13], whereas the devices with the TiO$_x$ layer were prepared by the conventional preheat-treatment method [14]. As a result, the initial performances of the two devices were (accidentally) almost identical. The devices with the TiO$_x$ layer exhibited significantly better air stability compared to the conventional device without the TiO$_x$ layer. The lifetime in air of the polymer solar cells was enhanced by the TiO$_x$ layer by two orders of magnitude.

The achievement of "air-stable" polymer devices implies that the promise of high performance "flexible" and "printable" devices can be realized with significantly enhanced device lifetimes. Since the TiO$_x$ layer is generated by a sol–gel process, all fabrication steps are carried out by processing the component materials from solution; that is, consistent with printing and coating technologies. The TiO$_x$ layer reduces the sensitivity to oxygen and water vapor to a point where simple barrier materials might be sufficient to enable the lifetimes required for printed, flexible, plastic electronics.

9.4
Efficient Polymer Solar Cells in the Tandem Architecture

The "tandem cell" architecture, a multilayer structure equivalent to two photovoltaic cells in series, offers a number of advantages. Because the two cells are in series, the open-circuit voltage is increased to the sum of the V_{oc} of the individual cells. The use of two semiconductors with different bandgaps enables absorption over a broad range of photon energies within the solar emission spectrum; the two cells typically utilize a wide bandgap semiconductor for the first cell and a smaller bandgap semiconductor for the second cell [38]. Since the electron–hole pairs generated by photons with energies greater than the energy gap rapidly relax to the respective band edges, the power conversion efficiency of the two cells in series is inherently better than that of a single cell made from the smaller bandgap material. Moreover, because of the low mobility of the charge carriers in the polymer–fullerene composites, an increase in the thickness of the active layer increases the internal resistance of the device, which reduces both the V_{oc} and FF [39]. Thus, the tandem cell architecture can have a higher optical density over a wider fraction of the solar emission spectrum than a single cell without increasing the internal resistance. The tandem cell architecture can therefore improve the light harvesting in polymer-based photovoltaic cells.

Polymer tandem cells have recently been successfully demonstrated, with each layer processed from solution, by using the solution-processed TiO$_x$ as the separation layer between the front cell and back cell. The performance of these initial tandem solar cells is summarized as follows: $J_{sc} = 7.8$ mA cm^{-2}, $V_{oc} = 1.24$ V, FF $= 0.67$, and $\eta_e = 6.5\%$.

The structure of the multilayer polymer tandem solar cell is shown in Figure 9.6a. The charge separation layer for the front cell is a bulk heterojunction composite of poly[2,6-(4,4-bis-(2-ethylhexyl)-4H-cyclopenta[2,1-b;3,4-b']dithiophene)-alt-4,7-(2,1,3-benzothiadiazole)] (PCPDTBT) and PCBM. The charge separation layer for the back cell is a bulk heterojunction composite of P3HT and [6,6]-phenyl-C$_{71}$ butyric acid methyl ester (PC$_{70}$BM). The two polymer–fullerene layers are separated by a transparent TiO$_x$ layer and a highly conductive hole transport layer, PEDOT:PSS (Baytron PH500). Electrons from the first cell combine with holes from the second cell at the TiO$_x$–PEDOT:PSS interface.

Cross-sectional images of the polymer tandem solar cells taken with high-resolution transmission electron microscopy (TEM) show the individual layers clearly; see Figure 9.6b [40]. Perhaps more important, the various interfaces are remarkably sharp; there is no interlayer mixing.

The energy level diagram shown in Figure 9.6b indicates the HOMO and LUMO energies of the individual component materials. The open-circuit voltage of the tandem cell is equal to the sum of the V_{oc} of each subcell, because the front and back cells are connected in series [41].

The molecular structures of PCPDTBT, P3HT, PCBM, and PC$_{70}$BM are shown in Figure 9.7a and the absorption spectrum of each material is shown in Figure 9.7b. The absorption bands of the P3HT and PCPDTBT complement each other, making these two materials appropriate for use in the two subcells of a spectrum splitting tandem cell device. The absorption spectrum of a film of the bulk heterojunction composite of each subcell and that of a bilayer film of PCPDTBT:PCBM and P3HT:PC$_{70}$BM are shown in Figure 9.7c. The absorption of the PCPDTBT:PCBM film is relatively weak in the visible spectral range but has two strong bands, one in the near-IR between 700 and 850 nm that arises from the interband π–π^* transition of the PCPDTBT [42] and one in the UV that arises primarily from the HOMO–LUMO transition of the PCBM. The absorption of the P3HT:PC$_{70}$BM film falls in the "hole" in the PCPDTBT:PCBM spectrum and covers the visible spectral range. The electronic absorption spectrum of the tandem structure can be described as a simple superposition of the absorption spectra of the two complementary composites. Therefore, the PEDOT:PSS and TiO$_x$ layers have negligible absorption in the tandem device structure.

Tandem cells generally use a wide bandgap material as the first charge separation layer and a narrow bandgap material as the second charge separation layer with a thinner front cell than back cell so that the photocurrents generated in each subcell are balanced. Because of the nonoptimum phase morphology of the PCPDTBT:PCBM composite, however, increasing the film thickness of this layer above 130 nm leads to

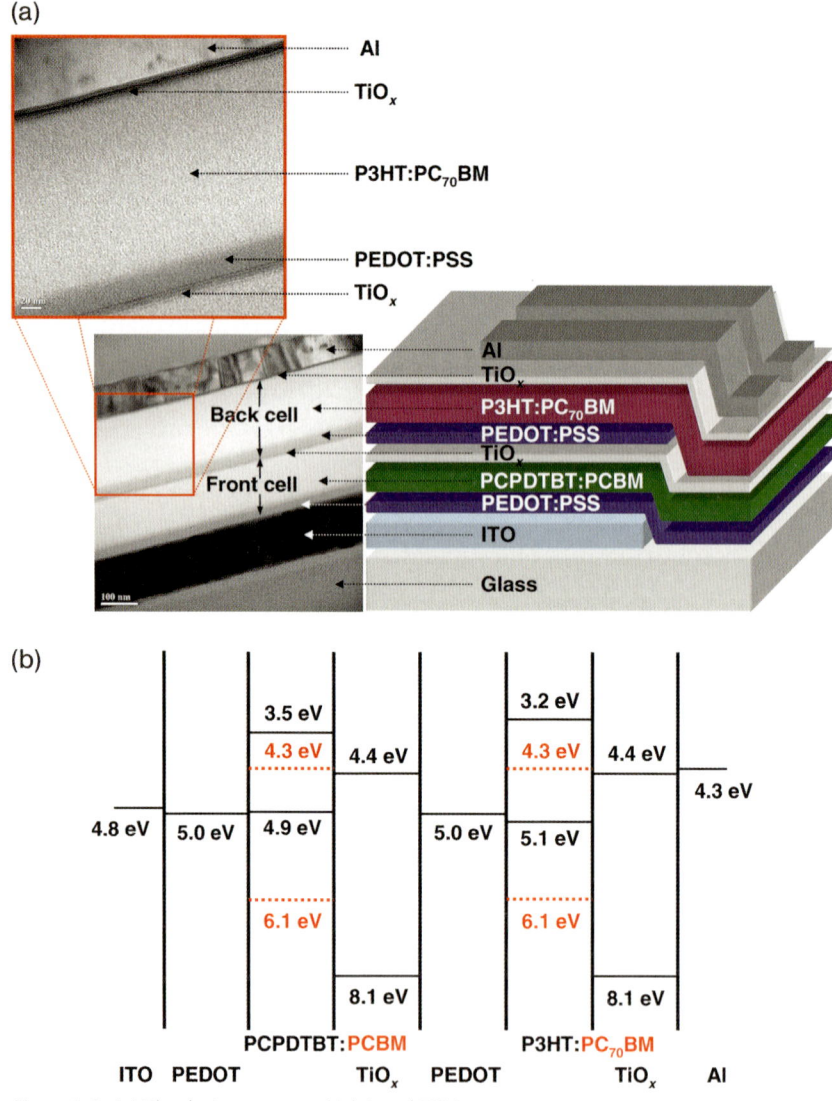

274 | *9 Titanium Oxide Films as Multifunctional Components*

Figure 9.6 (a) The device structure (right) and TEM cross-sectional image (left) of the polymer tandem solar cell. The scale bar is 100 nm in the lower image and 20 nm in the upper image. (b) Energy level diagram showing the highest occupied molecular orbital energies and the lowest unoccupied molecular orbital energies of each of the component materials.

reduced values for both J_{sc} and FF in single cells [42]. However, the J_{sc} of a P3HT:PC$_{70}$BM single cell increases as the film thickness increases up to 200 nm beyond which the FF is reduced. Because of these material characteristics, an "inverted tandem cell" structure was chosen with the low-bandgap bulk heterojunction composite

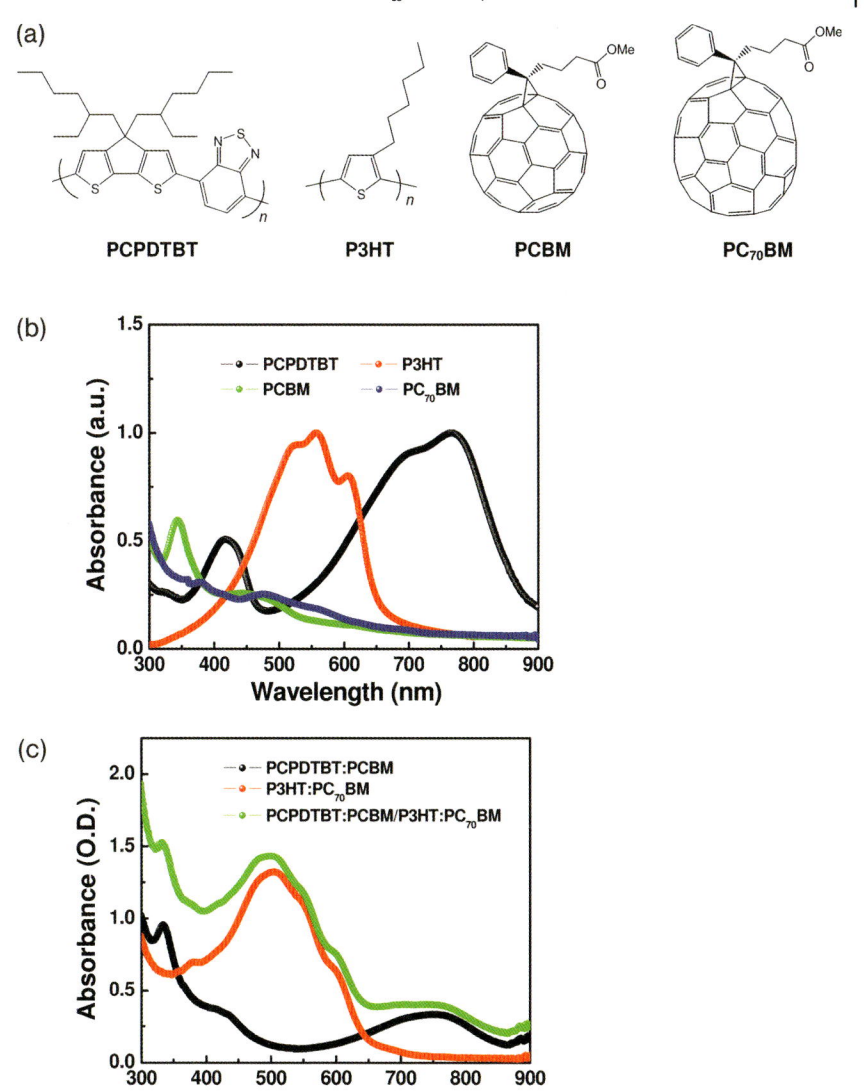

Figure 9.7 (a) Molecular structures of the active materials, PCPDTBT, P3HT, PCBM, and PC$_{70}$BM. (b) Absorption spectra of films of each material. The absorption band of the P3HT complements the absorption of PCPDTBT in visible range (a.u., arbitrary units). (c) Absorption spectra of a PCPDTBT:PCBM bulk heterojunction composite film, a P3HT:PC$_{70}$BM bulk heterojunction composite film, and a bilayer of the two as relevant to the tandem device structure.

(PCPDTBT:PCBM) as the front cell and the higher bandgap bulk heterojunction composite (P3HT:PC$_{70}$BM) as the back cell, as shown in Figure 9.6a.

The IPCE spectra of both the single cells and the tandem cell were measured, using a bias light on the tandem cell to confirm the series connection of the subcells and to

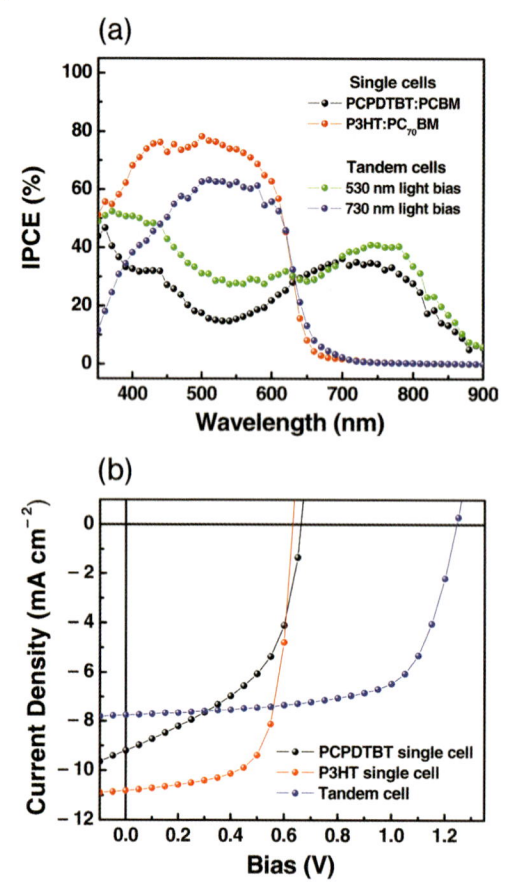

Figure 9.8 (a) IPCE spectra of single cells and a tandem cell with bias light. The IPCE measurements were carried out using modulation spectroscopy with a lock-in amplifier for the single cells and for the tandem cell without light bias. For the tandem cell measurements made with bias light, unmodulated monochromatic light with intensity of approximately 2 mW cm^{-2} was used. (b) J–V characteristics of single cells and tandem cell using PCPDTBT:PCBM and P3HT:PC$_{70}$BM composites under AM 1.5 G illumination from a calibrated solar simulator with irradiation intensity of 100 mW cm^{-2} are presented.

extend the data over the full spectral coverage as shown in Figure 9.8a [43,44]. Each single cell shows the known spectral response of its bulk heterojunction composite, the P3HT:PC$_{70}$BM composite gives a maximum IPCE of ~78% at 500 nm. This is somewhat lower than that shown in Figure 9.2a; as shown in Figure 9.6a, the TiO$_x$ layer that serves as the optical spacer in the tandem cell is too thin to enable the optimum redistribution of the light intensity.

The PCPDTBT:PCBM composite has two dominant peaks in the IPCE spectrum; approximately 35% at 750–800 nm and over 32% below 440 nm, in excellent agreement with the absorption spectra of the two composites (see Figure 9.7c).

When carrying out the measurements on the tandem cell, biasing the device with 530 nm blue light selectively excites the front cell and biasing with 730 nm red light selectively excites the back cell, showing that the device harvests photons from the UV to near IR and that each subcell functions individually [44].

The current density versus voltage (J–V) characteristics of single solar cells and the tandem solar cell respectively using PCPDTBT:PCBM and P3HT:PC$_{70}$BM composites under AM 1.5 G illumination from a calibrated solar simulator with an irradiation intensity of 100 mW cm^{-2} are shown in Figure 9.8b. The single devices show a typical photovoltaic response with device performance comparable with that reported in previous studies [14,42]; the PCPDTBT:PCBM single cell yields $J_{sc} = 9.2$ mA cm^{-2}, $V_{oc} = 0.66$ V, FF $= 0.50$, and $\eta_e = 3.0\%$, while the P3HT:PC$_{70}$BM single cell yields $J_{sc} = 10.8$ mA cm^{-2}, $V_{oc} = 0.63$ V, FF $= 0.69$, and $\eta_e = 4.7\%$.

With two subcells stacked in series, the current that is extracted from the tandem cell is determined by the current generated in either the front or back cell, whichever is smaller [44]. Accordingly, when there is greater carrier generation in either subcell, these excess charges cannot contribute to the photocurrent and so compensate for the built-in potential across that subcell. This compensation leads to a reduced V_{oc} in the tandem cell [45]. To optimize and balance the current in each subcell, all possible variations of the tandem cell architecture were fabricated by changing the order of the active materials, by varying the concentration and ratio of each component in the composite solutions, and by varying the thicknesses of the two bulk heterojunction materials. Because of the high extinction coefficient of the PCPDTBT:PCBM composite, the P3HT:PC$_{70}$BM back cell had the smaller J_{sc} of the two subcells and was thus the limiting cell [44]. The FF of the tandem cell can be very near the FF of the limiting cell. Thus, we used the P3HT:PC$_{70}$BM as the back cell to obtain a higher FF. Over 200 individual tandem cells were made to optimize the fabrication procedure and device architecture. This optimization led to the inverted tandem cell device depicted in Figure 9.1a. Using this inverted structure, more than 20 tandem cells were fabricated with efficiencies above 6.2%; typical performance parameters were as follows; $J_{sc} = 7.8$ mA cm^{-2}, $V_{oc} = 1.24$ V, FF $= 0.67$, and $\eta_e = 6.5\%$ (all the tandem cell devices were fabricated and their performance measured by J. Y. Kim). The J_{sc} in the tandem cell was consistent with the IPCE measurements since the photocurrent in the back cell from IPCE was 72% of that of P3HT:PC$_{70}$BM single cell, confirming that the back cell was the limiting cell for J_{sc} as well as FF.

9.5 Conclusions

The introduction of a thin layer (or layers) of titanium suboxide, TiO$_x$ into the structure by sol–gel chemistry has enabled a number of opportunities for improving the performance of bulk heterojunction photovoltaic cells. The TiO$_x$ layers serve six separate functions:

1. When deposited between the charge separation layer (e.g., comprised of a phase-separated mixture of poly(3-hexyl thiophene, P3HT, and a soluble:fullerene derivative) and the aluminum cathode, the TiO$_x$ layer functions as an optical spacer that redistributes the light intensity to optimize the efficiency of the back cell.

2. By introducing a TiO$_x$ layer between the charge-separating layer and the aluminum cathode, excellent air stability has been demonstrated. The TiO$_x$ layer acts as a shielding and scavenging layer preventing the intrusion of oxygen and humidity into the electronically active polymers, thereby improving the lifetime of unpackaged devices exposed to air by nearly two orders of magnitude.

3. The TiO$_x$ functions as an electron transport layer. Owing to the oxygen deficiency, the TiO$_x$ layer is n-type doped. As a result, the inclusion of the TiO$_x$ layer between the charge-separating layer and the aluminum cathode does not result in an increase in the series resistance. Moreover, since the lowest energy states at the bottom of the conduction band of TiO$_x$ are well matched to the Fermi energy of aluminum, there is facile electron transfer from the TiO$_x$ electron transport layer to the aluminum cathode.

4. The TiO$_x$ layer breaks the symmetry thereby creating the open-circuit voltage.

5. The TiO$_x$ functions as a hole-blocking layer since the top of the valence band of TiO$_x$ is sufficiently electronegative, 8.1 eV below the vacuum, to block holes.

6. The TiO$_x$ layer enables the fabrication of tandem cells. The transparent TiO$_x$ layer is used to separate and connect the front cell and the back cell. The TiO$_x$ layer serves as an electron transport and collecting layer for the first cell and as a stable foundation that enables the fabrication of the second cell to complete the tandem cell architecture.

The achievements summarized in this review emphasize the potential importance of the use of sol–gel processing to fabricate the TiO$_x$ layers in bulk hetrojunction solar cells comprising semiconducting polymers. The demonstration of polymer tandem cells in which each of the individual layers is processed from solution without significant interlayer mixing is a major step toward the achievement of high-efficiency solar cells that can be fabricated in large areas using low-cost printing and coating technologies. We anticipate that polymer tandem cells utilizing both new materials with fine-tuned absorption spectra and high mobility and the opportunity to increase the multilayer stack to create three cells in tandem will yield results approaching or even exceeding 10% efficiency. The use of TiO$_x$ as the separator, charge transport layer, hole-blocking layer, symmetry-breaking layer, and optical spacer in fully solution processable polymer tandem solar cells are important development steps toward large-scale commercialization.

Acknowledgments

The research was supported by Konarka Technologies (Lowell, MA), by the Air Force Office of Scientific Research, AFOSR, under FA9550-05-0139, by the Department of

Energy under DE-FG02-06ER46324, by the Heeger Center for Advanced Materials, Gwangju Institute of Science and Technology (Gwangju, Korea), and by the Ministry of Science and Technology of Korea under the International Cooperation Research Program (Global Research Laboratory Program with K. Lee and A. J. Heeger as Principal Investigators). The PCPDTBT, PCBM, P3HT, and PC$_{70}$BM materials were supplied for our use by Konarka Technologies.

References

1 Yu, G. and Heeger, A.J. (1995) *Journal of Applied Physics*, **78**, 4510.
2 Yu, G., Gao, J., Hummelen, J.C., Wudl, F. and Heeger, A.J. (1995) *Science*, **270**, 1789.
3 Halls, J.J.M., Pichler, K., Friend, R.H., Moratti, S.C. and Holmes, A.B. (1996) *Applied Physics Letters*, **68**, 3120.
4 Scharber, M.C., Wuhlbacher, D., Koppe, M., Denk, P., Waldauf, C., Heeger, A.J. and Brabec, C.J. (2006) *Advanced Materials*, **18**, 789.
5 Sariciftci, N.S., Smilowitz, L., Heeger, A.J. and Wudl, F. (1992) *Science*, **258**, 1474.
6 Hoppe, H. and Sariciftci, N.S. (2006) *Journal of Materials Chemistry*, **16**, 45.
7 Kraabel, B., McBranch, D., Sariciftci, N.S., Moses, D. and Heeger, A.J. (1994) *Physical Review B: Condensed Matter*, **50**, 18543.
8 Kraabel, B., Hummelen, J.C., Vacar, D., Moses, D., Sariciftci, N.S., Heeger, A.J. and Wudl, F. (1996) *Journal of Chemical Physics*, **104**, 4267.
9 Brabec, C.J., Zerza, G., Cerullo, G., De Silvestri, S., Luzzati, S., Hummelen, J.C. and Sariciftci, N.S. (2001) *Chemical Physics Letters*, **340**, 232.
10 Hwang, I.-W., Soci, C., Moses, D., Zhu, Z., Waller, D., Gaudiana, R., Brabec, C.J. and Heeger, A.J. (2007) *Advanced Materials*, **19**, 2307.
11 Sievers, D.W., Shrotriya, V. and Yang, Y. (2006) *Journal of Applied Physics*, **100**, 114509.
12 Dennler, G., Mozer, A.J., Juška, G., Pivrikas, A., Österbacka, R., Fuchsbauer, A. and Sariciftci, N.S. (2006) *Organic Electronics*, **7**, 229.
13 Ma, W., Yang, C., Gong, X., Lee, K. and Heeger, A.J. (2005) *Advanced Functional Materials*, **15**, 1617.
14 Kim, J.Y., Kim, S.H., Lee, H.-H., Lee, K., Ma, W., Gong, X. and Heeger, A.J. (2006) *Advanced Materials*, **18**, 572.
15 Pettersson, L.A.A., Roman, L.S. and Inganäs, O. (1999) *Journal of Applied Physics*, **86**, 487.
16 Stübinger, T. and Brütting, W. (2001) *Journal of Applied Physics*, **90**, 3632.
17 Hänsel, H., Zettl, H., Krausch, G., Kisselev, R., Thelakkat, M. and Schmidt, H.-W. (2003) *Advanced Materials*, **15**, 2056.
18 Snaith, H.J., Greenham, N.C. and Friend, R.H. (2004) *Advanced Materials*, **16**, 1640.
19 Melzer, C., Koop, E.J., Mihaletchi, V.D. and Blom, P.W.M. (2004) *Advanced Functional Materials*, **14**, 865.
20 O'Regan, B. and Grätzel, M. (1991) *Nature*, **353**, 737.
21 Bach, U., Lupo, D., Comte, P., Moser, J.E., Weissörtel, F., Salbeck, J., Spreitzer, H. and Grätzel, M. (1998) *Nature*, **395**, 583.
22 Arango, A.C., Johnson, L.R., Bliznyuk, V.N., Schlesinger, Z., Carter, S.A. and Hörhold, H.-H. (2000) *Advanced Materials*, **12**, 1689.
23 Breeze, A.J., Schlesinger, Z., Carter, S.A. and Brock, P.J. (2001) *Physical Review B: Condensed Matter*, **64**, 125205.
24 van Hal, P.A., Wienk, M.M., Kroon, J.M., Verhees, W.J.H., Slooff, L.H., van Gennip, W.J.H., Jonkheijm, P. and Janssen, R.A.J. (2003) *Advanced Materials*, **15**, 118.
25 Thelakkat, M., Schmitz, C. and Schmidt, H.-W. (2002) *Advanced Materials*, **14**, 577.

26 Brabec, C.J., Sariciftci, N.S. and Hummelen, J.C. (2001) *Advanced Functional Materials*, **11**, 15.
27 Brabec, C.J. (2004) *Solar Energy Materials and Solar Cells*, **83**, 273.
28 Shaheen, S.E., Brabec, C.J., Sariciftci, N.S., Padinger, F., Fromherz, T. and Hummelen, J.C. (2001) *Applied Physics Letters*, **78**, 841.
29 Padinger, F., Rittberger, R. and Sariciftci, N.S. (2003) *Advanced Functional Materials*, **13**, 85.
30 Lee, K., Chang, Y. and Kim, J.Y. (2003) *Thin Solid Films*, **423**, 131.
31 Burrows, P.E., Bulovic, V., Forrest, S.R., Sapochak, L.S., McCarty, D.M. and Thompson, M.E. (1994) *Applied Physics Letters*, **65**, 2922.
32 Burrows, P.E., Graff, G.L., Gross, M.E., Martin, P.M., Hall, M., Mast, E., Bonham, C., Bennet, W., Michalski, L., Weaver, M.S., Brown, J.J., Fogarty, D. and Sapochak, L.S. (2000) *Proceedings of SPIE*, **4105**, 75.
33 Weaver, M.S., Michalski, L.A., Rajan, K., Rothman, M.A., Silvernail, J.A., Brown, J.J., Burrows, P.E., Graff, G.L., Gross, M.E., Martin, P.M., Hall, M., Mast, E., Bohnam, C., Bennett, W. and Zurnhoff, M. (2002) *Applied Physics Letters*, **81**, 2929.
34 Chwang, A.B., Rothman, M.A., Mao, S.Y., Hewitt, R.H., Weaver, M.S., Silvernail, J.A., Rajan, K., Hack, M., Brown, J.J., Chu, X., Moro, L., Krajewski, T. and Rutherford, N. (2003) *Applied Physics Letters*, **83**, 413.
35 Moro, L., Krajeweski, T.A., Rutherford, N.M., Philips, O., Visser, R.J., Gross, M.E., Bennett, W.D. and Graff, G.L. (2004) *Proceedings of SPIE*, **5214**, 83.
36 Kwon, S.H., Paik, S.Y., Kwon, O.J. and Yoo, J.S. (2001) *Applied Physics Letters*, **79**, 4450.
37 Scherf, U. and List, E.J.W. (2002) *Advanced Materials*, **14**, 477.
38 Wanlass, M.W., Emery, K.A., Gessert, T.A., Horner, G.S., Osterwald, C.R. and Coutts, T.J. (1989) *Solar Cells*, **27**, 191.
39 Riedel, I. and Dyakonov, V. (2004) *Physica Status Solidi a: Applied Research*, **201**, 1332.
40 Steurman, D., Garcia, A. and Nguyen, T.-Q. (2008) *Advanced Materials*, **20**, 528.
41 Hiramoto, M., Suezaki, M. and Yokoyama, M. (1990) *Chemistry Letters*, **19**, 327.
42 Mühlbacher, D., Scharber, M., Morana, M., Zhu, Z., Waller, D., Gaudiana, R. and Brabec, C. (2006) *Advanced Materials*, **18**, 2884.
43 Dennler, G., Prall, H.-J., Koeppe, R., Egginger, M., Autengruber, R. and Sariciftci, N.S. (2006) *Applied Physics Letters*, **89**, 073502.
44 Burdick, J. and Glatfelter, T. (1986) *Solar Cells*, **18**, 301.
45 Hadipour, A., de Boer, B., Wildeman, J., Kooistra, F.B., Hummelen, J.C., Turbiez, M.G.R., Wienk, M.M., Janssen, R.A.J. and Blom, P.W.M. (2006) *Advanced Functional Materials*, **16**, 1897.

B
Bulk Heterojunction Solar Cells

10
Performance Improvement of Polymer: Fullerene Solar Cells Due to Balanced Charge Transport

L. Jan Anton Koster, Valentin D. Mihailetchi, Martijn Lenes, and Paul W.M. Blom

10.1
Introduction

An attractive way of producing energy is to harvest it directly from sunlight. The amount of energy that the Earth receives from the Sun is enormous: 1.75×10^{17} W. As the world energy consumption in 2003 amounted to 4.4×10^{20} J, Earth receives enough energy to fulfill the yearly world demand of energy in less than an hour. Not all of that energy reaches the Earth's surface due to absorption and scattering, however, and the photovoltaic conversion solar energy remains an important challenge. State-of-the-art inorganic solar cells have a record power conversion efficiency of close to 39% [1], whereas commercially available solar panels have a significantly lower efficiency of around 15–20%. Another approach to making solar cells is to use organic materials such as conjugated polymers. Solar cells based on thin polymer films are particularly attractive because of their ease of processing, mechanical flexibility, and potential for low cost fabrication of large areas. Additionally, their material properties can be tailored by modifying their chemical makeup, resulting in greater customization than traditional solar cells allow.

The field of organic photovoltaics dates back to 1959 when Kallman and Pope discovered that anthracene can be used to make a solar cell [2]. Their device produced a photovoltage of only 0.2 V and had an extremely low efficiency. Attempts to improve the efficiency of solar cells based on a single organic material (a so-called homojunction) were unsuccessful, mainly because of the low dielectric constant of organic materials (typically, the relative dielectric constant is 2–4). Due to this low dielectric constant, the probability of forming free charge carriers upon light absorption is very low. Instead strongly bound excitons are formed, with a binding energy of around 0.4 eV in the case of poly(*p*-phenylene vinylene) (PPV) [3–5]. Since these excitons are so strongly bound, the electric field in a photovoltaic device, which arises from the work function difference between the electrodes, is too weak to dissociate the excitons. A major advancement was realized by Tang who used two different materials, stacked in layers, to dissociate the excitons [6]. In this so-called heterojunction, an electron donor material (D) and an electron acceptor material (A) are brought together.

Organic Photovoltaics: Materials, Device Physics, and Manufacturing Technologies.
Edited by Christoph Brabec, Vladimir Dyakonov, and Ullrich Scherf
Copyright © 2008 WILEY-VCH Verlag GmbH & Co. KGaA, Weinheim
ISBN: 978-3-527-31675-5

By carefully matching these materials, electron transfer from the donor to the acceptor, or hole transfer from the acceptor to the donor, is energetically favored. In 1992, Sariciftci et al. demonstrated that ultrafast electron transfer takes place from a conjugated polymer to C_{60}, showing the great potential of fullerenes as acceptor materials [7]. To be dissociated, the excitons must be generated in proximity to the donor/acceptor interface, since the diffusion length is typically 5–7 nm [8–10]. This need limits the part of the active layer that contributes to the photocurrent to a very thin region near the donor/acceptor interface; excitons generated in the remainder of the device are lost. How can the problem of all excitons not reaching the donor/acceptor interface be overcome? In 1995 Yu et al. devised a solution [11]: by intimately mixing both components, the interfacial area is greatly increased and the distance that excitons have to travel to reach the interface is reduced. This device structure is called a bulk heterojunction (BHJ) and has been used extensively since its introduction in 1995. An important breakthrough in terms of power conversion efficiency was reached by Shaheen et al. who showed that the solvent used has a profound effect on the morphology and performance of BHJ solar cells [12]. By optimizing the device processing, an efficiency of 2.5% was obtained. State-of-the-art polymer/fullerene BHJ solar cells have an efficiency of more than 4% [13]. Although significant progress has been made, the efficiency of converting solar energy into electrical power obtained with plastic solar cells still does not warrant commercialization. To improve the efficiency of plastic solar cells, it is, therefore, crucial to understand what limits their performance.

The main steps in photovoltaic energy conversion by organic solar cells are depicted in Figure 10.1.

As a first step, we consider the simple case of a photoconductor with noninjecting contacts and an uniform electric field distribution. Goodman and Rose derived that, under the assumption of negligible recombination of charge carriers [14], the photocurrent J_{ph} is given by

$$J_{ph} = qGL, \tag{10.1}$$

with q the electron charge, G the generation rate of charge carriers, and L the thickness of the photoconductor. In this case all photogenerated charge carriers are simply extracted and the current density depends only on the generation rate G. In their derivation, Goodman and Rose took only drift of charge carriers into account and neglected the contribution of diffusion. Sokel and Hughes carried this analysis one step further by including diffusion of carriers, finding [15]

$$J_{ph} = qGL \left[\frac{\exp(V/V_t)+1}{\exp(V/V_t)-1} - 2\frac{V_t}{V} \right], \tag{10.2}$$

where V is the voltage drop across the active layer and V_t the thermal voltage, $V_t = kT/q$, where k is Boltzmann's constant and T is the temperature. For a BHJ device with a voltage V_0 already present from the different work functions of the contacts, the voltage drop amounts to $V_0 - V$. The result by Sokel and Hughes shows two regimes: a linear dependence of J_{ph} on voltage for small biases (regime I), while reducing

10.1 Introduction

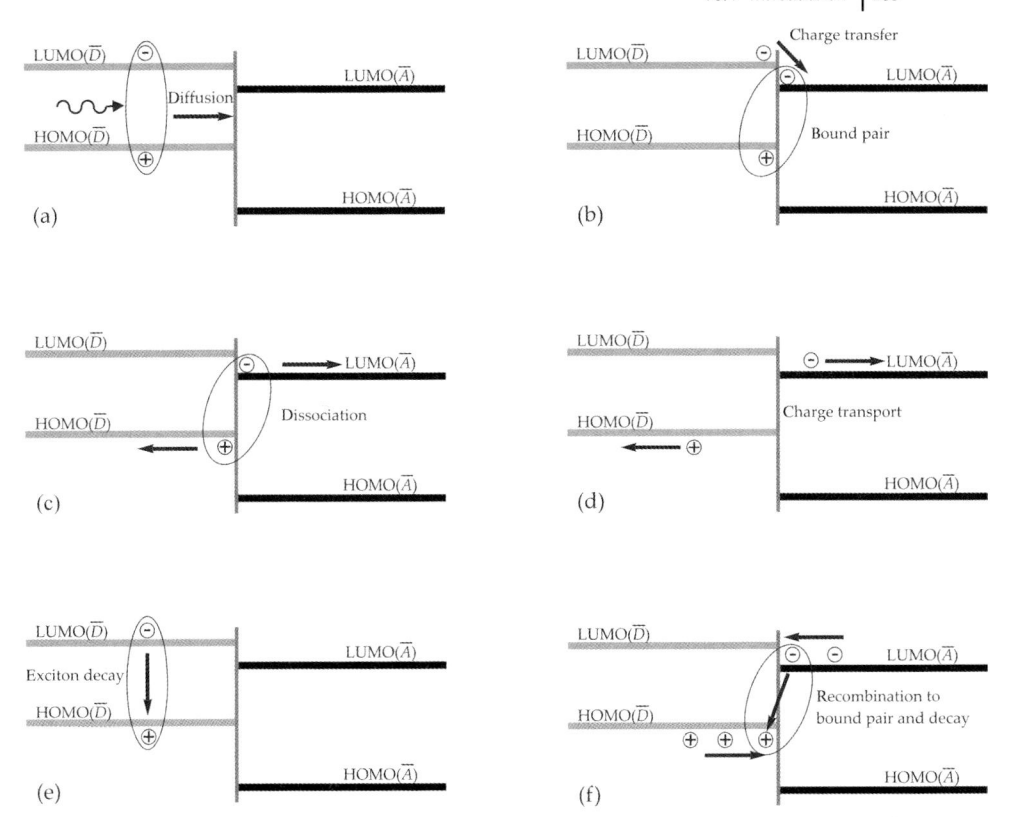

Figure 10.1 Organic photovoltaics in a nutshell: part (a) shows the process of light absorption by the polymer, yielding an exciton that has to diffuse to the donor/acceptor interface. If the exciton reaches this interface, electron transfer to the acceptor phase in energetically favored, as shown in (b), yielding a coulombically bound electron–hole pair. The dissociation of the electron–hole pair, either phonon or field assisted, produces free charge carriers, as depicted in (c). Finally, the free carriers have to be transported through their respective phases to the electrodes to be extracted (d). Exciton decay is one possible loss mechanism, see (e), while geminate recombination of the bound electron–hole pair and bimolecular recombination of free charge carriers (f) are two other possibilities.

to Equation 10.1 at moderately high bias (including short-circuit conditions) (regime II), see Figure 10.2a.

In the case of recombination losses, the extraction of photogenerated carriers is governed by the mean carrier drift length w, which is the mean distance a carrier travels before recombination occurs. When both the electron (w_n) and hole (w_p) drift lengths are larger than the active layer thickness, then the charges will readily flow out without distorting the field in the device, as shown in Figure 10.3a. However, for example, in the case where $w_n \gg w_p$ and $w_p < L$, there will be a net positive space charge near the anode, as shown in Figure 10.3b. For this situation, three regimes exist in the device: near the cathode, the electron density is much larger than the hole density; this is a small region (I). Next to this region, there exists a balance between

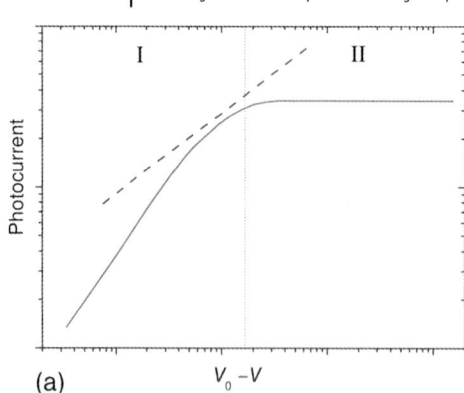

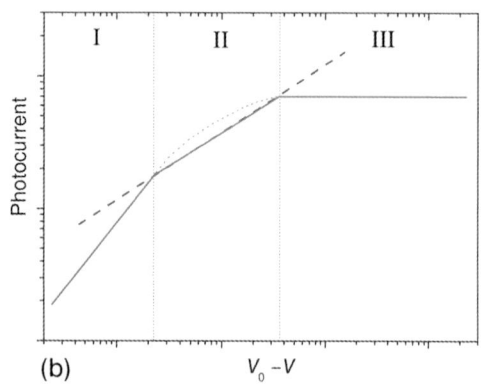

Figure 10.2 Schematic photocurrent versus voltage drop across the active layer for a device without (a) and with (b) space charge limitations. The dashed line represents the space-charge-limited photocurrent, given by Equation 10.2.

electron and hole density, yielding a neutral region (II). Near the anode, the holes dominate the device (III), resulting in a large net space charge and concomitant large voltage drop, as indicated in Figure 10.3b. The large field strength in region III facilitates the extraction of holes, ensuring that the extraction current of holes and electrons is equal. When the photocurrent is space-charge-limited (SCL), the following relation holds [14]:

$$J_{ph} \leq (qG)^{0.75} \left(\frac{9}{8}\varepsilon_0\varepsilon_r\mu_h\right)^{0.25} \cdot \sqrt{V}, \quad (10.3)$$

where μ is the mobility of the slowest carrier, holes in this case, and $\varepsilon_0\varepsilon_r$ the dielectric constant. Thus, fully space-charge-limited photocurrents are characterized by a square root dependence on voltage and are proportional to the incident light intensity I as $I^{0.75}$, irrespective of the amount of bimolecular recombination. In that case, as shown in Figure 10.2b, a third regime appears in the J_{ph}–V characteristics,

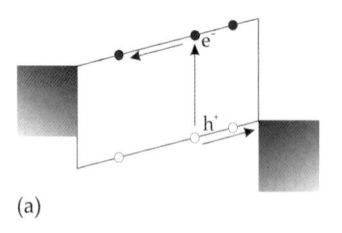

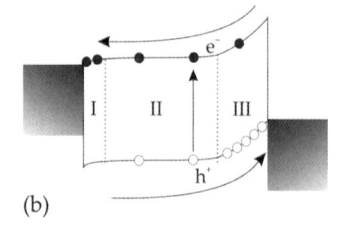

Figure 10.3 (a) Band diagram of a BHJ solar cell with balanced electron and hole mobilities; both types of charge carriers can readily flow out of the device and the field in the device is uniform. (b) Band diagram in the case of hole accumulation and concomitant space-charge-limited behavior. Near to the cathode, a small region dominated by electrons (I) exists next to a large neutral region (II), where electron and hole densities are comparable. Most of the potential drops across the hole accumulation layer (III) to facilitate the extraction of the slow holes.

thereby strongly limiting the fill factor (FF) of the solar cell. The occurrence of space-charge-limited photocurrents in BHJ solar cells based on [6,6]-phenyl C_{61}-butyric acid methyl ester (PCBM) and low mobility PPV derivatives has been demonstrated by Mihailetchi *et al.* [16].

10.2 MDMO-PPV:PCBM-Based Solar Cells

As shown in the previous section, in the case of a large difference in mean-free path for electrons and holes caused by, for example, a large difference in electron and hole mobility, the electric field in the device adjusts itself in such a way that the transport of the slowest carrier is enhanced. This results in a nonuniform field, since the charges of photogenerated electrons and holes do not cancel. Consequently, the slowest charge carrier will dominate the device because the faster carrier can leave the device much more easily. Since the hole mobility of neat poly(2-methoxy-5-(3′,7′-dimethoxyloctyloxy)-*p*-phenylene vinylene) (MDMO-PPV) was previously reported to be $5 \times 10^{-11}\,m^2\,V^{-1}\,s^{-1}$ [17], while an electron mobility of $2 \times 10^{-7}\,m^2\,V^{-1}\,s^{-1}$ was disclosed for [6,6]-phenyl C_{61}-butyric acid methyl ester (PCBM) [18], the charge transport in a heterojunction photovoltaic cell based on these materials is expected to be strongly unbalanced. However, from space-charge-limited conduction, admittance spectroscopy, and transient electroluminescence measurements, a hole mobility of $2 \times 10^{-8}\,m^2\,V^{-1}\,s^{-1}$ was found for the MDMO-PPV phase in the (1 : 4 wt%) blend at room temperature [19]. Consequently, the charge carrier transport in an MDMO-PPV:PCBM-based solar cell was much more balanced than previously assumed, which is a necessary requirement for the disclosed high fill factors of above 50%. This enhanced mobility is in agreement with the fact that to electrostatically allow the experimentally observed photocurrents, a hole mobility of at least $10^{-8}\,m^2\,V^{-1}\,s^{-1}$ is required. For lower mobilities the photocurrent is expected to be dominated by a nonuniform electric field and resulting space charge formation, as shown in Figure 10.3b. In that case a square root dependence of the photocurrent on voltage is expected. Figure 10.4 shows the current–voltage characteristics of a 120 nm thick MDMO-PPV:PCBM (1 : 4 by weight) BHJ solar cell. In this graph, the effective photocurrent density J_{ph}, obtained by subtracting the dark current from the current under illumination, is plotted as a function of effective applied voltage $V_0 - V$, where V_0 is the compensation voltage defined by $J_{ph}(V = V_0) = 0$ [20] In this way, $V_0 - V$ reflects the internal electric field in the device. It should be noted that $V_0 = 0.884$ V is very close to the open-circuit voltage (0.848 V). For low effective voltages $V_0 - V$, the photocurrent increases linearly with effective voltage and subsequently tends to saturate. Mihailetchi *et al.* [20] demonstrated that this low-voltage part can be described with an analytical model developed by Sokel and Hughes [15] for zero recombination, as indicated by the dashed line in Figure 10.4. The linear behavior at low effective voltage is the result of a direct competition between diffusion and drift currents. At higher effective voltage all free charge carriers are extracted for zero recombination and the photocurrent saturates to qGL.

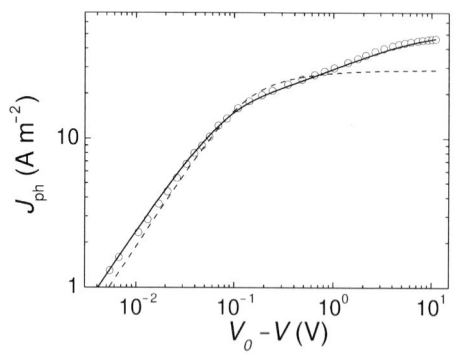

Figure 10.4 Photocurrent density J_{ph} as a function of effective applied voltage ($V_0 - V$). The symbols represent experimental data of MDMO-PPV/PCBM devices at room temperature. The solid line denotes a numerical simulation, while the dashed line represents the result of Sokel and Hughes.

The fact that the experimental photocurrent does not completely saturate at qGL but gradually increases for large effective voltages has been attributed to the field dependence of the generation rate G. The two parameters governing the field- and temperature-dependent generation rate $G(E,T)$ [21], the electron–hole pair distance a, and the decay rate k_f can be determined by equating the high field photocurrents to qGL. The value of a determines the field at which the dissociation efficiency fully saturates and hence a can be determined independent of k_f. By fitting the temperature dependence of the photocurrent at high reverse bias, $a = 1.3$ nm and $k_f = 2.5 \times 10^5$ s^{-1} are obtained [20]. Subsequently, these parameters were implemented in a numerical device model to describe the full voltage range [22]. It is evident from Figure 10.4 that the calculated photocurrent fits the experimental data over the entire voltage range. For comparison, in Figure 10.5 the experimental and calculated J_{ph} are also shown in a conventional linear plot focusing on the fourth quadrant.

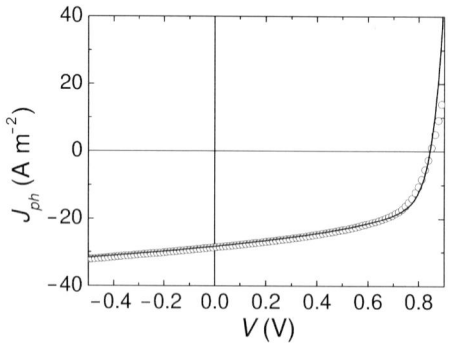

Figure 10.5 The current density under illumination of a MDMO-PPV:PCBM device (symbols) and the numerical result (line).

10.2 MDMO-PPV:PCBM-Based Solar Cells

The excellent agreement between experimental and calculated data now enables one to further analyze the losses in these devices in more detail.

A striking feature of these types of solar cells is that at the optimal device thickness of typically 100 nm, only 60% of the incident light is absorbed [12]. The absorption can be enhanced by increasing the thickness of the active layer. However, in spite of an increased absorption, the overall power conversion efficiency does not improve when the device thickness is increased beyond 100 nm. It is evident that a thickness increase is expected to also give rise to an enhanced charge recombination because of an increase in carrier drift length [23,24]. From a device point of view, the reduced performance with increasing thickness mainly originates from a decrease in the fill factor. As discussed above, for devices with a large difference in electron and hole mobility, a space-charge-limited photocurrent occurs at high intensity due to an unbalanced charge transport, described by Equation 10.3. It should be noted that Equation 10.3 does not depend on device thickness. On the contrary, for nonspace-charge-limited devices, as is the case for standard 100 nm MDMO-PPV:PCBM solar cells, the photocurrent density at short-circuit and reverse bias is closely approximated by $J_{ph} = qG(E,T)L$. Consequently, in this case, increasing the active layer thickness will generally result in a higher photocurrent due to an enhanced absorption. As a result, with increasing active layer thickness at some point the photocurrent will reach the (thickness independent) space charge limit given by Equation 10.3, and a transition will occur from a non-SCL, as shown in Figure 10.2a, to a SCL device as shown in Figure 10.2b. Such a transition will lead to a strong decrease in the fill factor, even when charge recombination does not play a role.

To investigate the effect of space charge formation, the photocurrents of devices with a thin (128 nm) and a thick (368 nm) active layer have been studied in more detail, including their illumination intensity dependence. Figure 10.6 shows the photocurrent density $J_{ph} = J_L - J_D$, where J_L and J_D are the current density under illumination and in dark, respectively, as a function of effective applied voltage $V_0 - V$ for both devices. Also shown is the predicted space charge limit using $\mu_h = 3 \times 10^{-8} \text{ m}^2 \text{ V}^{-1} \text{ s}^{-1}$ and $G = 1.9 \times 10^{27}$ and $0.9 \times 10^{27} \text{ m}^{-3} \text{ s}^{-1}$ for both devices. In the case of the 128 nm device, the photocurrent is still below the space charge limit and two regimes can be recognized, as also represented in Figure 10.2a. For voltages close to V_0, the photocurrent scales linearly with effective applied voltage due to a competition between drift and diffusion currents. As mentioned above, with increasing applied voltage ($V_0 - V > 0.1$ V) the photocurrent saturates to $J_{ph} = qG(E,T)L$. For the 368 nm device, however, the photocurrent intersects the predicted space charge limit and now three regimes appear, as indicated in Figure 10.2b: again, the photocurrent is linear for small applied voltages ($V_0 - V < 0.1$ V).

In the second regime (0.3 V $< V_0 - V <$ 0.7 V), the photocurrent now shows the typical square root dependence of an SCL photocurrent followed by a saturation of the photocurrent ($V_0 - V > 0.7$ V). It is evident that the occurrence of the space charge regime will have a strong effect on the fill factor of the 368 nm device. To further confirm the built-up of space charge in the thick devices, we investigated the dependence of the photocurrent J_{ph} on light intensity I, characterized by $J_{ph} \sim I^\alpha$. In Figure 10.6 the coefficient α is given for various effective voltages $V_0 - V$. For the

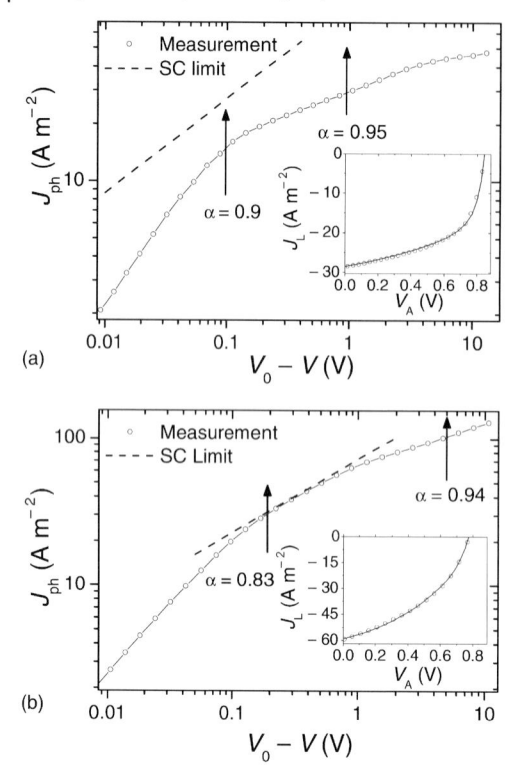

Figure 10.6 Experimental photocurrent density J_{ph} as a function of effective applied voltage $V_0 - V$ under $1\,kW\,m^{-2}$ illumination for a device consisting of a 128 nm active layer (a) and a 368 nm active layer (b). Circles indicate experimental data, solid line fit of the photocurrent and the dashed line the predicted space charge limit using Equation 10.3. The arrows indicate fits of the intensity dependence $J_{ph} \sim I^\alpha$. Inset: current under illumination J_L versus applied voltage V.

thin device α ranges from 0.9 in the linear regime to 0.95 in the saturated regime, indicating that almost no space charge effects occur. For the thick device $\alpha = 0.83$ at $V_0 - V = 0.2\,V$, approaching the theoretical value of $3/4$ for the pure space-charge-dominated regime [25]. Numerical simulations allow us to disentangle the various loss mechanisms recombination at maximum power point (MPP). First, the average dissociation rate <P> decreases for thicker devices as a result of the lower electric field in the device. At the MPP the dissociation efficiency drops from 51.5% for the 128 nm device to 40% for the 368 nm thick solar cell. Furthermore, the recombination losses at MPP increase from 14% for the thin device to 35% for the thick device. This shows that next to space charge formation also the reduced dissociation efficiency and increased recombination losses play a significant role in thick polymer solar cells. The main conclusion for MDMO-PPV:PCBM (1 : 4 wt%) based BHJ solar cells is that an electron mobility of $2 \times 10^{-7}\,m^2\,V^{-1}\,s^{-1}$ and a hole mobility of $2 \times 10^{-8}\,m^2\,V^{-1}\,s^{-1}$ is sufficient to prevent space charge formation in devices with a thickness of only 100 nm. However, the mobility difference of a factor of 10 limits the performance of devices of typically 300 nm, a thickness that is required to absorb most of the incoming photons, due to the formation of space charges in combination with an increased recombination. To further improve these thick solar cells, a better hole conductor is needed.

10.3
Annealed P3HT:PCBM-Based Solar Cells

It has been demonstrated by Padinger *et al.* that thermal annealing of devices based on blends of regioregular poly(3-hexylthiophene) (P3HT) and PCBM dramatically improves the external quantum efficiency of these cells [26]. It is well known that an enhanced degree of crystallinity can be induced in polythiophene films by thermal annealing. This controlled crystallization and orientation of polythiophene polymer chains can significantly improve the hole mobility. After annealing, an energy conversion efficiency as high as 3.5% has been reported. Besides this, a red shift of the optical absorption of P3HT inside the blend is observed, providing an improved overlap with the solar emission [27]. To determine the electron and hole mobilities in the P3HT:PCBM blend, we use SCL current–voltage measurements: by using suitable electrodes that either suppress the injection of electrons or holes, hole- or electron-only device are realized, respectively [19]. This technique has been applied to measure either the hole or electron current in blends of P3HT:PCBM as a function of the thermal annealing temperature of the spin-coated films. To fabricate the hole-only devices, palladium was evaporated as a top electrode on an indium tin oxide (ITO)/poly (3,4-ethylenedioxythiophene):poly(4-styrenesulfonate) (PEDOT:PSS)/P3HT:PCBM structure. The work function of PEDOT:PSS matches the highest occupied molecular orbital (HOMO) of P3HT at 4.9 eV, forming an ohmic contact for hole injection [28,29], whereas palladium strongly suppresses electron injection into PCBM due to the large mismatch between its work function and lowest unoccupied molecular orbital (LUMO) of PCBM. To suppress the hole injection into P3HT, the bottom contact must have a low work function. Recently, we have demonstrated that the work function of a noble metal (as silver) can be modified using a self-assembled monolayer (SAM) [30]. This technique works very well and successful electron-only devices were constructed for the P3HT:PCBM blends. Figure 10.7 shows the calculated zero-field mobility of electrons and holes in 50:50 wt% blends of P3HT:PCBM devices as a function of the annealing temperature. For comparison, the hole mobility of pristine P3HT, measured under the same experimental conditions, is also shown. It appears from Figure 10.7 that the hole mobility in pristine P3HT is hardly affected by thermal annealing, with a typical value of $(1.4–3.0) \times 10^{-8} \, m^2 \, V^{-1} \, s^{-1}$. This mobility was found to be completely field independent and fully consistent with the previously reported values for high molecular weight P3HT (as the one used here) [29]. In contrast, the hole mobility of P3HT in the blend is strongly affected by the

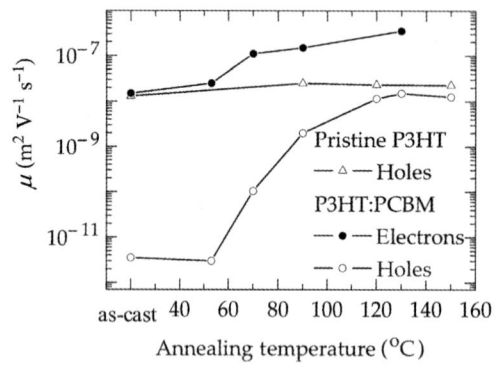

Figure 10.7 Electron and hole mobility in P3HT:PCBM blends as a function of annealing temperature, as well as the hole mobility in pristine P3HT.

presence of PCBM and it drops almost four orders of magnitude for an as-cast device. Upon annealing, however, the mobility starts to increase sharply with an onset at 50–60 °C, followed by saturation to approximately the value of the pristine polymer when the devices are annealed above 120 °C. Moreover, the electron mobility of PCBM in the blend is also affected by thermal annealing: For as-cast films, the electron mobility is $1-2 \times 10^{-8}\,\text{m}^2\,\text{V}^{-1}\,\text{s}^{-1}$, being typically a factor of 5000 higher than the hole mobility. As a result, the charge transport in as-cast films is strongly unbalanced and the current is fully dominated by the electrons.

As a next step, the effect of annealing on the photocurrent of P3HT:PCBM (1 : 1 wt%) solar cells is investigated: Figure 10.8 shows the experimental J_{ph} of P3HT:PCBM blends (50 : 50 wt%) in a double logarithm plot as a function of effective

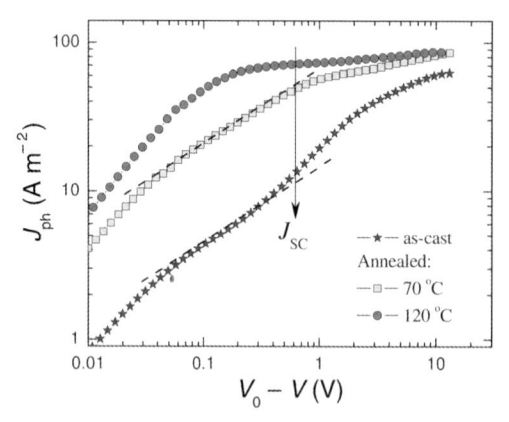

Figure 10.8 Experimental photocurrent (J_{ph}) versus effective applied voltage ($V_0 - V$) of the P3HT:PCBM devices at room temperature, for as-cast device and after thermally annealing of the photoactive layers (see the legend). The device thicknesses (L) are 96 nm and the arrow indicates the position of the short-circuit current (J_{sc}). The dashed lines represent the square root dependence of the J_{ph} on voltage.

applied voltage $(V_0 - V)$. The curves correspond to the different postproduction treatment as follows: as-cast thermally annealed at a temperature where the enhancement in hole mobility is maximized (120 °C), and annealed at lower temperature (70 °C). Thermal annealing was performed on complete devices, that is, with the photoactive layer between the electrodes, on the hot plate for a period of 4 min. It appears from Figure 10.8 that the photocurrent shows a strong enhancement after thermal annealing. For the completely annealed device (at 120 °C), the short-circuit current (J_{sc}) increases by a factor of 5, the FF by a factor of 2, and the overall enhancement of the efficiency is about one order of magnitude when compared with the device as-cast. For the device annealed at 70 °C, it is observed that for $V_0 - V < 0.03$ V, the J_{ph} shows linear dependence on voltage, which is caused by the opposite effect of drift and diffusion of charge carriers. Above 0.03 V, however, a square root dependence on voltage of the experimental J_{ph} is observed (dashed line), as is predicted for blends with a large difference in electron and hole mobilities [16]. At even larger voltages, the J_{ph} shows a clear transition to the saturation regime where it becomes limited by the field and temperature dependence of the dissociation of bound electron–hole pairs [20]. These results are distinctly different when the devices are annealed at higher temperature, where the electron and hole transport is more balanced. In that case, no square root dependence of J_{ph} is observed, as seen in the Figure 10.8 by the curve at 120 °C. The occurrence of SCL photocurrents for the devices annealed at only 70 °C has been further confirmed by investigations of the light intensity dependence [31].

These results now allow a true comparison between P3HT:PCBM (1:1) and MDMO-PPV:PCBM (1:4) blend devices. With respect to the charge transport, the P3HT-based devices have equal mobilities to those measured in MDMO-PPV:PCBM devices. Thus, the increased performance of the P3HT:PCBM solar cells does not originate from an enhanced charge transport, as is often assumed. The main difference, however, is that these identical charge transport properties are realized in blends with different polymer:PCBM weight fractions, namely (1:1) versus (1:4) for P3HT and MDMO-PPV, respectively. As a result, the larger volume fraction of absorbing material (P3HT), combined with more red-shifted absorption, enlarges the generation rate of charge carriers in 50:50 wt% P3HT:PCBM devices by more than a factor of 2, as compared to the 20:80 wt% MDMO-PPV:PCBM devices. Combining this with a higher separation efficiency of photogenerated bound electron–hole pairs under short-circuit conditions increases the J_{sc} with more than a factor of 2 for the P3HT-based devices. The most limiting factor of all P3HT-based devices remains, however, the V_{oc}, which is approximately 40% lower as compared to the V_{oc} of the MDMO-based devices. However, the increase in J_{sc} and FF make up for the loss in V_{oc} and, therefore, the power efficiencies of P3HT:PCBM cells are significantly higher. A main conclusion, however, is that also for these annealed P3HT:PCBM solar cells the mobility difference of a factor of 10 will limit the performance of devices with thicknesses exceeding 250 nm. As a result, to further improve the absorption by an increase in the active layer thickness without loss of fill factor also for these devices, the hole mobility needs to be further improved.

10.4
Slowly Dried P3HT:PCBM Solar Cells

Important progress was recently realized by Yang and coworkers, who demonstrated that the efficiency can exceed 4% by controlling the growth rate of the active layer [13]. Slowing down the drying process of the wet films leads to an enhanced self-organization, which is expected to enhance the hole transport in the P3HT. The charge transport properties in these slowly dried P3HT:PCBM blends had been investigated using time-of-flight (TOF) measurements. Electron and hole mobilities of $\mu_e = 7.7 \times 10^{-9}$ and $\mu_h = 5.1 \times 10^{-9}$ m^{2}V^{-1}s^{-1} were reported, respectively [13]. Remarkably, the reported mobility values for these slowly dried films with superior photovoltaic performance are much lower than the values reported for MDMO-PPV:PCBM and annealed P3HT:PCBM devices: As shown above, for MDMO-PPV:PCBM (1:4 wt%) values of $\mu_e = 2.0 \times 10^{-7}$ and $\mu_h = 1.4 \times 10^{-8}$ m^{2}V^{-1}s^{-1} have been found, and for P3HT:PCBM (1:1 wt%) using fast drying and annealing, similar values of $\mu_e = 3.0 \times 10^{-7}$ and $\mu_h = 1.5 \times 10^{-8}$ m^{2}V^{-1}s^{-1} have been measured. Therefore, the origin of the improved performance after slow drying is not clear. To further investigate the origin of this enhanced performance, we study the hole transport in blends that are spin-coated in chloroform and annealed at 110 °C for 4 min as well as blends that are spin-coated in *ortho*-dichlorobenzene (ODCB) and dried at room temperature in a closed Petri dish. To exclude contributions from the electron transport in the PCBM to the measured current, we used palladium top contacts. In Figure 10.9, the SCL hole-only currents are shown for the fast ($L = 220$ nm) and slowly dried ($L = 408$ nm) films.

For the fast dried and annealed film, the *J–V* characteristics are quadratic, as expected for an SCL current. The solid line is the calculated current employing a hole

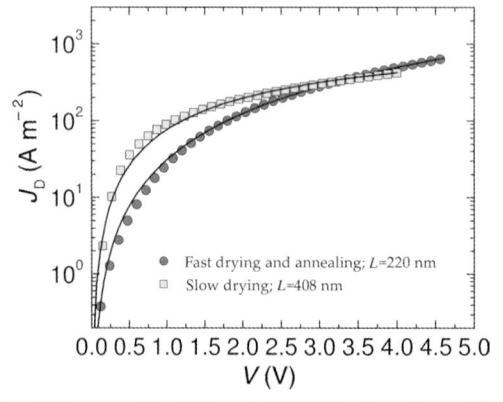

Figure 10.9 Experimental dark current densities (J_D) of the 50:50 wt% P3HT:PCBM blend devices, measured at room temperature in the hole-only device configuration. The symbols correspond to different drying conditions of the photoactive layer. The solid lines represent the fit using a model of single carrier SCL current.

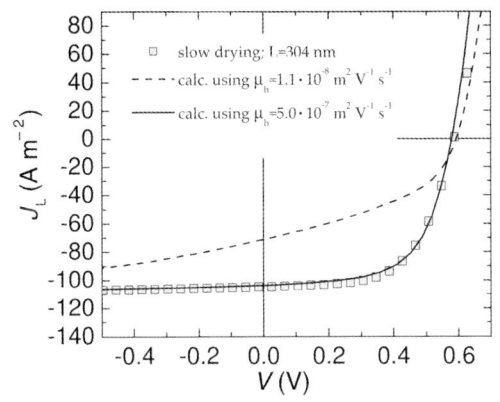

Figure 10.10 Experimental photocurrent (J_L) of a P3HT:PCBM blend solar cell device, prepared by the slow drying method of the photoactive layer (squares), together with the model calculation, using a hole mobility measured in the fast drying film (dashed line) and slow drying film (solid line).

mobility $\mu_h = 1.1 \times 10^{-8}\,m^2\,V^{-1}\,s^{-1}$ similar to the value reported before. For the slowly dried films, a mobility of $\mu_h = 5.0 \times 10^{-7}\,m^2\,V^{-1}\,s^{-1}$ is obtained. Thus, using slow drying, we observe that the hole mobility in the P3HT phase increases by a factor of 45 with respect to the annealed films of Figure 10.9 and compared to a previously reported value for annealed films it is 33 times higher [31]. In Figure 10.10, the photocurrent of the slowly dried device is modeled using the enhanced mobility of $\mu_h = 5.0 \times 10^{-7}\,m^2\,V^{-1}\,s^{-1}$ as input, while the other parameters were kept the same as for the case of the annealed devices: $\mu_e = 3.0 \times 10^{-7}\,m^2\,V^{-1}\,s^{-1}$, $a = 1.8\,nm$, and $k_f^{-1} = 7 \times 10^{-5}\,s$. Taking the enhanced mobility into account, the calculated photocurrent is in excellent agreement with the measurements. As a reference, the dashed line using the lower mobility of the annealed devices is also included in the plot. As expected, the increase in the mobility leads to a strong enhancement of the FF, going from 42 to 61%. This enhancement of FF together with the increased absorption in the thick film enhances the power efficiency from 3.1 to 3.7%. The role of the increased mobility is that the transition from non-SCL toward the SCL regime is extended to higher thickness. With a mobility of $5.0 \times 10^{-7}\,m^2\,V^{-1}\,s^{-1}$, the 304 nm device is still in the regime where space charge effects do not play a significant role. This is also confirmed by the linear intensity dependence of J_{sc}. As a result, these slowly dried P3HT:PCBM devices are the first plastic solar cells where the electron and hole transport is balanced.

10.5 Conclusions

The enhancement of the hole mobility in MDMO-PPV with two orders of magnitude upon blending with PCBM is the main reason for the achieved 2.5% efficiency in this type of cells. However, the remaining factor of 10 difference in electron and hole mobility leads to the formation of space charges in thick (>250 nm) devices. The resulting reduction of the fill factor then counteracts the increase in absorption. In annealed P3HT:PCBM solar cells, identical charge transport properties are obtained as compared to the MDMO-PPV:PCBM devices. The increase in performance toward 3.5%, however, results from the fact that these mobilities are obtained in a blend of 1 : 1 wt%, as compared to the 1 : 4 wt% for the MDMO-PPV:PCBM case. The larger polymer fraction in the blend, together with a red shift of the absorption, leads to an increase in the amount of absorbed light resulting in a higher efficiency. For slowly dried P3HT:PCBM devices, a balanced charge transport is obtained. As a result, for thick devices no significant space charge formation and reduction of the fill factor occur. The increased absorption then enhances the efficiency to typically 4%.

Acknowledgments

The authors especially acknowledge the contributions of Kees Hummelen, Bert de Boer, Jur Wildeman, Minte Mulder, Alex Sieval, and Edsger Smits to this work. These investigations were financially supported by the Dutch Ministries of EZ, O&W, and VROM through the EET program (EETK97115). The work of L. J. A. Koster forms part of the research program of the Dutch Polymer Institute (#323).

References

1 Green, M.A., Emery, K., King, D.L., Hishikawa, Y. and Warta, W. (2006) *Progress in Photovoltaics*, **14**, 455.
2 Kallmann, H. and Pope, M. (1959) *Journal of Chemical Physics*, **30**, 585.
3 Gomes da Costa, P. and Conwell, E.M. (1993) *Physical Review B: Condensed Matter*, **48**, 1993.
4 Marks, R.N., Halls, J.J.M., Bradley, D.D.C., Friend, R.H. and Holmes, A.B. (1994) *Journal of Physics of Condensed Matter*, **6**, 1379.
5 Barth, S. and Bässler, H. (1997) *Physical Review Letters*, **79**, 4445.
6 Tang, C.W. (1986) *Applied Physics Letters*, **48**, 183.
7 Sariciftci, N.S., Smilowitz, L., Heeger, A.J. and Wudl, F. (1992) *Science*, **258**, 1474.
8 Halls, J.J.M., Pichler, K., Friend, R.H., Moratti, S.C. and Holmes, A.B. (1996) *Applied Physics Letters*, **68**, 3120.
9 Markov, D.E., Tanase, C., Blom, P.W.M. and Wildeman, J. (2005) *Physical Review B: Condensed Matter*, **72**, 045217.
10 Markov, D.E., Amsterdam, E., Blom, P.W.M., Sieval, A.B. and Hummelen, J.C. (2005) *Journal of Physical Chemistry A*, **109**, 5266.
11 Yu, G., Gao, J., Hummelen, J.C., Wudl, F. and Heeger, A.J. (1995) *Science*, **270**, 1789.
12 Shaheen, S.E., Brabec, C.J., Sariciftci, N.S., Padinger, F., Fromherz, T. and Hummelen, J.C. (2001) *Applied Physics Letters*, **78**, 841.
13 Li, G., Shrotriya, V., Huang, J., Yao, Y., Moriarty, T., Emery, K. and Yang, Y. (2005) *Nature Materials*, **4**, 864.

14 Goodman, A.M. and Rose, A. (1971) *Journal of Applied Physics*, **42**, 2823.
15 Sokel, R. and Hughes, R.C. (1982) *Journal of Applied Physics*, **53**, 7414.
16 Mihailetchi, V.D., Wildeman, J. and Blom, P.W.M. (2005) *Physical Review Letters*, **94**, 126602.
17 Blom, P.W.M., de Jong, M.J.M. and Vleggaar, J.J.M. (1996) *Applied Physics Letters*, **68**, 3308.
18 Mihailetchi, V.D., van Duren, J.K.J., Blom, P.W.M., Hummelen, J.C., Janssen, R.A.J., Kroon, J.M., Rispens, M.T., Verhees, W.J.H. and Wienk, M.M. (2003) *Advanced Functional Materials*, **13**, 43.
19 Melzer, C., Koop, E., Mihailetchi, V.D. and Blom, P.W.M. (2003) *Advanced Functional Materials*, **14**, 865.
20 Mihailetchi, V.D., Koster, L.J.A., Hummelen, J.C. and Blom, P.W.M. (2004) *Physical Review Letters*, **93**, 216601.
21 Braun, C.L. (1984) *Journal of Chemical Physics*, **80**, 4157.
22 Koster, L.J.A., Smits, E.C.P., Mihailetchi, V.D. and Blom, P.W.M. (2005) *Physical Review B: Condensed Matter*, **72**, 085205.
23 Schilinsky, P., Waldauf, C., Hauch, J. and Brabec, C.J. (2004) *Journal of Applied Physics*, **95**, 2816.
24 Riedel, I. and Dyakonov, V. (2004) *Physica Status Solidi a: Applied Research*, **201**, 1332.
25 Lenes, M., Koster, L.J.A., Mihailetchi, V.D. and Blom, P.W.M. (2006) *Applied Physics Letters*, **88**, 243502.
26 Padinger, F., Rittberger, R.S. and Sariciftci, N.S. (2003) *Advanced Functional Materials*, **13**, 85.
27 Chirvase, D., Parisi, J., Hummelen, J.C. and Dyakonov, V. (2004) *Nanotechnology*, **15**, 1317.
28 Kim, Y., Choulis, S.A., Nelson, J., Bradley, D.D.C., Cook, S. and Durrant, J.R. (2005) *Applied Physics Letters*, **86**, 063502.
29 Goh, C., Kline, R.J., McGehee, M.D., Kadnikova, E.N. and Fréchet, J.M.J. (2005) *Applied Physics Letters*, **86**, 122110.
30 de Boer, B., Hadipour, A., Mandoc, M.M., van Woudenbergh, T. and Blom, P.W.M. (2005) *Advanced Materials*, **17**, 621.
31 Mihailetchi, V.D., Xie, H., de Boer, B., Koster, L.J.A. and Blom, P.W.M. (2006) *Advanced Functional Materials*, **16**, 599.

11
Morphology of Bulk Heterojunction Solar Cells

Joachim Loos, Svetlana van Bavel, and Xiaoniu Yang

11.1
Introduction

Organic electronics (OE) has the potential to become one of the major industries of the twenty-first century. R&D endeavors are focusing on continuous roll-to-roll printing of polymeric or organic compounds from solution – like newspapers – to produce flexible and lightweight devices at low cost. In particular, polymeric semiconductor-based solar cells (PSCs) are currently under investigation as potential low-cost devices for sustainable solar energy conversion. Because they are large-area electronic devices, readily processed polymeric semiconductors from solution have an enormous cost advantage over inorganic semiconductors. Further benefits are the low weight and flexibility of the resulting thin-film devices. Despite the progress made in the field, it is clear that PSCs are still in their early research and development stage. Several issues must be addressed before PSCs become practical devices. This includes further understanding of operation and stability of these cells and the control of the morphology formation – mainly morphology of the photoactive layer – which is directly linked to the performance of devices.

11.2
The Bulk Heterojunction of a Polymer Solar Cell

One of the main differences between inorganic and organic semiconductors is the magnitude of the exciton binding energy (exciton = bound electron–hole pair). In many inorganic semiconductors, the binding energy is small compared to the thermal energy at room temperature and therefore free charges are created under ambient conditions upon excitation across the bandgap [1]. An organic semiconductor, on the contrary, typically possesses an exciton binding energy that exceeds kT (roughly by more than an order of magnitude) [2]. As a consequence, excitons are formed upon excitation instead of free charges. This difference between inorganic

Organic Photovoltaics: Materials, Device Physics, and Manufacturing Technologies.
Edited by Christoph Brabec, Vladimir Dyakonov, and Ullrich Scherf
Copyright © 2008 WILEY-VCH Verlag GmbH & Co. KGaA, Weinheim
ISBN: 978-3-527-31675-5

and organic semiconductors is of critical importance in PSCs. While in conventional inorganic solar cells free charges are created upon light absorption, PSCs need an additional mechanism to dissociate the excitons.

A successful method to dissociate bound electron–hole pairs in organic semiconductors is the application of the so-called donor–acceptor interface. This interface is formed between two organic semiconductors with different valence and conduction bands, or equivalently dissimilar highest occupied molecular orbital (HOMO) and lowest unoccupied molecular orbital (LUMO) levels, respectively. The donor material is the material with the lowest ionization potential and the acceptor material the one with the largest electron affinity. If an exciton is created in the photoactive layer and reaches the donor–acceptor interface, the electron will be transferred to the acceptor material and the hole will recede in the donor. Afterwards, both charge carriers move to their respective electrodes when electrode materials are chosen with the right work functions.

The external quantum efficiency η_{EQE} of a photovoltaic cell based on exciton dissociation at a donor–acceptor interface, in which the donor material is the material with the lowest ionization potential and the acceptor material the one with the largest electron affinity, is $\eta_{EQE} = \eta_A \times \eta_{ED} \times \eta_{CC}$ [3], with the light absorption efficiency η_A, the exciton diffusion efficiency η_{ED}, which is the fraction of photogenerated excitons that reaches a donor–acceptor interface before recombining, and the carrier collection efficiency η_{CC}, which is the probability that a free carrier generated at a donor–acceptor interface by dissociation of an exciton reaches its corresponding electrode. Donor–acceptor interfaces can be very efficient in separating excitons: systems are known in which the forward reaction, the charge generation process, takes place on the femtosecond timescale, whereas the reverse reaction, the charge recombination step, occurs in the microsecond range [4]. The typical exciton diffusion length in organic semiconductors, and in particular in conjugated polymers, however, is limited to $\sim$10 nm [5–7]. Consequently, acceptor–donor interfaces have to be within this diffusion range for efficient exciton dissociation and free charge creation.

Independently, Yu et al. and Halls et al. have addressed the problem of limited exciton diffusion length by intermixing two conjugated polymers with different electron affinities [8,9], or a conjugated polymer with C_{60} molecules or their methanofullerene derivatives [10]. Because phase separation occurs between the two constituents, a large internal interface is created so that most excitons would be formed near the interface and dissociate at the interface. In the case of the polymer–polymer intermixed film, the approach has been found successful in the observation that the photoluminescence from each of the polymers was quenched. This implies that the excitons generated in one polymer within the intermixed film reach the interface with the other polymer and dissociate before recombining. This device structure, a so-called bulk heterojunction (Figure 11.1), provides a route by which nearly all photogenerated excitons in the film can be split into free charge carriers. At present, bulk heterojunction structures are the main candidates for high-efficiency polymeric solar cells. However, tailoring their morphology toward optimized performance is a challenge.

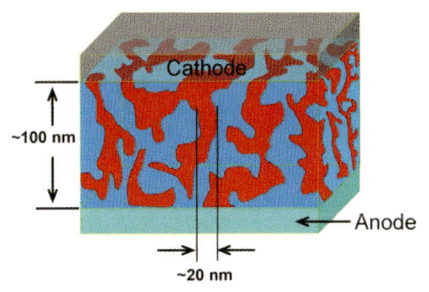

Figure 11.1 Schematic three-dimensional representation of a bulk heterojunction (electron donor and acceptor constituents in different colors) with top and back electrodes.

Several (morphological) requirements for the photoactive layer are summarized in Table 11.1. Since absorption spectrum range of the currently used donor materials is insufficient for optimum utilization of the incident light, photoinduced absorption of the cell should better fit the solar spectrum, which requires, for example, lower bandgap polymers to capture more near-infrared (NIR) photons. To enhance absorption capability, the photoactive layer should be thick enough to absorb more photons from the incident light, which needs layers with thickness of hundreds of nanometers; however, these layers still need controlled nanoscale phase separation between the donor and the acceptor material. During solidification of the constituents from the solution and thus forming a thin film, various approaches could be applied to tune the morphology of the thin blend film toward optimum organization, including both thermodynamic and kinetic aspects. The Flory–Huggins parameter χ

Table 11.1 Molecular and morphology requirements for the photoactive layer of a high-performance polymer solar cell.

Requirement	Influenced by Molecular architecture	Influenced by Morphology
Utilization of incident light	Molecule design to tune bandgap(s)	Layer thickness, roughness of interfaces
Exciton dissociation	Match of band properties between donor and acceptor	Max. interface
		Small acceptor–donor phases within exciton diffusion range
Charge transport	Molecule design with high charge carrier mobility	Short and continuous pathways to the electrodes
		Ordered (crystalline) transportation pathways
Processability	Optimized molecular architecture (stereoregularity, molecular weight, and branching)	Thermodynamics and kinetics of film formation and phase separation

between the constituents describes the main driving force of phase segregation in a blend from thermodynamics point of view, in which the ratio between the constituents and their solubility and conformation in the solution are the key aspects to determining the length scale of phase separation. The kinetic issues have significant influence on the morphology of the thin blend film obtained. For instance, by applying spin coating, a thin blend film with rather homogeneous morphology will be obtained while large-scale phase separation may undergo when using preparation methods like film casting, in which a more equilibrium organization is achieved during film formation [11]. Therefore, both thermodynamic aspects as well as kinetics determine the organization of the bulk-heterojunction photoactive layer in polymer solar cells, that is, they control the formation of a nanoscale interpenetrating network with, probably, crystalline order of both constituents [12,13].

In the following sections, we are going to describe typical preparation approaches and routes toward creating the desired morphology of the photoactive layer of a polymer solar cell. In particular, we discuss in detail the influence of constituents and solvent used, composition, and annealing treatments on morphology formation. Moreover, we highlight the importance of high-resolution microscopy techniques to provide the required information on the local nanometer-scale organization of the active layer as well as its local electrical properties. Finally, we demonstrate on the system P3HT/PCBM how we can apply successfully our toolbox to control the morphology of the photoactive layer.

11.3
Our Characterization Toolbox

11.3.1
Microscopy

The thin-film nature of the active layer with typical thickness of about 100 nm and the need for local morphology information have led to high-resolution microscopy techniques becoming the main investigation tools for morphology characterization. Transmission electron microscopy (TEM) including bright-filed observation and selected area electron diffraction (SAED) analysis, scanning electron microscopy (SEM) and scanning probe microscopy (SPM) – in particular atomic force microscopy (AFM) – have proven their versatility for a detailed characterization of the morphology of the active layer. The main difference between TEM on the one hand and SPM and SEM on the other is that TEM provides mainly morphological information of the lateral organization of thin-film samples by the acquisition of projections through the whole film (in transmission), whereas SPM and SEM probe the topography or phase demixing at the surface of such thin-film samples. On the basis of comprehensive microscopy studies, several morphology determining factors have been identified that will be described in Section 11.4. More details on sample preparation procedures and experimental setups can be found in the related references of each part of our study.

11.3.2
Characterization of Nanoscale Electrical Properties

In general, performance measurements of PSCs are carried out on operational devices having at least the size of square millimeters. On the contrary, the characteristic length scale determining the functional behavior of the active layer is on the order of 10 nm (exciton diffusion length) to about 100 nm (layer thickness). Moreover, it is believed that the local nanometer scale organization of nanostructures dominantly controls the electrical behavior of devices. Thus, it is necessary to obtain property data of nanostructures with nanometer resolution so as to be able to establish structure– property relations linking length scales from local nanostructures to large-scale devices.

In this respect, a very useful analytical tool is scanning probe microscopy and in particular the atomic force microscopy equipped with a conductive probe, the so-called conductive AFM (C-AFM) [14,15]. Because AFM uses the interaction force between the probe and the sample surface as a feedback signal, both topography and conductivity of the sample can be mapped independently. Theoretically, the resolution of C-AFM is as small as the tip-sample contact area, which can be less than 20 nm.

C-AFM is widely used for the characterization of electrical properties of organic semiconductors. For example, single crystals of sexithiophene have been studied [16], where the $I–V$ characteristics of the samples were measured. Several electrical parameters such as grain resistivity and tip-sample barrier height were determined from these data. In another study, the hole transport in thin films of poly[2-methoxy-5-(2′-ethyl-hexyloxy)-1,4-phenylene vinylene] (MEH-PPV) was investigated and the spatial current distribution and $I–V$ characteristics of the samples were discussed [17].

In our group, recently the first study on the spatial distribution of electrical properties of realistic bulk heterojunctions has been performed by applying C-AFM with a lateral resolution better than 20 nm [18]. For this study, poly[2-methoxy-5-(3,7-dimethyloctyloxy)-1,4-phenylenevinylene] blended with poly[oxa-1,4-phenylene-1,2-(1-cyanovinylene)-2-methoxy,5-(3,7-dimethyloctyloxy)-1,4-phenylene-1,2-(2-cyanovinylene)-1,4-phenylene] (MDMO-PPV/PCNEPV) was chosen (Figure 11.2). Measurements of the electrical current distribution over the sample surface were performed with an Au-coated tip. In such an experiment, the tip plays the role of the back electrode but with a much more localized contact area. A voltage was applied to the tip and the ITO front electrode was grounded (Figure 11.3). For conductive AFM measurements, the tip was kept in contact with the sample surface while the current through the tip was measured. C-AFM measurements of the same sample area were done several times and resulted in completely reproducible data. Subsequent analysis of the surface showed almost no destruction of the sample surface; only minor changes were detected from time to time.

A topography image and the corresponding current distribution measured at +8 and −8 V on the tip are shown in Figure 11.4. All images were acquired subsequently and thus some drift occurred. All pronounced domains in the topog-

304 | *11 Morphology of Bulk Heterojunction Solar Cells*

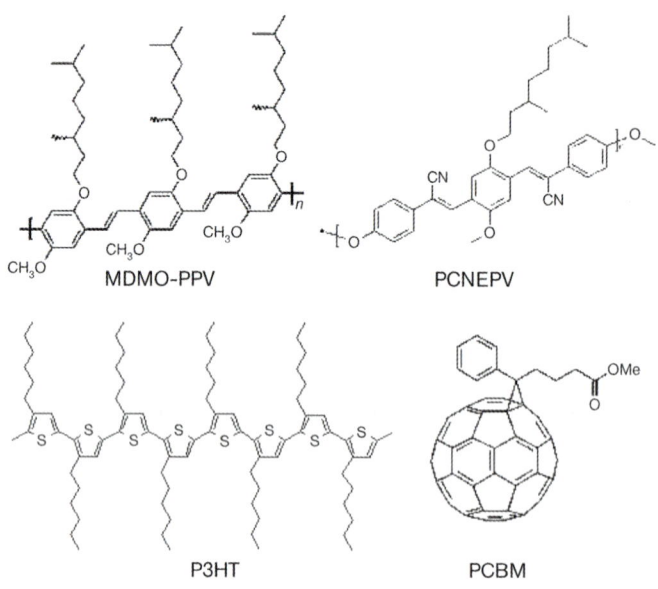

Figure 11.2 Chemical structures of some conjugated polymers and the fullerene derivative PCBM applied in polymer solar cells.

raphy image (Figure 11.4a) correlate with regions of minimal current in the C-AFM image (dark areas in Figure 11.4b).

From the corresponding energy level diagram (Figure 11.5) it follows that the difference between the HOMO level of MDMO-PPV and the Fermi level of both electrodes is rather small so that we expect ohmic contacts for a hole injection and strong energy barriers for electrons. Therefore, a hole-only current through the MDMO-PPV is expected for both polarities of voltage in an ITO/PEDOT:PSS/MDMO-PPV/Au-tip system. The energy difference between the HOMO and the

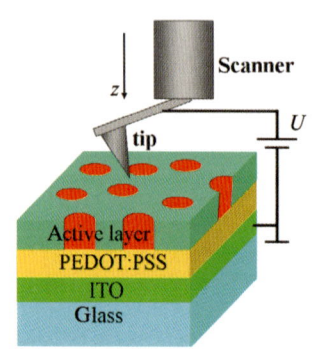

Figure 11.3 Scheme of the sample structure with segregated phases and the conductive AFM experimental setup, including scanner and conductive tip.

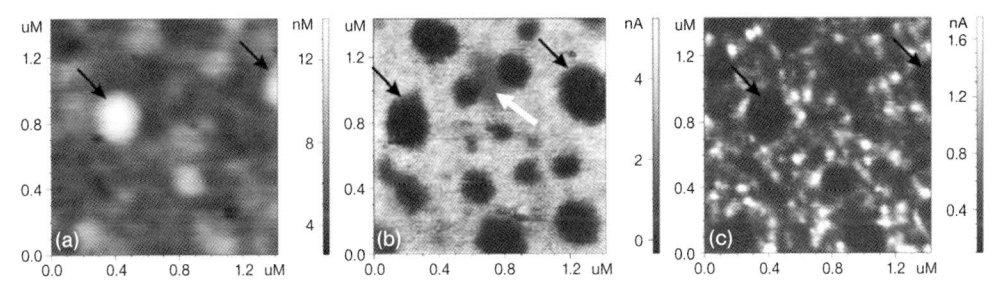

Figure 11.4 C-AFM images of the same area of a MDMO-PPV/PCNEPV active layer: (a) topography; (b) current distribution image with a positive bias at $U_{tip} = +8\,V$, the white arrow indicates a domain with reduced current; and (c) current distribution image with a negative bias at $U_{tip} = -8\,V$; black arrows indicate same domains for the reason of easy identification. (Reprinted with permission from Ref. [18]. Copyright 2006 Elsevier.)

LUMO of PCNEPV and the Fermi levels of both electrodes is about 1 eV, which means that a large barrier for electron injection exists in the structure ITO/PEDOT:PSS/PCNEPV/Au-tip (some changes of barrier heights are possible when contact between metal electrodes and organic material occurs [16,19]). Because the hole mobility of an n-type polymer is typically smaller than that of a p-type polymer, a hole-only current through the MDMO-PPV is larger than for PCNEPV in both bulk- and contact-limited regimes. Therefore, we assume that the observed contrast in Figure 11.4b is because of a hole current flowing through the MDMO-PPV-rich phase.

However, C-AFM measurement also shows regions with an intermediate current in between that of the MDMO-PPV matrix and the PCNEPV domains. A white arrow on Figure 11.4b marks one of these regions. These areas might be assigned as PCNEPV domains inside the active layer that are possibly covered by MDMO-PPV.

It is reported that the electrical contrast measured by C-AFM at the surface of samples depends on the sign of the voltage applied [16]. As shown in Figure 11.4c, the C-AFM measurements at negative bias on the tip showed drastic changes of the contrast in the current images compared to positive bias (Figure 11.4b). PCNEPV domains again showed only little current at low load; however, MDMO-PPV showed a heterogeneous spatial current distribution. These electrical heterogeneities indicate

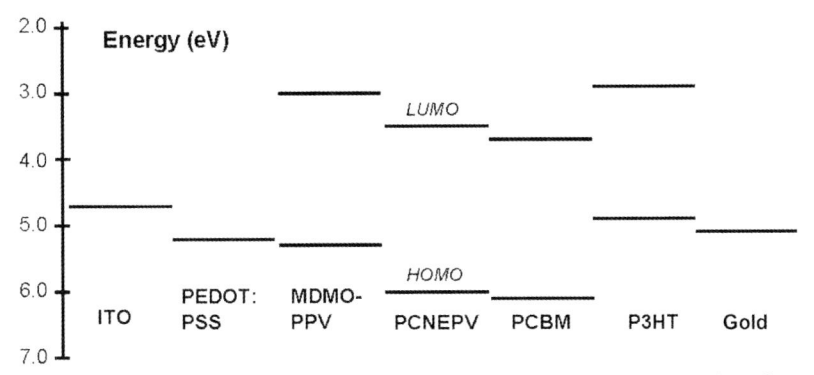

Figure 11.5 Energy level diagram for few materials commonly used in polymer solar cells.

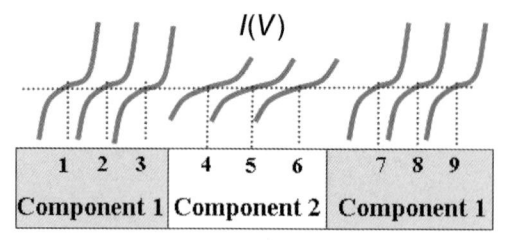

Figure 11.6 Schematic illustration of I–V measurements at nine consecutive points. The component 1 area represents the MDMO-PPV-rich matrix; component 2 area PCNEPV-rich domains.

small grains with a typical size of 20–50 nm, which differs by value of current. A similar structure was observed on MEH-PPV films [17]. In the case of MEH-PPV, the authors attribute these substructures to a special and very local organization of the film that could be caused by local crystallization of stereoregular parts of the MEH-PPV molecules or by impurities incorporated during the synthesis.

In addition to topography and current sensing analysis, current imaging spectroscopy was performed as well. The procedure of such measurements is similar to the so-called "force volume" technique [20], which implies measurements of the force–distance curve at each point of a scan to get complete information about the lateral distribution of mechanical properties at the surface. Here, we extend this method to measurements of electrical properties of the sample [21]. Current–voltage (I–V, for constant distance; always in contact) dependencies were collected at each point of a scan. The procedure of such measurements is shown schematically in Figure 11.6.

Figure 11.7 shows the typical I–V behavior of the MDMO-PPV/PCNEPV system and the corresponding C-AFM current distribution images for various biases when measured in inert atmosphere. The data were obtained by the application of the so-called I–V spectroscopy, which means that for a matrix of 128×128 points on the surface of the sample full I–V curves were measured for biases from -10 to $+10$ V. Three different types of I–V characteristics can be attained in the sample dependent on the location of the measurement. For domains of the electron acceptor compound PCNEPV, the current is always low and the general contrast of the current distribution images depends only on the bias applied. In the case of the electron donor matrix compound MDMO-PPV, two different I–V characteristics can be obtained showing almost the same behavior for positive bias but vary significantly for negative bias. The current distribution images of Figure 11.7 demonstrate this behavior and provide additional information about lateral sizes of the MDMO-PPV heterogeneities. The point resolution of the images is about 15 nm. Some heterogeneities can be recognized with sizes as small as about 20 nm. The C-AFM results obtained on a pure MDMO-PPV film are identical to that obtained on the matrix in the heterogeneous film MDMO-PPV/PCNEPV. The log I–log V plot of data obtained shows quadratic dependence of the current I on the voltage V measured on MDMO-PPV. This implies space-charge-limited current, that is, in agreement with I–V measurements on complete devices [22,23].

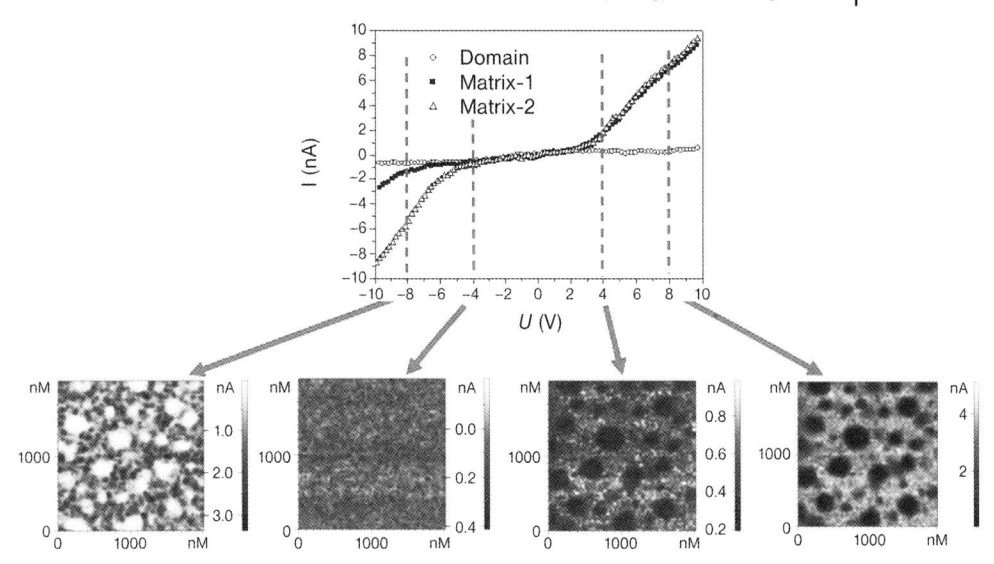

Figure 11.7 (Top) Typical *I–V* curves as measured for each point of the *I–V* spectroscopy scan (128 × 128 pixels) demonstrating the heterogeneous *I–V* characteristics of the MDMO-PPV matrix; (bottom) for four biases the corresponding current distribution images are shown demonstrating the obvious contrast between PCNEPV and MDMO-PPV for high bias as well as contrast within the MDMO-PPV matrix with heterogeneities as small as few tens of nanometers.

11.4
Morphology Determining Factors

11.4.1
Molecular Architecture

The chemical composition and local organization of the active layer (forced by the processing conditions applied) have been identified as important parameters for the performance of an organic solar cell based on the bulk heterojunction concept. However, monitoring the distinct details of the phase separation and the local organization on the nanometer scale of the functional blends is difficult. In a recent study, we have systematically varied the molecular architecture of the acceptor constituent PCNEPV and applied energy-filtered TEM (EFTEM) to identify phase separation in the system MDMO-PPV/PCNEPV by monitoring the nitrogen distribution within the active layer.

Figure 11.8a shows a zero-loss-filtered EFTEM image of sample 2 (Table 11.2). The corresponding nitrogen elemental distribution image of the same area visualizes the nitrogen distribution within the thin MDMO-PPV/PCNEPV film (Figure 11.8b). The signal-to-noise ratio is rather low; however, distinct regions showing higher nitrogen intensity can be identified. Especially, the bright domain in the central bottom part of the image, which has a size on the order of 80 nm, can be easily correlated with a similar dark-gray domain of the corresponding energy-filtered TEM

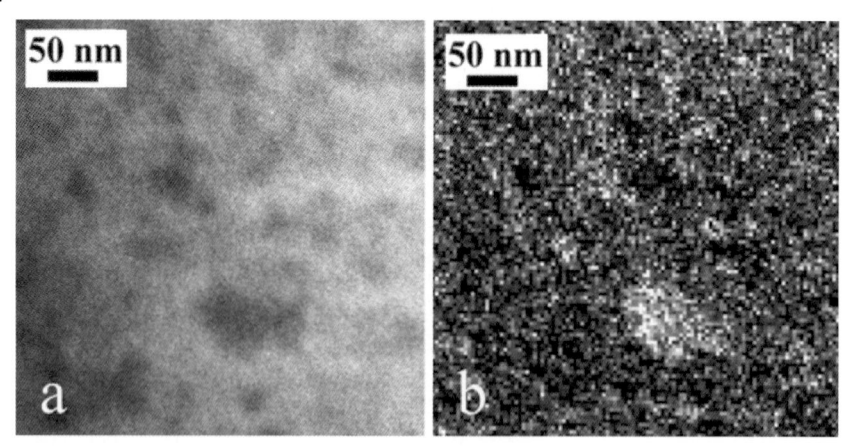

Figure 11.8 (a) Zero-loss filtered TEM and (b) corresponding N–K elemental map of the same area; the latter visualizes the nitrogen distribution within the thin MDMO-PPV/PCNEPV blend film (Table 11.2, sample 2).

image of Figure 11.8a. Keeping in mind that the nitrogen concentration in PCNEPV is only on the order of 5 wt% (or less for some derivatives), the result of the elemental mapping is very satisfying. More details on the imaging procedure applied can be found in Ref. [24].

Figure 11.9 shows zero-loss-filtered TEM images of the morphology of several functional polymer blend thin-film samples. For low molecular weight MDMO-PCNEPV (sample 1, Figure 11.9a), a homogeneous film without distinct phase separation can be observed. When the molecular weight was increased to a medium weight, small and homogeneously distributed PCNEPV domains with sizes of about 20–50 nm were observed in the MDMO-PPV matrix (sample 2, Figure 11.9b). In contrast, the use of high molecular weight MDMO-PCNEPV (sample 3, Figure 11.9c) or MEH-PCNEPV (sample 4, Figure 11.9d) results in a large-scale phase separation with relatively large PCNEPV domains (domain diameter 200–300 nm). In conclusion, we were able to demonstrate that tailoring the phase separation of functional blends is possible by systematic variation of the molecular architecture of their

Table 11.2 Specification of the acceptor polymer properties molecular weight M_w, polydispersity index PDI, glass transition temperature T_g, and resulting domain sizes.

Sample (acceptor material)	M_w (g mol^{-1})	PDI	T_g (°C)	Acceptor domain size (nm)
1 (MDMO-PCNEPV)	3500	1.7	65	<5
2 (MDMO-PCNEPV)	48 000	4	80	20–50
3 (MDMO-PCNEPV)	113 500	2.8	100	~200
4 (MEH-PCNEPV)	173 500	2.9	125	~300

The acceptor is a PCNEPV derivative (MDMO- or MEH-) blended with the donor polymer MDMO-PPV 1:1 by weight.

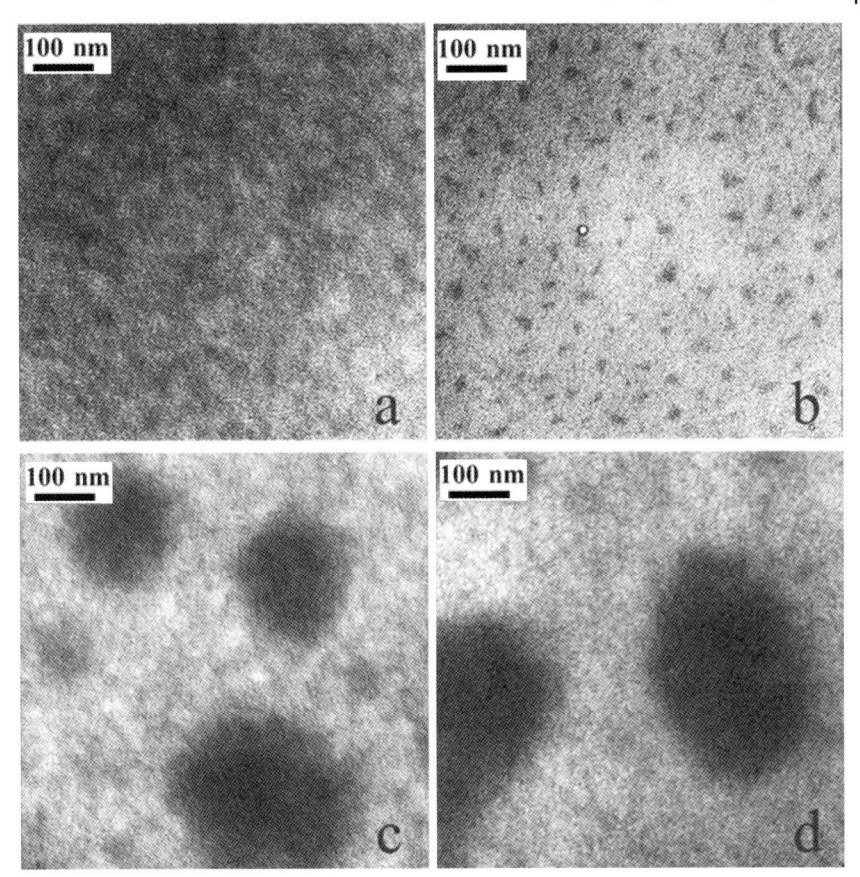

Figure 11.9 Zero-loss filtered TEM images of thin MDMO-PPV/PCNEPV blend film samples: (a) sample 1 with low molecular weight MDMO-PCNEPV acceptor, (b) sample 2 with medium molecular weight MDMO-PCNEPV acceptor, (c) sample 3 with high molecular weight MDMO-PCNEPV acceptor, and (d) sample 4 with MEH-PCNEPV acceptor.

constituents; the blend with PCNEPV of medium molecular weight shows optimum domain sizes of 20–50 nm and has the best *I–V* characteristics of all blends under investigation, too.

11.4.2
Solvents and Preparation Methods

The influence of solvent, constituent concentration, and preparation method used on the morphology formation in various functional bulk heterojunction systems has been studied in detail. Bright-field TEM (BF TEM) images reveal that phase segregation occurs in MDMO-PPV/PCBM (1:4 weight ratio, chemical structure of PCBM in Figure 11.2) films, spin-coated from toluene (Figure 11.10a) and chlorobenzene

310 | 11 Morphology of Bulk Heterojunction Solar Cells

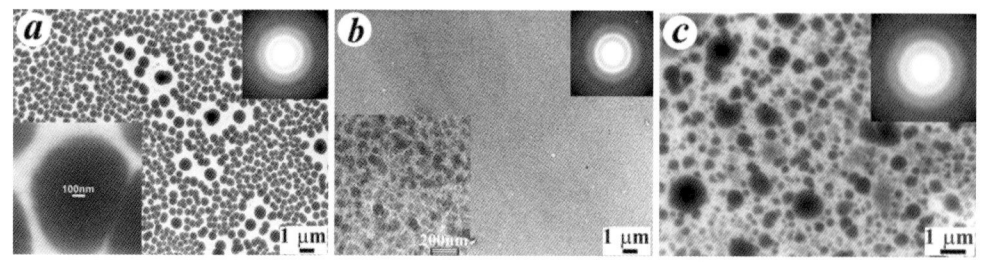

Figure 11.10 Bright-field TEM images of MDMO-PPV/PCBM films (1:4 wt ratio) prepared by spin coating from toluene (a), chlorobenzene (b), and by drop casting from chlorobenzene (c). The insets represent the corresponding SAED patterns. (Reprinted with permission from Ref. [11]. Copyright 2004 American Chemical Society.)

(Figure 11.10b). In the TEM images, the dark areas are attributed to PCBM-rich domains because the electron scattering density of PCBM is much higher than that of MDMO-PPV [25] and the thickness is rather homogeneous (rms roughness of ∼4 nm, as verified by atomic force microscopy) for the film prepared from chlorobenzene. The TEM images demonstrate a morphology in which PCBM-rich domains are dispersed in an MDMO-PPV-rich matrix [26]. The size of the PCBM-rich domains in the blend films, however, changes tremendously with the choice of solvent [26,27]. Using toluene (Figure 11.10a), the average size of the PCBM-rich domains is around 600 nm with a broad size distribution (roughly 350–1300 nm). In contrast, the size of PCBM clusters is quite small with about 80 nm when prepared from chlorobenzene solution. The solubility of both MDMO-PPV and PCBM in toluene is somewhat less than in chlorobenzene. This difference may explain the formation of different domain sizes.

Despite the difference in the length scale of phase separation, the Debye–Scherrer rings observed in the SAED patterns (insets in the micrographs of Figure 11.10) indicate that the crystalline structure of PCBM is identical in both films. The broad Debye–Scherrer rings with average d-spacing of 0.46, 0.31, and 0.21 nm result from the superposition of many single-crystal diffraction patterns originating from PCBM nanocrystals that are randomly distributed in the PCBM-rich domains. A more detailed discussion of the interpretation of the diffraction patterns of PCBM can be found in Ref. [28].

The dramatically different morphologies observed in Figure 11.10 rationalize the strongly different performance of photovoltaic devices fabricated using these solvents [26,27]. The energy conversion efficiency of photovoltaic devices prepared from toluene is approximately $\eta \sim 0.9\%$, while devices prepared using chlorobenzene give $\eta \sim 2.5\%$ [26]. Apart from possible changes in charge carrier mobility with the change of solvent, the fine phase separation obtained from chlorobenzene provides a larger interfacial area for excitons to dissociate and thereby (at least in part) explains the increased efficiency.

Thin polymer films prepared by spin coating are usually not in a thermodynamic equilibrium state, owing to the high rate of solvent evaporation associated with this method. Hence, changing the kinetics of solvent evaporation can influence the film morphology. For instance, drop casting reduces the rate of solvent evaporation compared to spin coating and thereby favors phase separation. Accordingly, the size of the PCBM-rich domains is dramatically increased when the MDMO-PPV/PCBM film is prepared by drop casting from (the same) chlorobenzene solution rather than by spin coating (Figure 11.10c). Drop casting also results in PCBM-rich domains that are composed of a substantial number of nanocrystals as shown by the SAED (inset in Figure 11.10c). Therefore, both the nature of the solvent and its evaporation rate are important parameters in determining the morphology of the active layer and thereby the performance of the devices.

11.4.3
Annealing

11.4.3.1 Probing the Morphology Stability

Another tool to influence the morphology of the active layer of polymer solar cells is the application of a controlled thermal treatment. One purpose of such treatment is to probe the long-term stability of the morphology. Improvement of the long-term stability of polymer solar cells in an ambient atmosphere is currently still a challenge. However, an acceptable lifetime is a key point for PSCs to compete with traditional photovoltaic technology and is a prerequisite for commercialization.

In general, the stability of PSCs is limited by two factors. One is the degradation of materials, in particular the conjugated polymers, upon being exposed to oxygen, water and/or UV radiation. The other limitation comes from the possible morphology instability of the photoactive layer during operation of devices at high temperature (exposed to sunlight, which means at least 60–80 °C are possible!).

Spin coating provides a simple and successful way to prepare films possessing homogeneous morphology within a relatively large area. In the case of the system MDMO-PPV/PCBM, the desired high rate of solvent evaporation suppresses phase separation. Figure 11.11 shows the schematic representation of the MDMO-PPV/PCBM film morphology, in which PCBM-rich domains are composed of PCBM nanocrystals; some PCBM nanocrystals can also be found in the PCBM-poor regions of the films. Hence, the spin-coated films are probably not in their equilibrium state and there is likely a strong thermodynamic driving force for the samples to reorganize toward the stable equilibrium state. This process will be accelerated at elevated temperatures. For PSCs, the morphological reorganization of the active layer with time or temperature may seriously affect the performance and long-term stability.

To investigate the effect of the annealing temperature on the phase separation, we annealed MDMO-PPV/PCBM films with 80 wt% PCBM at temperatures ranging from 60 to 120 °C. Figure 11.12 shows the corresponding bright-field TEM micrographs and electron diffraction patterns of PCBM single crystals grown under these conditions. Clearly, these PCBM crystals can be classified into two groups by their characteristic appearance. The crystals obtained from the annealing temperatures $\geq 80\,°C$ exhibit a prominent contrast to their background. However, the crystals obtained from lower annealing temperatures have large sizes in the lateral

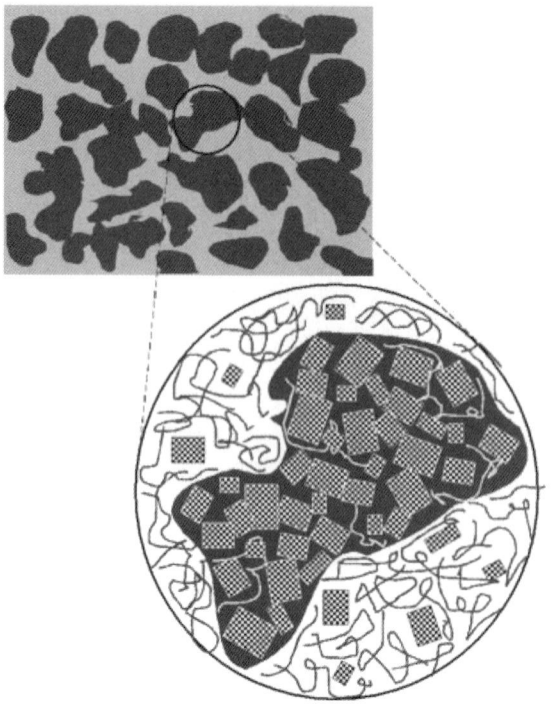

Figure 11.11 Schematic representation of the MDMO-PPV/PCBM film morphology, in which PCBM-rich domains are composed of PCBM nanocrystals; some PCBM nanocrystals can also be found in the PCBM-poor regions of the films. (Reprinted with permission from Ref. [11]. Copyright 2004 American Chemical Society.)

dimensions but seem to be thin and located at the surface. Since MDMO-PPV possesses a glass transition temperature (T_g) of around 80 °C, the diffusion of PCBM can benefit from the higher conformational dynamics of the polymer chains when the annealing temperatures are above this value. Contrarily, for annealing temperatures below T_g of the polymer matrix, the diffusion mobility of the PCBM molecules is hampered since the polymer matrix is approaching to a chain frozen state. Nevertheless, the diffusion of PCBM does happen at a temperature as low as 60 °C, although the mobility has drastically decreased and annealing times on the order of one week have to be applied before changes in the morphology could be detected.

To gain information on the evolution of the morphology of the thin blend film on the substrate, and in particular on the PCBM diffusion behavior, AFM operated in an intermittent-contact mode is used to acquire topography images during an annealing process [19]. Figure 11.13 shows a series of AFM topography images of MDMO-PPV/PCBM films recorded *in situ* upon annealing at 130 °C for different times. For the pristine film before any thermal treatment (Figure 11.13a), a homogeneous morphology within a relatively large area is observed at the scanning resolution we used.

11.4 Morphology Determining Factors | 313

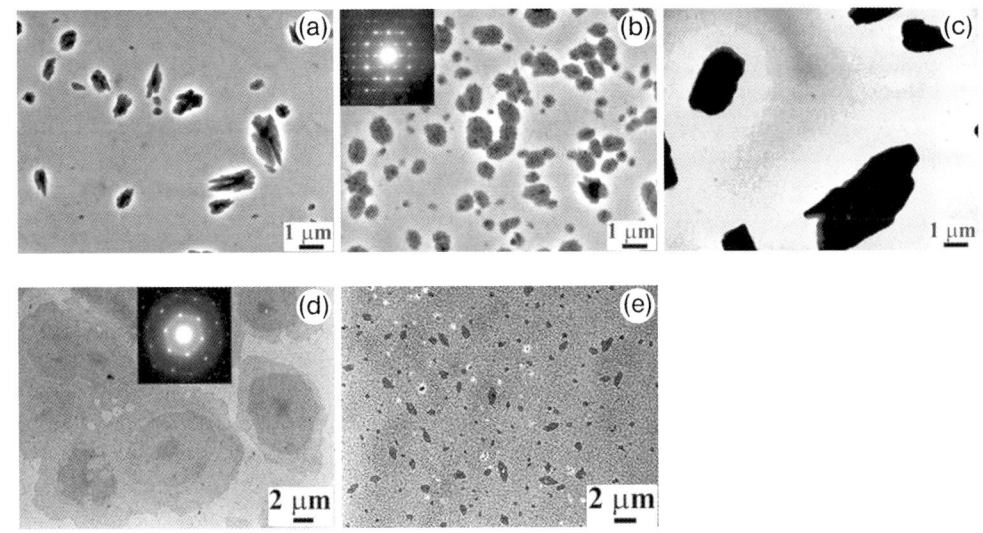

Figure 11.12 TEM bright-field images of MDMO-PPV/PCBM films (1:4 wt ratio) annealed at temperatures of (a) 120 °C for 50 min, (b) 100 °C for 8 h, (c) 80 °C for 25 h, (d) 70 °C for 80 h, and (e) 60 °C for 120 h. The SAED patterns in Figure 11.6b and d show typical diffraction patterns of PCBM single crystals. (Reprinted with permission from Ref. [11]. Copyright 2004 American Chemical Society.)

Upon annealing, PCBM single crystals grow up gradually with annealing time and stick out of the film plane (as shown in Figure 11.13b–f). Notably, in these AFM topography images, the bright domains are PCBM single crystals (marked as A in Figure 11.13d); and the dark areas (depletion zones, marked as B in Figure 11.13d) initially surrounding the PCBM crystals reflect thinner regions of the film, being composed of almost pure MDMO-PPV (i.e., depleted from PCBM). To acquire the exact growth kinetics for both the PCBM crystals and the depletion zones, volume quantification calculations were applied to topographic images from the composite film annealed for different times similar to the areas shown in Figure 11.13. However, this time the scan size was $100\,\mu m \times 100\,\mu m$ so that the detailed volume evolution of either the PCBM single crystals or depletion zones could be resolved. Since the calculations are carried out on the basis of quite large areas of the composite films, the results make statistical sense. More details on the procedures applied can be found in Ref. [29].

The principal reorganization scheme can be seen in Figure 11.14. At the very initial annealing time, as shown in Figure 11.14a, the volume amount of the film collapsed in the depletion area is smaller compared to that of the diffused PCBM inserted in the crystals. As annealing goes on, more and more PCBM is diffused toward the crystals, which ultimately causes a sudden collapse of large areas of the remaining MDMO-PPV matrix (Figure 11.14b). Ultimately almost the whole MDMO-PPV matrix film collapses however, the PCBM diffusion and crystal growth still continue

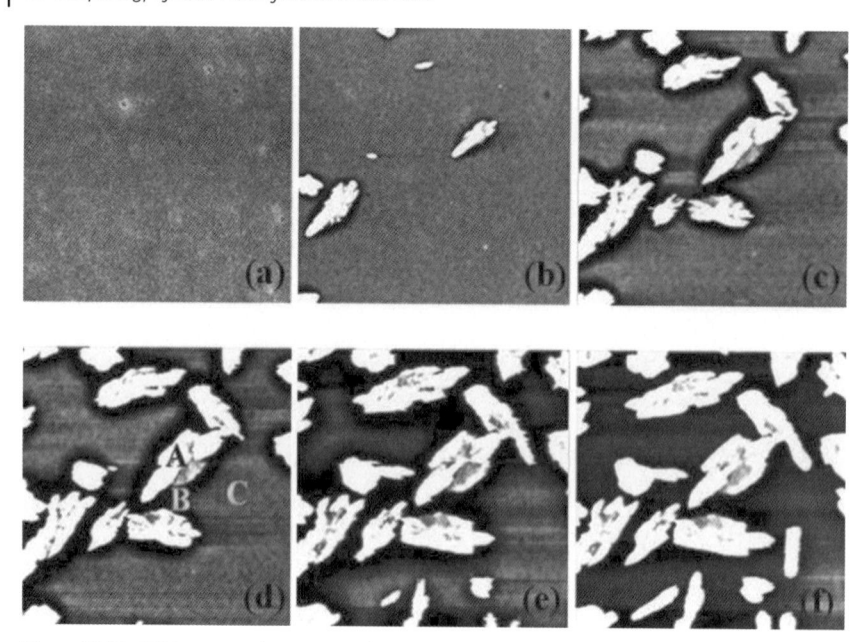

Figure 11.13 AFM topography images of MDMO-PPV/PCBM blend films (MDMO-PPV: PCBM = 1:4 by weight) *in situ* recorded upon annealing at 130 °C for (a) pristine film; (b) 12 min; (c) 22 min; (d) 27 min; (e) 38 min; and (f) 73 min. Scan size: 15 × 15 μm^2; height range (from peak to valley): 200 nm. The letters A, B, and C marked in Figure 11.8d represent the region where the PCBM nucleates and the crystal growths (A), the depletion zone that is formed owing to moving out of PCBM material toward the growing crystal (B), and the initial blend film still consisting of both MDMO-PPV and PCBM (C). (Reprinted with permission from Ref. [29]. Copyright 2004 American Chemical Society.)

(Figure 11.14c). Finally, the diffusion rate of PCBM within the whole film decreases, reaching its equilibrium state, as shown in Figure 11.14d.

The prominent morphology evolution of thin MDMO-PPV/PCBM films at elevated temperature is a typical phenomenon observed at various annealing temperatures (even as low as 60 °C), with different PCBM ratios in the film and under various spatial confinements. This morphological change of the film is ascribed to the diffusion of PCBM molecules within the MDMO-PPV matrix even at temperatures below the glass transition temperature of the MDMO-PPV matrix of about 80 °C and subsequent crystallization of PCBM molecules into large-scale crystals. However, for the high-performance polymer solar cell, the phase separation between electron donor and acceptor components should be controlled within a designed range to ensure a large interface for excitons to be dissociated efficiently. The large-scale phase separation enormously reduces the size of this interface area, causing significantly decreased performance or even leading to failure of the device. Therefore, large phase separation between donor and acceptor compounds should be prevented during both device fabrication and operation, particularly at elevated temperature.

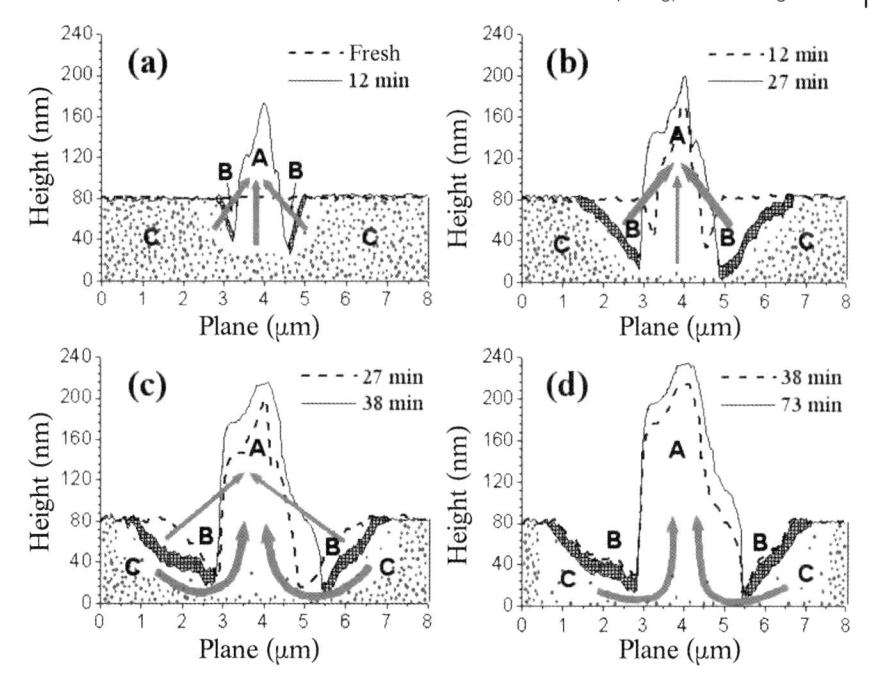

Figure 11.14 Schematic representations of the detailed morphology evolutions of thin MDMO-PPV/PCBM composite film upon thermal treatment. The dots in the profile represent PCBM molecules/nanocrystals and density of the dots represents the richness of PCBM; the diamond-outlined regions represent the depletion zones after PCBM material moved out for crystal growth, in which the density of diamond outlines represents the richness of MDMO-PPV. (sketches are similar to Figure 5 of Ref. [30].)

11.4.3.2 Morphology Control via Annealing

On the contrary, annealing of, for example, the P3HT/PCBM (chemical structure in Figure 11.2) system results in a significant improvement of its power conversion efficiency and in the stabilization of the morphology. After the pioneering work of Padinger *et al.* [30] and Waldauf *et al.* [31], in 2005, a series of studies dealing with the morphology-driven high performance of P3HT/PCBM-based polymer solar cells were published [13,32–35]. In all studies, a remarkable increase of the performance is observed after annealing the devices and power conversion efficiencies as high as 5.2% (using AM 1.5 testing conditions) are reported [35]. The influence of annealing and other parameters on morphology formation of the system P3HT/PCBM will be discussed in detail in Section 11.5 of this chapter.

11.4.4
Confinements

It has been shown that the dynamics of reorganization of a thin film upon thermal treatment is determined not only by the composition and organization of the film itself, but also by its local environment. Numerous experiments indicate that the mobility of polymer molecules in the vicinity of a surface or an interface is perturbed [36–38] and the extent to which they affect the mobility of polymer chains depends on the strength of their interactions with the surface/interface. At a free surface, there is a preferential aggregation of chain ends [39,40]. The presence of these chain ends will increase the free volume and consequently enhance the mobility of molecules at the surface. However, experimental studies have yielded contradictory results. For example, it is claimed that T_g is reduced at the regions close to the surface [41–44]; however, others have found that T_g of the free surface and the bulk is identical [45,46]. Anyway, these results suggest that the mobility of molecules in the vicinity of a free surface would either remain unchanged or increase. But when the thin film is covered and an impenetrable interface is created, the situation becomes more complicated. Strong attraction between the two films may retard the mobility of molecules and it results in a reduced diffusion coefficient [47]. In contrast, a weak interaction will hardly affect the mobility of molecules – and in the extreme case, even T_g can be decreased similar as in the case of a free surface [48].

For a ready-to-work PSC, the photoactive layer is actually sandwiched between two solid layers: a top metal layer of, for example, Al, LiF/Al, Ca, or Ba, and a bottom PEDOT:PSS-coated ITO/glass substrate. Therefore, the morphological evolution with time and the thermal stability of the active layer is strongly determined by its interaction with these two interfaces. Recently, we investigated the reorganization of photoactive thin MDMO-PPV/PCBM films upon annealing for different conditions of spatial confinements: no confinement, which corresponds to freestanding films; single-sided confinement, in which one side of the films is supported; and double-sided or sandwichlike confinement, in which the films are covered on both sides.

As discussed before, during spin coating of a solution of MDMO-PPV/PCBM in chlorobenzene, the high rate of solvent evaporation suppresses phase separation. Hence, the spin-coated films are probably not in their equilibrium state and there is likely a strong thermodynamic driving force to reorganize toward the stable equilibrium state. This process will be accelerated at elevated temperatures and is different for the three types of confinement investigated [49].

Figure 11.15 shows bright-field TEM images and corresponding SAED patterns of thin MDMO-PPV/PCBM film samples with 80 wt% PCBM. In the TEM images, the darker areas are attributed to PCBM-rich domains because the electron scattering density of PCBM is higher than that of MDMO-PPV [25]. The initial film morphology consists of PCBM nanocrystals homogeneously distributed in the MDMO-PPV matrix (Figure 11.15a).

The morphological evolution of the films upon annealing at 130 °C for different types of confinement is monitored. In the case of freestanding film samples, PCBM clusters are formed upon annealing that can be identified in the TEM image as dark areas in a gray MDMO-PPV/PCBM matrix. Notably, the brighter areas surrounding the PCBM clusters reflect thinner regions of the film, composed of almost pure MDMO-PPV (i.e., depleted from PCBM). We note that the dark PCBM clusters visualized in these images are single crystals, as evidenced from the corresponding SAED pattern (inset in Figure 11.15c). The PCBM single crystals develop gradually with annealing time and, particularly for these freestanding films, demonstrate a

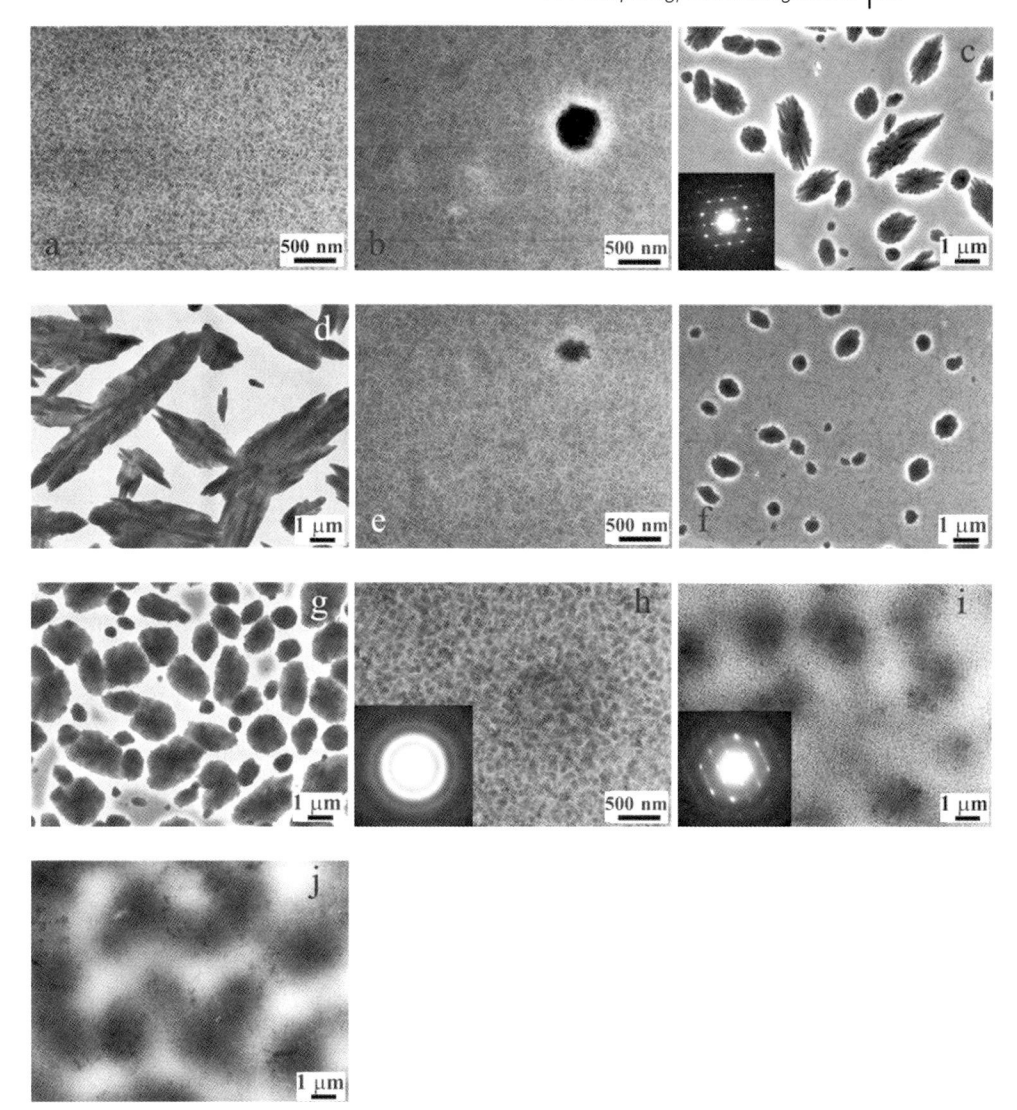

Figure 11.15 Bright-field TEM images demonstrating the formation of PCBM single crystals with time from 80 wt% MDMO-PPV/PCBM composite films upon annealing at 130 °C under different spatial confinements; (a) as spin-coated (fresh) sample; freestanding for (b) 10, (c) 20, and (d) 60 min; single-sided confined for (e) 10, (f) 20, and (g) 60 min; double-sided confined for (h) 20, (i) 38, and (j) 120 min. The insets are corresponding SAED patterns. For double-sided confined films, the metal cap was removed after annealing. (Reprinted with permission from Ref. [49]. Copyright 2004 American Chemical Society.)

pronounced and highly elongated shape (Figure 11.15d). Notably, the aspect ratio of the single crystals, which is related to the anisotropy of the crystal growth in the lateral dimensions, increases with annealing time from about 1.5 (almost circular) at initial stage to 4 after 60 min thermal treatment.

In case of a single-sided confinement, a somewhat similar evolution of the morphology of the thin MDMO-PPV/PCBM film can be followed. PCBM single crystals continuously grow during the annealing process (Figure 11.15e–g). However, when the thin-film samples are supported by a solid substrate (PEDOT/PSS), the PCBM crystals formed are smaller and less elongated as compared to freestanding film samples for the same annealing times. The aspect ratio of the crystals stays constant at approximately 1.5.

In contrast, the morphological evolution of samples in double-sided confinement follows a different route. For an annealing time of 20 min, the morphology of the samples seems to be unchanged (Figure 11.15h). However, the appearance of the PCBM-rich clusters becomes more prominent and the size of the clusters is increased from initially ∼80 to ∼120 nm. Corresponding SAED (Figure 11.15h inset) analysis has revealed that the clusters are still composed of PCBM nanocrystals.

When the annealing time is increased, further reorganization of the PCBM-rich domains can be observed. Relatively dark regions are emerging from the initially rather homogeneous film (Figure 11.15i and j), which seem to be PCBM crystals. Because the morphological appearance of these crystals is different when compared to the PCBM single crystals as seen in the case of freestanding or single-sided confined films, additional SAED analysis was performed. The diffraction pattern (Figure 11.15i inset) confirms that these dark regions are indeed PCBM single crystals possessing the same crystallographic structure as PCBM crystals formed for the other cases of confinement. Another feature of these crystals is that they have low contrast to the background and their contour is rather fuzzy. A possible reason is that these PCBM crystals are fairly small in the direction perpendicular to the film plane. However, the lateral size is quite similar compared to those formed from films annealed at single-sided confinement, but they possess almost circular shapes and show no preferred growth direction. It should be noted that the PCBM crystals are not surrounded by a bright halo that corresponds to the PCBM-depleted regions.

For all the confinement conditions, the large-scale crystallization of PCBM in the films is observed for the case of annealing temperatures above the glass transition temperature of bulk polymer. Elongated shapes of PCBM crystals are observed, especially in case of freestanding films. The crystal growth rate for this case is determined by the incorporation rate of PCBM molecules at the growth fronts instead of diffusion within the matrix. As more confinement is exerted on the composite film during annealing, the diffusion rate of PCBM molecules within the matrix is reduced and less elongated or even round PCBM crystals are obtained for single- and double-sided confined films, respectively.

As a successful and simple method to improve the performance of a device, thermal post-treatment should be performed under appropriate conditions. Notably temperature and time should be prudently chosen so that molecules/polymer chains in the photoactive layer can relax or reorganize efficiently enough and thus yield the optimum morphology with a maximum improvement for device performance. At the same time, annealing treatment should not exceed critical limitation and induce further change (damage) in the devices.

11.5
The P3HT/PCBM System: Nanoscale Morphology of an Efficient Bulk Heterojunction

In the last part of our study, we would like to demonstrate on the P3HT/PCBM system how to tailor and control the morphology of the photoactive layer toward the design of high-performance PSCs and to discuss why PSCs based on this system currently show the highest energy conversion efficiencies. To enhance charge transport within the interpenetrating networks (and thus reduce charge recombination), high charge carrier mobility for both holes and electrons is required. The general approach to enhance charge carrier transport in organic and polymer materials is to increase the mesoscopic order and crystallinity. A nanoscale interpenetrating network with crystalline order of both constituents is the desirable architecture for the active layer of PSCs [12]. Eventually, the electronic bandgaps of the materials in the photoactive layer should be tuned to harvest more light from the solar spectrum.

The high efficiency of devices based on P3HT/PCBM bulk heterojunctions can be related to the intrinsic properties of the two constituents. Regioregular P3HT self-organizes into a microcrystalline structure [50] and because of efficient interchain transport of charge carriers, the (hole) mobility in P3HT is high (up to $\sim 0.1\,\mathrm{cm}^2\,\mathrm{V}^{-1}\,\mathrm{s}^{-1}$) [51–53]. Moreover, in thin films, interchain interactions cause a redshift of the optical absorption of P3HT, which provides an improved overlap with the solar emission. The second component, PCBM, has an electron mobility of $2 \times 10^{-3}\,\mathrm{cm}^2\,\mathrm{V}^{-1}\,\mathrm{s}^{-1}$ [54]. The good solubility of PCBM in organic solvents allows the utilization of film deposition techniques requiring high-concentrated solutions. Also, PCBM can crystallize and control of nucleation and crystallization kinetics allows the adjustment of the crystal size [28]. However, continuous crystallization may result in single crystals with micrometer sizes.

During thin-film preparation, P3HT forms fibrillarlike crystals that build borders, which hamper the extensive diffusion of PCBM molecules and their large-scale crystallization. When applying the so-called solvent-assisted annealing procedure, these P3HT fibrils continuously crystallize during slow solvent evaporation and form the desired morphology in a single preparation step [55]. In the case of conventional spin-coating deposition with high solvent evaporation rates, a second annealing step is required to create a similar morphology of network forming high-crystalline P3HT nanowires.

Consequently, in the P3HT/PCBM system only small PCBM crystals can grow during the annealing treatment, whereas large crystals are formed in, for example, the MDMO-PPV/PCBM system. Furthermore, since the growth of P3HT is preferably in one direction, and finally leads to highly elongated fibrillarlike crystals, the increased crystallinity established by annealing does not significantly reduce the interface area with the electron acceptor PCBM. The percolation for charge carriers between the PCBM domains is established via PCBM nanocrystals that fill the space between the P3HT network and form a continuous pathway for electron transport.

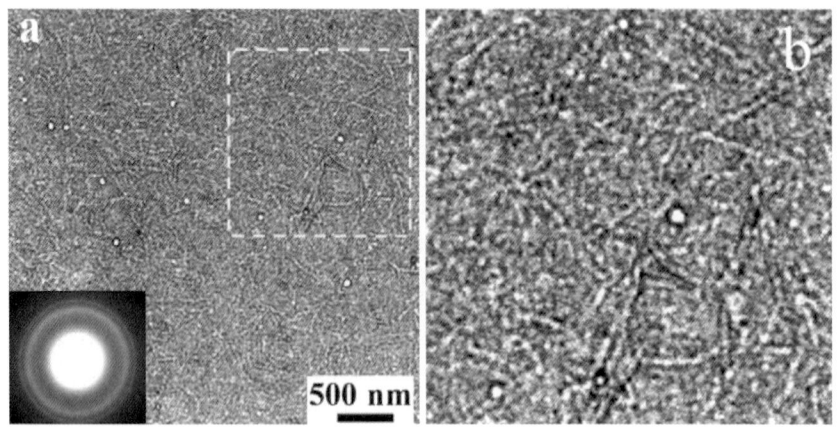

Figure 11.16 BF TEM images show the overview (a) and the zoom in (b) of the pristine P3HT/PCBM photoactive layer; the inset in (a) is the corresponding SAED pattern. (Reprinted with permission from Ref. [13]. Copyright 2004 American Chemical Society.)

Typical bright-field TEM images, recorded in slight underfocus conditions, of a pristine as-spin-coated P3HT/PCBM film are shown in Figure 11.16 [13]. Fibrillar P3HT crystals, which are relatively bright compared to the background, overlap with each other over the whole film. From the absence of other clearly detectable crystalline features, we infer that PCBM is rather homogeneously distributed in the matrix. The bright appearance of the P3HT crystals relative to a dark background is caused by the lower density of P3HT ($1.10\,g\,cm^{-3}$) compared to PCBM ($1.50\,g\,cm^{-3}$) [25]. The width of these fibrillar crystals is approximately 15 nm and their length can reach up to 500 nm. The P3HT crystals have a tendency to form a network, although not entirely connecting each other. The inset of Figure 11.16a shows the SAED pattern of this film. Two diffraction rings can be observed. The outer ring corresponds to a distance of 0.39 nm. This ring is attributed to the (0 2 0) reflection of P3HT crystals, which is associated with the typical π–π stacking distance of P3HT chains. A similar reflection with the same d-spacing has previously been observed for the SAED pattern of P3HT whiskers [56]. The crystallinity and the perfection of present P3HT crystals seem not very pronounced, as revealed by the low intensity of the reflection ring. The inner ring in the SAED pattern, corresponding to a d-spacing of 0.46 nm, appears even more diffuse and has been observed in pure PCBM layers [28]. We have previously shown that in pure PCBM layers the small size of PCBM nanocrystals and their dense stacking within the film in both lateral and vertical directions inhibit the formation of a pronounced bright-field contrast, whereas the SAED pattern reveals the crystalline order inside the film [28]. The inner ring is attributed to PCBM nanocrystals that are homogeneously dispersed throughout the film. Because P3HT apparently crystallizes faster than PCBM, small fibrillar P3HT crystals are formed while the crystallization of PCBM is almost suppressed and only nanometer-sized crystals are formed. The advantage of this morphology created

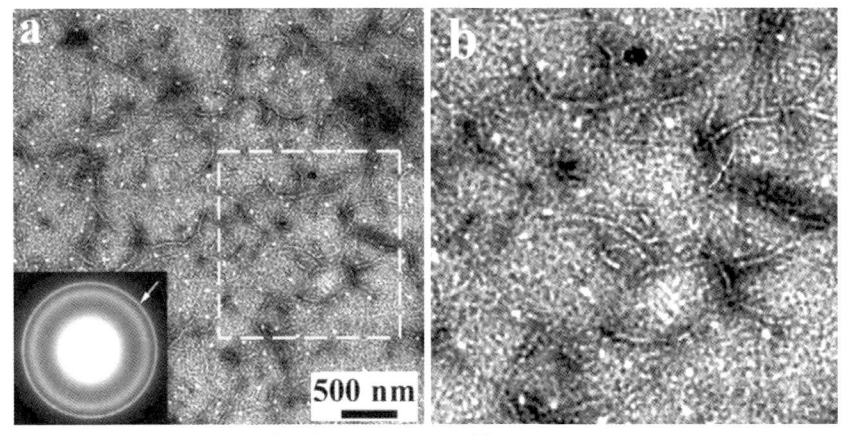

Figure 11.17 BF TEM images show the overview (a) and the zoom in (b) of the annealed P3HT/PCBM photoactive layer; the inset in (a) is the corresponding SAED pattern. (Reprinted with permission from Ref. [13]. Copyright 2004 American Chemical Society.)

already during conventional spin coating is that a homogeneous distribution of the two components on the nanoscale is achieved and that P3HT forms elongated crystals. Both features are essential for the success of the annealing step.

Figures 11.17 shows the BF TEM images of the composite film after controlled annealing at elevated temperature. The annealing was performed on a completed device, that is, with bottom and top contacts present that were later removed to obtain the images. The most pronounced feature in the TEM image of the annealed sample is the increased contrast and the appearance of bright fibrillar P3HT crystals throughout the entire film. The width of these crystals remains almost constant compared to the pristine composite film, but on average their length increases over 50%. We note that a conventional TEM image is a two-dimensional projection of the three-dimensional morphology in a thin film, which causes the missing of morphology information in a direction perpendicular to the film plane. Because there is no strong interaction between P3HT and the substrates, and bending of P3HT fibrils is clearly visible in Figure 11.17, we presume that the orientation of P3HT crystals in the film should be rather homogeneous, including the perpendicular direction. Recently, first results obtained by TEM tomography, a volume characterization technique with nanometer resolution, indicate the almost perfect 3D organization of the P3HT nanowires within the bulk heterojunction (Figure 11.18).

The increased crystallinity of P3HT after thermal treatment is evidenced by the increased intensity of the (0 2 0) reflection ring in the SAED pattern (inset of Figure 11.17a). In addition, various larger dark (PCBM-rich) areas can be observed that evidence an increased demixing between P3HT and PCBM. Furthermore, the intensity of the PCBM reflection rings in the SAED increases slightly. Despite the increased demixing, the image does not show any evidence of

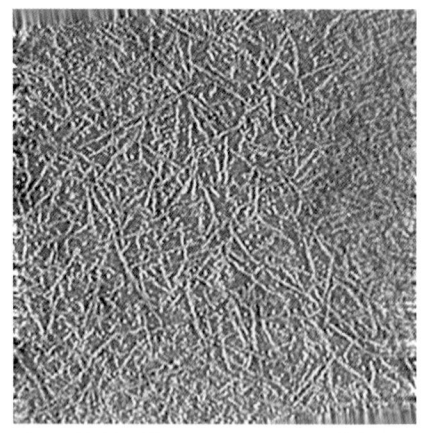

Figure 11.18 Volume reconstruction of the P3HT nanowire network in a P3HT/PCBM bulk heterojunction active layer: a 2D slice taken from a 3D data set. Size of the image is about 2 µm by 2 µm.

the large PCBM crystals (up to several microns) that were previously observed upon annealing of mixed MDMO-PPV/PCBM films in which demixing is much more pronounced [11].

In summary, these results point to a few important characteristics of the P3HT/PCBM bulk heterojunction: the crystallinity of P3HT is improved upon annealing and the demixing between the two components is increased, but large-scale phase separation is prevented. The resulting interpenetrating networks composed of P3HT crystals with high aspect ratio and aggregated nanocrystalline PCBM domains provide continuous pathways in the entire photoactive layer for efficient hole and electron transport (Figure 11.19). In general, the single-step preparation route via solvent-assisted annealing leads to similar morphologies as the conventional two-step spin coating and subsequent annealing procedure and results in the high energy conversion efficiencies.

11.6
Summary

Today, polymer solar cells with photoactive layers based on at least two constituents forming bulk heterojunctions become more and more attractive for commercial applications. In this respect, one key aspect toward high-performance devices is that over the years detailed knowledge on structure–property relations has been acquired. Besides the designed molecular architecture of the constituents, the electron donor and electron acceptor materials employed in the cell, and control on the nanoscale morphology formation within the photoactive layer have been identified to significantly contribute to the overall performance of polymer solar cells.

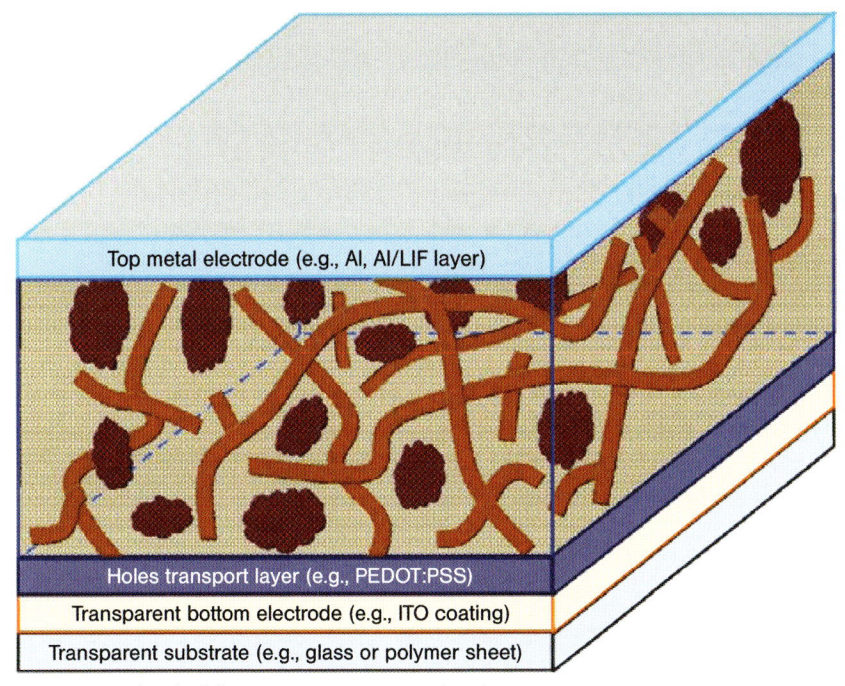

Figure 11.19 Sketch of the interpenetrating network in the active layer composed of P3HT and PCBM established after annealing.

Currently, by choosing appropriate solvent (or mixed solvents), optimized compound composition and correct preparation conditions, nanoscale phase separation between electron donor and acceptor constituents can be achieved, and continuous pathways for charge carriers to the electrodes can be provided. However, theoretical predication indicates polymer solar cells have the potential to reach an efficiency as high as 10%, twice the highest efficiency reported up to now. Such performance would not only make PSC commercially very attractive and comparable with amorphous silicon solar cells, but would also offer the strong advantage of easy processing. So, we constantly have to continue closing our knowledge gaps on structure–processing–property relations on the nanoscale to ultimately reach the highest performance of polymer solar cells. And certainly, as demonstrated by the present study, microscopy techniques as tools to explore the nanoworld have played and continue to play an important role in the research field of polymer solar cells.

Acknowledgments

We would like to thank Hanfang Zhong, Sasha Alexeev, and Erwan Sourty for their important assistance in the presented work. We thank René Janssen, Martijn Wienk, Jeroen van Duren, Sjoerd Veenstra, and Marc Koetse for fruitful discussions. Part of the work is embedded in the research program of the Dutch Polymer Institute

(DPI projects #326 and #524). Financial support was provided by the Dutch Science Organisation (NWO), the Royal Netherlands Academy of Arts and Sciences (KNAW), and the Chinese Ministry of Science and Technology (MOST) within the cooperation project 06CDP038. Xiaoniu Yang would like to thank the National Natural Science Foundation of China (Grant No. 20604029) and the "Hundred Talents Project" (initialization support) of the Chinese Academy of Sciences for financial support.

References

1 Bube, R.H. (1992) *Photoelectronic Properties of Semiconductors*, Cambridge University Press, Cambridge.
2 Pope, M. and Swenberg, C.E. (1999) *Electronic Processes in Organic Crystals and Polymers*, Oxford University Press, Oxford.
3 Peumans, P., Yakimov, A. and Forrest, S.R. (2003) *Journal of Applied Physiology*, **93**, 3693.
4 Smilowitz, L., Sariciftci, N.S., Wu, R., Gettinger, C., Heeger, A.J. and Wudl, F. (1993) *Physical Review B: Condensed Matter*, **47**, 13835.
5 Yoshino, K., Hong, Y.X., Muro, K., Kiyomatsu, S., Morita, S., Zakhidov, A.A., Noguchi, T. and Ohnishi, T. (1993) *Japanese Journal of Applied Physics, Part 2: Letters*, **32**, L357.
6 Halls, J.J.M., Pichler, K., Friend, R.H., Moratti, S.C. and Holmes, A.B. (1996) *Applied Physics Letters*, **68**, 3120.
7 Haugeneder, A., Neges, M., Kallinger, C., Spirkl, W., Lemmer, U. and Feldmann, J. (1999) *Physical Review B: Condensed Matter*, **59**, 15346.
8 Yu, G. and Heeger, A.J. (1995) *Journal of Applied Physiology*, **78**, 4510.
9 Halls, J.J.M., Walsh, C.A., Greenham, N.C., Marseglia, E.A., Friend, R.H., Moratti, S.C. and Holmes, A.B. (1995) *Nature*, **376**, 498.
10 Yu, G., Gao, J., Hummelen, J.C., Wudl, F. and Heeger, A.J. (1995) *Science*, **270**, 1789.
11 Yang, X., van Duren, J.K.J., Janssen, R.A.J., Michels, M.A.J. and Loos, J. (2004) *Macromolecules*, **37**, 2151.
12 Schmidt-Mende, L., Fechtenkötter, A., Müllen, K., Moons, E., Friend, R.H. and MacKenzie, J.D. (2001) *Science*, **293**, 1119.
13 Yang, X., Loos, J., Veenstra, S.C., Verhees, W.J.H., Wienk, M.M., Kroon, J.M., Michels, M.A.J. and Janssen, R.A.J. (2005) *Nano Letters*, **5**, 579.
14 Shafai, C., Thomson, D.J., Simard-Normandin, M., Mattiussi, G. and Scanlon, P. (1994) *Japanese Journal of Applied Physics, Part 2: Letters*, **64**, 342.
15 De Wolf, P., Snauwaert, J., Clarysse, T., Vandervorst, W. and Hellemans, L. (1995) *Applied Physics Letters*, **66**, 1530.
16 Kelley, T.W. and Frisbie, C.D. (2000) *Journal of Vacuum Science & Technology B: Microelectronics and Nanometer Structures*, **18**, 632.
17 Lin, H.-N., Lin, H.-L., Wang, S.-S., Yu, L.-S., Perng, G.-Y., Chen, S.-A. and Chen, S.-H. (2002) *Applied Physics Letters*, **81**, 2572.
18 Alexeev, A., Loos, J. and Koetse, M.M. (2006) *Ultramicroscopy*, **106**, 191.
19 Koch, N., Elschner, A., Schwartz, J. and Kahn, A. (2003) *Applied Physics Letters*, **82**, 2281.
20 Heinz, W.F. and Hoh, J.H. (1999) *Trends in Biotechnology*, **17**, 143.
21 Eyben, P., Xu, M., Duhayon, N., Clarysse, T., Callewaert, S. and Vandervorst, W. (2002) *Journal of Vacuum Science & Technology B: Microelectronics and Nanometer Structures*, **20**, 471.
22 Blom, P.W.M., de Jong, M.J.M. and Vleggaar, J.J.M. (1996) *Applied Physics Letters*, **68**, 3308.

23. Tanase, C., Blom, P.W.M. and de Leeuw, D.M. (2004) *Physical Review*, **B70**, 193202.
24. Loos, J., Yang, X., Koetse, M.M., Sweelssen, J., Schoo, H.F.M., Veenstra, S.C., Grogger, W., Kothleitner, G. and Hofer, F. (2005) *Journal of Applied Polymer Science*, **97**, 1001.
25. Bulle-Lieuwma, C.W.T., van Gennip, W.J.H., van Duren, J.K.J., Jonkheijm, P., Janssen, R.A.J. and Niemantsverdriet, J.W. (2003) *Applied Surface Science*, **203–204**, 547.
26. Martens, T., D'Haen, J., Munters, T., Beelen, Z., Goris, L., Manca, J., D'Olieslaeger, M., Vanderzande, D., De Schepper, L. and Andriessen, R. (2003) *Synthetic Metals*, **138**, 243.
27. Shaheen, S.E., Brabec, C.J., Sariciftci, N.S., Padinger, F., Fromherz, T. and Hummelen, J.C. (2001) *Applied Physics Letters*, **78**, 841.
28. Yang, X., van Duren, J.K.J., Rispens, M.T., Hummelen, J.C., Janssen, R.A.J., Michels, M.A.J. and Loos, J. (2004) *Advanced Materials*, **16**, 802.
29. Zhong, H., Yang, X., deWith, B. and Loos, J. (2006) *Macromolecules*, **39**, 218.
30. Padinger, F., Rittberger, R.S. and Sariciftci, N.S. (2003) *Advanced Functional Materials*, **13**, 85.
31. Waldauf, C., Schilinsky, P., Hauch, J. and Brabec, C.J. (2004) *Thin Solid Films*, **451–452**, 503.
32. Al-Ibrahim, M., Ambacher, O., Sensfuss, S. and Gobsch, G. (2005) *Applied Physics Letters*, **86**, 201120.
33. Reyes-Reyes, M., Kim, K. and Carrolla, D.L. (2005) *Applied Physics Letters*, **87**, 083506.
34. Ma, W., Yang, C., Gong, X., Lee, K. and Heeger, A.J. (2005) *Advanced Functional Materials*, **15**, 1617.
35. Reyes-Reyes, M., Kim, K., Dewald, J., López-Sandoval, R., Avadhanula, A., Curran, S. and Carroll, D.L. (2005) *Organic Letters*, **7**, 5749.
36. Forrest, J.A., Dalnoki-Veress, K. and Dutcher, J.R. (1997) *Physical Review E*, **56**, 5705.
37. Frank, B., Gast, A.P., Russell, T.P., Brown, H.R. and Hawker, C. (1996) *Macromolecules*, **29**, 6531.
38. Reiter, G. (1993) *Europhysics Letters*, **23**, 579.
39. Zhao, W., Zhao, X., Rafailovich, M.H., Sokolov, J., Komposto, R.J., Smith, J.J., Dozier, W.D., Mansfield, J. and Russell, T.P. (1993) *Macromolecules*, **26**, 561.
40. Mayes, A.M. (1994) *Macromolecules*, **27**, 3114.
41. Tanaka, K., Taura, A., Ge, S., Takahara, A. and Kajiyama, T. (1996) *Macromolecules*, **29**, 3040.
42. Forrest, J.A., Dalnoki-Veress, K., Stevens, J.R. and Dutcher, J.R. (1996) *Physical Review Letters*, **77**, 2002.
43. Toney, M.F., Russell, T.P., Logan, J.A., Kikuchi, H., Sands, J.M. and Kumar, S.K. (1995) *Nature*, **374**, 709.
44. Gusev, A.A. and Suter, U.W. (1993) *Journal of Chemical Physics*, **99**, 2228.
45. Liu, Y., Russell, T.P., Samant, M.G., Stöhr, J., Brown, H.R., Cossy-Favre, A. and Diaz, J. (1997) *Macromolecules*, **30**, 7768.
46. Ge, S., Pu, Y., Zhang, W., Rafailovich, M., Sokolov, J., Buenviaje, C., Buckmaster, R. and Overney, R.M. (2000) *Physical Review Letters*, **85**, 2340.
47. Wallace, W.E., van Zanten, J.H. and Wu, W.L. (1995) *Physical Review E*, **52**, 3329.
48. Orts, W.J., van Zanten, J.H., Wu, W.L. and Satija, S.K. (1993) *Physical Review Letters*, **71**, 867.
49. Yang, X., Alexeev, A., Michels, M.A.J. and Loos, J. (2005) *Macromolecules*, **38**, 4289.
50. Prosa, T.J., Winokur, M.J., Moulton, J., Smith, P. and Heeger, A.J. (1992) *Macromolecules*, **25**, 4364.
51. Bao, Z., Dodabalapur, A. and Lovinger, A. (1996) *Applied Physics Letters*, **69**, 4108.
52. Sirringhaus, H., Brown, P.J., Friend, R.H., Nielsen, M.M., Bechgaard, K., Langeveld-Voss, B.M.W., Spiering, A.J.H., Janssen, R.A.J., Meijer, E.W., Herwig, P. and de Leeuw, D.M. (1999) *Nature*, **401**, 685.
53. Sirringhaus, H., Tessler, N. and Friend, R.H. (1998) *Science*, **280**, 1741.

54 Mihailetchi, V.D., van Duren, J.K.J., Blom, P.W.M., Hummelen, J.C., Janssen, R.A.J., Kroon, J.M., Rispens, M.T., Verhees, W.J.H. and Wienk, M.M. (2003) *Advanced Functional Materials*, **13**, 43.

55 Li, G., Shrotriya, V., Huang, J.S., Yao, Y., Moriarty, T., Emery, K. and Yang, Y. (2005) *Nature Materials*, **4**, 864.

56 Ihn, K.J., Moulton, J. and Smith, P. (1993) *Journal of Polymer Science, Part B: Polymer Physics*, **31**, 735.

C
Hybrid Solar Cells

Organic Photovoltaics: Materials, Device Physics, and Manufacturing Technologies.
Edited by Christoph Brabec, Vladimir Dyakonov, and Ullrich Scherf
Copyright © 2008 WILEY-VCH Verlag GmbH & Co. KGaA, Weinheim
ISBN: 978-3-527-31675-5

12
TiO$_2$ Template/Polymer Solar Cells
Vignesh Gowrishankar, Brian E. Hardin, and Michael D. McGehee

12.1
Introduction

Templated titania (TiO$_2$) nanostructures in organic composite photovoltaics (OCPV) have a distinct advantage over organic bulk heterostructures – their morphology can be precisely controlled. Titania templates can be fabricated by several techniques with precise control of the continuous nanostructure. It is difficult to achieve the ideal device architecture by simply blending two materials together because the material properties of the constituents determine the domain size during phase separation. For example, variations in the side chain length of the polymer, weight percentage of materials, and solvent choices can lead to vastly different morphologies [1,2], which can severely affect exciton harvesting. Furthermore, blends may not contain completely bicontinuous pathways. If portions of a phase are not connected to an electrode, then charges can be trapped, inhibiting charge collection efficiency. Titania templates can be fabricated to have continuous pores of size of the order of the exciton diffusion length of a polymer, regardless of the polymers used, allowing for the possibility of complete exciton harvesting and charge collection.

In OCPVs, nanoporous or nanostructured titania films are fabricated first and then filled with a conjugated polymer [3–6]. It is not trivial to completely fill small pores (∼10 nm); however, there are a variety of pore-filling methods (see Section 12.5) and the titania surface (titania/polymer interface) can be modified, which has the potential to improve polymer wetting and charge transfer dynamics.

This chapter will focus on cells that use polymers designed to absorb light and transport charge, as opposed to cells where a dye is used to sensitize the titania and an organic semiconductor is only used to transport holes [7–11] or cells where the polymers are only used as sensitizers [12]. We first present the basic theory for OCPV device operations, following which we discuss material choices, fabrication techniques, and device performance for several types of titania-based OCPVs. We will then address an important issue with templated TiO$_2$ nanostructures, that is, pore filling and its effect on device efficiency, review OCPV models, and finally discuss the future

Organic Photovoltaics: Materials, Device Physics, and Manufacturing Technologies.
Edited by Christoph Brabec, Vladimir Dyakonov, and Ullrich Scherf
Copyright © 2008 WILEY-VCH Verlag GmbH & Co. KGaA, Weinheim
ISBN: 978-3-527-31675-5

outlook and potential techniques that can lead to highly efficient devices. We believe that nanostructured TiO$_2$ is an attractive approach to develop extremely efficient organic photovoltaics. Titania is both abundant and nontoxic, while also being well studied by the dye-sensitized solar cell community [7,13]. Future advancements in polymer solar cell technologies will benefit greatly from the TiO$_2$ scaffold, which provides greater processing flexibility and more polymer choices.

12.2
Basic Operation

The external quantum efficiency (EQE) is the number of electrons extracted from the device per photon absorbed and is described by the following equation:

$$\eta_{EQE} = \eta_{abs} \cdot \eta_{exharvest} \cdot \eta_{CT} \cdot \eta_{GS} \cdot \eta_{CC}.$$

Measuring EQE is a valuable method to both evaluate performance and understand the device operations of solar cells. The first three terms are the efficiencies of absorption (η_{abs}), exciton harvesting ($\eta_{exharvest}$), and charge transfer (η_{CT}) and are considered to be voltage independent. η_{EQE} is wavelength dependent because η_{abs} depends on the absorption coefficient, which varies by wavelength. The last two terms in the above equation corresponding to the efficiencies of geminate separation (η_{GS}) and of charge collection (η_{CC}) are the factors that lead to a voltage-dependent photocurrent. Figure 12.1a shows a schematic representation of some of these processes. We look at the correlation between polymer properties and these factors. A sufficiently thick layer of polymer with a high absorption coefficient would have large η_{abs}, while a lower bandgap would extend its ability to absorb to higher wavelengths. $\eta_{exharvest}$ mainly depends on the exciton diffusion length, which for most polymers is likely between 3 and 10 nm [14–17], although higher diffusion lengths have been reported [17,18]. Assuming 100% charge transfer efficiency (η_{CT}), which need not always be the case, a polymer (donor) layer of thickness equal to the exciton diffusion

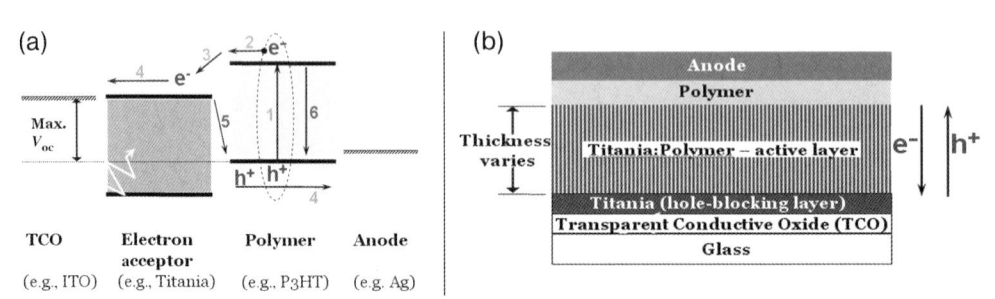

Figure 12.1 (a) Energy levels and maximum V_{oc} in a donor–acceptor (polymer–titania) system. Also shown are the various steps that give rise to photocurrent: light absorption (1) leads to the formation of excitons, which diffuse to the interface (2), and are split by electron transfer (3) before natural decay (6). After successful splitting, electrons and holes are transported (4) toward the cathode and anode, respectively, if they are able to avoid geminate recombination (5). (b) Generic physical structure of a titania-based organic composite photovoltaic device.

length would have 75% of the excitons generated within it, split by electron transfer to titania (acceptor) [15]. Upon charge transfer, the geminate electron–hole pair is still bound across the interface and must be pulled apart efficiently to yield a high η_{GS}. Finally, the holes (positive polarons) and electrons need to travel along the polymer and titania toward the electron-collecting cathode[1] and hole-collecting anode,[1] respectively. Low, morphology-dependent charge carrier mobilities in typical polymers (ranging from 10^{-1} to $10^{-7}\,cm^2\,V^{-1}\,s^{-1}$), and the possible danger of bimolecular recombination, can significantly lower η_{CC} (see Section 12.6.4).

For the titania/polymer solar cell, shown in Figure 12.1a, the maximum possible V_{oc} is considered to correspond to the difference between the conduction band of the titania and the HOMO (highest occupied molecular orbital) level of the polymer, for ohmic contacts. For nonohmic contacts the V_{oc} is less than this maximum and understanding it is more involved [19]. The V_{oc} is also affected by shunt pathways, dark current, and Fermi level pinning.

12.3
General Device Structure and Material Choices

12.3.1
General Device Structure

As shown in Figure 12.1b, titania/polymer solar cells consist of a transparent conducting oxide (TCO) electrode, a titania layer, a semiconducting polymer (e.g., poly(3-hexylthiophene) (P3HT) or poly(2-methoxy-5-(3′,7′-dimethoxyloctyloxy)-p-phenylene vinylene) (MDMO-PPV)), and a metallic electrode. For titania-based OCPV devices, an additional titania/polymer composite layer is present between the titania (hole-blocking) layer and polymer overlayer. The polarity of the device depends on the actual TCO and metal used; if the TCO is fluorine-doped tin oxide and the metal is Ag, the former would be the electron-collecting cathode and the latter the hole-collecting anode.

12.3.2
Transparent Conducting Oxide (Cathode)

The most widely used TCO is fluorine-doped tin oxide (SnO_2:F or FTO) and indium-tin oxide (ITO). It is important to note that although ITO generally has a lower sheet resistance and is smoother, its mechanical and electrical properties change when heated above 400 °C. This temperature is generally exceeded when preparing the titania and indium has been known to diffuse into TiO_2, thus making SnO_2:F the preferred TCO.

---

1) The nomenclature for the electrodes is consistent with that used by the electronics community for a diode.

12.3.3
Titania

Titania has a refractive index of about 2.5 and is largely nonabsorbing in the visible range (bandgap $\sim$3.2 eV). The anatase phase of titania is that which is usually used in solar cells. Titania is a strong electron acceptor, by virtue of its high electron affinity (conduction band at $\sim$4.2 eV below vacuum) with respect to the lowest unoccupied molecular orbital (LUMO) of almost all organic semiconductors [14,18,20,21]. It has a structure-dependent mobility ranging from 4–20 cm^2 V^{-1} s^{-1} (anatase crystals) to 10^{-6}–10^{-7} cm^2 V^{-1} s^{-1} (sintered TiO$_2$ nanoparticles) [22–25]. Titania is nontoxic and can be deposited in a variety of ways that will be described in Section 12.4. Furthermore, many molecules can be attached to the surface of titania [26] or other oxides (e.g., ZnO) [27], using carboxylic acid and phosphoric acid groups, making it suitable for donor–acceptor interface modification.

12.3.4
Semiconducting Polymer

Polyphenylene vinylene derivatives (e.g., OC$_{10}$C$_{10}$-PPV, OC$_1$C$_{10}$-PPV, MEH-PPV) and polythiophene derivatives (e.g., P3HT) are the most commonly used polymers for OCPVs because of their high absorptivity ($\sim 10^5$ cm^{-1}) and moderate mobilities. A discussion about mobility and the effects of pore filling on polymer mobility is presented in Section 12.6. The singlet exciton diffusion length varies from 2 to 10 nm depending on the polymer [14–18].

12.3.5
Anode

High work function metals (e.g., Au and Ag) are used as the anode. Au (5.1 eV) has a higher work function than Ag [28], but Ag contacts are partially oxidized, even in an inert environment, during evaporative deposition and subsequent annealing, which renders a work function range of 4.4–5.1 eV. Such electrodes are generally deposited by evaporation inside an inert environment, typically in a glove box.

12.4
Device Structures

This section will discuss past, present, and ideal device structures and their performance, including a brief discussion of ways to fabricate the various structures. Before looking at fabrication methods, it is worthwhile to discuss the desired properties and morphologies of an ideal nanostructure.

The nanostructure in Figure 12.2 is considered ideal because it has sufficiently small, straight pores and is thick enough to absorb most of the sunlight. The pore radius should be slightly less than the exciton diffusion length to enable maximum

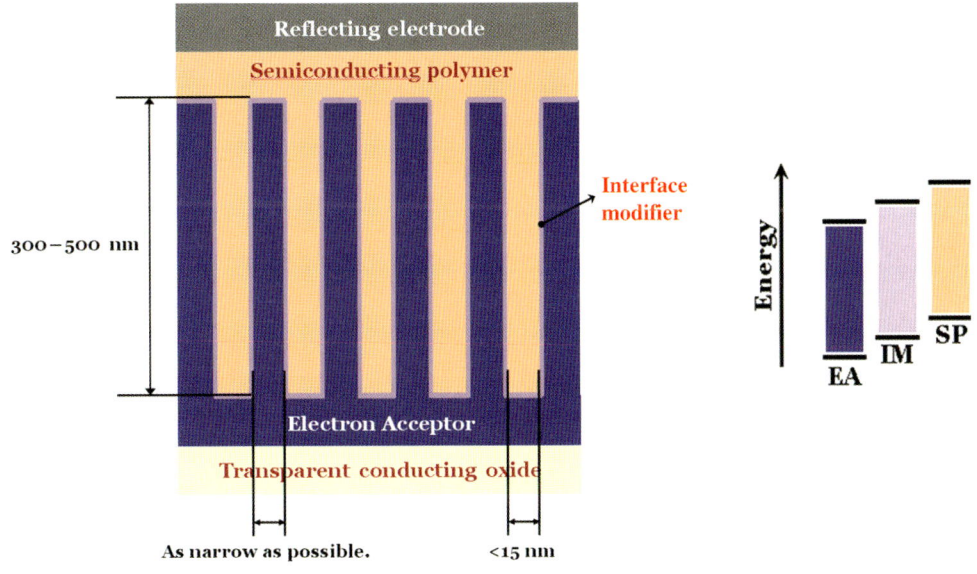

Figure 12.2 (Left) an ideal nanostructure with desirable dimensions and (right) suitable energy levels.

exciton harvesting. The thickness of the nanostructure should be 300–500 nm so that the infiltrated polymer can absorb most of the incident light. Nanostructures with a lower porosity (e.g., if the titania rods were thicker) would require thicker films to contain a similar amount of polymer. The pores (or channels) should be straight to provide the most direct and shortest path to the anode and cathode. Holes traverse the entire thickness of the nanostructure through the low-mobility polymer; therefore, thicker films may lead to high series resistance, recombination losses, and low η_{CC}. An ideal nanostructure should have as little titania as possible to enable maximum absorption for a given thickness. Since the titania nanostructure is fabricated first, the interface is amenable to interfacial modifications to maximize forward electron transfer and minimize recombination as well as promote polymer wetting. Finally, straight pores with smooth sidewalls would likely allow the polymer chains to align themselves preferentially in the vertical direction, thereby promoting exciton diffusion and charge transport (see Section 12.6). It would be additionally advantageous if the films were made at low cost on plastic substrates, which would require that the deposition and thermal treatments be done at modest temperatures (<500 °C).

Note that the figure is a cross-sectional depiction, whose three-dimensional form may consist of an array of vertical nanopillars in a matrix of the second semiconductor, or nanochannels (vertical slices) and nanoridges. Both structures are equally "ideal" for most of the part, perhaps differing slightly in porosity of the ultimate inorganic polymer composite. Of course, the inorganic of choice must possess suitable mobility and energy levels for exciton harvesting. We now discuss how titania/polymer cells have evolved while progressively striving to attain some of the above "ideal" qualities.

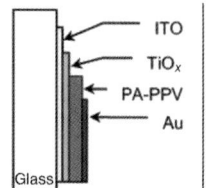

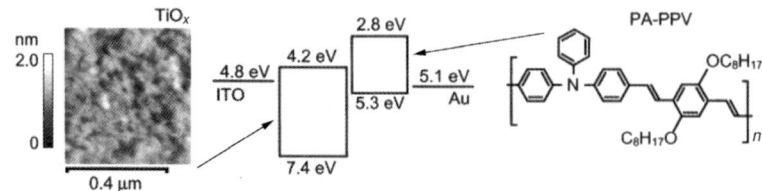

Figure 12.3 The ITO/TiO$_x$/PA-PPV/Au device structure, energy band diagram, and molecular structure of PA-PPV. An AFM image of the TiO$_x$ surface reveals surface-feature heights of less than 3 nm (reproduced from Ref. [18], with permission).

12.4.1
TiO$_2$/Polymer Bilayers

Bilayers represent the simplest structure consisting of a smooth, generally thin (10–100 nm) layer of titania underneath a flat film of polymer. Bilayers are routinely used to measure exciton diffusion lengths and evaluate the properties of new polymers in photovoltaic applications. Polymers with relatively large singlet exciton diffusion lengths can achieve moderate monochromatic external quantum efficiencies (EQEs) and power-conversion efficiencies [18]. It would be challenging to make extremely efficient bilayer cells over a wide range of wavelengths because singlet exciton diffusion lengths, typically 3–10 nm, are more than an order of magnitude less than absorption lengths,[2] generally 100–300 nm.

In bilayers, a compact titania layer is fabricated on top of the transparent conducting electrode using a sol–gel process with a titania precursor (e.g., titanium ethoxide). The first titania/polymer bilayer, described in Figure 12.3, used phenylamino-p-phenylenevinylene (PA-PPV), which had an intrinsic hole mobility of the order of $10^{-4}\,\text{cm}^2\,\text{V}^{-1}\,\text{s}^{-1}$ [18]. We have estimated the exciton diffusion length of basic PPV derivatives to be around 6 nm [15].

Power-conversion efficiencies over the solar spectrum are low because of low current densities due to poor exciton harvesting [18]. Several parameters can be optimized to "maximize" bilayer efficiency. It is possible to increase the J_{sc} by adjusting the thickness of the layers [15,17,29]. The back metal contact introduces interference effects, thus making it possible to concentrate the light intensity at the polymer/titania interface by adjusting the thickness of the individual layers. Also, reducing the polymer thickness lowers the series resistance and improves the fill factor. However, the interfacial area of the titania/polymer system must be increased to harvest more excitons and achieve considerably better power-conversion efficiencies.

---

2) Absorption length is defined as $1/\alpha$, where α is the absorption coefficient.

12.4.2
TiO$_2$ Nanoparticles/Polymer

TiO$_2$ nanoparticles have been used extensively to increase the interfacial surface area in dye-sensitized solar cells. Sintered TiO$_2$ particles can have a roughness factor of 100 µm^{-1} [30], which means that 1 µm thick film has a surface area 100 times greater than a flat TiO$_2$ layer of the same (projected) surface area. Because the dye attaches to the TiO$_2$ surface, the increase in surface area directly correlates to more dye adsorption and consequently higher light absorption and current density. Greater interfacial area for OCPVs improves exciton harvesting, because more excitons are generated within 5–10 nm of the polymer/titania interface. Sintered TiO$_2$ nanoparticles are reported to have an electron mobility of 10^{-6}–10^{-7} cm^2 V^{-1} s^{-1} [22,25].

Nanoparticle films are generally made by doctor blading a paste of titania nanocrystals and then sintering the particles together at 450–500 °C [3]. Typically, the pores have a diameter of approximately 20 nm, but these pores have an irregular shape and the films are many microns thick. Nanoporous titania films can also be made by spray pyrolysis [6], wherein droplets of titanium isopropoxide are formed with an ultrasonic nebulizer and then carried through a furnace with argon carrier gas. Anatase nanocrystals form as the droplets travel through the furnace. As these crystals exit the furnace, they are sprayed onto a substrate, which can be kept in air at room temperature. Since the nanocrystals are still hot when they hit the substrate, they tend to connect with each other. Spray pyrolysis does not require heating of the substrate and could potentially be used with plastic substrates. Another advantage of spray pyrolysis is that the film thickness can easily be varied, even in the submicrometer range. Various polymers were deposited inside the pores via spin casting, a method discussed in Section 12.6.

Carter *et al.* first made OCPVs from sintered TiO$_2$ nanoparticles, MEH-PPV, and gold contacts [3,31]. These devices had external quantum efficiencies at the peak absorption wavelength of 6%, three times greater than their titania/polymer bilayer cells. They attributed the enhancement in nanoporous titania films to the increased interfacial area between the two semiconductors that promoted exciton splitting. The exciton diffusion length in a pure film of MEH-PPV is reported to be 6 nm [32]. However, given the large fraction (>50%) of incident light that was absorbed in the device at the wavelength of maximum absorption of MEH-PPV, the fact that the EQE in the nanoporous devices only reached 6% indicates that either pore filling was not completely achieved by spin coating, excitons in the pores were not split prior to natural decay, or charge carriers were not able to escape the region of interpenetrating titania and polymer prior to bimolecular recombination.

There have also been several studies of PV cells made with nanocrystalline TiO$_2$ and polythiophene derivatives [5,33–37]. In some cases a ruthenium dye was attached to the TiO$_2$ before the polymer was infiltrated into the pores. The best devices made by Gebeyehu *et al.* have energy conversion efficiencies of 0.16% under simulated solar irradiation [5]. A study by Ravirajan *et al.* using polyfluorene derivatives and thin (100 nm) layers of nanocrystalline TiO$_2$ produced 14% EQE at low light intensity and 0.16% power efficiency under AM 1.5 G conditions [38]. External quantum efficiencies below 15% have been reported in nearly all of the reported bulk heterojunction PV cells made from nanocrystalline TiO_2 and conjugated polymers. The EQE and consequently the short-circuit current (J_{sc}) of these TiO_2–polymer PV cells are about an order of magnitude smaller than state-of-the-art polymer–fullerene cells [39,40]. The lower EQE and J_{sc} in these reported TiO_2–polymer cells may be due to a number of factors such as low red absorbance, low overall absorption due to low optical density, large pore size, or nanostructure spacing compared to exciton diffusion length, poor charge transport, and/or fast recombination. Systematic identification and analysis of the dominant loss mechanisms is difficult due to a wide variation in material properties and incomplete information about the extent of polymer infiltration [41,42].

To make better the dominant loss mechanisms, Ravirajan et al. studied the effects of different conjugated polymer optoelectronic properties on the photovoltaic performance of devices containing nanocrystalline TiO_2 and the conjugated polymers [42]. The device structures typically consisted of four layers – a dense titania layer (hole-blocking layer, HBL) on top of the transparent cathode, a porous titania layer (Figure 12.4a) that served to increase donor–acceptor interface area, a dip-coated polymer layer that made intimate contact with the porous titania, and a spin-coated polymer layer that made contact with the metal anode. Four different polymers were used, which were characterized in terms of their absorption, ionization potential, exciton diffusion lengths, and hole-transport properties.

The porous titania layer was filled by dip coating and while polymer penetration was found to vary among the polymers it was found to be sufficient. Photovoltaic cells, with the general structure ITO/dense titania (50 nm hole-blocking layer)/porous titania:polymer (100 nm)/polymer (50 nm)/Au, were fabricated and tested. As shown

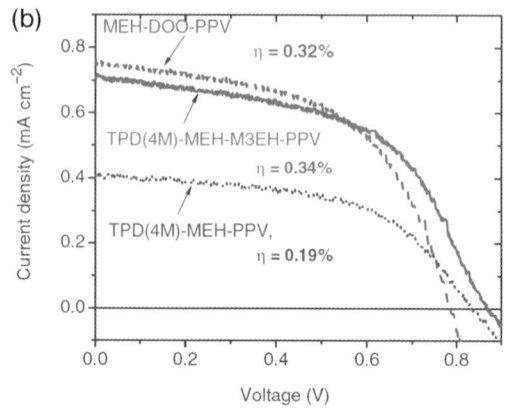

Figure 12.4 (a) Scanning electron micrograph of the surface of a mesoporous anatase film prepared from a hydrothermally processed TiO_2 colloid. The exposed surface planes have mainly {1 0 1} orientation (adapted from Ref. [30], with permission from Macmillan Publishers Ltd); (b) current density–voltage characteristics of ITO/ HBL (50 nm)/porous TiO_2 (100 nm)/polymer (50 nm, dip coated, and spin coated)/Au devices under simulated (100 MW cm^{-2}, AM 1.5) solar illumination. The polymers used were TPD(4M)-MEH-M3EH-PPV (dashed line), MEH-DOO-PPV (solid line), and TPD(4M)-MEH-PPV (dotted line) (reproduced from Ref. [42], with permission).

in Figure 12.4a, the device with the highest hole mobility (TPD(4M)-MEH-PPV) generated the lowest J_{sc}, primarily due to low absorption and poor sensitization of the porous titania. In these devices, it is clear that poor light-harvesting properties, and not insufficient hole mobility, are the primary cause of the low J_{sc}. The highest efficiency devices were those that used the polymer with intermediate light-harvesting and charge transport properties (TPD(4M)-MEH-M3EH-PPV); the device efficiency was 0.34% under AM 1.5 illumination. The fourth polymer that the authors used was F8T2 (not PPV based). Although this polymer has a high hole mobility and good red absorption, the device had lower J_{sc}, likely due to a low reported exciton diffusion length, thereby underscoring the importance of having large exciton diffusion lengths for obtaining large J_{sc}. As references, devices without the porous titania layer were fabricated for all four polymers. These devices showed efficiencies around 0.04%, which is at least a factor of 5 lower than when a porous titania layer is included, which increases donor–acceptor interfacial area and hence J_{sc}. A 20–40% improvement in device efficiency was achieved by introducing a layer of poly(3,4-ethylenedioxythiophene) poly(styrenesulfonate) (PEDOT:PSS) between the polymer and the evaporated gold. Their devices were significantly more efficient than titania/MEH-PPV (0.18%) and titania/F8T2 (0.17%) devices reported previously [31,38].

Although sintered nanoparticles increased the overall conversion efficiency of OCPVs, further refinements in nanostructure pore size and distribution are necessary to harvest even more excitons.

12.4.3
Mesoporous Titania/Polymer

Another approach for creating a titania nanostructure into which conjugated polymers are incorporated was reported by Coakley *et al.*, who made the first mesoporous titania films with relatively well ordered arrays of pores using evaporation induced self-assembly [43]. A high-resolution scanning electron microscopy (SEM) top-view image of a film made with this method is shown in Figure 12.5a.

The pore size is highly uniform, which is desirable for the efficient dissociation of excitons. Films made using this synthesis route have pores that form an interconnected network within the film, but it should be emphasized that these films do not have pores that go straight to the bottom of the film. It was shown that P3HT can be infiltrated into the pores of mesoporous TiO_2 very effectively by spinning a film of the polymer on top of the mesoporous film and then heating the sample to above the glass transition temperature of the polymer [43]. X-ray photoelectron spectroscopy (XPS) depth profiling was used to determine the degree of polymer infiltration into the 50–300 nm thick mesoporous TiO_2 film and it was found that the polymer chains could be infiltrated to the bottom of the TiO_2 film after heating at 200 °C for a few minutes. It is likely that the polymer undergoes a conformational entropy loss as it is infiltrated into the pores because the pore size in these films is smaller than the radius of gyration[3] of the polymer chain; hence, the authors hypothesize that a favorable

---

3) The radius of gyration is the average radial length of a coil of polymer in solution.

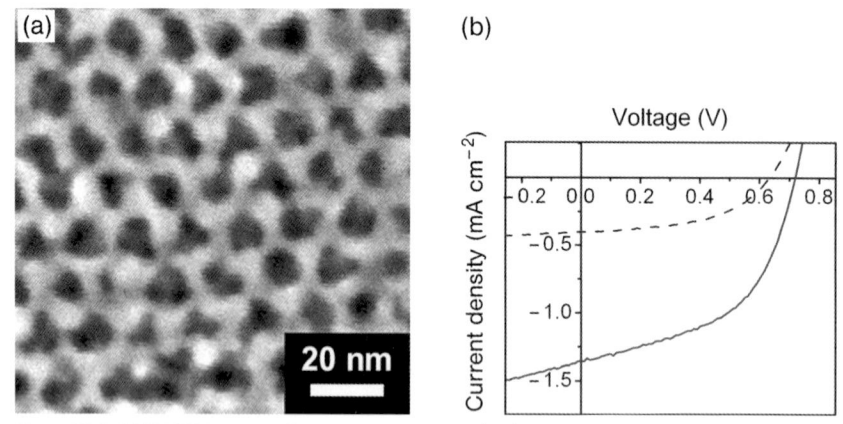

Figure 12.5 (a) HRSEM image of mesoporous titania after being calcined at 400 °C. Threefold symmetry of the pores can be seen here (reproduced from Ref. [43], with permission); (b) current–voltage curve for mesoporous TiO$_2$ (solid line) under 33 mW cm^{-2} 514 nm illumination. The current–voltage curve of a sample made using nonporous TiO$_2$ (dashed line) is included for comparison (reproduced from Ref. [4], with permission).

enthalpic interaction between the polymer chain and the TiO$_2$ surface drives the infiltration. Interestingly, despite the fact that the pore size in these films was only 8 nm, full photoluminescence quenching was not observed in the infiltrated P3HT depending on the infiltration conditions used. Also, both the absorption and photoluminescence spectra of the polymer in the pores following different infiltration conditions show a blue shift compared to the spectra of a neat film of P3HT, which suggests that the polymer chains are twisted and not π-stacked on each other. It is unclear whether the incomplete quenching resulted from reduced or just intrinsically low exciton diffusivity in the poorly packed infiltrated polymer chains or from inefficient electron transfer to the mesostructured titania [35]. OCPVs have been made with mesoporous TiO$_2$ and P3HT using SnO$_2$:F and silver as the cathode and anode, respectively [4]. In the optimized device geometry with the polymer chains infiltrated approximately 20–30 nm into the mesoporous TiO$_2$ film, 10% EQE under 514 nm illumination was obtained, a threefold improvement in EQE over devices made with nonporous TiO$_2$ (Figure 12.5b). The authors estimated that the power efficiency of an optimized PV cell would be 0.45% under AM 1.5 G illumination by integrating the EQE spectrum of the device. The EQE of cells made from these materials was found to drop if the conjugated polymer was infiltrated more deeply than 20–30 nm into the mesopores. This indicates that holes generated on P3HT chains infiltrated deep within the TiO$_2$ cannot escape from the film before undergoing back recombination and that the low hole mobility of the infiltrated P3HT chains limits the device performance. This is perhaps not too surprising given the fact that the P3HT chains take a highly tortuous pathway to the bottom of the film. This conclusion is also corroborated by the photoluminescence spectral data (Section 12.5.2) that indicate that the polymer chains are unable to π-stack in the highly confining pores of the TiO$_2$. Thus, straight pores and perhaps a larger pore diameter could be expected to enable good hole transport in these films, which will lead to improved EQE.

12.4.4
Ideal Nanostructures

At the beginning of Section 12.4, we discussed what qualities constitute an "ideal" nanostructure. There are several approaches to develop these nanostructures, which include nanoprinting, anodization, and self-assembly processes.

Goh et al. [44] created dense arrays of titania nanopores using nanoimprint lithography and embossing with diameters ranging from 35 to 65 nm (Figure 12.6a), much smaller than the >100 nm features previously reported. A nanoimprinting method was used, whereby a polymethylmethacrylate (PMMA) mold was first created from an anodic alumina template, which was dissolved away after the PMMA was infiltrated into it and cured. The PMMA chosen had sufficiently high modulus to facilitate replication of such small nanostructures. A thick PDMS backing layer enabled the easy handling of the patterned PMMA layer. A solution of titania was spin coated on the PMMA mold, which was then pressed onto different substrates, including FTO and ITO. After sufficient drying of the sol, the PMMA was itself dissolved leaving the titania nanopores, which were calcined at 450 °C for 6 h, resulting in anatase titania that was a good replica of the original anodic alumina template. The nonreusable nature of the master and mold allowed for the formidable

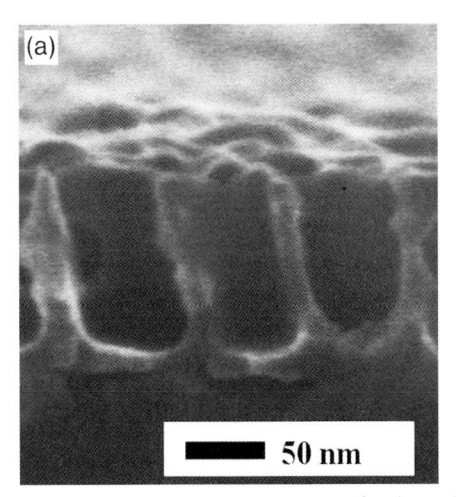

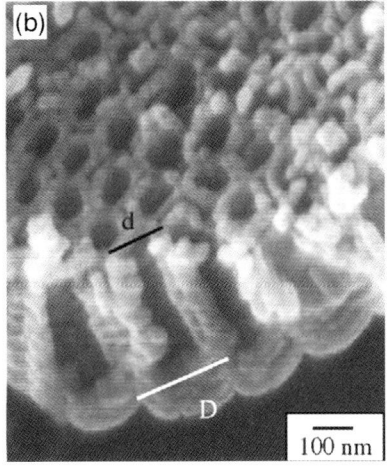

Figure 12.6 (a) Cross-sectional image of embossed titania after calcination at 450 °C (reproduced with permission from Ref. [44]. Copyright 2005 American Chemical Society); (b) FE-SEM images of HF-electrolyte anodized titania nanotube arrays anodized at 10 V for 20 min, then increasing the voltage at 1.0 V min^{-1} to 23 V, and finally kept at 23 V (reproduced from Ref. [46], with permission from Elsevier).

sub-100 nm spacing and periodicity – a characteristic that may or may not be conducive for large-area fabrication. Similar to the above work, Rolland et al. [45] have demonstrated sub-100 nm nanoimprint lithography using photocurable mold materials based on perfluoropolyethers (PFPE), reminiscent of "liquid Teflon" since they are liquid at room temperature (before curing) and are chemically resistant like Teflon.

Grimes et al. [46] have fabricated vertically oriented nanotubes of titania by direct anodization of titanium in fluoride-based electrolytes (Figure 12.6b). They are able to fabricate nanotube arrays of different shapes (cylindrical, tapered), pore size, length, and wall thickness by varying anodization parameters including electrolyte concentration, pH, voltage, and bath temperature. The nanotubes can be made sub-100 nm in diameter, the lower limit of the pore diameter being unclear. Several-micron long nanotubes have been used in dye-sensitized solar cells, which requires such large thicknesses of nanostructured titania to efficiently absorb most of the incident light. To this point, the anodized films have almost exclusively been used in dye-sensitized solar cells, where their relatively low surface roughness ($\sim 33\,\mu m^{-1}$) requires extremely thick films to absorb a sufficient amount of light. With polymer/titania devices, pore size is more important than total surface area and 300–500 nm thick films with 10–20 nm pores could be appropriate for use in titania–polymer solar cells.

Koganti et al. have fabricated a vertically aligned film of 2D hexagonally close-packed titania mesopores using a self-assembly process (Figure 12.7a) [47]. A titania precursor was mixed with a common surfactant pore template (P123). This solution was then spin coated onto a surface modified with a sacrificial layer of PEO-r-PPO random copolymer. This modified surface (with the random copolymer) is key to aligning the titania mesostructure in the vertical direction. When the titania mesostructure layer is deposited on top of the modified substrate with its top surface

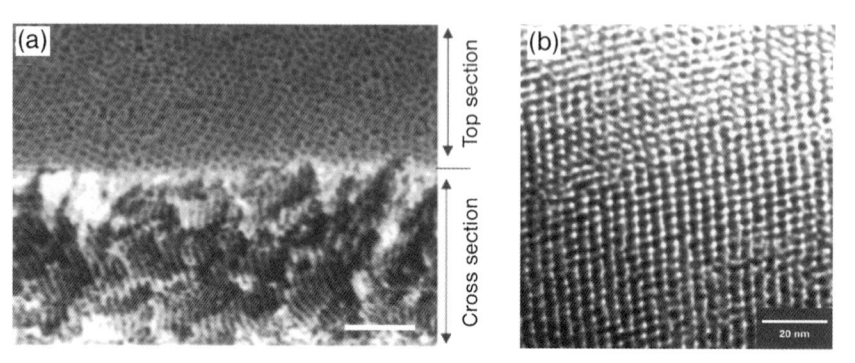

Figure 12.7 (a) SEM images of a fractured 200 nm thick titania film, prepared by confinement between two modified surfaces (scale bar = 100 nm) (reproduced with permission from Ref. [47]. Copyright 2006 American Chemical Society); (b) TEM image of a mesoporous titania thin film (reproduced from Ref. [48], with permission).

exposed to air, it is vertically aligned up to a critical thickness between 70 and 100 nm. When made thicker than this, the ability of the modified substrate to vertically align the mesostructure does not extend through the entire thickness of the mesostructure layer and some regions are not vertically aligned. However, by sandwiching the titania mesostructure layer between two modified surfaces, 200 nm thick films of titania mesopores have been aligned vertically. The vertical alignment of the titania mesopores has been confirmed using X-ray diffraction (XRD), grazing incidence small-angle X-ray scattering (GISAXS), scanning electron microscopy (SEM), and transmission electron microscopy (TEM).

Lee *et al.* have fabricated vertically oriented mesoporous structures in titania and silica by self-assembly (Figure 12.7b) [48,49]. By carefully controlling the humidity during different stages of the fabrication process and by using preformed titania nanoparticles as precursors for the mesostructure formation step, they are able to reproducibly create vertically oriented titania mesostructures (space group Pm-3m) by aligning the [1 1 1] direction perpendicular to the substrate [48]. These vertically oriented pores have a diameter that periodically increases and decreases along the length of the pore. Similar vertically oriented pores in silica have also been obtained by aligning the [1 1 1] direction of the mesostructure (space group $Pm\bar{3}m$) perpendicular to the substrate [49]. The alignment of these structures has been confirmed using XRD and TEM.

We have also created a dense array of nanopillars in silicon using block copolymer lithography [50]. Similar techniques could perhaps be used to obtain such structures in titania. Solar cells based on nanopillars (or nanowires as they are sometimes called) have been studied for ZnO/polymer systems [27,51,52].

Work is going on within the organic solar cell community to create suitable, "ideal" nanostructures that yield high-efficiency solar cells. The fabrication methods listed above are the first steps toward the realization of such a goal.

12.5
Pore Filling

Pore filling and how polymers pack in pores are extremely important in nanostructured polymeric devices. The degree of pore filling affects light absorption, exciton harvesting, and charge carrier injection efficiencies. Several techniques have been successfully used to fill pores in OCPVs: spin casting, melt infiltration, dip coating, and *in situ* polymerization.

12.5.1
Spin Casting

Arango *et al.* made polymer–titania PV cells by simply spin casting a solution of the polymer over a film of sintered titania nanocrystals [3]. They inferred that the polymer at least partially penetrated the titania since PV cells made with nanoporous titania were more efficient than ones made with solid titania. However, they did not report any measurements that revealed the extent of polymer penetration. Over the last couple of years, multiple research groups have tried to infiltrate various polymers into films of sintered titania nanocrystals by spin casting polymer solutions over the films or by dipping the films in polymer solutions and have found that only certain polymers go into the pores [35]. Huisman and coworkers introduced a simple method for determining whether or not a polymer reached the bottom of a film [6]. They spin-cast poly(3-octylthiophene) (P3OT) over several types of nanoporous films and then heated the films in air at 60 °C, which lowers the polymer's resistance, possibly by doping it. When they measured the current–voltage (I–V) characteristics of diodes made with these films, they observed linear I–V curves if the polymer penetrated all the way to the bottom electrode and nonlinear diodelike curves if the polymer did not reach the bottom. Using this test, they determined that spin casting P3OT was sufficient to penetrate up to 1 µm of a film made using 50 nm nanocrystals deposited by their spray pyrolysis method. They also made films of sintered 9 and 50 nm diameter titania nanocrystals. They found that P3OT could penetrate at least 2 µm into the film with 50 nm nanocrystals, but that it could not penetrate even 300 nm into the film with 9 nm nanocrystals. This confirms that pore size is an important variable for the infiltration process. For the films made by spray pyrolysis, they measured the absorbance of the film and concluded that only approximately 10% of the pore volume was filled with polymer. Given that the polymer reaches the bottom of the film, but only fills up a small fraction of the volume, we speculate that the polymer probably coats the pore walls.

12.5.2
Melt Infiltration and Dip Coating

Another method used for filling polymer into nanostructures is melt infiltration, which involves spin casting a polymer over a nanoporous film and then heating the film to soften or melt the polymer [43]. X-ray photoelectron spectroscopy depth profiling has been used to prove that regioregular poly(3-hexylthiophene) (P3HT) penetrates to the bottom of 200 nm thick mesoporous films of titania with 8 nm diameter pores when heated at temperatures in the range of 100–200 °C. At 200 °C, the infiltration only takes a few minutes. If shorter heating times are used, it is possible to infiltrate the polymer only part of the way into a film, which is useful for gaining insight into how the cells work [4]. It was also seen that when lower infiltration temperatures are used, less polymer is incorporated into the films. The most likely explanation for this observation is that the polymer coats the titania walls and that the coating is thicker at higher temperatures due to increased twisting of the P3HT chains. Blue shifts in the absorption and photoluminescence spectra with increasing infiltration temperature provide evidence for the twisted conformation of the chains. In Section 12.6, we will discuss the effect that the twisted chain conformation has on exciton diffusion and charge transport.

The dip coating process involves immersing TiO_2 nanostructures in a heated dilute polymer solution (e.g., 2 mg ml^{-1}) for several hours and then spinning an overlayer of polymer [38,53]. This technique was used to infiltrate 100 nm thick films composed

12.5.3
In situ Polymerization

Two in situ methods have been used to grow polymers inside pores: electropolymerization and surface-initiated growth. Liu et al. electropolymerized monomers such as 3-methyl thiophene (polymer is P3MeT) within 8 nm diameter pores with a thickness of 120 nm. The polymerization was initiated from the bottom, transparent conducting electrode, at the completion of which the films were immersed in an ammonia solution for 2 h to dedope the polymer [54]. A major problem with electropolymerization is that the polymer touches both electrodes, providing a pathway for dark current that runs in a direction that opposes the desired photocurrent. Photovoltaic cells made with this polymer inside unordered mesostructures had an overall power conversion efficiency of 0.23% at low intensity (3.3 mW cm^{-2}) 514 nm monochromatic illumination.

Recently, Zhang et al. have shown that it is possible to grow regiorandom P3HT inside pores by attaching a thiophene initiator on the TiO$_2$ surface and initiating polymerization through in situ oxidative coupling of 3-hexylthiophene monomers, as shown in Figure 12.8 [55]. This method successfully filled straight pores with diameters ranging from 10 to 100 nm. Optical absorption and photoluminescence quenching was greater than for films formed from immersion in toluene solution for several hours. Unfortunately, it is only possible to make regiorandom P3HT with a head–tail to head–head ratio (HT:HH) of 75:25 through this route. Regiorandom P3HT has an field effect transistor (FET) transistor mobility over three orders of magnitude lower than regioregular P3HT used in transistors [56], but this may not necessarily be indicative of its diode mobility, which has not been studied.

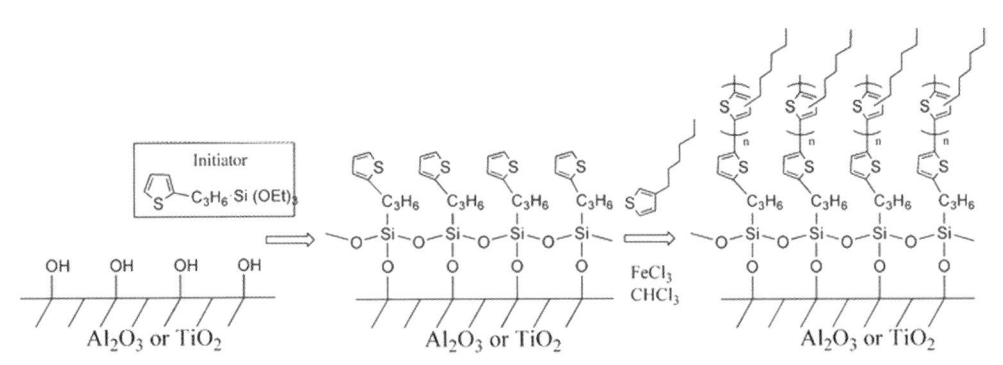

Figure 12.8 Schematic representation of the formation of a monolayer of thiophene initiator on an oxide surface and subsequent surface-initiated polymerization through in situ oxidative coupling of 3-hexylthiophene to fabricate regiorandom P3HT that is covalently attached to the substrate surface (reproduced from Ref. [55], with permission).

12.6
Effects of Pore Filling on Polymer Mobility and Exciton Harvesting

Polymer mobility plays a crucial role in determining the fill factor and J_{sc} of OCPVs; however, there is no general rule to understand the effects of pore sizes on the mobility of a given polymer, because hole transport is dependent on the specific structure of the given polymer film. This section will discuss the relative classifications of conjugated polymers, describe the relevant mobility measurement techniques, characterize the role of pore size for semicrystalline polymers, and explain the importance of hole transport in OCPVs.

12.6.1
Ordered Versus Disordered Polymers

Polymers can be roughly divided into two structure classes: amorphous/disordered (e.g., unsymmetric PPV derivatives like MEH-PPV) and semicrystalline/ordered (e.g., P3HT). Semicrystalline polymers generally have mobilities several orders of magnitude greater than disordered polymers due to better π–π stacking and crystal formation. These polymers exhibit extreme structure/property relations, meaning that the mobility is heavily linked to the degree of crystal formation and orientation. For these polymers, pore filling alone does not ensure high mobilities; in fact, the mobility of these films can vary by several orders of magnitude depending on the pore size (see Section 12.6.3).

Intrinsically disordered polymers (e.g., MEH-PPV derivatives) usually have unsymmetric side chains and are less affected by pore size because their mobilities are not dependent on crystallization. Table 12.1 lists the mobilities of several prominent polymers used in OCPVs.

12.6.2
Measurement Techniques

In neat[4] films, the polymer mobility is dependent on the measurement technique. A field effect transistor test measures the charge transport of the polymer in the direction parallel to the substrate; the maximum FET mobility for ordered polymers[5] is 0.6 and $\sim 10^{-4}$ $cm^2 V^{-1} s^{-1}$ for disordered polymers [40,57,58]. However, in solar cells, holes travel perpendicular to the substrate ("up" toward the electrode). This transport direction is measured in the diode configuration or via space charge limited current measurements and time-of-flight measurements [59]. Time-of-flight measurements cannot be taken accurately on thin films because the absorption depth ($\sim$200–300 nm) is not small compared to the film thickness as required. The space charge mobility of all polymers is lower than the FET mobility because the FET

---

[4] A neat film is defined as a film spun on a flat substrate.
[5] This polymer, poly(2,5-bis(3-alkylthiophen-2yl)thieno[3,2-b]thiophene (PBTTT), has yet to be incorporated into an organic/inorganic photovoltaic cell.

12.6 Effects of Pore Filling on Polymer Mobility and Exciton Harvesting

Table 12.1 Field effect transistor mobility and space charge (SC) mobility of commonly used polymers in OCPVs.

Structure	Polymer type	μ_{FET} (cm^2 V^{-1} s^{-1})	μ_{sc} (cm^2 V^{-1} s^{-1})	Reference
RR-P3HT	Ordered	1×10^{-1}	3×10^{-4}	[73]
BEH-PPV	Ordered	9×10^{-4}	1×10^{-5}	[40]
LC-PPV	Ordered	7×10^{-3}	2×10^{-6}	[40]
OC$_1$C$_{10}$-PPV	Disordered	4×10^{-4}	5×10^{-7}	[40]
MEH-PPV	Disordered	5×10^{-4}	2×10^{-7}	[40]

measurement uses a higher carrier concentration [60] and semicrystalline polymers are highly oriented at the buried interface. Higher carrier concentrations fill traps and consequently can increase the overall mobility of the polymer. The charge carrier density (hole density) can be around 10^{14}–10^{15} cm^{-3} for the diode configuration and 10^{18}–10^{20} cm^{-3} for the FET configuration. This difference is sufficient to explain the large change in mobility for disordered polymers, but not for semicrystalline films [60]. Semicrystalline polymers, such as P3HT, form highly oriented crystals at the polymer/dielectric interface where the current flows in thin-film transistors. The increased order, discussed below, can improve mobility by several orders of magnitude. It has also been shown that the high degree of order at the buried interface does not extend through the film [57]. The lack of order in the bulk is not a problem for FETs because charge is only transported in the first couple of nanometers at the interface. In the diode configuration, charge must traverse through the bulk and the interfacial ordering is much less influential. The space charge mobility measurement is the most relevant technique for measuring hole mobility in solar cells, because it measures the mobility in the direction through which holes travel in a solar cell while having similar carrier concentrations.

Semicrystalline polymers (e.g., PBTTT, P3HT, PQT) have a tendency to form crystals with the alkyl chains perpendicular to the substrate in the first couple of nanometers of the interface [57]. This is extremely beneficial for transistors, which transport charge parallel to the substrate and can take advantage of 2D charge transport along the chains and excellent π–π stacking in crystals. It has been hypothesized that the insulating alkyl chains can inhibit charge transport perpendicular to the substrate reducing hole mobility [61]. This anisotropy can be a detrimental for solar cells, which must either transport charge up through the alkyl chains or through chains that are aligned perpendicular to the substrate. For high polymer mobility in OCPVs, it would be ideal for the chains to lie perpendicular to the substrate; ideal nanostructures likely achieve this by the provision of a scaffold on which the polymers can align in such a way [62] (see Section 12.6.3).

12.6.3
Pore Size Effects for Semicrystalline Polymers

Table 12.1 shows that the mobility of regioregular P3HT can vary three orders of magnitude depending on the measurement; however, the space charge mobility can

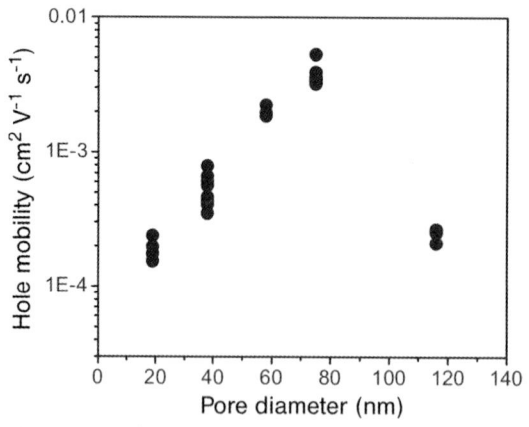

Figure 12.9 Hole mobility in P3HT infiltrated into anodic alumina as a function of pore diameter in the alumina (reproduced from Ref. [62], with permission).

vary significantly when chains are confined inside pores. Figure 12.9 shows the hole mobility of P3HT filled in various pore sizes; within 75 nm pores P3HT has a mobility that is 20 times greater than the space charge mobility of a neat film [62]. Conjugated polymer chains are much more polarizable along their chain and, consequently, absorption is stronger when the electric field is parallel to the chains. One can use optical techniques to determine the alignment of polymers in pores by rotating the sample and measuring the imaginary part of the refractive index. For the P3HT system, it was shown that polymer chains in pores were aligned much more perpendicularly to the substrate than in neat films, providing a plausible explanation for the increased mobility. Unfortunately, the space charge mobility decreases in smaller pores, even though polymer chains become more aligned in the vertical direction. It is likely that polymer chains do not pack as well on each other or that the pathways for current are disrupted in smaller pores.

12.6.4
Significance of Polymer Mobility in OCPVs

It is unclear what the ideal polymer mobility should be in titania-templated OCPVs. Clearly, in devices with extremely low hole mobility, space charge will limit device thickness. Higher hole mobilities will allow us to develop thicker cells that will have a stronger absorption and perhaps a broader absorption spectrum. It should be noted that although a 250 nm thick film of P3HT could absorb enough light to generate 14.0 mA cm^{-2}, a device with a 500 nm P3HT layer would produce 16.0 mA cm^{-2} (roughly 14% more) because of broadening of the absorption spectrum [63].

The relationship between hole mobility and charge carrier collection efficiency η_{CC} is less well defined. η_{CC} is dependent on bimolecular recombination at the donor–acceptor interface, whose magnitude is the product of the carrier concentrations and a recombination rate constant [64]. Improved charge carrier mobility increases the recombination rate constant. However, increasing the mobility also lowers the charge carrier density (as holes are transported more quickly to the anode). It has not yet been definitively shown whether lower or higher polymer mobility would reduce bimolecular recombination and enhance η_{CC}. Furthermore, this issue may depend on OCPV geometry and polymer morphology.

12.6.5
Pore Filling and Exciton Harvesting

Another important consideration is exciton diffusion to and splitting at the polymer/titania interface. It has been shown that there can be imperfect quenching in bilayers even when the polymer thickness is less than the diffusion length of the excitons [26]. Imperfect quenching can occur because of poor electronic coupling between the polymer chains and titania, thus implying poor η_{CT}. Only 68% of excitons are split at the interface inside titania mesoporous films [43], possibly because of reduced exciton diffusivity due to poor interchain coupling (π–π stacking), in addition to imperfect charge transfer. With the help of the scaffolding in an ideal nanostructure, when polymer chains align perpendicular to the substrate, the interchain coupling should likely be improved leading to better exciton diffusion, except when the pores are very small and the π–π stacking is disrupted [43,62]. Optimal alignment may also improve η_{CT} by promoting electronic coupling between the polymer and titania. Exciton splitting can be increased significantly by using interface modifiers (see Section 12.8).

The science of infiltrating conjugated polymers into nanopores is still in its infancy. Each research group working on this subject has used different polymers, titania films, and infiltration techniques. The results obtained so far suggest that the polymers' affinity for the titania surface, glass transition temperature, melting temperatures, and molecular weight are important variables, along with the titania pore size and shape. It is still not clear whether it is best to diffuse chains in solution, infiltrate them from a melt (or softened solid), or polymerize them directly in the pores. It is clear, however, that it is possible to incorporate high molecular weight polythiophene into pores as small as 8 nm in diameter with melt infiltration and electropolymerization. In the future, it will be important to determine how the various infiltration techniques affect the polymer chain packing and how this in turn affects absorption, exciton diffusion, electron transfer, and charge transport.

12.7
Organic Composite Photovoltaic Modeling

Intuitively, given the low exciton diffusion length of polymers, a bulk heterojunction architecture maximizing the interface area between the organic semiconductor and the second semiconductor is necessary for efficient power conversion. However, theoretical modeling of organic-based bulk heterojunction devices has not been extensively pursued. A couple of the more notable studies pertinent of titania-based OCPVs are discussed below. Other relevant modeling by Coakley *et al.* is discussed in Section 12.8.

Kannan et al. [65] model two geometries of bulk heterojunctions – polymer nanowires embedded in an inorganic matrix and the inverse, that is, inorganic nanowires surrounded by a matrix of the semiconducting polymer. The authors considered two limiting cases for exciton transport to the donor–acceptor interface: (a) the diffusion model, where the exciton diffusion length (mean free path) was smaller than the characteristic length scale of the bulk heterojunction and (b) the ballistic model, where the length scales were smaller than the exciton diffusion length and interpolated for cases in-between. A fundamental result, not unexpectedly, is that the exciton transport should be in the ballistic regime for maximized photocurrent and efficiency. The authors conclude that a nanostructure periodicity of the order of the exciton diffusion length is required for maximized exciton splitting, whereupon improved carrier transport has the greatest impact on charge extraction and photocurrent maximization. The authors also considered possible implication of quantum-size effects on the transport properties of the semiconductors.

Martin et al. [66] perform a conceptually similar analysis of the effect of the specific nanostructure geometry on the short-circuit current density. In contrast to the previous work, their model considers two limiting regimes of charge transport (not exciton transport). The authors' model system consists of an interpenetrating array of nanoridges and nanochannels of donor and acceptor sandwiched between thin layers of only acceptor and donor (on the outer sides of which are the electrodes). The effect of varying dimensions of this nanostructure on short-circuit current was computed. Increasing the length (or height) of the interpenetrating nanoridge–nanochannel region introduces two competing effects that directly affect the J_{sc}, that is, (a) increasing surface area causing greater charge injection (exciton splitting) and (b) increasing distance through which the charges need to travel, which, given the typically low polymer mobilities, promotes recombination via hole buildup and attraction of nearby electrons. Consequently, there exists an optimum channel length where the short-circuit current is maximized – between 200 and 300 nm for channel widths between 20 and 80 nm and polymer hole mobility of about $3 \times 10^{-4}\,cm^2\,V^{-1}\,s^{-1}$. The optimum thickness of the nanostructured region depends on factors that affect the two main competing effects (charge injection and recombination), that is, materials properties such as mobility, diffusivity, and recombination constant, as well as operating conditions such as light intensity.

12.8
Future Outlook

Currently, polymer–titania cells have achieved efficiencies of about 0.5%. If suitable modifications to the materials properties and architectures are made, then it is possible to achieve efficiencies greater than 10% [41]. We now discuss those suitable modifications that will lead to significant increase in efficiency.

To improve the short-circuit current from polymer–titania cells, it is important to absorb most of the light. To obtain EQE higher than 90% at wavelengths lower than the P3HT's absorption onset, 500 nm of P3HT is required [63]. Of course, the excitons generated within this film must be harvested. An ideal nanostructure, such as the one described in Section 12.4, would promote light absorption, exciton harvesting, charge transfer, and charge transport, as described earlier.

12.8.1
Low-Bandgap Polymers

However, with P3HT, just increasing EQE would be insufficient to attain 10% power-conversion efficiencies. As an illustration, we can assume 100% EQE for all photons above the 1.9 eV bandgap of P3HT, which yields a short-circuit current of approximately 17 mA cm^{-2}. Coupling this with a V_{oc} of 550 mV and a fill factor of 0.6 gives us an efficiency of 5.6% [41]. Reducing the bandgap of the semiconducting polymer will undoubtedly lead to significant improvement in efficiency. To first order with 100% EQE at wavelengths below the absorption onset, P3HT with a bandgap of 1.9 eV absorbs approximately 21% of the solar spectrum. But reducing the bandgap to 1.4 eV (thereby extending the absorption spectrum to about 900 nm) would mean that 46% of the solar photons are absorbed [41].

12.8.2
Polymer Engineering

Another avenue for improvement is increasing the V_{oc} by reducing the LUMO–LUMO gap between the donor and acceptor that aids in splitting excitons. The gap in polymer–titania cells is about 1 eV, which represents an energy loss. Coakley et al. [41] presented calculations of the maximum efficiency possible with a 1 eV LUMO–LUMO gap, as a function of the polymer bandgap, assuming 100% EQE for energies above the bandgap, fill factor of 1, and V_{oc} equal to the difference between titania-LUMO and polymer-HOMO. Although practically it is unlikely that such EQE, fill factor, and V_{oc} can be achieved, the maximum theoretical efficiency is an encouraging 15%, somewhat fortuitously occurring at a bandgap close to P3HT's 1.9 eV. The authors go on to calculate optimum bandgaps and, therefore, maximum efficiencies as a function of the LUMO–LUMO drop. From these plots, duplicated in Figure 12.10, it is clear that reducing the LUMO–LUMO drop is an option that has great potential for increasing efficiencies. One concern is that a lower LUMO–LUMO drop may reduce the efficiency of forward electron transfer. An encouraging report [67] demonstrates that electron transfer from a 1.6 eV bandgap polymer to phenyl C$_{61}$-butyric acid methyl ester (PCBM) is possible despite only a 0.3 eV drop. Assuming η_{CT} of 100% with a 0.3 eV drop and all previous assumptions, maximum power-conversion efficiencies of about 35% could be attained.

Adjusting the bandgap and LUMO–LUMO drop will affect other vital characteristics of the solar cell, such as the V_{oc}. V_{oc} is widely believed to be proportional to the acceptor-LUMO–polymer-HOMO gap, with the maximum attainable V_{oc} in a system being equal to the gap itself [19,68,69]. Reducing the donor (polymer) bandgap while simultaneously lowering the LUMO–LUMO gap allows the acceptor-LUMO–polymer-HOMO gap to be kept constant. Using an ideal nanostructure that enables

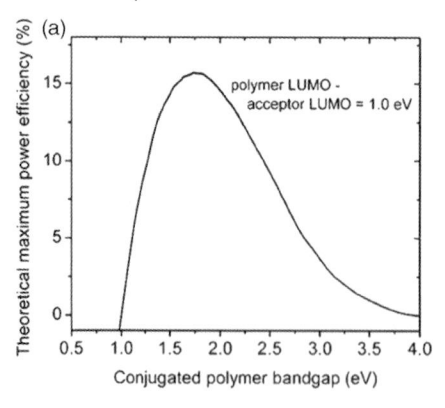

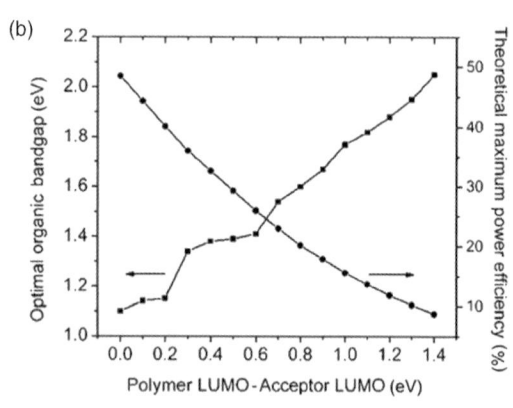

Figure 12.10 (a) Theoretical maximum power efficiency of a conjugated polymer-electron acceptor PV cell as a function of the polymer bandgap when 1 eV of energy is lost during electron transfer; (b) optimal polymer bandgap and theoretical maximum power efficiency as a function of the energy loss during electron transfer. The effect of the dark current on the theoretical maximum efficiency is not included in the calculations used to generate these plots (reproduced with permission from Ref. [41]. Copyright 2004 American Chemical Society).

interface modification allows more flexibility to independently change such parameters without adversely affecting other qualities such as polymer morphology.

12.8.3
Increasing Exciton Diffusion Lengths via Energy Transfer

Förster resonant energy transfer (RET), if suitably incorporated into solar cells, has the ability to increase the effective diffusion lengths of polymers, for example, from 3 to 27 nm [70]. RET, for the purposes of this text, can be thought of as the transfer of energy from an excited chromophore (exciton) in an energy donor to an energy acceptor. The strength of RET and, consequently, the distance over which RET is effective depends, among other things, on photoluminescence efficiency of the donor and overlap between donor emission and acceptor absorption spectra. Liu et al. [71] harnessed RET by placing an interfacial layer (PTPTB, an energy acceptor) between TiO_2 and P3HT (energy donor), thus harvesting more excitons and increasing photocurrent. By comparing the photoluminescence (PL) of P3HT on a thin (5 nm) layer of titania and on titania covered by 4 nm of PTPTB, it was seen that triple the excitons were quenched in the latter case because of PTPTBs ability to accept excitons via RET. RET is not possible between P3HT and titania because titania has a larger bandgap than P3HT, making it impossible for titania to act as an energy acceptor. On making solar cells incorporating a thin layer of PTPTB between titania and P3HT (i.e., titania/PTPTB/P3HT), a threefold improvement in the short-circuit current density over the titania/P3HT configuration was achieved (Figure 12.11). In the titania/polymer system, in the absence of RET, because polymers have a lower electron affinity than titania, exciton splitting occurs via electron transfer. However, in the presence of PTPTB, for the excitons transferred to it via RET to be split, it is

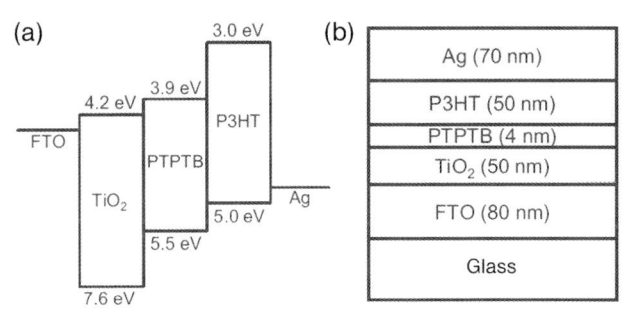

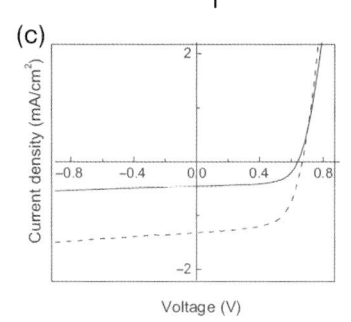

Figure 12.11 (a) Energy diagram of photovoltaic cells. P3HT energy levels were measured by cyclic voltammetry. (b) Device structure. (c) Current–voltage (I–V) characteristics of FTO/TiO$_2$/P3HT/Ag (solid) and FTO/TiO$_2$/PTPTB/P3HT/Ag (dash) cells tested under 100 mW cm^{-2} simulated air mass (AM) 1.5 spectrum. The thicknesses of titania, PTPTB, and P3HT are 50, 4, and 50, respectively (reproduced from Ref. [71], with permission).

essential for holes to jump back from the energy acceptor (PTPTB) to the energy donor (P3HT) (refer to Figure 12.11a for relevant energy levels), and for the electrons in PTPTB to transfer to titania for conduction [69]. RET has certain advantages that must not be overlooked: (a) it places the excitons right at the donor–acceptor interface, enabling the excitons to be split by hole transfer, (b) it is faster than exciton diffusion and can thus harvest the excitons before natural decay, and (c) it has the capability of minimizing trapping of excitons in low-energy sites of polymers, as well as possibly detrapping trapped excitons from such sites. Scully et al. [15] have studied the impact of RET on increasing the effective exciton diffusion length (i.e., the apparent exciton "diffusion" lengths were not RET considered) of polymers. By theoretically modeling the quenching of excitons, the authors find significant enhancements to the effective exciton diffusion length even with only moderately strong RET. The enhancement is more for lower exciton diffusion lengths.

The significance of enhanced effective diffusion lengths has significant implications for the nanostructures required. Clearly, a larger domain size will now suffice for complete exciton harvesting, which perhaps makes the fabrication of these nanostructures easier. Also, larger nanostructures would perhaps give more room for the polymer chains inside the nanostructure to align themselves in such a way as to maximize exciton diffusion, electron transfer (or hole transfer after RET), and charge carrier mobility, as described earlier.

12.8.4
Interface Modification

Another avenue for improvement is the use of interface modifiers to maximize forward electron transfer and minimize recombination. Ravirajan et al. [27] have reported the use of an amphiphilic dye, Z907, as an interface modifier and its manifold beneficial effects in nanostructured ZnO/P3HT cells. In cells with the dye, both the V_{oc} and the J_{sc} increase. The former is attributed to the suppression of recombination due to the physical barrier posed by the dye molecule, along with the suppression of shunt pathways between P3HT and the ITO through defects in the ZnO. The latter is attributed mainly due to improved wetting of the ZnO by the P3HT and consequent increase in the interfacial area. The use of another nonamphiphilic dye showed reduced recombination but no concomitant increase in the J_{sc}. Hence, the increase in J_{sc} was concluded not to be a direct consequence of the reduced recombination. Such a scheme of using interface modifiers to improve the output of the cell can be easily extended to titania/polymer systems, and especially to nanostructured composites, more so in the ideal configuration.

12.8.5
Conclusion

Nanostructured titania, especially in the ideal configuration, provides the essential scaffolding with fixed domain sizes, whose interface can be easily modified – this is the most important reason to develop nanostructured titania templates. Such a nanostructure enables great flexibility to fine-tune OCPVs by independently exploring polymer choices and device architectures, which among other things allows for large improvements in light absorption (e.g., by using a low-bandgap polymer), exciton harvesting, charge transport, and V_{oc} (e.g., by reducing the LUMO–LUMO difference between titania and the polymer by adjusting the polymer side chains or by using interface modifiers). Two broad avenues for significant progress lie in (a) creating the ideal nanostructure that enables maximum efficiency given a particular titania/polymer system, and (b) polymer engineering to enhance intrinsic properties such as bandgap, exciton diffusion lengths, energy levels, and mobility. In the case of titania/P3HT, harvesting all incident photons, perhaps by using nanostructures, coupled with reported V_{oc}s and fill factors achieves an efficiency of over 5%. Suitable polymer engineering can catapult efficiencies toward the Shockley–Queisser limit [72] in the future.

References

1 van Duren, J.K.J., Yang, X.N., Loos, J., Bulle-Lieuwma, C.W.T., Sieval, A.B., Hummelen, J.C. and Janssen, R.A.J. (2004) Relating the morphology of poly(p-phenylene vinylene)/methanofullerene blends to solar-cell performance. *Advanced Functional Materials*, **14**, 425–434.

2 Yang, X. and Loos, J. (2007) Toward high-performance polymer solar cells: The importance of morphology control. *Macromolecules*, **40**, 1353–1362.

3 Arango, A.C., Carter, S.A. and Brock, P.J. (1999) Charge transfer in photovoltaics consisting of interpenetrating networks of conjugated polymer and TiO_2 nanoparticles. *Applied Physics Letters*, **74**, 1698–1700.

4 Coakley, K.M. and McGehee, M.D. (2003) Photovoltaic cells made from conjugated polymers infiltrated into mesoporous titania. *Applied Physics Letters*, **83**, 3380–3382.

5 Gebeyehu, D., Brabec, C.J., Sariciftci, N.S., Vangeneugden, D., Kiebooms, R., Vanderzande, D., Kienberger, F. and Schindler, H. (2001) Hybrid solar cells

based on dye-sensitized nanoporous TiO$_2$ electrodes and conjugated polymers as hole transport materials. *Synthetic Metals*, **125**, 279–287.

6. Huisman, C.L., Goossens, A. and Schoonman, J. (2003) Aerosol synthesis of anatase titanium dioxide nanoparticles for hybrid solar cells. *Chemistry of Materials*, **15**, 4617–4624.

7. Bach, U., Lupo, D., Comte, P., Moser, J.E., Weissortel, F., Salbeck, J., Spreitzer, H. and Gratzel, M. (1998) Solid-state dye-sensitized mesoporous TiO$_2$ solar cells with high photon-to-electron conversion efficiencies. *Nature*, **395**, 583–585.

8. Kruger, J., Plass, R., Cevey, L., Piccirelli, M., Gratzel, M. and Bach, U. (2001) High efficiency solid-state photovoltaic device due to inhibition of interface charge recombination. *Applied Physics Letters*, **79**, 2085–2087.

9. Murakoshi, K., Kogure, R., Wada, Y. and Yanagida, S. (1997) Solid state dye-sensitized TiO$_2$ solar cell with polypyrrole as hole transport layer. *Chemistry Letters*, 471–472.

10. Smestad, G.P., Spiekermann, S., Kowalik, J., Grant, C.D., Schwartzberg, A.M., Zhang, J., Tolbert, L.M. and Moons, E. (2003) A technique to compare polythiophene solid-state dye sensitized TiO$_2$ solar cells to liquid junction devices. *Solar Energy Materials and Solar Cells*, **76**, 85–105.

11. Wang, P., Zakeeruddin, S.M., Moser, J.E., Nazeeruddin, M.K., Sekiguchi, T. and Gratzel, M. (2003) A stable quasi-solid-state dye-sensitized solar cell with an amphiphilic ruthenium sensitizer and polymer gel electrolyte. *Nature Materials*, **2**, 402–407.

12. Kim, Y.G., Walker, J., Samuelson, L.A. and Kumar, J. (2003) Efficient light harvesting polymers for nanocrystalline TiO$_2$ photovoltaic cells. *Nano Letters*, **3**, 523–525.

13. O'Regan, B. and Gratzel, M. (1991) Low-cost, high-efficiency solar cell based on dye-sensitized colloidal TiO$_2$ films. *Nature*, **353**, 737–740.

14. Savenije, T.J., Warman, J.M. and Goossens, A. (1998) Visible light sensitisation of titanium dioxide using a phenylene vinylene polymer. *Chemical Physics Letters*, **287**, 148–153.

15. Scully, S.R. and McGehee, M.D. (2006) Effects of optical interference and energy transfer on exciton diffusion length measurements in organic semiconductors. *Journal of Applied Physics*, **100**, 034907.

16. Theander, M., Yartsev, A., Zigmantas, D., Sundstrom, V., Mammo, W., Andersson, M.R. and Inganas, O. (2000) Photoluminescence quenching at a polythiophene/C$_{60}$ heterojunction. *Physical Review B: Condensed Matter*, **61**, 12957–12963.

17. Peumans, P., Yakimov, A. and Forrest, S.R. (2003) Small molecular weight organic thin-film photodetectors and solar cells. *Journal of Applied Physics*, **93**, 3693–3723.

18. Arango, A.C., Johnson, L.R., Bliznyuk, V.N., Schlesinger, Z., Carter, S.A. and Horhold, H.H. (2000) Efficient titanium oxide/conjugated polymer photovoltaics for solar energy conversion. *Advanced Materials*, **12**, 1689.

19. Mihailetchi, V.D., Blom, P.W.M., Hummelen, J.C. and Rispens, M.T. (2003) Cathode dependence of the open-circuit voltage of polymer:fullerene bulk heterojunction solar cells. *Journal of Applied Physics*, **94**, 6849–6854.

20. Anderson, N.A., Hao, E.C., Ai, X., Hastings, G. and Lian, T.Q. (2001) Ultrafast and long-lived photoinduced charge separation in MEH-PPV/nanoporous semiconductor thin film composites. *Chemical Physics Letters*, **347**, 304–310.

21. van Hal, P.A., Christiaans, M.P.T., Wienk, M.M., Kroon, J.M. and Janssen, R.A.J. (1999) Photoinduced electron transfer from conjugated polymers to TiO$_2$. *Journal of Physical Chemistry B*, **103**, 4352–4359.

22. Aduda, B.O., Ravirajan, P., Choy, K.L. and Nelson, J. (2004) Effect of morphology on electron drift mobility in porous TiO$_2$.

International Journal of Photoenergy, **6**, 141–147.

23 Forro, L., Chauvet, O., Emin, D., Zuppiroli, L., Berger, H. and Levy, F. (1994) High mobility n-type charge carriers in large single crystals of anatase (TiO$_2$). *Journal of Applied Physics*, **75**, 633–635.

24 Tang, H., Prasad, K., Sanjines, R., Schmid, P.E. and Levy, F. (1994) Electrical and optical properties of TiO$_2$ anatase thin films. *Journal of Applied Physics*, **75**, 2042–2047.

25 Xie, Z., Burlakov, V.M., Henry, B.M., Kirov, K.R., Smith, H.E., Grovenor, C.R.M., Assender, H.E., Briggs, G.A.D., Kano, M. and Tsukahara, Y. (2006) Intensity-dependent relaxation of photoconductivity in nanocrystalline titania thin films. *Physical Review B: Condensed Matter*, **73**, 113317.

26 Personal communication with Dr Chiatzuh Goh.

27 Ravirajan, P., Peiro, A.M., Nazeeruddin, M.K., Graetzel, M., Bradley, D.D.C., Durrant, J.R. and Nelson, J. (2006) Hybrid polymer/zinc oxide photovoltaic devices with vertically oriented ZnO nanorods and an amphiphilic molecular interface layer. *Journal of Physical Chemistry B*, **110**, 7635–7639.

28 Chelvayohan, M. and Mee, C.H.B. (1982) Work function measurements on (1 1 0), (1 0 0) and (1 1 1) surfaces of silver. *Journal of Physics C: Solid State Physics*, **15**, 2305–2312.

29 Persson, N.K., Arwin, H. and Inganas, O. (2005) Optical optimization of polyfluorene-fullerene blend photodiodes. *Journal of Applied Physics*, **97**, 034503–034508.

30 Gratzel, M. (2001) Photoelectrochemical cells. *Nature*, **414**, 338–344.

31 Breeze, A.J., Schlesinger, Z., Carter, S.A. and Brock, P.J. (2001) Charge transport in TiO$_2$/MEH-PPV polymer photovoltaics. *Physical Review B: Condensed Matter*, **64**, 125205–125209.

32 Markov, D.E., Amsterdam, E., Blom, P.W.M., Sieval, A.B. and Hummelen, J.C. (2005) Accurate measurement of the exciton diffusion length in a conjugated polymer using a heterostructure with a side-chain cross-linked fullerene layer. *Journal of Physical Chemistry A*, **109**, 5266–5274.

33 Gebeyehu, D., Brabec, C.J., Padinger, F., Fromherz, T., Spiekermann, S., Vlachopoulos, N., Kienberger, F., Schindler, H. and Sariciftci, N.S. (2001) Solid state dye-sensitized TiO$_2$ solar cells with poly(3-octylthiophene) as hole transport layer. *Synthetic Metals*, **121**, 1549–1550.

34 Gebeyehu, D., Brabec, C.J. and Sariciftci, N.S. (2002) Solid-state organic/inorganic hybrid solar cells based on conjugated polymers and dye-sensitized TiO$_2$ electrodes. *Thin Solid Films*, **403–404**, 271–274.

35 Luzzati, S., Basso, M., Catellani, M., Brabec, C.J., Gebeyehu, D. and Sariciftci, N.S. (2002) Photo-induced electron transfer from a dithieno thiophene-based polymer to TiO$_2$. *Thin Solid Films*, **403–404**, 52–56.

36 Huisman, C.L., Goossens, A. and Schoonman, J. (2003) Preparation of a nanostructured composite of titanium dioxide and polythiophene: A new route towards 3D heterojunction solar cells. *Synthetic Metals*, **138**, 237–241.

37 Spiekermann, S., Smestad, G., Kowalik, J., Tolbert, L.M. and Gratzel, M. (2001) Poly(4-undecyl-2,2-bithiophene) as a hole conductor in solid state dye sensitized titanium dioxide solar cells. *Synthetic Metals*, **121**, 1603–1604.

38 Ravirajan, P., Haque, S.A., Durrant, J.R., Poplavskyy, D., Bradley, D.D.C. and Nelson, J. (2004) Hybrid nanocrystalline TiO$_2$ solar cells with a fluorene-thiophene copolymer as a sensitizer and hole conductor. *Journal of Applied Physics*, **95**, 1473–1480.

39 Li, G., Shrotriya, V., Huang, J.S., Yao, Y., Moriarty, T., Emery, K. and Yang, Y. (2005) High-efficiency solution processable polymer photovoltaic cells by self-

organization of polymer blends. *Nature Materials*, **4**, 864–868.

40 Kemerink, M., van Duren, J.K.J., van Breemen, A.J.J.M., Wildeman, J., Wienk, M.M., Blom, P.W.M., Schoo, H.F.M. and Janssen, R.A.J. (2005) Substitution and preparation effects on the molecular-scale morphology of PPV films. *Macromolecules*, **38**, 7784–7792.

41 Coakley, K.M. and McGehee, M.D. (2004) Conjugated polymer photovoltaic cells. *Chemistry of Materials*, **16**, 4533–4542.

42 Ravirajan, P., Haque, S.A., Durrant, J.R., Bradley, D.D.C. and Nelson, J. (2005) The effect of polymer optoelectronic properties on the performance of multilayer hybrid polymer/TiO_2 solar cells. *Advanced Functional Materials*, **15**, 609–618.

43 Coakley, K.M., Liu, Y.X., McGehee, M.D., Frindell, K.L. and Stucky, G.D. (2003) Infiltrating semiconducting polymers into self-assembled mesoporous titania films for photovoltaic applications. *Advanced Functional Materials*, **13**, 301–306.

44 Goh, C., Coakley, K.M. and McGehee, M.D. (2005) Nanostructuring titania by embossing with polymer molds made from anodic alumina templates. *Nano Letters*, **5**, 1545–1549.

45 Rolland, J.P., Hagberg, E.C., Denison, G.M., Carter, K.R. and De Simone, J.M. (2004) High-resolution soft lithography: Enabling materials for nanotechnologies. *Angewandte Chemie – International Edition*, **43**, 5796–5799.

46 Grimes, C.A., Mor, G.K., Varghese, O.K., Paulose, M. and Shankar, K. (2006) A review on highly ordered, vertically oriented TiO_2 nanotube arrays: Fabrication, material properties, and solar energy applications. *Solar Energy Materials and Solar Cells*, **90**, 2011–2075.

47 Koganti, V.R., Dunphy, D., Gowrishankar, V., McGehee, M.D., Li, X.F., Wang, J. and Rankin, S.E. (2006) Generalized coating route to silica and titania films with orthogonally tilted cylindrical nanopore arrays. *Nano Letters*, **6**, 2567–2570.

48 Lee, U.H., Lee, H., Wen, S., Mho, S.I. and Kwon, Y.U. (2006) Mesoporous titania thin films with pseudo-cubic structure: synthetic studies and applications to nanomembranes and nanotemplates. *Microporous and Mesoporous Materials*, **88**, 48–55.

49 Lee, U.H., Lee, J.H., Jung, D.Y. and Kwon, Y.U. (2006) High-density arrays of platinum nanostructures and their hierarchical patterns. *Advanced Materials*, **18**, 2825.

50 Gowrishankar, V., Miller, N., McGehee, M.D., Misner, M.J., Ryu, D.Y., Russell, T.P., Drockenmuller, E. and Hawker, C.J. (2006) Fabrication of densely packed, well-ordered, high-aspect-ratio silicon nanopillars over large areas using block copolymer lithography. *Thin Solid Films*, **513**, 289–294.

51 Law, M., Greene, L.E., Johnson, J.C., Saykally, R. and Yang, P.D. (2005) Nanowire dye-sensitized solar cells. *Nature Materials*, **4**, 455–459.

52 Olson, D.C., Piris, J., Collins, R.T., Shaheen, S.E. and Ginley, D.S. (2006) Hybrid photovoltaic devices of polymer and ZnO nanofiber composites. *Thin Solid Films*, **496**, 26–29.

53 Ravirajan, P., Haque, S.A., Poplavskyy, D., Durrant, J.R., Bradley, D.D.C. and Nelson, J. (2004) Nanoporous TiO_2 solar cells sensitised with a fluorene-thiophene copolymer. *Thin Solid Films*, **451/452**, 624–629.

54 Liu, Y.X., Coakley, K.M. and McGehee, M.D. (2004) Electropolymerization of conjugated polymers in mesoporous titania for photovoltaic applications. *Proceedings of SPIE*, **5215**, 187–194.

55 Zhang, Y., Wang, C.W., Rothberg, L. and Ng, M.K. (2006) Surface-initiated growth of conjugated polymers for functionalization of electronically active nanoporous networks: Synthesis, structure and optical properties. *Journal of Materials Chemistry*, **16**, 3721–3725.

56 Bao, Z., Dodabalapur, A. and Lovinger, A.J. (1996) Soluble and processable

regioregular poly(3-hexylthiophene) for thin film field-effect transistor applications with high mobility. *Applied Physics Letters*, **69**, 4108–4110.

57 Goh, C., Kline, R.J., McGehee, M.D., Kadnikova, E.N. and Frechet, J.M.J. (2005) Molecular-weight-dependent mobilities in regioregular poly(3-hexyl-thiophene) diodes. *Applied Physics Letters*, **86**, 122110–122113.

58 Kline, R.J., McGehee, M.D. and Toney, M.F. (2006) Highly oriented crystals at the buried interface in polythiophene thin-film transistors. *Nature Materials*, **5**, 222–228.

59 McCulloch, I., Heeney, M., Bailey, C., Genevicius, K., Macdonald, I., Shkunov, M., Sparrowe, D., Tierney, S., Wagner, R., Zhang, W.M., Chabinyc, M.L., Kline, R.J., McGehee, M.D. and Toney, M.F. (2006) Liquid-crystalline semiconducting polymers with high charge-carrier mobility. *Nature Materials*, **5**, 328–333.

60 Hertel, D., Bassler, H., Scherf, U. and Horhold, H.H. (1999) Charge carrier transport in conjugated polymers. *Journal of Chemical Physics*, **110**, 9214–9222.

61 Tanase, C., Blom, P.W.M., de Leeuw, D.M. and Meijer, E.J. (2004) Charge carrier density dependence of the hole mobility in poly(p-phenylene vinylene). *Physica Status Solidi A*, **201**, 1236–1245.

62 Sirringhaus, H.C., Brown, P.J.C., Friend, R.H.C., Nielsen, M.M.C., Bechgaard, K.C., Langeveld-Voss, B.M.W.C., Spiering, A.J.H.C., Janssen, R.A.J.C., Meijer, E.W.C., Herwig, P.C. and de Leeuw, D.M.C. (1999) Two-dimensional charge transport in self-organized, high-mobility conjugated polymers. *Nature*, **401**, 685–688.

63 Coakley, K.M., Srinivasan, B.S., Ziebarth, J.M., Goh, C., Liu, Y.X. and McGehee, M.D. (2005) Enhanced hole mobility in regioregular polythiophene infiltrated in straight nanopores. *Advanced Functional Materials*, **15**, 1927–1932.

64 Personal communication with Shawn R. Scully.

65 Koster, L.J.A., Mihailetchi, V.D. and Blom, P.W.M. (2006) Bimolecular recombination in polymer/fullerene bulk heterojunction solar cells. *Applied Physics Letters*, **88**, 1–3.

66 Kannan, B., Castelino, K. and Majumdar, A. (2003) Design of nanostructured heterojunction polymer photovoltaic devices. *Nano Letters*, **3**, 1729–1733.

67 Martin, C.M., Burlakov, V.M. and Assender, H.E. (2006) Modeling charge transport in composite solar cells. *Solar Energy Materials and Solar Cells*, **90**, 900–915.

68 Brabec, C.J., Winder, C., Sariciftci, N.S., Hummelen, J.C., Dhanabalan, A., van Hal, P.A. and Janssen, R.A.J. (2002) A low-bandgap semiconducting polymer for photovoltaic devices and infrared emitting diodes. *Advanced Functional Materials*, **12**, 709–712.

69 Waldauf, C., Scharber, M.C., Schilinsky, P., Hauch, J.A. and Brabec, C.J. (2006) Physics of organic bulk heterojunction devices for photovoltaic applications. *Journal of Applied Physics*, **99**, 104503-1–104503-6.

70 Gowrishankar, V., Scully, S.R., McGehee, M.D., Wang, Q. and Branz, H.M. (2006) Exciton splitting and carrier transport across the amorphous-silicon/polymer solar cell interface. *Applied Physics Letters*, **89**, 252102-1.

71 Scully, S.R. and McGehee, M.D. (2007) Long-range resonant transfer for enhanced exciton harvesting, submitted.

72 Liu, Y.X., Summers, M.A., Edder, C., Frechet, J.M.J. and McGehee, M.D. (2005) Using resonance energy transfer to improve exciton harvesting in organic-inorganic hybrid photovoltaic cells. *Advanced Materials*, **17**, 2960.

73 Shockley, W. and Queisser, H.J. (1961) Detailed balance limit of efficiency of p–n junction solar cells. *Journal of Applied Physics*, **32**, 510–519.

13
Metal Oxide–Polymer Bulk Heterojunction Solar Cells
Waldo J.E. Beek, Martijn M. Wienk, and René A.J. Janssen

13.1
Introduction

In the past decade, enormous progress has been reported for photovoltaic devices using solution-processed p-type conjugated polymers, either in combination with polymers [1–3] or fullerenes [4–8] as the n-type material. The motivation for using two different semiconductors in the active layer originates in the fact that photoexcitation of organic semiconductors generally results in a strongly bound electron–hole pair, called an exciton [9,10], that is only effectively separated in free charges at an interface between a p-type (electron-donating) and n-type (electron-accepting) material. In polymer photovoltaic devices, the primary step after absorption of a photon is a photoinduced electron transfer at the interface of donor- and acceptor-type semiconducting materials, yielding a charge-separated state. This boosts the photogeneration of free charge carriers compared to the individual, pure materials. The exciton lifetime and diffusion length in organic semiconductors are limited by radiative and nonradiative decay. As a consequence only excitons generated in close vicinity of a p–n interface can give rise to charges. For poly(*p*-phenylene vinylene)s (PPVs), the exciton diffusion length is about 5–10 nm [11–13]. This implies that for effective charge separation the photogeneration of excitons should occur within a few nanometers from a p–n junction. For polythiophenes the exciton diffusion length is on the same order of magnitude, ∼5 nm [14]. These short exciton diffusion lengths impose an important condition for efficient charge generation. Anywhere in the active layer, the distance to the interface should be on the order of the exciton diffusion length. Despite their high absorption coefficients, exceeding $10^7\,\mathrm{m}^{-1}$, a 10 nm double layer of donor and acceptor materials would not be optically dense, allowing most photons to pass freely. The solution to this dilemma is elegantly simple. By creating a bulk heterojunction of the p- and n-type materials on a nanometer dimension, junctions throughout the bulk of the material are created that ensure quantitative dissociation of photogenerated excitons, irrespective of the thickness (Figure 13.1). To create a working photovoltaic device, the active

Organic Photovoltaics: Materials, Device Physics, and Manufacturing Technologies.
Edited by Christoph Brabec, Vladimir Dyakonov, and Ullrich Scherf
Copyright © 2008 WILEY-VCH Verlag GmbH & Co. KGaA, Weinheim
ISBN: 978-3-527-31675-5

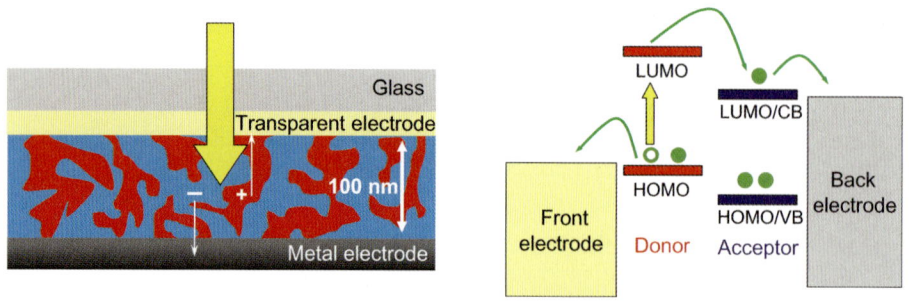

Figure 13.1 Schematic representation of a bulk heterojunction solar cell, showing the phase separation between donor (red) and acceptor (blue) materials and the schematic energy diagram involving the energy levels of the highest occupied and lowest unoccupied molecular orbitals (HOMO and LUMO) and valence and conduction bands (VB and CB).

layer is sandwiched between charge selective electrodes, one being transparent to light (Figure 13.1). Bulk heterojunction solar cells with efficiencies exceeding 5% have been obtained using a combination of a fullerene derivative phenyl C_{61}-butyric acid methyl ester (PCBM) and regioregular poly(3-hexylthiophene) (P3HT) [15–18].

In addition to these bulk heterojunction devices based on conjugated polymers, also the use of wide-bandgap semiconductor metal oxides shows great potential in dye-sensitized solar cells (DSSC) [19]. In a DSSC a nanoporous metal oxide, usually titania (TiO_2), covered with a monolayer of a dye is immersed in a liquid electrolyte. The nanoporous nature of the metal oxide creates a very large interface where dye molecules absorb. The primary step in a DSSC is that after absorbing a photon, the dye injects an electron injection into TiO_2. Subsequently, the electron is transported through the TiO_2 and the photooxidized dye is reduced by the electrolyte, which itself is regenerated at the counter electrode. The best DSSCs currently reach efficiencies of 11.04% [20,21]. A drawback of the traditional DSSC, hampering its wide use, seems the application of a liquid electrolyte. This liquid electrolyte is often related to limited thermostability of the cell and responsible for the corrosion of the Pt-covered counter electrode. For this reason alternatives for the liquid electrolyte are being developed, aiming at solid-state version of the DSSC [22–24]. Current "state of the art" quasi-solid-state dye-sensitized solar cells based on the iodide/triiodide redox couple reach stable and >6% efficient solar cells [25]. Also, the use of solvent-free dye-sensitized solar cell based on an ionic liquid electrolyte and using $SeCN^-/(SeCN)_3^-$ as the redox couple, replacing the iodide/triiodide redox couple, has been considered. This solar cell reaches measured AM 1.5 efficiencies of 8% [26].

With these developments, it is not surprising that the combination of organic or polymer p-type semiconductors with wide-bandgap n-type metal oxides attracts considerable interest for photovoltaic applications. In a metal oxide–polymer solar cell, the polymer takes care of absorption of light and injects an electron into the metal oxide. The photogenerated holes and electrons are the transported through the p-type polymer and the n-type metal oxide. When properly designed, these hybrid materials may have the processing properties of polymers, that is, allowing spin coating, doctor blading, or printing techniques to deposit thin films, have a high absorption coefficient in the visible range, and would benefit from the high charge carrier mobility for holes in the polymer and for electrons in the inorganic material.

Various approaches to hybrid polymer metal oxide solar cells can be envisioned and have been explored. Many of these aim at creating an intimately mixed bulk heterojunction nanostructure, ensuring short exciton diffusion ranges, and creating percolating pathways for both charge carriers to the relevant electrodes. Four conceptually different methods to create active layers can be distinguished (Figure 13.2):

(a) planar bilayer architectures;
(b) infiltrating polymers into nanoporous or nanostructured metal oxides;
(c) blending nanoparticles with conjugated polymers;
(d) *in situ* synthesis of the n-type semiconductor in the p-type polymer layer.

In this chapter we will discuss the most salient results that have been reported for each of these different cell architectures. In Table 13.1, we have made an attempt to collect the reported solar cell parameters and efficiencies reported for metal oxide–polymer hybrid solar cells.

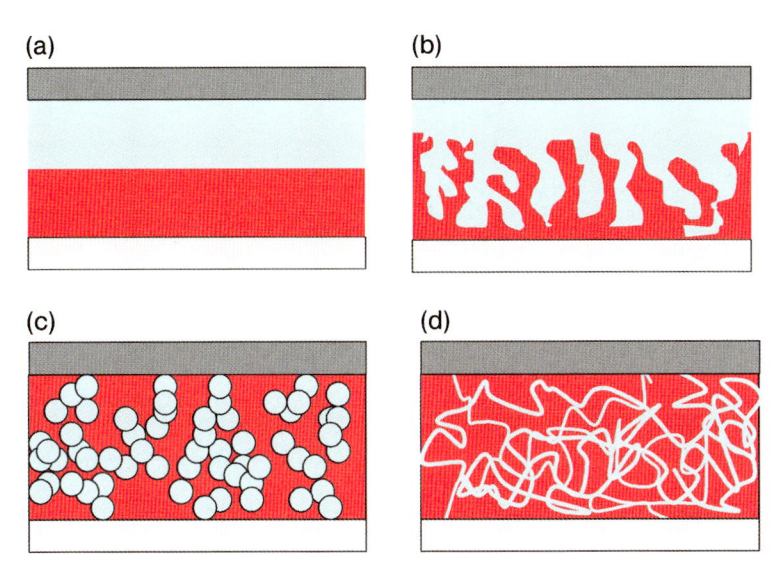

Figure 13.2 Device architectures for hybrid metal oxide–polymer solar cells: (a) bilayer, (b) filled nanoporous metal oxide, (c) polymer nanoparticle blend, and (d) *in situ* synthesized metal oxide network. In each drawing the polymer is represented in red and the metal oxide in cyan.

Table 13.1 Performance of hybrid polymer–inorganic photovoltaic cells.

Cell configuration	J_{sc} (mA cm^{-2})	V_{oc} (V)	FF	η (%)	I_L (mW cm^{-2})	EQE (%)	λ (nm)	Ref.
Planar metal oxide–polymer bilayers								
ITO/TiO$_2$/MEH-PPV/Hg	0.32	0.92	0.52	0.15	100	1	520	[13]
ITO/TiO$_x$/PA-PPV/Au	~0.5	0.85	0.52	~0.2	100	25	425	[34]
ITO/TiO$_2$/MEH-PPV/Au	0.24	0.70	0.42	0.07	100	3.3	500	[35]
ITO/TiO$_2$/M3EH-PPV/Au	1.2	0.65	0.40	0.40	80	7.5	480	[36]
ITO/TiO$_2$/MEH-PPV/Au	0.008	0.63	0.67	1.6	0.2^a	9.5	300	[37]
ITO/TiO$_2$/MEH-PPV/Au	0.33	0.87	0.49	0.13	100			[38]
ITO/TiO$_2$/MEH-PPV/Ag	0.4	0.74	0.25	0.07	100			[31]
ITO/TiO$_2$/MDMO-PPV/ PEDOT:PSS/Au	0.96	0.51	0.61	0.30	100	12	540	[39]
ITO/TiO$_2$/P3OT/Au	0.009	0.56	0.36	0.003	64			[42]
ITO/TiO$_2$/P3HT/Au	0.030	0.55			25			[46]
FTO/TiO$_2$/P3UBT/graphite	0.085	0.78	0.30	0.02	100			[44,45]
FTO/TiO$_2$/P3HT/Ag	0.68	0.66	0.60	0.79	39^b	4.2	514	[52]
FTO/TiO$_2$/PT-COOH/P3HT/Ag	2.06	0.52	0.40	1.10	39^b	12.6	514	[52]
FTO/TiO$_2$/P3HT/Ag	0.46	0.64	0.63	0.19	100			[55]
FTO/TiO$_2$/PTBTB/P3HT/Ag	1.33	0.67	0.63	0.56	100			[55]
ITO/TiO$_2$/MEH-PPV/Au	0.33	0.87	0.49	0.13	100			[37]
FTO/TiO$_2$/P3HT/Ag	0.85	0.60	0.67	0.34	100	~5	510	[54]
FTO/TiO$_2$(N3 + TBP)/P3HT/Ag	1.86	0.57	0.57	0.60	100	~14	510	[54]
ITO/Zn$_{0.75}$Mg$_{0.25}$O/P3HT/Ag	1.27	0.70	0.56	0.49	100			[59]
Nanoporous metal oxides filled with polymers								
ITO/TiO$_2$/MEH-PPV/Au	0.40	1.10	0.40	0.18	100			[35]
ITO/TiO$_2$/P3OT/Au	0.17	0.70	0.40	0.08	60			[42]
ITO/TiO$_2$/Ru-dye/P3OT/Au	0.33	0.64	0.44	0.16	60			[42]

ITO/TiO$_2$/F8BT/Au	0.40	0.97	0.44	0.16	100	13	440	[73,74]
ITO/TiO$_2$/MEH-PPV/PEDOT:PSS/Au	0.97	0.74	0.48	0.44	80			[75]
ITO/TiO$_2$/TPDs4Md-MEH-M3EH-PPV/PEDOT:PSS/Au	2.10	0.64	0.43	0.58	100	40	425	[77]
FTO/TiO$_2$/P3HT/Ag	1.22	0.72	0.51	0.45	100	10	514	[81]
ITO/TiO$_2$/MEH-PPV/Au	3.30	0.86	0.28	0.71	100	34	—	[83,84]
ITO/ZnO(pillars)/P3HT/Ag	2.20	0.475	0.56	0.53	100	17	540	[29]
ITO/ZnO(pillars)/P3HT:PCBM/Ag	10.0	0.475	0.43	2.0	100	57	515	[29]
ITO/ZnO(pillars)/Ru-dye/P3HT/PEDOT:PSS/Au	1.73	0.30	0.39	0.20	100	14	550	[89,90]
ITO/ZnO(pillars)/Ru-dye/TPD(4M)-MEH-M3EH-PPV/PEDOT:PSS/Au	~1.50	~0.30	~0.33	0.15	100			[90]
ITO/TiO$_2$(pillars)/MEH-PPV/Au	0.95	0.88	0.47	0.39	100	15	510	[37]
Blends of metal oxide nanoparticles and polymers								
ITO/TiO$_2$:MEH-PPV/Ca	~0.020	~1.00			2			[64]
ITO/PEDOT:PSS/TiO$_2$:P3HT/Al	2.76	0.44	0.40	0.42	100	15	475	[104,105]
FTO/TiO$_2$/P3OT/Au aerosol deposition	0.25	0.72	0.35	0.06	100	2.5	488	[108,109]
ITO/PEDOT:PSS/ZnO:MDMO-PPV/Al	2.40	0.81	0.59	1.6	71	40	510	[110]
ITO/PEDOT:PSS/ZnO:P3HT/Al	2.14	0.69	0.55	0.9	100	27	490	[119]
ITO/PEDOT:PSS/ZnO:APFO-3/Al	3.1	0.51	0.36	0.45	100	28	535	[126]
In situ metal oxide networks in polymers								
ITO/PEDOT:PSS/TiO$_x$:MDMO-PPV/LiF/Al	0.6	0.52	0.42	0.20	60	11	450	[128]

(*Continued*)

Table 13.1 (Continued)

Cell configuration	J_{sc} (mA cm^{-2})	V_{oc} (V)	FF	η (%)	I_L (mW cm^{-2})	EQE (%)	λ (nm)	Ref.
ITO/PEDOT:PSS/TiO$_x$: MDMO-PPV/LiF/Al	0.6	0.60	0.43	0.22	70	14	490	[129]
ITO/PEDOT:PSS/TiO$_x$: P3OT/LiF/Al	0.7	0.45	0.41	0.17	70	10	490	[129]
ITO/PEDOT:PSS/ZnO: MDMO-PPV/Al	2.0	1.14	0.42	1.10	90	27	520	[132]
ITO/PEDOT:PSS/ZnO: P3HT/Sm/Al	3.5	0.83	0.50	1.40	100	26	500	[135]

Short-circuit current (J_{sc}), open-circuit voltage (V_{oc}), fill factor (FF), power conversion efficiency (η), light intensity (I_L), and external quantum efficiency (EQE) at wavelength λ. When comparing currents and efficiencies, care must be taken that light conditions may vary significantly even for entries that are nominally similar, due to spectral differences.
[a]Monochromatic light of 500 nm.
[b]Monochromatic light of 514 nm.

13.2
Planar Metal Oxide–Polymer Bilayer Cells

Covering a planar layer of a wide-bandgap inorganic semiconductor as TiO_2 with a thin polymer film is the most direct approach to hybrid photovoltaic cells. The simple device architecture is advantageous in the sense that photogenerated charge carriers can easily be transported to the electrodes and that the necessity of creating percolation pathways for charge carriers is easily met in this layout. On the contrary, the relatively small interfacial area between polymer and inorganic semiconductors may limit the charge carrier generation, especially when the exciton diffusion length is small. Because of this fundamental limitation, the progress – in terms of power conversion efficiency – for hybrid polymer–metal oxide bilayers has been modest since the first report [13]. Nevertheless, the device architecture structure has received considerable interest and allows for investigating fundamental scientific issues with respect to charge generation, interface processes, and device operation. Most of the planar metal oxide–polymer devices employ TiO_2 for charge generation. Planar zinc oxide (ZnO) layers have been used for charge collection in inverted polymer:fullerene solar cells [27–30], but bilayer polymer/ZnO cells where charge generation occurs predominantly in the organic layer have only recently been reported [31,32]. In the following two sections, we therefore focus mainly on TiO_2 layers and discuss the progress using poly(p-phenylene vinylene)s and polythiophenes.

13.2.1
Metal Oxide–Poly(p-Phenylene Vinylene)

The photophysical properties of thin films of poly[2-methoxy-5-(2′-ethylhexyloxy)-1,4-phenylene vinylene] (MEH-PPV) on planar TiO_2 substrates were first investigated in 1998 by Savenije et al. [13]. Evidence for dissociation of excitons at the MEH-PPV/TiO_2 interface and electron injection into the conduction band of TiO_2 was clearly established from current–voltage characteristics for TiO_2/MEH-PPV bilayers sandwiched between a transparent indium-tin oxide (ITO) electrode and a mercury (Hg) drop contact. The cells gave a short-circuit current (J_{sc}) of 0.32 mA cm^{-2} and open-circuit voltage (V_{oc}) of 0.92 V and a fill factor (FF) of 0.52 under AM 1.5 white-light conditions (100 mW cm^{-2}), resulting in an energy conversion efficiency (η) of 0.15%. The external quantum efficiency (EQE) reached 1% for light with a wavelength of 520 nm. More detailed investigations using time-resolved microwave conductivity (TRMC) on charge separation at the interface of a smooth film of anatase TiO_2 (about 80 nm) and spin-coated layers of MEH-PPV (20–140 nm thick), with nanosecond photoexcitation of the polymer at 544 nm, revealed that the quantum efficiency of electron injection into the TiO_2 per incident photon is about 6% for thicknesses in the range of 30–50 nm of the polymer layer [33]. The charge separation was found to persist well into the microsecond region, which is advantageous because it allows for photogenerated charges to be collected.

A considerable improvement in terms of performance was reported by Carter et al. for bilayer devices containing a transparent planar TiO$_x$ layer and a conjugated polymer based on a phenylamino-p-phenylene vinylene (PA-PPV) repeating unit [34]. The electron-rich amino groups in PA-PPV enhance the hole mobility of the polymer. When positioned between TiO$_x$ and Au, an 80 nm layer of PA-PPV gave an EQE of 25% at 440 nm and afforded $V_{oc} = 0.85$ V and FF = 0.52 under 100 mW cm^{-2} white-light illumination. Similar bilayer devices made from MEH-PPV on planar TiO$_2$ gave rather low efficiencies of $\eta \leq 0.1\%$ ($J_{sc} = 0.24$ mA cm^{-2}, $V_{oc} = 0.7$ V, FF = 0.42) with a maximum EQE = 3.3% for a 65 nm thick layer [35], consistent with the results of Savenije et al. [13,33]. Replacing MEH-PPV with the better hole transporting poly[2,5-dimethoxy-1,4-phenylene-1,2-ethenylene-2-methoxy-5-(2-ethylhexyloxy)(1,4-phenylene-1,2-ethenylene)] (M3EH-PPV) in a bilayer device provided $\eta = 0.4\%$ and EQE 7.5% [36]. Spin casting a mixture of M3EH-PPV and (poly[oxa-1,4-phenylene-1,2-(1-cyano)ethenylene-2,5-dioctyloxy-1,4-phenylene-1,2-(2-cyano)ethenylene-1,4-phenylene]) (PCNEPV) on top of the TiO$_2$ layer increases this efficiency to 0.75% ($J_{sc} = 3.3$ mA cm^{-2}, $V_{oc} = 0.65$ V, FF = 0.28, EQE = 23.5% at 480 nm) [36]. In this example, the charge separation is likely to be dominated by the bulk heterojunction created by p-type M3EH-PPV and the n-type PCNEPV.

Similar cells and results using PPV derivatives have now been reported by several others [31,37–39]. For example, Fan et al. fabricated a planar heterojunction photovoltaic cell using sol–gel TiO$_2$ anatase and MEH-PPV with a peak power conversion efficiency of 1.6% when illuminated with monochromatic light at 500 nm at low intensity (0.20 mW cm^{-2}) [37]. Krebs et al. used various different metal oxides prepared via sol–gel methods and found that none of the oxides tested (Nb$_2$O$_5$, ZnO, CeO$_2$, and CeO$_2$-TiO$_2$) surpassed the performance of TiO$_2$ layers ($J_{sc} = 0.39$ mA cm^{-2}, $V_{oc} = 0.7$ V, FF = 0.25, $\eta = 0.07\%$) in an ITO/metal oxide/MEH-PPV/Ag device structure under identical conditions [31]. Photocurrents were generally low although they increased somewhat with time (except for ZnO). Slooff et al. have reported an estimated AM 1.5 efficiency of 0.3% ($J_{sc} = 0.96$ mA cm^{-2}, $V_{oc} = 0.508$ V, FF = 0.61) for planar bilayer devices constructed from poly(2-methoxy-5-(3′,7′-dimethoxyloctyloxy)-p-phenylene vinylene) (MDMO-PPV) and TiO$_2$ using a cell configuration in which the TiO$_2$/MDMO-PPV bilayer is sandwiched between an ITO bottom electrode and a poly(3,4-ethylenedioxythiophene):poly(styrenesulfonate) (PEDOT:PSS)/Au top electrode. [39].

The influence of oxygen on fluorine-doped tin oxide (FTO)/TiO$_2$/MDMO-PPV/Hg cells has been studied by Goossens et al. [40,41]. During the illumination of these cells, the presence of oxygen causes an increase in the dark reverse and forward currents when the cell is held at open-circuit voltage. This effect does not occur when the cell is short-circuited. The results were explained by considering that under open-circuit conditions the photoexcited electron is not transferred to TiO$_2$, but remains in the LUMO of MDMO-PPV and interacts with available oxygen to form O$_2^-$ and a charge transfer complex, increasing the acceptor density and conductivity. Under short-circuit conditions, the photoexcited electrons injected into TiO$_2$, are removed from the interface by the electric field at the junction, and do not cause oxygen doping.

13.2.2
Metal Oxide–Polythiophene

Early studies on polythiophene derivatives such as P3HT in combination with TiO_2 revealed that the working devices could be obtained, but that photocurrents and efficiencies were relatively small. In a study that focused on solid-state dye-sensitized TiO_2 solar cells with poly(3-octylthiophene) (P3OT) as hole transport layer, Sariciftci et al. showed that under 64 mW cm^{-2} white-light illumination, a planar TiO_2/P3OT control cell gives $J_{sc} = 9\,\mu A\,cm^{-2}$, $V_{oc} = 0.56$ V, and FF $= 0.36$ [42]. Similar bilayer solar cells constructed from poly(3-undecyl-2,2'-bithiophene) (P3UBT) and planar TiO_2 were reported by Smestad et al. [43]. The best initial performance of FTO/TiO_2/P3UBT/graphite devices gave $J_{sc} = 55\,\mu A\,cm^{-2}$ and V_{oc} 0.7 V under 100 mW cm^{-2} AM 1.5 illumination. Interestingly, the photocurrent and voltage improved over time to $J_{sc} = 85\,\mu A\,cm^{-2}$, $V_{oc} = 0.78$ V, FF $= 0.30$, and η 0.02% [44,45]. Kim et al. have described a comparison of ITO/TiO_2/P3HT/Au devices made with regiorandom and regioregular P3HT at different temperatures [46]. For regiorandom P3HT, V_{oc} decreased from ∼0.9 to 0.55 V going from 140 to 300 K, while J_{sc} increased from less than 1 to ∼20 $\mu A\,cm^{-2}$ (under 25 mW cm^{-2} AM 1.5 illumination). For regioregular P3HT, $J_{sc} \approx 30\,\mu A\,cm^{-2}$ was almost constant between 200 and 295 K.

Goossens et al. studied the photophysics of polythiophenes on TiO_2 layers [47]. Contrary to their expectation, they observed a significantly increased photoluminescence of a 5 nm P3OT layer on TiO_2 compared to glass. The authors ascribe the enhancement to the interaction of long-lived positive polarons (P3OT radical cations created by electron transfer to TiO_2) with subsequently generated excitons. The exact nature of this interaction is not clear, but the authors postulated that a fraction of the nonemissive triplet excitons is converted into emissive singlet excitons. This is an interesting idea, but it has been established that positive polarons effectively quench singlet excitons in conjugated polymers [48–51]. Moreover, excitation of the triplet exciton to a singlet exciton in P3HT requires about ∼0.7 eV.

Savenije et al. studied the charge separation efficiency in polythiophene-sensitized P3HT/TiO_2 bilayers using flash photolysis TRMC [14]. Following a 3 ns pulse, the transient photoconductivity was monitored from nanoseconds to milliseconds. Interfacial charge separation and electron injection were found to persist well into the millisecond domain. The action spectrum of the photoconductivity followed the photon attenuation spectrum with a maximum wavelength of 540 nm. By studying both front-side and back-side illumination for different P3HT layer thicknesses and comparing the results to direct bandgap excitation of TiO_2, the efficiency of charge separation per incident photon (IPCSE) could be determined. At 540 nm, the IPCSE reached a maximum of 0.8% for 10 nm P3HT films. In principle, the IPCSE may be taken as an upper limit for the EQE of a similar cell for the same composition and excitation wavelength. In this respect the value of 0.8% seems rather low and actually considerably less than the EQE $= 4.6\%$ that was subsequently reported by Fréchet and McGehee for a TiO_2/P3HT cell [52]. The experiment, however, demonstrates that reaching high efficiencies in bilayer devices is a challenging task.

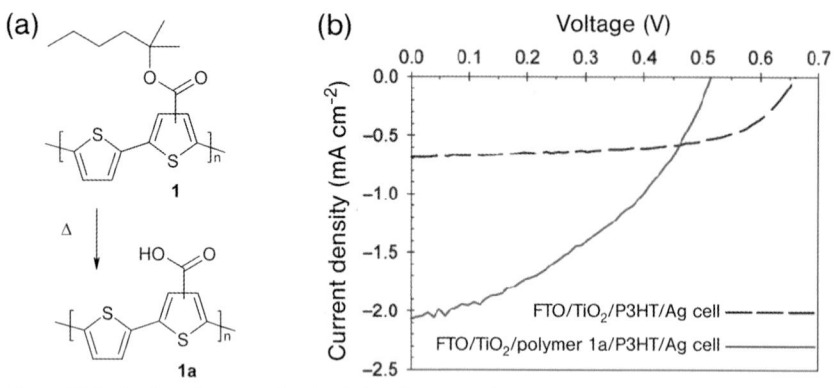

Figure 13.3 (a) Ester-functionalized polythiophene **1** and its thermal conversion into insoluble acid-functionalized polymer **1a** (b) *I–V* curves of FTO/TiO$_2$/polymer **1a**/P3HT/Ag cell (solid line) and FTO/TiO$_2$/P3HT/Ag cell (dashed line) under 39 mW cm^{-2} 514 nm illumination. Reprinted with permission from Ref. [52]. Copyright 2004 American Chemical Society.

Fréchet *et al.* have used a polythiophene derivative (Figure 13.3a) with a thermally removable solubilizing group to increase the performance of TiO$_2$/polymer solar cells [52]. The structure of the ester-functionalized polythiophene was based on several considerations: (1) the tertiary ester can be cleaved at relatively low temperatures; (2) the branched side chain increases solubility, which facilitates synthesis, purification, and deposition; (3) the carboxylic groups that remain after thermal cleavage not only allow tuning of the energy levels but also enhance the interface interaction with TiO$_2$; and (4) after conversion the acid functionalized polythiophene is insoluble, which allows making multilayer devices.

For device fabrication, a thin layer (5 nm) of the ester-functionalized polythiophene **1** was deposited from solution on TiO$_2$ and converted into the acid-functionalized derivative **1a** at 210 °C for 45 min. Subsequently, P3HT was deposited and the device was completed by evaporating 80 nm Ag. The FTO/TiO$_2$/polymer **1a**/P3HT/Ag device gave enhanced photoresponse compared to a control cell (FTO/TiO$_2$/P3HT/Ag) (Figure 13.3b). Under monochromatic (514 nm) illumination the EQE of this cell is 12.6%, compared to 4.2% for the control cell. The enhanced photocurrent has been ascribed to advantageous properties of polymer **1a**, that is, a higher chromophore density, an enhanced exciton diffusion length because of a shorter interchain distance, and chelation of the carboxylic acid groups to the TiO$_2$. For the FTO/TiO$_2$/polymer **1a**/P3HT/Ag cell, the power conversion efficiency at 39 mW cm^{-2} of 514 nm illumination is about 1.10%. Compared to the control cell, this represents a tripling of the photocurrent but a simultaneous loss of V_{oc} and FF. The effect of carboxylic acid group on the V_{oc} has been studied in more detail by McGehee *et al.* [53]. Kelvin probe measurements revealed that the decrease in the open-circuit voltage originates from interfacial dipoles. These dipoles create a reduction in the band offset between the conduction band of TiO$_2$ and HOMO of the polymer interface, resulting in the loss of V_{oc}. In a subsequent study they investigated the effects of surface modification of TiO$_2$ in hybrid TiO$_2$/regioregular P3HT photovoltaic cells. By employing a series of *para*-substituted benzoic acids with varying dipoles and a series of multiply substituted benzene carboxylic acids, the energy offset at the TiO$_2$/polymer interface and thus the open-circuit voltage could be tuned systematically by 0.25 V [54]. The saturated photocurrent of TiO$_2$/P3HT devices exhibited more than a twofold enhancement when molecular modifiers with large electron affinity were employed.

A similar multipolymer layer structure has been used in combination with resonance energy transfer to improve exciton harvesting in organic hybrid photovoltaic cells by McGehee *et al.* [55]. In this cell a thin layer (4 nm) of a low-bandgap polymer (poly(N-dodecyl-2,5-bis(2′-thienyl)pyrrole-2,1,3-benzothiadiazole, PTPTB) is inserted between P3HT (50 nm) and TiO$_2$ (50 nm) to triple the efficiency (Figure 13.4). The rationale of this interesting idea is that the thicker P3HT layer can efficiently transfer excitation energy via a Förster mechanism to the thin low-bandgap PTBTB layer that subsequently injects an electron into TiO$_2$, followed by hole injection into P3HT. Indeed the PL of P3HT on a thin TiO$_2$ layers was strongly quenched when a PTBTB interlayer was used. The use of Förster transfer to increase the distance over which an exciton can be transferred beyond the normal diffusion length has been studied by the same group recently, including optical interference effects in multilayer configurations [56,57]. Clear evidence for the effect was obtained from the study of the performance of solar cells. A control device of FTO/TiO$_2$/P3HT/Ag without the PTBTB interlayer provides $J_{sc} = 0.46$ mA cm^{-2}, $V_{oc} = 0.64$ V, FF = 0.63, and an overall power conversion efficiency of $\eta = 0.19\%$ under simulated AM 1.5 100 mW cm^{-2} solar illumination. The device with PTBTB was three times better, with $J_{sc} = 1.33$ mA cm^{-2} $V_{oc} = 0.67$ V, FF = 0.63, and $\eta = 0.56\%$ [55]. Importantly, the spectral response of the device corresponds to the absorption spectrum of P3HT, showing that excitons initially generated in P3HT are effectively transferred to PTBTB.

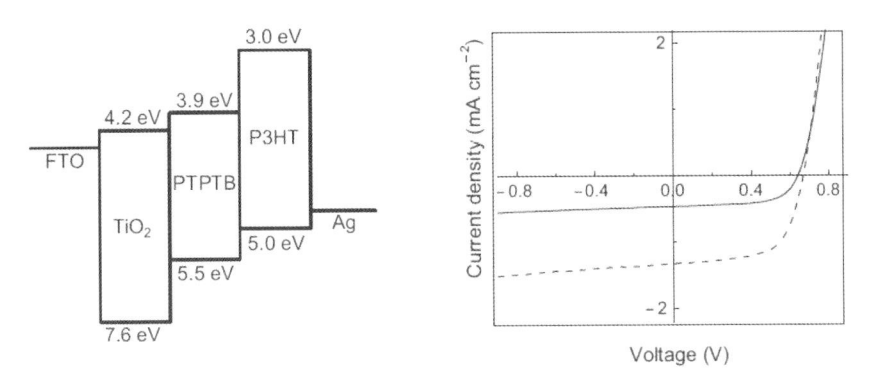

Figure 13.4 Energy levels and performance of FTO/TiO$_2$/PTBTB/P3HT/Ag devices. The J–V characteristics measured under 100 mW cm^{-2} AM 1.5 illumination with PTBTB (solid line) and without (PTBTB) (dashed line). Reprinted with permission from Ref. [55]. Copyright 2005 Wiley-VCH.

With respect to this PTPTB/P3HT cell, it is interesting to note that Kroeze and Savenije have studied the photoinduced charge separation in a double layer of the low-bandgap PTPTB spin coated onto a smooth layer of SnO_2 or TiO_2, using the TRMC technique [58]. The TRMC technique allows intrinsic properties such as the charge separation efficiency to be measured without the necessity of applying electrodes. It was found that SnO_2 that has a high electron affinity (low-lying conduction band) could successfully be applied as the electron acceptor in combination with PTPTB that has relatively low excited state energy level. Ultimately, the major advantage of using such a low-bandgap material is the improved overlap with the solar energy spectrum. As a result of the lower lying SnO_2 conduction band edge and the increased driving force for photoinduced charge transfer, the charge separation efficiency per incident photon is increased by a factor of 30 as compared to the TiO_2 acceptor. Hence, SnO_2/low-bandgap polymer combinations form an interesting candidate for application in photovoltaic devices.

Olson et al. have recently reported a simple method to systematically tune the band offset in a π-conjugated polymer–metal oxide hybrid donor–acceptor system in order to maximize the V_{oc} [59]. Substitution of magnesium into a zinc oxide acceptor (ZnMgO) reduces the band offset and results in a substantial increase in the V_{oc} of planar $P3HT/Zn_{1-x}Mg_xO$ devices. The V_{oc} increases from 0.5 V at $x = 0$ up to values in excess of 0.9 V for $x = 0.35$. A concomitant increase in overall device efficiency is seen as x increases from 0 to 0.25, with a maximum power conversion efficiency of $\eta = 0.49\%$ obtained at $x = 0.25$ ($J_{sc} = 1.27$ mA cm^{-2}, $V_{oc} = 0.70$ V, and $FF = 0.56$). Beyond $x = 0.25$, the efficiency decreases because of increased series resistance in the device.

13.3
Filling Nanoporous and Nanostructured Metal Oxides with Conjugated Polymers

With respect to increasing the interface area between the p-type and n-type materials, infiltrating a nanoporous semiconductor electrode with a conjugated polymer is an attractive option. It is a logical extension of the solid-state DSSC concept in which the semiconducting polymer takes care of light absorption and hole transport, replacing both the dye and electrolyte. Furthermore, hybrid bulk heterojunctions based on well-ordered inorganic semiconductor nanostructures such as vertical pillars, straight pores, or inverted opal structures are a promising approach to create effective photoactive layers for solar cells. By controlling the dimensions into the range of the exciton diffusion length, charge generation can be optimized while efficient charge transport can be ensured by creating short, straight direct pathways for both charge carriers [60].

13.3.1
Polymers in Nanoporous TiO_2

The large surface area of nanoporous TiO_2 electrodes is highly advantageous for charge generation in hybrid bulk heterojunction with a conjugated polymer.

Consequently, the idea of replacing the dye and liquid electrolyte of a DSSC by a semiconducting polymer that takes care of light absorption, electron injection, and hole transport has been explored in recent years. In contrast to many polymer:fullerene and polymer:polymer bulk heterojunctions, the stability of the morphology of a filled nanoporous metal oxide would be high. The main challenge that has been encountered in this approach is efficient filling of the pores. This affects the solar energy conversion efficiencies that slowly approach 1%.

Van Hal et al. performed photoinduced absorption (PIA) and photoinduced ESR experiments to investigate the electron transfer from poly(p-phenylene vinylene)s and polythiophenes to nanoporous TiO_2 [61]. The PIA spectra of the conjugated polymers on nanoporous TiO_2 recorded at 80 K showed two strong bands at ~0.5 and ~1.5 eV, characteristic of cation radicals (polarons) generated on the polymer chains. Photoinduced ESR experiments confirmed this result by the appearance of ESR signals at $g = 2.0022$–2.0026 due to photogenerated spins on the polymers, whose intensity could be modulated by changing the light intensity. A complex range of lifetimes for the charge-separated state was observed. In PIA spectroscopy a decay mechanism with a steady-state lifetime of 2–10 ms gave an important contribution but other, slower processes occurred as well. The complex decay behavior has been attributed to the inhomogeneous nature of the polymer/nanoporous TiO_2 interface, giving rise to a range of trapping depths. Photoinduced ESR at 130 K shows that recombination occurs with time constants of 10 s and longer. Importantly, only a partial quenching of the polymer photoluminescence was observed when the polymer infiltrated in the nanoporous TiO_2. The residual photoluminescence was attributed to polymer chains that are at a longer distance from the interface with TiO_2 than the exciton diffusion range. This will result in fluorescence, even when an ultrafast forward electron transfer is possible in regions of improved contact. Luzzati et al. confirmed these results for a poly(dithenothiophenedioxide) (PTOX) derivative [62]. The optical absorption spectra of this polymer showed a remarkable interaction with the TiO_2 as compared to other alkyl-substituted thiophene-based conjugated polymers. Moreover, the PL quenching and the PIA spectra evidenced that TiO_2 acts as an efficient electron acceptor toward PTOX in the excited state.

Lian et al. used femtosecond infrared transient absorption spectroscopy to study the separation and recombination dynamics of the electron transfer of photoexcited MEH-PPV to nanoporous SnO_2 and TiO_2 films [63]. The forward electron transfer is indeed ultrafast with time constants of 800 fs (SnO_2) and <100 fs (TiO_2). Despite a lower level of its conduction band, the injection rate for SnO_2 is less than for TiO_2, demonstrating that the rate is determined not just by the band offsets. SnO_2/MEH-PPV showed no carrier recombination with the 1 ns, but a significant signal decay was observed for TiO_2/MEH-PPV. In both systems, some charges persisted from microseconds to seconds.

One of the first attempts to make solar cells by infiltrating a conjugated polymer into nanoporous TiO_2 has been reported by Carter et al. [64]. Sintered nanocrystalline TiO_2 layers were made by coating an ITO/glass substrate with a viscous TiO_2 water solution and subsequent annealing at 500 °C to fuse the particles. This resulted in 4–6 μm thick layers with TiO_2 particles of ~80 nm and a pore diameter of 20 nm. The pores were filled with MEH-PPV by spin coating. The best ITO/TiO$_2$/MEH-PPV/Au cells gave $J_{sc} = 40\,\mu\text{A cm}^{-2}$, $V_{oc} = 0.7$ V, FF = 0.43 at 2 mW cm^{-2} light intensity over the absorption region of the polymer. Subsequent devices made using 200 nm thick sintered TiO$_2$ nanoparticle layers with a surface roughness of up to 100 nm showed noticeable improvements and provided $J_{sc} = 0.4$ mA cm^{-2}, $V_{oc} = 1.1$ V, FF = 0.40, at incident light intensity near 100 mW cm^{-2}, giving a power conversion efficiency of 0.18% [35]. The improvement can possibly be rationalized by the fact that a thinner TiO$_2$ film (100–200 nm) allows for a better transport of the photogenerated holes and allows filling the porous material better than in the case of a TiO$_2$ film that is several micrometers thick.

Sariciftci et al. studied sintered nanocrystalline TiO$_2$ layers in combination with P3OT. Sandwiched between an ITO or FTO front electrode and an Au back electrode, these devices provided $J_{sc} = 0.17$ mA cm^{-2}, $V_{oc} = 0.70$ V, and FF = 0.40 under 60 mW cm^{-2} white-light illumination [42]. The performance doubles to $J_{sc} = 0.33$ mA cm^{-2}, $V_{oc} = 0.64$ V, and FF = 0.44 when a sensitizer layer of cis-bis(isothiocyanato)bis(2,2'-bipyridyl-4,4'-dicarboxylato)ruthenium(II) bistetrabutylammonium was inserted between the TiO$_2$ and P3OT. In this last device the conjugated polymer acts as a solid-state hole conductor, replacing the liquid electrolyte of standard DSSCs. Conjugated polymers have attracted more attention as hole conductor in solid-state ruthenium dye-sensitized solar cells [65–67], but given the fact that solid-state dye-sensitized solar cells are a vast research area in itself [68,69], these cells will not be discussed here. Similarly, we excluded liquid dye-sensitized solar cells in which the dye is a conjugated oligomer [70] or polymer [71,72] from this overview.

Nelson et al. have studied photovoltaic devices consisting of poly(9,9'-dioctylfluorene-co-bithiophene) (F8BT) and nanocrystalline TiO$_2$ [73,74]. In this system, efficient photoinduced charge transfer occurs while charge recombination is relatively slow (~100 μs–10 ms). In the best ITO/TiO$_2$/F8BT/Au device, the EQE maximized at 13% at a wavelength of 440 nm. In these devices a dense "hole-blocking" TiO$_2$ layer, about 50 nm thick, was introduced between the ITO and the nanoporous TiO$_2$ to prevent direct contact between the polymer and the substrate. Under simulated AM 1.5 illumination, the device produced $J_{sc} = 0.40$ mA cm^{-2}, $V_{oc} = 0.92$ V, FF = 0.44, and $\eta = 0.16\%$. The authors proposed that the performance of these devices is limited by the energy step at the polymer/metal interface.

Consistently, Kim et al. were the first to show that by inserting a conducting PEDOT:PSS layer between the Au back contact and the polymer layer, the device efficiency of nanoporous TiO$_2$ cells in combination with MEH-PPV can be significantly enhanced [75]. The PEDOT-modified device (ITO/TiO$_2$/MEH-PPV/PEDOT:PSS/Au) cell has $J_{sc} = 0.97$ mA cm^{-2}, $V_{oc} = 0.74$ V, FF = 0.48, and $\eta = 0.44\%$ at 80 mW cm^{-2} AM 1.5 illumination. Without PEDOT, the fill factor was very small (FF = 0.16). The enhancing effect of PEDOT:PSS was explained by a better ohmic contact.

In optimizing the power conversion efficiency of hybrid solar cells based on nanostructured TiO$_2$ and a MEH-PPV-based conjugated polymer, Ravirajan and Nelson et al. also established that the charge collection efficiency and device performance was enhanced by introducing a PEDOT:PSS layer under the gold electrode as the hole collector [76,77]. Several possible reasons for the improvement in J_{sc} resulting from insertion of the PEDOT layer were proposed: (1) PEDOT may cause a chemical doping of the polymer that reduces the contact resistance between the polymer and metal contact; (2) the PEDOT layer may protect the polymer film from damage during evaporation of the Au electrode; and (3) PEDOT improves collection by minimizing the energy step between the polymer and top contact. Maximum device performance was obtained for a 100 nm thick active layer, giving an EQE of 40% at the polymer's maximum absorption wavelength and $J_{sc} = 2.10$ mA cm^{-2}, $V_{oc} = 0.64$ V, FF $= 0.43$, and $\eta = 0.58\%$ at AM 1.5 conditions 100 mW cm^{-2} and 1 sun [77].

The results above show that filling the porous material with (conjugated) polymers or other hole-conducting materials is seriously hampered by the densely agglomerated TiO$_2$. Long polymer strands will find it more difficult to infiltrate into a porous network than smaller strands [78]. This subject has been studied in detail by Heeger and Bartholomew [78]. By combining surface analysis, imaging, and depth profiling, they have studied the infiltration of regioregular P3HT into nanoporous TiO$_2$ networks. A very low incorporation of the polymer was found (0.5%), even for highly porous ($\approx$65%) networks. Strategies to increase the incorporation via heat treatment and surface derivatization and the use of low molecular weight polymer resulted in a high level of 22% P3HT of the total volume. These authors concluded that if semiconducting polymers can be incorporated at high level, efficient hybrid cells will result [78]. Circumventing filling problems by *in situ* polymerization of polythiophene in the pores of the TiO$_2$ layer has led to moderate photovoltaic effects so far [79] and this effect is dependent on the doping level of the polythiophene [80].

13.3.2
Filling Structured Inorganic Semiconductors with Polymers

Creating structured inorganic semiconductors with a predefined morphology may be useful to resolve the problems encountered in filling the random nanoporous semiconductors described in the previous sections. Nanostructuring the inorganic material allows for optimizing dimensions, small enough for all excitons to reach the junction, but still allowing a simple and effective filling of the electrodes. A controlled height of the structures will also lead to an improved hole transport and collection efficiency at the top electrode. Since most conjugated polymers still have rather low hole mobilities, the height of the structures, and therefore the maximal achievable increase in contact area and light absorption, will likely limit the device. Ultimately, major improvements, therefore, can be expected to come from polymers with significantly higher hole mobilities, an expectation that actually holds for all hybrid solar cells described in this chapter.

13.3.2.1 Structured Porous TiO$_2$
McGehee *et al.* have reported an elegant example of nanostructured TiO$_2$ [81,82]. Mesoporous TiO$_2$ films were synthesized using the structure-directing properties of block copolymers. Using titanium(IV) tetraethoxide as the titania precursor and a

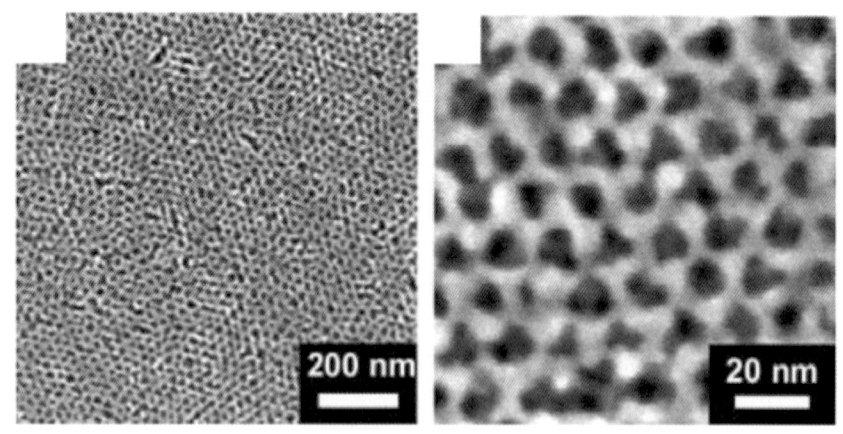

Figure 13.5 Mesoporous TiO$_2$ films made using the structure-directing properties of block copolymers; after the TiO$_2$ network is formed, the polymers are removed and the TiO$_2$ is crystallized by calcination at 400 °C. Reprinted with permission from Ref. [82]. Copyright 2003 Wiley-VCH.

Pluronic poly(ethylene oxide)–poly(propylene oxide)–poly(ethylene oxide) triblock copolymer (P123, average composition HO-PEO$_{20}$-PPO$_{70}$-PEO$_{20}$-OH) as a template, TiO$_2$ films were formed that have a regular and open structure with pores of approximately 10 nm (Figure 13.5), which have been filled with regioregular P3HT. Infiltration of the mesoporous TiO$_2$ has been performed by heat treatment of a spin-cast polymer film. External quantum efficiencies of EQE = 10% have been obtained for cells that have a FTO front electrode, an Ag back electrode, and a 30 nm P3HT layer. By combining the measured V_{oc} = 0.72 V and FF = 0.51 with the AM 1.5 integrated spectral response, a power conversion efficiency of η = 0.45% was estimated, corresponding to J_{sc} = 1.22 mA cm^{-2} at 100 mW cm^{-2}.

Djurišić et al. followed an approach similar to McGehee et al. to create a TiO$_2$ interconnected network structure for photovoltaic applications using a polystyrene block–polyethylene oxide diblock copolymer (PS-b-PEO) as the templating agent and MEH-PPV as the light-absorbing and hole-transporting polymer. The pore size of the structure could be controlled by the amount of Ti precursor provided. The best heterojunction solar cells provided J_{sc} = 3.3 mA cm^{-2}, V_{oc} = 0.86 V, FF = 0.28, and η = 0.71% under AM 1.5 solar illumination and used a 20 nm compact TiO$_2$ layer between the porous TiO$_2$ and the ITO electrode. The maximum EQE for optimum MEH-PPV thickness was 34% in these ITO/TiO$_2$/MEH-PPV/Au devices [83,84].

13.3.2.2 Oriented Nanorods

Ultimately, the ideal bulk heterojunction structure would be one where ordered nanosized channels or pillars of both the inorganic material and the polymer are present throughout the bulk of the film (Figure 13.6) [85,86]. When the typical dimensions on the order of the exciton diffusion length can be realized, high charge generation efficiency and good charge transport can possibly be combined.

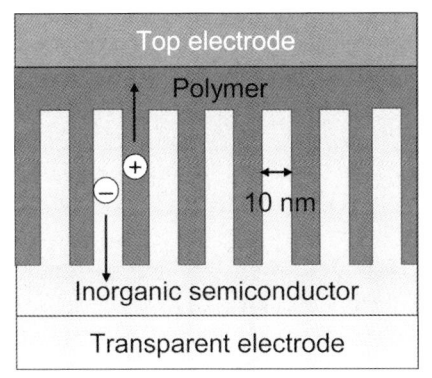

Figure 13.6 Schematic representation of the ideal bulk heterojunction hybrid solar cell in which an ordered inorganic semiconductor is infiltrated with a conjugated polymer and typical dimensions are on the order of the exciton diffusion length.

Nanocrystalline ZnO can adopt a wide range of interesting morphologies [87], and hence fabrication of well-defined ZnO nanopillars is an appealing approach for hybrid polymer solar cells [88], especially when the dimensions of the rods and the spacing in between can be controlled to create a structure with a high-interface contact area, but still allowing a proper filling of the voids with polymer. The actual realization of hybrid solar cells involving ZnO nanopillars has been reported by two groups recently [29,89,90].

Olson *et al.* have grown ZnO nanofibers hydrothermally from a ZnO nucleation layer on a conducting ITO/glass substrate (Figure 13.7) [29]. Then P3HT was spin coated from solution, followed by thermal annealing at 200 °C for 1 min before completing the device with an Ag top electrode. The best device gave a significant photovoltaic effect with $J_{sc} = 2.2\,\text{mA}\,\text{cm}^{-2}$, $V_{oc} = 0.475\,\text{V}$, and $FF = 0.56$, providing an efficiency of $\eta = 0.53\%$ with a maximum EQE of 17% at 520 nm. Incorporation of

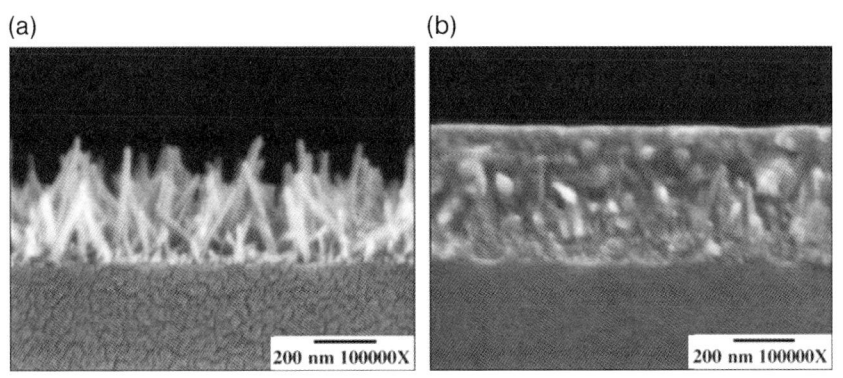

Figure 13.7 (a) SEM image of a glass/ZnO nucleation layer/ZnO nanofiber structure. (b) SEM image of P3HT intercalated into the ZnO nanofiber structure. Reprinted with permission from Ref. [29]. Copyright 2006 Elsevier.

PCBM as an additional acceptor in the blend gave an increase in performance to 2.03%. In this case, however, the active layer is the P3HT:PCBM blend, which is known to give very high conversion efficiencies of 4–5% [15–18]. The power efficiency in the ZnO-fiber/P3HT devices seems to be limited by the 0.2 V reduction in V_{oc} compared to ZnO nanoparticle cells (see Section 13.4) and by the spacing between the ZnO nanofibers, which is on the order of 100 nm and therefore substantially larger than the exciton diffusion length in P3HT [29].

Nelson et al. have performed an extensive study on the different solution chemical routes for preparing columnar ZnO structures on a dense ZnO backing layer that act as a seed-growth layer for the rods in relation to hybrid photovoltaic devices [90]. They established that the growth of ZnO nanorod arrays depends on the morphological and structural characteristics of the seed layers. Different polymers (high hole mobility MEH-PPV and P3HT) were tested in devices. A unique feature of this approach is that the ZnO nanorods were covered with an amphiphilic dye cis-RuLL'(SCN)$_2$ (L = 4,4'-dicarboxylic acid-2,2'-bipyridine, L' = 4,4'-dinonyl-2,2'-bipyridine), referred to as Z907, before applying P3HT by spin coating [89]. The rationale for this approach is that the dye helps in wetting the ZnO surface by the polymer and that the dye can assist in the interfacial electron transfer from the polymer to the ZnO. For the top contact a layer of PEDOT:PSS was spin coated, followed by deposition of Au. The PEDOT:PSS improves the smoothness of the top surface and the electrode quality. The best ZnO nanorod:P3HT device using the Z907 interfacial layer gave $J_{sc} = 1.73$ mA cm^{-2}, $V_{oc} = 0.30$ V, and FF = 0.39, providing an efficiency of $\eta = 0.2\%$ [90], with a maximum EQE = 14% at 550 nm [89]. Devices with the high hole mobility MEH-PPV derivative showed an overall efficiency of 0.15%. Without the Z907, dye or the PEDOT:PSS showed large leakage currents and very low V_{oc}.

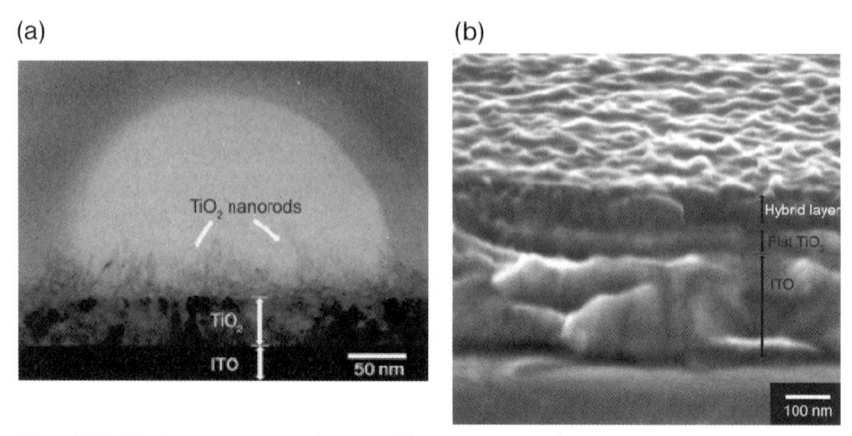

Figure 13.8 (a) TEM cross-sectional image of the nanostructured TiO$_2$ substrate; (b) FE-SEM cross-sectional images of the MEH-PPV films deposited on the nanostructured TiO$_2$ substrates. Reprinted with permission from Ref. [38]. Copyright 2006 American Chemical Society.

On the basis of low-temperature sol–gel synthesis of anatase TiO$_2$ nanorods in a reversed micelle solution, Hashimoto et al. succeeded in depositing an array of TiO$_2$ nanostructure assemblies with a height of ∼40 nm selectively on a TiO$_2$ substrate that was immersed in planar TiO$_2$ substrate [38]. Cross-sectional transmission electron microscopy (TEM) showed that each nanostructure is an assembly of TiO$_2$ nanorods with a diameter and length of 4 and 40 nm, respectively (Figure 13.8). The fabricated array of TiO$_2$ rods was applied to construct photovoltaic devices in combination with MEH-PPV. The best devices showed $J_{sc} = 0.95$ mA cm^{-2}, $V_{oc} = 0.88$ V, and FF $= 0.47$, providing an efficiency of $\eta = 0.39\%$, which is a considerable improvement compared to a planar TiO$_2$/MEH-PPV control device ($J_{sc} = 0.33$ mA cm^{-2}, $V_{oc} = 0.87$ V, FF $= 0.49$, and $\eta = 0.13\%$). The EQE of the nanostructure TiO$_2$/MEH-PPV maximized at 15% at a wavelength of 510 nm.

13.4
Nanoparticle–Polymer Hybrid Solar Cells

By mixing nanoparticles of inorganic semiconductors with conjugated polymers, p–n junctions can be created throughout the thin film that give rise to efficient charge generation from photogenerated excitons. The approach is in essence simple and applicable to large areas. The hybridization of inorganic nanoparticles and conjugated polymers potentially combines all the advantages of inorganic and polymeric semiconductors. Polymers allow for simple, large-scale deposition techniques and, hence, combining them with soluble inorganic nanoparticles into bulk heterojunction layers opens a versatile route toward hybrid photovoltaic devices [91]. The challenge, however, is to create sufficient particle–particle contacts to sustain charge transport. Moreover, it is crucial to use a solvent that allows processing of the inorganic and the organic materials at the same time, which may not be trivial because of the different nature of the two components. This may require modification of the inorganic material or conjugated polymer.

The use of inorganic nanoparticles in combination with conjugated polymers originates from the seminal work of Greenham and Alivisatos using CdSe as the inorganic semiconductor [92,93]. By using elongated rods and tetrahedrally branched nanoparticles (tetrapods) that enhance the transport of electrons, efficiencies close to 3% under AM 1.5 conditions have now been obtained [94–99]. Crucial for the good performance is exchanging the surfactants of the nanocrystals by volatile molecules, for example, replacing trioctylphosphine oxide by pyridine. Pyridine evaporates during film processing to enable electrical contact between the individual nanocrystals and between the nanocrystals and the polymer. Nevertheless, it is difficult to control the morphology and dispersion of nanocrystals within the polymer in these layers. In the polymer:CdSe cells, both components contribute to light absorption and charge generation.

Transparent semiconducting metal oxides such as TiO$_2$ and ZnO are promising materials to be used as the electron-accepting material in combination with p-type conjugated polymers. These metal oxides are not considered to be toxic and are available in abundance. Hence, it is not surprising that blends of metal oxide nanoparticles with conjugated polymers have been explored for use in solar cells.

13.4.1
TiO$_2$ Nanoparticles

First examples on blends from TiO$_2$ nanoparticles and PPV have shown moderate external quantum efficiencies and short-circuit currents of tens of microamperes. In one example, PPV was obtained by thermal conversion of a methanol soluble precursor mixed with 20 nm TiO$_2$ nanocrystals and spin casting on fluorinated tin oxide (SnO$_2$:F) [100,101]. The resulting SnO$_2$:F/TiO$_2$:PPV/Al devices with 20 wt% TiO$_2$ gave $J_{sc} = 25$ µA cm^{-2} and $V_{oc} = 0.65$ V, with EQE = 2% in the absorption range of PPV. Later, TEM and PL studies on similarly prepared TiO$_2$:PPV blends revealed large phase separation between the TiO$_2$ crystals and PPV [102]. In another attempt to create bulk heterojunction solar cells, MEH-PPV was mixed with TiO$_2$ nanoparticles dispersed in p-xylene and cast into thin films [103]. For ITO/nc-TiO$_2$:MEH-PPV/Ca devices prepared in this way, containing 70 wt% TiO$_2$, gave $J_{sc} \approx 20$ µA cm^{-2} and $V_{oc} \approx 1$ V.

These examples show that it is not straightforward to make bulk heterojunction solar cells by blending TiO$_2$ nanoparticles with conjugated polymers; poor mixing of the polymer and the nanoparticles leads to agglomerates of nanoparticles in the blend, which severely limited the performance of the first photovoltaic devices of this type.

The synthesis of TiO$_2$ nanoparticle powders mainly occurs in water- or alcohol-based media. Transferring these nanoparticle powders into organic solvents regularly leads to aggregate formation and therefore a low miscibility with conjugated polymers. For this reason the application of TiO$_2$ nanoparticle powders in the bulk heterojunctions is often limited. Only recently it was shown that 20–40 nm TiO$_2$ nanoparticles and conjugated polymers can be blended from common organic solvents. Kwong et al. demonstrated solar cells with an efficiency of approximately $\eta = 0.42\%$ [104,105], providing the first account, showing that the ideas advanced by Alivisatos and Greenham for CdSe also work for TiO$_2$ nanoparticles. The most efficient devices incorporated 60 wt% TiO$_2$ and a ~100 nm thick TiO$_2$:P3HT film, spin coated from xylene. The solutions for fabricating the films were prepared in a heated ultrasonic bath (at 50 °C) to increase solubility of P3HT. The films were baked in a vacuum oven for 24 h at 110 °C. The completed ITO/PEDOT:PSS/nc-TiO$_2$:P3HT/Al devices gave $J_{sc} = 2.759$ mA cm^{-2}, $V_{oc} = 0.44$ V, FF = 0.396, and $\eta = 0.424\%$ under AM 1 spectral conditions at 100 mW cm^{-2}. The maximum EQE amounted to 15% at 475 nm. For blend compositions with less than 40% or more than 70% TiO$_2$, the performance was significantly less, due to extensive recombination (at low concentration) or poor film quality (at high concentration).

Agostiano et al. have performed an extensive optical and photoelectrochemical study of blended systems composed of organic-capped TiO$_2$ crystals with a spherical ($d \sim 5$ nm) or rodlike ($d \sim 3$–4 nm, $l = 25$–30 nm) morphology and MEH-PPV [106,107]. PL quenching experiments indicated that photoinduced charge separation occurs at the interface for both oleic acid and n-tetradecylphosphonic acid capping layers.

Photoelectrochemical measurements in three-electrode cell employing an ITO working electrode covered with the 1nc-TiO$_2$:MEH-PPV blends provided photocurrents that were higher for the spherical particles than for the rodlike counterparts, and also increased when replacing the bulky oleic acid with the phosphonic acid capping layer.

An interesting alternative method to develop hybrid solar cells using an aerosol technique has been proposed by Huisman et al. [108,109]. In this technique, ultrasonically formed droplets of titanium(IV) isopropoxide are pyrolyzed to deposit thin films of nanosized anatase TiO$_2$ particles. This method eliminates the need for a sintering step after deposition. Using these films cells could be constructed by spin coating P3OT on top. Devices with a 1 μm thick active layer give $J_{sc} = 0.25$ mA cm^{-2}, $V_{oc} = 0.72$ V, FF $= 0.35$, and $\eta = 0.06\%$ under 100 mW cm^{-2} white-light illumination and EQE $= 2.5\%$ at 488 nm. Devices prepared by adding P3OT during the deposition of the porous showed a similar performance.

13.4.2
ZnO Nanoparticles

Beek et al. chose to use ZnO instead of TiO$_2$ or CdSe [110]. ZnO is a nontoxic, crystalline material and at room temperature the stable phase is a wurtzite-type crystal. In contrast to TiO$_2$, crystalline ZnO is already formed at temperatures as low as 4 °C [111]. ZnO has a direct bandgap at 3.2 eV, a conduction band level of –4.4 V versus vacuum [112], and an electron mobility in the bulk exceeding 100 cm^2 V^{-1} s^{-1} [113,114]. Crystalline ZnO nanoparticles (nc-ZnO) are soluble in organic solvents and can be mixed with conjugated polymers without the use of surfactants [110]. Figure 13.9a shows a transmission electron microscopy image of crystalline ZnO nanoparticles of approximately 4.9 nm in diameter (according to UV absorption onset). X-ray analysis confirmed the presence of a wurtzite type ZnO crystal (Figure 13.9b) [115].

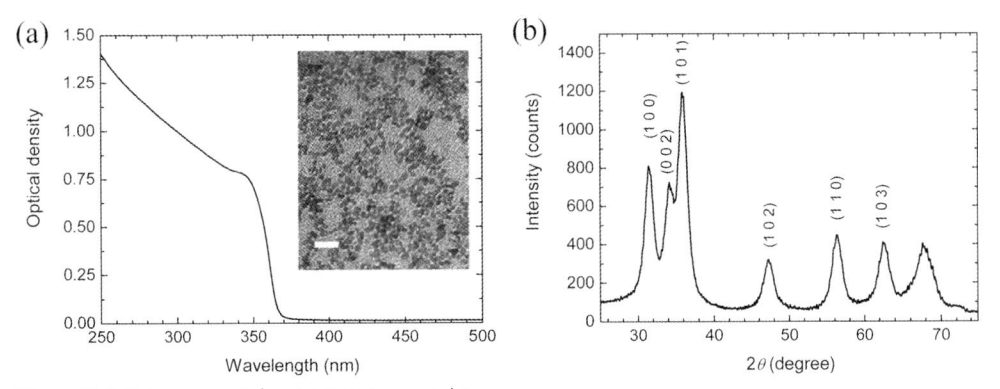

Figure 13.9 ZnO nanoparticles. (a) TEM image (white bar $= 20$ nm) and UV–Vis absorption spectrum in a chloroform:methanol (v:v $= 90:10$) mixture. (b) Powder X-ray diffraction pattern. Reprinted in part with permission from Ref. [115]. Copyright 2005 American Chemical Society.

13.4.2.1 Photophysics of Nanocrystalline ZnO–Polymer Blends

Spin casting a mixed solution of these nc-ZnO particles and a conjugated polymer (MDMO-PPV) from a common solvent mixture (chlorobenzene:methanol 95 : 5 v/v) can be used to prepare thin bulk heterojunction films. PIA spectroscopy performed on these films (Figure 13.10) gave direct spectral evidence that under illumination electron transfer occurred from the polymer to the ZnO nanoparticles. The PIA spectrum (solid line) of the nc-ZnO:MDMO-PPV blend exhibits the characteristic absorption bands of the polymer radical cation at 0.4 and 1.3 eV [116]. Compared to the spectrum of TiO$_2$:MDMO-PPV (dashed line), the increased intensity of the 0.4 eV band indicates the presence of negatively charged nc-ZnO [117]. The presence of the radical cation on the MDMO-PPV and the electron on the ZnO nanoparticle proves that after photoexcitation the electron is transferred from the MDMO-PPV to the ZnO, leaving a hole on the conjugated polymer. Pump-probe spectroscopy (Figure 13.9b) revealed that the photoinduced charge transfer occurs within a picosecond, similar to the ultrafast electron injection (<300 fs) reported by Bauer et al. [118] for electron injection in a dye-sensitized solar cell utilizing Ru (dcbpy)$_2$(NSC)$_2$ dyes on nanoparticulate ZnO electrodes. The occurrence of fast electron injection into the ZnO and the formation of long-lived charges on the polymer and the ZnO show that this combination of materials has promising properties for application in photovoltaic devices.

In a subsequent study, composite films of P3HT and nc-ZnO, spin-cast from chloroform, have been characterized with absorption and photoluminescence spectroscopy [119]. The UV–Vis spectra of these blends (Figure 13.11a) show that the relative contribution of the ZnO absorbance (at 350 nm) increases compared to the π–π^* band of P3HT in the 400–650 nm region with the increase in the

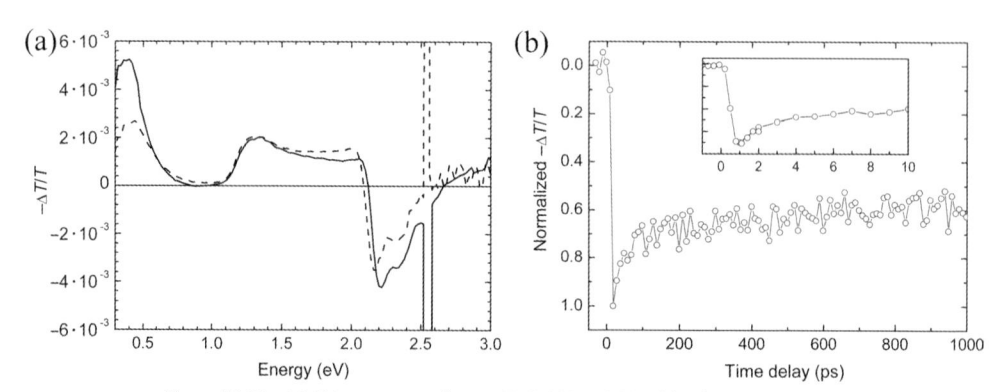

Figure 13.10 (a) PIA spectrum of an nc-ZnO:MDMO-PPV blend on quartz (solid line) and of MDMO-PPV on TiO$_2$ (dashed line). (b) Time-resolved pump-probe spectroscopy, monitoring the intensity of the radical cation band at 0.56 eV, after excitation at 2.43 eV (510 nm). The inset shows the transient absorption at short time delays. Reprinted with permission from Ref. [110]. Copyright 2004 Wiley-VCH.

Figure 13.11 (a) Absorbance spectra of nc-ZnO: P3HT blends with 15 vol% nc-ZnO (○), with 26 vol% nc-ZnO (▲) and 42 vol% nc-ZnO (□) compared to pristine P3HT (■). The inset shows the effect of the addition of these amounts of nc-ZnO on the photoluminescence of P3HT. (b) PIA spectrum of an nc-ZnO:P3HT blend (26 vol% ZnO) on quartz (solid line), measured at 80 K with modulated (275 Hz) excitation at 2.54 eV (488 nm). The pump power was 25 mW with a beam diameter of 2 mm. The dashed line gives the PIA spectrum of a P3HT film. Reprinted with permission from Ref. [119]. Copyright 2006 Wiley-VCH.

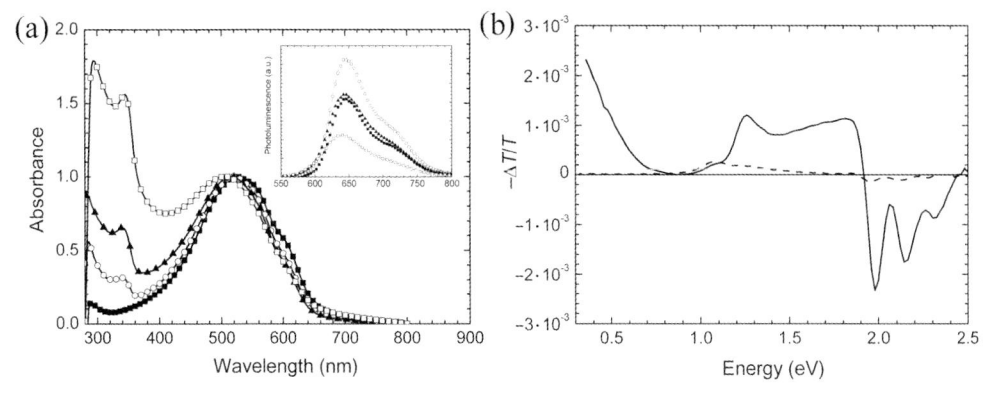

concentration of ZnO in the blend. Simultaneously, a slight blue shift of the P3HT band and a concomitant loss of the (weak) vibronic structures have been observed. The shift has been attributed to a loss of P3HT polymer chain stacking and conformational disorder, caused by the mixing with the ZnO nanoparticles. With the increase in ZnO concentration, the photoluminescence intensity of the blend increased initially compared to the intensity of pristine P3HT (Figure 13.11a, inset), and only after addition of more ZnO (>26 vol%), the photoluminescence eventually dropped below the initial intensity. The initial increase in photoluminescence intensity is unexpected because a photoinduced electron transfer between P3HT and ZnO would rather result photoluminescence quenching. However, the photoluminescence quantum yield of P3HT is low and known to be sensitive to the degree of chain order. Hence, the changes in PL intensity with ZnO concentration were rationalized by a competition between an increase in intensity due to more disordered P3HT chains and a decrease in intensity due to photoinduced electron transfer to the ZnO.

Photoinduced charge separation between P3HT and nc-ZnO was confirmed by PIA spectroscopy (Figure 13.11b). The PIA spectrum of an nc-ZnO:P3HT blend exhibits a vibronically resolved photobleaching band (1.98, 2.15, and 2.41 eV) of the neutral P3HT and the characteristic absorption bands of the P3HT radical cation around 0.4 eV and 1.26 eV [61]. The intensity ratio of these bands is different from the usual ones and we attribute the higher intensity of the 0.4 eV band to electrons injected into the nc-ZnO [117]. In addition to the spectral features of photoinduced charges in P3HT and ZnO, a small signal at 1.06 eV is observed for the blend. This band is also present in the PIA spectra of pure P3HT (dashed line) due to the triplet–triplet absorption of P3HT. Its presence indicates that not all absorbed photons give rise to charges, consistent with the partial photoluminescence quenching (Figure 13.10a) and indicating the presence of relatively large P3HT domains.

Savenije et al. have studied the photogeneration and decay of charge carriers in blend films of ZnO nanoparticles and MDMO-PPV and P3HT by means of TRMC experiments [120]. Excitation of the polymer was found to lead to a long-lived transient photoconductance signal, due charge formation at the nc-ZnO:polymer interface. The signal has been attributed to be mainly due to excess electrons in ZnO. Increasing the weight fraction of nc-ZnO in the blends leads to a higher photoconductance via the generation of more charge carriers.

13.4.2.2 Photovoltaic Properties of nc-ZnO–Polymer Blends

Hybrid bulk heterojunction photovoltaic devices have been made using an nc-ZnO MDMO-PPV blend, sandwiched between PEDOT:PSS/ITO as front electrode and Al as back electrode (Figure 13.12) [110].

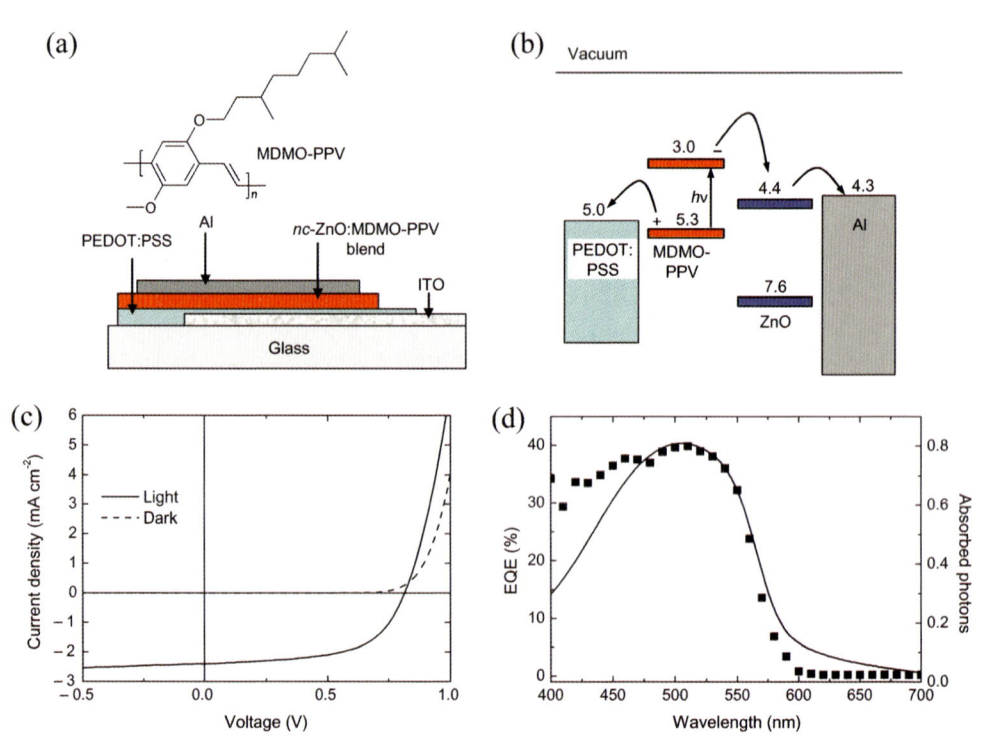

Figure 13.12 (a) Solar cell consisting of an 80 nm thick nc-ZnO:MDMO-PPV blend, sandwiched between ITO/PEDOT:PSS (80 nm) and Al (100 nm) electrodes. (b) Schematic energy level diagram for the cell with energy levels in eV relative to vacuum. (c) Current density–voltage (J–V) characteristic of an nc-ZnO:MDMO-PPV cell (67 wt% nc-ZnO) in the dark and under illumination with white light from a tungsten halogen lamp estimated at 0.71 sun equivalent intensity. (d) EQE as a function of the wavelength of monochromatic irradiation, coplotted with the fraction of photons absorbed by the blend. Reprinted in part with permission from Ref. [110]. Copyright 2004 Wiley-VCH.

The EQE of the hybrid nc-ZnO:MDMO-PPV solar cell with 26 vol% ZnO (Figure 13.12d) has been measured with white-light bias illumination to create conditions resembling operation under solar illumination. The EQE spectrum reaches a value of 40% at the absorption maximum of MDMO-PPV. Integration of the EQE with the solar spectrum (AM 1.5 standard, normalized to 100 mW cm^{-2}) affords an estimate of the short-circuit current density of $J_{sc} = 3.3$ mA cm^{-2} under AM 1.5 (1 sun) conditions. J–V measurements measured with white-light illumination (equivalent to AM 1.5 at 71 mW cm^{-2}) are shown in Figure 13.12c. Under these conditions $J_{sc} = 2.40$ mA cm^{-2} has been obtained, together with $V_{oc} = 0.814$ V and FF $= 0.59$. At this light intensity the power conversion efficiency is $\eta = 1.6\%$.

The effect of the ZnO concentration on the photovoltaic performance is significant. Hybrid nc-ZnO:MDMO-PPV devices with different amounts of nc-ZnO show different J–V characteristics under illumination [115]. The effects of the ZnO concentration and layer thickness on the maximum power point (MPP) under white-light illumination are shown in Figure 13.13. The maximum performance is observed for devices containing 26–35 vol% ZnO and a layer thickness between 100 and 150 nm.

Blom et al. recently characterized the transport of electrons and holes in nc-ZnO:MDMO-PPV layers by selectively suppressing the injection of one of the charge carriers through the use of either high or low work function electrodes [121]. Both hole and electron currents were found to be space charge limited. The hole mobility in the polymer phase of MDMO-PPV/nc-ZnO (1 : 2 by weight) was found to

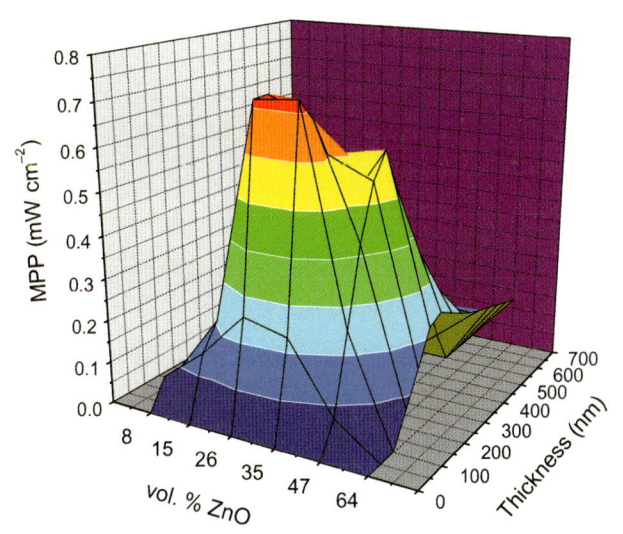

Figure 13.13 The maximum power point of nc-ZnO:MDMO-PPV solar cells, as a function of vol% ZnO and the thickness of the active layer. Reprinted with permission from Ref. [115]. Copyright 2005 American Chemical Society.

be equal to the mobility in pristine MDMO-PPV (5.5×10^{-6} cm^2 V^{-1} s^{-1}), whereas the electron mobility amounts to 2.8×10^{-5} cm^2 V^{-1} s^{-1}. In contrast to MDMO-PPV: PCBM solar cells where the hole mobility increases by more than two orders of magnitude upon addition of 70 vol% PCBM, the hole mobility in the hybrid layer is not affected by the presence of nc-ZnO. This explains the lower efficiency of the nc-ZnO: MDMO-PPV system compared to MDMO-PPV:PCBM cells that reach 2.5% efficiency [5].

Regioregular P3HT is an interesting material to apply in nc-ZnO hybrid polymer solar cells. P3HT can have a high hole mobility, up to ~0.1 cm^2 V^{-1} s^1 in field effect transistors [122,123]. Compared to MDMO-PPV, the improved hole mobility, together with a more "red-shifted" absorbance of the material, might lead to an improved performance of photovoltaic devices. Beek et al. have demonstrated that in combination with nc-ZnO, solution-processed bulk heterojunction solar cells can be obtained [119]. Thermal annealing of the nc-ZnO: P3HT blend further enhanced the photovoltaic performance [119]. Figure 13.14 shows the effects of the annealing step on the photovoltaic properties of nc-ZnO: P3HT solar cells. The effects of annealing are very obvious in blends with low amounts of nc-ZnO, that is, the increase in short-circuit current density and the open-circuit voltage, leading to a pronounced increase in efficiency. In some respect these solar cells behave very similar to P3HT:PCBM solar cells, where the performance also improves with annealing [17,124]. The improvement in cell efficiency in both types of cells mainly results from an increased, almost doubled, short-circuit current density, attributed to an increased crystallinity and hence higher charge carrier mobility.

Figure 13.14a indicates that the largest effect of annealing is found in blends with low amounts of ZnO, that is, the highest amount of polymer. This shows that the annealing step mainly influences P3HT phase and therefore likely has a beneficial effect on the hole transport properties of this phase [119,125]. The J–V characteristics of the optimized nc-ZnO:P3HT cell under white-light illumination using a tungsten–halogen lamp (75 mW cm^{-2}) are shown in Figure 13.13b. Under these conditions $J_{sc} = 2.14$ mA cm^{-2}, $V_{oc} = 0.685$ V, and FF = 0.55, giving an overall power conversion efficiency of $\eta = 0.9\%$. Compared to $\eta = 1.6\%$ of the nc-ZnO:MDMO-PPV, the lower value for nc-ZnO:P3HT seems unexpected, given the higher (pristine) hole mobility of the P3HT and the red-shifted absorbance. However, difficulties during processing and film formation indicated that the processing conditions and the blend morphology play a significant role.

Recently, Greenham et al. described the use of poly(2,7-(9,9-dioctylfluorene)-alt-5,5-(4′,7′-di-2-thienyl-2′,1′,3′-benzothiadiazole)) (APFO-3) as an electron donor blended with ZnO nanoparticles [126]. The cells gave $J_{sc} = 3.1$ mA cm^{-2}, V_{oc} 0.51 V, FF = 0.36, and $\eta = 0.45\%$. The EQE maximized at 28% for 535 nm light. The performance of these devices was limited by the nonoptimal morphology of the films, which influences not only the degree of exciton separation (seen from the incomplete PL quenching) but also the transport of charges out of the device.

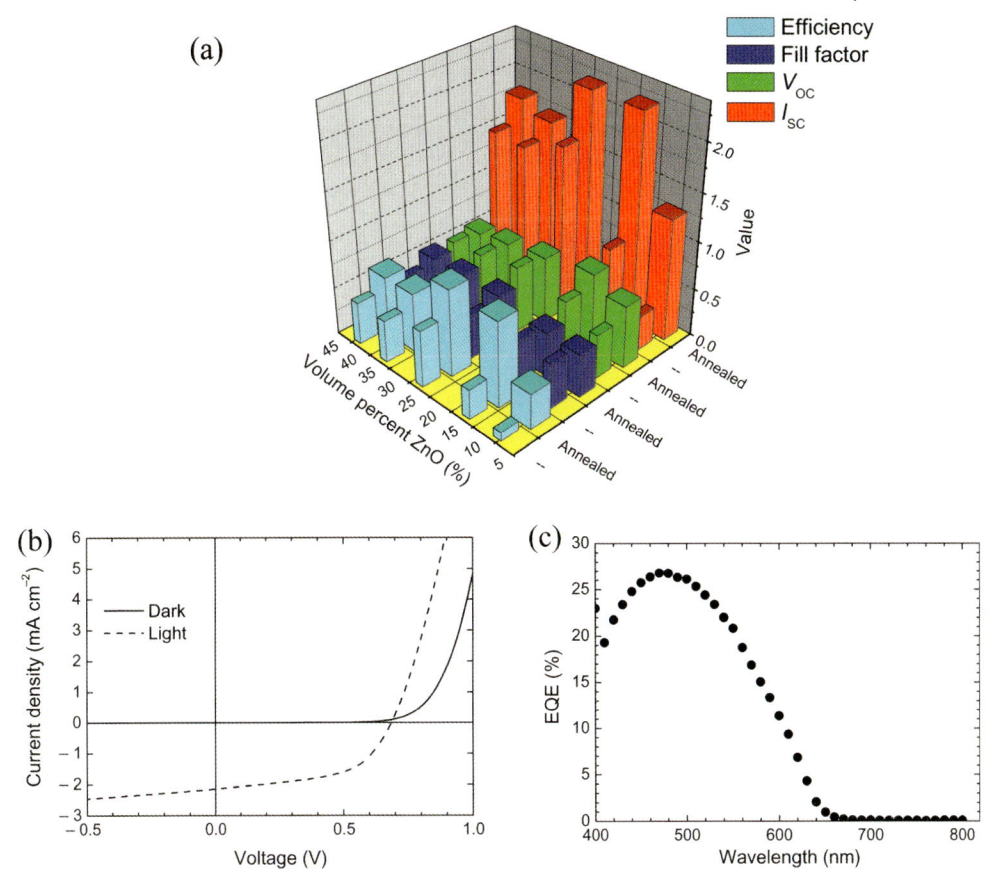

Figure 13.14 (a) Performance of nc-ZnO:P3HT solar cells as a function of thermal annealing composition. (b) Current density–voltage (J–V) characteristics of an nc-ZnO:P3HT blend (with 26 vol% nc-ZnO) in dark and under illumination with white light from a tungsten halogen lamp (75 mW cm^{-2}) (c) EQE as a function of the wavelength of monochromatic irradiation. Reprinted in part with permission from Ref. [119]. Copyright 2006 Wiley-VCH.

13.4.2.3 Morphology of nc-ZnO:Polymer Blends

For bulk heterojunction solar cells, the phase separation between the two components is of utmost importance because it determines the effectiveness of charge generation and charge transport [124,127]. Charge generation requires a high surface to volume ratio, whereas charge transport requires the presence of pathways consisting of almost pure phases of the starting compounds, formed by phase separation. Transmission electron microscopy of a thin (50 nm) film (Figure 13.15) containing 26 vol% nc-ZnO in MDMO-PPV revealed phase separation on nanometer scale. Dark nc-ZnO regions and bright polymer regions, some larger than 10 nm, can be seen. The occurrence of pure polymer domains, larger than the exciton diffusion length, shows that not all excitons can dissociate into photogenerated charges,

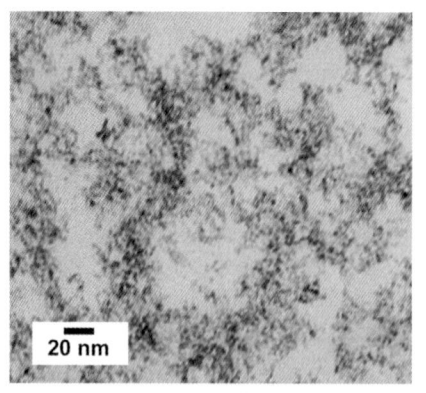

Figure 13.15 TEM image of a 50 nm thin film of an nc-ZnO:MDMO-PPV blend with 26 vol% ZnO. Reprinted with permission from Ref. [115]. Copyright 2005 American Chemical Society.

lowering the charge generation efficiency. From photoluminescence quenching data, it appeared that a maximum of 85% of the absorbed photons gave rise to charge separation [110].

The TEM image seems to indicate that the desired morphology is present; however, the phase-separated morphology was more difficult to observe with tapping mode atomic force microscopy (AFM) (Figure 13.16). When mixing the ZnO nanoparticles with MDMO-PPV or P3HT spherical aggregates and clusters that

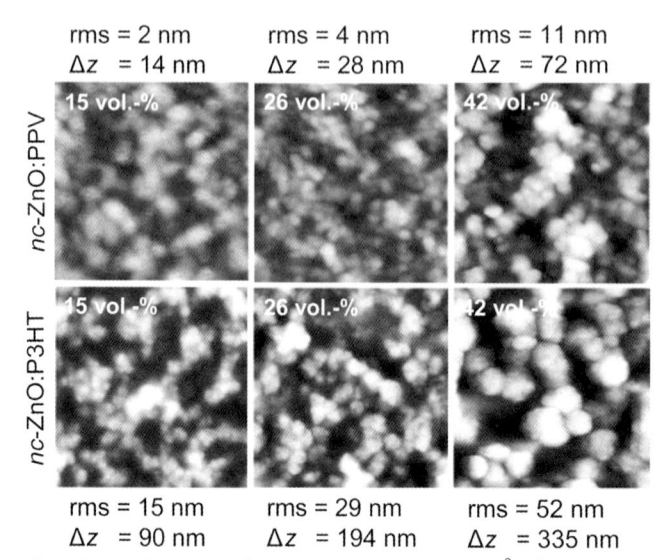

Figure 13.16 Tapping mode AFM pictures (500 × 500 nm^2) of nc-ZnO:MDMO-PPV (top) and nc-ZnO:P3HT (bottom) blends for different compositions. Reprinted in part with permission from Ref. [115]. Copyright 2005 American Chemical Society.

appeared thereof, the spherical structures appeared to consist of a mixture of polymer and ZnO, but no clear phase separation has been observed at the surface of the *nc*-ZnO:MDMO-PPV film.

Closer examination of the AFM images revealed that there is a significant difference in morphology between the P3HT and MDMO-PPV-based blends. The MDMO-PPV-based blends appeared much smoother and the spherical structures and aggregates were much smaller. This is a likely explanation for the differences in photovoltaic effect. The roughness of the *nc*-ZnO:P3HT films increased dramatically when the amount of ZnO in the blend increased. The likelihood of shunted films increases, which limits the photovoltaic performance that can be reached. The *nc*-ZnO:MDMO-PPV blends also showed an increase in roughness with ZnO addition, but much less. This allows a further loading of the *nc*-ZnO:MDMO-PPV films with sufficient ZnO nanoparticles to improve charge generation and separation. Figures 13.13 and 13.14 show that the optimum volume percent of ZnO in the *nc*-ZnO:P3HT cells is slightly lower than that in the case of the *nc*-ZnO:MDMO-PPV cells. There seems further evidence that it is the morphology that limits the performance of the *nc*-ZnO:P3HT devices.

13.5
Metal Oxide Networks and Conjugated Polymers

In the previous sections it was shown that both mixing conjugated polymers with inorganic nanoparticles and filling of (nanostructured) pores still pose processing challenges with respect to reaching higher efficiencies. For the nanoparticles, a difficult issue is to find conditions that allow efficient mixing of the particles with the polymer and create good electrical contacts between the particles to obtain a percolating pathway. The hybrid polymer nanoporous inorganic materials face the problem that the small (~20 nm) pores that are required to ensure that the majority of photoexcitations in the polymer reach the interface with inorganic semiconductor are difficult to fill efficiently.

In an attempt to overcome these disadvantages, Van Hal *et al.* introduced the use of a molecular precursor for the inorganic semiconductor to create a metal oxide network inside the polymer film [128]. Because the metal oxide precursor is readily soluble in organic media, it can be cast into a thin film together with the polymer from solution. Upon exposure to moisture from air, the precursor is converted into a metal oxide semiconductor within the polymer layer, ensuring intimate mixing and concomitantly efficient photoinduced charge generation.

This approach differs in an essential way from the previous methods, in the sense that the inorganic material is made only after film formation. It promises to be a very simple procedure for making hybrid bulk heterojunctions and can make use of well-established sol–gel chemistry. Similar to the two other methods, the challenge is to control morphology of the blend and the crystalline nature of the inorganic phase after conversion, as these likely affect the electron transport in the material. One drawback of this method is that the applied precursor or sol–gel chemistry should be compatible with the use of conjugated polymers, limiting it to the application of organic media and at low temperatures.

13.5.1
In situ Blends Based on TiO$_x$

The validity of this approach was first demonstrated using titanium(IV) isopropoxide (Ti(*i*-PrO)$_4$) as a precursor for TiO$_2$ in combination with MDMO-PPV as the conjugated polymer [128]. To fabricate the hybrid bulk heterojunction, Ti(*i*-PrO)$_4$ was added to a solution of MDMO-PPV in dry tetrahydrofuran (THF) prior to spin casting the mixture into a thin solid film. The Ti(*i*-PrO)$_4$ precursor was subsequently converted "*in situ*" into TiO$_2$ by exposure of the cast film to moisture from the air to promote hydrolysis and a consecutive high vacuum treatment to promote condensation of the formed titanium hydroxide into titania. As a consequence of the formation of a TiO$_2$ network, the bulk heterojunction film became resistant to scratching and could no longer be wiped off the substrate or dissolved in organic solvents. X-ray photoelectron spectroscopy (XPS) of the *in situ* TiO$_2$:MDMO-PPV indicated that the conversion of Ti(*i*-PrO)$_4$ into TiO$_2$ in the bulk heterojunction was at least 65% under these conditions [128]. The characteristic distance of this *in situ* TiO$_x$:MDMO-PPV bulk heterojunction was in the nanometer range, as inferred from AFM studies (Figure 13.17) [128].

The PL at 580 nm of the *in situ* TiO$_x$:MDMO-PPV bulk heterojunction was significantly quenched compared to that of a pristine MDMO-PPV film. The PL intensity decreased with the increasing amount of TiO$_2$ and was quenched by a factor of 19 in the 50% blend (Figure 13.18a), implying a fast deactivation of the MDMO-

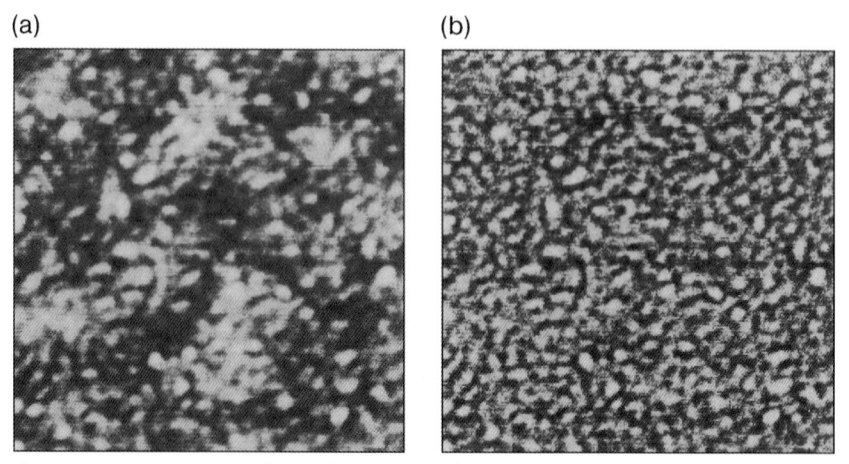

Figure 13.17 (a) Tapping mode AFM height image of a 1 : 1 (v/v) *in situ* TiO$_x$:MDMO-PPV bulk heterojunction (2 × 2 µm^2, z-range = 10 nm) created by the *in situ* method. (b) Corresponding AFM phase image (z-range = 7.5°) . Reprinted with permission from Ref. [128]. Copyright 2003 Wiley-VCH.

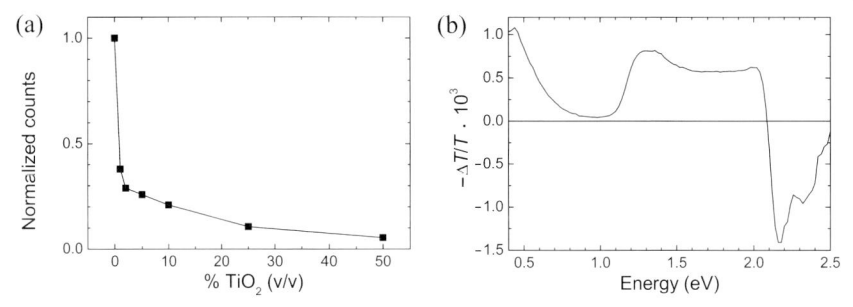

Figure 13.18 (a) Photoluminescence intensity of *in situ* TiO$_x$:MDMO-PPV bulk heterojunctions with varying amounts of TiO$_2$ with excitation at λ = 488 nm. (b) PIA spectrum of 1 : 1 (v/v) *in situ* TiO$_x$:MDMO-PPV bulk heterojunction. Reprinted with permission from Ref. [128]. Copyright 2003 Wiley-VCH.

PPV singlet excited state. The significantly reduced PL intensity corroborates with the fine-scale phase separation inferred from AFM (Figure 13.17) and indicates an efficient formation of photoinduced charges. The PIA spectrum of *in situ* TiO$_x$:MDMO-PPV recorded at 80 K exhibits two absorptions at 0.42 and 1.32 eV and a bleaching signal at 2.18 eV with a shoulder at 2.32 eV (Figure 13.18b) and gives direct spectral evidence of the formation of photogenerated charges. The absorption bands are due to the two dipole-allowed transitions of the radical cation of MDMO-PPV, while the two bleaching signals are attributed to the MDMO-PPV ground state [61].

The spectral response of an ITO/PEDOT:PSS/*in situ* TiO$_x$:MDMO-PPV/LiF/Al solar cell prepared using this simple approach gave an EQE = 11% at 450 nm (Figure 13.19a) for a 4 : 1 (v/v) blend. J–V measurements on these cells under illumination with a halogen lamp (∼0.7 sun intensity) revealed a moderate performance with $J_{sc} = 0.6$ mA cm^{-2}, $V_{oc} = 0.520$ V, and FF = 0.42 (Figure 13.19b), representing an

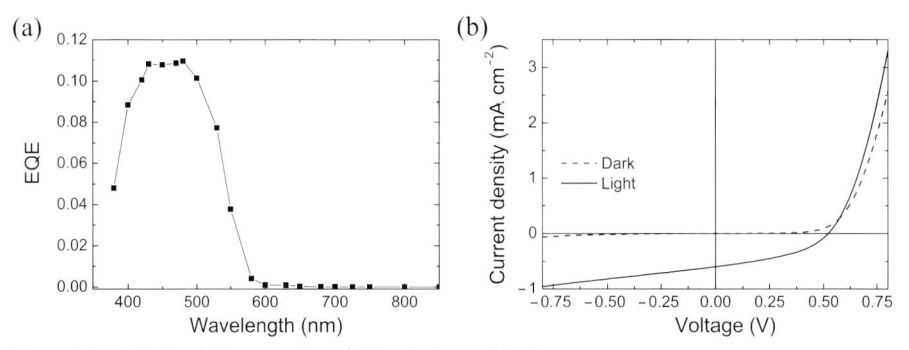

Figure 13.19 Photovoltaic properties of ITO/PEDOT:PSS/*in situ* TiO$_x$:MDMO-PPV/LiF/Al devices 4 : 1 (v/v); (a) external quantum efficiency; (b) J–V curve in the dark (dashed line) and under illumination (∼70 mW cm^{-2}) (solid line). Reprinted with permission from Ref. [128]. Copyright 2003 Wiley-VCH.

energy conversion efficiency of $\eta = 0.2\%$. Integration of the measured spectral response of the cell with the AM 1.5 solar spectrum affords an estimate for the short-circuit current density under AM 1.5 conditions (1000 W m^{-2}) of $J_{sc} = 0.87$ mA cm^{-2} [128]. In a subsequent study, Slooff et al. confirmed these results for in situ TiO$_x$:MDMO-PPV ($J_{sc} = 0.6$ mA cm^{-2}, $V_{oc} = 0.60$ V, FF = 0.43, and $\eta = 0.22\%$ at 70 mW cm^{-2}) and showed that the procedure is also applicable to polythiophene derivatives [129]. For an in situ TiO$_x$:P3OT active layer, they found $J_{sc} = 0.7$ mA cm^{-2}, $V_{oc} = 0.45$ V, FF = 0.41, and $\eta = 0.17\%$ at 70 mW cm^{-2}.

Wang et al. followed a similar approach using a triethoxysilane-containing poly(3-nonylthiophene) (P3NT) and Ti(i-PrO)$_4$ to make hybrid in situ TiO$_x$/P3NT material [130]. The presence of the silane group effectively prevented the macroscopic aggregation of TiO$_2$ during the sol–gel reaction, leading to more uniform distribution of TiO$_2$ throughout the sample and a better PL quenching. This promising material, however, has not been tested in solar cells so far.

While the in situ TiO$_x$ synthesis has various potential advantages in terms of creating intimate mixing, the main drawback of the use of in situ prepared TiO$_x$ is that the resulting inorganic phase is essentially amorphous. Crystallization of TiO$_2$ would require high temperatures (>350 °C [131]). Hence, it is likely that the presence of an amorphous TiO$_x$ phase, rather than a crystalline network of TiO$_2$, limits the charge transport in photovoltaic devices based on this method.

13.5.2
In situ Blends Based on ZnO

Due to the poorly defined amorphous nature of the formed TiO$_x$ phase, the power conversion efficiency of the resulting photovoltaic cells described in the previous section remained rather low at about 0.2%. Improvement can be expected from crystallization of TiO$_2$, but the high temperatures (>350 °C [131]) required for crystallization are not compatible with the presence of the semiconducting polymer. ZnO, on the contrary, is known to crystallize at much lower temperatures [111]. To overcome the limitations encountered in the in situ synthesis of TiO$_2$, Beek et al. introduced in situ ZnO:polymer bulk heterojunction solar cells [132]. The active layer was prepared by spin coating a solution containing an organozinc compound and a conjugated polymer, followed by thermal annealing at moderate temperature. This affords a crystalline ZnO network in the polymer phase. The resulting photovoltaic devices exhibit a significantly increased power conversion efficiency compared to the amorphous in situ TiO$_2$:polymer cells.

The new method to make crystalline ZnO films starts from diethylzinc. Like other sol–gel processes, the conversion of diethylzinc into ZnO requires two steps: hydrolysis and condensation. These reactions initiate very rapidly when diethylzinc is exposed to air. To obtain smooth ZnO films, the hydrolysis and condensation reactions need to be moderated. This has been achieved by the addition of THF to the solution. THF stabilizes diethylzinc by coordination to the zinc atom. Room temperature conversion of diethylzinc does not result in the direct formation of ZnO, as evidenced from the lack of absorption above 250 nm in the absorption spectrum of a pristine spin-coated film. However, upon heating to 110 °C the absorption onset shifted toward 375 nm, indicating that ZnO is formed [132]. The ZnO formed via this route appears to be crystalline by powder X-ray diffraction. The observed peak positions correspond very well to the values expected for crystalline ZnO [132,133]. The line broadening that was present in the XRD spectrum indicated the formation of 6 ± 1 nm small domains of ZnO.

Because of the moderate conversion temperature, the diethylzinc route is compatible with conjugated polymers. Composite films on glass were made by spin coating solutions of diethylzinc and MDMO-PPV in chlorobenzene/THF/toluene mixtures. Because water is one of the reagents, the relative humidity is a critical processing parameter. The best results were obtained when spin coating was performed in air under relative humidity of 40%, followed by aging for 15 min and annealing at 110 °C for 30 min in nitrogen at 40% relative humidity. The UV–Vis absorption spectrum revealed the presence of ZnO at $\lambda < 360$ nm (Figure 13.20a). At the same time a distinct 26 nm blue shift of the MDMO-PPV absorbance was observed, indicating partial degradation of the polymer [132].

The photoluminescence of *in situ* ZnO:MDMO-PPV films was significantly quenched compared to a pristine annealed MDMO-PPV film, consistent with an electron transfer in the blend from the singlet-excited state of the polymer to ZnO precursor. The residual photoluminescence amounted to ~20% and was attributed to excitons that cannot reach a junction between MDMO-PPV and ZnO [132]. Hence, up to 80% of the absorbed photons may result in charge formation. Evidence for long-lived charges was provided by PIA measurements. The PIA spectrum of an annealed *in situ* ZnO:MDMO-PPV film shows the photobleaching band of neutral MDMO-PPV at 2.2 eV and the characteristic absorption bands of the polymer radical cation around 0.4 and 1.5 eV (Figure 13.20b) [132]. The band at 1.5 eV also contained a residual signal from the MDMO-PPV triplet excited state absorption. Some triplet formation via intersystem crossing is likely to occur because the competing intrinsic

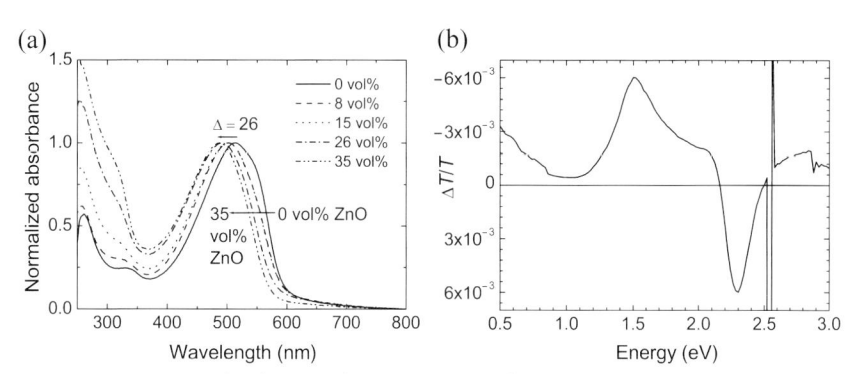

Figure 13.20 (a) Normalized UV–Vis absorption spectra of *in situ* ZnO:MDMO-PPV films after annealing at 110 °C. (b) PIA spectrum of an *in situ* ZnO:MDMO-PPV blend on quartz measured at 80 K with modulated (275 Hz) excitation at 2.54 eV (488 nm). Reprinted with permission from Ref. [132]. Copyright 2005 Wiley-VCH.

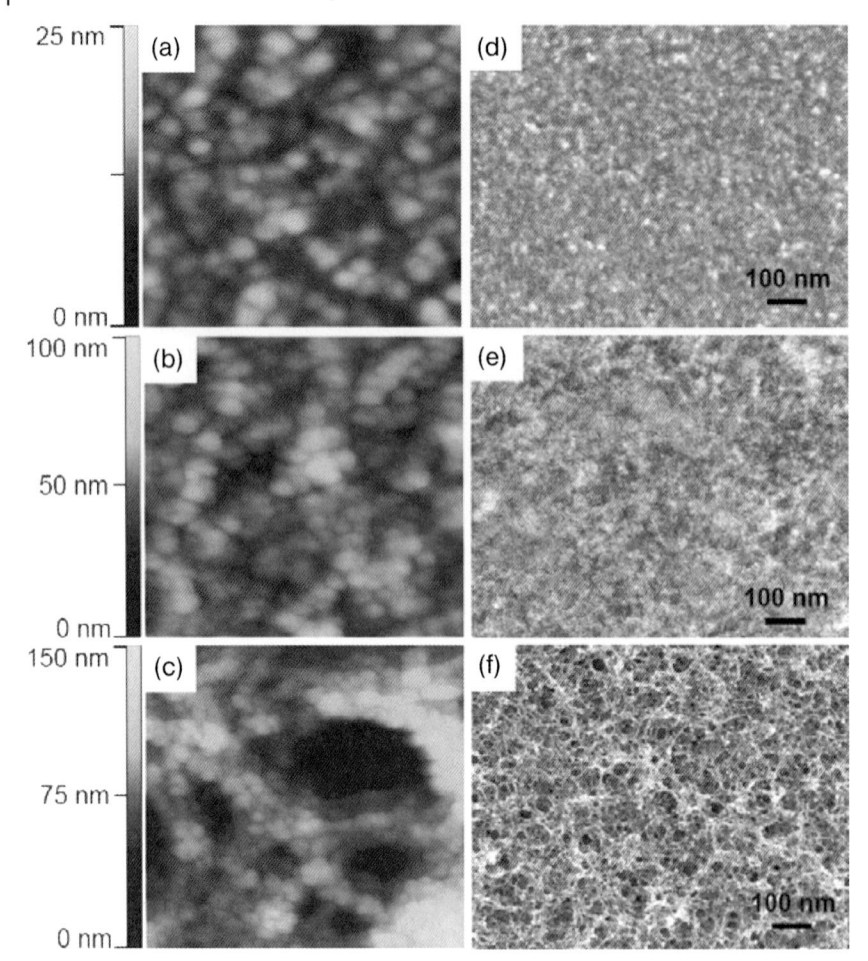

Figure 13.21 (a–c) AFM tapping mode images of *in situ*-ZnO:MDMO-PPV blends, image size is 500 × 500 nm^2: (a) 8 vol% ZnO (height scale = 25 nm, Δh (peak-to-peak roughness) = 7 nm); (b) 26 vol% ZnO (height scale = 100 nm, Δh = 40 nm); (c) 47 vol% ZnO (height scale = 150 nm, Δh = 100 nm). (d–f) SEM images: (d) 15 vol% ZnO; (e) 26 vol% ZnO; (e and f) 35 vol% ZnO. Reprinted with permission from Ref. [132]. Copyright 2005 Wiley-VCH.

process, photoluminescence of the polymer, was not completely quenched. No distinct contribution of ZnO$^-$ to the PIA spectrum was observed, but this absorption may be obscured by the low energy absorption of the cation radical. The formation of charge carriers in *in situ* ZnO:MDMO-PPV bulk heterojunctions has also been studied by Savenije *et al.* by the electrodeless TRMC technique [134]. Photoexcitation of the polymer at 510 nm leads to a long-lived (up to 10 µs) conductance signal, indicating efficient charge separation from the polymer to ZnO.

The morphology of the *in situ* ZnO:MDMO-PPV films has been examined with tapping mode AFM and SEM. With AFM conglomerates of spherical aggregates have been identified. The roughness of the film increases significantly with increasing ZnO content (Figure 13.21a–c). A similar conclusion can be drawn from the SEM images (Figure 13.21d–f). The increased roughness at higher ZnO content leads to formation of large pores, as observed with AFM and SEM (Figure 13.21c and f). However, the increased roughness and larger pores does not seem to change the morphology on a smaller scale, as the same spherical aggregates are observed with AFM.

Photovoltaic devices were fabricated by spin coating, aging, and annealing the photoactive layer under controlled relative humidity on glass substrates covered with ITO and PEDOT:PSS and completed by deposition of an Al top electrode. Under white-light illumination (90 mW cm^{-2}), the hybrid device gives $J_{sc} = 2.0$ mA cm^{-2}, $V_{oc} = 1.14$ V, and FF $= 0.42$ (Figure 13.22). The EQE reached to 27% at 520 nm and integrating the spectral response with the AM 1.5 spectrum gave an estimate for $J_{sc} = 2.3$ mA cm^{-2} and a calculated power conversion efficiency of $\eta = 1.1\%$. This efficiency is five times higher than for optimized cells based on TiO$_2$ precursors [128] and demonstrates the effectiveness of crystalline ZnO as n-type material in hybrid bulk heterojunction solar cells. Moreover, such photovoltaic effect is consistent with the formation of a bicontinuous network of both materials. It is important to note that an unannealed *in situ* ZnO:MDMO-PPV blend shows only a very small photovoltaic effect with $J_{sc} = 0.05$ mA cm^{-2}, $V_{oc} = 1.15$ V, and FF $= 0.18$. This shows the necessity of the annealing step and obtaining crystalline ZnO for preparing well-performing photovoltaic devices.

The degradation of MDMO-PPV during the processing of the layers with diethylzinc has been studied in more detail by Blom *et al.* [135]. Apart from the blue shift of the absorption spectrum, the hole transport through the polymer deteriorates dramatically, indicating a reduction in the conjugation length of the polymer backbone. To prevent polymer degradation through the breaking of the *trans*-vinyl bonds, regioregular P3HT is introduced as the electron donor. This system of P3HT and diethylzinc as a precursor ZnO reveals an unchanged UV–Vis absorption profile and zero-field hole mobility with respect to the pristine polymer, as well as an improved photovoltaic performance with an estimated power conversion

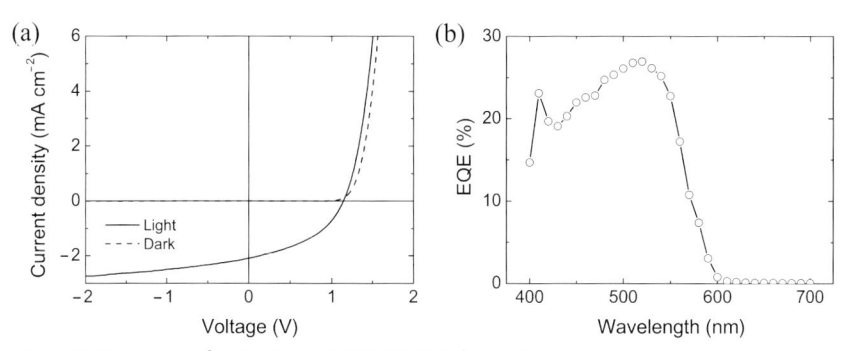

Figure 13.22 (a) J–V of an *in situ* ZnO:MDMO-PPV photovoltaic device (15 vol% ZnO). (b) EQE of the device. Reprinted with permission from Ref. [132]. Copyright 2005 Wiley-VCH.

efficiency of 1.4% under AM 1.5 conditions (J_{sc} of 3.5 mA cm^{-2}, V_{oc} = 0.83 V, and FF = 0.50).

13.6
Conclusions and Outlook

The field of hybrid metal oxide–polymer solar cells has attracted strong interest and progressed enormously over the past decade. An overview of the different cell architectures and their performance as published up to middle of 2007 is collected in Table 13.1. The best planar devices so far are FTO/TiO$_2$/P3HT/Ag cells in which the TiO$_2$ interface is modified with PTPTB or the N3 dye reach and η = 0.60% [54,55]. For filling nanoporous or nanostructured TiO$_2$, the highest reported efficiency is η = 0.71% for an ITO/TIO$_2$/MEH-PPV/Au cell in which the TiO$_2$ layer was created using a polystyrene block–polyethylene oxide diblock copolymer (PS-b-PEO) as the templating agent [83]. By using nanoparticles or a precursor method, efficiencies well above η = 1.0% have been reached. ZnO nanoparticles blended in MDMO-PPV give η = 1.6% [110] and an *in situ* formed ZnO network in P3HT, η = 1.4% [135]. The latter two efficiencies are about a factor of two less than the best values for blends of CdSe nanoparticles and conjugated polymers, where η = 2.4–2.8% was reported for cells based on P3HT and CdSe tetrapods [97].

Compared to the best polymer/fullerene cells with η = 5%, hybrid cells still face some important challenges in closing the gap. One of the most important questions to be solved is the creation of the bulk heterojunction itself. The different nature of the organic and inorganic materials shows that creating an intimate blend with two fully percolating pathways is not trivial, and it is our opinion that this remains the most important hurdle to be taken. It will be up to the imagination and creativity of researchers to come up with novel techniques and approaches to solve this issue. This will pave the way for further improving the cells by optimizing the optical bandgap of the polymer to the solar emission and by minimizing voltages losses at electrodes and in the actual charge transfer step.

Finally, it should be noted that apart from being part of the active layer, transparent inorganic layers are emerging as additional layers in improving the efficiency of polymer/fullerene solar cells. In this respect Krausch *et al.* [136] and more recently Heeger and coauthors [137] have used a TiO$_2$ layer as an optical spacer between the active layer and the Al back electrode. The optical spacer affects the spatial distribution of the optical field strength in the devices and causes an increased absorption of light. Krausch *et al.* have shown that the optical effect alone cannot explain the increased performance and concluded that the TiO$_2$ layer also acts as a charge transfer interface. For ITO/PEDOT:PSS/P3HT:PCBM/TiO$_x$/Al cells, the introduction of the TiO$_x$ layer has resulted in a more than twofold improvement of the power conversion efficiency compared to a similarly prepared device without the TiO$_x$ layer.

Likewise, White *et al.* developed inverted organic photovoltaic devices based on a blend of P3HT and PCBM by inserting a solution-processed ZnO interlayer between the ITO electrode and the active layer using Ag as a hole-collecting back contact [30].

Efficient electron extraction through the ZnO and hole extraction through the Ag, with minimal loss in open-circuit potential, were observed with a certified power conversion efficiency of 2.58%. The inverted architecture removes the need for the use of PEDOT:PSS as an ITO modifier and for the use of a low work function metal as the back contact in the device.

Hybrid materials have also contributed to the development of double and triple junction solar cells. Gilot *et al.* [138] have presented a first example of solution-processed multiple junction solar cells, incorporating polymer:fullerene bulk heterojunctions as active layers and a hybrid recombination layer. The hybrid recombination layer, deposited between the active layers, was fabricated by spin coating ZnO nanoparticles, followed by spin coating neutral pH PEDOT from water, and short UV illumination of the completed device. The key advantage of this procedure is that each step does not affect the integrity of previously deposited layers. The open-circuit voltage (V_{oc}) for double and triple junction solar cells is close to the sum of the V_{oc}s of individual cells. This simple procedure to create multijunction devices will enable increasing the overall efficiency when active materials with different optical bandgaps are used.

More recently Heeger and coauthors demonstrated a similar principle to make tandem solar cells with a current world record efficiency of $\eta = 6.5\%$ [139]. They used a TiO$_x$/PEDOT recombination layer and two complementary active layers that were made of a low-bandgap polymer and P3HT in combination with fullerene derivatives.

In conclusion, hybrid polymer–inorganic photovoltaic cells are one of the various strategies that are currently explored for creating renewable sources of sustainable energy. While the progress in the basic understanding of their operation principles and in the actual performance of these devices has been very significant, it is evident that breakthroughs will be needed before we can start thinking of using these devices for large-scale energy production. Yet, the importance of this subject for society and the interesting opportunities for fundamental research make exploring this fascinating area worth all the efforts.

References

1 Halls, J.J.M., Walsh, C.A., Greenham, N.C., Marseglia, E.A., Friend, R.H., Moratti, S.C. and Holmes, A.B. (1995) *Nature*, **395**, 498.

2 Veenstra, S.C., Verhees, W.J.H., Kroon, J.M., Koetse, M.M., Sweelssen, J., Bastiaansen, J.J.A.M., Schoo, H.F.M., Yang, X., Alexeev, A., Loos, J., Schubert, U.S. and Wienk, M.M. (2004) *Chemistry of Materials*, **16**, 2503.

3 Koetse, M.M., Sweelssen, J., Hoekerd, K.T., Schoo, H.F.M., Veenstra, S.C., Kroon, J.M., Yang, X. and Loos, J. (2006) *Applied Physics Letters*, **88**, 083504.

4 Yu, G., Gao, J., Hummelen, J.C., Wudl, F. and Heeger, A.J. (1995) *Science*, **270**, 1789.

5 Shaheen, S.E., Brabec, C.J., Sariciftci, N.S., Padinger, F., Fromherz, T. and Hummelen, J.C. (2001) *Applied Physics Letters*, **78**, 841.

6 Schilinsky, P., Waldauf, C. and Brabec, C.J. (2002) *Applied Physics Letters*, **81**, 3885.

7 Wienk, M.M., Kroon, J.M., Verhees, W.J.H., Knol, J., Hummelen, J.C., van

Hal, P.A. and Janssen, R.A.J. (2003) *Angewandte Chemie – International Edition*, **42**, 3371.

8   Peet, J., Kim, J.Y., Coates, N.E., Ma, W.L., Moses, D., Heeger, A.J. and Bazan, G.C. (2007) *Nature Materials*, **6**, 497.

9   Schweitzer, B. and Bässler, H. (2000) *Synthetic Metals*, **109**, 1.

10 Sariciftci, N.S.(ed.) (1998) *Primary Photoexcitations in Conjugated Polymers: Molecular Exciton Versus Semiconductor Band Model*, World Scientific Publishers, Singapore.

11 Markov, D.E., Amsterdam, E., Blom, P.W.M., Sieval, A.B. and Hummelen, J.C. (2005) *Journal of Physical Chemistry A*, **109**, 5266.

12 Halls, J.J.M., Pichler, K., Friend, R.H., Moratti, S.C. and Holmes, A.B. (1996) *Applied Physics Letters*, **68**, 3120.

13 Savenije, T.J., Warman, J.M. and Goossens, A. (1998) *Chemical Physics Letters*, **287**, 148.

14 Kroeze, J.E., Savenije, T.J., Vermeulen, M.J.W. and Warman, J.M. (2003) *Journal of Physical Chemistry B*, **107**, 7696.

15 Ma, W., Yang, C., Gong, X., Lee, K. and Heeger, A.J. (2005) *Advanced Functional Materials*, **15**, 1617.

16 Brabec, C.J. (2004) *Solar Energy Materials and Solar Cells*, **83**, 273.

17 Padinger, F., Rittberger, R.S. and Sariciftci, N.S. (2003) *Advanced Functional Materials*, **13**, 85.

18 Li, G., Shrotriya, V., Huang, J., Yao, Y., Moriarty, T., Emery, K. and Yang, Y. (2005) *Nature Materials*, **4**, 864.

19 O'Regan, B. and Grätzel, M. (1991) *Nature*, **353**, 737.

20 Nazeeruddin, M.K., Kay, A., Rodicio, I., Humphry-Baker, R., Muller, E., Liska, P., Vlachopoulos, N. and Grätzel, M. (1993) *Journal of the American Chemical Society*, **115**, 6382.

21 Grätzel, M. (2004) *Journal of Photochemistry and Photobiology A*, **164**, 3.

22 Bach, U., Lupo, D., Comte, P., Moser, J.-E., Weissörtel, F., Salbeck, J., Spreitzer, H. and Grätzel, M. (1998) *Nature*, **395**, 583.

23 Li, B., Wang, L., Kang, B., Wang, P. and Qiu, Y. (2006) *Solar Energy Materials and Solar Cells*, **90**, 549.

24 Grätzel, M. (2005) *MRS Bulletin*, **30**, 23.

25 Wang, P., Zakeeruddin, S.M., Moser, J.E., Nazeeruddin, M.K., Sekiguchi, T. and Grätzel, M. (2003) *Nature Materials*, **2**, 402.

26 Wang, P., Zakeeruddin, S.M., Moser, J.-E., Humphry-Baker, R. and Grätzel, M. (2004) *Journal of the American Chemical Society*, **126**, 7164.

27 Umeda, T., Shirakawa, T., Fujii, A. and Yoshino, K. (2003) *Japanese Journal of Applied Physics, Part 2: Letters*, **42**, P L1475.

28 Shirakawa, T., Umeda, T., Hashimoto, Y., Fujii, A. and Yoshino, K. (2004) *Journal of Physics D: Applied Physics*, **37**, 847.

29 Olson, D.C., Piris, J., Collins, R.T., Shaheen, S.E. and Ginley, D.S. (2005) *Thin Solid Films*, **496**, 26.

30 White, M.S., Olson, D.C., Shaheen, S.E., Kopidakis, N. and Ginley, D.S. (2006) *Applied Physics Letters*, **89**, 143517.

31 Lira-Cantu, M. and Krebs, F. (2006) *Solar Energy Materials and Solar Cells*, **90**, 2076.

32 Olson, D.C., Shaheen, S.E., White, M.S., Mitchell, W.J., van Hest, M.F.A.M., Collins, R.T. and Ginley, D.S. (2007) *Advanced Functional Materials*, **17**, 264.

33 Savenije, T.J., Vermeulen, M.J.W., de Haas, M.P. and Warman, J.M. (2000) *Solar Energy Materials and Solar Cells*, **61**, 9.

34 Arango, A.C., Johnson, L.R., Bliznyuk, V.N., Schlesinger, Z., Carter, S.A. and Hörhold, H.H. (2000) *Advanced Materials*, **12**, 1689.

35 Breeze, A.J., Schlesinger, Z., Carter, S.A. and Brock, P.J. (2001) *Physical Review B: Condensed Matter*, **64**, 125205–125211.

36 Breeze, A.J., Schlesinger, Z., Carter, S.A., Tillmann, H. and Hörhold, H.H. (2004) *Solar Energy Materials and Solar Cells*, **83**, 263.

37 Fan, Q., McQuillin, B., Bradley, D.D.C., Whitelegg, S. and Seddon, A.B. (2001) *Chemical Physics Letters*, **347**, 325.
38 Wei, Q., Hirota, K., Tajima, K. and Hashimoto, K. (2006) *Chemistry of Materials*, **18**, 5080.
39 Slooff, L.H., Kroon, J.M., Loos, J., Koetse, M.M. and Sweelssen, J. (2005) *Advanced Functional Materials*, **15**, 689.
40 van der Zanden, B., Goossens, A. and Schoonman, J. (2001) *Synthetic Metals*, **121**, 1601.
41 van der Zanden, B. and Goossens, A. (2003) *Journal of Applied Physics*, **94**, 6959.
42 Gebeyehu, D., Brabec, C.J., Padinger, F., Fromherz, T., Spiekermann, S., Vlachopoulos, N., Kienberger, F., Schindler, H. and Sariciftci, N.S. (2001) *Synthetic Metals*, **121**, 1549.
43 Roberson, L.B., Poggi, M.A., Kowalik, J., Smestad, G.P., Bottomley, L.A. and Tolbert, L.M. (2004) *Coordination Chemistry Reviews*, **248**, 1491.
44 Grant, C.D., Swartzberg, A.M., Smestad, G.P., Kowalik, J., Tolbert, L.M. and Zhang, J.Z. (2003) *Synthetic Metals*, **132**, 197.
45 Grant, C.D., Schwartzberg, A.M., Smestad, G.P., Kowalik, J., Tolbert, L.M. and Zhang, J.Z. (2002) *Journal of Electroanalytical Chemistry*, **522**, 40.
46 Song, M.Y., Kim, J.K., Kim, K.-J. and Kim, D.Y. (2003) *Synthetic Metals*, **137**, 1389.
47 van der Zanden, B., van de Krol, R., Schoonman, J. and Goossens, A. (2004) *Applied Physics Letters*, **84**, 2539.
48 Dyreklev, P., Inganäs, O., Paloheimo, J. and Stubb, H. (1992) *Journal of Applied Physics*, **71**, 2816.
49 List, E.J.W., Kim, C.H., Naik, A.K., Scherf, U., Leising, G., Graupner, W. and Shinar, J. (2001) *Physical Review B: Condensed Matter*, **64**, 155204.
50 Scheblykin, I., Zoriniants, G., Hofkens, J., de Feyter, S., van der Auweraer, M. and de Schryver, F.C. (2003) *ChemPhysChem*, **4**, 260.
51 Yu, J., Song, N.W., McNeill, J.D. and Barbara, P.F. (2004) *Israel Journal of Chemistry*, **44**, 127.
52 Liu, J., Kadnikova, E.N., Liu, Y., McGehee, M.D. and Fréchet, J.M.J. (2004) *Journal of the American Chemical Society*, **126**, 9486.
53 Liu, Y., Scully, S.R., McGehee, M.D., Liu, J., Luscombe, C.K., Fréchet, J.M.J., Shaheen, S.E. and Ginley, D.S. (2006) *Journal of Physical Chemistry B*, **110**, 3257.
54 Goh, C., Scully, S.R. and McGehee, M.D. (2007) *Journal of Applied Physics*, **101**, 114503.
55 Liu, Y., Summers, M.A., Edder, C., Fréchet, J.M.J. and McGehee, M.D. (2005) *Advanced Materials*, **17**, 2960.
56 Liu, Y., Summers, M.A., Scully, S.R. and McGehee, M.D. (2006) *Journal of Applied Physics*, **99**, 093521.
57 Scully, S.R. and McGehee, M.D. (2006) *Journal of Applied Physics*, **100**, 034907.
58 Kroeze, J.E. and Savenije, T.J. (2004) *Thin Solid Films*, **451–452**, 54.
59 Olson, D.C., Shaheen, S.E., White, M.S., Mitchell, W.J., van Hest, M.F.A.M., Collins, R.T. and Ginley, D.S. (2007) *Advanced Functional Materials*, **17**, 264.
60 Coakley, K.M., Liu, Y., Goh, C. and McGehee, M.D. (2005) *MRS Bulletin*, **30**, 37.
61 van Hal, P.A., Christiaans, M.P.T., Wienk, M.M., Kroon, J.M. and Janssen, R.A.J. (1999) *Journal of Physical Chemistry B*, **103**, 4352.
62 Luzzati, S., Basso, M., Catellani, M., Brabec, C.J., Gebeyehu, D. and Sariciftci, N.S. (2002) *Thin Solid Films*, **403–404**, 52.
63 Anderson, N.A., Hao, E., Ai, X., Hastings, G. and Lian, T. (2001) *Chemical Physics Letters*, **347**, 304.
64 Arango, A.C., Carter, S.A. and Brock, P.J. (1999) *Applied Physics Letters*, **74**, 1698.
65 Gebeyehu, D., Brabec, C.J., Sariciftci, N.S., Vangeneugden, D., Kiebooms, R., Vanderzande, D., Kienberger, F. and Schindler, H. (2001) *Synthetic Metals*, **121**, 1549.
66 Spiekermann, S., Smestad, G., Kowalik, J., Tolbert, L.M. and Grätzel, M. (2001) *Synthetic Metals*, **121**, 1603.

67 Smestad, G.P., Spiekermann, S., Kowalik, J., Grant, C.D., Swartzberg, A.M., Zhang, J., Tolbert, L.M. and Moons, E. (2003) *Solar Energy Materials and Solar Cells*, **76**, 85.
68 Bach, U. (2003) *Encyclopedia of Electrochemistry*, **6**, 475.
69 Nogueira, A.F., Longo, C. and De Paoli, M.A. (2004) *Coordination Chemistry Reviews*, **248**, 1455.
70 van Hal, P.A., Wienk, M.M., Kroon, J.M. and Janssen, R.A.J. (2003) *Journal of Materials Chemistry*, **13**, 1054.
71 Senadreea, G.K.R., Nakamura, K., Kitamura, T., Wada, Y. and Yanagida, S. (2003) *Applied Physics Letters*, **83**, 5470.
72 Kim, Y.G., Walker, J., Samuelson, L.A. and Kumar, J. (2003) *Nano Letters*, **3**, 533.
73 Ravirajan, P., Haque, S.A., Poplavskyy, D., Durrant, J.R., Bradley, D.D.C. and Nelson, J. (2004) *Thin Solid Films*, **451–452**, 624.
74 Ravirajan, P., Haque, S.A., Durrant, J.R., Poplavskyy, D., Bradley, D.D.C. and Nelson, J. (2004) *Journal of Applied Physics*, **95**, 1473.
75 Song, M.Y., Kim, J.K., Kim, K.-J. and Kim, D.Y. (2003) *Synthetic Metals*, **137**, 1387.
76 Ravirajan, P., Haque, S.A., Durrant, J.R., Bradley, D.D.C. and Nelson, J. (2005) *Advanced Functional Materials*, **15**, 609.
77 Ravirajan, P., Bradley, D.D.C., Nelson, J., Haque, S.A., Durrant, J.R., Smit, H.J.P. and Kroon, J.M. (2005) *Applied Physics Letters*, **86**, 143101.
78 Bartholomew, G.O. and Heeger, A.J. (2005) *Advanced Functional Materials*, **15**, 677.
79 Huisman, C.L., Huijser, H., Donker, H., Schoonman, J. and Goossens, A. (2004) *Macromolecules*, **37**, 5557.
80 Kajihara, K., Tanaka, K., Hirao, K. and Soga, N. (1997) *Japanese Journal of Applied Physics*, **36**, 5537.
81 Coakley, K. and McGehee, M.D. (2003) *Applied Physics Letters*, **83**, 3380.
82 Coakley, K., Liu, Y., McGehee, M.D., Frindell, K.L. and Stucky, G.D. (2003) *Advanced Functional Materials*, **13**, 301.
83 Wang, H., Oey, C.C., Djurišić, A.B., Xie, M.H., Leung, Y.H., Man, K.K.Y., Chan, W.K., Pandey, A., Nunzi, J.-M. and Chui, P.C. (2005) *Applied Physics Letters*, **87**, 023507.
84 Oey, C.C., Djurišić, A.B., Wang, H., Man, K.K.Y., Chan, W.K., Xie, M.H., Leung, Y.H., Pandey, A., Nunzi, J.M. and Chui, P.C. (2006) *Nanotechnology*, **17**, 706.
85 Salafsky, J.S. (2001) *Solid-State Electronics*, **45**, 53.
86 Kannan, B., Castelino, K. and Majaumdar, A. (2003) *Nano Letters*, **3**, 1729.
87 Wang, Z.L. (2004) *Materials Today*, **7**, 26.
88 Greene, L.E., Law, M., Tan, D.H., Montano, M., Goldberger, J., Somorjai, G. and Yang, P. (2005) *Nano Letters*, **5**, 1231.
89 Ravirajan, P., Peiró, A.M., Nazeeruddin, M.K., Grätzel, M., Bradley, D.D.C., Durrant, J.R. and Nelson, J. (2006) *Journal of Physical Chemistry B*, **110**, 7635.
90 Peiró, A.M., Ravirajan, P., Govender, K., Boyle, D.S., O'Brien, P., Bradley, D.D.C., Nelson, J. and Durrant, J.R. (2006) *Journal of Materials Chemistry*, **16**, 2088.
91 Milliron, D.J., Gur, I. and Alivisatos, A.P. (2005) *MRS Bulletin*, **30**, 41.
92 Greenham, N.C., Peng, X. and Alivisatos, A.P. (1996) *Physical Review B: Condensed Matter*, **54**, 17628.
93 Greenham, N.C., Peng, X. and Alivisatos, A.P. (1997) *Synthetic Metals*, **84**, 545.
94 Huynh, W.U., Peng, X. and Alivisatos, A.P. (1999) *Advanced Materials*, **11**, 923.
95 Liu, J., Tanaka, T., Sivula, K., Alivisatos, A.P. and Fréchet, J.M.J. (2004) *Journal of the American Chemical Society*, **126**, 6550.
96 Sun, B., Marx, E. and Greenham, N.C. (2003) *Nano Letters*, **3**, 961.
97 Sun, B., Snaith, H.J., Dhoot, A.S., Westenhoff, S. and Greenham, N.C. (2005) *Journal of Applied Physics*, **97**, 014941.
98 Wang, P., Abrusci, A., Wong, H.M.P., Svensson, M., Andersson, M.R. and Greenham, N.C. (2006) *Nano Letters*, **6**, 1789.
99 Sun, B. and Greenham, N.C. (2006) *Physical Chemistry Chemical Physics*, **8**, 3557.

100 Salafsky, J.S., Lubberhuizen, W.H. and Schropp, R.E.I. (1998) *Chemical Physics Letters*, **290**, 297.

101 Salafsky, J.S. (1999) *Physical Review B: Condensed Matter*, **59**, 10885.

102 Zhang, J., Wang, B., Ju, X., Liu, T. and Hu, T. (2001) *Polymer*, **42**, 3697.

103 Carter, S.A., Scott, J.C. and Brock, P.J. (1997) *Applied Physics Letters*, **71**, 1145.

104 Kwong, C.Y., Djurišić, A.B., Chui, P.C., Cheng, K.W. and Chan, W.K. (2004) *Chemical Physics Letters*, **384**, 372.

105 Kwong, C.Y., Choy, W.C.H., Djurišić, A.B., Chui, P.C., Cheng, K.W. and Chan, W.K. (2004) *Nanotechnology*, **15**, 1156.

106 Petrella, A., Tamborra, M., Cozzoli, P.D., Curri, M.L., Striccoli, M., Cosma, P., Farinola, G.M., Babudri, N.F. and Agostiano, A. (2004) *Thin Solid Films*, **451–452**, 64.

107 Petrella, A., Tamborra, M., Curri, M.L., Cosma, P., Striccoli, M., Cozzoli, P.D. and Agostiano, A. (2005) *Journal of Physical Chemistry B*, **109**, 1554.

108 Huisman, C.L., Goossens, A. and Schoonman, J. (2003) *Synthetic Metals*, **138**, 237.

109 Huisman, C.L., Goossens, A. and Schoonman, J. (2003) *Chemistry of Materials*, **15**, 4617.

110 Beek, W.J.E., Wienk, M.M. and Janssen, R.A.J. (2004) *Advanced Materials*, **16**, 1009.

111 Meulenkamp, E.A. (1998) *Journal of Physical Chemistry B*, **102**, 5566.

112 Hagfeldt, A. and Grätzel, M. (1995) *Chemical Reviews*, **95**, 49.

113 Hutson, A.R. (1957) *Physical Review*, **108**, 222.

114 Hunger, R., Iwata, K., Fons, P., Yamada, A., Matsubara, K., Niki, S., Nakahara, K. and Takasu, H. (2001) *Materials Research Society Symposium Proceedings*, **668**, H2.8.1.

115 Beek, W.J.E., Wienk, M.M., Kemerink, M., Yang, X. and Janssen, R.A.J. (2005) *Journal of Physical Chemistry B*, **109**, 9505.

116 Wei, X., Vardeny, Z.V., Sariciftci, N.S. and Heeger, A.J. (1996) *Physical Review B: Condensed Matter*, **53**, 2187.

117 Shim, M. and Guyot-Sionnest, P. (2001) *Journal of the American Chemical Society*, **123**, 11651.

118 Bauer, C., Boschloo, G., Mukhtar, E. and Hagfeldt, A. (2001) *Journal of Physical Chemistry B*, **105**, 5585.

119 Beek, W.J.E., Wienk, M.M. and Janssen, R.A.J. (2006) *Advanced Functional Materials*, **16**, 1112.

120 Quist, P.A.C., Beek, W.J.E., Wienk, M.M., Janssen, R.A.J., Savenije, T. and Siebbeles, L.D.A. (2006) *Journal of Physical Chemistry B*, **110**, 10315.

121 Koster, L.J.A., van Strien, W.J., Beek, W.J.E. and Blom, P.W.M. (2007) *Advanced Functional Materials*, **17**, 1297.

122 Bao, Z., Dodabalapur, A. and Lovinger, A. (1996) *Applied Physics Letters*, **69**, 4108.

123 Sirringhaus, H., Brown, P.J., Friend, R.H., Nielsen, M.M., Bechgaard, K., Langeveld-Voss, B.M.W., Spiering, A.J.H., Janssen, R.A.J., Meijer, E.W., Herwig, P.T. and de Leeuw, D.M. (1999) *Nature*, **401**, 685.

124 Yang, X., Veenstra, S.C., Verhees, W.J.H., Wienk, M.M., Janssen, R.A.J., Kroon, J.M., Michels, M.A.J. and Loos, J. (2005) *Nano Letters*, **5**, 579.

125 Mihailetchi, V.D., Xie, H., De Boer, B., Popescu, L.M., Hummelen, J.C., Blom, P.W.M. and Koster, L.J.A. (2006) *Applied Physics Letters*, **89**, 012107.

126 Wong, H.M.P., Wang, P., Abrusci, A., Svensson, M., Andersson, M.R. and Greenham, N.C. (2007) *Journal of Physical Chemistry C*, **111**, 5244.

127 van Duren, J.K.J., Yang, X.N., Loos, J., Bulle-Lieuwma, C.W.T., Sieval, A.B., Hummelen, J.C. and Janssen, R.A.J. (2004) *Advanced Functional Materials*, **14**, 425.

128 van Hal, P.A., Wienk, M.M., Kroon, J.M., Verhees, W.J.H., Slooff, L.H., van Gennip, W.J.H., Jonkheijm, P. and Janssen, R.A.J. (2003) *Advanced Materials*, **15**, 118.

129 Slooff, L.H., Wienk, M.M. and Kroon, J.M. (2004) *Thin Solid Films*, **451–452**, 634.

130 Wang, L., Ji, J.S., Lin, Y.J. and Rwei, S.P. (2005) *Synthetic Metals*, **155**, 677.

131 Okuya, M., Nakade, K. and Kaneko, S. (2002) *Solar Energy Materials and Solar Cells*, **70**, 425.
132 Beek, W.J.E., Slooff, L.H., Wienk, M.M., Kroon, J.M. and Janssen, R.A.J. (2005) *Advanced Functional Materials*, **15**, 1703.
133 Albertsson, J., Abrahams, S.C. and Kvick, A. (1983) *Acta Crystallographica Section B*, **45**, 34.
134 Quist, P.A., Slooff, L.H., Donker, H., Kroon, J.M., Savenije, T.J. and Siebbeles, L.D.A. (2005) *Superlattices and Microstructures*, **38**, 308.
135 Moet, D.J.D., Koster, L.J.A., de Boer, B. and Blom, P.W.M. (2007) *Chemistry of Materials*, **19**, 5856.
136 Hänsel, H., Zettl, H., Krausch, G., Kisselev, R., Thelakkat, M. and Schmidt, H.W. (2003) *Advanced Materials*, **15**, 2056.
137 Kim, J.Y., Kim, S.H., Lee, H.H., Lee, K., Ma, W., Gong, X. and Heeger, A.J. (2006) *Advanced Materials*, **18**, 572.
138 Gilot, J., Wienk, M.M. and Janssen, R.A.J. (2007) *Applied Physics Letters*, **90**, 143512.
139 Kim, J.Y., Lee, K., Coates, N.E., Moses, D., Nguyen, T.Q., Dante, M. and Heeger, A.J. (2007) *Science*, **317**, 222.

III
Technology

A
Electrodes

14
High-Performance Electrodes for Organic Photovoltaics
Cecilia Guillén and José Herrero

14.1
Introduction

The electrodes or contacts are important components in any photosensitive optoelectronic device. The term electrode or contact refers to layers that provide a medium for delivering photogenerated power to an external circuit or providing a bias voltage to the device. Thus, the electrodes or contacts provide the interface between the photoactive regions of the optoelectronic device and a wire, lead or other means for transporting the charge carriers to or from the external circuit. Sometimes in the term are also included other charge transfer layers that deliver charge carriers from one subsection of the device to the adjacent subsection. An ideal electrode must be highly conductive, stable chemically and electrochemically, and should not cause the degradation of the active materials contacted in the device. In addition, in a photovoltaic cell, it is desirable to allow the maximum amount of ambient solar radiation from the exterior to be admitted to the photoactive interior region. This indicates that at least one of the electrical contacts should be minimally absorbing and minimally reflecting of the incident solar radiation; namely, it should be a transparent conductive electrode.

The electrodes or contacts used for the polymer, molecular, and dye-sensitized solar cells that have shown the highest efficiencies [1–3] are typical metal or metal oxide layers, adding sometimes conducting polymer films in combination with the inorganic electrode materials or substituting them [2,4]. Virtually, all organic solar cells fabricated today use a conductive metal layer (usually Al, Ag, or Au) as rear electrode, and a conductive transparent metal oxide film to provide the frontal electrode exposed to the solar radiation. The most widely used conductive transparent metal oxides are indium-tin oxide (ITO) (In_2O_3:Sn) and fluorine-doped tin oxide (FTO) (SnO_2:F), which are highly doped degenerate semiconductors with optical bandgap values above 3 eV that render them transparent to wavelengths greater than approximately 400 nm [5]. Conducting polymers are especially attractive for application in all-organic devices. One such polymer that has been studied extensively

Organic Photovoltaics: Materials, Device Physics, and Manufacturing Technologies.
Edited by Christoph Brabec, Vladimir Dyakonov, and Ullrich Scherf
Copyright © 2008 WILEY-VCH Verlag GmbH & Co. KGaA, Weinheim
ISBN: 978-3-527-31675-5

is poly(3,4-ethylenedioxythiophene) (PEDOT), which has an optical bandgap of around 1.7 eV, giving a transmissive sky-blue color in the oxidized state [6].

It is well known that cell series resistance is an important factor in improved photovoltaic performance and design. Reduction of its magnitude is necessary to continuous device improvement, especially for devices with scaling-up areas and/or exposed to high light intensities. In general, high conductive electrodes are needed to reduce the series resistance, but the optimal electrode resistance value depends on the specific device materials and their configuration. For metal-free phthalocyanine solar cells with small areas and low illumination levels, the series resistance has been found increasing and the power conversion efficiency decreasing as the In electrode thickness was decreased below a critical value around 70 nm, but no further performance improvement was achieved by increasing the In electrode thickness beyond when the organic semiconductor bulk resistivity became the dominant contributor [7]. Otherwise, for double-heterostructure copper-phthalocyanine/C_{60} photovoltaic cells, the dependence of the series resistance on the device area suggests the dominance of the bulk resistance of the ITO electrode as a limiting factor in practical cell efficiencies at areas above 0.01 cm^2 [8].

Besides its electrical resistance, the choice of the best electrode material is determined by its work function and the organic photovoltaic device configuration, because it can be desirable for the contact interface to contribute to the net charge separating action or, contrarily, it presents the smallest possible barrier to carrier transport. Thus, for the basic Schottky-type cell, constituted by a single organic photoconductive material sandwiched between a pair of contacts, a high work function material (Pt, Au, ITO, or PEDOT) is commonly used for n-type photoconductors and a metal with a low work function (Ag, Al, or In) is used for p-type photoconductors. In this way, a rectifying junction is built at the electrode–photoconductor interface [9,10]. Although it is possible for charge separation to occur at both interfaces and be additive, it is generally preferable to have all charge separation occurring at one interface. This has been achieved by choosing for the other electrode a metal providing little or no barrier to carrier transport [10,11]. For these Schottky-type cells, the open-circuit photovoltage is attributed to the barrier height at the electrode–photoconductor interface [9] or the difference in work function of the two electrode materials [11], depending on the depletion width extension throughout the photoactive material thickness. In any case, the adjustment of the electrode work function is needed to improve the solar cell performance.

In organic bilayer heterojunction devices, made by electron and hole accepting polymers, a scaling of the open-circuit voltage with electrode work function difference has also been observed, with an additional contribution from the active layer within the device depending on light intensity [12]. In organic bulk heterojunction devices, prepared by blending electron donor and acceptor, a linear correlation of the open-circuit voltage with the reduction potential of the acceptor or with the oxidation potential of the donor has been reported for fullerene derivatives with varying acceptor strength [13] and for different donor polythiophene-conjugated polymers [14]. The maximum photovoltage for these cases is governed by the bulk material properties. Such correlation is expected in the case of nonrectifying or ohmic contacts, when the interfaces at the electrodes have the smallest possible barrier. This is usually achieved by using a high work function material for the hole-collecting contact (positive electrode) and a low work function material for the electron-collecting contact (negative electrode). In the case of nonohmic contacts, when the interface at the electrodes has a significant barrier, the open-circuit voltage can also be modified by varying the electrode work function. Thus, for polymer:fullerene bulk heterojunction solar cells, a total variation of more than 0.5 V of the open-circuit voltage was observed by variation of the negative electrode work function [15]. The open-circuit voltage increases linearly until pinning of the electrode Fermi level close to the reduction potential of the fullerene is reached [13,15].

Attention has mainly been focused on the effect of the electrode work function value on the open-circuit voltage of the cells. However, for organic bulk heterojunction photovoltaic devices, it has been observed that a decrease in the work function of the metal top electrode leads to an enhancement not only of the open-circuit voltage, but also of the short-circuit current, and the power conversion efficiency [15,16]. It has been demonstrated that the photocurrent obtained from a polymer:fullerene bulk heterojunction with different metal electrodes shows a universal behavior when scaled against the effective voltage across the device [16]. Indeed, model calculations confirm that the dependence of the photocurrent on the effective voltage is responsible for the observed variation in performance of each different electrode. Consequently, by selecting the most suitable electrode material for each specific organic cell, the open-circuit voltage can be maximized and the efficiency will be enhanced accordingly.

It is important to note that interaction between the organic semiconductor and the electrode can also affect strongly the photovoltaic device characteristics, which can be related to the presence of interface dipoles that generate an electrostatic potential close to the electrodes. Such interface dipoles have a deep impact on the built-in potential and consequently on the open-circuit voltage of organic composite photovoltaic cells [17]. This implies that the choice of contact material for optimal device performance should be based on a consideration of the specific semiconductor–electrode interaction.

In the following sections, those electrode material properties and interactions that can be interesting in relation with their application to organic photovoltaic devices will be analyzed separately for the metal, metal oxide, and conducting polymers typically utilized. Finally, the application of multilayer electrodes developed by combining different materials will be analyzed.

14.2
Metal Electrodes

14.2.1
Metal Properties

The work function value is a fundamental property to take into account for the metal electrode selection, as has already been pointed out in Section 14.1. For monocrystalline metals, dependence of the work function upon crystal face has been studied in detail and it has been proved that the work function varies from one crystal face to another owing to changes in the effective mass of the electrons when the direction of motion varies [18]. This can also apply to polycrystalline structures, where any stress or change in the size of the lattice unit cell can give rise to a change in the material work function [18]. Besides, the metal surface conditions (roughness, oxidation, or atomic species adsorption) can substantially affect the value of the work function [18,19].

The metal electrode should be selected according to the solar cell configuration and the energy band diagram corresponding to the active materials involved. Figure 14.1 represents the energy diagram for typical donor and acceptor semiconductors (Figure 14.1a) and for the most commonly used metal electrodes (Figure 14.1b), for which average work function values have been taken from the experimental data reported for different polycrystalline and single-crystal specimens [20]. For the adjustment of the electrode work function and other properties, metal alloys have also been investigated. Among the various kinds of alloys acting as negative electrodes, Mg:Ag alloy thin films have been extensively used because of their low tunneling barrier and low work function [21,22].

Moreover, the electrode or contact should provide a minimum electrical resistance. The electrical resistivity of several pure metals is given in Table 14.1. For metal thin films, different resistivity values have been found, depending on the film thickness and other preparation conditions, which are generally superior to the value of the electrical resistivity of the pure metal in bulk [23–26]. Such higher thin-film resistivity is because of increased surface scattering, as the film thickness becomes smaller than the intrinsic electron mean free path (this is known as the size effect) and scattering owing to a distribution of planar potentials (the grain boundaries effect) that depend on the thickness and other experimental preparation conditions such as the film

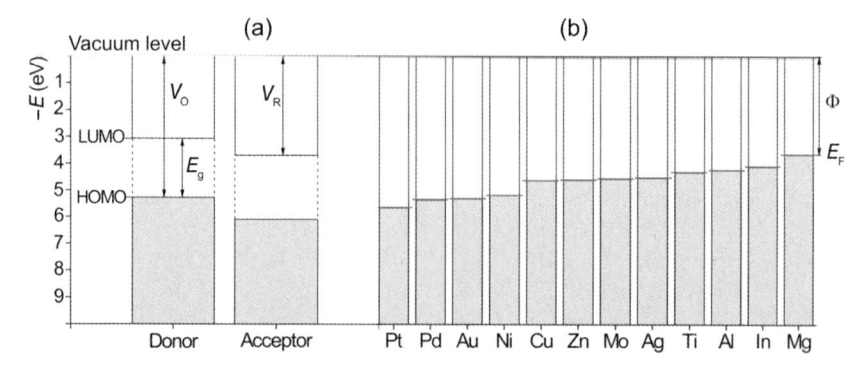

Figure 14.1 Energy band diagram corresponding to (a) typical donor or acceptor organic semiconductors and (b) several metal electrodes. Here LUMO refers to the lowest unoccupied molecular orbital, HOMO to the highest occupied molecular orbital, E_g is the bandgap energy, V_O the oxidation potential of the donor, V_R the reduction potential of the acceptor, E_F is the Fermi level energy, and Φ the metal work function for which average values have been calculated from the reported experimental data [20].

Table 14.1 Electrical resistivity values reported for several bulk metals [20].

Metal	Resistivity ($\mu\Omega$ cm) at 20 °C
Ag	1.59
Cu	1.72
Au	2.44
Al	2.82
Mg	4.6
W	5.6
Mo	5.7
Zn	5.8
Ni	7.8
In	8.0
Pt	10
Pd	11
Sn	11.5
Cr	12.6
Ta	15.5
Ti	39

deposition rate or the growth temperature [27]. In general, the resistivity of metal thin films approaches close to the bulk metal value for film thickness above 100 nm [24,26], whereas the film optical transmittance decreases and the reflectance increases with such thickness increment [23,24].

Highly conducting Ag and Al layers with 80–100 nm thickness are commonly used as rear electrodes in organic solar cells [2,8,16]. When the active layer thicknesses are comparable with the penetration depth of the incident light, the interference effects because of the superposition of the incident light with that backreflected from the rear electrode can improve the optical field intensity within the cell [2,28,29]. For photovoltaic devices based on the CuPc/C_{60} heterojunction, the presence of a highly reflecting Al electrode causes strong optical interference effects, increasing the field intensity, and hence absorption, in the CuPc layer by a factor of ~4 [2]. The optimal device layer structure was determined by pushing the CuPc/C_{60} interface into regions of high optical field intensity. Taking into account that the maximum optical field intensity is placed at a distance $\sim\lambda/(4n)$ from the metal electrode, an organic material with a large index of refraction (n) over the visible wavelength range (400 nm < λ < 800 nm) is needed to allow acceptor thickness reduction making it possible to take advantage of the interference effect to improve the device efficiency.

When semitransparent frontal electrodes are required, the metal film thickness should be reduced. Metallic films that are sufficiently thin compared to the wavelength of light can be optically transparent; and if the film is continuous on the nanoscale over macroscopic distances, then the electrical resistivity of the film can remain usefully low as well. There is an increasing interest in the development of ultrathin metallic films, with very low surface roughness, that act like more efficient electrodes in photovoltaic cells and in other optoelectronic devices. In this sense, conductive metal films with thickness as low as 2 nm, providing optical transmittance as large as 80%, have been obtained by several techniques such as filtered vacuum arc plasma deposition [23] or ion-beam sputtering [24]. Such ultrathin transparent metal contacts can provide large optical transmittance in the ultraviolet part of the spectrum, providing the potential for increased photovoltaic conversion efficiency and can be deposited at room temperature without heating, allowing fabrication on low-cost, lightweight, and flexible polymeric substrates [30,31].

Below a threshold value of the film thickness that is highly dependent on the deposition parameters and the substrate characteristics, the film becomes discontinuous. It has been shown experimentally [30] that the optical absorption in discontinuous metal layers is very high, the maximum of the absorption coinciding with the percolation threshold. When the electrons are trapped in individual islands and cannot move freely, there is an increase in the possibility for the excitation of surface plasmons by the incoming radiation and thus the absorption. At the percolation threshold, the maximum number of metal atoms in isolated islands is present and therefore the maximum of absorption takes place. Significant enhancement of the photovoltaic conversion efficiency of Schottky-type organic devices by the incorporation of small metal clusters has been reported [32,33]. The enhancement is explained in terms of resonant light absorption in the metal cluster, which is accompanied by a strengthened electric field in the vicinity of the particle, thus enhancing absorption of the organic dye film [32]. Besides, it has been shown that an excited plasmon in a metal cluster is also capable to emit an electron directly in a preferential direction if the particles are placed inside an oriented electrical field like the one existing in the depletion layer of a Schottky junction. As a result a primary photocurrent is observed in a spectral region without any direct absorption in the organic film [33].

In stacked tandem photovoltaic devices, the use of discontinuous metal clusters for connecting the heterojunction bilayer subcells has allowed to improve the efficiency over that achieved by using continuous semitransparent metal contact layers [34]. In the operation of a dual-heterojunction device, a balance of the photoresponse of each subcell in the stack is essential to achieve high efficiency. Ultrathin Ag layers (~0.5 nm) deposited as nanoscale clusters have been able to provide efficient recombination sites for the unpaired charges that are photogenerated in the interior of the device, being used as efficient internal floating electrodes for the fabrication of multiple stacked organic cells with increased open-circuit voltage and conversion efficiency [34,35].

14.2.2
Metal/Organic Semiconductor Interactions

In metal/organic thin-film structures, two types of interfaces can be obtained depending on the preparation procedure: the metal-on-organic interface and the organic-on-metal interface. When Al atoms [36], as well as other type of metals [37], are deposited on conjugated materials, the formation of covalent bonds between metal and carbon atoms changes the hybridization of those carbon atoms from sp2 to sp3, and thus, the π-conjugation is broken at these sites. The metal–C bond leads to a redistribution of the charge density. Often, metal atoms become positively charged and electron density is transferred to the organic part; consequently, interface dipoles are formed at the metal-on-organic interface. Besides, the crystalline structure of metal layers is found to strongly depend on the underlying organic substrate. Correspondingly, work functions are found to be different by more than 200 meV in agreement with the crystalline orientation of the metal films [38]. During metal deposition, a significant diffusion into the organic layer has also been observed, which strongly depends on the metal and the order of deposition [39]. Thus, it has been found that Mg doping occurs during the deposition of Mg top electrodes onto C_{60} films and that the conversion efficiency of Schottky-type [40] and heterojunction [41] solar cells is markedly improved by the doping. The large enhancement of the short-circuit current density is attributed to a reduction of series resistance, which is mainly a reduction of the bulk resistance of C_{60} film by Mg doping [41].

Other studies [42–44] have demonstrated that an interface dipole can also appear when depositing the organic material on a metal surface, which can modify the energy of the electronic levels in the organic material and the work function of the metal surface. It has been found that passivating the metal surfaces with oxygen prevents the formation of strong metal–C bond [45]. Besides, several results clearly show the importance of atmosphere for interface formation and related electrical properties. Band bending leading to Fermi level alignment is not achieved for some metal/organic systems prepared and measured under ultrahigh vacuum, whereas it may be obtained under ambient atmosphere, possibly owing to the modification of the interface or doping into the organic layer [39]. In Schottky-type merocyanine solar cells, higher efficiencies have been achieved when there is a very thin interfacial oxide layer between the barrier forming metal and merocyanine [46]. In addition, the electrical properties of phthalocyanine thin films are known to critically depend upon the ambient conditions and the oxygen adsorption is responsible for lower hole mobility and higher charge carrier concentrations [47]. Thus, the exposure of the metal/organic system to an oxidizing atmosphere can modify the conductivity of the organic material and/or create a metal oxide layer at the interface. Both factors should be taking into account to determine the rectification properties of the device [48].

It is interesting to note that the interface dipole created by chemisorbing molecular species on the metal surface has been proposed as a route to control the charge injection in organic-based optoelectronic devices. With molecules carrying a high dipole moment oriented perpendicular to the metal surface, it has been observed that the total interface dipole can be rather easily tuned by changing chemically the magnitude of the molecular dipole moment [49]. The direction of charge transfer can be estimated from the properties of the isolated components (the bare metal and the isolated molecule) that are accessible experimentally: the work function of the bare metal and the ionization potential and electron affinity of the molecule [43]. This liberates the device builder from the limitation of contact material choice based entirely on substrate work function, as the interaction can be modified by the choice of organic functional groups and/or substrate surface composition. Figure 14.2 illustrates the influence of interface dipole formation on the energy band diagram.

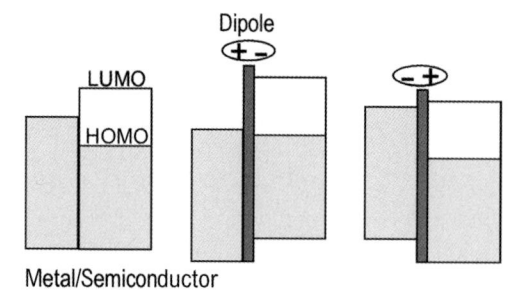

Figure 14.2 Sketch of the impact of the formation of an interface dipole on the electronic levels at the metal/organic semiconductor interface.

One important degradation process for organic solar cells involves oxygen incorporation and its diffusion during the cell operation [50–52]. Conjugated polymers are particularly susceptible to photodegradation induced by oxygen and moisture [50]. Moreover, the metals typically chosen for the low work function electrode are Al and Ca, which maximize the open-circuit voltage of the solar cells but rapidly undergo oxidation when exposed to air. Mixing of conjugated polymers with fullerenes can decrease the degradation effect in the dark air storage, but illumination accelerates the oxidation/degradation processes [50–52]. In photovoltaic devices based on conjugated polymer materials with Al electrodes, it has been demonstrated that oxygen species derived from the atmosphere diffuse through aluminum electrode during illumination and react with the constituents of the active layers, giving rise to increased oxygen content throughout the active region, especially near the aluminum interface [53]. Besides, aluminum species diffuse into the organic layer and react with molecular oxygen. This indicates that two simultaneous degradation mechanisms take place: the direct photooxidation of the active layer constituents and photoreduction of the active layer constituents and subsequent oxidation. Thus, to increase the lifetime of electrodes in organic devices two means are proposed: the preparation of oxygen-tight electrodes and the introduction of a barrier layer that does not react or photoreduce in the presence of aluminum while allowing for the transport of electrons across the barrier.

Recent progresses on high-performance organic solar cells include a strategy of incorporating a very thin layer of LiF (<2 nm) at the interface between the photoactive layer and the aluminum electrode, thus allowing to increase the power conversion efficiency [54,55]. Although LiF/Al electrodes are already widely used for enhancing the efficiency of organic light emitting diodes, the underlying mechanisms are still under investigation. Several factors have been suggested thus far, including the protection of the organic layer during Al deposition and dissociation of the LiF giving a low work function contact [56] or doping the organic layer [57]. For organic solar cells, the formation of a dipole moment across the junction, owing either to the orientation of LiF or chemical reactions leading to a charge transfer across the interface, is suggested as an efficiency enhancement mechanism [54]. The influence of LiF on the photovoltaic performance is larger in devices fabricated with organic

14.3
Metal Oxide Electrodes

14.3.1
Metal Oxide Properties

Thin metal oxide films that are both transparent and electrically conductive have been studied extensively as a result of their wide range of technological applications, particularly as electrodes for photovoltaic devices [58]. These transparent and conductive oxides (TCO) are wide-bandgap (in general >3 eV) semiconductors that show metal-like electrical characteristics when doped to high levels to become degenerate. Historically, most research to develop TCO thin films as transparent electrodes has been conducted using two basic families of n-type semiconductors: metal oxides with tetrahedrally coordinated cations (as ZnO) and metal oxides with octahedrally coordinated cations (CdO, In_2O_3, SnO_2) [59], including the related binary and ternary compounds. The recent discoveries of p-type oxides with linearly coordinated cations ($CuAlO_2$, Cu_2SrO_2, and related compounds) [60] and UV light-activated n-type oxides with cage structures ($12CaO \cdot 7Al_2O_3$) [61] represent enabling materials for novel technological applications, but are still not suitable for use as practical transparent electrodes. The energy diagrams corresponding to several binary and ternary metal oxides now available for thin-film transparent electrodes are shown in Figure 14.3, for which experimental data of the intrinsic bandgap energy and work function parameters were taken from the literature [62]. The multicomponent oxides composed of combinations of binary and/or ternary compounds allow a fine adjustment of their properties by altering the components' proportion [63]. It should be noted that the values showed

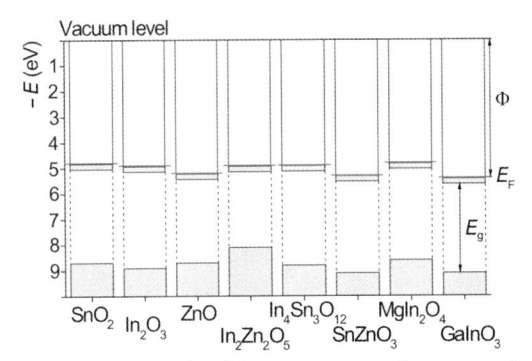

Figure 14.3 Energy band diagram corresponding to several metal oxide electrodes, where the energy gap (E_g) and the work function (Φ) have been represented according to the reported experimental data [62].

in Figure 14.3 are only approximate because for degenerate semiconductors the bandgap energy increases and the work function decreases with increasing carrier concentration. This is owing to the Fermi energy variation and the bandgap widening by the Burstein–Moss effect, a blocking of the lowest states of the conduction band by excess electrons [58]. Significant work function variations have also been observed depending on the surface cleaning method [64,65].

One advantage of using binary compounds, as compared to using ternary or multicomponent oxides, is the relative ease of controlling the chemical composition in film deposition. Compounds most extensively used for practical applications are ZnO, In_2O_3, and SnO_2. In each case, there are studies both of the pure oxides, where the conduction electrons result from oxygen vacancies acting as donors, and of mixed oxides such as ITO, FTO, or ZnO:Al (aluminum-doped zinc oxide (AZO)), where the metal atom of higher valency can form a donor state by substituting an atom of the host metal. ITO thin films with thicknesses in the range of a few hundred nanometers are typically used as transparent conducting electrodes for high-performance organic solar cells [2,8,29,34,35], although FTO and AZO layers with similar characteristics have also been used with good results [3,66–68]. For these and other applications, it is interesting to note that high-quality metal oxide electrodes can be obtained at low-preparation temperatures, which allow minimizing interdiffusion processes on previously deposited materials and utilizing plastics for replacing the conventional glass substrates [69,70].

About the electrical conductivity of the primary TCOs, an intrinsic limit has been established [71]. To preserve charge neutrality, there must be a sufficient number of positive charges to balance out the negatively charged free electrons. There are the ionized donor impurities from which the free electrons were originally produced. The Coulomb interaction between these impurities and the free electrons provides a source of scattering that is intrinsic to the doped material, which therefore sets a lower limit to the resistivity that can be achieved, regardless of scattering mechanisms as neutral impurities, grain boundaries or other forms of structural disorders depend on the precise details of the preparation procedure. The resistivity because of the ionized impurity scattering has been calculated to give a lower limit to the attainable resistivity [71] and is represented in Figure 14.4 together with some experimental data for ITO thin films in different processing conditions [72].

In the TCO layers, carrier concentration increase leads not only to a lower resistivity but also to a degradation of the optical transmission. This occurs because the free electrons reflect radiation below the plasma frequency and this frequency is an increasing function of the carrier concentration. Consequently, when the electron concentration is increased, the metal oxide film begins to reflect in the visible region as well as in the infrared. Thus, the requirement for visible transparency restricts the carrier concentration in TCOs to $\sim 2 \times 10^{21}$ cm^{-3} and the conductivity limit to $\sim 2.5 \times 10^4$ S cm^{-1} [71]. In practice, commercially available FTO or ITO does not exceed a conductivity of 10^4 S cm^{-1}. However, processing techniques have been refined in the labs to improve crystal quality, thus limiting the mobility losses caused by defects and grain boundaries. These scattering mechanisms can be eliminated in the limit of a perfect single crystal. Heteroepitaxial ITO on yttria-stabilized zirconia

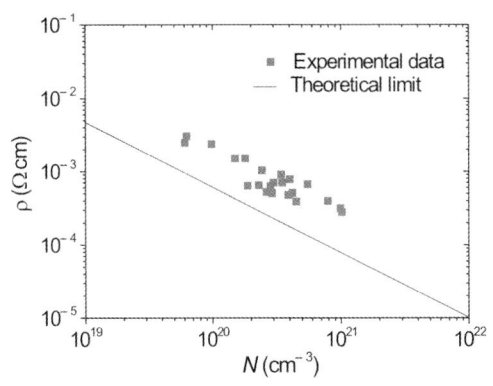

Figure 14.4 Representation of the electrical resistivity as a function of the carrier concentration for metal oxide electrodes. The linear plot is the lower limit imposed by the ionized impurity scattering [71]. The experimental data are unpublished results corresponding to ITO thin films at different processing conditions [72].

showed electrical conductivity of 1.3×10^4 S cm^{-1} that was in good agreement with the predicted performance limit at the measured carrier concentration of 1.9×10^{21} cm^{-3}, with a mobility of 42 cm^2 (V s)$^{-1}$ and transmittance in the visible range above 85% [73]. The ordered microstructure of ITO thin films crystallized under the action of a temperature gradient allowed to achieve even higher conductivity of 2.3×10^4 S cm^{-1}, with a carrier concentration of 1.4×10^{21} cm^{-3} and mobility of 103 cm^2 (V s)$^{-1}$ [74]. Such high mobility values allow to increase the conductivity without sacrificing the transparency.

Regarding the chemical stability, AZO films are more easily etched by both acid and alkaline solutions than ITO films [75]. This fact makes difficult to apply to AZO the presently used photolithography process with wet treatments. On the contrary, for applications involving reducing atmospheres at high temperatures and hydrogen plasma, AZO film properties remain more stable than those of ITO and TO films. Hydrogen treatments for ITO and TO films led to chemical reduction with coloration and then vaporization of the corresponding metals, whereas AZO films exhibited no coloring because of the simultaneous development of reduction and vaporization [76]. In contrast, for use in oxidizing atmospheres at high temperatures, ITO and TO films are preferred to AZO layers, because Zn is more chemically active in an oxidizing atmosphere than either Sn or In, and controlling the Zn oxidation is much more difficult [63]. Additionally, ITO thin films have shown excellent resistance to damp-heat conditions [77].

14.3.2
Metal Oxide/Organic Semiconductor Interactions

Although metal oxide electrodes (ITO, AZO, FTO) are the most widely used transparent conductors in the photovoltaic industry, there are some unique requirements from the viewpoint of organic solar cells that differ from photovoltaic devices based on inorganic semiconductors. Several problems related to the metal oxide/organic layer system have been pointed out, especially the adjustment of the work function, passivation of defects, and smoothening of the surface. It has been found that the surfaces of these polar hydrophilic oxides are often chemically incompatible with nonpolar organic thin films, leading to delamination of the organic layers and high series resistances [78]. Besides, the film roughness is very important when the electrode is used as substrate, because its surface morphology is directly transferred to the functional organic layers and uneven interface is not desirable for the efficiency and stability of the solar cell [79]. Organic device performance has been found to strongly depend on the surface roughness of the structure, which can be modified by changing the cell configuration or the preparation sequence [79,80].

Strategies have been developed to increase the wettability of the metal oxide surface by the nonpolar organic thin films, to tune the effective work function of this material through chemisorption or covalent attachment of polar organic functional groups, and to increase the electrode stability through the introduction of buffer layers at the metal oxide surface [81–83]. The efficiency of polymer-based photovoltaic devices is greatly improved when a buffer layer is introduced between the ITO electrode and the active polymer layer. Conducting polymers (PEDOT:PSS (polystyrenesulfonic acid)) as well as transition metal oxides (V_2O_5, MoO_3) have been demonstrated as efficient buffer layer materials, not only preventing unwanted chemical reaction or dipole formation between the ITO and the active layer but also increasing the efficiency of organic solar cells [80,83]. On the contrary, the surface of ITO electrodes has been electronically modified by a chemically bonded monolayer of organic molecules or a single layer of adsorbed acid ions, such as illustrated in Figure 14.5, to attain a uniform surface with covalent organic–inorganic bonds or to create a suitable

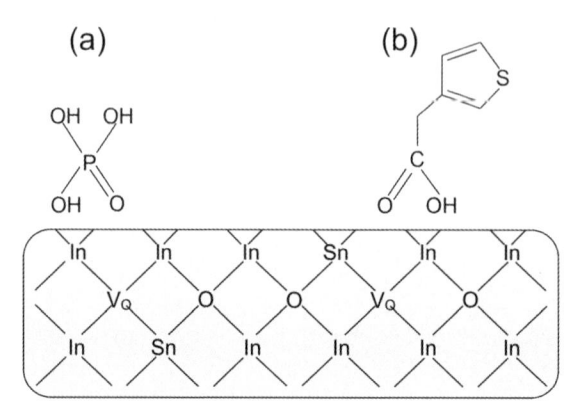

Figure 14.5 Schematic view of an ITO electrode (constituted by In_2O_3 doped with Sn and intrinsic oxygen vacancies V_O) that has been modified by (a) phosphoric acid, creating a dipole layer on the ITO surface, or (b) chemisorption of thiophene acetic acid, placing the polar functional groups at the ITO interface and exposing the less polar thiophene functionality.

interface dipole. Treatments with phosphoric acid and periodic acid both have led to an increase in work function of about 0.7 eV with a good homogeneity [82]. A chemically bonded monolayer of 8-hydroxy-quinoline (8-HQ) has also been useful to achieve a uniform work function different from untreated ITO films. With such treatments by acids or a chemically bonded monolayer, the efficiency of organic solar cells based on phthalocyanine/C_{60} heterojunction has been improved by approximately 40% compared to that obtained with untreated ITO [82]. Modification of the ITO surface with electroactive small molecules such as $Fc(COOH)_2$ and 3-thiophene acetic acid, has also promoted a better wettability of the organic layers to the polar ITO surface and an enhanced electrical contact (lower series resistance) that resulted in a significant improvement in the photovoltaic response of multilayer CuPc/C_{60}/BCP solar cells [81]. Similar surface modification of metal oxide layers by creation of interface dipoles has been applied to enhanced carrier injection in organic light-emitting diodes or electroluminescent devices, for which SnO_2 surface has been modified with functional groups such as −COOH and −COCl, whereas ITO has been chemically modified with H-, Cl-, and CF_3-terminated benzoyl chlorides [84].

For polymer light-emitting diodes, in which the electroluminescent polymer is deposited on a metal oxide electrode, it has been pointed out that oxidation of the polymer by oxygen diffusing out of the metal oxide limits the device lifetime [85]. In this sense, it is interesting to note that several surface treatments commonly used to increase the metal oxide work function (such as UV-ozone or oxygen-plasma treatment [64,65]) induce a peroxidic electrode surface that favors the migration of oxygen atoms into the organic active region [86]. Thus, overlong treatments should be avoided because they increase the incorporation and diffusion of oxygen atoms, apart from inducing a significant increase of the electrode roughness and resistivity [65]. When the electrode conditioning is optimized, no substantial influence on the device lifetime should be expected. Thus, from the study of the decay of bulk heterojunctions formed with MEH-PPV (poly(p-phenylene vinylene)) and PCBM in a film sandwiched between ITO and aluminum electrodes, it has been found that ITO electrode conditioning can produce a small increase in the overall current density but no significant changes in the decay parameters. This implies that the interface between the active layer and ITO does not play a significant role in the degradation of the performance of the photovoltaic cell under illumination, although it does, however, improve the overall performance of the device [51].

14.4
Conducting Polymer Electrodes

14.4.1
Conducting Polymer Properties

Conducting polymers represent a promising class of materials for the use as electrodes, in particular in view of all-organic devices. Common classes of organic conductive polymers include poly(acetylene)s, poly(pyrrole)s, poly(aniline)s, and poly (thiophene)s [87]. There are two primary methods of doping a conductive polymer, both through an oxidation–reduction process. The first method is chemical doping that involves exposing a polymer to an oxidant or reductant. The second is electrochemical doping in which a polymer-coated working electrode is suspended in an electrolyte solution where the application of a suitable electric potential causes a charge to enter the polymer in the form of electron addition (n-doping) or removal (p-doping). Besides, polymers may also be self-doped when associated with a protonic solvent such as water or an alcohol.

Among the various organic conducting polymers, PEDOT is the most widely studied polymer for photovoltaic applications because of its unique properties such as high electrical conductivity, almost transparent thin film in the oxidized state, and excellent stability at ambient and elevated temperatures [88]. PEDOT has in its oxidized form a good hole conductivity but is as such unstable in water. The polystyrenesulfonic acid acts as a charge-compensating counterion to stabilize the p-doped conducting polymer and forms a processable waterborne dispersion with PEDOT. The proposed mechanism of the doping reaction involving the acid is protonation at a carbon site involved in the delocalized π-electron system, thus introducing a net positive charge (hole). Values of conductivity are generally lower for this type of doping than for oxidative doping. The commercially available polyelectrolyte complex, PEDOT:PSS, can be easily spin-coated resulting in a highly transparent polymer film, with resistivity and work function values that depend on the PEDOT:PSS ratio [89,90]. Several articles describe the adjustment of the work function of PEDOT:PSS. For instance, it is possible to exchange the protons by addition of NaOH or CsOH to the dispersion and to modify the material work function through the exchange of protons by alkali metal ions. In this way, the work function changed from 5.1 to 4.0 eV with increasing alkali surface concentration [90]. Moreover, the work function can be modified by chemical doping or dedoping through the addition of oxidizing or reducing agents or by an electrochemical treatment [91,92].

For some applications, it is desirable to avoid the use of water or the presence or excess of an acidic polyelectrolyte such as PSS. Recently, dispersions of PEDOT block copolymers with poly(ethylene glycol) (PEG) have also become commercially available. Because the colloidal stabilization is achieved by a neutral molecule rather than an anionic or acidic species such as PSS, these copolymers do not contain an excess ionic species and the dopant of the PEDOT segments can be chosen from a variety of anions, including perchlorate (PC) or para-toluenesulfonate (PTS) anions. The structure of such organic compounds is represented in Figure 14.6. The work function values reported for the PTS-doped PEDOT:PEG copolymer (4.19 eV) and the PC-doped PEDOT:PEG copolymer (4.33 eV) do scale well with conductivity and doping level trends [93]. In this sense, Figure 14.7 shows the energy band diagrams corresponding to different doping levels. From a device view, PEDOT:PEG doped with PTS yields a comparable conductivity to the PEDOT:PSS but without the corrosive PSS or water dispersion requirements. The availability of organic dispersions of PEDOT will give rise to hydrophobic layers and will allow to increase the life of the devices as the presence of small water traces limits the device life.

14.4 Conducting Polymer Electrodes

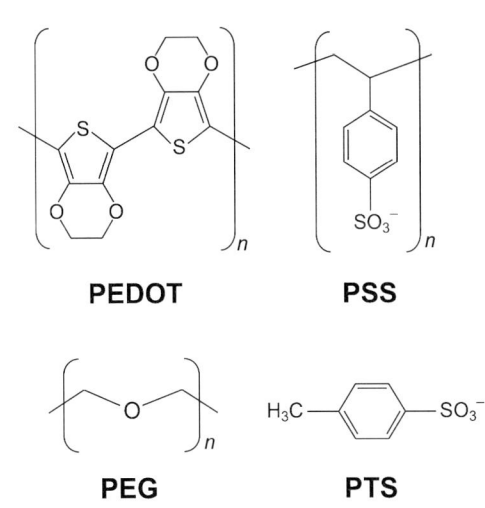

Figure 14.6 Structure of PEDOT, PSS, PEG, and PTS. Transparent and conducting polymer electrodes can be obtained from ionic stabilized PEDOT:PSS blends (where PSS acts also as dopant) or from sterically stabilized PEDOT:PEG copolymers (adding dopants such as PTS).

The electrical conductivity of PEDOT:PSS films is typically below $10\,\mathrm{S\,cm^{-1}}$. Owing to such limited conductivity with respect to metal or metal oxide electrodes, PEDOT:PSS is usually utilized in combination with ITO [2,13–16] or with an underlying metallic grid [94,95]. PEDOT:PSS has also been successfully used to fabricate small-area devices with organic materials that exhibit high carrier mobilities that can minimize the overall cell series resistance and offset losses because of the high resistivity of the electrode material [96]. However, to broaden the use of the polymer electrode in photovoltaic devices, higher values of conductivities are needed.

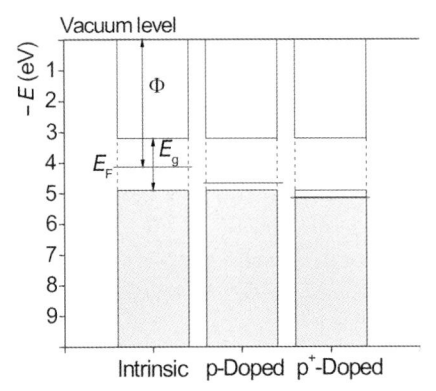

Figure 14.7 Energy band diagram corresponding to PEDOT-based electrodes with different doping, which makes to change the Fermi level energy (E_F) and the resulting work function value (Φ).

Electrochemical polymerization is widely used to get high conductivity, but the method is restricted to depositing the polymer only on conducting substrates. Recently, several groups are working to improve the conductivity of PEDOT using different approaches. Significant improvement of the conductivity of the commercially available PEDOT:PSS up to 84 S cm^{-1}, with 190 nm thickness and 81% optical transmittance, has been achieved by mixing the polymer solution with sorbitol as a secondary dopant [97]. Similar works using other secondary dopants such as dimethyl sulfoxide (DMSO), N,N'-dimethyl formamide (DMF), tetrahydrofuran (THF), or 2,2'-thiodiethanol also notably improved the conductivity of PEDOT:PSS in thin films up to 98 S cm^{-1} and the optical transmission to 84% [98,99]. Others have used the chemical oxidative polymerization of PEDOT from its monomer (EDOT) and a methanol-substituted derivative (EDOT-CH$_2$OH). By optimizing the ratio of the monomer, the oxidant (iron(III)p-toluenesulfonate (Fe(OTS)$_3$), a weak base (imidazole), and the solvent, further enhancement of the conductivity up to 900 S cm^{-1} and 82% transparency have been achieved [100]. Another approach is to use the vapor-phase polymerization (VPP). The base-inhibited VPP method has given the highest conductivity of PEDOT, exceeding 1000 S cm^{-1} [101]. In this method, a surface covered with ferric p-toluenesulfonate as oxidant mixed with a volatile base (pyridine) is exposed to EDOT vapors. The base is added to suppress an acid-initiated polymerization of EDOT, leading to a product with little or no conjugation. The product of the base-inhibited VPP is confirmed to be virtually identical to PEDOT obtained by wet chemical oxidation. Polymer solar cells with VPP PEDOT and PEDOT:PSS doped with sorbitol showed an order of increase in power conversion efficiency under sunlight illumination conditions compared to devices with electrodes made of PEDOT:PSS [97].

14.4.2
Conducting Polymer/Organic Semiconductor Interactions

It has been noted that doped conducting polymer films can interact with semiconducting polymer layers in complex ways. Clear evidences for such interactions have been presented for the case of PPV formed on top of PEDOT:PSS via the sulfonium precursor route, which involves high process temperatures [4]. The most direct interaction between doped polymer and semiconducting polymer is the movement of the dopant (in this case, the strong polymeric acid PSS) from the former to the latter, giving rise to partial doping of the semiconducting polymer, much lower than for the PEDOT owing to the higher value of ionization potential for PPV. It is found that even a slight doping of the PPV facilitates the carrier transport and could improve exciton dissociation in the PEDOT:PSS/PPV devices. When PEDOT:PSS is deposited on top of the PPV precursor prior to the thermal conversion, the PEDOT:PSS reaches or approaches the interface between the PPV and substrate in some areas at which the PPV is thinner, resulting in a very large interfacial area between the two polymers. The large interpenetrating interface leads to a larger portion of the PPV film being exposed to PSS, the increased doping improving the carrier transport, and also the collection [4]. On the contrary, in polymer solar cells with electrodes constructed from different forms of PEDOT:PSS, variations in open-circuit voltage, short-circuit current, and fill factor were observed and attributed to the slight segregation and rearrangement of the insulating layer PSS on top of PEDOT, which altered the charge injection at the interface between electrode and photoactive layer [102].

PEDOT:PSS is known as a hygroscopic material [103] and, therefore, there is some possibility that PEDOT:PSS electrode can adsorb water during exposure to a humid atmosphere, giving rise to some device degradation. In this sense, stability studies of unencapsulated photovoltaic devices based on blends of MDMO-PPV and PCBM with and without a PEDOT:PSS layer indicate that the degradation after air exposure is associated with water absorption into the PEDOT:PSS layer [104]. Charge transport measurements showed that water on PEDOT:PSS increases the resistivity of the PEDOT:PSS–blend layer interface, but it does not affect the charge mobilities of the active layer. This degradation of PEDOT:PSS layer appears to be spatially inhomogeneous, associated with the formation of insulating patches resulting in the loss of device current and therefore device efficiency [104]. The use of such hygroscopic materials increases the requirement of water barrier layers for the encapsulation of organic solar cells [105].

From the study of solar cells with an $Al/C_{60}/C_{12}$-PSV/PEDOT:PSS geometry, it has been found that PEDOT:PSS is involved in other degradation mechanism through the formation of particles in the device during storage. The formation of these particles is proposed to follow one of the two possible mechanisms: (i) PSS migrates from the PEDOT:PSS layer to the C_{12}-PSV/C_{60} interface where it reacts with partially oxidized C_{12}-PSV; (ii) some of the PSS is desulfonated and subsequently reacts with unreacted PSS forming an adduct, which is susceptible to oxidation resulting in the formation of the particles [52]. In any case, the necessary means to increase the lifetime of the organic devices involve the introduction of layers that are more resistant to oxygen and that are not involved in alternative processes such as particle formation.

14.5
Multilayer Electrodes

Owing to the numerous requirements that an electrode must fulfill as far as conductivity, transparency, stability, and so on are concerned, it is sometimes difficult to find a material that satisfies them in optimal way. Therefore, for certain specific applications, the optimization of the contact can require the development of structures of multiple layers that allow to reach optical and electrical characteristics globally superior to the attainable ones with a single-layer material.

Double layer metal electrodes of the type Al/Au, Ca/Au, or Ca/Ag have been used in semitransparent devices with very good results [106,107]. Fitting the thickness of the individual layers and the stack, it is possible to achieve a maximum value of transmittance and additionally to control the spectral region to reach the maximum. Besides, the extraction of carriers in the contact can be improved by adjusting the work function in the electrode region next to the active material. Surface roughness and other morphological characteristics can also be modified by introducing an additional layer. In this way, a uniform transmission over 70% through the visible spectral range and a sheet resistance as low as 12 Ω sq^{-1} have been obtained with a Ca (10 nm)/Ag (10 nm) bilayer structure [106]. The 10 nm thick layer of Ag on glass showed a strong absorption at 480 nm and much higher sheet resistance because it nucleated in small isolated islands. However, when the Ca layer was introduced, it wetted the glass surface and allowed the Ag layer to be continuous, making the electrode more transparent and conductive.

Double-layer metal oxide structures containing tin-doped In_2O_3 (ITO) in combination with aluminum-doped ZnO or indium-doped CdO (CIO) have shown improved characteristics compared to the corresponding single layers on glass substrates [108,109]. For both AZO/ITO and CIO/ITO structures, the overall improved optical and electrical characteristics are related to changes in the structure and morphology of the stack with regard to the individual films on glass substrate. The CIO/ITO bilayers have shown a sheet resistance of 5.6 Ω sq^{-1} and average visible transmittance of 87%, with a low overall indium content of 16 at%, being interesting for large-area applications [108]. In other works, the introduction of a thin ZnO layer between the ITO electrode and the active organic layer is reported to improve the power conversion efficiency of heterojunction photovoltaic devices [110,111], owing mainly to the enhanced carrier extraction achieved by the introduction of the ZnO.

Combination of metal oxide layers (mostly ITO) with metal or conducting polymer films is also widely used in organic solar cells. The work function of ITO is well matched to the highest occupied molecular orbital (HOMO) energy of a large number of hole-transporting organic materials. Despite such suitable matching, in heterojunction solar cells fabricated with ITO contact to hydrazone-substituted polysiloxane (PSX-Hz) p-type material, it has been found [112] that ITO bandgap structure can be responsible for photocurrent losses in the spectral region of high PSX-Hz absorption. Photocurrent losses are related to the lack of effective charge generation near the ITO interface, owing to the charge separation limits imposed by the large energy gap of the ITO electrode just below the Fermi level. In this sense, the introduction of a thin coating of metal or conducting polymer on the ITO layer can improve the photoresponse [112]. Compound electrodes of ITO/Ag, ITO/PEDOT:PSS, and ZnO/PEDOT:PSS have effectively allowed to improve the performance of different organic photovoltaic devices [68,80,113]. Sheet resistance, optical transmittance, and surface roughness of the ITO/Ag electrode can be adjusted through the individual layer thickness [113,114]. When PEDOT:PSS is introduced between the metal oxide and the active organic layer, the increase in the efficiency has also been related to an improvement in the morphological and chemical properties of the region in contact with the organic material [68,80,83]. Besides, the introduction of a PEDOT interlayer between the metal and the active organic material has been useful to reduce the series resistance and improve the overall conversion efficiency [115].

Other structures such as ITO/Ag/ITO have demonstrated their capacity to fit the maximum of optical transmission (above 80%) to 550 nm with thicknesses of ITO of 30–50 nm, and simultaneously to reach low sheet resistance (below 20 Ω sq^{-1}) with a thickness of Ag of $\sim$8 nm [116–118], showing on plastic substrates better mechanical and electrical properties than electrodes constituted by a single ITO layer [119]. It is important to note that for the multilayer structures such as ITO/Ag/ITO continuous and uniform coverage is required for the ITO films but not for the metal interlayer that can be discontinuous allowing to increase the global transmission. Moreover, the total ITO thickness used for the multilayer electrodes is lower than that required to achieve the same electrical resistivity with a single ITO layer, and thus the use of multilayer electrodes can provide a significant material saving. Recently, there have also been developed structures ZnO/Ag/ZnO [120] and ZnO/Al [121] with characteristics superior to individual layers.

14.6 Conclusions

It has been pointed out that both the fundamental material properties and the interactions with the active organic semiconductor are essential to achieve high-performance electrodes. In this sense, several examples have been given to illustrate the influence of the work function value or the interface dipole formation on the device photovoltage and efficiency. Besides, increased reflectance in the back contact or higher transmittance in that exposed to the light can contribute to a significant photocurrent and efficiency enhancement. In addition, the development of homogeneous and low-resistivity electrodes is required to go to scaling-up areas and/or to high light intensities, where the series resistance becomes an important limiting factor.

It has been found that recognized metal, metal oxide, or conducting polymer materials can in general fulfill the basic requirements for high-performance electrodes, taking into account that the best material choice depends on the specific device configuration and active semiconductor selection. Sometimes, the optimization of the contact can require the development of structures of multiple layers that allow to reach optical, electrical, and chemical characteristics globally superior to the attainable ones with a single-layer material.

References

1 Spanggaard, H. and Krebs, F.C. (2004) *Solar Energy Materials and Solar Cells*, **83**, 125–146.
2 Peumans, P., Yakimov, A. and Forrest, S.R. (2003) *Journal of Applied Physics*, **93**, 3693–3723.
3 Schmidt-Mende, L., Zakeeruddin, S.M. and Grätzel, M. (2005) *Applied Physics Letters*, **86**, 013504.
4 Arias, A.C., Granström, M., Thomas, D.S., Petritsch, K. and Friend, R.H. (1999) *Physical Review B: Condensed Matter*, **60**, 1854–1860.
5 Manifacier, J.C. (1982) *Thin Solid Films*, **90**, 297–308.
6 Pei, Q., Zuccarello, G., Ahlskog, M. and Inganäs, O. (1994) *Polymer*, **35**, 1347–1351.
7 Loutfy, R.O., Sharp, J.H., Hsiao, C.K. and Ho, R. (1981) *Journal of Applied Physics*, **52**, 5218–5230.
8 Xue, J., Uchida, S., Rand, B.P. and Forrest, S.R. (2004) *Applied Physics Letters*, **84**, 3013–3015.
9 Ghosh, A.K., Morel, D.L., Feng, T., Shaw, R.F. and Rowe, C.A. (1974)

Journal of Applied Physics, **45**, 230–236.

10 Sharma, G.D., Gupta, S.K. and Roy, M.S. (1997) *Journal of Physics and Chemistry of Solids*, **58**, 195–205.

11 Ghosh, A.K. and Feng, T. (1973) *Journal of Applied Physics*, **44**, 2781–2788.

12 Ramsdale, C.M., Barker, J.A., Arias, A.C., MacKenzie, J.D., Friend, R.H. and Greenham, N.C. (2002) *Journal of Applied Physics*, **92**, 4266–4270.

13 Brabec, C.J., Cravino, A., Meissner, D., Sariciftci, N.S., Rispens, M.T., Sanchez, L., Hummelen, J.C. and Fromherz, T. (2002) *Thin Solid Films*, **403–404**, 368–372.

14 Gadisa, A., Svensson, M., Andersson, M.R. and Inganäs, O. (2004) *Applied Physics Letters*, **84**, 1609–1611.

15 Mihailetchi, V.D., Blom, P.W.M., Hummelen, J.C. and Rispens, M.T. (2003) *Journal of Applied Physics*, **94**, 6849–6854.

16 Mihailetchi, V.D., Koster, L.J.A. and Blom, P.W.M. (2004) *Applied Physics Letters*, **85**, 970–972.

17 Melzer, C., Krasnikov, V.V. and Hadziioannou, G. (2003) *Applied Physics Letters*, **82**, 3101–3103.

18 Brodie, I. (1995) *Physical Review B: Condensed Matter*, **51**, 13660–13668.

19 Akbi, M. and Lefort, A. (1998) *Journal of Physics D: Applied Physics*, **31**, 1301–1308.

20 Lide, D.R.(ed.) (1997) *Handbook of Chemistry and Physics*, CRC Press.

21 Choo, D.C., Im, H.C., Lee, D.U., Kim, T.W., Han, J.W. and Choi, E.H. (2005) *Solid State Communications*, **136**, 365–368.

22 Taima, T., Chikamatsu, M., Yoshida, Y., Saito, K. and Yase, K. (2004) *IEICE Transactions on Electronics*, **E87C**, 2045–2048.

23 Avrekh, M., Thibadeau, B.M., Monteiro, O.R. and Brown, I.G. (1999) *Review of Scientific Instruments*, **70**, 4328–4330.

24 Klauk, H., Huang, J.R., Nichols, J.A. and Jackson, T.N. (2000) *Thin Solid Films*, **366**, 272–278.

25 Guillén, C. and Herrero, J. (2003) *Journal of Materials Processing Technology*, **143–144**, 144–147.

26 Lim, J.W. and Isshiki, M. (2006) *Journal of Applied Physics*, **99**, 094909.

27 Mayadas, A.F. and Shatzkes, M. (1970) *Physical Review B: Condensed Matter*, **1**, 1382–1389.

28 Stubinger, T. and Brutting, W. (2001) *Journal of Applied Physics*, **90**, 3632–3641.

29 Peumans, P. and Forrest, S.R. (2001) *Applied Physics Letters*, **79**, 126–128.

30 Charton, C. and Fahland, M. (2003) *Surface & Coatings Technology*, **174–175**, 181–186.

31 Charton, C. and Fahland, M. (2004) *Thin Solid Films*, **449**, 100–104.

32 Stenzel, O., Stendal, A., Voigtsberger, K. and von Borczyskowski, C. (1995) *Solar Energy Materials and Solar Cells*, **37**, 337–348.

33 Westphalen, M., Kreibig, U., Rostalski, J., Lüth, H. and Meissner, D. (2000) *Solar Energy Materials and Solar Cells*, **61**, 97–105.

34 Yakimov, A. and Forrest, S.R. (2002) *Applied Physics Letters*, **80**, 1667–1669.

35 Xue, J., Uchida, S., Rand, B.P. and Forrest, S.R. (2004) *Applied Physics Letters*, **85**, 5757–5759.

36 Konstadinidis, K., Papadimitrakopoulos, F., Galvin, M. and Opila, R.L. (1995) *Journal of Applied Physics*, **77**, 5642–5646.

37 Ettedgui, E., Hsieh, B.R. and Gao, Y. (1997) *Polymers for Advanced Technologies*, **8**, 408–416.

38 Kampen, T.U., Das, A., Park, S., Hoyer, W. and Zahn, D.R.T. (2004) *Applied Surface Science*, **234**, 333–340.

39 Seki, K., Hayashi, N., Oji, H., Ito, E., Ouchi, Y. and Ishii, H. (2001) *Thin Solid Films*, **393**, 298–303.

40 Taima, T., Chikamatsu, M., Bera, R.N., Yoshida, Y., Saito, K. and Yase, K. (2004) *Journal of Physical Chemistry B*, **108**, 1–3.

41 Chikamatsu, M., Taima, T., Yoshida, Y., Saito, K. and Yase, K. (2004) *Applied Physics Letters*, **84**, 127–129.

References

42 Yan, L., Watkins, N.J., Zorba, S., Gao, Y. and Tang, C.W. (2002) *Applied Physics Letters*, **81**, 2752–2754.

43 Crispin, X. (2004) *Solar Energy Materials and Solar Cells*, **83**, 147–168.

44 De Renzi, V., Rousseau, R., Marchetto, D., Biagi, R., Scandolo, S. and del Pennino, U. (2005) *Physical Review Letters*, **95**, 046804.

45 Koller, G., Blyth, R.I.R., Sardar, S.A., Netzer, F.P. and Ramsey, M.G. (2000) *Applied Physics Letters*, **76**, 927–929.

46 Ghosh, A.K. and Feng, T. (1978) *Journal of Applied Physics*, **49**, 5982–5989.

47 Saleh, A.M., Hassan, A.K. and Gould, R.D. (2003) *Journal of Physics and Chemistry of Solids*, **64**, 1297–1303.

48 Shafai, T.S. and Anthopoulos, T.D. (2001) *Thin Solid Films*, **398–399**, 361–367.

49 Campbell, I.H., Kress, J.D., Martin, R.L., Smith, D.L., Barashkov, N.N. and Ferraris, J.P. (1997) *Applied Physics Letters*, **71**, 3528–3530.

50 Neugebauer, H., Brabec, C., Hummelen, J.C. and Sariciftci, N.S. (2000) *Solar Energy Materials and Solar Cells*, **61**, 35–42.

51 Krebs, F.C., Carlé, J.E., Cruys-Bagger, N., Andersen, M., Lilliedal, M.R., Hammond, M.A. and Hvidt, S. (2005) *Solar Energy Materials and Solar Cells*, **86**, 499–516.

52 Norrman, K., Larsen, N.B. and Krebs, F.C. (2006) *Solar Energy Materials and Solar Cells*, **90**, 2793–2814.

53 Norrman, K. and Krebs, F.C. (2006) *Solar Energy Materials and Solar Cells*, **90**, 213–227.

54 Brabec, C.J., Shaheen, S.E., Winder, C., Sariciftci, N.S. and Denk, P. (2002) *Applied Physics Letters*, **80**, 1288–1290.

55 Wen, F.S., Li, W.L., Liu, Z. and Wei, H.Z. (2006) *Materials Chemistry and Physics*, **95**, 94–98.

56 Jonsson, S.K.M., Carlegrim, E., Zhang, F., Salaneck, W.R. and Fahlman, M. (2005) *Japanese Journal of Applied Physics*, **44**, 3695–3701.

57 Parthasarathy, G., Shen, C., Kahn, A. and Forrest, S.R. (2001) *Journal of Applied Physics*, **89**, 4986–4992.

58 Chopra, K.L., Major, S. and Pandya, D.K. (1983) *Thin Solid Films*, **102**, 1–46.

59 Mason, T.O., Gonzalez, G.B., Kammler, D.R., Mansourian-Hadavi, N. and Ingram, B.J. (2002) *Thin Solid Films*, **411**, 106–114.

60 Robertson, J., Peacock, P.W., Towler, M.D. and Needs, R. (2002) *Thin Solid Films*, **411**, 96–100.

61 Bertoni, M.I., Mason, T.O., Medvedeva, J.E., Freeman, A.J., Poeppelmeier, K.R. and Delley, B. (2005) *Journal of Applied Physics*, **97**, 103713.

62 Minami, T. (1999) *Journal of Vacuum Science & Technology A: Vacuum Surfaces and Films*, **17**, 1765–1772.

63 Minami, T. (2005) *Semiconductor Science and Technology*, **20**, S35–S44.

64 Sugiyama, K., Ishii, H., Ouchi, Y. and Seki, K. (2000) *Journal of Applied Physics*, **87**, 295–298.

65 Chen, S.-H. (2005) *Journal of Applied Physics*, **97**, 073713.

66 Yang, F. and Forrest, S.R. (2006) *Advanced Materials*, **18**, 2018.

67 Qiao, Q., Beck, J., Lumpkin, R., Pretko, J. and Mcleskey, J.T. (2006) *Solar Energy Materials and Solar Cells*, **90**, 1034–1040.

68 Bernède, J.C., Derouiche, H. and Djara, V. (2005) *Solar Energy Materials and Solar Cells*, **87**, 261–270.

69 Guillén, C. and Herrero, J. (2005) *Thin Solid Films*, **480–481**, 129–132.

70 Guillén, C. and Herrero, J. (2006) *Thin Solid Films*, **515**, 640–643.

71 Bellingham, J.R., Phillips, W.A. and Adkins, C.J. (1992) *Journal of Materials Science Letters*, **11**, 263–265.

72 Guillén, C. and Herrero, J. (2006) *Vacuum*, **80**, 615–620.

73 Ohta, H., Orita, M., Hirano, M., Tanji, H., Kawazoe, H. and Hosono, H. (2000) *Applied Physics Letters*, **76**, 2740–2742.

74 Rauf, I.A. (1996) *Journal of Applied Physics*, **79**, 4057–4065.

75 Minami, T., Suzuki, S. and Miyata, T. (2001) *Thin Solid Films*, **398–399**, 53–58.

76 Minami, T., Sato, H., Nanto, H. and Takata, S. (1989) *Thin Solid Films*, **176**, 277–282.
77 Guillén, C. and Herrero, J. (2006) *Surface & Coatings Technology*, **201**, 309–312.
78 Kim, J.S., Friend, R.H. and Cacialli, F. (1999) *Journal of Applied Physics*, **86**, 2774–2778.
79 Djara, V. and Bernède, J.C. (2005) *Thin Solid Films*, **493**, 273–277.
80 Nüesch, F., Tornare, G., Zuppiroli, L., Meng, F., Chen, K. and Tian, H. (2005) *Solar Energy Materials and Solar Cells*, **87**, 817–824.
81 Armstrong, N.R., Carter, C., Donley, C., Simmonds, A., Lee, P., Brumbach, M., Kippelen, B., Domercq, B. and Yoo, S. (2003) *Thin Solid Films*, **445**, 342–352.
82 Johnev, B., Vogel, M., Fostiropoulos, K., Mertesacker, B., Rusu, M., Lux-Steiner, M.-C. and Weidinger, A. (2005) *Thin Solid Films*, **488**, 270–273.
83 Shrotriya, V., Li, G., Yao, Y., Chu, C.-W. and Yang, Y. (2006) *Applied Physics Letters*, **88**, 073508.
84 Ganzorig, C., Kwak, K.-J., Yagi, K. and Fujihira, M. (2001) *Applied Physics Letters*, **79**, 272–274.
85 Scott, J.C., Kaufman, J.H., Brock, P.J., DiPietro, R., Salem, J. and Goitia, J.A. (1996) *Journal of Applied Physics*, **79**, 2745–2751.
86 Lee, K.H., Jang, H.W., Kim, K.B., Tak, Y.H. and Lee, J.L. (2004) *Journal of Applied Physics*, **95**, 586–590.
87 Brédas, J.L., Thémans, B., Fripiat, J.G., André, J.M. and Chance, R.R. (1984) *Physical Review B: Condensed Matter*, **29**, 6761–6773.
88 Groenendaal, L., Jonas, F., Freitag, D., Pielartzik, H. and Reynolds, J.R. (2000) *Advanced Materials*, **12**, 481–494.
89 Koch, N. and Vollmer, A. (2006) *Applied Physics Letters*, **89**, 162107.
90 Weijtens, C.H.L., van Elsbergen, V., de Kok, M.M. and de Winter, S.H.P.M. (2005) *Organic Electronics*, **6**, 97–104.
91 de Jong, M.P., Denier van der Gon, A.W., Crispin, X., Osikowicz, W., Salaneck, W.R. and Groenendaal, L. (2003) *Journal of Chemical Physics*, **118**, 6495–6502.
92 Petr, A., Zhang, F., Peisert, H., Knupfer, M. and Dunsch, L. (2004) *Chemical Physics Letters*, **385**, 140–143.
93 Sapp, S., Luebben, S., Losovyj, Ya.B., Jeppson, P., Schulz, D.L. and Caruso, A.N. (2006) *Applied Physics Letters*, **88**, 152107.
94 Aernouts, T., Vanlaeke, P., Geens, W., Poortmans, J., Heremans, P., Borghs, S., Mertens, R., Andriessen, R. and Leenders, L. (2004) *Thin Solid Films*, **451–452**, 22–25.
95 Niggemann, M., Glatthaar, M., Lewer, P., Müller, C., Wagner, J. and Gombert, A. (2006) *Thin Solid Films*, **511–512**, 628–633.
96 Kushto, G.P., Kim, W. and Kafafi, Z.H. (2005) *Applied Physics Letters*, **86**, 093502.
97 Admassie, S., Zhang, F., Manoj, A.G., Svensson, M., Andersson, M.R. and Inganäs, O. (2006) *Solar Energy Materials and Solar Cells*, **90**, 133–141.
98 Kim, J.Y., Jung, J.H., Lee, D.E. and Joo, J. (2002) *Synthetic Metals*, **126**, 311–316.
99 Martin, B.D., Nikolov, N., Pollack, S.K., Saprigin, A., Shashidhar, R., Zhang, F. and Heiney, P.A. (2004) *Synthetic Metals*, **142**, 187–193.
100 Ha, Y.-H., Nikolov, N., Pollack, S.K., Mastrangelo, J., Martin, B.D. and Shashidhar, R. (2004) *Advanced Functional Materials*, **14**, 615–622.
101 Winther-Jensen, B. and West, K. (2004) *Macromolecules*, **37**, 4538–4543.
102 Zhang, F.L., Gadisa, A., Inganäs, O., Svensson, M. and Andersson, M.R. (2004) *Applied Physics Letters*, **84**, 3906–3908.
103 Winter, I., Reese, C., Hormes, J., Heywang, G. and Jonas, F. (1995) *Chemical Physics*, **194**, 207–213.
104 Kawano, K., Pacios, R., Poplavskyy, D., Nelson, J., Bradley, D.D.C. and Durrant, J.R. (2006) *Solar Energy Materials and Solar Cells*, **90**, 3520–3530.

105 Dennler, G., Lungenschmied, C., Neugebauer, H., Sariciftci, N.S., Latrèche, M., Czeremuszkin, G. and Wertheimer, M.R. (2006) *Thin Solid Films*, **511–512**, 349–353.

106 Lee, C.J., Pode, R.B., Moon, D.G. and Han, J.I. (2004) *Thin Solid Films*, **467**, 201–208.

107 Shrotriya, V., Hsing-En Wu, E., Li, G., Yao, Y. and Yang, Y. (2006) *Applied Physics Letters*, **88**, 064104.

108 Yang, Y., Wang, L., Yan, H., Jin, S., Marks, T.J. and Li, S. (2006) *Applied Physics Letters*, **89**, 051116.

109 Herrero, J. and Guillén, C. (2004) *Thin Solid Films*, **451–452**, 630–633.

110 Umeda, T., Shirakawa, T., Fujii, A. and Yoshino, K. (2003) *Japanese Journal of Applied Physics*, **42**, L1475–L1477.

111 White, M.S., Olson, D.C., Shaheen, S.E., Kopidakis, N. and Ginley, D.S. (2006) *Applied Physics Letters*, **89**, 143517.

112 Jahng, W.S., Francis, A.H., Moon, H., Nanos, J.I. and Curtis, M.D. (2006) *Applied Physics Letters*, **88**, 093504.

113 Bailey-Salzman, R.F., Rand, B.P. and Forrest, S.R. (2006) *Applied Physics Letters*, **88**, 233502.

114 Hao, X.T., Zhu, F.R., Ong, K.S. and Tan, L.W. (2006) *Semiconductor Science and Technology*, **21**, 19–24.

115 Song, M.Y., Kim, K.J. and Kim, D.Y. (2005) *Solar Energy Materials and Solar Cells*, **85**, 31–39.

116 Fahland, M., Karlsson, P. and Charton, C. (2001) *Thin Solid Films*, **392**, 334–337.

117 Jung, Y.S., Choi, Y.W., Lee, H.C. and Lee, D.W. (2003) *Thin Solid Films*, **440**, 278–284.

118 Bertran, E., Corbella, C., Vives, M., Pinyol, A., Person, C. and Porqueras, I. (2003) *Solid State Ionics*, **165**, 139–148.

119 Lewis, J., Grego, S., Chalamala, B., Vick, E. and Temple, D. (2004) *Applied Physics Letters*, **85**, 3450–3452.

120 Sahu, D.R., Lin, S.Y. and Huang, J.L. (2006) *Applied Surface Science*, **252**, 7509–7514.

121 Hu, Y.M., Lin, C.W. and Huang, J.C.A. (2006) *Thin Solid Films*, **497**, 130–134.

15
Reel-to-Reel Processing of Highly Conductive Metal Oxides
Matthias Fahland

15.1
Introduction

The technology of reel-to-reel processing is of high importance for organic solar cells. There are two major reasons for this statement.

First, the absorber material is a polymer-based composite. So its material properties are very similar to the properties of plastic substrates. This is a big advantage on the way to stable flexible solar cells.

Secondly, the efficiency of organic solar cells is presently below that of other cell types. So they are designated for occupying the low-price segment of the solar cell market, that is, they need a very cost-effective production process.

Because of these two peculiarities, the continuous reel-to-reel process is a production principle of high importance for the successful market introduction of organic solar cells. This technology works with flexible materials and has been proven to achieve outstanding low production costs.

The special advantages are primarily based on the easy handling of the polymer film rolls. The following example may serve as an illustration of this fact. The pilot roll coater *coFlex* 600 at FEP site in Dresden (see Figure 15.4, Section 15.4) can be fitted with rolls of 400 mm outer diameter and a width of 650 mm. The weight of such a roll of polymer material is approximately 100 kg. Therefore, it can easily be handled and transported. Assuming a 50 µm film thickness, there is a total area of 1380 m^2 (one side) on such a roll.

On the contrary, the same area of glass (4 mm thickness) would weigh more than 13 ton. Obviously, flexible polymer materials have a huge handling advantage over other substrate materials. This is the key to low production costs.

Another important factor is the usage of a cheap substrate material. Polyethylene terephthalate (PET, brands: DupontTejin: Melinex, Mitsubishi: Hostaphan) possesses an optimum combination of low-price and acceptable properties. That is why this chapter will primarily focus on this material. The range of several important properties of PET is summarized in Table 15.1.

Organic Photovoltaics: Materials, Device Physics, and Manufacturing Technologies.
Edited by Christoph Brabec, Vladimir Dyakonov, and Ullrich Scherf
Copyright © 2008 WILEY-VCH Verlag GmbH & Co. KGaA, Weinheim
ISBN: 978-3-527-31675-5

15 Reel-to-Reel Processing of Highly Conductive Metal Oxides

Table 15.1 Typical properties of PET films [1].

Property	Value	Property	Value
Maximum processing temperature	100–150 °C	Modulus	490–510 kg mm^{-2}
Density	1.39 g cm^{-3}	Refractive index	1.65 (at 550 nm)
Shrinkage	0.5–0.9% (at 105 °C)	Water content	Up to 0.8%

Especially, the low processing temperature is a challenge for the deposition of transparent electrodes (which will be discussed in Section 15.2). That is why other substrate materials such as polyimide (PI) or polyethylene naphthalate (PEN) are frequently used for scientific purposes. However, the high price of these substrates would be a serious obstacle for the market introduction.

The deposition technology of highly conductive oxides in reel-to-reel machines has a history of about 30 years. In 1978, Sierracin, Inc. (USA) first introduced a sputter roll coater for the deposition of indium-tin oxide on PET film [2]. The coated material was intended for transparent electrodes used in an X-ray imaging technology. The rapid growth of mobile communication at the beginning of the 1990s was a driving force for extending R&D and production capacities.

Today, the solar cell industry can benefit from these developments for mobile communication. However, there are some important differences that need to be considered. Transparent electrodes for various kinds of displays are normally optimized with respect to the wavelength dependence of the sensitivity of the human eye. The transparent electrode of a solar cell must be adapted to the wavelength dependence of the quantum efficiency of the absorber material. In the case of organic solar cells, the difference is only small. Nonetheless, it needs to be considered for achieving an optimum cell performance.

A second difference can be found in the price restrictions that are harder in the case of solar cells compared to displays.

Sheet resistance R_{sh} is the most important property of transparent electrodes. It is defined as the electrical resistance between two opposite sides of a quadratic sample piece of any size (Figure 15.1).

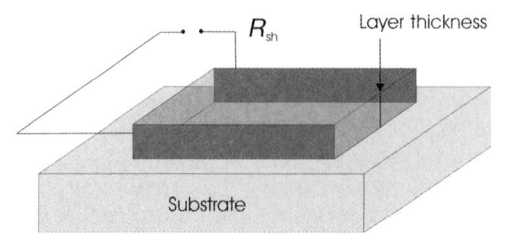

Figure 15.1 Schematic concerning the definition of the sheet resistance R_{sh}.

Table 15.2 Typical requirements for transparent electrodes on flexible polymer substrates.

Application	Required sheet resistance Ω_{squ}
Antistatic layers	500–2000
Electrodes for touch screens	400–700
Electrodes for light emitting films (inorganic electroluminescence)	20–120
Electrodes for organic electroluminescence (OLED)	20
Electrode for solar cells	8–80
Flexible electrochromics	1–10

In case of a single layer electrode, one yields by a simple calculation

$$R_{sh} = \rho/d, \tag{15.1}$$

ρ being the specific resistivity and d the layer thickness.

Various devices require transparent electrodes with different values of the sheet resistance. Table 15.2 summarizes different applications on flexible polymer substrates.

One can see that solar cells cover a wide range. Generally, the sheet resistance is in series with the internal resistance of the absorber material and therefore limits the short-circuit current of the cell. This effect is more pronounced in large-scale cells compared to small-area cells. Additionally, it is important whether the solar cell has a substrate or a superstrate configuration. The lower resistances in Table 15.2 are required for the superstrate configuration. In these cells the transparent electrode is deposited directly on the substrate. The absorber layer and the opaque counter electrode are deposited on top afterward. In contrast to that, the structure of the substrate configuration is vice versa. For these devices, the deposition of the transparent electrode is the last step. This opens the possibility to apply a metallic grid on top of the cell, which reduces the requirements to the sheet resistance.

Organic solar cells normally have the superstrate configuration. This avoids the application of plasma processes after the deposition of the sensitive absorber material. Owing to this, a sheet resistance below $20\,\Omega_{squ}$ is required.

In the following, both the materials and the technologies for the manufacturing of such electrodes are discussed.

15.2
Materials

In Table 15.2 one can see that the sheet resistance requirements for transparent electrodes cover three orders of magnitude. Naturally, it is impossible that only one material or one layer stack can provide the solution in all cases. Nonetheless, indium oxide, doped with 10% tin oxide (in short, ITO), is the dominant transparent conducting oxide (TCO) material. This is due to the extraordinary combination of electrical and optical properties. A specific resistivity as low as $0.44 \times 10^{-4}\,\Omega\,cm$ had been obtained by combining electron beam evaporation with an intended substrate temperature gradient [3]. The material has an intrinsic bandgap of 3.75 eV, which makes it transparent in the visible spectral range [4]. A resistivity below $2 \times 10^{-4}\,\Omega\,cm$ has been obtained by sputtering at elevated substrate temperatures [5]. The major drawback of ITO is the high material price of indium. This material is only available to a limited extent. Approximately 400 ton are produced annually. More than 70% of the production is used for thin-film electrodes, mainly for flat panel displays. The high demand and the limited resources resulted in a considerable price increase. The indium price shifted from the lowest value of $60 kg^{-1} in 2002 to over $1000 kg^{-1} in 2005.

Because of this, the solar industry is very much focused on alterative materials. The most interesting one is zinc oxide, doped with 2% aluminum oxide (in short, ZAO). Sputtered at optimum conditions, this material has excellent optical properties and is only slightly worse in the electrical properties compared to ITO [6,7]. Additionally, it has some specific benefits for the solar industry. This is mainly the possibility to form a textured surface for silicon-based solar cells [8]. The price perspective is promising because the raw materials are cheap and available without limitation.

Compared to the solar industry in general, the situation with respect to alternative TCO materials is slightly different in the case of organic solar cells. Both ITO and ZAO have their optimum performance when they are deposited on substrates maintained at temperatures 200 °C and above. Room temperature deposition, as is required for PET, results mainly in a decline of the electrical performance. The consequence of low-temperature deposition can be seen in Figure 15.2.

As one can see the electrical properties of both ITO and ZAO become worse when the material is deposited onto substrates at room temperature. The decrease in

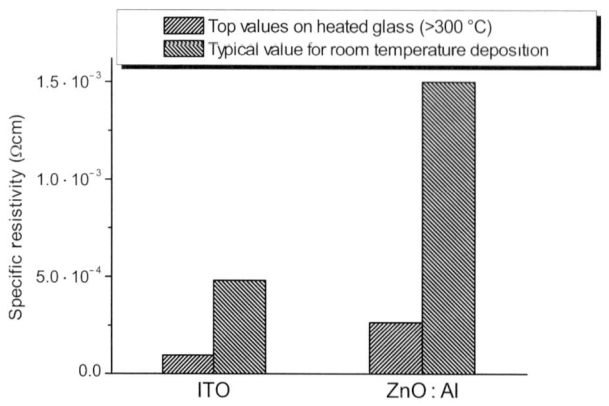

Figure 15.2 Difference between room temperature deposition and elevated temperature deposition for the specific resistivity of ITO and ZAO. The values are taken from own measurements and from [7,9,10].

the conductivity is especially pronounced for ZAO. More than a micrometer thickness of this material would be necessary to achieve acceptable sheet resistance values for organic solar cells. This is too much for an economic production and for preserving a sufficient flexibility of the final device.

ITO shows an increase in the specific resistivity that is moderate compared to ZAO. That is why ITO still plays an important role in the R&D work for flexible solar cells. However, for seizing a high-volume market, it is necessary either to improve the deposition technology or to look for a replacement of single layer transparent oxides. Both approaches will be discussed in the following chapters.

15.3
Deposition Technology

Transparent conductive oxides have been prepared by various deposition technologies. Among them are chemical vapor deposition [11], pulsed laser deposition [12], RF sputtering [13,14], DC sputtering [15], mid-frequency DMS sputtering [16], spray pyrolysis [17], and e-beam evaporation [3].

Among all these technologies, sputtering is the most appropriate one for large-scale applications. Various modifications of this technology have been developed over the past decades. First of all, the different approaches can be distinguished by the type of target. Today, the usage of ceramic targets is widely used in the industry and can easily be applied to sputter roll-coating machines. The ceramic targets have a chemical composition that is very close to the desired layer material. The key of this technology is the fact that they still possess an electrical conductivity that is high enough for applying conventional DC sputtering. Therefore, the process is easy to handle. It is necessary to add a small amount of oxygen to the argon atmosphere for the fine-tuning of the optimum layer property. This technology has achieved a very high maturity for ITO due to its extensive application in the display industry. The solar panel industry has adopted this technology, mainly for depositing ZAO.

As it was outlined in the previous chapter, both ITO and ZAO require elevated substrate temperatures (more than 200 °C) for showing the optimum performance. The preferred PET substrates cannot be heated up to such temperatures. According to formula (15.1) the increased specific resistivity needs to be compensated by an increased layer thickness to maintain the necessary sheet resistance. This is boosting the material costs and can therefore be regarded as a serious problem for market acceptance of flexible electronic devices. Several attempts have been made to overcome this difficulty.

Improved performance could be achieved by double magnetron arrangements [16,18] as they can be seen in Figure 15.3. Such an arrangement can be run with deposition rates of more than 100 nm on a film moving with 1 m min^{-1} (dynamic deposition rate $100 \text{ nm m min}^{-1}$).

Superimposing RF power to the DC voltage of the sputtering process is another way to reduce the specific resistivity. This approach had been reported to both

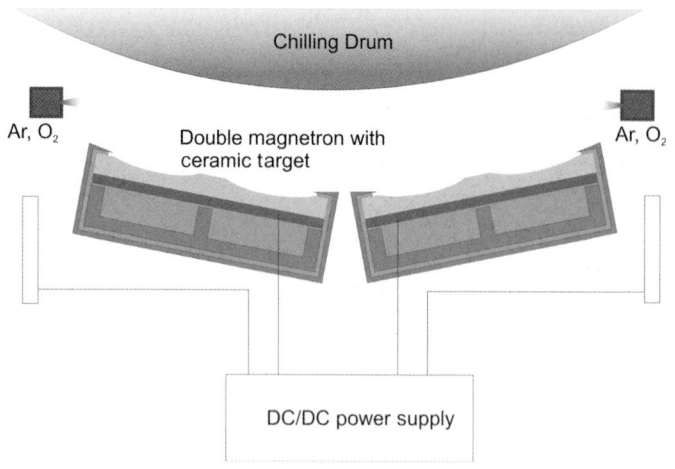

Figure 15.3 Double magnetron arrangement with DC–DC-powering for ITO deposition from ceramic targets.

ITO [19] and ZAO [20]. An improvement of up to 25% could be achieved in the case of ITO without significant loss of rate [19]. However, this approach has not been demonstrated in the temperature range for PET deposition, that is, between 20 and 80 °C.

In Ref. [21], it was reported that a low-energy ion bombardment provided an improved specific resistivity for ZAO. Adding new dopants such as molybdenum to the ITO lattice is another possibility for reducing the specific resistivity [22]. These attempts are partly already used for production processes. All of them are a topic of intense R&D work worldwide.

Improving the magnetron and the sputter technology is another way for reducing the material cost. The standard solution for DC sputtering of ceramic targets is the usage of planar magnetrons. However, they have limited target utilization due to the formation of an erosion groove. This difficulty can be overcome by the use of rotatable magnetrons. These magnetrons show a target utilization of more than 85% [23]. The layer properties for ITO [23] and ZAO [24] have been proven to be the same or even better than those for their planar counterparts.

A second important way of reducing the material cost is the use of metallic targets instead of ceramic targets. Thus, material costs can drop down to 20% of the original value in case of planar magnetrons. However, it is necessary to run a reactive sputtering process with a much higher oxygen partial pressure. The technology has to deal with the typical hysteresis behavior of reactive sputtering [25–27]. The oxygen partial pressure of the sputtering atmosphere needs to be regulated by a feedback control loop. Regardless of these difficulties, the technology has successfully been demonstrated for both ZAO [7] and ITO [27].

The use of both rotatable magnetrons and the sputtering from metal alloy targets are promising ways for cost reduction in TCO deposition.

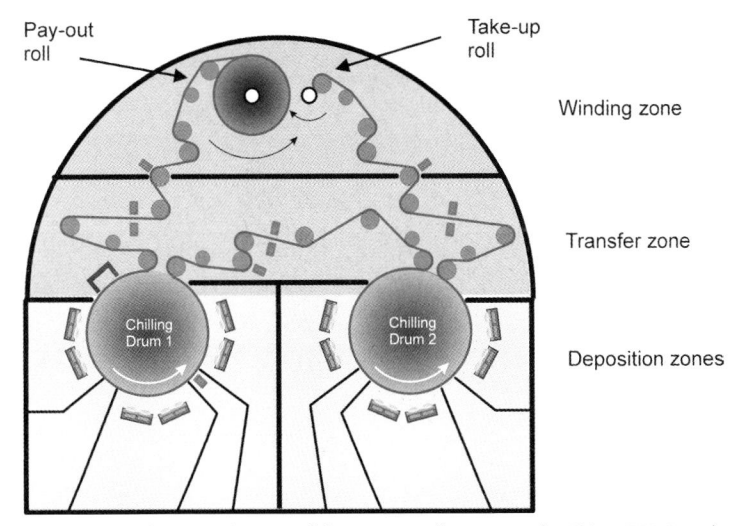

Figure 15.4 Schematic drawing of the sputter roll coater *coFlex* 600 at FEP Dresden.

15.4 Equipment

As it was already outlined in Section 15.1 that roll coaters are the preferred machines for the deposition of flexible transparent electrodes. The basic principle of such equipment can be seen in Figure 15.4.

The coating machine is divided into the winding zone, the film transfer zone, and various deposition zones. A reel with the substrate is installed on the payout roller. After evacuating the vessel the film is unwinded, transferred through the winding system, and finally rewinded on the take-up roll. On the way through the machine, it passes several deposition, treatment, and measurement zones.

The deposition zones are usually located around the chilling drums. These drums maintain the substrate temperature within a tolerable range during the heat load of the deposition process.

Roll-coating machines, as shown in Figure 15.4, have been developed since the 1950s. They are mainly used for evaporation purposes. Evaporation of aluminum from electrically heated boats is the most common deposition technique. The coated material is preferably used in the packaging industry. Another large volume production is the deposition of metal layers for capacitors. These two applications sum up to several million square meter of coated material per year. On the contrary, the market volume of the transparent electrodes on polymer film is quite small. However, the driving forces behind that application are the electronic and the solar industries. Both industries are growing rapidly. This makes the technology interesting and has led to an increased competition among manufacturers. Presently two companies in Europe have set up a large-scale production of one or more products listed in Table 15.2: Southwall Europe GmbH (Germany) and ISF/Bekaert (Belgium). Other companies exist mainly in the United States and in Asia. Owing to the present boom of the solar industry, the installation of new facilities can be expected.

In the following, the most important features of the roll-coating equipment will be discussed.

15.4.1
Vacuum System

The common sputter processes require a pressure range between 0.1 and 1 Pa. The gas atmosphere is mainly composed of argon. The background pressure plays an important role for a consistent deposition of conducting oxides. Several authors have found that the specific resistivity of ITO is dropping when the background pressure is increased from 1×10^{-5} to 1×10^{-3} Pa [28,29]. However, the author of this article observed an increase in the specific resistivity when the background pressure exceeds 3×10^{-3} Pa (Figure 15.5).

Figure 15.5 shows the typical U-shaped dependence of the sheet resistance on the oxygen flow rate for the ITO deposition. The two curves are measured at different background pressure. The oxygen that is added to the sputtering gas for fine-tuning of the layer properties is partly replaced by oxygen originating from water vapor. Additionally, hydrogen is incorporated into the layer structure. On the basis of this result and the literature, one can draw the conclusion that a water vapor pressure around 1×10^{-3} Pa is obviously the optimum for ITO deposition.

Achieving this background pressure within a short time is not an easy task to fulfill. The main difficulty consists in the outgassing behavior of the polymer film. Compared to other types of vacuum equipment, two additional gas sources become active when the film is unwinded. Trapped air is set free and water is outgassed from the bulk polymer film. The first component can easily be pumped away in the winding zone. The second component is the most serious problem for roll coaters.

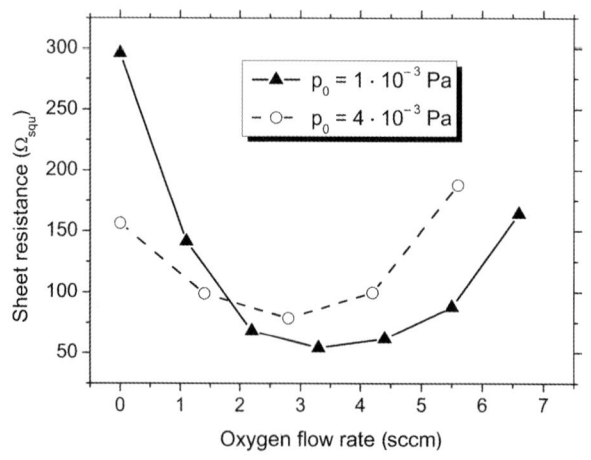

Figure 15.5 Dependence of the sheet resistance on the oxygen flow rate for two different levels of background pressure.

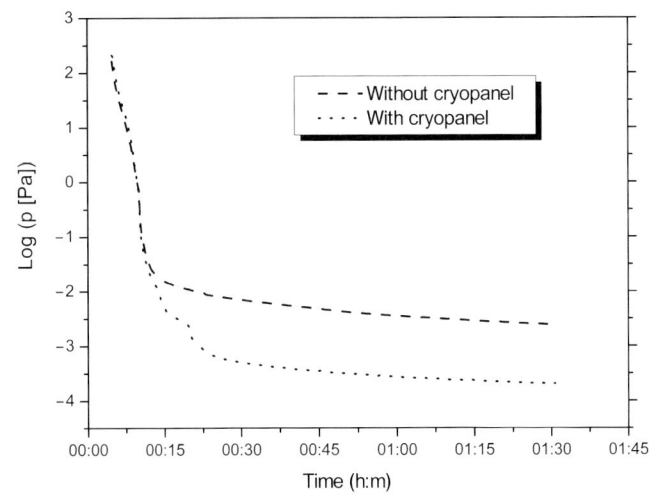

Figure 15.6 Pump-down curve of a typical sputter roll coater with and without the use of a cryopanel (e.g., *coFlex* 600).

The polymer film is exposed to vacuum only for the moment when it is unwinded. This means that the deposition starts within less than a minute after the film is in "contact" with vacuum for the first time. The water vapor atmosphere can only be kept in a tolerable range if pumps with high throughput are installed. Typically, the task is solved by a combination of turbomolecular pumps and cryopanels. The first is maintaining the argon gas pressure. The second is removing excess water from the vessel. The difference in the pump-down curve with and without cryopanel can be seen in Figure 15.6.

Additionally, plasma treatment and/or the deposition of a barrier layer reduce the activation of water molecules during the deposition process. This is a part of the technological know-how that needs to be adapted for all equipment.

15.4.2
Winding System

The basic functions of the winding system are the appropriate transport of the film, the chilling of the material during the deposition, and maintaining the roll quality at the rewinder.

A very important part of the winding system is the chilling drum. Depending on the number of deposition processes, there can be one or more chilling drums in one roll coater. They have to ensure that the film temperature is kept within the specified range. Owing to the sputtering process, a considerable heat load arrives at the substrate. It consists of heat of condensation, the energy of the sputtered particles, radiation, and energetic plasma particles. It can sum up to several tens of electron volt per deposited atom [30].

Only 10% of the heat is transferred by direct contact between the polymer web and the chilling drum surface. The remaining 90% is transferred by the gas that is trapped between the drum and the backside of the film. This is mainly water that was adsorbed on the surface of the film. The gas pressure between the film and the drum can reach values of more than 1000 Pa [31]. The exact heat transfer coefficient is difficult to estimate because it strongly depends on the geometry of the trapping. However, several models predict that the heat transfer is increased when the gap distance is minimized. Therefore, the requirements to the surface finish and cleanliness of the drums are very strict. A mean roughness of 0.2–0.5 µm is a typical value. Combined with the film roughness, one yields 0.5–10 µm gap width. This provides heat transfer coefficients of up to $1000\,\mathrm{W\,m^{-2}\,K^{-1}}$ [31]. Normally, these values are sufficient for the sputtering of conducting oxides.

15.4.3
Inline Measurement System

An inline measurement of both the optical and the electrical coating results is necessary for assuring the quality of the product. Naturally, the electrical measurement has the priority for transparent electrodes. There are two types of measurement units used in roll coaters. The first type is a contact mode device. This device consists of one or more electrically insulated rollers in the winding system. The rollers must be placed at locations where the coated side is touching the roller surface. In a simple configuration, the resistance between two rollers or the resistance between one roller and the ground potential is measured. A more sophisticated system consists of several independent segments and allows a crossweb measurement of the sheet resistance. The drawback of this type of measurement is the high contact resistance between the film and the roller and between the roller and the sliding contact. Because of these contact resistances, an offset is measured generally and the signal has a noise in the range of 10%. Because of that, the contact mode setup can only be applied for the sheet resistance range above $50\,\Omega_{squ}$.

The preferred measurement units of transparent electrodes for organic photovoltaics are contactless measurement devices. They consist of a source of electromagnetic radiation and a receiver. The coated film is moved between these two parts. The receiver measures the attenuation of the signal. This is a measure of the sheet resistance of the transparent electrode. The applicability of this device is well proven in the range below $20\,\Omega_{squ}$. In this range the noise is very low and small shifts during the deposition process can be determined with high accuracy. However, the calibration to the absolute value needs to be done by a conventional off-line four-point probe.

The electrical measurement is often combined with an inline system for the measurement of reflection and transmission. Such a device can determine the layer thickness, assuming the knowledge of the right dispersion function of the layer material.

Normally, it is essential to know the optical behavior between 350 and 800 nm. A great variety of detectors and lamp systems exist for this spectral range. Usually, the light is guided to the measurement point by an optical fiber. The generated measurement signal is transported to the detector by an optical fiber as well. Some difficulty is included in the reflectivity measurement. Usually, black rollers are applied to suppress fake illumination through the transparent substrate.

The setup for the measurement of transmission and reflection is relatively easy to handle. More sophisticated devices use ellipsometric data or work in the UV spectral range. However, in most cases this is not necessary for the purpose of production of transparent electrodes.

15.5
Alternative Approaches

The preceding chapters demonstrated that single layers with transparent conductive oxides do not provide the ideal solution for solar cells in the superstrate setup. It is rather difficult to achieve the necessary low sheet resistance at an acceptable price.

On the basis of this fact, various approaches had been developed for overcoming this difficulty. The most matured one is the replacement of the single ITO or ZAO layer by a so-called IMI-stack. IMI stands for a system of at least three layers built up in the structure insulator–metal–insulator. The key element for the sheet resistance is the metal layer.

This metal layer must be an excellent electrical conductor and should be thin enough for providing a sufficient transmittance. The transmittance in the visible spectral range is still increased by embedding the metal into two nonabsorbing high-index layers (Figure 15.7).

Silver or gold or alloys of these metals can be used for IMI layers. In this case, the thin metal determines the sheet resistance of the stack entirely. Thin silver layers on plastic substrates can reach specific resistivity values as low as $0.5 \times 10^{-5}\,\Omega\,cm$ [32]. Even if ITO is used as a high index embedding material, the influence on the sheet resistance can be neglected [33]. The typical characteristics of IMI systems are shown in Figure 15.8.

The diagram shows both the transmittance and the sheet resistance of an ITO–Ag–ITO stack, which is dependent on the silver thickness. There is a maximum in transmittance at a silver thickness of approximately 8 nm. The transmittance in the maximum is above 80%, which is quite sufficient for many applications. The

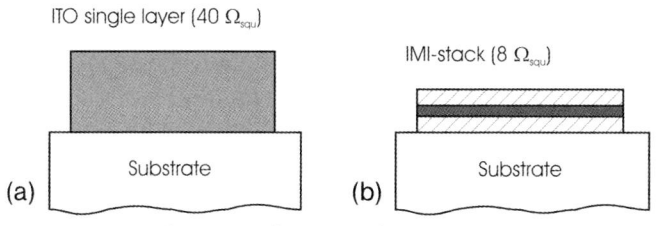

Figure 15.7 Typical structure of an IMI-stack (b) compared to a 150 nm ITO single layer (a). The structure of the layers is drawn to scale.

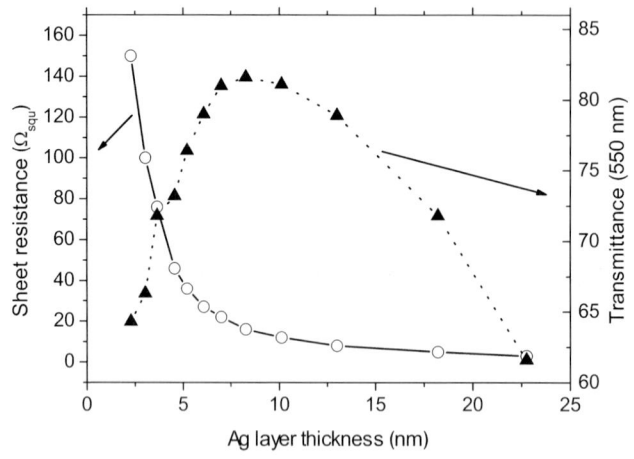

Figure 15.8 Dependence of the sheet resistance and the transmittance on the silver thickness for an ITO–Ag–ITO stack.

sheet resistance is in the range 10–20 Ω_{squ}. This value is already fairly below a typical single layer electrode and is thus opening the way for several applications shown in Table 15.2. The total thickness of whole layer stack is between 80 and 90 nm. This can effectively be deposited by sputtering in a roll-coating machine.

Figure 15.8 reveals a typical characteristic for IMI systems. Layers thinner than 8 nm show a decreased transmittance. This behavior is due to plasmon resonances occurring in discontinuous thin silver films. Very thin films are forming islands that are only weakly connected to each other. Both the optical and electrical properties of such films are not suitable for an application as transparent electrodes. In most cases, the minimum layer thickness for the formation of a continuous film is best for transparent electrodes.

A very serious problem for IMI systems is the stability against environmental stress. Although the metal layer is formed from a noble metal, there are mechanisms of degradation. The most important one is the agglomeration. This behavior is especially pronounced in very thin silver layers. The whole system has a general tendency to minimize its free enthalpy. The global minimum, that is, the stable condition, is corresponding to an agglomerated silver layer (this is true in the silver thickness range that is typical for IMI stacks). This situation can be avoided by preventing the silver from rearranging. One way is embedding the silver into suitable materials including a moisture barrier. Another possibility is the use of gold instead of silver or doping of silver with other metals like palladium. The advantage of IMI stacks with respect to layer thickness can be seen in Figure 15.9.

Another important advantage over single layer TCO materials was found by Lewis et al. [34]. The authors studied the bending behavior of the polymer films with different conducting coatings. The IMI system turned out to be superior after a series of bending around a 6 mm tube. This is a very important feature for any type of flexible electronic device.

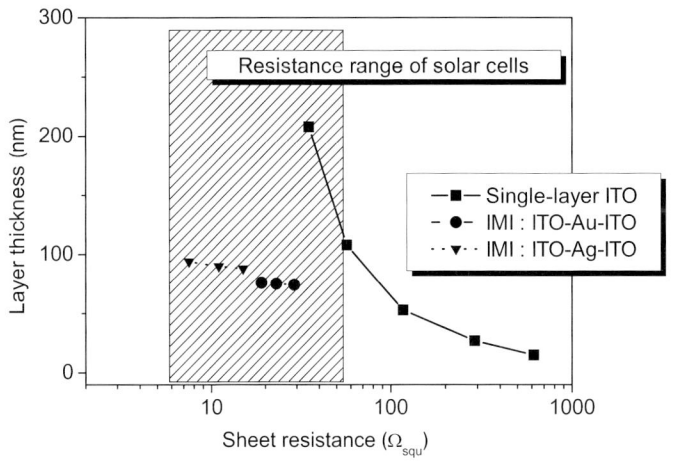

Figure 15.9 Necessary layer thickness and corresponding resistance range for different types of transparent electrodes.

Besides the use of IMI systems, presently extensive R&D work is put into the carbon nanotubes (CNT). These nanomaterials have fascinating properties, among which is a very high electrical conductivity along the tube. There are attempts from several research groups to suspend CNTs in a host material and form a transparent composite layer. Electrodes with sheet resistances below $100\,\Omega_{squ.}$ could be achieved [35]. The transmittance is not yet sufficient for the needs of solar cells. At the moment this technology is still on a laboratory level. Nonetheless, this development is very promising for a future solution.

References

1 Product information DupontTejinFilms.
2 Kittler, C. and Ritchie, I.T. (1982) Continuous coating of indium tin oxide onto large area flexible substrates, SPIE. *Optical Thin Films*, **325**, P 61.
3 Rauf, I.A. (1996) Structure and properties of tin-doped indium oxide thin films prepared by reactive electron-beam evaporation with a zone-confining arrangement. *Journal of Applied Physics*, **79**, P 4057.
4 Qiao, Z. (2003) Fabrication and study of ITO thin films prepared by magnetron sputtering, Doctoral Theses, Duisburg-Essen.
5 Calnan, S., Upadhyaya, H.M., Du, H.L., Thwaites, M.J. and Tiwari, A.N. (2006) Effects of substrate temperature on indium tin oxide films deposited using high target utilisation sputtering. 21st European Photovoltaic Solar Energy Conference, Dresden, Germany.
6 Malkomes, N., Vergöhl, M. and Szyszka, B. (2001) Properties of aluminum-doped zinc oxide films deposited by high rate mid-frequency reactive magnetron sputtering. *Journal of Vacuum Science & Technology A: Vacuum Surfaces and Films*, **19** (2), 414–419.
7 Szyszka, B. (1999) Transparent and conductive aluminum doped zinc oxide

films prepared by mid-frequency reactive magnetron sputtering. *Thin Solid Films*, **351**, 164–169.

8 Huepkes, J., Rech, B., Calnan, S., Kluth, O., Zastrow, U., Siekmann, H. and Wuttig, M. (2006) Material study on reactively sputtered zinc oxide for thin film silicon solar cells. *Thin Solid Films*, **502**, 286–291.

9 Schmidt, N.W., Totushek, T.S., Kimes, W.A., Callender, D.R. and Doyle, J.R. (2004) Effect of substrate temperature and near substrate plasma density on the properties of aluminum doped zinc oxide. *Journal of Applied Physics*, **94** (9), 5514–5521.

10 Cormia, R.L., Fenn, J.B., Memarian, H. and Ringer, G. (1998) Roll-to-roll coating of indium tin oxide - a status report. 41st Annual Technical Conference Proceedings of Society of Vacuum Coaters, pp. 452–457.

11 Chandrasekhar, R. and Choy, K.L. (2001) Innovative and cost-effective synthesis of indium tin oxide films. *Thin Solid Films*, **398–399**, 59–64.

12 Savu, R. and Joanni, E. (2006) Low-temperature, self-nucleated growth of indium-tin oxide nanostructures by pulsed laser deposition on amorphous substrates. *Scripta Materialia*, **55** (11), 979–981.

13 Singh, V., Saswat, B. and Kumar, S. (2005) Low temperature deposition of indium tin oxide (ITO) films on plastic substrates. Materials, Integration and Technology for Monolithic Instruments, March 29–30, San Francisco, CA, USA, Materials Research Society, Warrendale, PA, USA, pp. 139–144.

14 Kim, D.-H., Park, M.-R. and Lee, G.-H. (2006) Preparation of high quality ITO films on a plastic substrate using RF magnetron sputtering. *Surface and Coatings Technology*, **201** (3–4), 927–931.

15 Song, P.K., Shigesato, Y., Yasui, I., Ow-Yang, C.W. and Paine, D.C. (1998) Study on the crystallinity of tin-doped indium oxide films deposited by dc magentron sputtering. *Japanese Journal of Applied Physics*, **37**, 1870–1876.

16 Fahland, M., Charton, C. and Klein, A. (2002) Deposition technology of transparent conductive coatings on PET foils. 45th Annual Technical Conference Proceedings of Society of Vacuum Coaters, pp. 487–491.

17 Vasu, V. and Subrahmanyam, A. (1990) Reaction kinetics of the formation of indium tin oxide films grown by spray pyrolysis. *Thin Solid Films*, **193–194**, 696–703.

18 May, C. and Struempfel, J. (1999) ITO coating by reactive magnetron sputtering–comparison of properties from DC and MF processing. *Thin Solid Films*, **351**, 48–52.

19 Bender, M., Trube, J. and Stollenwerk, J. (1999) Characterization of RF/dc magnetron discharge for the sputter deposition of transparent and highly conductive ITO films. *Applied Physics*, **A69**, 397–401.

20 Cebulla, R., Wendt, R. and Ellmer, K. (1998) Al-doped zinc oxide films deposited by simultaneous rf and dc excitation of a magentron plasma. *Journal of Applied Physics*, **83** (2), 1087–1095.

21 Ellmer, K. (2007) Magnetron sputtering of transparent conductive zinc oxide: relation between the sputtering parameters and the electronic properties. *Journal of Physics D: Applied Physics*, **33**, R17–R32.

22 van Hest, M.F.A.M., Dabney, M.S., Perkins, J.D. and Ginley, D.S. (2006) High-mobility molybdenum doped indium oxide. *Thin Solid Films*, **496**, 70–74.

23 De Bosscher, W., Delrue, H., Van Holsbeke, J., Matthews, S. and Blondeel, A. (2005) Rotating cylindrical ITO targets for large area coating. 48th Annual Technical Conference Proceedings of Society of Vacuum Coaters, pp. 111–115.

24 Hagentstroem, H., Milde, F. and Struempfel, J. (2006) Advanced processes and tools for architectural glass coating. 6th International Conference on Coating on Glass and Plastics, Proceedings, pp. 55–57.

25 Safi, I. (2000) Recent aspects concerning DC reactive magnetron sputtering of thin films: a review. *Surface and Coatings Technology*, **127**, 203–219.

26 Berg, S. and Nyberg, T. (2005) Fundamental understanding and modelling of reactive sputtering processes. *Thin Solid Films*, **476**, 215–230.

27 Frach, P., Glöß, D., Goedicke, K., Fahland, M. and Gnehr, W.-M. (2003) High rate deposition of insulating TiO_2 and conducting ITO films for optical and display applications. *Thin Solid Films*, **445**, 251–258.

28 Lee, B., Yi, C., Kim, I. and Lee, S. (1997) Effect of base pressure in sputter deposition on characteristics of indium tin oxide thin films. *Materials Research Society Symposium Proceedings*, **424**, 335–340.

29 Ishibashi, S., Higuchi, Y., Ota, Y. and Nakamura, K. (1990) Low resistivity indium tin oxide for transparent conducting films. I. Effect of introducing H_2O and H_2 gas during dc magnetron sputtering. *Journal of Vacuum Science & Technology A: Vacuum Surfaces and Films*, **8** (4), 1399–1402.

30 Druesedau, T., Bock, T., John, T. and Klabunde, F. (1999) Energy transfer into the growing film during sputter deposition: an investigation by calorimetric measurements and Monte Carlo simulations. *Journal of Vacuum Science & Technology A: Vacuum Surfaces and Films*, **17** (5), 2896–2905.

31 Roehrig, M., Bright, C. and Evans, D. (2000) Vacuum heat transfer models for web substrates: review of theory and experimental heat transfer data. 43rd Annual Technical Conference Proceedings of Society of Vacuum Coaters, pp. 335–341.

32 Charton, C. and Fahland, M. (2004) Electrical properties of Ag films on polyethylene terephthalate deposited by magnetron sputtering. *Thin Solid Films*, **449**, 100–104.

33 Fahland, M., Karlsson, P. and Charton, C. (2001) Low resistivity transparent electrodes for displays on polymer substrates. *Thin Solid Films*, **392**, 334–337.

34 Lewis, J., Grego, S., Chalama, B., Vick, E. and Temple, D. (2004) Electromechanics of a highly flexible transparent conductor for display applications. 47th Annual Technical Conference Proceedings of Society of Vacuum Coaters, pp. 129–133.

35 Gruner, G. (2006) Two-dimensional carbon nanotube networks: a transparent electronic material. *Materials Research Society Symposium Proceedings*, **905E**, 0905-DD06-05.1–0905-DD06-05.11.

16
Novel Electrode Structures for Organic Photovoltaic Devices
Michael Niggemann and Andreas Gombert

16.1
Introduction

In general, it is accepted that organic solar cells have to be manufactured using reel-to-reel processing to utilize the advantages of this new type of solar cell. These cells can be processed by dry or wet deposition at relatively moderate temperatures. In practice, polymer films function well as flexible substrates. Thus, the processing technologies of microreplication and perforation that have become well-established procedures for polymer films can be applied to these solar cells to optimize the efficiency and lower costs. Using microreplication and perforation technologies, the standard planar device architecture of organic solar cells can be transferred to a three-dimensional (3D) device architecture.

The basic requirements on electrode structures are as follows:

(a) They must allow an efficient extraction of charge carriers. Although primarily determined by the organic semiconductor–electrode interface, this can also be influenced by the electrode geometry with respect to the photoactive bulk.
(b) They have to guarantee an efficient illumination of the semiconductor.
(c) The effective sheet resistance has to be sufficiently low to minimize ohmic losses.

These three requirements are, in principle, fulfilled by planar electrode systems for organic solar cells comprised of a transparent indium-tin oxide (ITO) electrode, which is coated with (PEDOT:PSS) poly(3,4-ethylenedioxythiophene:poly(styrenesulfonate) serving as anode and an evaporated highly reflective metal cathode. However, there is one major obstacle faced in this standard electrode system hindering the ambitious low-cost goal for organic solar cells: the ITO film. ITO is a well-known transparent conductive oxide (TCO) with a superior figure of merit, but indium suffers from a shortage and its limited resources make it rather expensive. It is also the standard TCO used for flat panel displays. Thus, costs that fit the requirements of photovoltaic applications in the future cannot be realistically expected. In photovoltaics, it is a common approach to use a metal grid to collect the current with comparably low shadowing loss. But, to produce such a metal grid

Organic Photovoltaics: Materials, Device Physics, and Manufacturing Technologies.
Edited by Christoph Brabec, Vladimir Dyakonov, and Ullrich Scherf
Copyright © 2008 WILEY-VCH Verlag GmbH & Co. KGaA, Weinheim
ISBN: 978-3-527-31675-5

with adequate dimensions either is technologically elaborate (e.g., lithographic structuring) or suffers from the low conductivity of finely printed silver lines. (The latter often needs thermal treatment to improve the conductivity and has thicknesses in the micrometer range, inferring problems with thin-film formation of the active layers.) Therefore, two possible solutions exist: (1) to develop new TCOs as substitutes for ITO or (2) to use novel 3D electrode configurations as described in the following. All requirements for optimizing the electrode structure are set by the device efficiency, lifetime, and cost. Light trapping plays an important role in optimizing efficiency. This topic can be addressed with a 3D electrode or device structures. Furthermore, nanostructured electrodes open up new applications for organic solar cells.

Three-dimensional microstructured and nanostructured substrates can be manufactured by microreplication on large areas at low costs. The first step is producing the master structures. This can be done for larger structures (typical dimensions >10 µm) by using ultra precision machining and for smaller structures by applying lithographic methods. In the latter case, however, upscaling is a severe problem. The only lithographic method suitable for very fine pitch structures on large areas is interference lithography. It offers the possibility to generate periodic surface relief structures in a large variety of shapes and dimensions from subwavelength nanostructures (200 nm grating period) up to several tens of micrometers on large areas up to 1 m^2. The interference pattern is recorded by a thin photoresist layer. In a subsequent developing process, the intensity profile of the interference pattern is transferred into a three-dimensional surface relief. The photoresist structures can be replicated into nickel stamps for subsequent microreplication into a suitable substrate material. Several generations of nickel stamps from one master structure provide many embossing tools and numerous replicas making the whole process very cost efficient. Roller UV embossing of microstructures is an already established technology as an industrial process (e.g., Autotype Ltd, Wantage, England). This method allows a very good reproducibility and very high resolution and is well suited for mass production as the setup is realized in a reel-to-reel process. Both nanoelectrode and microprism types of substrates, which will be presented in the following sections, were produced by these processes. The metallic nanoelectrodes were prepared by oblique evaporation on 3D nanostructured surfaces using the self-shadowing effect of 3D structures (see Figure 16.2) assisted by sacrificial layers. Thus, sophisticated electrode structures can be produced without using costly lithographic steps.

16.2
Buried Nanoelectrodes

Buried nanoelectrodes are vertically orientated metallic electrodes embedded in the photoactive layer of the solar cell. Different electrode configurations of planar electrodes and nanoelectrodes can be imagined. Asymmetric electrodes are obtained by combining a planar electrode with a transparent array of vertical nanoelectrodes [1,2].

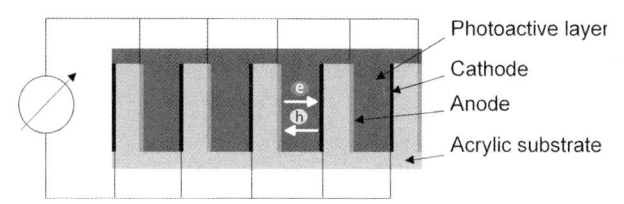

Figure 16.1 Cross section of an organic photovoltaic cell based on the concept of interdigital buried nanoelectrodes. Both electrodes are vertically orientated forming an array of interdigital electrodes.

The substitution of both planar electrodes by buried nanoelectrodes results in an interdigital electrode setup (see Figure 16.1). The manufacture and the application of this architecture for organic photovoltaic devices will be discussed here in detail. The two electrodes with their specific work functions are deposited on the side walls of transparent nanolamellae. The width of the cavities between the electrodes measures approximately half the period of the nanostructure. Utilizing interference lithography, structure periods in the subwavelength range can be achieved. A 720 nm period structure will be presented here, but structures with periods as low as 250 nm, resulting in electrode spacings of 125 nm, have been successfully realized at Fraunhofer ISE.

These dimensions are very attractive from the perspective of organic electronic devices, as the thickness of the semiconductor layers, limited by the relatively low charge carrier mobilities, is within this range. In principle, the geometry of vertically oriented nanoelectrodes follows the requirement of separating the light absorption from the charge carrier transport. Hence, the effective thickness of the absorber can be increased while keeping the distance of the electrodes small enough for an efficient charge extraction. However, technological limitations with respect to the aspect ratio of the structures have to be considered. In contrast to the planar solar cell architecture based on an ITO substrate, both electrodes are deposited prior to the application of the photoactive film. This can be advantageous for the device processing, because no vacuum deposition step is required after the deposition of the organic compounds. Another feature is the absence of a highly reflecting back electrode. This makes semitransparent applications possible. As the dimensions of these structures range within the wavelength of light, near-field optics plays an important role. Strong interactions of the light in terms of high absorption in transversal magnetic (TM) polarization are observed in previous simulations [2].

16.2.1
Experimental

The substrate for buried nanoelectrodes is made by replication of holographically originated structures into an acrylic UV curable photopolymer. The microstructure has a lamellae period of 720 nm, a depth of approximately 400 nm, and a cavity width of 400 nm. The anode and cathode can be realized by oblique evaporation of different metals resulting in an interconnection at the lamellae tips. The proper separation of

444 | *16 Novel Electrode Structures for Organic Photovoltaic Devices*

these two electrodes is challenging from a technological point of view and is performed in a lift-off procedure. The process sequence is illustrated in Figure 16.2. Prior to the deposition of the metal electrodes, a sacrificial layer is evaporated under steep opposing angles so that only a capping is formed on the tips of the lamellae.

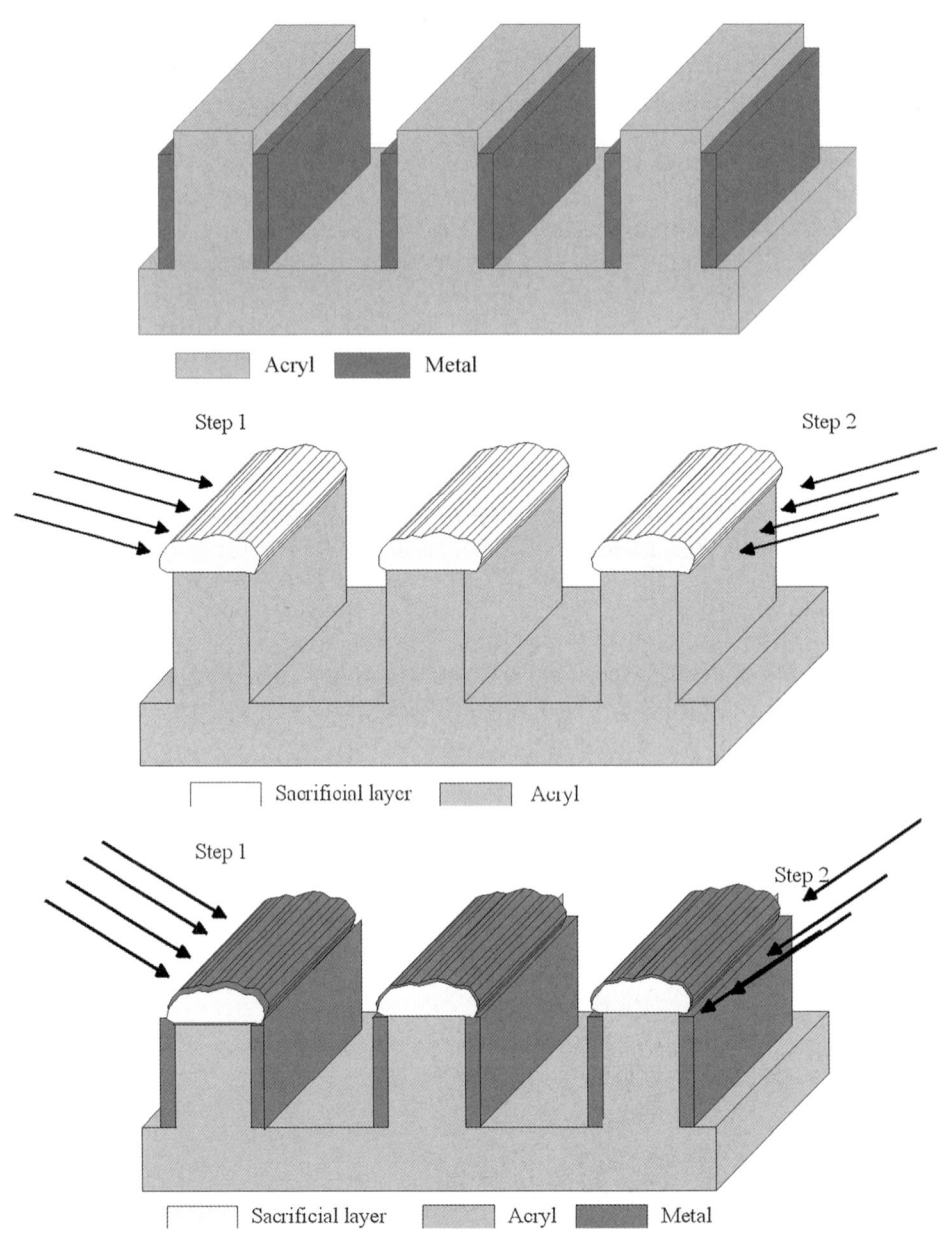

Figure 16.2 Nanoelectrode preparation by oblique evaporation and subsequent lift-off procedure.

Subsequently, the vertical metal electrodes are evaporated under a certain angle in such a way that only the vertical regions and the tips of the structure are coated. The areas of the respective electrodes are defined by shadow masks during evaporation under inclined angles, whereas the interconnection of the gridlines is made by evaporation of a stripe perpendicular to the lamellae from normal incidence. The electrode materials have to provide good contacts for the organic semiconductor. One prerequisite is matching the electrode work function to the respective quasi-Fermi potentials of the semiconductor. Investigations of inverting the layer sequence in organic solar cells have shown that titanium is suited as a predeposited electron contact for the acceptor material 6,6-phenyl C_{61}-butyric acid methyl ester (PCBM) [3]. The conductivity of the titanium electrode was enhanced by an underlying aluminum layer.

Initial investigations of organic photovoltaic devices on the basis of buried nanoelectrodes were made with gold as anode, which forms an ohmic contact with the hole transport level highest occupied molecular orbital (HOMO) of poly-3-hexylthiophene (P3HT). The electrical characterization of the interdigital nanoelectrode structure prior to filling with the photoactive material resulted in a high shunt resistance of $2.9 \times 10^5 \, \Omega \, cm^2$. This is sufficiently high in comparison to the shunt resistance of $1540 \, \Omega \, cm^2$ determined from a dark current–voltage characteristic of a P3HT:PCBM organic solar cell [4]. Scanning electron microscopy (SEM) images of nanoelectrodes deposited on an acrylic substrate before and after filling with the photoactive material are shown in Figures 16.3 and 16.4, respectively.

The current–voltage characteristic of the nanoelectrode structure filled with the photoactive composite is shown in Figure 16.5. The high shunt resistance photovoltaic cell has a small fill factor (FF) of 25%, which can be explained by the semiconductor–electrode contacts rather than by a low shunt resistance owing to the direct contact of the electrodes. The low rectification can be attributed to an insufficient selectivity of the gold electrode for the respective charge carriers. Both types of charge carriers can be injected into the gold electrode. This is confirmed by investigations on planar solar cells using gold as electron collecting material [5].

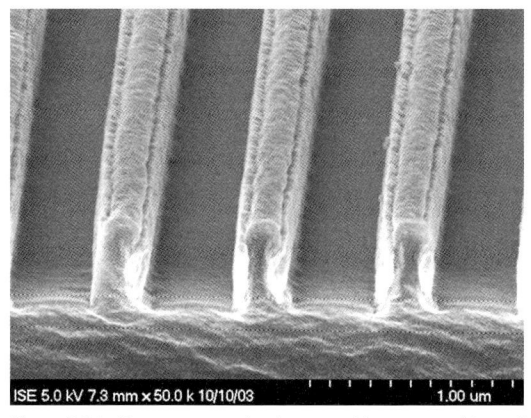

Figure 16.3 Nanostructured substrate with separated interdigital nanoelectrodes.

446 *16 Novel Electrode Structures for Organic Photovoltaic Devices*

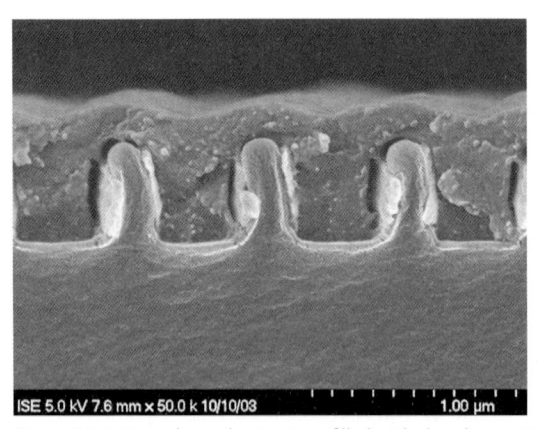

Figure 16.4 Nanoelectrode structure filled with the photoactive material.

A detailed investigation of interface dipole layers between C_{60} and Au is presented by Veenstra et al. [6]. A shift of the lowest unoccupied molecular orbital (LUMO) of the C_{60} derivative to the Fermi level of the metal electrode was observed.

Obviously, the hole contact needs to be optimized to allow an efficient charge collection. Suitable anode materials are currently under investigation. Promising candidates are metal oxides such as V_2O_5 and MoO_3 [7].

Buried nanoelectrodes are still at an early stage of investigation. The approach to collect the charge carriers by nanoelectrodes widens the possibilities to create organic optoelectronic devices possessing properties that cannot be achieved with the planar electrode geometry such as

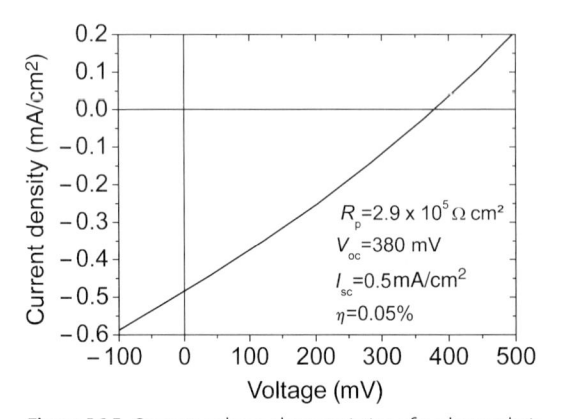

Figure 16.5 Current–voltage characteristics of a photovoltaic device based on interdigital buried nanoelectrodes. The parallel resistance prior to the filling of the cavities with photoactive material was $R_p = 2.9 \times 10^5\,\Omega\,cm^2$ (FF = 25%, $I_{sc} = 0.5\,mA\,cm^{-2}$, $V_{oc} = 380\,mV$, $P_{in} = 100\,mW\,cm^{-2}$, $\eta = 0.05\%$).

(a) Both electrodes can be made of nontransparent metal. This widens the range of available electrode materials and may allow a better matching of the electrode work function to the organic semiconductor. The disadvantages of some transparent electrode materials, such as the diffusion of indium from transparent ITO electrodes into the organic layer as observed in the case of organic light emitting diodes (OLEDs) or the high price of ITO, can be circumvented [8,9].

(b) Although light absorption has been observed in vertically oriented nanoelectrodes, a significantly lower sheet resistance accompanied by a high transparency for optimized electrode structures is expected.

(c) The deposition of the photoactive material from solution is made after the deposition of electrodes onto the substrate. This can be advantageous for future device production. Further, no final vacuum procedure, as necessary for organic solar cells in standard geometry, is required.

(d) The interdigital electrode geometry is very versatile. It is already being applied in organic field effect transistors and is very common in sensor applications. The outstanding property of buried nanoelectrodes with a vertical orientation and a spacing as low as 125 nm even for heteroelectrodes can lead to a significant improvement of these devices. Therefore, nanoelectrodes can serve as a versatile platform for novel organic-electronic devices and circuits.

16.3
Organic Photovoltaic Devices on Functional Microprism Substrates

Using substrates with a three-dimensional surface topography in the micrometer scale offers the possibility to increase the light absorption within the photoactive layers of a solar cell.

Here, a cell architecture for organic solar cells based on a microprism substrate is presented. In principle, the geometry can be described by a folded planar solar cell (see Figure 16.6). This setup comprises two advantages in comparison to the standard organic solar cell. First, the ITO electrode is substituted by a highly

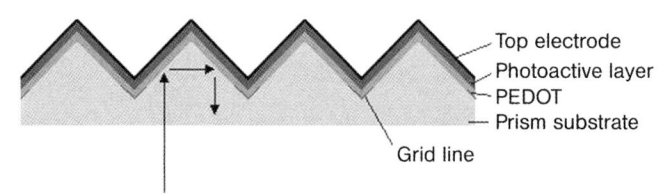

Figure 16.6 Schematic cross section of an organic microprism solar cell. The transparent microstructured polymer substrate is made by microreplication. Metal gridlines are deposited in the valley of the structures. The transparent PEDOT anode, the photoactive layer, and the evaporated cathode follow the shape of the microstructure.

conductive polymer layer with a supporting metal grid located in the valley of the structure. Second, the microprism structure contributes to an increased light absorption owing to a twofold reflection of the incident light. The microprism structure can be originated by micromachining or by interference lithography. The cost efficient substrate generation by microreplication is common to both approaches. An optimal shape and dimension of the three-dimensional device architecture has to be derived from modeling supported by experiments. The influence of the incident angle and the twofold reflection is investigated by optical modeling. These calculations allow the favorable shape of the structure to be estimated. The dimensions of the structure are primarily determined by the conductivity of the transparent electrode and the widths of the metal gridlines. These calculations are discussed in the following sections. For a further optimization of the structure, the experimental constraints, for example, the thin-film formation on corrugated surfaces, have to be considered.

16.3.1
Optical Simulations

The influence of the incident angle and twofold reflections on the light absorption in the photoactive layer will be addressed here. In particular, photoactive materials with different spectral extinction coefficients will be compared. The calculations of the absorption in distinct layers of the solar cell were performed by rigorous coupled wave analysis (RCWA), a method which was implemented for describing near-field phenomena at corrugated interfaces [10–13]. Another approach especially suited for thin-film devices is the transfer matrix formalism [14].

Each compound of an organic solar cell has to be described by its optical constants. Ellipsometry is the most widely used method for their determination. The crystallinity of some photoactive compounds, the interaction with the substrate, and the influence of the deposition process can strongly affect the optical constants. Optical anisotropy can be observed for some photoactive blends [15–18]. For reasons of simplicity, optical isotropy is assumed in the optical modeling described here. Nevertheless, this effect has to be kept in mind while evaluating the calculation results. The absorption properties of a bulk-heterojunction absorber layer are predominantly determined by the molecular structure of the single components, the packing in the thin film, and the ratio of the donor and acceptor compounds. The process of crystallization can be influenced by the annealing of the photoactive layers or by using solvents with high boiling points. This phenomenon can be observed for the P3HT:PCBM donor–acceptor system [15,19,20]. Other organic semiconductors do not tend to crystallize owing to their complex molecular structure. As will be shown in the following, the interdependence between the optical constants (refractive indices and extinction coefficients) of the photoactive layer and the three-dimensional geometry of the device can strongly affect the light harvesting. Therefore, we compare three different types of photoactive layers. One based on a poly(2-methoxy-5-(3′,7′-dimethyloctyloxy)-1,4-phenylene vinylene) (MDMO-PPV): PCBM blend with a weight ratio of 1 : 4 [14] and two blends based on P3HT:PCBM

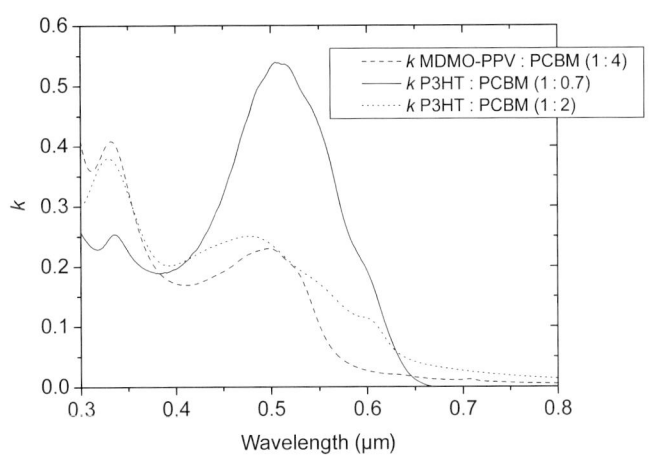

Figure 16.7 Extinction coefficients of three photoactive composites.

with different weight ratios of 1 : 2 and 1 : 0.7. The extinction coefficients are shown in Figure 16.7.

The local maximum values of the extinction coefficient at 330 nm and 480–520 nm can be attributed to the acceptor (in this case PCBM) and the donor components, respectively. Compared to P3HT-based photoactive layers, a scaling of the absorption coefficient with the weight fraction is obvious. With respect to the solar spectrum, an efficient photon harvesting at the visible and near infrared, and therefore a high extinction coefficient in this wavelength range, is desirable. Nevertheless, composites that do not tend to form crystalline phases because of their complex molecular structure of the donor component require a large fraction of 75–80% (by weight) PCBM to allow an efficient charge carrier transport [21,22]. In contrast, planar P3HT:PCBM-based photoactive layers that form crystalline phases [15] show the best performance with strongly reduced fraction of PCBM of approximately 40% (by weight) [23,24].

Optical constants for the PEDOT:PSS layer were determined for the formulation AI4083 purchased from H. C. Starck and may deviate slightly from the optical constants of the highly conductive PEDOT CPP 105 D, which was used in these devices. The thickness of the PEDOT:PSS layer was 100 nm. Both polarizations of the incident light – transversal electric (TE) and TM were considered. The optics of the thin-film system which is deposited on the prismatic structure is investigated by simulating the effect of inclined incident light onto the thin-film system. In the second step, a second absorption process under 45° is considered. As a consequence, the influence of the rounded tip of the holographic structure on the light absorption will not be considered here.

The light is incident from a semiinfinite nonabsorbing medium with a refractive index of 1.5. The light absorption by the substrate is considered by a 200 μm thick layer described by the dielectric function of glass. The thickness of the photoactive layer and the incident angle are varied from 60 to 190 nm and from 0° (normal incidence) to 85°, respectively.

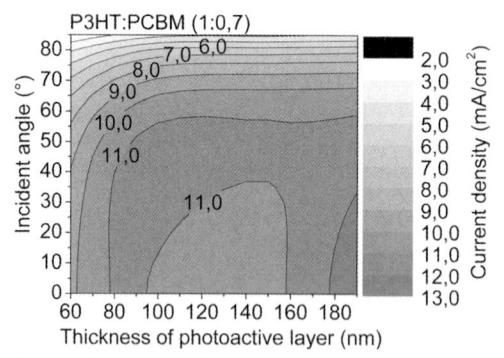

Figure 16.8 Calculated current density depending on the thickness of the photoactive layer of P3HT:PCBM (1:0.7) and the incident angle of the solar irradiation (AM 1.5, 100 mW cm^{-2}) assuming an internal quantum efficiency of unity.

In a first approximation, the short-circuit current density is calculated from the spectral absorption in the blend layer and the solar spectrum. To estimate the maximum current for the given system, an internal quantum efficiency (IQE) of unity is assumed. Calculations of the short-circuit current for the photoactive layer containing the highest fraction of P3HT (P3HT:PCBM 1:0.7) are shown in Figure 16.8.

For a normal incident angle (0°), the typical effect of a stepwise increase in the photocurrent with increasing film thickness and the occurrence of a local minimum for a thin-film system can be observed. This observation can be explained by interference. A local maximum can be observed at approximately 85 nm. An inclined incident angle does not lead to a significant increase in the photocurrent. An enhanced light absorption can be expected only in the range between 100 and 150 nm thick films by increasing the incident angle up to 45°. The situation changes for the photoactive layer with a lower fraction of the donor component (P3HT:PCBM 1:2) (see Figure 16.9) and therefore with a lower extinction coefficient in the visible range of the spectrum (see Figure 16.7).

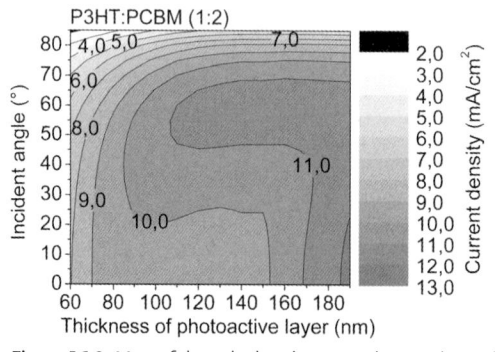

Figure 16.9 Map of the calculated current density depending on the thickness of the P3HT:PCBM (1:2) photoactive layer and the incident angle of the solar irradiation (AM 1.5, 100 mW cm^{-2}) assuming an internal quantum efficiency of unity.

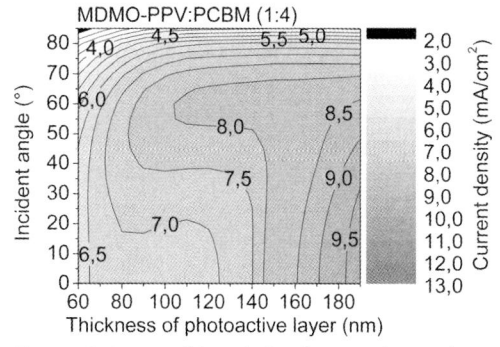

16.3 Organic Photovoltaic Devices on Functional Microprism Substrates | 451

Figure 16.10 Map of the calculated current density depending on the thickness of the MDMO-PPV:PCBM (1:4) photoactive layer and the incident angle of the solar irradiation (AM 1.5, 100 mW cm^{-2}) assuming an internal quantum efficiency of unity.

Up to a film thickness of approximately 160 nm, the current density can be increased by increasing the incident angle up to 55°. This is very beneficial for devices with limited charge carrier mobility and is valid, in particular, for the model donor–acceptor system consisting of MDMO-PPV and PCBM. A weight ratio of 1 : 4, and therefore a low extinction coefficient, produced devices that performed the best [25]. The shape of the calculated photocurrent map (see Figure 16.10) is similar to that given in Figure 16.9. Again, a higher photocurrent is obtained for a film thickness up to 120 nm by increasing the incident angle up to 60°.

The microprism structure provides the possibility of a second absorption process. Assuming normal incident light on a 90°-prism structure, two reflections at 45° will occur. Different scenarios for the spectral absorption of a 120 nm thick photoactive layer of P3HT:PCBM (1 : 2) are shown in Figure 16.11.

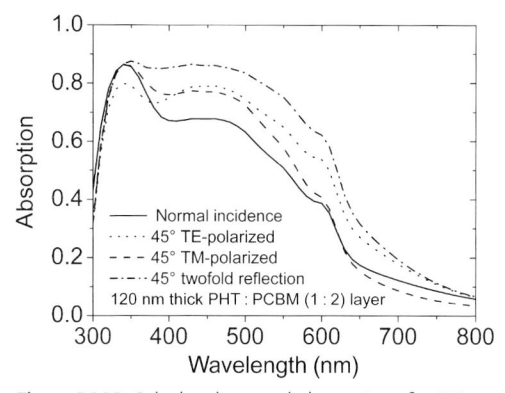

Figure 16.11 Calculated spectral absorption of a 120 nm thick P3HT:PCBM (1:2) photoactive layer under consideration of different illumination scenarios of the solar irradiation (AM 1.5, 100 mW cm^{-2}) assuming an internal quantum efficiency of unity.

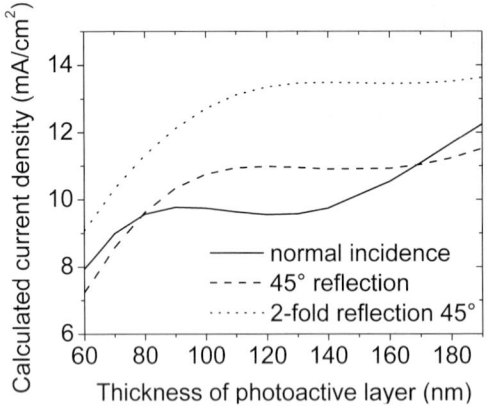

Figure 16.12 Calculated maximum current density (internal quantum efficiency of unity anticipated) for P3HT:PCBM (1:2) photoactive layer depending on the film thickness under consideration of different illumination scenarios of the solar irradiation (AM 1.5, 100 mW cm^{-2}) assuming an internal quantum efficiency of unity.

The different gain in absorption for TE- and TM-polarized light with respect to the normal incidence is because of the additional light absorption in the metal electrode for TM polarization. A further increase of the spectral absorption can be observed by considering the second absorption process (only the mean value of both polarizations is shown here). Figure 16.12 shows the calculated short circuit with respect to the thickness of the photoactive layer assuming solar irradiation under 100 mW cm^{-2} and an internal quantum efficiency of 1. A gain of 15–40% in the short-circuit current can be derived by comparing the case of normal incidence on a 120 nm thick photoactive layer with a single and a twofold reflection under 45°, respectively.

16.3.2
Dimensioning of the Microstructure

The dimensions of the microstructure strongly affect the efficiency of the solar cell, as the period of the microprism structure defines the distance of the gridlines. Ohmic losses have to be minimized by a sufficiently small grid distance, whereas the ratio between the lattice distance of the grid and the width of the conducting lines should be as large as possible to minimize the shadowing effect. In addition, the metal gridlines have to be sufficiently conductive so that the series resistance is minimized. The contributions of both the aluminum top electrode and the gold grid to the series resistance are neglected. The current–voltage characteristics of an infinitesimal small elementary cell with $V_{oc} = 600$ mV, $I_{sc} = 15$ mA cm^{-2}, and FF $= 0.55$ were taken as input parameters for the calculations. Such an elementary cell has an efficiency of approximately 5%. In order to determine the optimum lattice distance for a given type of structure, calculations were made for different gridline widths ranging from 2 up

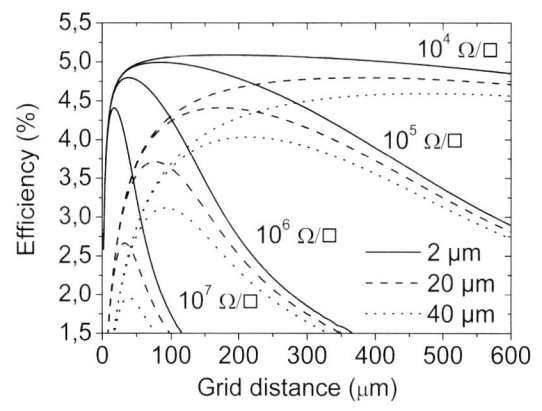

Figure 16.13 Calculation of the energy conversion efficiency of a microprism solar cell depending on the grid distance for different sheet resistances of the PEDOT layer and various widths of the gridlines.

to 40 μm and for a sheet resistance varying between 10^4 and 10^7 Ω/square (see Figure 16.13).

The two competing loss mechanisms – ohmic losses and losses by shadowing – determine the characteristics of the calculations. The steepness of the rising efficiency with increasing grid distance is strongly affected by the shadowing losses caused by the gridlines, whereas the decaying efficiency at larger grid distances is determined by the PEDOT:PSS (PEDOT CPP 105 D by H. C. Starck) sheet resistance and therefore by the ohmic losses. The geometry of the microprism structure is considered in the calculations by an effectively reduced current density caused by the tilted absorbing surface.

It is intuitive that the efficiency of the cell is weakly affected by the grid distance for small gridline widths and a small sheet resistance of the PEDOT. To minimize absorption losses in the PEDOT layer, the thickness should range between 80 and 120 nm. A strong variation of the PEDOT sheet resistance between approximately 10^4 and 10^5 Ω/square was measured for film thicknesses in this range. Assuming a conservative value of 10^5 Ω/square and a realistic width of the gold grid of 2 μm for structures (with a period on the order of 20 μm) generated by interference lithography, a suited grid distance of 22–215 μm (5% deviation from peak efficiency) can be extracted from the calculations shown in Figure 16.13.

This value corresponds to a structure period of 16–152 μm (the grid distance is calculated from the path lengths on the prismatic surface). Larger structure periods, for example, micromachined structures on the order of 100 μm allow the generation of approximately 20 μm wide gridlines. A lower sheet resistance of 10^4 Ω/square would be favorable. Allowing a 5% decrease in efficiency from an optimum period at 400 μm (96% of the elementary cell efficiency), the structure period can range between 150 and 650 μm. Scaling the suitable structure period over approximately one order of magnitude provides further possibilities to develop appropriate coating processes on these corrugated surfaces.

454 *16 Novel Electrode Structures for Organic Photovoltaic Devices*

Figure 16.14 Cross section of a microprism substrate with a structure period of 100 μm. The structure was replicated from a micromachined masterstructure (scanning electron microscopy image).

16.3.3
Experimental

Different types of structured substrates originated by micromachining and by interference lithography have been investigated. The micromachined structures had a period of 100 μm and relatively sharp tips (see Figure 16.14), whereas structures originated by interference lithography had a period of 20 μm and rounded tips (see Figure 16.15).

The metal grid was made by a combination of evaporation processes under inclined incident angles using the microstructure as a self-aligning shadow mask similar to the process described for the fabrication of nanoelectrodes in Section 16.2.

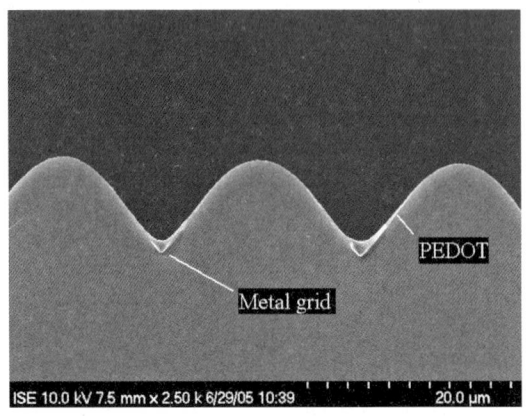

Figure 16.15 Cross section of a holographic microprism substrate with a structure period of 20 μm carrying a microgrid and coated with the polymer anode (scanning electron microscopy image).

High-quality microstructures with a small defect density are required to generate several centimeters long metal gridlines without interruptions. The minimum width of uniform gridlines scales with the period of the microstructure. The shape of the microstructure can have an additional influence. A gold grid with a mean width of 22.5 µm and a thickness of 100 nm resulting in an effective sheet resistance of 0.86 Ω/square was obtained for the micromachined (100 µm period) substrate, whereas 2–3 µm wide gold gridlines with a thickness of 50 nm and an effective sheet resistance of 10 Ω/square were reached for the holographic microstructure (20 µm period). Compared to the sheet resistance of ITO on foils of 50–60 Ω/square (10–15 Ω/square on glass), these low resistivities of the microgrid will lead to a further reduction of the losses for solar cell modules. Broader cell stripes can be realized and thus shadowing losses upon series circuitry of the cell stripes can be decreased.

A challenging task is the deposition of sufficiently homogeneous thin films from solution on these corrugated surfaces. The homogeneity of the film thickness is determined by the complex interplay of the shape and size of the microstructure, the rheological properties of the solutions, and the coating process and parameters. In general, an increased film thickness in the valley of the structures can be anticipated. This can cause absorption losses in the PEDOT layer and recombination losses in the photoactive layer, but it can also be advantageous by suppressing the electrical shunts between gridlines and the evaporated top electrode.

According to the optical simulations for a P3HT:PCBM (1:2) shown in Figure 16.12 a photoactive layer thickness ranging between 100 and 150 nm is favorable. Initial coating experiments were carried out with spin coating. A film thickness less than 200 nm was obtained on 80% of the surface of the micromachined substrates. A sufficiently thin PEDOT layer can be anticipated on 80% of the surface.

Solar cells based on micromachined prismatic structures suffered from electrical shunt defects predominately at the sharp tips of the structures [26]. However, this is not a principal drawback of micromachined structures as rounded tips could be manufactured with appropriate tools. Holographic microprisms have rounded tips that allow a better wetting of the photoactive layer in this region. As a consequence, the probability for electrical shunts is reduced. A homogeneous coating of the rounded tips with a film thickness of approximately 100 nm was observed. However, a film thickness exceeding 400 nm can be observed in the valley of the structure. Experimental details can be found in Refs [26,27]. An improvement of the homogeneity is expected by further optimizing the viscosity and spin coating parameters.

The current–voltage characteristics of organic solar cells built up on a micromachined and a holographic microprism substrate are shown in Figure 16.16. The photoactive layers consist of a P3HT:PCBM blend.

The low fill factor (0.35) of the cell based on the micromachined substrate can be traced back to electrical shunts predominately at the tips of the structure. This has been verified by shunt mapping using thermography and by electrical isolation of the tips. A higher observed short-circuit current and an enhanced light absorption of the whole device with respect to a planar reference cell based on ITO give a first indication for an enhanced light absorption [26]. However, owing to the variations of the film

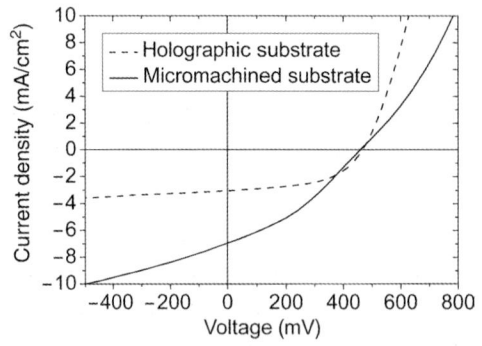

Figure 16.16 Current–voltage characteristics of organic solar cells on microprism substrates originated by holography and micromachining (photoactive layer: P3HT:PCBM).

thickness, such comparisons have to be carried out on a statistical basis. In contrast, the fill factor of the cell based on the holographic substrate is significantly higher (FF = 0.52) indicating a reduction of the shunts owing to a better wetting of the rounded tip. The differential resistances $R_s = 3.8\,\Omega\,\text{cm}^2$ at 1.0 V and $R_p = 840\,\Omega\,\text{cm}^2$ (under illumination) at 0 V were extracted from the current–voltage characteristics. These values are comparable with characteristic data for planar reference cells. Nevertheless, the short-circuit current and the open-circuit voltage are low. As the measured film thickness (100 nm at the tips) should be sufficient for the generation of significantly higher currents, a negative chemical interaction with the substrate might be one explanation. Additionally, the thin-film deposition needs to be further optimized.

Finally, we can draw the following conclusions: Using optical modeling, we could identify and quantify two effects that contribute to light harvesting in the photoactive layer – an inclined incident angle and a twofold reflection of the light. It turns out that the gain in light absorption caused by the inclined incident angle is significantly influenced by the absorption coefficient of the photoactive layer. Lower absorption coefficients result in a significantly higher gain in light absorption at inclined incident angles. In general, multiple reflections are favorable to increase the light absorption.

These results have to be evaluated with respect to the synthesis of novel photoactive compounds. The focus of ongoing research activities are to optimize the bandgap of the donor and acceptor components as well as to optimize the energy offset between the LUMO levels of the donor and acceptor components in order to reduce the energy loss caused by the charge carrier transfer. These effects can be achieved by incorporating additional functional groups into the polymer chain leading to copolymers [17,18,21,22]. As a consequence, the formation of crystalline phases together with an enhancement of the light absorption and an increased charge carrier mobility is suppressed. Additionally it is observed that a large fraction of the acceptor component is required for an efficient charge separation. Therefore, light trapping by microprism structures can be one route to compensate for these disadvantages by benefiting from the reduced bandgap and increased open-circuit voltage.

16.4
Anode Wrap-Through Organic Solar Cell

Low material costs and an efficient production were the main drivers for the development of the wrap-through organic solar cell. The costs are lowered predominately by excluding the ITO electrode and by an efficient production by incorporating reel-to-reel manufacturing processes.

Anode wrap-through organic solar cells allow the use of transparent electrodes with a significantly lower conductivity such as PEDOT:PSS in the range of several hundred Ω/squares. The current is collected over sufficiently small areas reduce the ohmic losses. Instead of a metal grid that can serve as a current-collecting structure, a polymeric hole contact is led through vias in the device to the backside of the perforated substrate, where a metal layer supports the transparent hole contact by enhancing the effective sheet conductance. In this way, a scalable parallel circuitry is formed without the need for a metal grid (see Figure 16.17). In contrast to the most widely used standard solar cell architecture, the organic solar cell is built on a thin plastic substrate with inverted layer sequence (see Figure 16.19), that is, starting with the metallic electron contact [3]. Next, the active absorber layer is applied, followed by the PEDOT:PSS layer, which forms the transparent hole contact.

A series connection of single cells to make a solar cell module with the required current–voltage characteristics can also be achieved monolithically on the substrate by connecting the electron contact of one cell stripe on the front side of the substrate and the metallic back contact of the neighbored cell stripe through additional vias in the substrate (Figure 16.18).

Two aspects of this solar cell architecture related to the electrode structures will be discussed in detail as follows: (1) the inversion of the layer sequence as a prerequisite for the presented architecture and (2) the dimensioning of the perforation.

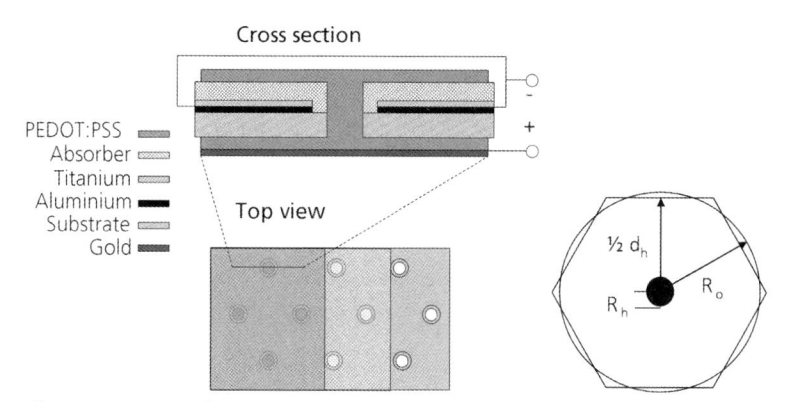

Figure 16.17 Top: Schematic drawing of the hole contact wrap-through concept. The organic solar cell is built up on a plastic substrate with inverted layer sequence [3]. The PEDOT:PSS hole contact is led through vias to the backside of the substrate, forming a scalable parallel circuitry. Bottom: geometry used for the dimensioning of the wrap-through devices.

Monolithic parallel and serial circuitry

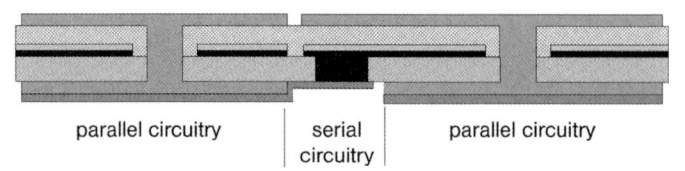

Figure 16.18 Monolithic serial circuitry based on the wrap-through concept. In addition to the parallel circuitry (see Figure 16.17), a serial circuitry can also be included. Therefore, the metallic electron contact of one cell segment and the metallic back contact of the neighbored cell segment are connected through additional vias in the substrate.

16.4.1
Organic Solar Cell with Inverted Layer Sequence

The most widely used organic solar cell architecture (electrode structure) is shown in Figure 16.19. The devices are fabricated on ITO-coated substrates. Between the photoactive layer and the ITO, a layer of poly(ethylene dioxythiophene) doped with poly(styrene sulfonate) is deposited. ITO/PEDOT:PSS forms the anode, which collects the holes. The cathode, collecting the electrons, is formed by thermally evaporated aluminum or calcium on top of the photoactive layer.

The inversion of the layer sequence cannot be done by simply inverting the deposition sequence of the electrodes. Aluminum is not suitable for a bottom electrode as its native oxide forms a barrier for the charge injection. The formulation

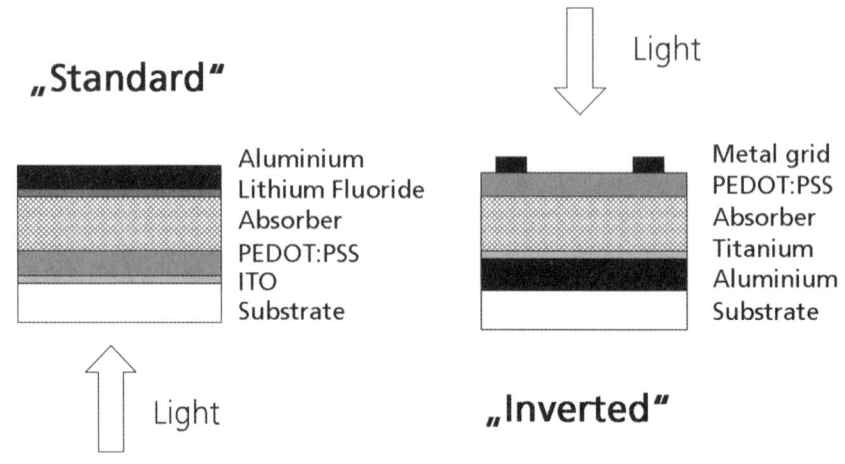

Figure 16.19 Comparison of the widely used layer sequence on ITO/PEDOT:PSS electrode (left) and inverted layer sequence (right), where the ITO is replaced by a metal grid for small area devices. In the next step, the metal grid is replaced by the wrap-through circuitry as shown in Figure 16.18.

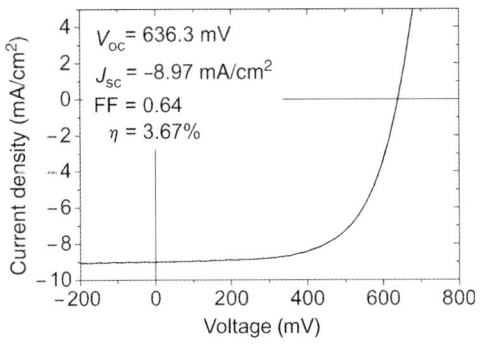

Figure 16.20 Current–voltage curve of an inverted bulk-heterojunction solar cell measured under 1000 W m^{-2} AM 1.5 solar spectrum on a Steuernagel solar simulator calibrated with a Si reference solar cell and using a mismatch factor to account for spectral mismatch (active area 1 cm^2).

[Graph shows: V_{oc} = 636.3 mV, J_{sc} = -8.97 mA/cm^2, FF = 0.64, η = 3.67%]

of the PEDOT:PSS anode has to be altered to allow for a good adhesion on the photoactive layer. Additionally, a low series resistance is required for solar cell architectures with current collecting structures such as metal grids or the presented wrap-through concept. Titanium supported by a highly conductive aluminum layer underneath is a suitable substitute for the aluminum cathode [3]. An alternative electrode is a solution-processed titanium oxide [29]. The inverted layer sequence device is shown in Figure 16.19. The sheet resistance of the polymer anode is supported by a metal grid.

Open-circuit voltage (630 mV) and fill factor (0.65) of the inverted solar cell are comparable with the state-of-the-art devices with ITO/PEDOT:PSS electrode and the standard layer sequence (see Figure 16.20). The lower short-circuit current density (8.6 mA cm^{-2}) as compared to 10–12 mA cm^{-2} for the state-of-the-art devices can be at least partially explained by the shadowing loss (~10%) induced by the metal grid and a smaller reflectivity of the titanium electrode in comparison with aluminum [24,30,31].

16.4.2
Calculation of Optimal Device Geometry for the Wrap-Through Device

The hole distance and diameter have to be optimized to maximize the solar cell efficiency [32,33]. Two competing loss mechanisms can be identified: ohmic losses and area losses. Ohmic losses increase with increasing distance of the holes. Area losses arise from the fraction of the hole surface area, which represents a photovoltaic inactive area.

The module efficiency is calculated using a numerical model. First, the hexagonal unit cell of the module pattern is approximated by a circular unit cell (see Figure 16.17, right hand side). Utilizing the radial symmetry, a one-dimensional model is sufficient to calculate the losses for this circular cell. The via distance in a hexagonal pattern is then calculated by using a scaling factor, taking into account a hexagon with the same area as the circle used for the calculation. The device area used to calculate the current densities is the total area of the circle/hexagon, whereas the integration for current collection is taken from the outer radius R_o to the radius of the via hole R_h, as the via does not contribute to current generation. In this way, the area loss arising from the via is included. The following differential equations are to be solved:

$$\frac{dV}{dr}(r) = -\frac{\rho}{2\pi r}I(r),$$

$$\frac{dI}{dr}(r) = -2\pi r j(V(r)),$$

$$V(R_o) = V_0,$$

$$I(R_o) = 0,$$

where $V(r)$ is the voltage at radius r and ρ denotes the sheet resistance of the PEDOT:PSS layer; $I(r)$ is the lateral current flowing in the PEDOT:PSS layer toward the hole, whereas $j(V(r))$ is the current density delivered by the solar cell segments of infinitesimal size at the radius r. R_o denotes the outer radius of the circular cell. The boundary condition $V(R_o) = V_0$ has to be chosen such that the electrical power of the device $P = V(R_h)I(R_h)$ is maximized, where R_h is the radius of the via hole, that is, the solar cell module works at the maximum power point. Varying R_o and finding the maximum power point for each R_o by varying $V(R_o)$ leads to curves as shown in Figure 16.21.

Taking the measured current–voltage curve of a small area inverted device with an efficiency of 3.4% [33] as input (a higher efficiency of 3.67% has already been reached, see Figure 16.20), the efficiency of wrap-through modules was calculated as a function of the via hole distance. A via hole diameter of 200 μm can be easily realized in reel-to-reel production (e.g., hot needle or laser perforation). One can obtain nearly the same efficiency for the module as for the small area device using a readily

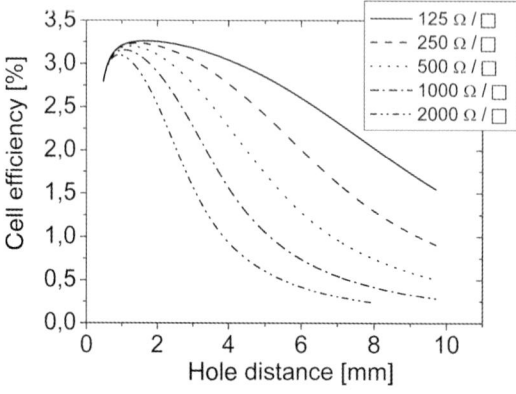

Figure 16.21 Calculation of the device efficiency upon variation of the via hole distance for different sheet resistances (200 μm hole diameter).

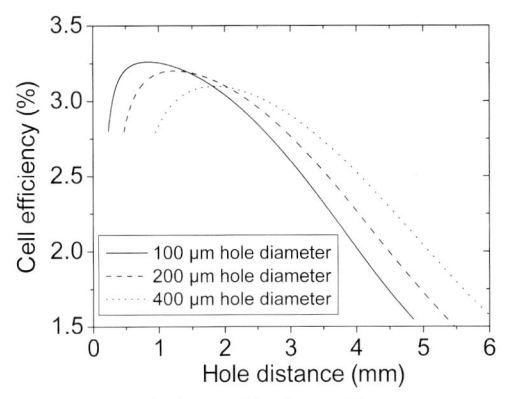

Figure 16.22 Calculation of the device efficiency upon variation of the different via hole diameters (500 Ω/square sheet resistance).

available transparent hole contact material with 500 Ω/square sheet resistance. For better conductive materials, the maximum obtainable module efficiency increases slightly, but the more striking effect is that the via hole distance becomes less critical for obtaining high efficiencies (see Figure 16.21). Varying the via hole diameter while keeping the sheet resistance constant (500 Ω/square) induces different area losses, leading to an increased maximally obtainable module efficiency for small via holes and a reduced value for larger via holes (see Figure 16.22).

A nontrivial effect is that for smaller via holes and same via hole distance, the efficiency is smaller at comparably large via hole distances. Trivially, one would expect that for the same via hole distance and smaller via hole diameter, the efficiency is higher owing to the larger active area (which is in fact true for comparably small via hole distances). But, for large via hole distances the loss is dominated by the ohmic losses. The reduced efficiency for smaller via holes is, therefore, because of the fact that the main ohmic loss occurs close to the via hole where the current is high and the cross-sectional area ($2\pi R_h d_{PEDOT:PSS}$) of the conductive sheet is small. For a smaller via hole, this cross-sectional area is even smaller, so that the ohmic loss for the same hole distance is higher for a smaller via hole as compared to a larger one. Nevertheless, a smaller via hole leads to an overall higher efficiency, if the via hole distance is correctly chosen.

16.4.3
Performance of Wrap-Through Devices

The current–voltage curve of the wrap-through organic solar cell on an area of 2 cm^2 and an image of the respective device are shown in Figures 16.23 and 16.24. Experimental details can be found in Ref. [32]. The short-circuit current scales with device area as expected, where the short-circuit current density and the inherently area-independent quantities V_{oc} and FF are independent of the cell size.

The FF is not yet as high as expected. This is owing to an additional series resistances in the via holes, because the PEDOT:PSS layer does not fill the via holes

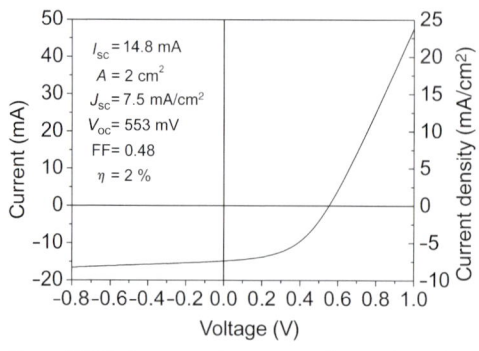

Figure 16.23 Current–voltage curve of wrap-through solar cells with an active area of 2 cm² (1 mm hole distance). The short-circuit current is scaling with area, while V_{oc}, J_{sc}, FF, and η are nearly unaffected.

well because of the low solid content of the PEDOT:PSS solution. As the liquid evaporates, the film in the via hole shrinks and often detaches from the via hole boundary. Additionally, there might also be shunts that reduce the FF as well. Nevertheless the series resistance appears to be the more severe limitation, as the current density under forward bias is about 10 times lower than for small area devices. A shunt by contrast should increase the current flow in the forward direction and significantly lower V_{oc}.

The connection of single cells to make a solar cell module can be done through perforations as well. For reel-to-reel production, this should be producible monolithically on the substrate. The polymer anode and the metal layers need to be separated to define the area of the single cell. This can be done manually on the lab scale. For future production, either printing techniques [34] or coating techniques

Figure 16.24 Wrap-through organic solar cell (without encapsulation).

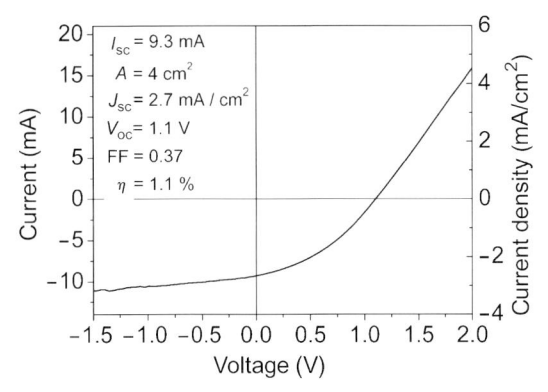

Figure 16.25 Current–voltage curve of a wrap-through solar cell with two cell segments of approximately 2 cm² active area each connected in series monolithically. The photovoltage is doubled as expected for a serial circuitry. The short-circuit current density is calculated using the complete area of the device (including the circuitry), such that it should be ideally half the value of a single cell element. In this case one would expect approximately 7 mA cm^{-2} for the single segment, such that the module would be expected to have a short-circuit current density of 3.5 mA cm^{-2}, if the loss owing to the circuitry would be negligible.

combined with a subsequent structuring, for example, laserscribing [35], need to be established. The series interconnection of the cell was performed through additional perforations, which were filled by conductive silver ink. Figure 16.25 shows the current–voltage curve of the first device manufactured with such a monolithic serial circuitry of two wrap-through cell segments of nearly 2 cm² each. The serial circuitry works in principle, as the open-circuit voltage is doubled. The short-circuit current density is lower than expected, but it is still acceptable when the area loss induced by the manual PEDOT:PSS structuring is accounted for. The fill factor is significantly lower for this device as compared to the devices with only parallel circuitry, suggesting again severe series resistance and shunts in at least one cell segment.

16.5
Summary

Extending the planar cell architecture of organic solar cells to three-dimensional device architecture widens the capacity for further optimization and development of these novel organic photovoltaic devices. To meet the requirements for the commercialization of organic solar cells, all aspects of an overall device optimization – solar cell efficiency, stability, and costs – have to be considered. For the presented device architectures – based on buried nanoelectrodes, microprisms, and the wrap-through solar cell – cost-efficient manufacturing technologies, such as microreplication and perforation technologies, are utilized and the ITO electrode is substituted by less expensive materials. Buried nanoelectrodes are a very versatile approach. They have the potential to increase the solar cell efficiency through efficient charge carrier extraction and can also serve as a platform for novel organic-electronic devices and circuits. Compared to the presented cell architectures, buried nanoelectrodes are still at an early stage of development. Important topics for elaboration are the development of suitable electrodes, the optimization of the device geometry on the basis of optical modeling, and the development of reel-to-reel evaporation processes.

The microprism solar cell architecture utilizes the light trapping effect of inclined incident light in the thin film and a twofold reflection. From optical modeling, an increased gain in light absorption can be predicted for weakly absorbing photoactive layers. This makes the application of the microprism cell architecture for amorphous copolymer-acceptor bulk heterojunctions with optimized energy levels promising. The coating of these corrugated surfaces needs to be further improved before the predicted increase in the solar cell efficiency with respect to the planar solar cell architecture can be demonstrated.

The wrap-through concept serves as a suitable module for ITO-free reel-to-reel and therefore cost-efficient, large area organic solar cell modules. First devices with active areas of 2–4 cm^2 have been manufactured reproducibly with decent power conversion efficiencies of up to 2% (1000 W m^{-2} AM 1.5) for parallel wrap-through circuitry. The first device with additional serial circuitry has shown a power conversion efficiency of 1.1%. Besides, optimizing the devices for high efficiencies, the next steps toward commercialization will be the further upscaling of the devices and the encapsulation of the ITO-free, thin, flexible solar cell with suitable barriers to protect the organic semiconductors and the metallic electrode against oxygen and water vapor.

Acknowledgments

This work was supported by the Bundesministerium für Bildung und Forschung (BMBF) under contract number 01SF0119 and by an internal research project of the Fraunhofer Institute for Solar Energy Systems (ISE).

We thank the following people for their contributions on the development of the novel electrode structures for organic photovoltaic devices.

The work on nanoelectrode structures was supported by Daniel Ruf and Claas Müller from the Institute of Microsystem Technology (IMTEK) at the Albert Ludwigs University Freiburg im Breisgau and by Benedikt Bläsi from Fraunhofer ISE.

Special thanks to Autotype Ltd, Wantage, England for the replication of the nanostructured and microstructured substrates.

We thank Philippe Lalanne from Laboratoire Charles Fabry de L'Institut d'Optique, Orsay Cedex, France for the simulation tool and Harald Hoppe from TU Ilmenau for the determination of the optical constants of the MDMO-PPV:PCBM blend. The development of the wrap-through solar cell was carried out by Birger Zimmermann. Markus Glatthaar initially developed the inverted organic solar cell and carried out the electrical modeling of the module circuitry. Many thanks to Andreas Georg, Andreas Hinsch, Kristian Silvester-Hvid, Peter Lewer, and Moritz K. Riede for their contribution to the work presented in this chapter.

References

1. Niggemann, M., Glatthaar, M., Gombert, A., Hinsch, A. and Wittwer, V. (2004) Diffraction gratings and buried nano-electrodes – architectures for organic solar cells. *Thin Solid Films*, **451**–452, 619–623.
2. Niggemann, M. (2005) Fundamental investigations on periodic nano- and microstructured organic solar cells, PhD thesis, Albert-Ludwigs-Universitaet Freiburg im Breisgau, Germany.
3. Glatthaar, M., Niggemann, M., Zimmermann, B., Lewer, P., Riede, M., Hinsch, A. and Luther, J. (2005) Organic solar cells using inverted layer sequence. *Thin Solid Films*, **491** (1–2), 298–300.
4. Schilinsky, P., Waldauf, C., Hauch, J. and Brabec, C. (2004) Simulation of light intensity dependent current characteristics of polymer solar cells. *Journal of Applied Physics*, **95** (5), 2816–2819.
5. Zimmermann, B., Glatthaar, M., Niggemann, M., Riede, M. and Hinsch, A. (2005) Electroabsorption studies of organic bulk-heterojunction solar cells. *Thin Solid Films*, **493** (1–2), 170–174.
6. Veenstra, S., Heeres, A., Hadziioannou, G., Sawatzky, G. and Jonkman, H. (2002) On interface dipole layers between C-60 and Ag or Au. *Applied Physics A: Materials Science & Processing*, **75** (6), 661–666.
7. Shrotriya, V., Li, G., Yao, Y., Chu, C.W. and Yang, Y. (2006) Transition metal oxides as the buffer layer for polymer photovoltaic cells. *Applied Physics Letters*, **88** (7), P 073508.
8. de Jong, M.P., Simons, D.P.L., Reijme, M.A., van IJzendoorn, L.J., van der Gon, A.W.D., de Voigt, M.J.A., Brongersma, H.H. and Gymer, R.W. (2000) Indium diffusion in model polymer light-emitting diodes. *Synthetic Metals*, **110** (1), 1–6.
9. Kim, J., Granstrom, M., Friend, R., Johansson, N., Salaneck, W., Daik, R., Feast, W. and Cacialli, F. (1998) Indium-tin oxide treatments for single- and double-layer polymeric light-emitting diodes: the relation between the anode physical, chemical, and morphological properties and the device performance. *Journal of Applied Physics*, **84** (12), 6859–6870.
10. Moharam, M.G. and Gaylord, T.K. (1982) Diffraction analysis of dielectric surface-relief gratings. *Journal of the Optical Society of America*, **72** (10), 1385–1392.
11. Moharam, M.G. and Gaylord, T.K. (1981) Rigorous coupled-wave analysis of planar-grating diffraction. *Journal of the Optical Society of America*, **71** (7), 811–818.
12. Lalanne, P. and Morris, G.M. (1996) Highly improved convergence of the coupled-wave method for TM polarization. *Journal of the Optical Society of America A: Optics Image, Science and Vision*, **13** (4), 779–784.
13. Lalanne, P. and Jurek, M.P. (1998) Computation of the near-field pattern with the coupled-wave method for transverse magnetic polarization. *Journal of Modern Optics*, **45** (7), 1357–1374.
14. Hoppe, H., Arnold, N., Meissner, D. and Sariciftci, N. (2003) Modeling of optical absorption in conjugated polymer/fullerene bulk-heterojunction plastic solar cells. *Thin Solid Films*, **451**–452, 589–592.
15. Erb, T., Zhokhavets, U., Gobsch, G., Raleva, S., Stuhn, B., Schilinsky, P., Waldauf, C. and Brabec, C.J. (2005) Correlation between structural and optical properties of composite polymer/fullerene films for organic solar cells. *Advanced Functional Materials*, **15** (7), 1193–1196.
16. Nguyen, T.Q., Martel, R., Bushey, M., Avouris, P., Carlsen, A., Nuckolls, C. and Brus, L. (2007) Self-assembly of 1-D organic semiconductor nanostructures. *Physical Chemistry Chemical Physics*, **9** (13), 1515–1532.
17. Persson, N., Sun, M., Kjellberg, P., Pullerits, T. and Inganas, O. (2005) Optical properties of low band gap alternating copolyfluorenes for photovoltaic devices. *The Journal of Chemical Physics*, **123** (20), P 204718.

18 Persson, E., Wang, X., Admassie, S., Inganäs, O. and Andersson, M. (2006) An alternating low band-gap polyfluorene for optoelectronic devices. *Polymer*, **47** (12), 4261–4268.

19 Li, G., Shrotriya, V., Yao, Y. and Yang, Y. (2005) Investigation of annealing effects and film thickness dependence of polymer solar cells based on poly (3-hexylthiophene). *Journal of Applied Physics*, **98** (4), 043704-1–043704-5.

20 Yamanari, T., Taima, T., Hara, K. and Saito, K. (2006) Investigation of optimum conditions for high-efficiency organic thin-film solar cells based on polymer blends. *Journal of Photochemistry and Photobiology A: Chemistry* **182** (3), 269–272.

21 Wienk, M.M., Struijk, M.P. and Janssen, R.A. (2006) Low band gap polymer bulk heterojunction solar cells. *Chemical Physics Letters*, **422** (4–6), 488–491.

22 Muhlbacher, D., Scharber, M., Morana, M., Zhu, Z.G., Waller, D., Gaudiana, R. and Brabec, C. (2006) High photovoltaic performance of a low-bandgap polymer. *Advanced Materials*, **18** (21), P 2884.

23 Kim, Y., Choulis, S., Nelson, J., Bradley, D., Cook, S. and Durrant, J. (2005) Composition and annealing effects in polythiophene/fullerene solar cells. *Journal of Materials Science*, **40** (6), 1371–1376.

24 Reyes-Reyes, M., Kim, K., Dewald, J., Lopez-Sandoval, R., Avadhanula, A., Curran, S. and Carroll, D. (2005) Meso-structure formation for enhanced organic photovoltaic cells. *Organic Letters*, **7** (26), 5749–5752.

25 Shaheen, S.E., Brabec, C.J., Sariciftci, N.S., Padinger, F., Fromherz, T. and Hummelen, J.C. (2001) 2.5% efficient organic plastic solar cells. *Applied Physics Letter*, **78**, 841–843.

26 Niggemann, M., Glatthaar, M., Lewer, P., Muller, C., Wagner, J. and Gombert, A. (2006) Functional microprism substrate for organic solar cells. *Thin Solid Films*, **511–512**, 628–633.

27 Niggemann, M., Ruf, D., Blasi, B., Glatthaar, M., Riede, M., Muller, C., Zimmermann, B. and Gombert, A. (2005) Functional substrates for flexible organic photovoltaic cells. *Proceedings of the SPIE*, **5938**, 1–9.

28 Scharber, M., Muhlbacher, D., Koppe, M., Denk, P., Waldauf, C., Heeger, A. and Brabec, C. (2006) Design rules for donors in bulk-heterojunction solar cells – towards 10% energy-conversion efficiency. *Advanced Materials*, **18** (6), 789–794.

29 Waldauf, C., Morana, M., Denk, P., Schilinsky, P., Coakley, K., Choulis, S.A. and Brabec, C.J. (2006) Highly efficient inverted organic photovoltaics using solution based titanium oxide as electron selective contact. *Applied Physics Letters*, **89** (23), P 233517.

30 Ma, W.L., Yang, C.Y., Gong, X., Lee, K. and Heeger, A.J. (2005) Thermally stable, efficient polymer solar cells with nanoscale control of the interpenetrating network morphology. *Advanced Functional Materials*, **15** (10), 1617–1622.

31 Vanlaeke, P., Vanhoyland, G., Aernouts, T., Cheyns, D., Deibel, C., Manca, J., Heremans, P. and Poortmans, J. (2006) Polythiophene based bulk heterojunction solar cells: morphology and its implications. *Thin Solid Films*, **511–512**, 358–361.

32 Zimmermann, B., Glatthaar, M., Niggemann, M., Riede, M.K., Hinsch, A. and Gombert, A. (2007) ITO-free wrap through organic solar cells – a module concept for cost-efficient reel-to-reel production. *Solar Energy Materials and Solar Cells*, **91** (5), 374–378.

33 Glatthaar, M. (2007) Zur funktionsweise organischer solarzellen auf basis interpenetrierender donator/akzeptor-Netzwerke, PhD thesis, Albert-Ludwigs-Universitaet Freiburg im Breisgau, Germany.

34 Aernouts, T., Vanlaeke, P., Geens, W., Poortmans, J., Heremans, P., Borghs, S., Mertens, R., Andriessen, R. and Leenders, L. (2004) Printable anodes for flexible organic solar cell modules. *Thin Solid Films*, **451–452**, 22–25.

35 Dennler, G., Lungenschmied, C., Neugebauer, H., Sariciftci, N.S. and Labouret, A. (2005) Flexible, conjugated polymer-fullerene-based bulk-heterojunction solar cells: basics, encapsulation, and integration. *Journal of Materials Research*, **20** (12), 3224–3233.

B
Packaging

17
Flexible Substrates Requirements for Organic Photovoltaics
William A. MacDonald

17.1
Introduction

There is currently considerable interest in photovoltaics and growing interest in organic photovoltaics (o-PV) on a flexible substrate as it opens new PV cell design opportunities because of their light weight, mechanical flexibility, and semitransparency. In addition, the conjugated polymers being developed for organic cells can be made soluble in common solvents and hence can be deposited by solution processing and printing technologies. This opens up the possibility for low-cost mass production by roll-to-roll (R2R) manufacturing. There currently is considerable activity in developing flexible electronic devices based on organic semiconductors and organic light-emitting polymers [1]. This emerging technology area is slightly more mature than the corresponding flexible o-PV area, and the substrate requirements for flexible electronics [1–3] are in many ways similar to the property set, which is likely to be required for o-PV devices. This property set for flexible substrates for the PV industry is less well defined at this stage, but the PV industry will be able to "piggyback" on the learning that has been developed within the flexible electronic programs. This review will concentrate on biaxially oriented polyester films and will also discuss the properties of these materials in the context of the likely property set requirements for o-PV devices.

17.2
Polyester Substrates

Polyethylene terephthalate (PET), for example, DuPont Teijin Films Melinex® polyester film, and polyethylene naphthalate (PEN), for example, DuPont Teijin Films Teonex® polyester film, are biaxially oriented semicrystalline films [4]. The difference in chemical structure between PET and PEN is shown in Figure 17.1.

The substitution of the phenyl ring of PET by the naphthalene double ring of PEN has very little effect on the melting point (T_m), which increases by only a few degree

Organic Photovoltaics: Materials, Device Physics, and Manufacturing Technologies.
Edited by Christoph Brabec, Vladimir Dyakonov, and Ullrich Scherf
Copyright © 2008 WILEY-VCH Verlag GmbH & Co. KGaA, Weinheim
ISBN: 978-3-527-31675-5

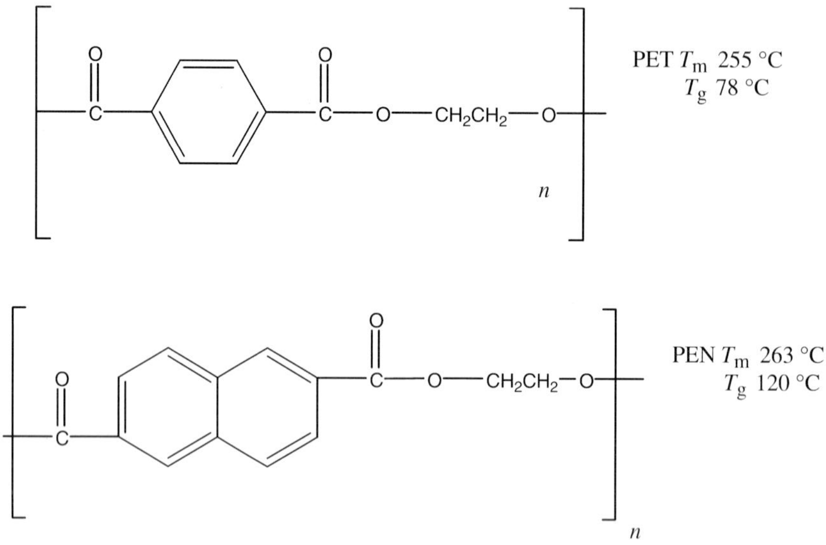

Figure 17.1 Chemical structure of PET and PEN films.

PET T_m 255 °C
T_g 78 °C

PEN T_m 263 °C
T_g 120 °C

centigrades. However, there is a significant effect on the glass transition temperature (T_g) (T_g is the temperature at which the polymer changes from a rigid glass to a rubber or polymer melt, and the polymer molecules start to show significant mobility), which increases from 78 °C for PET to 120 °C for PEN [1–3]. PET and PEN films are prepared by a process whereby the amorphous cast is drawn in both the machine and transverse directions. The biaxially oriented film is then heat set to crystallize the film [5,6].

The success of the polyester film in general application is derived from the properties of the basic polymer coupled with the manufacturing process of biaxial orientation and heat setting, as described above. These properties include high mechanical strength, good resistance to a wide range of chemicals and solvents, low water absorption, excellent dielectric properties, good dimensional stability, and good thermal resistance in terms of shrinkage and degradation of the polymer chains. Fillers can be incorporated into the polymer to change the surface topography and opacity of the film. The film surface can also be altered by the use of pretreatments to give a further range of properties, including enhanced adhesion to a wide range of inks, lacquers, and adhesives. These basic properties have resulted in PET films being used in a wide range of applications, ranging from magnetic media and photographic applications, where optical properties and excellent cleanliness are of paramount importance, to electronics applications such as flexible circuitry and touch switches, where thermal stability is the key. More demanding polyester film markets that exploit the higher performance and benefits of PEN include magnetic media for high-density data storage and electronic circuitry for hydrolysis-resistant automotive wiring. The property set described above provides the basis to meet the demands of the o-PV market.

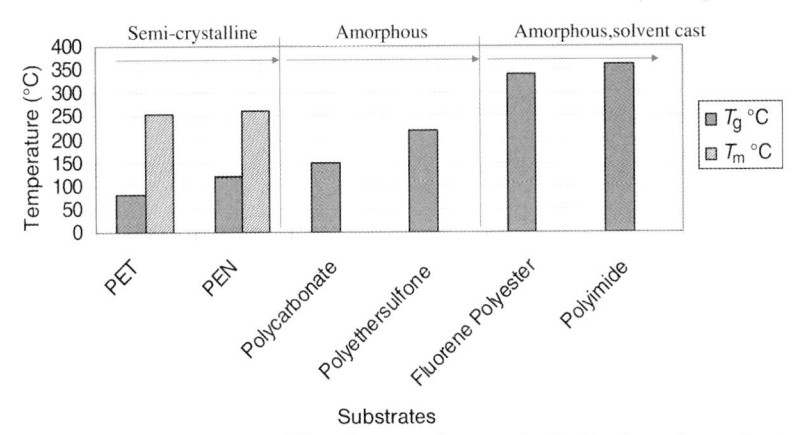

Figure 17.2 Glass transition of film substrates of interest for flexible electronics applications.

It is interesting to contrast these films with other materials considered for flexible PV devices. Taking organic-based films first, possible candidate films are shown in Figure 17.2, which lists the substrates in terms of increasing glass transition (T_g) [1–3,7].

The polymers can be further categorized into films that are semicrystalline (PET and PEN as mentioned above), amorphous and thermoplastic, and amorphous, but solvent cast. Polymers with T_g higher than 140 °C that are semicrystalline tend to generally have melting points that are too high to allow the polymers to be melt processed without significant degradation – PEN is the highest performance material available as a biaxially oriented semicrystalline film. The next category are polymers that are thermoplastic but noncrystalline, and these range from polycarbonate (PC), for example, Teijins PURE-ACE® [8] and GEs Lexan® [9] with a T_g of ~150 °C, to polyethersulfone (PES), for example, Sumitomo Bakelite's Sumilite® [10] with a T_g of ~220 °C. Although thermoplastic, these polymers may also be solvent cast to give high optical clarity. The third category are high T_g materials that cannot be melt processed, such as polyimide (PI), for example, DuPont's Kapton® [11].

In addition to polymer films, there is also a considerable activity in developing PV devices on stainless steel. The advantages and disadvantages of polyesters films relative to the other polymer films and stainless steel are discussed in Section 17.3.8.

17.3
Properties of Base Substrates

17.3.1
Optical Properties

The clarity of the substrate is not necessarily important for the base substrate, but the top encapsulant layer needs to be transparent. Clear plastic substrates such as polyester typically have total light transmission (TLT) of >85% over 400–800 nm coupled with a haze of less than 0.7%. This should be acceptable for PV devices.

17.3.2
Thermal Properties

Dimensional and thermal stability and reproducibility are critical to be able to withstand the high temperatures of deposition of conductive coatings or any barrier coatings and the curing of dielectric coatings; to ensure the registration of the different layers in the final device; and for the multilayer device to be able to withstand thermal cycling. This is manifested in two ways. First is the shrinkage a film undergoes when is heated and then cooled to the starting temperature. Low levels of shrinkage are desired to make accurate alignments on the substrates after each thermal processing step. The dimensional stability of both PET and PEN films can be enhanced by a heat stabilization process, where the internal strain in the film is relaxed by exposure to high temperature while under minimum line tension [1–3]. Both films are dimensionally reproducible up to the temperature at which they are heat stabilized, and this ability to undergo a heat stabilization process above the T_g is unique to the biaxially oriented crystalline films. This is illustrated in Figure 17.3, which shows the shrinkage versus temperature for heat-stabilized PET and PEN. This clearly shows the higher shrinkage performance of PEN relative to PET.

The T_g process does not define the upper processing temperature of crystalline films, unlike for the amorphous polymers. For PET, this is typically 150 °C and for PEN it is 180–220 °C. Below these temperatures both films exhibit shrinkages of the order 200–500 ppm, respectively.

The second factor that affects the dimensional reproducibility is the natural expansion of the film as the temperature is cycled, as measured by the coefficient of linear thermal expansion (CLTE). (The CLTE defines an increase in length along one given axis; it is often abbreviated as CTE.) A low CLTE typically <20 ppm/°C is

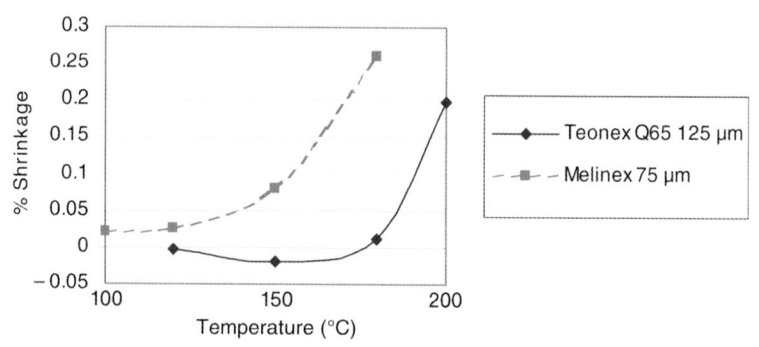

Figure 17.3 Shrinkage versus temperature for PET and PEN.

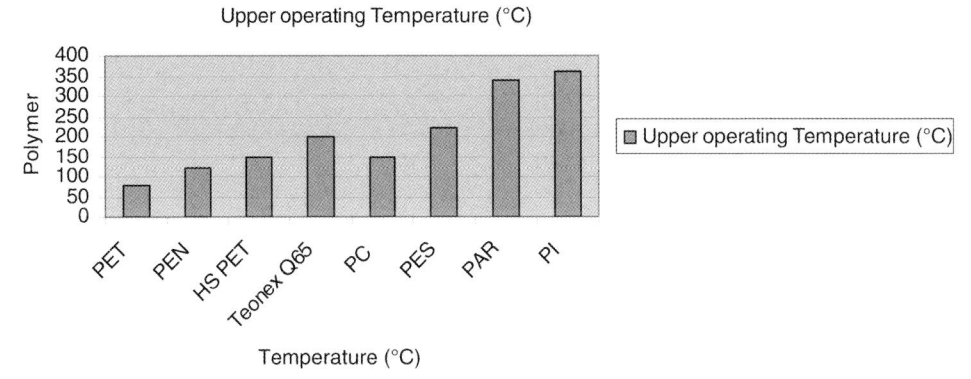

Figure 17.4 Upper processing temperature of film substrates of interest for o-PV.

desirable to match the thermal expansion of the base film to the layers that are subsequently deposited. A mismatch in thermal expansion means that the deposited layers become strained and cracked under thermal cycling. In the temperature range from room temperature up to the T_g, the typical CTLE of PEN is 18–20 ppm and PET 20–25 ppm (but note that the T_g of PEN is 40 °C higher than that of PET). Above the T_g, the natural expansion of PET and PEN films that have not been stabilized is dominated by the shrinkage that the films undergo as the internal strains in the film relax, as discussed above. However, the heat-stabilized films discussed above show only a very small increase in CTLE, in the temperature range from T_g to the temperature at which they were heat stabilized. This contrasts favorably with the quoted coefficient of expansion of amorphous polymers that is typically 50 ppm C^{-1} [12] below the T_g, but can increase by a factor of 3 above the T_g.

In addition to dimensional stability, another important factor to be considered is the upper processing temperature (T_{max}) at which a film can be used. Although as outlined above, the T_g does not define T_{max} with the semicrystalline polymers, it largely does with the amorphous polymers. Figure 17.4 shows the upper operating temperature if the effect of heat stabilization is taken into account

17.3.3
Solvent Resistance

A wide range of solvents and chemicals can potentially be used when laying down the organic solar cells, depending on the processing steps involved. Amorphous polymers in general have poor solvent resistance compared to semicrystalline polymers (Table 17.1).

This deficiency is overcome by the application of a hard coat to the amorphous resins that significantly improves the solvent and chemical resistance to solvents such as NMP, IPA, acetone, methanol, THF, ethyl acetate, 98% sulfuric acid, glacial acetic acid, and 30% hydrogen peroxide, and saturated bases such as sodium hydroxide [12]. With polyethylene terephthalate and polyethylene naphthalate films, a hard coat is not required to give solvent resistance.

17 Flexible Substrates Requirements for Organic Photovoltaics

Table 17.1 Table of solvent resistance.

	Unit	Q65	PET	PC	PES
Ketone	Acetone	Good	Good	(Fair)	(NG)
	MEK	Good	Good	(Fair)	(—)
Alcohol	Methanol	Good	Good	(Good)	(—)
	Ethanol	Good	Good	(Good)	(Good)
	Isopropanol	Good	Good	(Good)	(—)
	Butanol	Good	Good	(Good)	(—)
Ester	Ethyl acetate	Good	Good	(Fair)	(Good)
Hydrocarbon	Formalin	Good	Good	(NG)	(—)
	Tetrachloroethane	Good	Good	(—)	(Good)
Acid	10% HCl	Good	Good	(—)	(Good)
	10% HNO_3	Good	Good	(—)	(Good)
	10% H_2SO_4	Good	Good	(—)	(Good)
	Acetic acid	Good	Good	(—)	(—)
Alkali	10% NaOH	Good	Fair	(—)	(Good)

In addition to this residual shrinkage, the processing environment and, in effect, the prevailing humidity must be taken into account, as even crystalline Teonex®Q65 will absorb up to 1500 ppm of moisture under ambient conditions, yet readily loose this at higher processing temperatures. With the knowledge of the solubility level of moisture in PEN film and its rate of diffusion as a function of temperature, it is possible to model the impact of such environmental conditions on volumetric changes in the film [2]. Figure 17.5 shows the moisture loss against time as film of 125 m is heated and held at 90, 100, 120, and 150 °C at an RH of 40% at 20 °C.

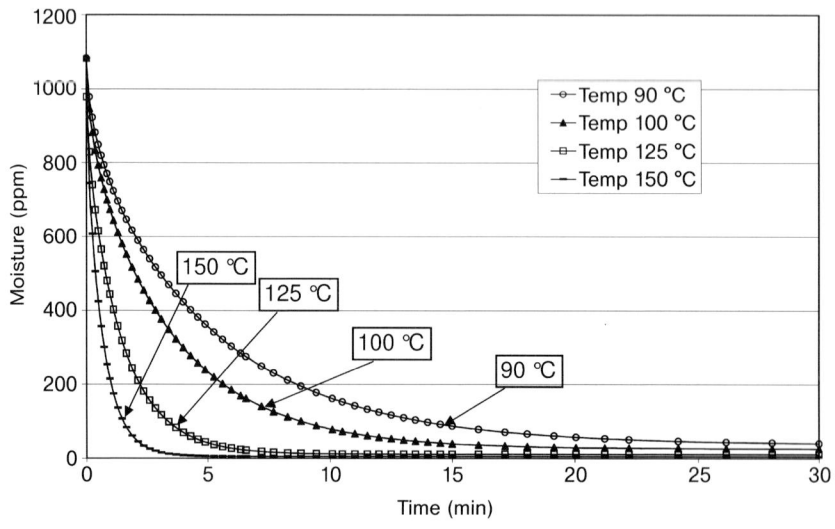

Figure 17.5 Moisture loss versus time on heating to different temperatures at RH of 40%/20 °C.

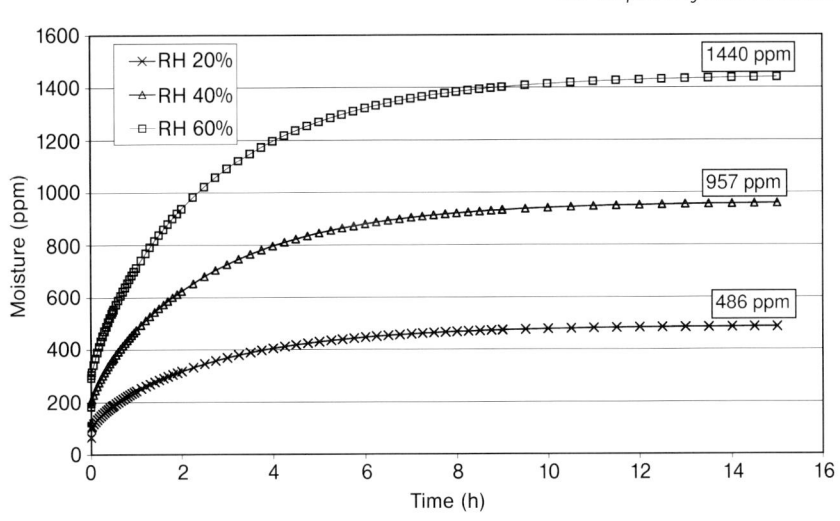

Figure 17.6 Effect of RH on moisture pickup at 23 °C.

At 150 °C, the film takes 6 min to reach an equilibrium moisture level of 5 ppm, but at 90 °C the film takes 30 min to reach an equilibrium level of 40 ppm.

Upon removal from the heated environment back into ambient conditions, the film initially picks up to 50% of its equilibrium moisture level while cooling within a few minutes, but the remaining moisture to reach equilibrium is picked up at a much slower rate. Films can typically take over 12 h to reach equilibrium. Figure 17.6 shows the impact of RH on this process and indicates that final (equilibrium) moisture levels will change significantly with varying RH.

At an RH of 20%, the film reaches an equilibrium level of approximately 500 ppm but at 60% RH an equilibrium level of approximately 1500 ppm.

Measurements in our laboratory indicate that the film increases in length in each axis by approximately 45 ppm per 100 ppm of moisture absorbed. This is what one would predict for such small changes in dimensions from a simple volume calculation. Taking into account the variation in moisture content that can result depending upon temperature, RH, and film thickness, which can be of the order of several hundreds of ppm, it can be seen that uncontrolled moisture pickup will have a significant influence on dimensional reproducibility. This will be particularly significant at lower processing temperatures where there is minimal film shrinkage, but where the film takes a longer time to reach the equilibrium moisture level than the time taken to carry out the processing step at a given temperature, the film ideally should be allowed to equilibrate before any registration points are established.

By carrying out modeling studies of the type described above, the impact of this moisture change once understood can be minimized.

PET and PEN films have an inherent advantage over the amorphous polymer films, being both semicrystalline and biaxially oriented, and typically will absorb approximately 1400 ppm of moisture at equilibrium (the exact figure depending upon temperature and relative humidity). Polyethersulfone and polyimide films are particularly problematic and will absorb over 1% moisture at equilibrium.

The residual shrinkage of Teonex®Q65 after 30 min at 150 °C is of the order 500 ppm but can be reduced to below 200 ppm or better by careful process control. Dimensional reproducibility down to 100 ppm has been demonstrated in the filed of a-Si active matrix backplanes on flexible substrates with Teonex®Q65 [13].

17.3.4
Surface Quality

The surface quality of the film is essential to minimize the presence of any surface defects or debris that could protrude through subsequent barrier or o-PV layers. Surface quality can be divided into the following:

(i) *Inherent surface smoothness:* PET and PEN have an inherent surface smoothness of less than 1 nm, in terms of both Ra and Rq using white-light interferometric methods of measurement. However, the extent of the surface smoothness depends upon the propensity of "internal" contaminants such as catalyst residues, polymer degradants, and so on. Typically, over a 1 cm sample area, these internal contaminants exhibit a range of surface peaks less than 0.2 μm high. The frequency of such surface peaks (Rp values) can be shown in the graph below in Figure 17.7.

(ii) *Surface cleanliness:* this is dominated by dust and surface scratching. These defects can range in size up to tens of microns both laterally and vertically and are unavoidable in film that is handled and slit in a nonclean room environment.

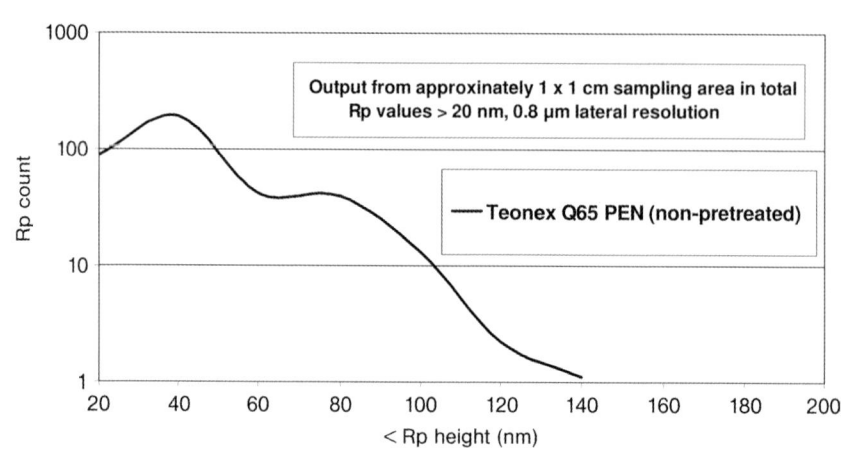

Figure 17.7 Extreme surface peak "Rp" (highest point) frequency distribution for TeonexQ65 125 μm. The graph comprises numerous measurements, which in total make $1 \times 1 \, cm^2$ sampling area. For each measurement the highest point (Rp value) is recorded and the following distribution observed. The particles are predominantly below 140 nm. The predominance of these particles can be influenced by the polymer recipe and by process control to minimize degradation.

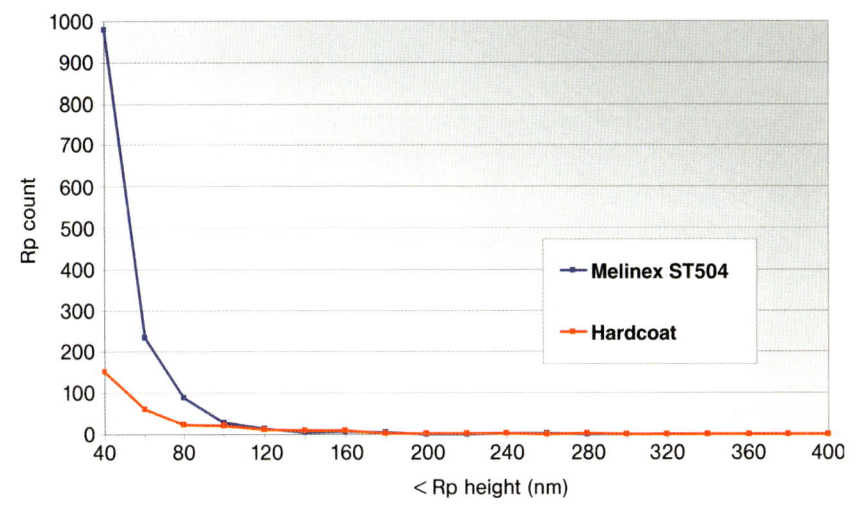

Figure 17.8 Extreme surface peak Rp (all high points >25 nm) – frequency distribution comparison, for Melinex ST504 nonpretreated surface and hard coat upon its pretreated surface.

A proportion of the dust particles can be removed by some form of surface cleaning, but to eliminate these types of defects, it is necessary to apply a planarizing coating that is ideally coated in a clean room environment [1–3,6]. These coatings have the dual function of providing not only a surface smoothness but also a certain surface hardness to prevent scratching of the surface during film processing. Figure 17.8 shows a typical reduction in particle size achieved when a planarizing coating is added. Work is going on within DuPont Teijin Films to reduce the peak count further.

17.3.5
Mechanical Properties

Production of glass-based displays currently involves moving batches of glass between the different processing stages. The mechanical difference, in particular stiffness, between rigid glass and flexible substrates will likely require very different methods of processing. PET and PEN films are inherently stiffer than amorphous films with their Young's modulus being typically three times higher – an artifact of being semicrystalline and biaxially oriented [15]. Young's modulus is thickness independent and does not indicate how stiffness will change with thickness. Another way to express this is by defining the rigidity by the equation below:

$$D = \frac{Et^3}{12(1-n)},$$

where E is the tensile or Young's modulus, t is the thickness, and n is Poisson's ratio (0.3–0.4).

Table 17.2 Comparative rigidity of amorphous and semicrystalline films.

Material	Thickness (µm)	Rigidity (nm × 10^{-4})	Rigidity relative to 125 µm amorphous film
Amorphous	125	5	1
Amorphous	200	20	4
TeonexQ65	125	15	3
TeonexQ65	200	61	12

Assuming a Young's modulus of 2 GPa for amorphous films and 6 GPa for PEN films, it can be seen that (Table 17.2) 200 µm polyethylene naphthalate film is four times more rigid than a 125 µm polyethylene naphthalate film and 12 times more rigid than a 125 µm amorphous film. This stiffness may prove to be an advantage in a batch-based display manufacturing process [14]. The extra rigidity of the 200 versus 125 µm PEN film has also been found to be an advantage in applications such as a-Si TFT backplanes, where the thicker and more rigid film is better able to withstand the stress induced when coating inorganic layers onto polyester film.

One of the main drivers in moving to plastic substrates is that it opens up the possibility of roll-to-roll processing and the process and economic advantages that this brings. Under these conditions, a winding tension will clearly be present and polymer film substrates with low moduli will be susceptible to internal deformation, particularly at elevated process temperatures if the tension in the film exceeds the stiffness of the film. Figure 17.9 shows a comparison between polyethylene terephthalate and polyethylene naphthalate films. Similarly, if the film is restrained on a rigid carrier during a heating cycle, the film may be susceptible to internal deformation that will manifest itself as shrinkage upon subsequent reheating.

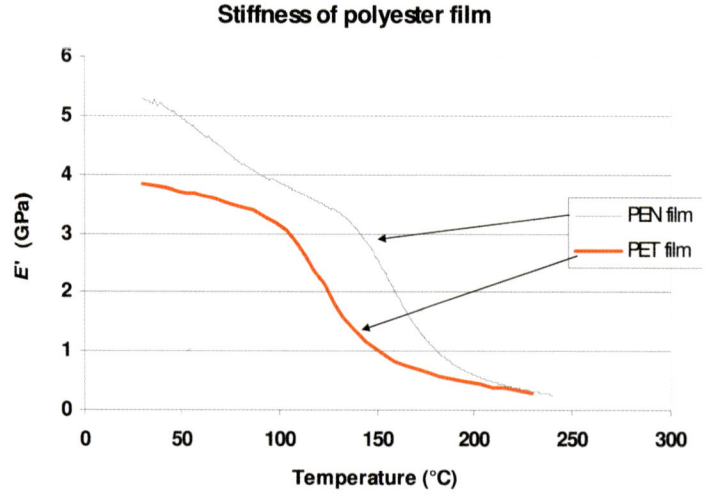

Figure 17.9 Change in storage modulus of PET and PEN with temperature.

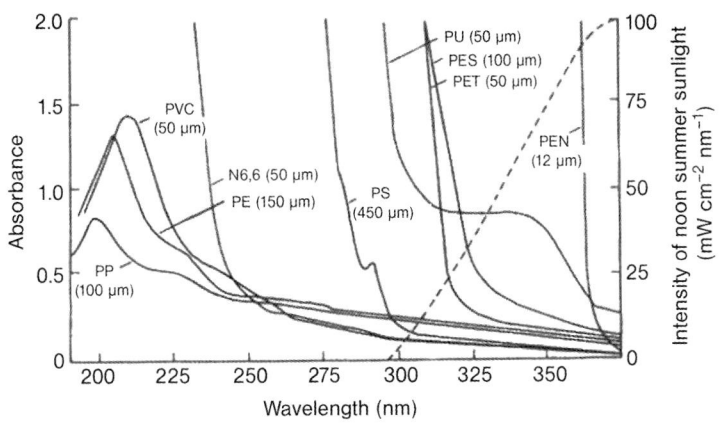

Figure 17.10 UV absorption spectra of films of type A polymers – polypropylene (PP), polyethylene (PE), poly(vinyl chloride) (PVC), nylon 6,6 (N6,6), and polystyrene (PS) – and type B polymers – PES, PET, PEN, and polyurethane (based on 4,4′-diphenylmethane diisocyanate) (PU) – compared with the UV spectral energy distribution of natural sunlight (- - -). The thicknesses of the films are shown in parentheses [15].

The storage modulus E' is recorded using dynamic mechanical thermal analysis, and as the temperature is increased, the stiffness of both materials is seen to fall. However, in the region 120–160 °C, PEN is significantly stiffer and stronger, with a modulus almost twice as that of PET.

17.3.6
UV Stability

Clearly, in an application area where the cells are exposed to sunlight, the UV stability of the substrate is an issue. Photoreaction mechanisms are classified into two main types depending on the mode of light absorption, which itself is governed by the cutoff point for terrestrial sunlight (wavelengths >290 nm). The first class of mechanism (class A) is based upon the absorption of light through impurity chromophores that are present in the polymer. Polyolefins are examples of type A.

The second class of mechanism (class B) is based on the direct absorption of solar radiation by units or groups that form a part of the basic chemical constitution of the polymer. PET and PEN (and aromatic polyesters in general) exhibit type B behavior. This effect is shown in the Figure 17.10 below, which shows the absorption spectra of a number of polymers compared with that of sunlight. It can be seen from this that PET and PEN are prone to the effects of solar radiation as they exhibit strong absorption in the near-UV region. PET photodegrades over the range 290–350 nm with the production of acid end groups at a maximum close to 315 nm [15].

The main reports on PET are the works of Day and Wiles [16,17]. These authors showed [16] that there is a significant overlap of the UV absorption spectrum of PET with the emission spectra of solar simulating devices (xenon and carbon arc lamps). Degradation, measured by color change, MW decrease, tensile strength loss, and formation of new –COOH end groups, was significantly faster in xenon than in carbon arc light because xenon lamp gives greater irradiation at wavelengths below 315 nm. However, they also showed that light with wavelengths above 350 nm can cause degradation on prolonged exposure. These data were confirmed by a later work [17] that also showed that the major sensitivity of PET is to light with wavelengths below 315 nm, which causes rapid loss of properties, but that prolonged exposure through a filter cutting off all wavelengths below 320 nm also leads to degradation, with formation of –COOH end groups and fluorescent products.

The strongly absorbing nature of the naphthalene structure in the UV leads to an absorption tail of PEN up to 400 nm (well into the near-UV). PEN shows a greater tendency to photoyellow than PET on exposure to sunlight due to chromophores formed during photooxidation attributed to naphthalene-related derivatives. Although PEN photoyellows are faster than PET, the rate of chain scission is higher for PET. PEN therefore maintains its mechanical integrity to a greater extent than PET on photodegradation. Due to the highly absorbing nature of PEN and the low oxygen permeability, the degradation in PEN is limited to the very outer surface of the film to a depth of approximately 10 µm [18]. This oxidized layer acts as a protective UV barrier layer in PEN.

The effective stabilization of any polymer against UV-induced degradation typically uses a mixture of additives, among which the most common are the so-called UV absorbers, UVA. Of these, the main classes used in polymers have been the o-hydroxybenzophenones, the hydroxybenzotriazoles, and the hydroxyphenyltriazenes [19].

All of these stabilizers are designed to function by absorption of radiation over the activating range for degradation (wavelength = 290–400 nm) and its conversion into low energy radiation (heat). Their functionality derives from internal hydrogen bonding between the chromophore and phenolic OH group; in the excited state the transfer of the H atom to the chromophore requires loss of aromaticity and results in a return to the ground state with conversion of electronic into vibrational energy.

Figure 17.11 shows how that transmission spectra of PET is altered by the addition of increasing levels of UV stabilizer.

Figure 17.12 shows how the addition of a UV stabilizer has a significant effect on the retention of mechanical properties of PET as a measure by ultimate tensile strength (UTS) on accelerated aging. (Aging at 10 000 h in a weatherometer equates to 3 years/1000 h for Northern Europe, 2 years/1000 h for Southern Europe, and 1 year/1000 h for Australia.)

17.3.7
Barrier

One of the key properties of any substrate that replaces glass will be to offer glasslike barrier properties. The materials used in organic PV cells are sensitive to moisture and oxygen. Conjugated polymers like poly(phenylene vinylene) (PPV), for example, are known to be unstable in air and undergo detrimental photodegradation reactions

% Transmission of non-UV versus UV absorbing films

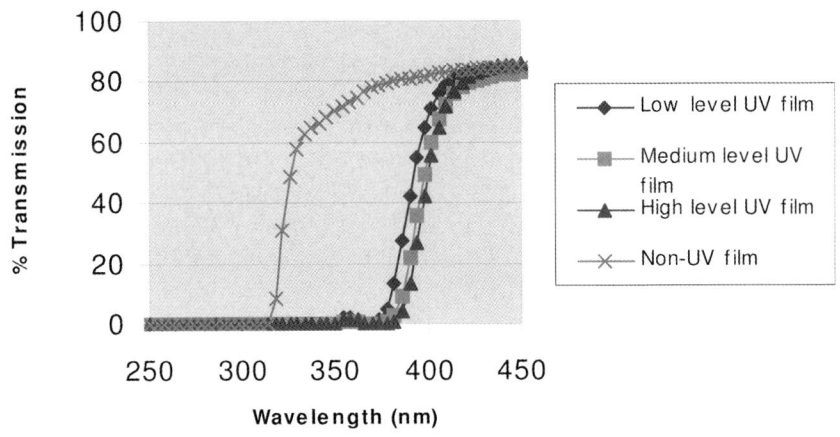

Figure 17.11 Transmission spectra of PET with increasing levels of UV stabilizer.

induced by oxygen and moisture that react with the double bonds, hence breaking the conjugation and reducing the efficiency of the active layer. In addition, the top electrode materials are usually chosen from low work function metals such as Al and Ca that undergo severe oxidation when exposed to air, leading to the formation of insulating thin oxide barriers that hinder electrical conduction and collection of charge carriers [20].

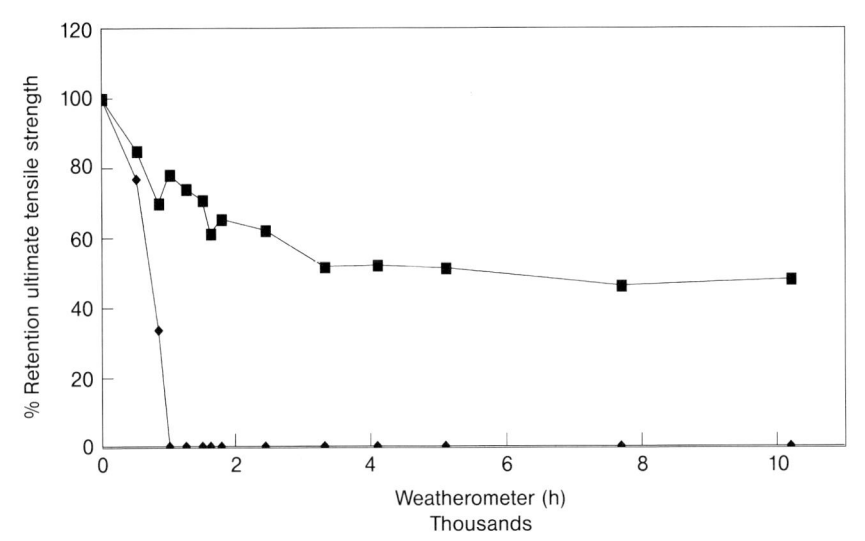

Figure 17.12 Change in UTS versus hours in a weatherometer for PET (triangle) and UV-stabilized PET (squares).

The plastic substrates that are under consideration as flexible substrates typically have barriers of the order of $1-10\,\mathrm{g\,m^{-2}\,day^{-1}}$ for water vapor transmission rate and of about $1-10\,\mathrm{ml\,m^{-2}\,day^{-1}}$ for oxygen transmission rate. It is generally accepted that OLED displays will require water vapor transmission rates of about $10^{-6}\,\mathrm{g\,m^{-2}\,day^{-1}}$ and oxygen transmission rates of about $10^{-3}-10^{-5}\,\mathrm{ml\,m^{-2}\,day^{-1}}$ [21].

The organic materials used in o-PV cells are generally more stable than the OLED materials and are less sensitive to "dark spots" caused by pinhole decay, which visually detract from a display performance and therefore may not require the transmission rates predicted for OLED displays. However, when exposed to sunlight these cells may experience temperatures above 50 °C. This has to be taken into account when designing encapsulation solutions for these devices.

Barrier coatings on films have been extensively researched for a number of years [22–30] because of the necessity to prolong the shelf life of packaged food products by reducing water and oxygen ingress, and metal and dielectric layers are now routinely coated on polymers for this purpose with great success. However, the permeability specification for o-PV cells is still several orders of magnitude lower than that for food packaging, and for the top layer it is of course not possible to use opaque metal films.

In principle, a perfect layer of silica only a few nanometers thick should reduce the diffusion of water and oxygen to acceptable levels (just as the 2–3 nm native oxide layer on silicon wafers protects the wafer surface from further oxidation). The problem in practice is that thin-film coatings can have defects that provide easy paths for water and oxygen molecules to penetrate [24]. Also, surface defects on the polymer substrates cause pinholes in the final coating and lead to catastrophic failure of barrier properties. However, even clean, pinhole-free coatings are not perfect. In the particular case of vacuum-deposited thin-film barrier coatings, this is due to the tendency of such films to show columnar growth and to exhibit densities less than the bulk material [28]. Water and oxygen molecules can then diffuse through the nanospacings between columns [31]. On the relatively rough surfaces of normally processed polymer sheet, this effect is enhanced. Heating substrates to elevated temperatures can help reduce columnar growth but this option is not open for all polymer substrates.

Symmorphix [32,33] has developed a single barrier coating but based on a dense inorganic coating of aluminosilicate. They have shown that by continually improving the surface of the base film via different cleaning techniques, it is possible to achieve barrier films with permeability beyond the Mocon limit [33].

Although good progress is being made with single-layer inorganic coatings, it is difficult to completely eliminate defects over the surface area of a device size film and there are concerns that OLED-type barrier properties are unlikely to be achieved by this route. An alternative route to achieving high-performance barrier is based on vacuum-deposited organic–inorganic multilayers [31,34–37], and several organizations are currently developing optically transparent multilayer barrier coatings for flexible substrates based on this approach. The improved barrier performance is achieved by using the organic coating to provide a smoothing layer and to reduce mechanical damage, hence reducing damage caused by surface defects to the inorganic layer, and then repeating this alternating process that allows the organic layer to decouple the defects from each successive inorganic layer. This multilayer geometry provides a long tortuous diffusion path, and the barrier performance is due primarily to a lag time effect attributed to this rather than low steady-state permeability, that is, the barrier works by delaying water vapor ingress rather than by preventing it. Vitex Systems [38,39] is one of the leading protagonists of this technology, and results demonstrated at external conferences show continually improving barrier performance, as evidenced by the calcium test, and indicate that a 5-dyad barrier film (a dyad being one organic–inorganic layer) capable of yielding the required performance for OLED applications can be achieved. A simpler dyad stack may provide acceptable barrier performance for o-PV cells with obvious cost benefits.

One of the potential disadvantages of the multilayer stack approach is that this type of structure may suffer from poor adhesion and delamination through the thermal cycles involved in device manufacture. In order to overcome this, one approach adopted by General Electric Global Research Center has been to use plasma-enhanced chemical vapor deposition to lay down a graded single hybrid layer of organic–inorganic coating based on SiO_xN_y and SiO_xC_y [40].

The barrier developments described above have been targeted at OLED-type devices and, as discussed earlier the barrier requirements of these devices are more demanding than that required for o-PV cells. At this stage there is less information available on lifetime of flexible o-PV cells, but it is likely that the barrier films being developed for OLED displays may well be already good enough for o-PV cells and in fact simpler barrier structures may be viable.

Typically, the barrier films described above have light transmission >80%, which is likely to be adequate for PV device cell operation.

Although excellent results are now being achieved generally with batch-based processes, the next major step in the development process will be to achieve the same barrier performance on barrier film generated in a roll-to-roll process.

17.3.8
Summary of the Key Properties of Base Substrates

The main properties of heat-stabilized PET and PEN relevant to flexible electronics are summarized in Figure 17.13.

Unstabilized PET and PEN films exhibit the same property set apart from the shrinkage and upper processing temperature that for these films is largely indicated by T_g. As discussed in this chapter, several factors contribute to the dimensional stability of PET and PEN:

(i) off-line stabilization;

(ii) control of mechanical stress through exploitation of film thickness and control of tension during processing;

(iii) control of processing environment to minimize the effect of moisture and feedback to DTF by end users, indicating that if these factors are controlled, shrinkages below 100 ppm at 150 °C can be achieved with TeonexQ65.

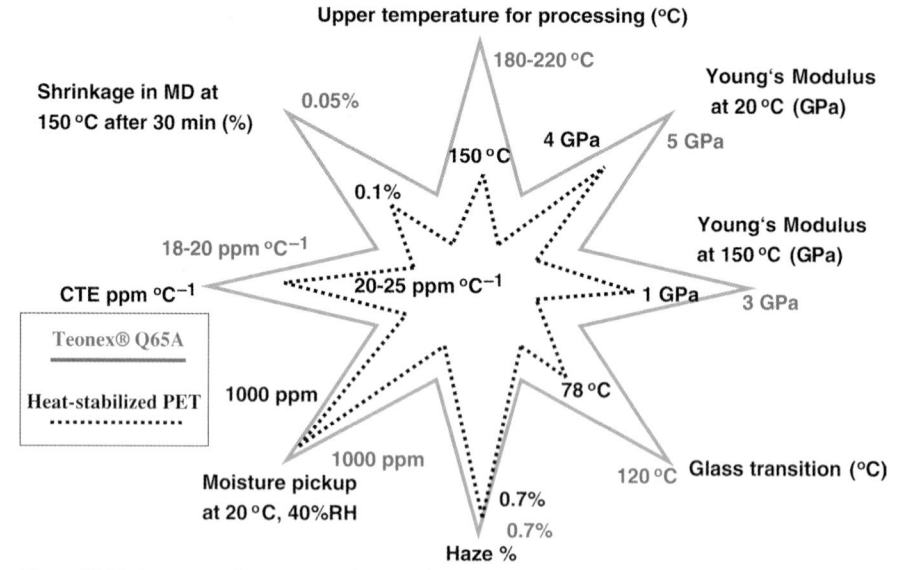

Figure 17.13 Summary of properties of PET and PEN relevant to organic PV.

Table 17.3 contrasts the property set of different substrates under consideration for flexible electronics and the table remains relevant to o-PV.

These properties have been discussed in previous papers [1–3]. This table shows that both heat-stabilized PET (e.g., Melinex ST506) and heat-stabilized PEN (TeonexQ65A) have an excellent balance of the key properties required for o-PV. TeonexQ65A has a higher temperature performance than Melinex (Figure 17.12).

Table 17.3 Comparison of the key properties relevant to flexible electronics.

Property	Heat-stabilized PEN	Heat-stabilized PET	Polycarbonate	Polyethersulfone	Polyimide
CTE (−55 to 85 °C) (ppm °C^{-1})	✓✓	✓✓	✓	✓	✓✓
%Transmission (400–700 nm)	✓✓	✓✓	✓✓✓	✓✓	×
Water absorption (%)	✓✓	✓✓	✓	×	×
Young's modulus (GPa)	✓✓	✓✓	✓	✓	✓
Tensile strength (MPa)	✓✓	✓✓	✓	✓	✓
Solvent resistance	✓✓	✓✓	×	×	✓✓
Upper operating temperature	✓✓	✓	✓	✓✓	✓✓✓

Flexible stainless steel substrates are used for large-scale solar modules, and products have been fabricated using roll-to-roll processing for inorganic solar cells. Stainless steel offers some obvious advantages over plastic substrates namely high thermal stability, good dimensional stability, excellent barrier properties, and good UV stability. However, commercially available stainless steel has poor surface quality, and for o-PV cells the surface must be planarized. In addition, an insulating layer may also have to be applied to provide electrical passivation, and this layer also acts as a barrier to inorganic ions that can migrate from the stainless steel at elevated processing temperature and contaminate the active organic components. The presence of these coatings can also lead to warpage due to a mismatch in CTE. Stainless steel is not transparent and a transparent flexible barrier material will still be required to allow exposure of the organic solar cell to sunlight, and this will to an extent negate the advantage of stainless steel. The radius of curvature of stainless steel is higher than plastic film, and stainless steel is prone to "kinking" on impact due to its lower yield strain relative to plastic film. Stainless steel with a density of 7.9 g cm^{-3} is considerably heavier than, for example, PEN with a density of 1.36 g cm^{-3}. Given that one of the expected benefits of o-PV is a high power to weight ratio, this would be a problem. Stainless steel has obvious advantages with inorganic solar cells where high-temperature processing is required, but with flexible o-PV cells that require lower processing temperatures and high barrier on both sides, the advantages are less obvious and coated stainless steel is possibly overspecified and too expensive for the application.

17.4
Concluding Remarks

The basic property set of PET and PEN films, their availability at commercial scale coupled with the new developments discussed in this chapter, leads polyester films to be the leading substrate candidates for o-PV cells. The substrate requirements for o-PV largely mirror those required for flexible displays, but with the specific difference that UV stability is an additional requirement. There will be a requirement for additional barrier coatings, but it is likely that the o-PV developers will be able to take advantage of the advances in barrier technology that have been made for OLED displays.

Acknowledgments

I would like to thank Karl Rakos, Kieran Looney, Robert Eveson, Raymond Adam, and Duncan MacKerron of DuPont Teijin Films for their contributions.

References

1 MacDonald, W.A., Rollins, K., MacKerron, D., Rakos, K., Eveson, R., Rustin, R.A. and Hashimoto, K. (2005) Engineered films for display technologies, in *Flexible Flat Panel Displays* (ed. G. Crawford), John Wiley & Sons, Ltd, Chapter 2.

2 MacDonald, W.A. (2006) *Organic Electronics: Materials, Manufacturing and Applications* (ed. H. Klauk), Wiley-VCH Verlag GmbH, Chapter 7.
3 MacDonald, W.A. (2004) Engineered films for display application. *Journal of Materials Chemistry*, **14**, 4.
4 Melinex, Teonex, DuPont Teijin Films, Luxembourg, S.A., P.O. Box 1681, L-1016, Luxembourg.
5 MacDonald, W.A., Mackerron, D.H. and Brooks, D.W. (2002) PET film and sheet, in *PET Packaging Technology* (ed. D.W. Brookes), Academic Press, Sheffield.
6 MacDonald, W.A. (2003) Polyester film, *Encyclopedia Polymer Science & Technology*, 3rd edn, John Wiley & Sons, Inc.
7 MacDonald, B.A., Rollins, K., Eveson, R., Rakos, K., Rustin, B.A. and Handa, M. (2003) New developments in polyester films for flexible electronics. *Materials Research Society Symposium Proceedings*, **769**, paper H9.3.
8 PURE-ACE, Teijin Chemicals Ltd, 1–2–2 Uchisaiwai-cho, Chiyoda-ku, Tokyo, Japan.
9 Lexan, GE Plastics, One Plastics Ave, Pittsfield, MA 01201, USA.
10 Sumilite, Sumitomo Bakelite Co. Ltd., Ten-Nouzu Parkside Blgd., 5-8,2-Chome, Higashi-Shinagawa, Shinagawa-Ku; Tokyo; 140-0002, Japan.
11 Kapton, DuPont High Performance Films, P.O. Box 89, Route 23, South and DuPont Road, Circleville, OH 43113, USA.
12 Angiolini, S., Avidano, M., Bracco, R., Barlocco, C., Young, N.G., Trainor, M. and Zhao, X.-M. (2003) High performance plastic substrates for active matrix flexible FPD, Society For Information Display, Digest of Technical Papers, pp. 1325–1327.
13 Sarma, K.R., Chanley, C., Dodd, S., Roush, J., Schmidt, J., Srdanov, G., Stevenson, M., Wessel, R., Innocenzo, J., Yu, G., O'Regan, M., Macdonald, W.A., Eveson, R., Long, K., Gleskova, H., Wagner, S. and Sturm, J.C. (2003) Active matrix OLED using 150 °C a-Si TFT backplane built on flexible plastic substrate. Proceedings from SPIE Aerosense, Technologies and Systems for Defense and Security, April 22–25.
14 MacDonald, W.A., Rollins, K., MacKerron, D., Eveson, R., Rustin, R.A., Rakos, K. and Handa, M. (2004) Society For Information Display, Digest of Technical Papers, pp. 420–423.
15 Allen, N.S. (1994) *Trends in Polymer Science*, **2**, 366, and references contained therein.
16 Days, M. and Wiles, D.M. (1972) *Journal of Applied Polymer Science*, **16** (1), 175.
17 Days, M. and Wiles, D.M. (1972) *Journal of Applied Polymer Science*, **16** (1), 191.
18 Scheirs, J. and Gardette, J.-L. (1997) *Polymer Degradation and Stability*, **56**, 339.
19 Gugumus, F. (2001) *Plastic Additives Handbook*, 5th edn (ed. H. Zweifel), Hanser Gardner, Chapter 2.
20 Dennler, G., Lungenschmied, C., Neugebauer, H. and Sariciftci, N.S. (2005) *Journal of Materials Research*, **20**, 3224.
21 Lewis, J.S. and Weaver, M.S. (2004) *IEEE Journal of Selected Topics in Quantum Electronics*, **10**, 45.
22 Prins, W. and Hermans, J.J. (1959) *Journal of Physical Chemistry*, **63**, 716.
23 Jamieson, E.H.H. and Windle, A.H. (1983) *Journal of Material Science*, **18**, 64–80.
24 Rossi, G. and Nulman, M. (1993) *Journal of Applied Physics*, **74**, 5471.
25 Philips, R.W., Markantes, T.M. and LeGallee, C. (1993) 36th Annual Technical Conference Proceedings of the Society of Vacuum Coaters, p. 293.
26 Amberg_Schwab, S., Hoffmann, M., Bader, H. and Gessler, M. (1998) *Journal of Sol–Gel Science and Technology*, **1/2**, 141.
27 Moosheimer, U. and Langowski, H.-C. (1999) 42nd Annual Technical Conference Proceedings of the Society of Vacuum Coaters, p. 408.
28 Henry, B.M., Erlat, A.G., Grovenor, C.R.M., Deng, C.S., Briggs, G.A.D., Miyamoto, T., Noguchi, N., Niijima, T. and Tsukahara, Y. (2001) 44th Annual Technical Conference Proceedings of the Society of Vacuum Coaters, p. 469.

29. Smith, A.W., Copeland, N., Gerrard, D. and Nicholas, D. (2002) 45th Annual Technical Conference Proceedings of the Society of Vacuum Coaters, p. 525.
30. Henry, B.M., Erlat, A.G., Grovenor, C.R.M., Briggs, G.A.D. and Tsukahara, Y. (2002) 45th Annual Technical Conference Proceedings of the Society of Vacuum Coaters, p. 503.
31. Langowski, H.-C., Melzer, A. and Schubert, D. (2002) 45th Annual Technical Conference Proceedings of the Society of Vacuum Coaters, p. 471.
32. Symmorphix Inc., 1278 Reamwood Avenue, Sunnyvale, CA 94089-2233.
33. Stickney, B. (2006) USDC Flexible Display Conference, Phoenix.
34. Affinato, J.D., Gross, M.E., Coronado, C.A., Graff, G.L., Greenwell, E.N. and Martin, P.M. (1996) *Thin Solid Films*, **290–291**, 63–67.
35. Bright, I.C. (2000) US Patent 20020022156A1.
36. Graff, G.L., Gross, M.E., Shi, M.K., Hall, M.G., Martin, P.M. and Mast, E.S. (2001) WO 01/81649 A1.
37. Graff, G.L., Gross, M.E., Shi, M.K., Hall, M.G., Martin, P.M. and Mast, E.S. (2001) WO 01/82389 A1.
38. Graff, G.L., Burrows, P.E., Williford, R.E. and Praino, R. (2005) *Flexible Flat Panel Displays* (ed. G. Crawford), John Wiley & Sons, Ltd, Chapter 4.
39. Barix, Vitex Systems Inc., 3047 Orchard parkway, San Jose, CA 95134.
40. Yan, M., Kim, T.W., Erlat, A.G., Pellow, M., Foust, D.F., Liu, J., Schaepkens, M., Heller, C.M., McConnelee, P.A., Feist, T.P. and Duggal, A.R. (2005) *Proceedings of the IEEE*, **93**, 1468.

18
Barrier Films for Photovoltaics Applications
Lorenza Moro and Robert Jan Visser

18.1
Introduction

Significant amount of research and development efforts have been focused in improving the photovoltaics (PV) efficiency and/or reducing the production cost of the PV cell. This has brought to the development of new concepts for solar cells. Thin-film concepts for solar cells have been proposed based on silicon and other inorganic or organic semiconductors as a method to save material cost. For terrestrial applications, other materials as an alternative to silicon with a potential of sufficient efficiency and low cost are CdTe, CIGS, and, at an earlier stage of development, organic materials.

Organic photovoltaics (OPV) based on semiconducting polymers has the potential to provide a low-cost solution for solar power conversion [1–3]. Up to now the main work has focused on improvement of the light conversion efficiency under AM 1.5 conditions [4,5]. An efficiency of 5% is regarded as the critical efficiency necessary for commercialization [6]. Cells with such efficiency had been previously realized by Konarka [7]. The most impressive recent results reported are a "tandem cell" with power-conversion efficiencies of more than 6% at illuminations of $200\,\text{mW}\,\text{cm}^{-2}$ by Heeger and coworkers [8] and single layer organic solar cells with 5.4% power efficiency reported by Plextronics and certified by the National Renewable Energy Laboratory (NREL) [9]. Production methods based on web coating with either vacuum or printing methods on rigid and flexible substrates have been studied for OPV with the double scope of decreasing production and installation costs. The most economical production techniques are continuous roll-to-roll printing techniques [6]. The use of these techniques requires processing on flexible substrates, and similar to organic light emitting devices (OLEDs), the use of barrier layers will be a requirement in order to achieve lifetime of 3–5 years, which is believed to be necessary for a successful product [6]. This paper focuses on describing what are the requirements for barrier in encapsulation and packaging of OPV cells with reference to the most common degradation mechanisms of cells and barriers. A review of available barrier

Organic Photovoltaics: Materials, Device Physics, and Manufacturing Technologies.
Edited by Christoph Brabec, Vladimir Dyakonov, and Ullrich Scherf
Copyright © 2008 WILEY-VCH Verlag GmbH & Co. KGaA, Weinheim
ISBN: 978-3-527-31675-5

technologies will be given and examples of applications of an ultralow permeation barrier to OPV packaging will be presented.

18.2
Requirements for OPV Environmental Barriers

Long-term stability is an essential issue for any commercial PV devices. Solar modules for terrestrial application based on bulk and thin-film inorganic PV materials such as wafer-based silicon, crystalline and amorphous silicon, thin-film silicon, chalcogenides, and II–VI semiconductors to be used for energy generation in the PV industry will have to last 30 years in the field. At the current stage of development of OPV, the lifetime of the solar cells is limited by the inherent instability of the organic photovoltaic absorber. This fact, together with the limited 5% conversion efficiency makes OPV targeting commercial markets as disposable photovoltaics with a target lifetime of weeks with applications to clocks, price tags, small sensors, portable energy generation systems with a lifetime less than 3 years for cell phones, PDA, computers, and, probably in the next future, small power generation systems with a lifetime of less than 5 years.

To meet commercial success in these markets with relaxed lifetime and requirements other factors such as lightweight, flexibility, and very low cost becomes indispensable. In the following sections, we will try to define the specifically critical issues for environmental barriers enabling the successful commercial deployment of OPV in the market.

Organic solar cells are extremely sensitive to oxygen and moisture exposure [6]. Degradation mechanisms induced by oxidation will be briefly described in the next section. This environmental sensitivity of the cells dictates the need for substrates and superstrates with moisture and oxygen permeation in the range 10^{-6}–10^{-4} $g\,m^{-2}\,day^{-1}$. These requirements are comparable to the requirements of organic light emitting devices where the oxidation of a *single monolayer* at the reactive interface between the cathode and the organic layers leads to localized dimming of the display and, with more oxidation, to a complete loss of the ability to inject current, that is, the formation of a so-called "black spot".

Glass substrates and/or superstrates are not an option as an environmental barrier for OPV since lightweight and flexibility are among its most attractive features.

Plastic films with barrier coating or with inherent barrier proprieties are available for food and pharmaceutical packaging. While these types of films meet the weight and flexibility requirements, their barrier propriety is insufficient. Figure 18.1 describes the barrier performance of different types of engineered plastic films in comparison with the requirement of different applications in photovoltaics. Plastic films used in general packaging have permeation in the range of 1–10 $g\,m^{-2}\,day^{-1}$. The environmental barrier performance required for specialized food or medical packaging is not better than $10^{-2}\,g\,m^{-2}\,day^{-1}$ and is in general given by specialized polymeric material, as for example Aclar by Honeywell, or by the deposition of a single inorganic barrier layer on other plastic films. Performance in this range

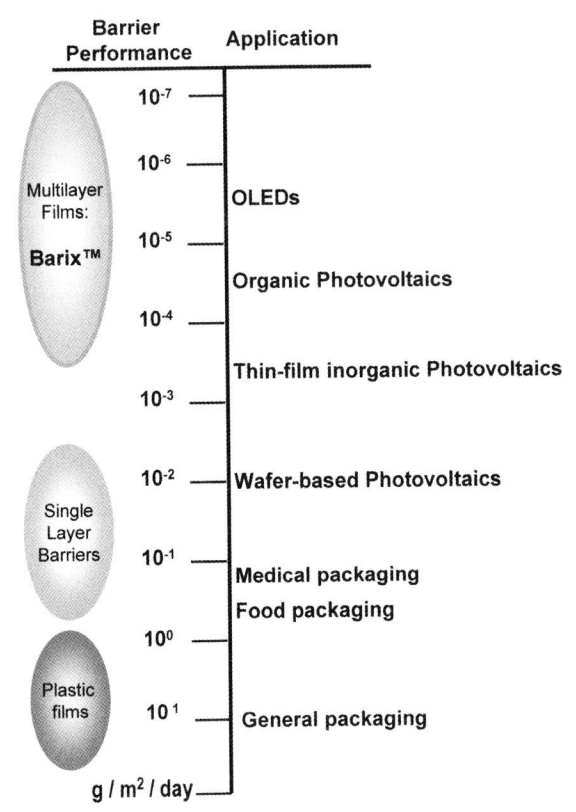

Figure 18.1 Barrier performance of engineered plastic films and typical applications.

is suitable for silicon-based PV. Permeation lower than 10^{-3}–10^{-4} g m^{-2} day^{-1} can be achieved only by the deposition of multilayer barriers on plastic and is sufficient for thin-film inorganic PV. OPV has permeation requirements comparable to those of OLEDs (5×10^{-5}–10^{-6} g m^{-2} day^{-1}) that can be achieved only by a few selected multilayer barrier technologies.

The barrier coating on plastic film for food/pharmaceutical applications is often realized by depositing aluminum layer(s) and, therefore, is not transparent. Transparency of more than 85% to the solar spectrum is another essential attribute for solar cells packaging films to be used on the cell side exposed to light. In addition, antireflective characteristics would be desirable. The side not exposed to the light may be reflective, although it should be nonconductive.

The engineered plastic films to be used as substrates must be able to sustain the deposition of a transparent conductor that is in most cases indium-tin oxide (ITO) deposited by sputtering. The layers constituting an organic solar cell are extremely thin (200–500 nm). To avoid shorts between the electrodes, the roughness of substrates and of the barrier encapsulation coating must be very smooth, ideally in the nanometer range and with peaks less than 40 nm, that is, about 1/10 of the total absorber film thickness.

The barrier must be environmentally resistant, with specific regard to stability to UV light and weathering. Its mechanical properties include good adhesion, robustness to breaking and impact, and antiscratch protection.

The two possible approaches to OPV packaging include direct encapsulation by depositing the barrier directly on the cell or lamination of a plastic film precoated with the barrier to a substrate where the device has been fabricated. The best method may be chosen accordingly for the optimized production process. However, if direct encapsulation is the choice, the barrier deposition process must be compatible with the device itself. This requires a deposition process at low temperature (less than 120 °C), with no damage from radiation or chemicals that may be used in the barrier deposition. Mechanical (low induced stress) and chemical compatibility between the device and the deposited layer(s) is also required. These last two requirements also apply to the lamination process.

As for any other component used in technologies related to solar energy, a major problem in the choice of the encapsulation is to be cost competitive. This is one of the reasons that ethylene vinyl acetate (EVA) was chosen for use as an encapsulant in traditional Si-based PV modules, rather than other more expensive polymers known to have better properties. The cost of the materials used and production processes must be held to a minimum. This requires proper choice of multilayered stacks of superstrates, substrates, and coating and lamination processes. These must be made from inexpensive, durable, and easily processed materials. An indicative target cost for solar cell encapsulation is less than $10 \, m^{-2}$.

18.3
Degradation Mechanisms of OPV Cells

Extensive studies of the degradation mechanisms of OPV cells have been carried on by several groups and several degradation mechanisms have been identified [10–13].

The first mechanism is the degradation of the polymeric materials induced by the UV exposure in air [10]. Conjugated polymers such as poly(p-phenylene vinylene) (PPV) are unstable in air. Oxygen atoms bind to the vinyl bonds, breaking the conjugation and forming carbonyl groups [10]. The degradation is mitigated, but not suppressed, when the conjugated polymers are mixed with fullerene. The oxidation reduces the light absorption in the range of visible spectrum and degrades the charge transport characteristics of the polymer blend by lowering the charge-carrier mobility of the polymer and creating a higher density of deep traps. Both mechanisms reduce the current of short circuit, I_{sc}, of the solar cells.

A second degradation mechanism is associated with exposure of the cells to oxygen and moisture in the dark [11]. One effect of it is the oxidation of the low work function materials commonly used in the standard OPV architectures so as to achieve selective carrier injection at the interface between cathode and light-absorbing layer. The degradation of the organic material by direct reaction with the Al cathode has also been postulated [12]. By both mechanisms, cells degrade through loss of active area, similar to the loss of active area in OLEDs due to growth of dark spots. The collection efficiency is reduced and contact resistance at the interface increases. Exposure of solar cells based on semiconductive polymers to moisture in the dark may result in water absorption into the poly(3,4-ethylenedioxythiophene) poly(styrenesulfonate) (PEDOT:PSS) layer. In Ref. [12], Krebs et al. explain the loss of I_{sc}, and the consequent loss of efficiency, through higher resistivity at the anode/cell interface induced by the formation of localized and spatially inhomogeneous insulating areas at the PEDOT:PSS/blend interface.

By a longer time exposure to oxygen and moisture, actual mechanical disintegration may occur either by an electrochemical process that leads to delamination of the electrode and/or segregation of the donor and acceptor components of the bulk heterojunction structure.

Since the different degradation mechanisms are initiated by oxidation and water, absorption barriers with permeation lower than $10^{-4}\,g\,m^{-2}\,day^{-1}$ are required for achieving cell lifetime of at least 5000 h at room temperature (5000 h being considered the minimum lifetime for any, even disposable, commercial device) [6].

Considering a packaging technology for OPV cells, the potential degradation of cell performance induced by deleterious effects of the direct encapsulation or barrier lamination processes needs to be addressed in the selection of processes and materials. Few examples of degradation induced by packaging have been reported in literature. For example, Sariciftici et al. [14] report 20% reduction in I_{sc} in the first 50 h after lamination of the cells with a high-performance barrier. The authors explain the phenomenon either by trapping of residual oxygen or formation of by-products in the reaction of the two epoxy glue components used in the lamination. Glue induced degradations are known to occur in glass encapsulation OLED devices.

Extending the observations made by Park et al. [15] reporting on encapsulation of OLED devices by alumina films deposited by atomic layer deposition (ALD), it is expected that the direct deposition of AlO_x by ALD on OPV cells would also lead to the reduction of the active area. In the cited paper the problem was minimized by the adoption of a parylene buffer layer between barrier layer and device.

The present authors [16] reported 20% I_{sc} decrease measured after direct plasma exposure of organic photovoltaics devices based on semiconducting polymers in encapsulation by the Barix multilayer barrier. We also reported that, similar to OLEDs [17], the detrimental effect to OPV cells can be avoided by using a Vitex proprietary protective layer. These experimental results will be further discussed in the following paragraph.

The potential degradation of packaged OPV performance that may come from the presence of the barrier itself must also be considered. Deleterious interactions may be induced by the physical or chemical long-term incompatibility between the materials used in the barrier and the OPV cells. Mechanical stresses and weathering of the barrier polymeric materials induced by irradiance (especially by UV radiation), temperature and thermal cycles, exposure to atmospheric gases (O_2, O_3, CO_2), and pollutants (gases, mists, and particulates) are examples of these long-term effects. To avoid this type of degradation, the first prerequisite is that the bulk properties of the substrate, superstrates, thin-film coating, and other materials be stable. For example, photothermally induced degradation can be the predominant factor for polymeric- or organic-based materials used in the packaging. The activation spectrum depends on the bond strengths in the polymer and is sensitive to the incident UV wavelengths. Interface reactions must also be addressed with particular attention given to the intrinsic higher reactivity and diffusivity of species located there and the potential of macroscopic failures due to loss of adhesion between layers.

18.4
Current Approaches to Oxygen and Moisture Barriers

In the most conventional approach to solar cell packaging, the cell is sandwiched between two sheets of glass using EVA as cell encapsulant. This technology, mainly developed for wafer-based PV, has evolved more recently. Especially for thin-film PV applications, the rigid glass sheets are being substituted with polymeric sheets to reduce weight, breakage, and possibly cost. Alternatively the backplane of the solar cell is replaced by a thin metal foil with the same objectives in mind. Among the polymeric materials used as back sheet, the most widely used is Tedlar [(polyvinyl fluoride (PVF)], very often in the form of a trilaminate sheet Tedlar/polyethylene terephthalate (PET)/Tedlar. Unfortunately, these polymeric front and back sheets considered for thin-film inorganic solar modules have poor barrier proprieties and are, therefore, not suitable for OPV. The thin metal foil is obviously a good barrier material, but needs to be polished and planarized and then covered by a pinhole-free passivation layer to prevent electrical contact between the anode and the metal backplane. This is obviously not a very easy solution either.

Jorgensen *et al.* conducted an extensive comparative study focused on solar packaging materials and modeled the rates of moisture ingress into photovoltaic modules [18]. The study looked at the moisture transport properties of several engineered polymeric films and their interfacial adhesion to glass as a function of damp-heat test at 85 °C/85% RH. The comparison was based on corrosion of Al layers deposited on glass. In this study oriented to silicon-based photovoltaics, breathable sheets were also considered since they are acceptable for silicon-based PV. In this approach, the outer face of the sheet is in equilibrium with the environment. Moisture can penetrate in the modules, but it will go back to equilibrium quite rapidly (time <1 week), possibly removing harmful species created by corrosion and material decomposition. This solution is not suitable for OPV cells that degrade in *few hours* when exposed to oxygen or moisture. The modeling effort showed that the water vapor transport rate (WVTR) of impermeable sheets must be at least less than $10^{-4}\,\mathrm{g\,m^{-2}\,day^{-1}}$ so that the water will not penetrate to the central part of the module over the lifetime of the panel. In the experimental study the permeation of commercial monolithic sheets such as PET, polyethylene naphthalate (PEN), PVF, and polychlorotrifluoroethylene (PCTFE) and single-layer laminate sheets (PVF, PET/SiO_2, or Al) were compared with experimental sheets made of PET/PEN substrates coated with single or multilayer barriers. All the multilayer barriers had WVTR less than $1\,\mathrm{g\,m^{-2}\,day^{-1}}$ at 85 °C and below MOCON detection limit even at 60 °C. They were transparent with the exception of a PET/Al/PET laminate. The transparent multilayer barriers investigated were based on the Barix technology initially proposed for application to OLED devices at Pacific Northwest National Laboratories (PNNL) and developed and commercialized by Vitex Systems, Inc. (San Jose, CA) [17]. This technology and its promising results in the photovoltaics field will be extensively discussed in Section 18.5.

Most of the literature and the published data for the high-performance barriers (WVTR < 10^{-4} g m^{-2} day^{-1}) required by OPV are presented in the contest of OLED technology.

Single layer barriers can achieve low permeation. However, in general, thin-film barriers of single inorganic layers on plastic have not been able to demonstrate their capabilities in a technically and economically feasible way. The requirements of the layers of being totally pinhole and crack free over very large (>1 m^2) surface areas, low stress and high robustness while being deposited at low temperatures well below 100 °C, and transparent, if to be used as frontsheet, have proven to be very difficult to meet. Early attempts to solve this problem with single layer oxides or nitrides, while obtaining some success on small areas, failed in the scalability to large area substrates because of the presence of particles, crack, and defects in the layer and residual stress.

Among the deposition techniques used for coating inorganic barrier layer on plastic, the most commonly used are reactive sputtering [17,19], SiO$_2$ or SiN$_x$ plasma enhanced chemical vapor deposition (PECVD) [14,20,21,23], SiN$_x$ by "Hot-wire"-chemical vapor deposition (CVD) (also called "Cat"-CVD) [22], and Al$_2$O$_3$ ALD [23–26]. All the studies report improved barrier performance with the deposition of a polymeric smoothing layer prior to the deposition of the inorganic barrier layer. PECVD or sputtered films can reach typical permeation rate in the range of $1-10^{-1}$ g m^{-2} day^{-1} with film thickness of several micrometers. ALD can achieve lower permeation rate ($10^{-1}-10^{-2}$ g m^{-2} day^{-1}) with thinner (30–50 nm) layers, but because of low deposition rate it is costly and has been proved to be used only on small area (30 × 40 cm^2). Cat-CVD has been scaled to several square meters, but only in a discreet sheet configuration. Symmorphix [19] has demonstrated a physical vapor deposited (PVD) single-layer aluminosilicate barrier on polycarbonate (PC) and PEN films. The most recent data show that a permeation rate in the $10^{-5}-10^{-6}$ g m^{-2} day^{-1} range was achieved for a 150–200 nm single layer barrier on PEN film. For a bending radius of 12 mm, no cracking was observed. However, the maximum size of the substrate is 600 mm × 700 mm.

Most developers have transitioned to multilayer structures to improve yield and manufacturability. Among the barrier multilayer approach for photovoltaic packaging, the Fraunhofer ISC and partners [27] have developed a barrier multilayer on PET composed by (four) multiple layers of alternating PVD-SiO$_x$ and ORMOCER, a sol–gel organic/inorganic composite material manufactured in a two-step process. The reported permeation is below 1×10^{-2} g m^{-2} day^{-1}. Sariciftci et al. [14,27] have reported an optically transparent multilayer barrier made of five alternating inorganic/organic layers of SiO$_x$ and PECVD-deposited organosilicon deposited on PEN. The thickness of each layer is 100 nm for a total of 500 nm. The permeation was below the MOCON detection limit (<5×10^{-3} g m^{-2} day^{-1}) and survived the "Ca test" for 1000 h at 85 °C, 85% RH. Organic solar cells encapsulated with this barrier survived

3000 h, the same as glass encapsulated cells. The total area of the cells was 56×30 mm^2. Another type of multilayer PECVD barrier on plastic (polycarbonate) with WVTR below $1 \times 10^{-5}\,\text{g m}^{-2}\,\text{day}^{-1}$ is developed by Duggal et al. from GE Research [20]. The barrier is described as graded ultrabarrier fabricated by two base PECVD processes: an inorganic and an organic process. This barrier was tested on OLED pixels that reached 144 h at 60 °C/85% RH with no black spots. Dow Corning has also presented results on graded barrier coatings of SiO_xC_y (organic zone)/SiC alloy for application to OLED displays and solid-state lighting [28].

The multilayer solution proposed by Vitex is an organic and inorganic stack called Barix [17,29–31]. It consists of thicker polymer layers alternated by thin layers of oxide or nitride. The Barix barrier can be directly deposited on devices to provide an ultralow permeation thin-film encapsulation or can be deposited on plastic films by using a roll-to-roll web coater creating an engineered plastic film that can be used as a lightweight flexible substrate or laminated on devices as packaging material. The Vitex Barix multilayer has been shown to meet telecommunication and automotive application specifications for OLED displays [16,17,31–33]. Among the different ultrabarrier technologies for OLED application, the Barix technology is the only one now for having been commercialized. Recently, the roll-to roll barrier technology on plastic per display application has been licensed to a major plastic film supplier and is being scaled up for manufacturing web coaters. The technology has also been commercialized for direct encapsulation of display by the sale of three tools and technology licenses to major OLED display manufacturers with the tools being used for mass production. Although the multilayer technology proposed by Vitex has been specifically developed for OLED display applications, it has been demonstrated that the basic principles to achieve permeation rated in the range 1×10^{-4}–$1 \times 10^{-6}\,\text{g m}^{-2}\,\text{day}^{-1}$ satisfy the requirements for PV inorganic and organic solar cells [16,34].

18.5
Barix Multilayer Technology

A cross section of the Barix multilayer system of organic and inorganic layers is shown in Figure 18.2. The thin films of oxide serve as the barrier layers to oxygen and water. The multilayer provides redundancy. Since the remaining defects in the inorganic layers are few, farther apart, and not connected, a very long diffusion path for water and oxygen molecules from the external surface to the substrate results [17,35]. Therefore, the main effect of the multilayer is increasing the lag time between exposing the top layer to water vapor and the water molecules arriving at the interface between the OLED and the Barix encapsulation layer.

Theoretical modeling of the permeation through this type of multilayer structure has been done by several groups [36–38]. In the extensive model developed by Graff et al. [35], the main findings are that (i) high-quality inorganic films coupled with a multilayer architecture are necessary to achieve OLED barrier requirements (large spacing between defects); (ii) lag time (transient diffusion), not steady state flux,

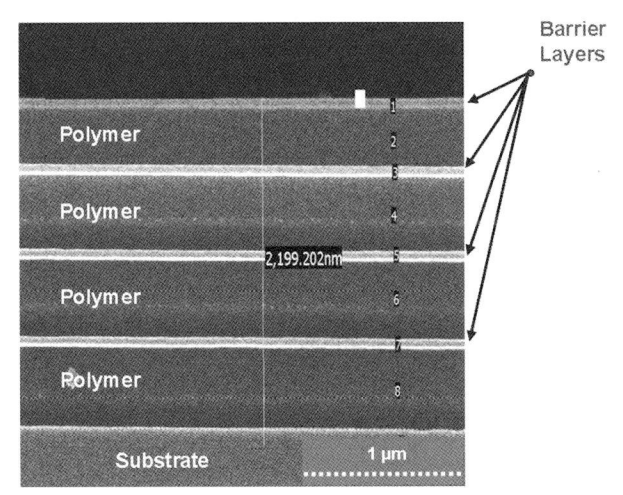

Figure 18.2 SEM cross section of a four-dyad Barix barrier.

dominates gas permeation in these multilayer thin-film systems; (iii) consideration of steady state alone is not sufficient to describe and predict the performance of multilayer barrier films; the transient regime must be considered. Similar conclusions were independently reached by Paetzold *et al.* by using Monte Carlo simulations [36] and Greener *et al.* by using the laminate diffusion equation for diffusion [37]. In all these theoretical simulations, it has been assumed that organic layers and substrates follow the Fick's law.

Details of the apparatus used for the deposition of the Barix barrier are described elsewhere [16,17,29–33]. The deposition of the Barix multilayer stack for direct encapsulation or for the deposition of barrier on discrete plastic foils is done by moving the device(s) or substrate(s) back and forth between the different stations. Similarly, in the experimental web coater, only one station for oxide deposition and one station for polymer deposition and curing are present. Each layer of the stack is coated on the 30-cm wide web by sequentially depositing on the entire roll of plastic film, while the film is rewinded on a temporary roller.

Alumina layers with typical thickness between 10 and several hundred nanometers are deposited by DC reactive sputtering of a metallic Al (99.999%) target. Polymer layers are deposited in vacuum by a deposition technique that involves the flash evaporation of liquid monomers. The monomer vapors are condensed on the sample surface in liquid phase and cross-linked by UV radiation to form a solid film. A proprietary blend of acrylates monomers (Barix Resin System) optimized for encapsulation of OLED devices is used to give a highly cross-linked acrylic organic film [39,40]. Typical thicknesses of the organic films vary between 0.25 and several micrometers. What is really unique about this process is that the organic phase is deposited as a liquid: the film is very smooth (<2 Å variation) locally and also has extremely good planarizing properties over high topographical structures. So while the local flatness creates an ideal surface for growing an almost defect-free inorganic layer, the liquid takes care of covering topography.

The Barix barrier properties depend on the number of organic/inorganic layers deposited (each pair may also be called dyad). The initial validation of the technology has been done by using five to six dyad barriers, but recently the number has been reduced to two to three (see below). The Barix multilayer has been shown to meet telecommunication application specifications requiring WVTR lower than 10^{-6} g m^{-2} day^{-1} for a wide variety of OLED displays: passive and active matrix displays, bottom, top, and transparent displays and working equally well for small molecule, polymer, and phosphorescent OLEDs [17,40–46].

Barix-coated plastic film can be used as transparent substrates and/or superstrates for highly sensitive moisture and oxygen devices like OPV cells, or Barix multilayers can be directly deposited on the devices as a thin-film encapsulation. Both these packaging options are compatible with the use of metal foil as a substrate.

Direct Barix encapsulation has been applied with success to inorganic PV cells [34]. Figure 18.3 shows some results of accelerated lifetime testing at 60 °C/90% RH of CIGS (a) and CdTe (b) solar cells. While the efficiency of bare CIGS solar cells drops to 50% of the initial value after 100 h of lifetime testing, the efficiency of the Barix encapsulated cells stays constant throughout the duration of the test. The encapsulated CdTe solar cells present a 20% drop in efficiency after 100 h of testing and remains constant thereafter. However, the initial drop is not related to moisture/oxygen permeation but rather to other temperature-accelerated degradation mechanisms not related to barrier performance as shown by the identical behavior of cells aged in damp or dry atmosphere (see Figure 18.3b).

Typical barrier performance and champion data of the Barix multilayer measured using the Ca test [47] are reported in Table 18.1. The permeation performance of the Barix barrier realized by direct encapsulation of Ca coupons on glass is typically below 10^{-6} g m^{-2} day^{-1}. Accelerated testing at 85 °C/85% RH for more than 1000 h has shown that 10^{-7} g m^{-2} day^{-1} can be achieved. Permeation in the range of 10^{-6} g m^{-2} day^{-1} has been demonstrated by discrete plastic sheets coated on a linear tool. In this case the multilayer barrier is first deposited on plastic films (PET or PEN), then a series of Ca coupons are deposited and directly encapsulated. Figure 18.4 shows the pristine status of the Ca coupons encapsulated in this way after aging for 1000 h at 60 °C/90% RH. All the samples made on the roll-to-roll machine pass the Mocon test (WVTR $<5 \times 10^{-3}$ g m^{-2} day^{-1}) for both two dyad- and five dyad-coated films, but in the Ca test pinholes start to appear after 100 h at 60 °C/90% RH indicating the presence of local defects in the barrier and a barrier performance more than $10 \sim 100$ time worse than the linear tool. Careful analyses of the major failure modes and dedicated experiments have shown that in the web coater the oxide–roller interaction damages the oxide by creating particles that degrade the barrier causing the calcium film to oxidize faster. This and other mechanisms related to poor handling and repeated winding of the web on the experimental roll-to-roll coater explain the difference in performance. The proper design of the production roll-to-roll coater can remove the difference.

The multilayer barrier is transparent ($T > 90\%$) in the visible range as shown in Figure 18.5 by the light transmittance spectrum of a stack coated on plastic. Computer simulation of the optical properties of the stack have demonstrated that

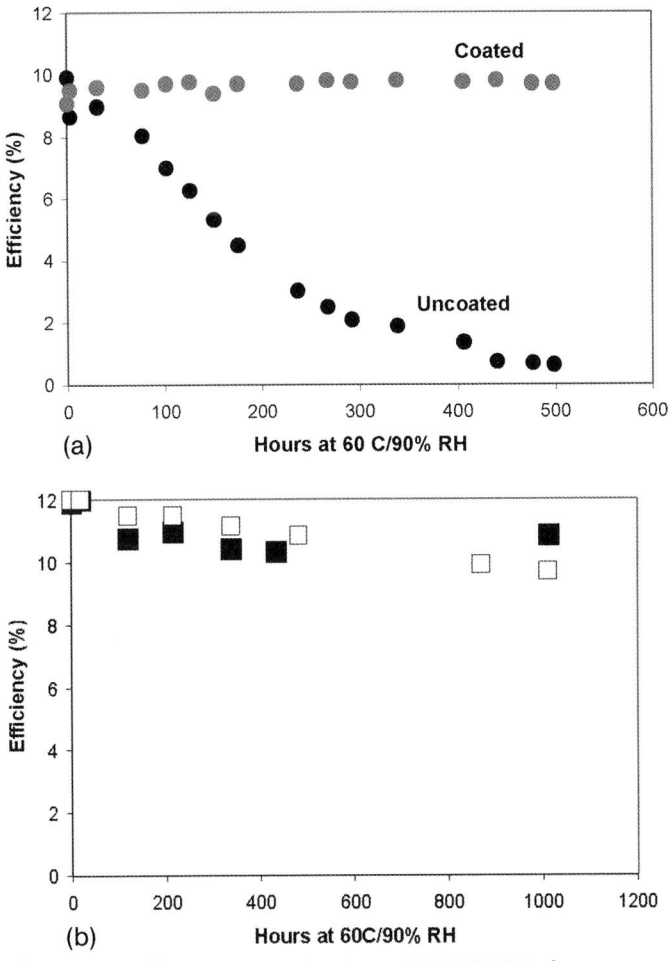

Figure 18.3 (a) Efficiency versus aging time at 60 °C/90% RH of CIGS bare (dark dots) and Barix encapsulated (light dot) solar cells (courtesy of PNNL). (b) Efficiency versus aging time of Barix encapsulated CdTe cells: 60 °C dry heat (dark dots) and 60 °C/90% RH (light dot) solar cells (courtesy of PNNL).

Table 18.1 Typical and champion permeation values of Barix multilayers.

Samples	Typical performance[a]	Best results[a]
Barrier on glass	$<10^{-6}$	$<10^{-7a}$
Barrier on plastic[b]	$<10^{-6}$	
Barrier on plastic roll-to-roll	$<10^{-4}$	5×10^{-6}

[a]Values in g m^{-2} day^{-1}.
[b]Sheet coating.

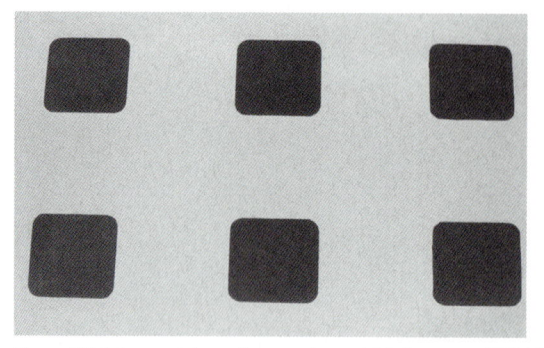

Figure 18.4 Image taken after accelerated lifetime test (60 °C/ 90% RH) of Ca coupons deposited on a Barix/PEN substrate and directly encapsulated. The substrate size is 200 mm × 200 mm.

by properly selecting the thickness of the layers and the refractive index of the inorganic layer, the optical properties of the barrier could be optimized so as to enhance the antireflective property for the light of interest.

The plastic substrates engineered with the Barix barrier are flexible. Table 18.2 reports the result of a bending test for PET substrates. The pass–fail criteria are defined by the appearance of cracks and defects that eventually lead to the deterioration of the barrier performance measured by the Ca test. Up to 1000 flexes can be sustained with no loss of performance for curvature radius between 25 and 50 mm (see Figure 18.6).

The barrier on plastic can withstand ITO deposition as well as patterning and annealing processes with no loss of performance [48].

In direct encapsulation of OPV cells, the device sensitivity to the barrier deposition process must be considered. In applying the Barix encapsulation to OLEDs, it had

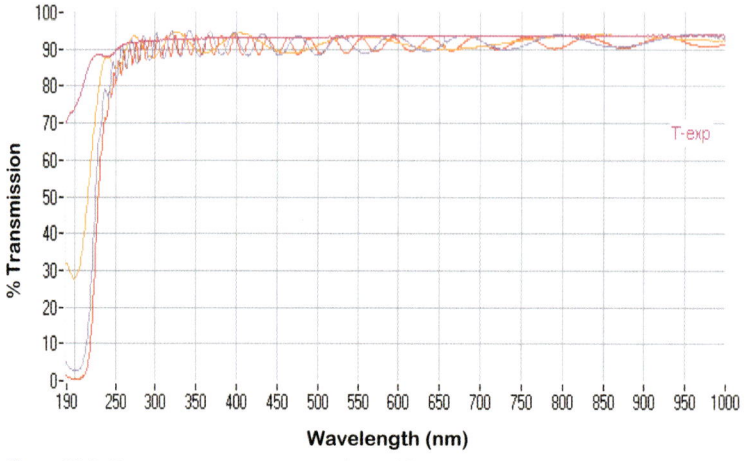

Figure 18.5 Transmission spectrum in the visible range of a Barix barrier on PET substrates.

Table 18.2 Results of bending test for Barix multilayer on plastic.

Radius (mm)	Sample	Pass/fail
6	Barrier on PET	Fail
8	ITO on PET	Fail
12	Barrier on PET	Pass 100
		Fail 1000
25–50	Barrier on PET	Pass 1000 flexes

been seen that some types of OLED devices are sensitive to the plasma process, but this detrimental effect can be avoided by using a Vitex proprietary protective layer [49] or by laminating a Barix-coated plastic film on top of the solar cell. Similar investigations of sensitivity to the direct encapsulation process have been carried on organic solar cells based on photoactive blends of conjugated polymers deposited on glass [50]. The solar cells were fully characterized after different steps of the encapsulation process and after full encapsulation. Three different encapsulation processes, SOP (Standard Operating Procedure), A, and B, were compared. Process A included the deposition of a protective layer at the interface cell/encapsulation, while process B included the protective layer plus some modification in the layers deposition conditions. Since the cells were measured and transported in sealed stainless steel container in a N_2 atmosphere to the encapsulation tool, reference samples were included to monitor the effect of the transportation. The main cell performance parameters normalized to the values measured prior to shipping are reported in Table 18.3. All devices showed no change in open-circuit voltage (V_{oc}). However, the short-circuit current (I_{sc}) decreased following transportation, oxide deposition, and full encapsulation with the standard process (SOP). The full encapsulation was repeated twice after different shipping for checking reproducibility. As a consequence of the drop in I_{sc}, the efficiency also dropped. By modifying the

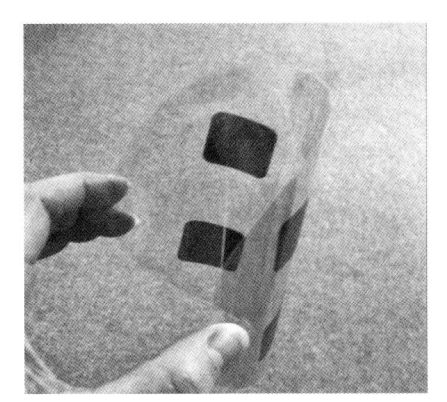

Figure 18.6 Bended plastic sheet with Ca coupons deposited on a Barix/PEN substrate and directly encapsulated. The substrate size is 200 mm × 200 mm.

Table 18.3 Variation of I_{sc}, V_{oc}, and efficiency for OPV cells after different process steps used in Barix encapsulation.

Process	FF	V_{oc}	I_{sc}	Efficiency
Transportation	0.85	0.98	0.85	0.70
UV exposure	0.98	1.00	0.90	0.87
Oxide deposition	1.00	0.95	0.81	0.80
Full barrier SOP deposition (1)	0.80	0.90	0.80	0.58
Full barrier SOP deposition (2)	0.90	0.96	0.88	0.74
Full barrier deposition (process A)	0.94	0.96	1.00	0.88
Full barrier deposition (process B)	0.93	0.98	0.94	0.90

All variables are normalized to the value measured before processing.

encapsulation process with the introduction of the protective layer (process A) and, additionally, by changing some of the deposition conditions (process B), the performance loss can be avoided. While the reduction in I_{sc} for the witness cells (i.e., only shipped, no processing) suggests that some of the performance loss can be attributed to loss of active area due to oxidation during transportation, it is conceivable that this type of OPV cells are sensitive to plasma damage similar to the OLED devices. The results obtained with the modified processes also support this interpretation and demonstrate that this type of deterioration can be avoided by the protective layer.

Similar type of OPV cells fabricated on glass substrates were packaged by direct encapsulation using process B and by lamination using PET films with a five dyad Barix barrier deposited on the prototype roll-to-roll coater [16]. By storing bare and Barix encapsulated cells in a dry box for more than 800 h and comparing the variation of the cell parameters, it was possible to verify that no long-term deterioration is introduced by the encapsulation process. When exposed to a climate chamber (30 °C/90% RH followed by 40 °C/90% RH), the lifetime of the encapsulated cells was 100 times longer than for bare cells, although the efficiency dropped to less than 50% in about 200 h. Once again, the main deterioration was seen in I_{sc} and related to loss of active area. The main reason for the poor barrier performance was attributed to the excessive number of particles introduced by handling of the samples in measuring and transportation. The laminated cells had a lifetime of more 5000 h before dropping the efficiency below 50% of the original vales. Analogous to the directly encapsulated cells, the drop in efficiency is due to a drop in I_{sc}. The measured lifetime is in agreement with the predicted value calculated by using WVTR equal to 5×10^{-5} g m^{-2} day^{-1}. This permeation value had been measured in a simultaneous and parallel characterization of calcium coupons. This permeation value agrees with the typical barrier performance of films coated in the experimental roll-to-roll coater. The typical barrier performance of multilayers deposited on plastic on discrete plastic sheets (see Table 18.1) would already yield in solar cells with lifetime acceptable for a product.

The capability of the Barix process for mass manufacturing has also been demonstrated. The main improvements toward scalability and manufacturability readiness have been the reduction in the number of dyads (2–3 days) maintaining the high-performance yield (>95%) [31,33]. The improved Barix process with less dyads was made possible by improvement in materials and process conditions. The proprietary blend of acrylate monomers was specifically formulated to meet the requirements of a good barrier on OLED devices. Hardware, metal targets, and process conditions used for reactive sputtering of the alumina layer were improved. The quality of the inorganic layer was improved, yielding a film with less intrinsic defects deposited with tighter uniformity on the plate and thickness/composition control through the target life (±10%). The reduction in added particles also had a large significance. A clear correlation has been proven between barrier failures on OLED and particle density [51].

Analysis of the Barix barrier deposition process had demonstrated that the oxide deposition process is the rate-limiting step in the overall process. To meet the throughput requirements for larger substrates and web coating, a much faster oxide process has been developed. Faster deposition rates were obtained by changing both the cathode configuration and the parameters of the deposition process. The actual limitations to the development of a fast process are determined by finding a process that does not damage the devices and/or the barrier layers. In developing an oxide process capable of meeting throughput requirements, the following criteria were met: (i) no damage to the electroluminescence characteristic of the OLED devices, (ii) no damage to the organic layers in the barrier, (iii) same moisture and oxygen barrier performance as the "slower" process, and (iv) no loss of device yield or faster black spot growth than the previous process.

A complete discussion of the interacting mechanisms leading to degradation of the overall barrier performance on devices is beyond the scope of this paper. As an example of a cause of barrier performance degradation occurring in changing the sputtering conditions, nonoptimized hardware may introduce particles or lateral nonuniformities in the thickness of the deposited oxide and these results in defective barrier stacks with worse performance.

By a modification in the sputter conditions, a process has been developed that is five times faster then the previous standard process. The barrier performance on Ca coupons (and OLED devices) of the fast process successfully compared to the standard process [40–43]. A 95% yield of good pixels after 500 h aging at 60 °C/90% RH was achieved on a sample set of 180 devices. The 5% yield loss is due to occasional shorts and independent of the encapsulation process, since it is also observed for glass lid encapsulated test samples. These results are comparable with previously published Barix encapsulation results using the standard process. Equivalent/better barrier performance of faster and standard process upon aging at 85 °C/85% RH was also obtained. For roll-to-roll machines able to coat 12, 24, and 60″ wide web, coating speeds in the range of 2–3 m min^{-1} are possible. An important consideration in changing the configuration and condition of sputtering oxide deposition is the potential damage to the devices induced by plasma processing. By using OLED devices we have proved that the protective layer is effective in avoiding the previously reported [17,42,43,52] short- and long-term damage induced by plasma as it is for the standard process. While the manufacturability readiness has been proved by using Ca test and OLED devices, the similarity in barrier requirements and process sensitivity between OLEDs and OPV cells described before suggests that its extension for OPV applications is justified.

The estimated cost of the Barix barrier coated by a web coater on a plastic film is $5–50 m^{-2} with the ultimate cost depending on the cost of the basic film. In a fully integrated OPV production process on roll-to-roll coater, direct encapsulation also would be possible on the same web coater.

18.6
Conclusions

In the last few years, research and development in photovoltaics has dramatically increased, pushed by the worldwide need for renewable energy sources. In the near future, among the new approaches to solar power harvesting, the use of organic solar cells may be an option for applications where efficiency is not crucial, provided that the lifetime of the cells can be extended to several thousand hours with fabrication costs kept low. The extreme sensitivity of OVP to moisture and oxygen makes this a challenge for the development and application of an encapsulation technology able to meet the requirements of low weight, flexibility, and low cost with a permeation rate well below 10^{-5} g m^{-2} day^{-1}.

Traditional PV packaging solutions are not adequate. The adoption of multilayer barrier technologies deposited on plastic substrates or by direct device encapsulation seems the most viable option.

We have reviewed here several multilayer barriers with water permeation below 10^{-4}–10^{-6} g m^{-2} day^{-1}. We propose the Barix multilayer coating as a technology that on OLED, on plastic or glass substrates applications, has demonstrated permeation below 10^{-6} g m^{-2} day^{-1} with available process and equipment for low-cost mass manufacturing.

The preliminary applications of Barix barrier coatings to inorganic and organic PV have extended the lifetime of the cells to several thousand hours (at RT) with no evident damage to the cells. While further research and development will be necessary so that Barix can meet all the application-specific requirements, Barix barrier deposited by roll-to-roll on a web seems to offer a low-cost solution to produce flexible OPV.

Acknowledgments

The authors warmly recognize N. Rutherford's contribution to the development of the Barix encapsulation and to the scientific and technical understanding of the barrier requirements for plastic electronics and OPV that are the subject of this manuscript. The indispensable and creative contribution of the Xi Chu, T. Krajewski, C. Hutchinson, S. Kapoor, T. Ramos, H. Hirayama, and all the Vitex team to the deployment of the Barix technology is also acknowledged. The great support of Chyi-San Suen in contact with customers and economic evaluations is also recognized. We also thank the PNNL Staff whose work appears in this presentation. Konarka Technologies, and in particular C. J. Brabec and J. A. Hauch, have stimulated the initial applications of Barix to OPV and collaborated to preliminary results. Identified (Samsung SDI, Philips Mobile Display Systems, Universal Display Corporation, MED) and unidentified Vitex collaborators have contributed to the Barix development by supplying OLED test pixels. The collaboration of industrial partners such as DuPont Teijin Films, Techni-Met, Inc., Tokki Corporation, and ANS is also recognized.

Part of the work on the development of Barix and its applications to flexible substrate and encapsulation of plastic displays was funded by USDC contracts RFP98-37 and RFP01-63 and subcontract no. 070102.10 to UDC by ARL DAAD19-02-2-0019.

References

1 Sariciftci, N.S., Smilowitz, L., Heeger, A.J. and Wudl, F. (1992) Photoinduced electron transfer from a conducting polymer to buckminsterfullerene. *Science*, **258**, 1474–1476.

2 Halls, J.J.M., Walsh, C.A., Greenham, N.C., Marseglia, E.A., Friend, R.H., Moratti, S.C. and Holmes, A.B. (1995) Efficient photodiodes from interpenetrating polymer networks. *Nature*, **376**, 498–500.

3 Yu, G., Gao, J., Hummelen, J.C., Wudl, F. and Heeger, A.J. (1995) Polymer photovoltaic cells: enhanced efficiencies via a network of internal donor-acceptor heterojunctions. *Science*, **270**, 1789–1791.

4 Schilinsky, P., Waldauf, C. and Brabec, C.J. (2002) Recombination and loss analysis in polythiophene based bulk heterojunction photodetectors. *Applied Physics Letters*, **81**, 3885–3887.

5 Waldauf, C., Schilinsky, P., Hauch, J. and Brabec, C.J. (2004) Material and device concepts for organic photovoltaics: towards competitive efficiencies. *Thin Solid Films*, **451–452**, 503–507.

6 Brabec, C.J., Hauch, J.A., Schilinsky, P. and Waldauf, C. (2005) Production aspects of organic photovoltaics and their impact on the commercialization of devices. *MRS Bulletin*, **30**, 50–52.

7 NREL certificate for Konarka solar cell E8-6, July 18, 2005.

8 Kim, J.Y., Lee, K., Coates, N.E., Moses, D., Nguyen, T.-Q., Dante, M. and Heeger, A.J. (2007) Efficient tandem polymer solar cells fabricated by all-solution processing. *Science*, **317**, 222–225.

9 Plextronics, Press Release, August 9, 2007, Pittsburgh, PA.

10 Pacios, R., Chatten, A.J., Kawano, K., Durrant, J.R., Bradley, D.D.C. and Nelson, J. (2006) Effects of photo-oxidation on the performance of poly{2-methoxy-5-(3′,7′-dimethyloctyloxy)-1,4-phenylene vinylene}:{6,6}-phenyl C_{61}-butyric acid methyl ester solar cells. *Advanced Functional Materials*, **16**, 2117–2126.

11 Kawano, K., Pacios, R., Poplavskyy, D., Nelson, J., Bradley, D.D.C. and Durrant, J.R. (2006) Degradation of organic solar cells due to air exposure. *Solar Energy Materials and Solar Cells*, **90**, 3520–3530.

12 Krebs, F.C., Carle', J.E., Cruys-Bagger, N., Andersen, M., Lilliedal, M.L., Hammond, M.A. and Hvidt, S. (2005) Lifetimes of organic photovoltaics: photochemistry, atmosphere effects and barrier layers in

ITO-MEHPPV:PCBM-aluminium devices. *Solar Energy Materials and Solar Cells*, **86**, 499–516.
13 Norrman, K. and Krebs, F.C. (2006) Chemical degradation mechanisms of organic photovolatics studied by TOF-SIMS and isotopic labeling. Proceedings of SPIE, Organic Photovoltaics VI, vol. 5938 (eds Z.H. Kafafi and P.A. Lane), 59380D, 10.1117/12.613433.
14 Denner, G., Lungenschmied, C., Neugebauer, H. and Sariciftci, N.S. (2005) Flexible, conjugated polymer-fullerene-based bulk-heterojunction solar cells: basics, encapsulation, and integration. *Journal of Materials Research*, **20**, 3224–3233.
15 Park, S.-H.K., Oh, J., Hwang, C.-S., Lee, J.-I., Yang, Y.S., Chu, H.Y. and Kang, K.-Y. (2005) Ultra thin film encapsulation of organic light emitting diode on a plastic substrate. *ETRI Journal*, **27**, 545–550.
16 Moro, L., Rutherford, N.M., Visser, R.J., Hauch, J.A., Klepek, C., Denk, P., Schilinsky, P. and Brabec, C.J. (2006) Barix multilayer barrier technology for organic solar cells. Proceedings of SPIE Organic Photovoltaics VII, vol. 6334 (eds Z.H. Kafafi and P.A. Lane), 63340M, 10.1117/12.687185.
17 Moro, L., Krajewski, T.A., Rutherford, N.M., Philips, O. and Visser, R.J. (2004) Process and design of a multilayer thin film encapsulation of passive matrix OLED displays. Proceedings of SPIE Organic Light Emitting Devices VII, vol. 5214 (eds Z.H. Kafafi and P.A. Lane), pp. 83–93.
18 Jorgensen, G.J., Terwillinger, K.M., DeCueto, J.A., Glick, S.H., Kempe, M.D., Pakow, J.W., Pern, F.J. and McMahon, T.J. (2006) Moisture transport, adhesion, and corrosion protection of PV module packaging materials. *Solar Energy Materials and Solar Cells*, **90**, 2739–2775.
19 Pakbaz, H. (2004) Thin film barrier properties of high performance dielectric materials: an update. Presented at OSC (Organic Semiconductors Conference), September 27–29, Cambridge, UK.

20 Kim, T.W., Yan, M., Erlat, A.G., McConnelee, P.A., Pellow, M., Deluca, J., Feist, T.P., Duggal, A.R. and Schaepkens, M. (2005) Transparent hybrid inorganic/organic barrier coatings for plastic light-emitting diode substrates. *Journal of Vacuum Science & Technology A: Vacuum Surfaces and Films*, **23**, 971–977.
21 Hemerik, M., van Erven, R., Yang, J., van Rijswijk, T., Winters, R. and van Rens, B. (2006) Lifetime of thin film encapsulation and its impact on OLED device performance. Society for Information Display (SID) 2006 International Symposium – Digest of Technical Papers, vol. XXXVII, pp. 1571–1574.
22 Matsumura, H. (2001) Summary of research in NEDO Cat-CVD project in Japan. *Thin Solid Films*, **395**, 1–11.
23 Park, S.-H.K., Oh, J., Hwang, C.-S., Lee, J.-I., Yang, Y.S. and Chu, H.Y. (2005) Ultra thin film encapsulation of an OLED by ALD. *Electrochemical and Solid-State Letters*, **8**, H21–H23.
24 Groner, M.D., George, S.M., McLean, R.S. and Carcia, P.F. (2006) Diffusion barriers on polymers using Al_2O_3 atomic layer deposition. *Applied Physics Letters*, **88**, P 051907.
25 Ghosh, A.P., Gerenser, L.J., Jarman, C.M. and Fornalik, J.E. (2005) Thin-film encapsulation of organic light-emitting devices. *Applied Physics Letters*, **86**, P 223503.
26 HIPROLOCO Project, European Commission Project (ENK5-CT-2000-00325) (2006).
27 Denner, G., Lungenschmied, C., Neugebauer, H., Sariciftci, N.S., Latreche, M., Czeremuszkin, G. and Wertheimer, M.R. (2006) A new encapsulation solution for flexible organic solar cells. *Thin Solid Films*, **349**, 511–512.
28 IR http://www.netl.doe.gov/ssl/portfolio-07/current-organic/ThinFilmPackagingSolutions.htm.
29 Affinito, J.D., Gross, M.E., Coronado, C.A., Graff, G.L., Greenwell, E.N. and Martin,

P.M. (1996) A new method for fabricating transparent barrier layers. *Thin Solid Films*, **290–291**, 63–67.

30. Burrows, P.E., Graff, G.L., Gross, M.E., Martin, P.M., Happ, M., Mast, E., Bonham, C., Bennett, W., Michalski, L., Weaver, M., Brown, J.J., Fogarty, D. and Sapochak, L.S. (2000) Gas permeation and lifetime tests on polymer-based barrier coatings. *Proceedings of SPIE – The International Society for Optical Engineering*, **4105**, 75–84.

31. Moro, L., Chu, X., Hirayama, H., Krajewski, T. and Visser, R.J. A mass manufacturing process for Barix encapsulation of OLED displays: a reduced number of dyads, higher throughput and 1.5 mm edge seal. In IMID/IDMC '06 Digest, August 22–25, 2006, Daegu, South Korea, pp. 754–758.

32. Rutherford, N. (2005) Flexible displays – a low cost substrate/encapsulation packaging solution. Proceedings of USDC Flexible Displays & Microelectronics Conference, February 1, Phoenix.

33. Chu, X., Moro, L., Rosenblum, M. and Visser, R.J. (2007) Barix thin film encapsulation of OLED's manufacturing aspects and high temperature performance. To be presented at IMID/IDMC, August 30, 2007, Daegu, Korea.

34. Olsen, L.C., Kundu, S.N., Bonham, C.C. and Gross, M. (2005) Barrier coatings for CIGSS and CdTe cells. Proceedings of 31st IEEE Photovoltaic Specialists Conference, pp. 327–330.

35. Graff, G.L., Williford, R.E. and Burrows, P.E. (2004) Mechanisms of vapor permeation through multilayer barrier films: lag time versus equilibrium permeation. *Journal of Applied Physics*, **96**, 1840–1849.

36. Paetzold, R., Henseler, D., Heuser, K., Cesari, V., Sarfert, W., Wittmann, G. and Winnacker, A. (2004) High-sensitivity permeation measurements on flexible OLED substrates. Proceedings of SPIE – The International Society for Optical Engineering Organic Light-Emitting Materials and Devices VII, vol. 5214 (eds Z.H. Kafafi and P.A. Lane), p. 14.

37. Greener, J., Ng, K.C., Vaeth, K.M. and Smith, T.M. (2006) The barrier problem in flexible OLED displays: a modeling approach. Society for Information Display (SID) 2006 International Symposium – Digest of Technical Papers, vol. XXXVII, pp. 916–919.

38. Moro, L., Krajewski, T.A., Ramos, T., Rutherford, N.M., Chu, X., Hirayama, H. and Visser, R.J. (2006) UV curable layers in barrier films for flexible electronics. Presented at RadTech 2006 – UV & EB Technology Expo & Conference, Chicago, IL, USA.

39. US Patent Application 2007/0281174 A1, December 6, 2007.

40. Visser, R.J. (2003) Presented at 3rd International Display Manufacturing Conference, IDMC 2003 Conference, February 18–21, Taipei, Taiwan.

41. Visser, R.J. (2004) Barix multi-layer as thin film encapsulation of organic light emitting diodes, in *Organic Electroluminescence Materials and Technologies* (ed. Y. Sato), CMC Books, pp. 141–152.

42. Moro, L., Krajewski, T.A., Rutherford, N.M., Philips, O., Visser, R.J., Gross, M., Bennett, W.D. and Graff, G. (2004) Integrated encapsulation of bottom and top emission OLED displays. Presented at SPIE – The International Society for Optical Engineering, Organic Light-Emitting Materials and Devices VIII, Denver, CO, USA.

43. Moro, L., Krajewski, T.A., Ramos, T., Rutherford, N.M., Chu, X. and Visser, R.J. (2005) Integrated encapsulation of flexible electronics. Presented at SPIE – The International Society for Optical Engineering, Organic Light-Emitting Materials and Devices IX, San Diego, CA, USA.

44. Chwang, A.B., Rothman, M.A., Mao, S.Y., Hewitt, R.H., Weaver, M.S., Silvernail, J.A., Rajan, K., Hack, M., Brown, J.J., Chu, X., Moro, L., Krajewski, T. and Rutherford, N. 2003, Thin film encapsulated

flexible organic electroluminescent displays. Applied Physics Letters, 83, 413-415.

45 Chwang, A., Hewitt, R., Urbanik, K., Silvernail, J., Rajan, K., Hack, M., Brown, J., Lu, J.P., Shih, C., Ho, J., Street, R., Ramos, T., Moro, L., Rutherford, N., Tognoni, K., Anderson, B. and Huffman, D. (2006) Full color 100 dpi AMOLED displays on flexible stainless steel substrates. Society for Information Display (SID) 2006 International Symposium – Digest of Technical Papers, vol. XXXVII, pp. 1858–1861.

46 Jin, D.U., Jeong, J.K., Shin, H.S., Kim, M.K., Ahn, T.K., Kwon, S.Y., Kwack, J.H., Kim, T.W., Mo, Y.G. and Chung, H.K. (2006) 5.6 inch flexible full color top emission AMOLED display on stainless steel foil. Society for Information Display (SID) 2006 International Symposium – Digest of Technical Papers, vol. XXXVII, pp. 1855–1857.

47 Nisato, G., Kuilder, M., Bouten, P., Moro, L., Philips, O. and Rutherford, N. (2003) Thin film encapsulation for OLEDs: evaluation of multi-layer barriers using the Ca test. Society for Information Display (SID) 2003 International Symposium – Digest of Technical Papers, vol. XXXIV, p. 88.

48 Ramos, T. (2006) Progress toward early-to-market flexible OLEDs: packaging of low-resolution plastic and high-resolution foil-based displays. Presented at Flexible Displays and Microelectronics Conference, Phoenix, AZ, USA.

49 US Patent Application 2006/0216951-A1 (2006).

50 Rutherford, N., Moro, L. and Visser, R., unpublished results.

51 Chu, X., Moro, L. and Visser, R. (2004) Analysis of failure modes of multilayer thin film encapsulation of OLED devices and Ca films. Proceedings of IDW 2004, pp. 1011–1012.

52 Kim, H.K., Kim, D.G., Lee, K.S., HuH, M.S., Jeong, S.H., Kim, K.I., Kim, H., Han, D.W. and Kwong, J.H. (2004) Plasma damage-free deposition of Al cathode on organic light-emitting devices by using mirror shape target sputtering. Applied Physics Letters, 85, 4295–4297.

C
Production

19
Roll-to-Roll Processing of Thin-Film Organic Semiconductors
Arved C. Hübler and Heiko Kempa

19.1
Introduction

Roll-to-roll processing of functional materials can be divided into gas-phase (vacuum), solid-state (transfer), and liquid-state deposition techniques (Figure 19.1). Depending on the concrete process, these methods result in either patterned or nonpatterned films. For the case of solid-state processes, this depends on whether the pristine solid film, which is usually prepared by one of the other techniques, is patterned or not. For the case of vacuum deposition, patterned (additive) deposition can be realized, for example, by means of shadow masks. For the case of liquid-state deposition, methods that directly result in patterned films are frequently summarized under the term "printing," although several techniques have been developed, which do not belong to the traditional printing methods. The same holds for deposition onto prepatterned substrates; however, printing methods can be subsequently employed in order to create the patterned film. Methods that result in nonpatterned films are commonly summarized under the term "coating". Subsequent (subtractive) patterning steps can be applied, which are preferably carried out in a continuous roll-to-roll process as well.

The application of roll-to-roll processing of functional materials usually has the following objectives: (1) covering large substrate areas with a single process step and (2) reducing the fabrication costs. Roll-to-roll vacuum deposition of inorganic as well as organic materials has become a routine technique especially for coatings in the packaging industry [1]. It can be employed, for example, for the preparation of organic light-emitting diodes [2]. However, the reduction of fabrication costs as well as the ability to process large areas is limited by the low process speed and by the necessity of pumping large volumes and locking in and out the roll substrate. Similar arguments apply to solid-state transfer methods, which nevertheless can be employed in order to prepare functional layers [3] and complete electronic devices [4].

Therefore, liquid-state processing, unlike the usage of flexible substrates, is not a strict prerequisite for roll-to-roll processing. However, the application of high-volume liquid-state coating and printing technologies enables the exploitation of advantages of

Organic Photovoltaics: Materials, Device Physics, and Manufacturing Technologies.
Edited by Christoph Brabec, Vladimir Dyakonov, and Ullrich Scherf
Copyright © 2008 WILEY-VCH Verlag GmbH & Co. KGaA, Weinheim
ISBN: 978-3-527-31675-55

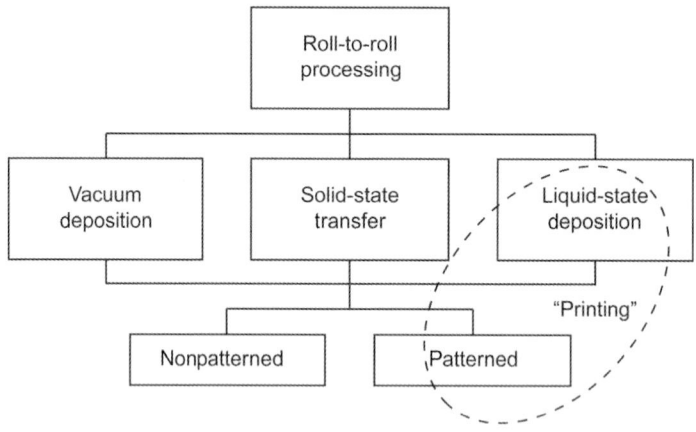

Figure 19.1 Roll-to-roll processing methods.

roll-to-roll deposition techniques. In what follows, we will focus on liquid-state roll-to-roll processing methods. In terms of materials, we will concentrate on organic conductors and semiconductors, and devices that are based on these materials, namely organic field effect transitors (OFETs), organic light-emitting diodes (OLEDs), and organic photovoltaic cells (OPVCs). In terms of processes, we will distinguish among coating (i.e., nonpatterned deposition), printing, and other patterning methods. The chapter is organized as follows: in Section 19.2, a brief overview on coating techniques in connection with organic electronics is given. In Section 19.3, methods for patterned deposition including printing techniques are discussed. In Section 19.4, examples for complete roll-to-roll processes for the fabrication of organic electronics are reviewed. Finally, Section 19.5 reviews the progress in roll-to-roll fabrication of organic photovoltaic cells. Section 19.6 provides some conclusions.

19.2
Coating

There is a large number of various coating methods (Figure 19.2), most of which were employed for the preparation of organic electronic devices. By selecting an appropriate method and adjusting the process parameters, a broad variety of materials can be processed and a wide range of desired film properties is accessed. However, for a specific material, the properties of deposited films can significantly depend on the chosen method and process parameters.

Surin et al. [5] have studied the influence of the coating technique and of the solvent on the performance of OFETs based on a widely used polymer semiconductor, poly (3-hexylthiophene) (P3HT). The semiconductor was deposited from different solvents by means of spin coating, dip coating, and drop casting. The semiconducting layer exhibited strongly different morphologies for the different combinations (Figure 19.3a–d). Consequently, the field effect mobility turned out to depend on the deposition method for one and the same solvent and vice versa (Figure 19.3e).

19.2 Coating

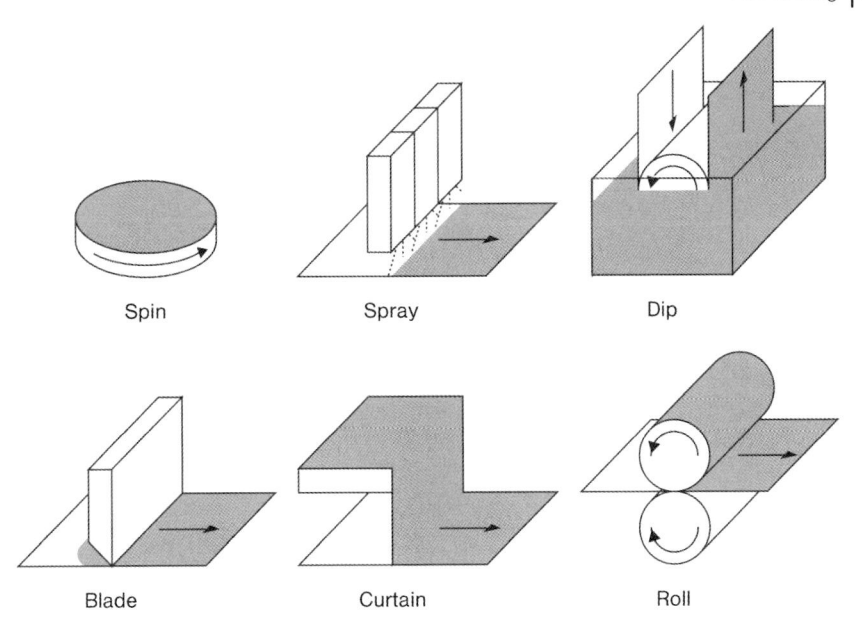

Figure 19.2 Selected coating techniques.

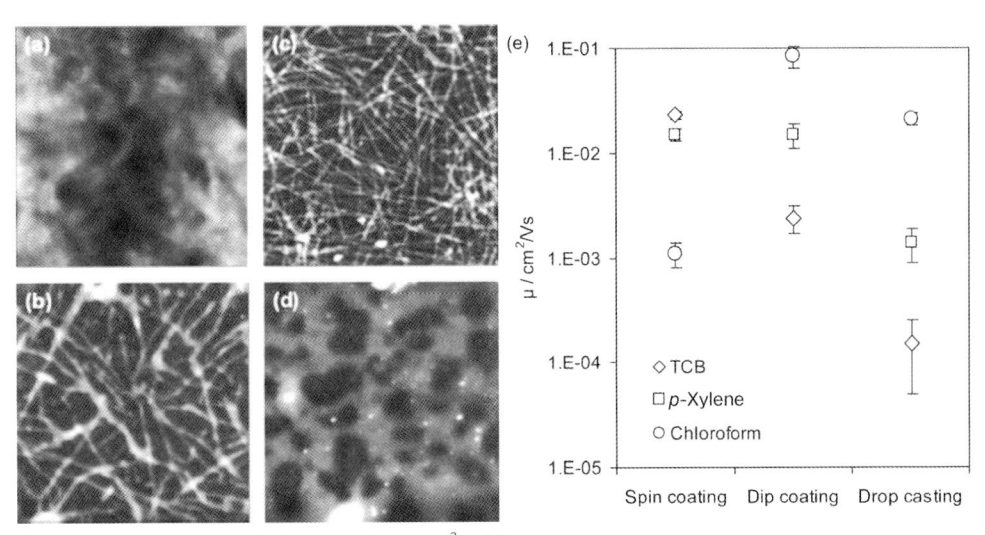

Figure 19.3 Tapping-mode AFM images ($2 \times 2\,\mu m^2$) of films of P3HT deposited by means of drop casting (a) from 1,2,4-trichlorobenzene (TCB), (b) from p-xylene, by means of spin coating, (c) from p-xylene, (d) from chloroform; (e) field effect mobilities for different coating methods and solvents (images and data taken from Ref. [5]).

Although not being roll-to-roll compatible, spin coating is the most common coating method in organic electronics. It is often used as a reference method, because it is applicable to almost all soluble materials with only minor adjustments of process parameters [6]. Therefore, it is well suited for comparative material studies. Furthermore, it represents an easily setup method and is considered to provide quasi-ideal films. For example, spin-coated films of a soluble pentacene derivative exhibit field effect mobility similar to drop-casted films and close to the intrinsic mobility limit [7]. However, the results of Surin et al. [5] suggest that careful process adjustment is necessary in order to obtain perfect films. The same holds for drop casting, which is often used as the simplest method for preliminary material tests and for materials that are difficult to process [8]. In addition, it is used as laboratory model process for spray coating and inkjet printing [9].

Roll-to-roll compatible methods, especially blade coating, are employed where large-scale processes and commercialization of organic semiconductor applications are targeted [10,11]. Thin semiconductor films and dielectrics of controlled thickness are obtained in continuous processes with high throughput. Roll-to-roll coating of organic conductors has become a routine technique and foils coated with poly (3, 4-ethylene dioxythiophene):poly(styrene sulphonate) (PEDOT:PSS) or polyaniline (PANI) are commercially available. Furthermore, coating methods for these materials are employed in order to obtain films for subsequent subtractive patterning [10–12]. Spray coating is a well-suited method for nonpatterned deposition of organic semiconductors, especially when low viscosity of the solutions makes the application of printing methods complicated. This technique has, for example, been used for the preparation of the active layer of organic photovoltaic cells from poly-(2-methoxy-5-(2′-ethylhexoxy)-1,4-phenylenevinylene):l-(3-methoxycarbonyl)-propyl-l-phenyl-(6,6)-C_{61} (MEH-PPV:PCBM) [13]. Spray coating of semiconductors and dielectrics has been used for the fabrication of printed OFETs in inline processes, before today's gravure printing methods were available [14]. In this case, spray and spin coating result in high-quality films with field effect mobilities that were achieved only recently by means of gravure printing [15]. Finally, due to its similarity with printing, roll coating is rarely applied to electronically functional materials, because patterned deposition by means of printing is almost always favorable. However, similar requirements in terms of materials properties and process conditions apply in order to obtain reasonable film quality [16].

19.3
Patterning

Techniques for patterning of thin films are divided into additive and subtractive methods. In the latter case, a nonpatterned (coated) film is patterned in a separate process step, whereas in the case of additive methods, coating and patterning are done in one and same process step, that is, the desired pattern is directly deposited. In the case of subtractive processes, additive methods can be used to transfer the pattern to the coated film. Besides conventional patterning techniques that are used on a large scale in microelectronics industry, such as photolithography and scanning beam lithography, there are several novel methods that are especially applied to the patterning of organic conductors and semiconductors, including stamping and printing techniques [17].

Techniques that employ the principle of stamping are widely used in order to prepare patterned films of organic conductors and semiconductors. They exist in various forms, the most common ones being microcontact printing (μCP) and nanoimprint lithography (NIL). While μCP uses soft stamps and typically achieves minimum feature sizes in the μm range, with NIL sub-μm structures can be prepared using hard stamps. At least μCP was claimed to be roll-to-roll compatible [18], although this has not yet been demonstrated. The potential of NIL of being roll-to-roll compatible is yet unclear. In typical subtractive μCP processes [19,20], an etch resist pattern is transferred by means of a soft stamp to a film, which is supposed to be patterned and which is subsequently etched. Thereby, for example, gold electrode structures for OFETs can be obtained. Additive μCP processes have been applied in order to directly deposit patterned films of organic conductors on rigid [21] and foil [22] substrates. Alternatively, seed layers were prepared on foil substrates by means of μCP, which allowed for subsequent patterned metallization [23] or growth of vacuum-deposited organic semiconductors [24]. In NIL, typically the desired pattern is molded into an etch resist layer followed by direct etching [25] or lift-off processes [26].

Pad printing as a rather unconventional printing method has been employed in a roll-to-roll process for the fabrication of OFET circuits [10]. Ink from a pattern engraved into a planar plate was transferred to a flexible pad and printed onto the substrate. Apart from this, printing techniques most commonly used for the fabrication of organic electronics can be distinguished by the principles of ink transfer to the substrate and ink separation between printing and nonprinting areas, respectively (Figure 19.4). The basic principle of ink transfer and separation deter-

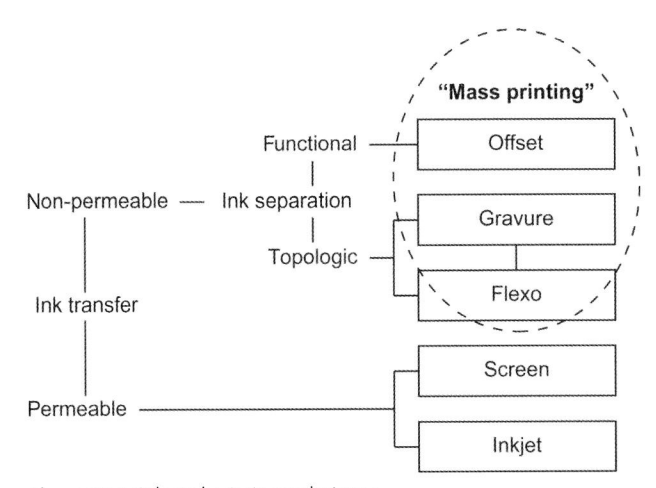

Figure 19.4 Selected printing techniques.

Table 19.1 Characteristic properties of printing methods.

	Inkjet	Screen	Flexo	Offset	Gravure
Resolution (μm)	>50	>100	>40	>20	>20
Layer thickness (μm)	0.3–20	3–15	0.8–2.5	0.5–2	0.8–8
Ink viscosity (Pa s)	0.001–0.04	0.5–50	0.05–0.5	30–100	0.05–0.2
Maximum throughput ($m^2 s^{-1}$)	>50	>100	>50 000	>50 000	>100 000

mines the overall setup of the respective machinery, ink supply, as well as processable materials and achievable properties of deposited films. In Table 19.1, some characteristic boundary conditions for individual printing methods are listed. Offset, gravure, and flexographic printing are typical rotary printing methods, where ink is transferred from a physical printing form in a continuous high-speed process. The throughput of these methods is several orders of magnitude higher compared to any other patterning methods; therefore, they are usually summarized under the term "mass printing." Inkjet and screen-printing techniques are typically used as sheetfed methods; however, they also can be laid out as webfed (roll-to-roll) methods. In the case of mass-printing methods, both variations exist.

In screen printing, a stencil is patterned by closing selected areas of a screen. Upon imprint, ink is forced through open meshes by means of a blade. Screen printing is widely used in conventional electronic fabrication wherever comparatively thick and coarse layers are needed, for example, in printed circuit board and antenna fabrication. In the context of organic electronics, screen printing is most commonly employed not only for inorganic metal inks and dielectric layers [11,26], but also for semiconductors in organic photovoltaics [27] and even complete OFETs [28].

Inkjet printing is probably the most widely used printing method in organic electronics and technological principles were frequently reviewed (see, for instance, [29]). Two basic approaches exist for drop creation and acceleration: continuous and drop-on-demand systems. In the former case, a continuous liquid jet through a nozzle is broken into drops by means of an acoustic wave generated by a piezoelectric device. Ejected droplets are then electrically charged and deflected according to the desired pattern. In drop-on-demand systems, droplets are discontinuously created by acoustic (mostly for nonwater-based inks) or heat (mostly for water-based inks) pulses when they are needed. Complete OFETs have been prepared by means of inkjet printing [30]. The large variety of materials processed in this experiment demonstrates the flexibility of the method. However, the achieved resolution (transistor channel length $L = 50$ μm) shows its limitations. Additionally, covering large areas and achieving competitive throughputs with inkjet are rather difficult.

In order to overcome the limitations in terms of resolution, several methods for printing on prestructured substrates have been developed. In this case, the patterning step precedes ink deposition, and techniques other than printing are involved. For example, hydrophobic mesas defined by means of photolithography served as a separator between inkjet printed source and drain electrodes from PEDOT:PSS [31], resulting in a resolution of $L = 5$ μm. Even higher resolution of $L < 1$ μm was achieved by inkjet printing PEDOT:PSS into trenches in a polymer layer that were prepared by a method similar to NIL [32]. The highest resolutions were obtained from self-alignment processes [33,34]. Nevertheless, inkjet printing turns out to be a very versatile method for the fabrication of various electronic structures. It has been successfully employed for the preparation of vertical interconnects by printing defined amounts of solvents in order to open holes in dielectric layers [35,36]. Furthermore, frontplanes [28] and backplanes [37] of OLED displays, integrated circuits [36], and OPVCs [38] as well as other devices have been prepared using inkjet printing. The high expectations in terms of reduction in fabrication costs that are connected with the development of organic electronics usually implicitly or explicitly refer to productivities and throughputs that can be achieved by using mass-printing methods [39]. However, there are few research groups that work on the application of mass printing for the fabrication of electronics. This is probably due to the extensive know-how and comparatively high effort necessary for setting up mass-printing machines. Additionally, required properties of deposited films largely differ depending on whether electronic devices or conventional printing products are supposed to be printed [28]. Consequently, setting up an electronic printing process requires significant modification of traditional processes. Fortunately, mass-printing technology is in a mature state of development and several distinct methods with variations can be chosen from [40]. When ink separation is due to topological differentiation, the printing parts of the form can be situated below or above the nonprinting parts. These two principles are most commonly realized in gravure and flexographic printing, respectively. The former letter printing, which is a variation of the latter principle, is nowadays widely replaced by offset printing. In offset printing, ink separation is not due to the surface topology of the printing form (i.e., printing and nonprinting parts are (nearly) on the same height level), but due to a differentiation of surface energies.

In flexographic printing, the printing form is made of a material of certain flexibility, leading to the broadening of the image, which is difficult to control and significantly reduces the achievable resolution and layer quality. However, flexographic printing offers the possibility to deposit layers with comparatively high and variable thickness up to several microns. Furthermore, with this method a relatively broad range of ink viscosities is accessible. Due to these properties, flexographic printing has been used for printing inorganic [41] and organic [42] conductors. In the latter case, PANI has been printed on foil substrates at a resolution of 200 μm with carefully adjusted ink formulation (Figure 19.5a). However, a trade-off among resolution, conductivity, roughness, and adhesion of the deposited films depending on the solid content of the ink was observed.

In gravure printing, printing parts are laid out as patterns of small cells with variable geometry. Upon inking, cells are filled and extra ink is removed by means of a blade. Imprint is done at high speed and under high pressure in order to remove the ink from the cells. Restrictions in terms of ink formulation are somewhat relaxed; variations can be compensated within limits by control over the process. Gravure printing produces good film qualities at high resolutions,

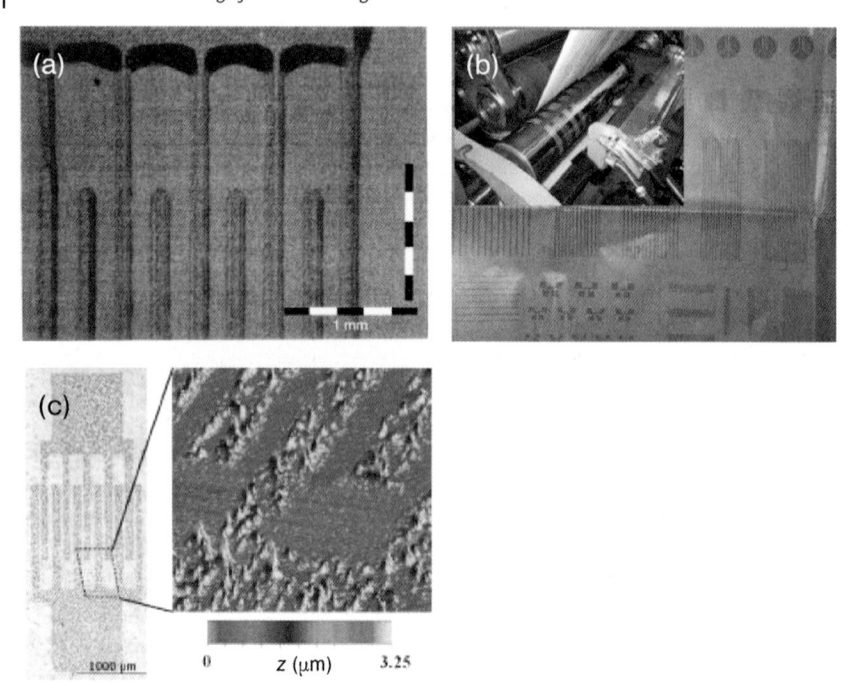

Figure 19.5 Examples of printed organic conductors: (a) flexographic printed PANI, (b) gravure printed PANI, (c) offset printed PEDOT:PSS (images reproduced from (a) Ref. [42], (b) Ref. [44], (c) Ref. [47]).

however, at the expense of complicated and cost-intensive printing form fabrication. For highest resolution, traditional cell patterns cannot be used; instead; the desired pattern is directly engraved into the form. In this case modification of the inking process is necessary in order to assure a uniform inking of the printing areas. Although not yet extensively used, gravure printing is an attractive method for high-volume, patterned deposition of quality-sensitive functional films such as inorganic [43] and organic [44] conductors (Figure 19.5b) and semiconductor–dielectric interfaces in OFETs based on various organic semiconductors [15,45]. Field effect mobility obtained from gravure printed semiconductor–dielectric interfaces was comparable to the one from spin-coated layers, thereby proving the high film quality [15].

Offset printing shares the high resolution and good film quality with gravure printing and exceeds it in terms of edge sharpness. Additionally, offset printing, on the one hand, offers the advantage of a fast and inexpensive printing form preparation. On the other hand, it has the most restricted requirements for ink formulation, because the principle of ink separation requires carefully adjusted, relatively high viscosity and surface tension of the ink. Most often, imprint is done not directly from the printing form but to a transfer cylinder, which then prints the image on the substrate (therefore the term "offset"). However, direct imprint is also possible; then the slightly inconsistent term "direct offset" is used. Frequently, the process involves oil-based inks and printing forms, where the printing parts are hydrophobic and the nonprinting parts are hydrophilic, and which are continuously wetted; this method is referred to as "wet offset." For water-based inks also a "dry offset" variation with appropriate printing forms exists. Offset printing has been used for the deposition of inorganic [46] and organic [47] conductors. In the latter case, a water-based PEDOT:PSS dispersion was printed by means of direct dry offset on a foil substrate with a resolution of 100 µm (Figure 19.5c). However, the films showed a characteristic nonideal morphology resulting from a specific hydromechanical phenomenon known as "viscous fingering," which occurs upon splitting of the ink film between printing form and substrate [48].

To enhance layer quality and resolution of printed patterns, several modifications of printing methods are under development. For example, an approach similar to prepatterning of the substrate in inkjet printing methods has been developed, which is, however, compatible with mass printing and runs at high process speed (Figure 19.6a). The surface of a foil substrate is activated by a conventional corona device as it is found in many printing machines. Subsequently, a special printing form comes into contact with the activated surfaces, however, without conducting any ink. Thus, selected parts of the foil surface are deactivated according to the desired pattern. Subsequent coating of the substrate with an organic conductor solution affects only the deactivated areas due to the large difference in surface energies. From this process thin and smooth films are obtained (Figure 19.6b) with a potentially enhanced resolution compared to conventional printing.

As can be seen from the given examples, mass printing and other printing methods are able to process all the layers necessary for thin film, namely organics-based, electronic devices. Therefore, the main challenge for the realization of low-cost, large area roll-to-roll fabrication of electronics consists of the integration of various methods into comprehensive, continuous processes that can be potentially transferred to industrial production.

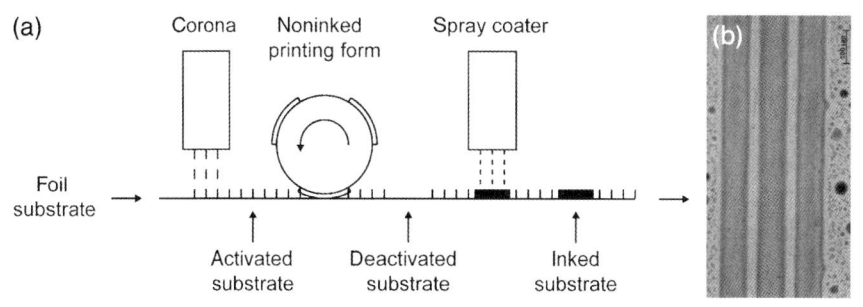

Figure 19.6 (a) Corona structuring technique derived from mass-printing technology (see the text for details). (b) Electrode structures from PEDOT:PSS prepared by means of this method.

19.4
Roll-to-Roll Processes

We will now discuss three different approaches for roll-to-roll processes for the fabrication of organic electronics in order to illustrate challenges that occur upon setting up such a process. All of these approaches were focused on fabrication of OFET circuits. In all cases, OFETs were realized in the so-called top-gate configuration, that is, source and drain electrodes are primarily created on the substrate. The distance between these electrodes (the channel length L) largely determines the current in the transistor; therefore, the highest resolution is required in this layer. However, resolution is promoted when this critical layer is prepared on the pristine substrate. Subsequently, semiconductor, dielectric, and gate electrode are deposited. Ring oscillators are widely used as demonstration object to show the capability of preparing integrated OFET circuits [49]. In these simple circuits, an uneven number of inverter stages, each consisting of two transistors with different channel widths W, are connected in series. When the output signal of the last inverter stage is coupled back into the first one and the circuit is supplied with a sufficient DC voltage, a sinusoidal AC signal is delivered. Ring oscillators are useful as demonstrators, because they include the preparation of several uniform transistors and vertical interconnects between them. Furthermore, the frequency of the signal is given by $f = f_i/2i$, where f_i is the switching frequency of a single inverter stage and i the number of inverter stages with f_i approximately given by

$$f_i = \frac{\mu V_{DD}}{2\pi L^2},$$

where μ is the average field effect mobility, V_{DD} the supply voltage, and L the transistor channel length. Therefore, by measuring the frequency, technological progress in terms of μ and L can be directly determined.

PolyIC [10] has been using two methods, which are roll-to-roll compatible or even replaceable by mass-printing methods: pad printing and blade coating (Figure 19.7). Source and drain electrodes were prepared by a process similar to lithography, however, under the involvement of pad printing. First, poly(ethylene-terephtalate) (PET) foils were coated with PANI from a toluene-based solution using a commercial blade coater. An area of 12 × 15 cm^2 could be covered by this method. An etch resist was pad printed with the desired pattern onto the PANI-coated foil, which is then etched. The resulting source and drain electrode structures had a channel length of 20–60 µm. Vertical interconnects were prepared by printing bars from carbon ink, which exceed the subsequently bladed semiconductor and the dielectric in height. Different semiconductors, among them P3HT, and different dielectrics, among them poly(methyl methacrylate) (PMMA) were used. One important criterion for these materials was the low surface tension of the dielectric solution in comparison with the surface energy of the dry semiconductor film, in order to ensure good wetting when coating the dielectric. Finally, gate electrodes were pad printed from carbon ink, including connections with the vertical interconnects. With this process, a ring oscillator circuit was prepared, which oscillated with a frequency of 0.86 Hz at a

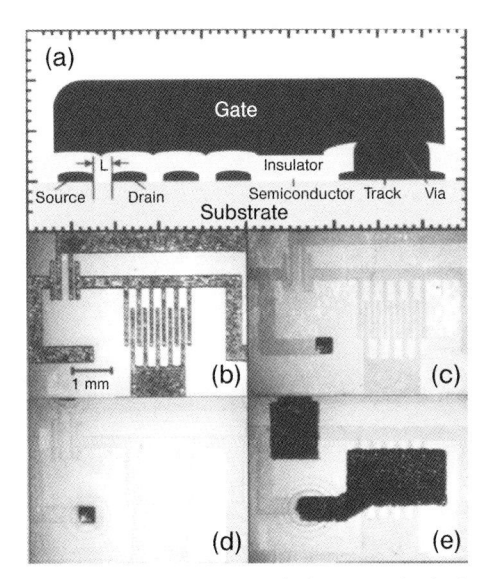

Figure 19.7 OFET preparation by the approach of PolyIC: (a) cross section showing the transistor configuration, (b) etch resist pad printed on a blade-coated PANI film, (c) etched source and drain electrodes with pad printed vertical interconnect, (d) blade-coated semiconductor and dielectric, and (e) complete OFET with pad-printed gate electrodes and vertical interconnects (image reproduced from Ref. [10]).

supply voltage of 90 V. However, using gold electrodes on foil substrate, which were patterned by means of conventional photolithography, spin-coated semiconductor, and dielectric and vertical interconnects defined by photolithography as well, a transistor channel length of 10 μm and a ring oscillator frequency of 106 kHz at 80 V were obtained [50]. These results clearly show the performance loss that is typically observed when conventional fabrication methods are replaced by fast roll-to-roll compatible ones. Nevertheless, the potential of the approach used here to be scaled into an industrial process remains to be demonstrated.

Fraunhofer IZM Munich [11] has been following the approach of integrating photolithography with liquid-state processing into a roll-to-roll process for the fabrication of organic integrated circuits (Figure 19.8). Copper-metallized PET foils were patterned using an especially set up roll-to-roll capable photolithography unit. Although this process step was carried out in a roll-to-roll fashion, it was not continuous, because the web had to be stopped for processing individual batches. Therefore, source and drain electrode structures with a resolution of 10 to 20 μm were obtained. The polymer semiconductor, a poly(arylamin) (PAA), and the dielectric were supposed to be prepared by means of blade coating in a roll-to-roll process. However, yield turned out to be very low for this process step; therefore, spin coating was used for the preparation of these layers. Subsequently, gate electrodes were screen printed from PEDOT:PSS or PANI formulations, followed by a screen printed

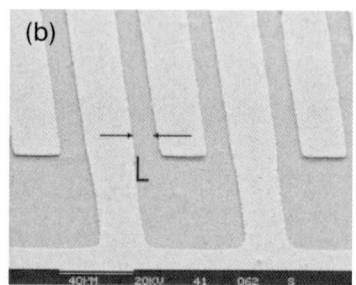

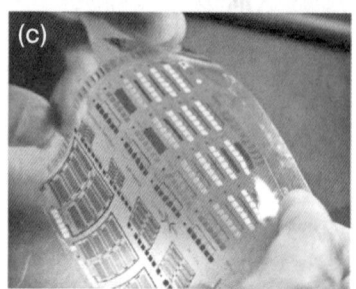

Figure 19.8 Preparation of OFET circuits by the approach of Fraunhofer IZM Munich: (a) electrode structures made from copper-metallized foils using a roll-to-roll photolithography process, (b) SEM image of source and drain electrodes, and (c) completely processed integrated circuits, including ring oscillators (images reproduced from Ref. [11]).

insulation layer and an interconnect layer, which was either an organic conductor or a silver ink and was screen printed as well. However, no details were reported on how vertical interconnects in the semiconductor and dielectric layers were prepared. Presumably, a solvent was applied to selected spots before printing of the interconnect layer. However, complete ring oscillator circuits were prepared, oscillating with a frequency of 5 Hz, although no operation voltage was reported. Nevertheless, the approach of integrating various fabrication methods, including photolithography, coating, and potentially also printing, into a single roll-to-roll process is clearly attractive in view of different functionalities that potentially can be integrated into an organic electronic system. Such a process is not necessarily limited to OFET fabrication, but could be extended to sensors, displays, solar cells, and so forth.

An attempt to fabricate integrated OFET circuits solely using mass-printing methods has been made by the Institute for Print and Media Technology at Chemnitz University of Technology (pmTUC) in collaboration with BASF and Bell-Lucent [45]. All process steps (Figure 19.9) have been carried out on an especially built up laboratory-printing machine, which nevertheless is assumed to model industrial printing processes, namely the high process speed of the order of $1\,\mathrm{m\,s^{-1}}$. The circuits were based on source and drain electrodes, which were offset printed on

Figure 19.9 Preparation of OFET circuits by the approach of pmTUC: (a) offset printing of source and drain electrodes, (b) gravure printing of the polymer semiconductor, (c) flexographic printing of the insulating layer, (d) flexographic printing of gate electrodes, (d) offset printed electrode structures for ring oscillators.

PET foil from a formulation of PEDOT:PSS with a resolution of 100 μm. The semiconductor, poly(9,9-dioctyl-fluorene-*co*-bithiophene) (F8T2), and the dielectric was gravure printed due to the high requirements in terms of layer quality. However, due to the aforementioned rough morphology of the underlying electrodes, no sufficient insulation of the gate electrode was achieved. Therefore, an additional insulating layer was deposited on top. To maintain a sufficient gate capacitance, a Bariumtitanate-based ink with a high dielectric constant was used. The main challenge turned out to print this highly polar material on top of the underlying nonpolar film. This was achieved by using an especially adjusted flexographic printing process, making use of the capability of working at low pressure. Finally, gate electrodes were printed from silver ink by means of a similar flexographic printing process. As all underlying layers were additively patterned, forming of the vertical interconnects was straightforward upon printing of the gate layer, due to the high film thickness achievable with flexographic printing. By means of this process, ring oscillator circuits could be fabricated, which were oscillating with a frequency of about 4 Hz at 80 V.

As can be seen, all of the mass-printing methods have been employed in order to deposit the necessary layers. However, this was mainly determined by the rough structure of the first electrode layer. On the contrary, gravure printing was shown to be capable of producing thin and smooth electrodes with high resolution [15,44]. Therefore, a printing process is currently under development, where the preparation of source and drain electrodes eliminates the necessity for an additional insulating layer and thereby the necessity for using other methods than gravure. It is hoped that such a comprehensive gravure printing approach will lead to better film quality, simplified processes, and improved circuit performance.

The given examples illustrate the potential of various roll-to-roll approaches for the fabrication of low-cost, in this case OFET-based, electronics on flexible substrates. However, there are still several challenges to be mastered, before mass production of organic electronic devices by means of roll-to-roll processes comes into sight. The main one of these is the enhancement of film quality and resolution while maintaining the cost advantage and productivity of roll-to-roll methods. Other important aspects under development are reliability and yield issues, as well as long-term stability of the processes in question.

19.5
OPVC Fabrication

Although not yet extensively used, roll-to-roll processes are clearly an attractive option for the fabrication of OPVCs [51]. However, potential and requirements are slightly different compared to OFET-based circuitry: large area and high film quality are more important than high resolution and integration density. Nevertheless, flexible substrates and liquid processable materials are prerequisites as well, in order to make use of the advantages of roll-to-roll fabrication. Important steps into this direction have been made in the past [52].

One of these steps was the preparation of an OPVC on a flexible substrate by Brabec et al. [53]. Poly(2-methoxy-5-(3′,7′-dimethyloctyloxy)-1,4-phenylene vinylene):l-(3-methoxycarbonyl)-propyl-I-phenyl-(6,6)-C_{61} (MDMO:PCBM) was deposited from solution on an indium-tin oxide (ITO)-coated foil and resulted in a remarkably high efficiency around 1.2%. Comparison was made with smaller sized cells on ITO-covered glass substrates, which had an efficiency of around 2.3%. Furthermore, an experiment with blending polystyrene (PS) into the active layer solution was carried out. PS is a typical additive used to enhance the viscosity of functional inks for optimized processing. Small amounts of PS (below 10%) turned out to not significantly affect the efficiency.

A PEDOT:PSS layer, which is usually spin coated on top of the ITO cathode, is known to improve hole injection. However, various coating methods have also been used for deposition of the active layer. Shaheen et al. [54] have studied the influence of the solvent on the efficiency of OPVCs based on spin-coated layers of MDMO:PCBM. Films deposited from chlorobenzene resulted in an efficiency of 2.5%, while those from toluene resulted in 0.9%. Similar to the aforementioned results of Surin et al. [5] for the case of a polymer semiconductor as the active layer in an OFET, this could be attributed to different film morphologies (Figure 19.10). Evaporative spray coating as deposition method for MEH-PPV:PCBM has been employed by Ishikawa et al. [55]. An efficiency of 0.63% was obtained from a dilute tetrahydrofurane (THF) solution, as compared to 0.87% from a spin-coated chlorobenzene solution.

Several authors reported the application of printing methods for the deposition of the active layer. Shaheen et al. [27] have used screen printing for the preparation of

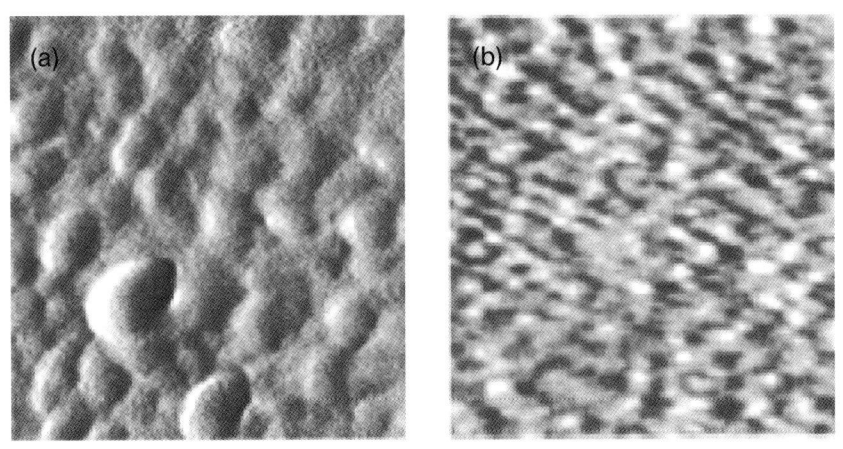

Figure 19.10 AFM images of spin-coated films of MDMO:PCBM: (a) from toluene solution (efficiency 0.9%), (b) from chlorobenzene solution (efficiency 2.5%) (images reproduced from Ref. [54]).

MDMO:PCBM films and obtained an efficiency of 4.3%, which exceeds the one of most vacuum-deposited films. Inkjet printing has been used by Shah and Wallace [38] for the deposition of P3HT:PCBM as the active layer and for the PEDOT:PSS layer; however, an extremely low efficiency was obtained.

Although similar strategies for process optimization apply to OFETs and OPVCs [5,54], the encouraging results for OPVCs prepared on flexible substrates [53] and using printing methods [27] suggest that roll-to-roll fabrication might be even more beneficial and easier to realize than for OFETs. However, the lack of liquid processable anode materials so far prevents the implementation of completely solution-based processes. Recently, a polymer alternative for inorganic cathode materials was reported [56]. Vapor-phase polymerized PEDOT:PSS was used instead of aluminum in an OPVC. It is, however, yet unclear if this approach will lead to a soluble cathode material.

19.6
Conclusions

Roll-to-roll fabrication of organic electronics based on OFETs, OLEDs, OPVCs, and other devices offers great potential for reducing fabrication costs and realizing large, flexible area applications. Significant technological developments could be observed in the past in order to implement flexible substrates and liquid-state processing, both in coating and patterning methods for the fabrication of electronics. However, there are still considerable challenges on the way to industrial roll-to-roll processes. Namely, the performance of roll-to-roll fabricated OFET-based devices needs to be improved by enhancing layer quality and resolution of coating and patterning techniques. Fortunately, there is still room for process optimization, modification, and innovation. Several successful approaches have been demonstrated on a laboratory scale and are to be tested in a production environment. In the case of OPVCs, beside the problem of developing a soluble cathode material, fundamental requirements (flexible substrate, liquid processing) have been successfully fulfilled. Therefore, the way has been paved for the development of large-scale roll-to-roll processes.

References

1 Bishop, C. (2006) *Vacuum Deposition onto Webs, Films and Foils*, William Andrew Publishing.
2 Thompson, M.J. (2004) *Solid State Technology*, **47**, 5.
3 Blanchet, G.B. et al. (2003) *Applied Physics Letters*, **82**, 463
4 Hines, D.R. et al. (2007) *Journal of Applied Physics*, **101**, 024503.
5 Surin, M. et al. (2004) *Journal of Applied Physics*, **100**, 033712.
6 Yoon, M.-H. et al. (2006) *Journal of the American Chemical Society*, **128**, 12851.
7 Park, S.K. et al. (2005) *International Electron Devices Meeting Technical Digest '05*, 105.
8 Lee, W.H. et al. (2007) *Applied Physics Letters*, **90**, 132106.
9 Ai, Y. et al. (2004) *Thin Film Solids*, **450**, 312.
10 Knobloch, A. et al. (2004) *Journal of Applied Physics*, **96**, 2286.
11 Bock, K. et al. (2005) *Proceedings of IEEE*, **93**, 1400.
12 Hohnholz, D. et al. (2005) *Advanced Functional Materials*, **15**, 51.
13 Ishikawa, T. et al. (2004) *Applied Physics Letters*, **84**, 2424.
14 Hübler, A. et al. (2004) unpublished.
15 Kempa, H. et al. (2006) Proceedings of Organic Electronics Conference, Frankfurt/M. (Germany).
16 Carvalho, M.S. and Scriven, L.E. (1997) *Journal of Fluid Mechanics*, **339**, 143.
17 Gates, B.D. et al. (2005) *Chemical Reviews*, **105**, 1171.
18 Rogers, J.A. et al. (1999) *Advanced Materials*, **11**, 741.
19 Leufgen, M. et al. (2004) *Applied Physics Letters*, **84**, 1582.
20 Li, D. and Guo, L.J. (2006) *Applied Physics Letters*, **88**, 063513.
21 Cosseddu, P. and Bonfiglio, A. (2006) *Applied Physics Letters*, **88**, 023506.
22 Zschieschang, U. et al. (2005) *Advanced Materials*, **15**, 1147.
23 Briseno, A.L. et al. (2006) *Nature*, **444**, 913.
24 Guo, L.J. (2007) *Advanced Materials*, **19**, 495.
25 Leising, G. et al. (2006) *Microelectronic Engineering*, **83**, 831.
26 Bao, Z. et al. (1997) *Chemistry of Materials*, **9**, 1299.
27 Shaheen, S.E. et al. (2001) *Applied Physics Letters*, **79**, 2996.
28 Holdcroft, S. (2001) *Advanced Materials*, **13**, 1753.
29 de Gans, B.-J. et al. (2004) *Advanced Materials*, **16**, 203.
30 Subramanian, V. et al. (2005) *Proceedings of IEEE*, **93**, 1330.
31 Wang, J.Z. et al. (2004) *Synthetic Metals*, **146**, 287.
32 Wang, J.Z. et al. (2006) *Applied Physics Letters*, **88**, 133502.
33 Sele, C.W. et al. (2005) *Advanced Materials*, **17**, 997.
34 Zhao, N. et al. (2007) *Journal of Applied Physics*, **101**, 064513.
35 Kawase, T. et al. (2003) *Advanced Materials*, **13**, 1601.
36 Sirringhaus, H. et al. (2000) *Science*, **290**, 2123.
37 Arias, A.C. et al. (2004) *Applied Physics Letters*, **85**, 3304.

38 Shah, V.G. and Wallace, D.B. (2004) Proceedings of IMAPS, Long Beach, USA.
39 Sheats, J.R. (2004) *Journal of Materials Research*, **19**, 1974.
40 Kipphan, H.(ed.) (2001) *Handbook of Print Media*, Springer.
41 Siden, J. *et al.* (2005) Proceedings of Polytronic, Wroclaw, Poland.
42 Mäkelä, T. *et al.* (2005) *Synthetic Metals*, **153**, 285.
43 Leppavuori, S. *et al.* (1994) *Sensors and Actuators*, **41–42**, 593.
44 Mäkelä, T. *et al.* (2003) *Synthetic Metals*, **135**, 41.
45 Huebler, A. *et al.* (2007) JT Organic Electronics, in press.
46 Harrey, P.M. *et al.* (2002) *Sensors and Actuators B: Chemical*, **87**, 226.
47 Zielke, D. *et al.* (2005) *Applied Physics Letters*, **87**, 123580.
48 Reuter, K. *et al.* (2007) *Progress in Organic Coatings*, **58**, 312.
49 Brown, A.R. *et al.* (1997) *Synthetic Metals*, **88**, 37.
50 Fix, W. *et al.* (2002) *Applied Physics Letters*, **81**, 1735.
51 Chopra, K.L. *et al.* (2004) *Progress in Photovoltaics Research and Applications*, **12**, 69.
52 Hoppe, H. and Sariftci, N.S. (2004) *Journal of Materials Research*, **19**, 1924.
53 Brabec, C.J. *et al.* (1999) *Synthetic Metals*, **102**, 861.
54 Shaheen, S.E. *et al.* (2001) *Applied Physics Letters*, **78**, 841.
55 Ishikawa, T. *et al.* (2004) *Applied Physics Letters*, **84**, 2424.
56 Gadisa, A. *et al.* (2006) *Synthetic Metals*, **156**, 1102.

20
Socio-Economic Impact of Low-Cost PV Technologies
Gilles Dennler and Christoph J. Brabec

20.1
Introduction

20.1.1
The Energy Supply

According to the first law of thermodynamics, any human activity requires an energy input. The industrial revolution and the development of the capitalist economy induced a world that does rely entirely on energy supply. The two oil crises of the 1970s illustrated the weakness of the economy of the member countries of the Organization for Economic Cooperation and Development (OECD) toward uncontrolled energy imports. Figure 20.1 displays the world total primary energy supply (TPES) growth from 1850 to 2100 [1]. This figure clearly shows that the pace of the growth of the TPES changed significantly after the Second Word War. In 2004, the TPES was evaluated to about 470 EJ (10^{18} J) or 11.2 gigatons oil equivalent (Gtoe) (see Table 20.1 for the conversion to different units) [2,3]. While the transport activities account for about 25% of the total energy consumption, the industrial activities represent about 27%, the agriculture, commercial, public service and residential activities about 38%, and the nonenergy activities (fossil fuel feedstock and material) 10% [2]. Most of the previsions forecast a future growth of the TPES between 1.5 and 2% per year [1]. This trend is mostly driven by the expected growth of the overall population (6 billion in 2004; 12 billion expected in 2100) and the increase in the level of life and access to consumption for an increasing part of the population. For example, while the fleet of vehicles in OECD countries is expected to rise by 33% by 2030, the growth is believed to approach 270% in the rest of the world [4,5].

The shares of the energy source in 2004 are illustrated in Figure 20.2. Fossil fuels alone (oil and gas) represent about 55% of the overall TPES, whereas coal still represents almost 25%. The so-called *new renewables*, namely, geothermal, wind, solar, and modern biomass (mostly biocarburant) totalize only 3%. These values, being averages generated by considering the planet whole consumption, might

Organic Photovoltaics: Materials, Device Physics, and Manufacturing Technologies.
Edited by Christoph Brabec, Vladimir Dyakonov, and Ullrich Scherf
Copyright © 2008 WILEY-VCH Verlag GmbH & Co. KGaA, Weinheim
ISBN: 978-3-527-31675-5

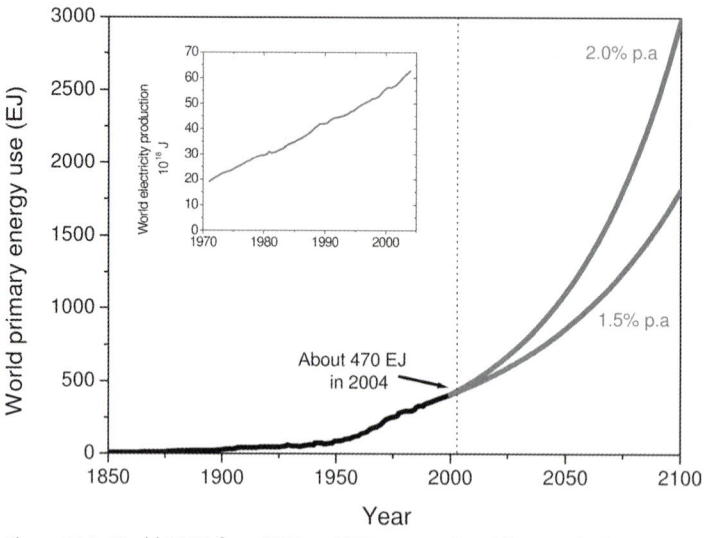

532 | 20 Socio-Economic Impact of Low-Cost PV Technologies

Figure 20.1 World TPES from 1850 to 2100, reported and forecast [1,2].

Table 20.1 General conversion factors for energy units.

From	10⁹ tons of oil equivalent (Gtoe)	Exajoule (10^{18} J)	Terawatt-hour (TWh)
	To		
10^9 tons of oil equivalent (Gtoe)	1	42	11 676
Exajoule (10^{18} J)	0.024	1	278
Terawatt-hour (TWh)	8.5×10^{-5}	0.0036	1

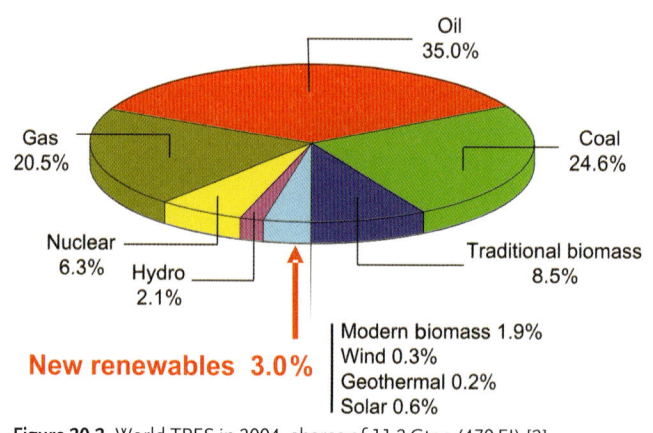

Figure 20.2 World TPES in 2004, shares of 11.2 Gtoe (470 EJ) [2].

change from region to region. Indeed, the geographical dependence of the energy consumption and sources of energy are severely influenced not only by the gross domestic product (GDP) of the region considered, but also by the climate, the availability of resources, the history, the way of life, and the strategic and political choices. Figure 20.3 gives an overview of the disparities that can be observed around the globe. Contrary to the common belief, Iceland is one of the OECD countries that consumes more energy per capita, that is, 11.94 toe, than any other country. This is mainly explained by the toughness of the climate of the country. The next two countries are Canada and USA, whose TPES per capita are 8.42 and 7.91 toe, respectively. Most of the West European countries have TPES close to Japan's, about 4 toe capita^{-1}. But these values drop down significantly by moving east or south. China's TPES is about 1.24 toe capita^{-1}, more than twice of India's (0.53 toe capita^{-1}). The lowest values are found in Central and South America (Paraguay 0.67; Bolivia 0.55; Peru 0.48; Haiti 0.24), in the sub-Saharan Africa (Mozambique 0.44; Togo 0.45; Ghana 0.39; Senegal 0.24), and in Asia (Vietnam 0.61; Bangladesh 0.16). Figure 20.3 indicates as well the amount of electricity consumption per capita and the ratio of electricity versus the TPES for the chosen countries. In the OECD, this ratio most of the time hovers between 15 and 20%. However, for developing countries, this ratio decreases significantly: 10% in India and China; 1.6% for Togo; 1.0% for Haiti, with only 30 kWh consumed per year per person. This is 1000 times less than in Iceland and about 500 times less than in the USA. This shows that electricity consumption is much more discriminative and representative of the GDP/capita than TPES, as TPES of Haiti, for example, is 46 times smaller than in Iceland and 30 times smaller than in the USA. In other words, nowadays access to electricity

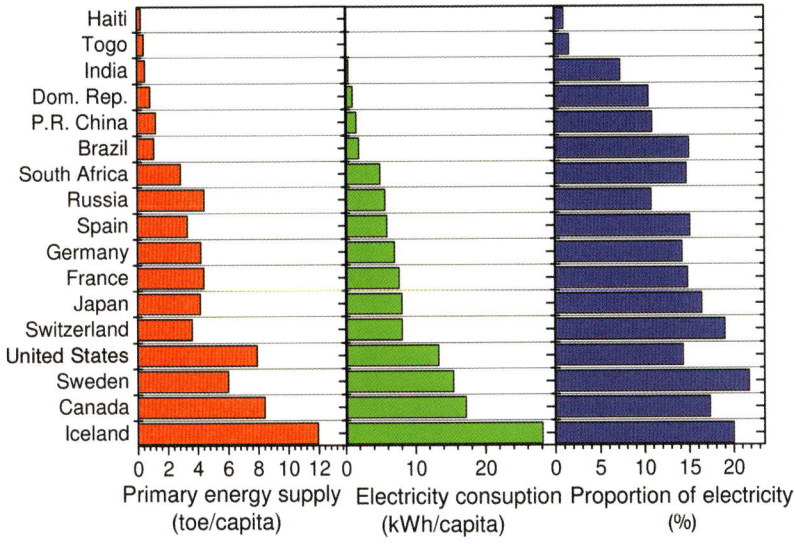

Figure 20.3 TPES per capita, total electricity consumption per capita, and electricity shares in the TPES in some selected countries [2].

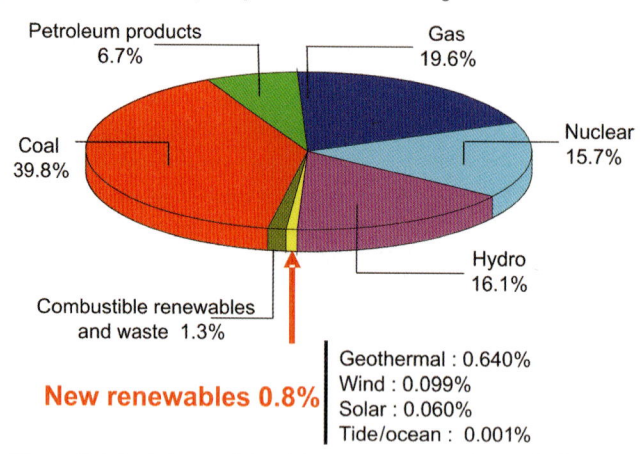

Figure 20.4 Fuel shares of electricity generation in 2004 (total of 17 450 TWh) [2].

is reserved to the richest countries mainly because of the intrinsic nature of this energy, which requires a costly infrastructure and which is not transportable.

Out of the 470 EJ supplied worldwide in 2004, 13.4% was directly created as or transformed in electricity (63 EJ or 17 450 TWh). Figure 20.4 shows the worldwide average fuel shares of electricity generation. It appears that coal is the major source of energy employed and that nuclear power represents almost 16% of the overall electricity produced. However, it is very interesting and highly educative to observe the fuel shares of electricity generation by region (Figure 20.5). It reveals, as explained above, that each region has a very specific fuel share scheme, driven by several factors. In the Middle East and North Africa, about 90% of the electricity is produced by fossil fuels, whereas in Asia and Sub-Saharan Africa more than 63% of the electricity comes from coal. On the contrary, Latin America produces its electricity mainly from hydrodynamics, whereas the OECD countries show a quite balanced production.

20.1.2
The Oil Shortage

The world's oil resources resulted from two brief epochs of extreme global warming, about 90 and 150 million years ago, respectively; that is, during the Jurassic era. During these periods, an excessive growth of algae and similar microorganisms was induced in waters by abnormally high temperature. Those algae gathered in stagnant rifts, which formed as the continents moved apart. Chemical reaction converted this organic debris into oil once they were buried by sediments and heated by burial. The natural gas appeared when plant remains and ordinary oil were overheated by excessive burial [6]. Once formed, the oil and gas began to migrate upward to zones of lesser pressure. In some cases, it simply dissipated. In others, it encountered a porous and permeable carrier bed, such as sandstone, through which it could move. Where folded or faulted, the carrier bed provided traps in which the oil and gas accumulated. Only some of them were large enough to support oilfields. The quality

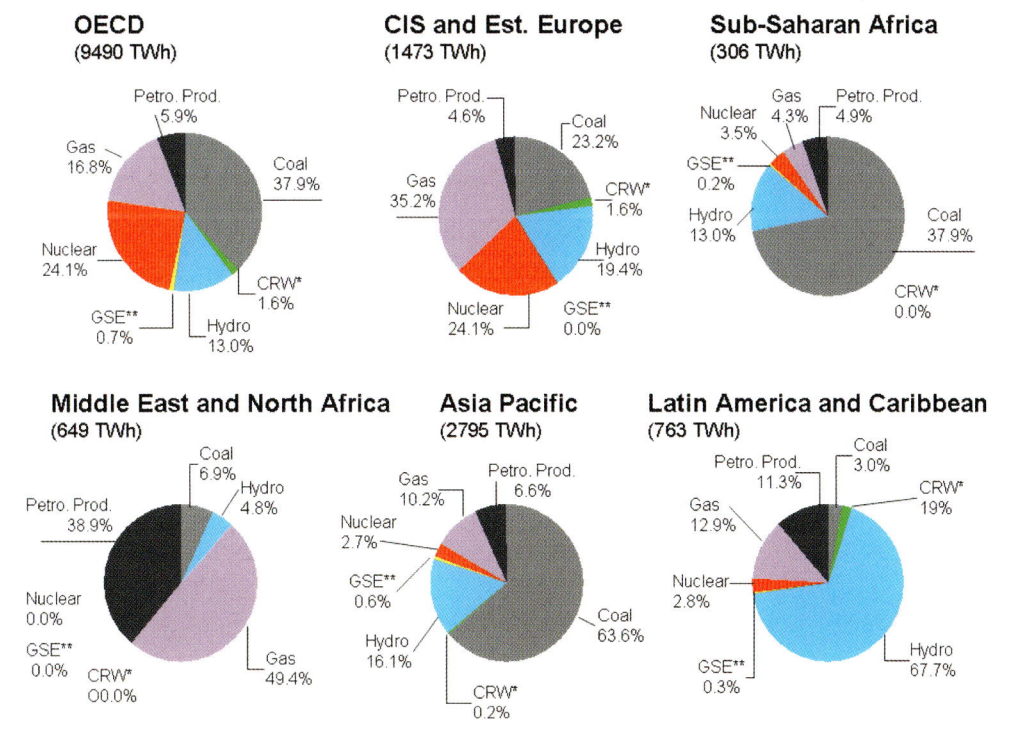

Figure 20.5 Shares of electricity production in various regions, by energy sources for 2001 [3] (nb: CRW* = combustible renewable and wastes; GSE** = geothermal, solar, etc.).

of reservoir ranges widely from, for example, a pure sandstone with plenty of porespace to one in which the spaces are clogged by fire-grained material. It may be variously thick and homogeneous, thin, or made up of alternating beds of different composition. On average, only about 35% of the oil in a reservoir is producible, the rest being held immovable by capillary forces or physical constrictions in the rock. Lastly, the reservoir needs to be capped by an impermeable layer, such as clay or rock salt, to prevent the oil and gas from escaping. However, no seal has perfect integrity, as much oil, and even more gas, has escaped over geological time.

Owing to their formation history, context, and location, there are several types of oil and gas each having their own characteristics, depletion profile, and cost. Some are easy and cheap to produce, while some other are costly and require complicated processes. The vast majority of oil and gas available till now is called *conventional* and is extractable with usual well techniques. The other main categories are

- heavy oil,
- deep water oil and gas (deeper than 500 m),
- polar oil,
- natural gas liquids,
- nonconventional gas.

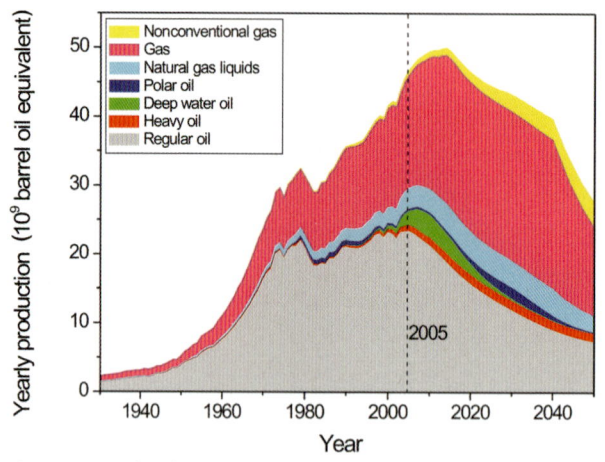

Figure 20.6 Oil and gas production and depletion [6].

The *conventional* oil and gas have contributed most to-date and will dominate all supply far into the future. Accordingly, it determines the peak of all production. The other categories will be important after peak by ameliorating the decline, but they have a minimal impact on the peak itself. This is substantiated by Figure 20.6, which makes it clear that conventional oil and gas are the main resources and control the peaks and trends.

The geophysics M. K. Hubbert proposed in 1956 that the fossil fuel production in a region does follow a symmetrical bell shape [7]. Detailed studies of site exploitation have shown that in most of the cases the Hubbert law is verified [8]. Thus, if the consumption of fuel is kept constant, the peak of the Hubbert graph indicates the end of the first half of the lifetime of the site. Figure 20.7 shows the world oil production

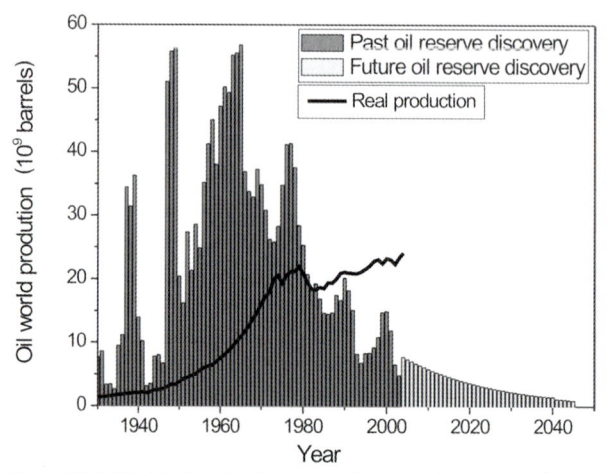

Figure 20.7 World oil production, past oil reserve discovery, and estimated future oil reserve discovery [6].

versus the proven reserves and the expected discoveries. If the Hubbert law is verified on the planet scale, this graph suggests that the peak is close to appear (between 2005 and 2015, depending on the sources [9]), announcing the end of the first half of the age of oil. It should be kept in mind that extensive debates are currently going on the quality of the estimation of the reserves. These ones are usually classified into three appellations, namely, *proven* reserves, *probable* reserves, and *possible* reserve with respective indexes of confidence ranging from "reasonably certain" to "having a chance of being developed under favorable circumstances." But even the proven reserves (total estimated to about 1720×10^9 barrels) are difficult to assess as they are the objects of severe political, economical, and strategic stresses. However, in essence, the depletion of fossil fuel started with the very first barrel extracted. Although the last barrel is still very far in the future, the time of cheap, easy to extract oil is fading off. The need of diversification of energy sources is, therefore, getting obvious to ensure long-term economic and social stability.

20.1.3
The Global Warming

The amount of energy reaching the top of Earth's atmosphere each second on a surface area of $1\,m^2$ facing the Sun during daytime is about $1370\,W$ and the amount of energy per square meter per second averaged over the entire planet is one-quarter of this value [10] (see Figure 20.8). About 30% of the sunlight that reaches the top of the atmosphere is reflected back to space. Roughly two thirds of this reflectivity is caused by clouds and small particles in the atmosphere known as "aerosols." The energy that is not reflected back to space is absorbed by the Earth's surface and atmosphere. This amount is approximately $240\,W\,m^{-2}$. To balance the incoming energy, the Earth itself must radiate, on average, the same amount of energy back to space. The Earth does this by emitting outgoing longwave radiation. To emit $240\,W\,m^{-2}$, a surface would have to have a temperature of around $-19\,°C$. This is much colder than the conditions that actually exist at the Earth's surface (the global mean surface temperature is about $14\,°C$). Instead, the required $-19\,°C$ is found at an altitude about $5\,km$ above the Earth's surface. The reason the Earth's surface is so warm is the presence of greenhouse gases (GHGs), which reflect back to Earth the longwave radiation coming from the surface. As it can be seen in Figure 20.8, the contribution of the greenhouse effect is a significant part of the Earth's thermal budget. The most important greenhouse gases are water vapor and carbon dioxide (CO_2). The two most abundant constituents of the atmosphere – nitrogen and oxygen – have no such effect. Clouds, on the contrary, do exert the same blanketing effect as that of the greenhouse gases. However, this effect is balanced by their reflectivity. The capability of gas to reflect long wavelength radiation is called the radiative force (RF), expressed in $W\,m^{-2}$. Currently, the three gases with a larger RF are CO_2, nitrous oxide (N_2O), and methane (CH_4). They are all mainly anthropogenic gases, related to the human activity, although a small part of them originates by natural processes:

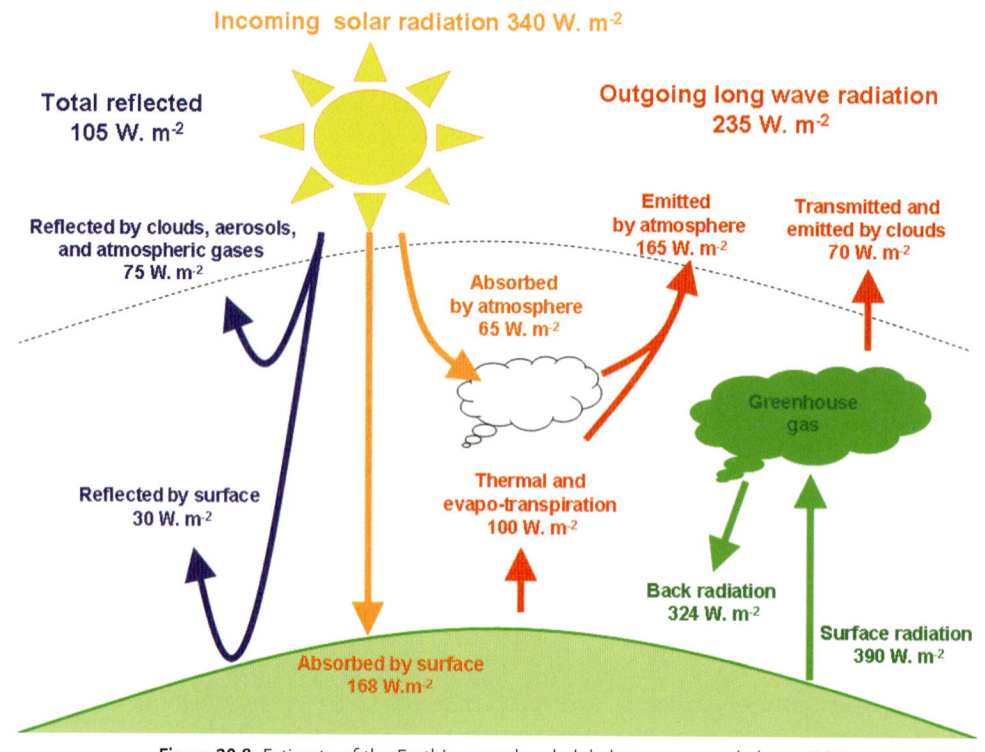

Figure 20.8 Estimate of the Earth's annual and global mean energy balance [10].

- CO_2 (RF ≈ 1.66 W m^{-2} in 2005) is mainly released by burning fossil fuels. It is generated by deforestation (fires) and decay of plant matters, as well.
- CH_4 (RF ≈ 0.46 W m^{-2} in 2005) emission is a result of human activities related to agriculture, natural gas distribution, and landfills. But methane is also released from natural processes that occur, for example, in wetlands.
- N_2O (RF ≈ 0.16 W m^{-2} in 2005) is released by the use of fertilizers in modern agriculture and fossil fuel burning. But it is also induced by some natural processes in the soil and ocean.

Figure 20.9 illustrates the growth of the concentration of these three gases in the atmosphere during the past 2000 years. Interestingly, their concentration started to increase significantly with the beginning of the industrial era. Since then, their abundance has kept on growing at an impressive pace. The International Panel on Climate Change (IPCC) has recently compared the growth of the concentration of these gases with some climate indicators [10]. It has been finally concluded and agreed by more than 2500 scientists that the increase of the average temperature of Earth (13.6 °C in 1860; 14.4 °C in 2005), the increase of the average sea level (150 mm since 1860), and the decrease of the average thickness of polar ice are related to the increasing RF of the anthropogenic gases [10].

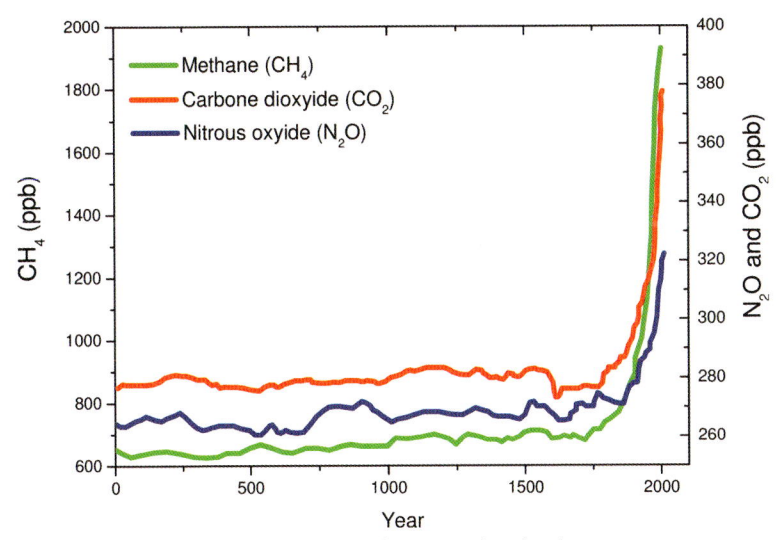

Figure 20.9 Atmospheric concentrations of important long-lived greenhouse gases over the last 2000 years [10].

Although limiting the emission of GHGs and sequestrating them are one of the priorities of policy makers, yet the industrial world that depends mainly on fossil fuels shows quite a strong reluctance in this regard. The overwhelming majority of electricity sources used nowadays are strong generators of GHGs, as shown in Figure 20.5. This is especially true for developing countries where electricity generation is mainly based on coal.

20.1.4
Renewable Energies

As explained above, the intrinsic limitation of fossil and mineral fuels arises from the fact that their process of creation was over thousands of years ago. Thus, their depletion has, by definition, started with the first joule used. Moreover, the conversion of these resources into genuine energy does release a significant amount of GHGs, which endanger the equilibrium of the Earth's climate and therefore the very existence of life – human, animal, and vegetal.

One approach to overcome both of these issues consists in the so-called *renewable energies* (RE) that are based on limitless resources and that generate most of the time significantly less GHGs than usual energy sources. The first historical RE and still more employed one is the usage of the biomass in the form of woody, nonwoody, and processed wastes. As shown in Figure 20.2, it represents about 9% of the current world energy supply. However, it has to be pointed out that this RE is not necessarily used in a sustainable manner [1]. The second most popular RE is the hydropower generating either electricity or pure mechanical force. But beyond these two traditional cases, some *new renewable energies* (NRE) are being developed and exploited. Before studying in detail the very NRE related to this book, that is, the photovoltaic (PV) approach, we would like to describe briefly three of the most preeminent NREs related to the production of electricity. Of course, this short list is not exhaustive and does not comprise some exciting and promising approaches such as the usage of Sun-driven fuel cells, new biomass-based liquid fuels, or the extraction of energy out of the ocean, the tides, and the waves:

- *Wind*: It is generally accepted that wind resources can be exploited in regions where the wind power density is at least 300–400 W m^{-2} at 50 m ($\approx$6.4–7 m s^{-1} or 23–25 km h^{-1}) or 150–200 W m^{-2} at 10 m ($\approx$5.1–5.6 m s^{-1} or 18–20 km h^{-1}), constituting the regions of class 3 and above in the widely used US classification of wind resources. However, technological advances might reduce this minimum limit. It has been estimated that about 23% of the Earth land area fulfills this requirement, representing a gross electric potential of 500 000 TWh [11]. Out of this, the World Energy Council has estimated that only 4% can be realistically exploited because of the noise and visibility nuisance as well as space requirement [12]. This brings the overall potential down to 20 000 TWh, what is still more than the overall world electricity consumption in 2004 (Figure 20.3). Offshore wind resources have been identified as even much larger than the onshore ones. To be economically viable, however, they have to be close to the coasts. In 2004, about 82 TWh of electricity based on wind power has been produced worldwide [13] with a total of about 48 GW installed production capacity. Since 1971, wind electricity generation has witness a growth rate of more than 30% per annum. The common size of wind turbines has reached the MW range (maximum 4–6 MW available), up from about 30 kW in the mid-1970s [14], with an energy conversion efficiency of about 30%. The average price of wind electricity ranges from 5 to 10 cents kWh^{-1} depending on the location.

- *Geothermal*: The geothermal energy can be defined as the heat energy stored within the Earth's crust. It is distributed between the constituent host rock and the natural fluid that is contained in its fractures and pores. Geothermal energy is exploited either directly as heat source or indirectly as electricity source. On average, the Earth's temperature increases by about 3 °C for every 100 m in depth, though this value is highly variable. To extract thermal energy economically, one must drill to depths where the rock temperatures are sufficiently high to justify investment in any heat-mining project. For generating electricity, drilling to rock temperatures in excess of 150–200 °C is usually necessary, whereas for many space or process heating applications, lower temperatures would be acceptable, such as 100–120 °C. Furthermore, geothermal energy can have a significant impact even at temperatures below 50 °C: Geothermal heat pumps provide an important example of how a low-grade thermal energy, available at shallow depths from 2 to 200 m, leads to substantial energy savings in the heating and cooling of buildings. The impressive worldwide geothermal resources have been estimated to about 140 000 000 EJ based on a 5 km drilling depth [15]. But new technologies allowing access to depth beyond 10 km suggest a potentially much larger reserves [16]. In 2004, about 56 TWh of electricity was produced by geothermal energy [13] with a total of more than 10 GW installed production capacity. Geothermal electricity generation has seen a growth rate between 5 and 10% per annum and the average price of geothermal electricity ranges from 5 to 10 cents kWh^{-1}, depending on the location [13,16].

- *Solar-thermal*: The total solar energy striking the Earth's surface is estimated to about 3 600 000 EJ year^{-1}, that is, 7700 times the current world annual primary energy consumption. This fact makes it the largest available energy source on Earth, ultimately much larger than the geothermal one since it is renewed constantly. According to the authors [17,18], between 50 000 and 1600 EJ is realistically exploitable, the latter being still 26 times the current world annual electricity consumption. The concentration solar power (CSP) technologies aim at collecting the solar heat, concentrating it between 50 and 50 000 times and then converting it into electricity. The three major approaches employed are (i) the parabolic trough-shaped mirror reflectors concentrating sunlight onto receiver tubes heating a thermal transfer fluid – this technology is the most mature one; (ii) the tower receiving a 2D concentrating sunlight reflected from numerous movable mirrors (heliostats) and offering a high Carnot conversion efficiency; (iii) parabolic dish-shaped reflectors concentrating the sunlight on to a small engine or turbine at the focal point [19]. The largest program run so far is the trough solar electric energy system (SEGS) built in the California Mojave desert between 1984 and 1990, totaling 354 MW. This plant occupies about 2 ha MW^{-1}, for a final electricity cost of about 0.16 cents kWh^{-1} [20]. Europe's first commercial concentrating power plant was inaugurated end of March 2007 in Andalusia, Spain. It consists of 624 heliostats having each 120 m^2 focusing the sunlight on to a 115 m tall tower, which contains the 10 MW steam turbine. According to the International Energy Agency (IEA), the CSP generated about 50% of the total world Sun-based electricity in 2004 [20]. But, the growth rate of this technology is expected to remain lower than that of PV because of some intrinsic limitations such as the need of an annually high insulation (ranging from 2000 to 2500 kWh m^{-2} to be competitive) and a minimum surface area, which runs into several tens of hectares per megawatt, as this technology demands a high concentration to be efficient and depends on a strong economy of scale. However, IEA, Greenpeace, and the European Solar Thermal Power Industry Association agree that CSP may represent several tens of GW worldwide by 2020 and the US Department of Energy (USDOE) expects CSP to soon provide electricity at a cost lower than 4 cents kWh^{-1} [21].

20.2 Photovoltaic Energy

20.2.1 World Market

During the last 5 years, the annual world PV cell and module production has experienced a growth rate above 40%, making this activity the fastest growing

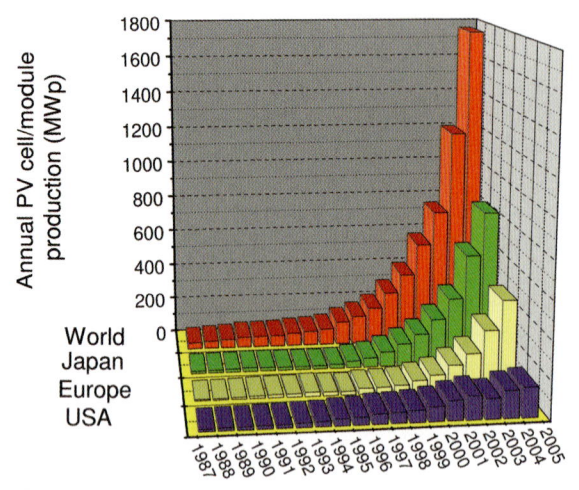

Figure 20.10 World PV cell/module from 1990 to 2005 [22].

renewable sources of energy, as well as one of the most promising industrial sectors, attracting a large amount of capital investments. In 2005, the photovoltaic industry delivered about 1760 MWp [22], representing an overall market estimated at about €9 billion. As shown in Figure 20.10, this growth has been observed not only in Japan but also in the United States and Europe. Besides, most Japanese, US and European roadmaps all seem to agree that this trend is going to last at least for the next 30 years [22].

At the end of 2005, the overall worldwide cumulative installed PV capacity reached 3.7 GWp [23]. The capacity installed in Europe increased about six times between 2001 and 2005, to reach almost 1.8 GWp. More than 80% of this capacity is located in Germany (1.5 GWp). However, Luxembourg still leads the world in terms of installed power per capita capacity at 52.4 Wp capita^{-1}, followed by Germany (17.32 Wp capita^{-1}) and Japan (11.13 Wp capita^{-1}). In Japan, the total installed capacity increased in 2005 by 8.3% to reach 1.42 GWp. Figure 20.11 illustrates the growth of installed capacity in several countries, showing the exponential growth in Germany. It has been estimated that the worldwide growth rate should remain the same within the next 25 years [22]. Thus, in 2030 a significant capacity of about 920 GWp should be operational around the globe, shared mostly among Europe, North America, and Japan. This growth is the equivalent of more than 600 state-of-the-art nuclear power plants.

Out of the 1760 MWp produced in 2005, Germany was the largest market with 600 MWp, accounting for more than 93% of the total 25-member European market, followed by Japan with 291 MWp and the United States with 110 MWp, where California and New Jersey cumulate more than 90% of the total installed national PV. Despite the fact that Europe's production increased by 50% in 2005 with 470 MWp, the fastest growth of the German market makes Europe a net PV cell and module importer. Japan's export-oriented PV production increased by 65%, totaling 528 MWp, out of which 390 MWp was exported to the European market. The world

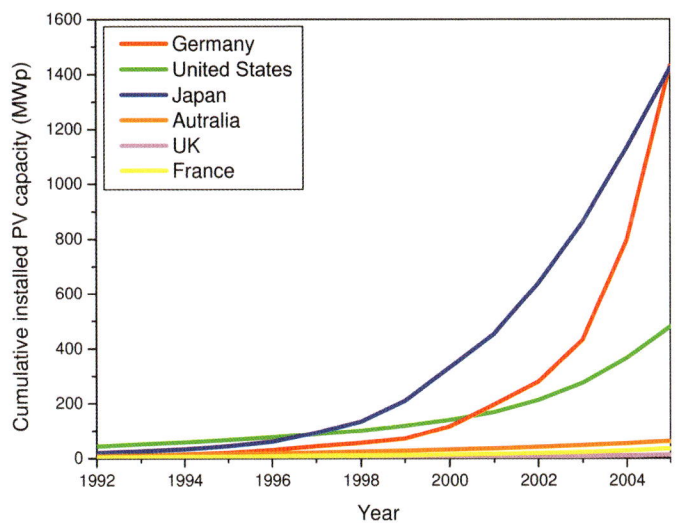

Figure 20.11 Evolution of the cumulative PV capacity for several countries, from 1992 to 2005 [23].

market share of PV produced in Japan was about 47%, experiencing a small decrease (of 2%) compared to the previous year. This reduction was mainly owing to the entrance in the business of Taiwan and the People's Republic of China (PRC), which almost tripled their production of 2004 (75 MWp) to reach 210 MWp in 2005, that is, significantly larger than that of the United States (154 MWp in 2005). This trend is partly driven by the Chinese government's decision to replace 10% of the total primary energy consumed in the country by renewable energy and also by the large demand abroad; for instance, in 2005, 90% of the PRC's PV production was exported.

20.2.2
Technologies

In 2005, more than 90% of the PV cells produced were manufactured from silicon wafers (c-Si), the rest from the thin-film technology based on amorphous silicon (a-Si), cadmium–indium–gallium–selenide (CIGS), and cadmium telluride (CdTe). During 2007, the usual module efficiency ranged between 10 and 15%, for an average retail price of €5 Wp^{-1} [24] and with the lowest reported module costs for CdTe at below $2 Wp^{-1} [25]. The c-Si cells were typically 250 μm thick and had been reduced during the last 2 years to below 200 μm and below a weight 10 g Wp^{-1}. Interestingly, the current c-Si cell price is driven mostly by the reduced availability of silicon as against the growing demand. This situation is expected to change within the next few years as some major silicon providers are investing in new plants [22]. To be operative, a module requires some additional components comprising electronic parts, support structures, and potentially electricity storage: the sum of these additional features is called the balance of system (BOS). For grid connected PV, the BOS accounts for 20–30% of the total system cost, whereas for standalone installations, this can go up to 70%.

During the last few years, the silicon-based PV cell price decreased by 5% each year, or by 20% each time the total capacity installed was doubled. In that respect, European Photovoltaic Industry Association (EPIA) expects to see module prices to drop to €2–3 Wp^{-1} around 2010. According to EPIA roadmaps, this cost reduction will be triggered mostly by a reduction not only in the Si feedstock price, once the shortage is resolved, but also in the cell thickness (150 µm) and an increase in the device efficiency (more than 17%).

Although the roadmaps of EPIA and the UN ensure that Si technology will dominate the market at least for the next 20 years, they both forecast a significant increase of the market share of the thin film and new technologies such as dye-sensitized solar cells and organic bulk-heterojunction solar cells. These new technologies are expected to follow the path of a-Si, which today is a rooftop technology with a soon-by annual production capacity of 500 MWp year^{-1}. Like a-Si in the previous years, the new technologies will not likely compete with c-Si on the rooftop market within the first few years of their launch. More realistically, they target the so-called consumer and off-grid market, which was worth about ∼300 MWp in 2006 (15–20% of the estimated annual production of 2 GWp). It is worth noting that since 1990, organic solar cells represented 27% of the worldwide patent applications [26].

20.2.3
Political Incentives

As it can be seen in Figure 20.12, the share of on-grid and off-grid installed PV systems has drastically changed during the past 15 years. While in 1992 more than 70% of the world PV capacity was off-grid, this type of PV usage went down to about

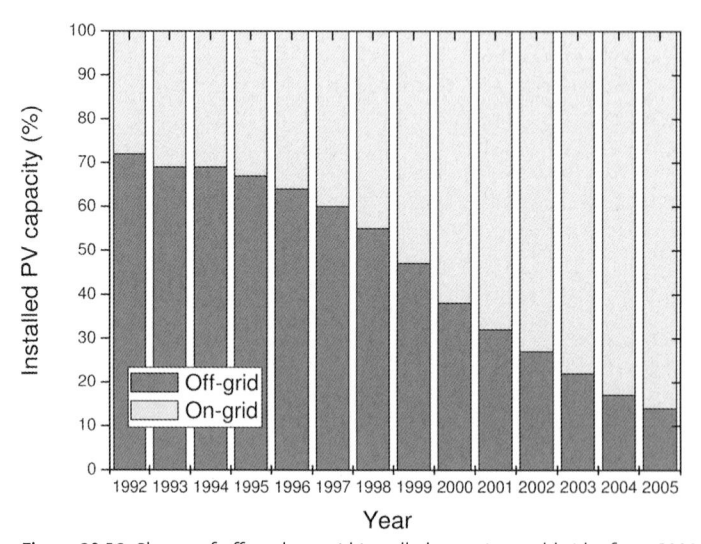

Figure 20.12 Shares of off- and on-grid installed capacity worldwide, from 1992 to 2005 [23].

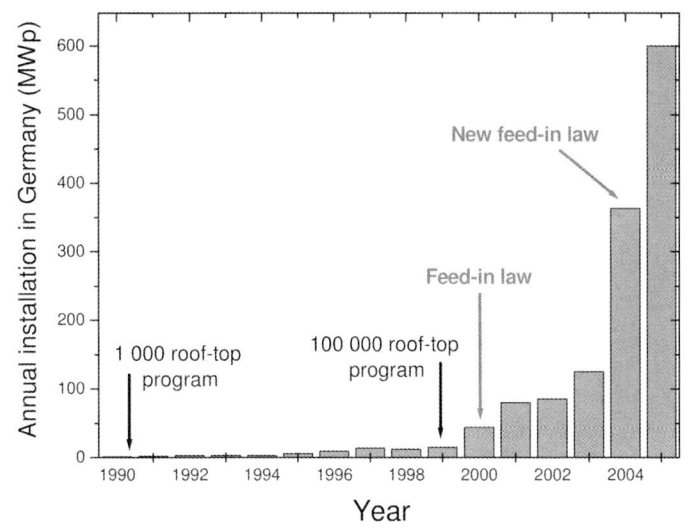

Figure 20.13 Annual installation in Germany (MWp) in the period 1990–2005. The different government incentives are indicated (from EPIA).

14% in 2005. In countries such as Germany and Japan, the current on-grid capacity is as high as 98 and 93%, respectively. However, in Canada, for example, off-grid systems account still for more than 93%. This observation can be partially explained by the geography of this country, mostly comprised of remote and isolated areas not connected to any grid as yet. But this vast difference can be explained by another argument also. PV, alike all new energy resources, needs the long-term commitment of governments to ensure a sustained development. This is especially true for this technologically advanced approach, as it is for nuclear energy. In that respect, the will of a state to promote a renewable source of energy does dictate its growth. Figure 20.13 illustrates this statement, showing that the evolution of the German PV market was triggered by some governmental decisions. The launch of the "100 000 rooftop program" and the following first feed-in law in 2000 induced a widespread demand. The reevaluation of the feed-in law in 2004 renewed and accelerated the trend.

Currently, about 15 of the 25 European states have passed some bills to encourage both the consumers and the industries to purchase and install PV modules. These initiatives follow the recommendations of the European Union White Paper, "Energy for the Future: Renewables Sources of Energy" [27], and the Green Paper, "Towards a European Strategy for the Security of Energy Supply" [28]. Some examples are listed below:

- *Germany:* For 20 years, a feed-in tariff of €0.406 kWh^{-1} for free-standing systems and of €0.518 kWh^{-1} < 30 kWp and ≈€0.49 kWh^{-1} > 30 kWp for systems on buildings and sound barriers. An additional bonus of €0.05 kWh^{-1} is allocated for facade- integrated PV.

- *Italy:* For 20 years, a feed-in tariff of €0.445 kWh^{-1} < 20 kWp, 20 kWp < €0.46 kWh^{-1} < 50 kWp, and €0.49 kWh^{-1} > 50 kWp.
- *Spain:* A feed-in tariff of €0.44 kWh^{-1} < 100 kWp (i.e., 575% of the average electricity price) and €0.23 kWh^{-1} > 100 kWp.
- *France:* A feed-in tariff of €0.30 kWh^{-1} for 20 years. An additional bonus of €0.25 kWh^{-1} is allocated for building integrated PV. Moreover, 50% of the investment costs are tax deductible.

This list is of course not exhaustive, and some additional details can be found in the reference [22]. This document provides a description of the policy incentives in the United States and Japan that are slightly more complicated as they operate at several geographical (states, county, and municipality).

An interesting point to note is that beyond securing the national primary energy supply, state incentives have other indirect positive socio-economical impacts. In Germany, for example, the number of employees in the PV sector was evaluated to about 30 000 in 2005 [29], that is, 75% of the total workforce employed in this sector in Europe. This number is expected to double by 2010. Moreover, the turnover of the German PV industry was about €3 billion in 2005 and 70% of the added value remained in the country. According to the EPIA, new production facilities create about 20 new jobs per MWp per year of production capacity, while providing about 30 additional jobs in the wholesale, retail, installation, and maintenance service sector. In other words, even in countries that do not produce PV cell/modules but import them, a significant amount of economic activity is created.

20.2.4
Potential of PV

The overall potential of PV as a global renewable energy source is related to the total energy available on Earth from solar radiation. As explained above, this quantity is simply huge. But before installing a PV panel at a certain location, a detailed study of the local sunlight insulation is necessary to assess the expected returns on investment. Figure 20.14 gives an overview of the yearly global irradiance (Ir) at optimal angle for the NorthAfrica/Europe/Middle East region [30]. It can be observed that this irradiance varies from about 1000 Wh m^{-2} in northern Europe to 1600 Wh m^{-2} in southern Europe and shows values beyond 2200 Wh m^{-2} in the southern part of the Mediterranean region. The very question that needs to be answered relates to the electrical energy that can be extracted. The yield of a system (Yf), expressed in kWh kWp^{-1}, is related to the irradiance by the so-called performance ratio (PR), according to the following equation: PR = Yf/Ir. For typical crystalline silicon based modules with a module efficiency of about 10%, the average PR value reported for the geographical regions described above is close to 0.8 [31,32]. In other words, 800 kWh kWp^{-1} can be generated yearly in northern Europe, 1300 kWh kWp^{-1} in southern Europe, and beyond 1750 kWh kWp^{-1} in North Africa. Assuming a module price of €5 Wp^{-1}, a 1 kWp system installed in middle Europe would be paid back by the

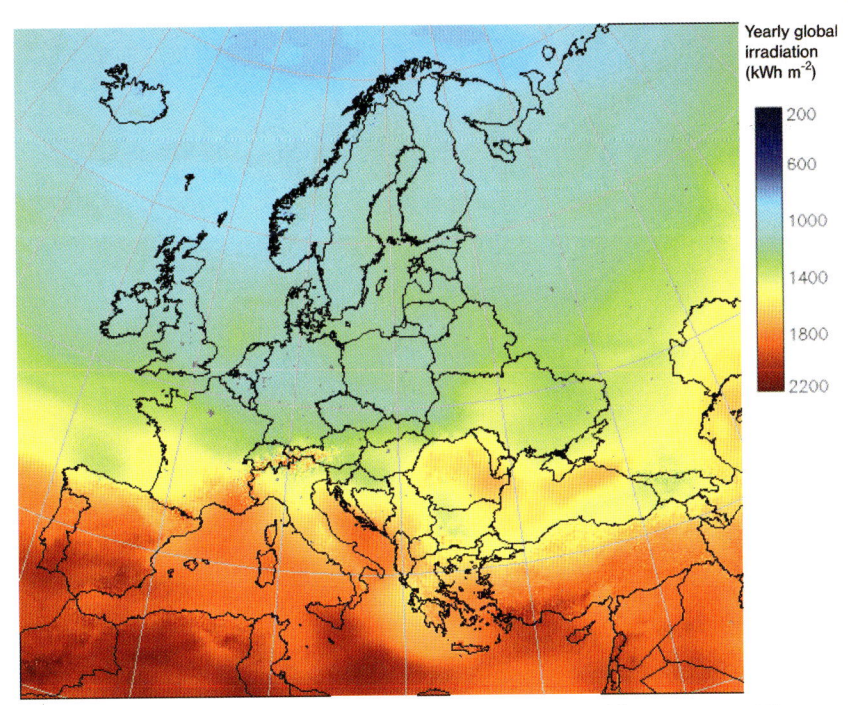

Figure 20.14 Yearly global irradiation in the North Africa/Europe/Middle East region [30].

feed-in tariff within <12 years in Germany and <10 years in France, Italy, and Spain. Knowing that most of the PV module providers offer a 25-year guaranty, the investment is almost free of any risk.

Another interesting economic aspect that deserves attention is the sticking correlation of the electricity demand on the market and the sunlight illumination. Figure 20.15 shows the price of the MWh negotiated on the Electricity European Exchange (EEX) on Tuesday June 12, 2007. It appears that the price increases significantly during the working hours, between 6:00 a.m. and 18:00 p.m. at 11:00 a.m. the price was about 18 times higher than at 3:00 a.m. This phenomenon is easily explained by the fact that most of the human activities follow the sunlight, starting with the sunrise and fading off by night; about 57% of the total electricity is consumed by agricultural, commercial, and residential activities [2] (20% for the lighting of houses and offices). The line in Figure 20.15 illustrates the typical solar irradiance versus time in middle Europe. It does nicely follow the evolution of the price of the electricity, suggesting that PV is a perfect electricity source for any direct human activities. This statement was substantiated during the heat wave of July 2006, when the peak price paid at the EEX exceeded the feed-in tariff paid in Germany.

Finally, we would like to point out two advantages of the PV approach that are somehow related to the solar irradiance and that widen the potential of this energetic

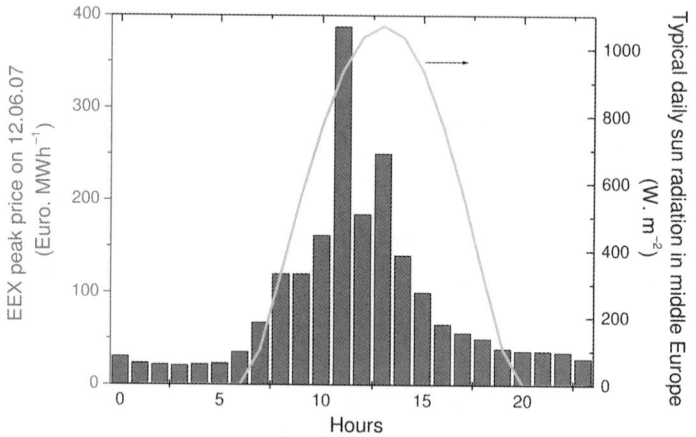

Figure 20.15 Price of peak electricity negotiated at EEX on Tuesday, June 12, 2007 and typical solar radiation in middle Europe during the same period.

solution. The first one relates to the emission of GHG, which is significantly less for PV than for any fossil-fuel-based approach. Although the emissions related to c-Si PV cells manufacturing are not currently completely negligible, they are expected to decrease with the usage of GHG treatment equipment within PV cell plants. Therefore, this renewable energy source has only a very small burden on the environment. The second point pertains to the relationship between access to energy and poverty. The United Nations has reported that poor people spend much of their income on energy, more than a third of household expenditures in some countries [33]. It is estimated that worldwide there are 2.4 billion people (more than one third of humanity) who rely on wood, charcoal, and dung as their principal source of energy for cooking and heating. At least 1.6 billion people have no access to electricity in their homes and consequently are without means for electric lighting, mechanical power, and telecommunications. Moreover, this lack of electricity hampers the development of modern industry and weakens the provision of public services such as public lighting, education, and health care. Although people from both rural and urban areas suffer from a lack of access to modern energy services, those in rural areas are especially deprived. It is estimated that four out of five people in rural areas of the developing world, mainly South Asia and Sub-Saharan Africa, live without electricity. And according to the IEA, these figures will remain largely unchanged in 2015 unless new policies are adopted to expand investment in rural energy infrastructure. In this respect, PV module installation appears as a very attractive solution in these countries where large yearly solar irradiance correlates with extreme poverty. Indeed, the PV approach can cover a range of power of more than three orders of magnitude, namely, from kWp to MWp. It can be directly used in individual households, without the necessity of a grid, which is often not installed in poor countries. Moreover, the lifetime guaranty of more than 25 years can allow a certain sustainability of the development.

20.3
Organic Photovoltaics and its Potential as a Low-Cost PV Technology

Do we really need another PV technology?
This question is a highly popular and frequently used killer argument once the public realizes the costs and the timeframe for the development of a new photovoltaic technology: for the Gen 1 and Gen 2 technologies, the timeframe was ~20–30 years and development costs were significantly >$1 billion. The mono-Si took longer and was more expensive because it had to open up and drive the whole field of PV, pioneering the suppliers, the machine vendors, the customers as well as the government and environment institutions.

Today's community has to answer the question: Is it worth to invest a billion dollars into a new field that may or may not work out in some 20 years? Or is it more beneficial to invest this amount of money into the expansion of given production capabilities, knowing that one billion dollars is the equivalent of about half a GW production capacity for conventional solar cells.

The answer is trivial: Our society needs as many and as diversified PV technologies as possible. Photovoltaic is expected to become a sustainable source of terawatt (TW) energy. However, terawatt production needs hundreds of thousands of tons of raw materials and ten thousands of square kilometers of area. Relying on one technology, with one set of raw materials, is too much of a risk and incompatible with the concepts of sustainable volume production. The last few years have taught us that raw materials can become rare and precious overnight. The cost increase in feedstock Si was artificially caused by mismanagement within the PV community and took some 5 years to fix. More serious are the limitations expected for metals and other process materials such as indium, copper, silver, and so on. The indium's price rose by a factor of 50 within the last few years and those of the copper's by a factor of 3–5. Today, no one can predict how material shortage or raw material costs are going to affect the individual PV technologies. If PV industry wants to become a sustainable energy supplier in the future, a diversified technology and multiple technology platforms are a must.

20.3.1
The Costs of PV

So far, PV is undoubtedly a success story. It took some 50 years to develop the technology and a suitable business model to install and operate PV. After these long years of building up, the PV community is now being rewarded by margins significantly over 10%. Such high margins are unusual in strongly cost-driven sectors such as renewable energy; however, these margins are expected to come down once the initial hype around it as a profitable business fades away. The annual production capacities are well beyond 1 GWp year^{-1} and the cumulative installed PV capacity will overcome the 10 GWp limit before 2010. Growth rates are still in the 30–40% regime, and everyone involved in this business would confirm PV has finally arrived on the market. So what are our future expectations for photovoltaics?

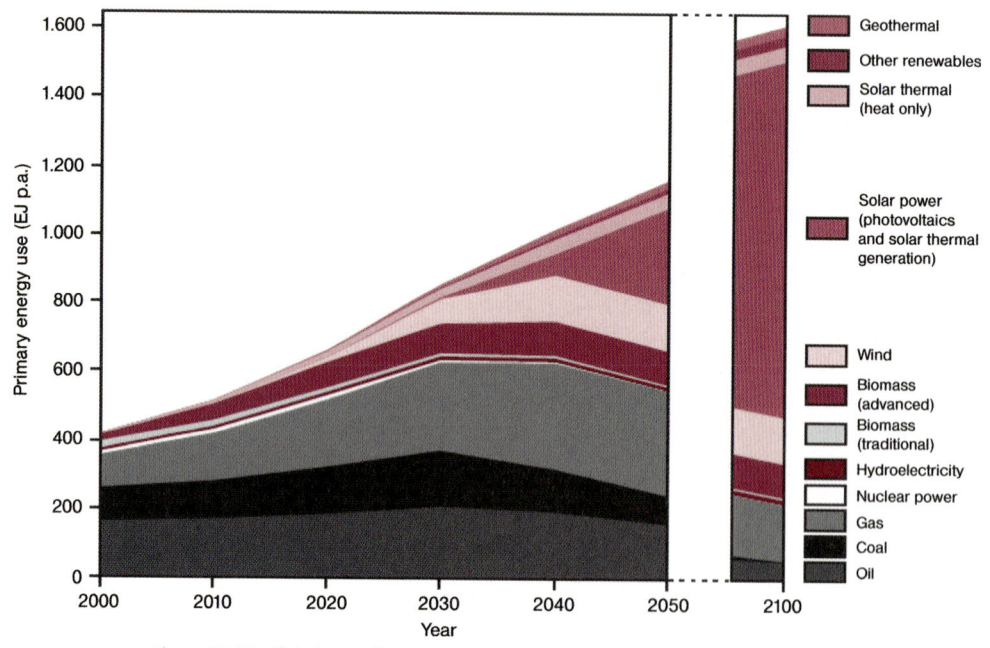

Figure 20.16 Global annual energy use in terajoules, listed for the various energy resources. (According to the scientific advisory committee of the German government for global environmental changes. Available at www.wbgu.de)

The expectations for photovoltaics are huge. Figure 20.16 shows one possible scenario for the energy consumption over the next 50 years. This particular study, performed and published by the scientific advisory committee for the German government for global environmental changes, concludes that the fossil fuels and energy sources will peak within the next 25 years, followed by a slow decline. With the fossil fuel reserves going down, the energy demand will have to be met with renewable energies, and most of the studies conclude that solar energy will take the leadership there. What makes the solar energy different compared to other renewable energy sources such as hydro or wind? The answer was given in Section 20.1.4, where the dimension of mankind's demand for primary energy sources was discussed. From all the renewable energies, only solar has the practical capacity to guarantee mankind's sustainable energy supply. Depending on the different studies, the energy supply is expected to at least double till 2050, and quadruple till 2100. We will be lacking approximately 500 EJ (∼140 000 TWh) till 2050 and approximately 1500 EJ (420 000 TWh) till 2100. If the bigger part of that energy is to be supplied by solar energy, we need peak capacities of up to 140 TW by 2050 and some 420 TW by 2100 (assuming an average 1000 h of Sun). Compared to the ∼5–10 GW production capacity in 2010, such a volume does require a constant growth rate of 15–25% for the next 40 years, resulting in an annual production capacity in the five to low double-digit TWp. Assuming that the investment into Si manufacturing is as low as US$$10^6$ MWp^{-1}, the solar energy program requires an investment of US$$10^{12}$–$10^{13}$ into manufacturing plants. If the investment into the PV-related costs (vendors, mounting, installation, conversion, etc.) is taken into account, the overall investment would increase by a factor of 3–5.

If photovoltaics become a major contributor as the world's primary, nonCO_2, energy supplier as outlined above, the PV industry will evolve into one of the world's largest industries. The biggest challenge of PV is not to make a profitable business, but to supply terawatt capacities at reasonable prices. Consequently, Hoffert [34], Nate Lewis [35] or recently Ken Zweibel from the National Renewable Energy Laboratory (NREL) named the problem "the Terawatt challenge" [36]. In detail, the terawatt challenge is based on the projection that the world's need for non-CO_2 energy would be approximately 10–20 TW by 2050. There is a lot of uncertainty behind this estimate, since a considerable part of this energy is projected to reduce and stabilize our household CO_2. CO_2 sequestration is one of the current slogans intended to bring CO_2 in our atmosphere under control. At the current stage, it is fair to state that CO_2 sequestration is an early, although potentially a promising, concept with the details still to be planned out. Depending on the type of sequestration, renewable energies play a more or less relevant role. Conventional sequestration, that is, capturing CO_2 during burning of coal and subsequent piping and storing underground, will stretch out the time for the fossil fuels, delaying but not reducing the need for renewable energies. Biomass produced CO_2 sequestration, in which coal and biomass are used to produce liquid fuels, could actually be adopted to remove CO_2 from the atmosphere. If biomass CO_2 sequestration works out, other renewable energy sources would be of less importance and the required capacity of other renewable sources would only be a fraction of the volumes outlined above. With all the information available today, and with all the uncertainty behind this information, it is not possible to choose one against the other. But independent of the detailed scenario to come, a huge available capacity for photovoltaic energy will be of benefit – for the security of our energy supply as well as our household CO_2.

Which areas have to be addressed to prepare photovoltaics for the terawatt challenge? In addition to the financial and cost estimates given above, the terawatt challenge will also face associated issues such as material availability, land area needs, and energy storage systems. Solar energy can easily meet the global energy needs from a resource point of view. The question is what does it take to make it happen?

- Cost is certainly the biggest driver.
- Availability of raw materials.
- Scalable, high-volume and low-cost production processes.
- Available area is crucial.

- Energy storage systems (such as hydrogen, bio fuels, and solar-produced hydrocarbons, as well as classical energy storage systems such as batteries, caps, hydro, and thermal).
- Distribution and grid efficiency for the energy mix.
- Centralized versus decentralized energy production.

Let us now take a look at the organic photovoltaic (OPV) technologies and see where they can contribute to the photovoltaic market and, consequently, to the terawatt challenge. There are a number of distinct features that make organic PV technologies attractive compared to other technologies and, apparently, give OPV a stable competitive advantage:

1. Thanks to the monolithic roll-to-roll manufacturing process, OPV modules (like any other low-temperature printable technology) can be produced at significantly *lower unit costs* than any other solar technology. This is owing to the use of organic semiconductors that can be deposited through printing technologies.

2. The use of films as a substrate leads to *flexible and lightweight modules* that enable completely *new form factors* and *novel application categories*. Current inorganic technologies still struggle to give full performance when processed on flexible substrates, but breakthrough developments are expected in this field. It is expected that a-Si will be the first technology to become fully commercialized on PET substrates.

3. The lower initial investments for print-based solar installations pave the way for *new business models and funding mechanisms*.

4. The compatibility to printing and printing-related technologies would allow what is called a "fables production" concept. Printed organic solar cells will be produced by already existing printing and coating facilities, which are available worldwide with huge capacities. With the photographic film, the magnetic tapes as well as CD and CD-R business going down, the available printing capacities worldwide will rise rapidly. No or little specific investment will be necessary to set up a production process in such printing lines. No current inorganic technology can compete printed PV technologies in this respect.

5. The *energy balance* of organic solar cells will be significantly better than the balance of any other solar technology. While this is a positive and encouraging fact, energy payback times of the inorganic technologies at present are also that low (between 2 and 8 years, depending on the technology and the region of installation), so that even shorter payback times do not give further competitive advantage, except for business models working with shorter amortization times than 20 years and consequently shorter module lifetimes. For a positive energy balance, the module lifetime needs to be longer than the energy payback time.

Some additional thoughts around these five arguments:

  ad 1. *Lower unit costs.* The initial investments for an OPV manufacturing facility are comparatively small because the technology is designed into a well-established printing technology. It is expected worldwide that printed technologies will meet production cost of less than €1 Wp^{-1} early after commercialization. Even if adjusted to a shorter lifetime, this is a significant change.

By contrast, crystalline silicon-based technologies are fabricated in a batch process from small wafers of silicon semiconductor material. The production requires substantial investments in manufacturing facilities. For example, between 1999 and 2003, Shell publicly announced an investment of more than US$34 million in a manufacturing facility for polycrystalline PV in Gelsenkirchen, Germany, with a capacity of 25 MWp p.a. Therefore, investment costs need to be brought down. The easiest way to do so is by achieving economy of scale. If more and more players set up bigger and bigger production lines, the investment costs will come down automatically. However, depending on the detailed process, the productivity of Si- or wafer-based lines will remain in the 10–1000 m^2 h^{-1} regime, whereas the one of printing lines is in the 10 000–100 000 m^2 h^{-1} regime.

ad 2. *Flexible and lightweight modules enable new form factors.* Printed solar cells will be, by definition, light and flexible and would be realized even as semitransparent modules. This last attribute will allow completely new applications such as PV windows.

Crystalline silicon modules are rigid, heavy, and difficult to customize to individual design needs. Amorphous silicon PV on the contrary can be applied on foils and films and has been commercialized as a flexible product already.

ad 3. *New business and funding mechanism.* Printed PV installations result in reduced capital costs and therefore provide the opportunity of using new funding approaches. This is especially important in developing countries in which PV installations are mostly financed with loans. An important problem of all silicon-based solar installations is the relatively high initial investment. If the period of use is not sufficient to redeem this upfront payment, the installation is far too expensive.

ad 4. *Fables production concept.* Imagine printing of solar cells in one of the huge printing lines operating all around the world. The fastest printing and coating lines worldwide run on a web speed of up to 10 m s^{-1}, at widths of 2–3 m. This results into an incredible productivity of 50 000–100 000 m^2 h. Other interesting aspects are the product cycle times in the printing industry. This industry is experienced to produce different patterns and products each day. Layout and architecture modifications of solar modules, that is, the width, length, and number of stripes could be done with only few days of preparation.

ad 5. *Energy balance.* Renewable energies are judged by their energy balance, that is, the relation of the energy used during production of the solar module compared to the energy it produces during its lifetime. Owing to the low-temperature production processes, printed PV is expected to need only a few weeks to attain a positive energy balance. The energy balance of silicon is now at around 4–8 years, and the thin-film technologies have between 2 and 4 years. Printed PV has a serious advantageous ecological profile among the renewable energies.

In the following sections, we will strictly focus on the cost discussion. Bringing the costs of photovoltaic produced energy down is the biggest challenge to fully exploit this resource. Thus, it is imperative for any new thin-film PV technology to (a) analyze the costs and (b) understand whether this new technique can manage explosive growth (scalable production). These two issues will answer the question whether a new technology can contribute to the terawatt challenge or not.

20.3.1.1 Conclusion

The intrinsic advantages of printed PV throw up the opportunity for serious business. Its realization depends very much on the technological profile that can be achieved. As the profile will significantly change over time, selection of the right market entry strategy is of great importance.

In the short and medium term, the characteristics of printed organic technologies will allow the penetration of specific market segments in which lifetime and efficiency requirements are met. The development of these segments alone will allow any company in that field to build a profitable business. When the organic printed technologies succeed to realize their full potential, they have a fair chance to impact the whole solar energy industry. And that is essential. The long-term vision of any PV technology must be to contribute significantly to global energy production.

20.3.2
The Costs of OPV: BOM and BOS

Costs in photovoltaics business are typically separated in terms of the costs for the module and the costs for the installation. The balance for the module costs is frequently called BOM (balance of modules). The balance of the installation costs is called *BOS*. The BOM typically contains

- all *material* costs, including material waste and process materials;
- all *production* costs, including capex, depreciation of machinery, maintenance, and the production yield;
- all *overhead* costs, including R&D, marketing, sales, and so on.

The BOS typically contains

- all *area related* costs, including rent, mounting hardware, racks, shipping, and installation;
- all *energy related* costs, including cables, converters, shipping, installation, servicing.

There is little doubt that low-temperature, roll-to-roll printing of solar cells on plastic sheets can manage explosive growth better than most other production processes in terms of cost, volume, capacity and scale-up. As such, OPV is expected to master the "volume supply" hurdle with brilliance. The BOM of OPV, and here especially the material costs, will be analyzed in greater detail. For the BOM-related production and overhead costs, we simply draw an analogy with comparable multilayer printed large volume products, such as housings for consumer electronics (i.e., mobile phones), optical decor films, or high-quality optical films for packaging. Typical production and overhead costs for such products are in the range of €1–10 m^{-2}, with <€5 m^{-2} as a reasonable expectation at larger volume.

An excellent summary on BOS costs for thin-film PV was recently given in a review by Zweibel [36]. Depending on the application (e.g., power PV, residential rooftop or commercial flat roof) and volume, he found BOS costs between €100 and €30. These numbers will be used for the energy cost calculations in Section 20.3.3.

20.3.2.1 BOM of OPV

The material costs for OPV will significantly change over time and with produced volume. A typical cost analysis has to work with economy of scale, assuming specific volumes over time driving the material costs down. However, a generic look at the BOM cost situation of OPV at present is already quite helpful at this early stage and allows an educated guess on the upper and lower boundaries for the material costs. In the following section, we will analyze the individual cost drivers of OPV. This section does not intend to give a detailed cost analysis, that is, by the way, impossible before large-scale commercialization takes place. But a high level cost analysis, using costs of comparable but already available components, does help to identify maximum as well as minimum cost ranges for a product.

A simple example helps to clarify the logic behind this approach: the cost estimate for the barrier will be based on the cost situation for the currently available food packaging barriers. Today's packaging films, opaque and semitransparent ones, are produced in huge volumes from high-throughput production machines, with roll widths as high as 4 m and roll lengths of up to 48 000 m. The assumptions are that the packaging materials for OPV will be produced in similar volumes, on comparable machines.

The cost situation is summarized in Tables 20.2 and 20.3. We distinguish between the direct material related and the indirect production and overhead-related costs. Even at this early stage of pre-commercialization, it becomes obvious that OPV costs are outstanding low. There is strong encouragement that OPV has the potential to become an outstanding cheap PV technology. Nevertheless, requesting that the BOM should be below €50 m^{-2} even at low volumes, there are a few cost points that need to be carefully monitored over time. The three big cost drivers are the semitransparent electrode, the packaging material, and the semiconductor. Behind each of these components is a huge parameter space, urging optimization of the architectural design. Some of the urgent questions from the product design are, for example, whether it is justifiable to use a twice as expensive semiconductor for a relative efficiency increase, say, of 5–6%. Another challenge that may be answered by a precise cost model: is it profitable to prolong the lifetime of a product, say, for 3–5

Table 20.2 Potential direct cost structure for a OPV volume product.

Layer	Typical materials	Price €m^{-2} or €m^{-2} equivalent	Thickness of layer in μm	% of area per meter square	Cost targets for individual components €m^{-2}
Bottom package (barrier, hard coat, UV filter)	SiO$_x$-coated barrier	1–10	25–100	100	<€5 m^{-2}
Adhesives	Epoxy, PSA, hotmelts	1–3	1–10	90	<€2 m^{-2}
Substrate	PET	2–4	25–150	100	<€2 m^{-2}
Transparent bottom electrode	ZnO/ITO/ nanotubes, ...	5–20	0.1–0.5	90	<€10 m^{-2}
Electron blocker	PEDOT/PANI	1–5	0.1–0.5	90	<€5 m^{-2}
Active Layer (first semiconductor)	p-type	5–25	0.1–0.5	90	€5–25 m^{-2}
Active layer (second semiconductor)	n-type	2–10			€2–10 m^{-2}
Hole blocker	For example, LiF, ...	1–5	0.01–0.1	90	€1 m^{-2}
Opaque top electrode	Metals, carbon, organic ...	2–4	1	90	<€5 m^{-2}
Bus bars	Printed metal	2–5	1–10	5–10	<€1 m^{-2}
Adhesive	Epoxy, PSA, hotmelts, ...	1–3	1–10	90	<€2 m^{-2}
Top package (barrier)	PET/Al laminate	1–10	25–100	100	<€5 m^{-2}
Total costs		∼€25–100 m^{-2}			<€50 m^{-2}

Table 20.3 Potential indirect (production costs + overhead) cost structure for an OPV volume product.

Components/process	Price (euro) per meter square likely to be less than
Process chemicals	<€1 m^{-2}
Other consumables	<€1/m^{-2}
Utilities	<€1 m^{-2}
Maintenance	<€1 m^{-2}
Rent	<€1 m^{-2}
Labor	<€1 m^{-2}
Total	<∼€5 m^{-2}

OPV device architecture
(simplified)

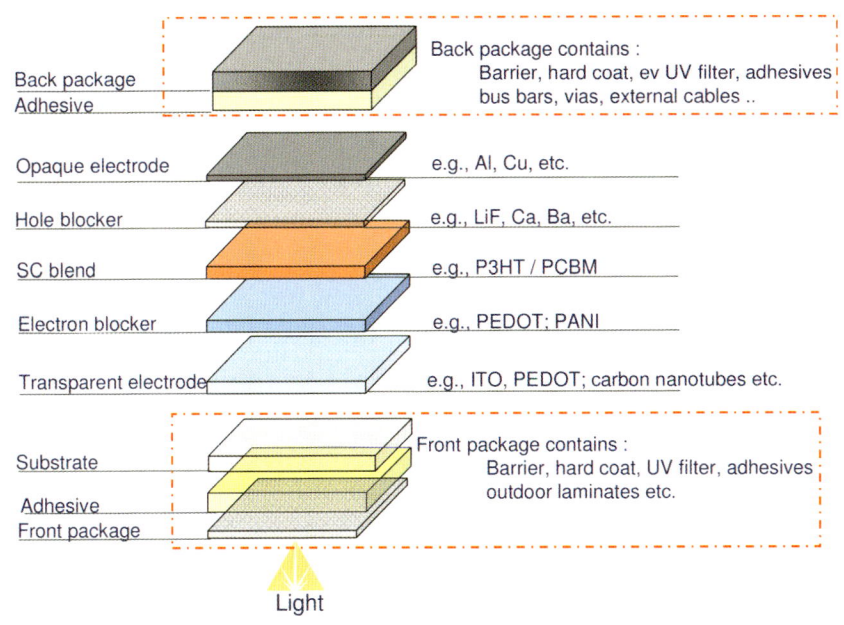

Figure 20.17 Typical cross section and material set for an OPV cell.

years by using a 10 times more expensive barrier? These and similar questions can be answered by a cost model, which should be an essential tool for product design. The rather large tolerances for the individual material items (see Table 20.2) are less of a problem for such a fundamental analysis; they rather offer an opportunity to understand the right product optimization scenarios.

Summarizing, OPV has a direct cost potential between a €25 and €100 m^{-2}, whereas the €30–60 m^{-2} regime appears as a very reasonable, fast to realize, cost scenario at rather low volumes. Indirect costs of <€10 m^{-2}, but more likely even <€5 m^{-2} are reasonable at these volumes.

20.3.3
Cost Model for OPV: Representative for any Low-Cost and Low-Performance Technology

After understanding the direct and indirect contributions to the BOM, one can analyze the energy production costs of the technology. Such an analysis can be conducted with significant ease by working with BOM and BOS costs only, neglecting the single contributions to each of these cost items. Section 20.3.2 worked out the reasonable BOM and BOS cost regimes for the OPV technology, which are used as entrance parameters for the energy production cost calculations in this chapter.

The focus of this analysis is on to develop an understanding to which extent a shorter product lifetime can be compensated by significantly lower costs. As a model scenario, we will calculate the costs for a rooftop residential PV installation. The following assumptions are taken:

- The installation is designed for a 1 MW annual capacity (equivalent to a 1 KW peak installation).
- The lifetime for the installation is 25 years.
- The BOS costs are varied between 40 and €100 m^{-2} costs.
- The BOM costs are varied between 10 and €100 m^{-2} costs.
- The module lifetime is varied between 3 and 10 years.
- The module efficiency is varied between 3 and 10%.

The competitive position of an organic printed PV technology in the on-grid market depends on its cost–performance profile. To compare OPV modules with silicon-based installations, a lifetime-adjusted calculation needs to be made. The concept behind the lifetime adjustment is based on two assumptions:

- A module with shorter lifetime can be exchanged. A cost calculation for a 25-year lifetime installation can take a shorter lifetime module into account by a further investment at a later time. A module with 5 years of lifetime needs to be replaced four times within the 25-year life cycle.
- The replacement installation entail costs for the new modules plus an installation fee.
- Discounting future investments: all future replacements and investment costs necessary to replace the shorter lifetime PV modules are discounted by 7% to come with the net present value.

The assumption to discount future investments is a major assumption. An alternative scenario is to finance the total costs for a 25 yr installation by a credit. In that case interest rates need to be added to the total costs for the installation. We have chosen to work with the discount model, since it is clearly the more attractive model for a solar technology with shorter lifetime. The differences in final costs between these two models can be up to a factor of 2.

One broad benchmark for any new PV technology is the expected price of installations in €Wp^{-1}. A price of €3–4 Wp^{-1} is translated into a €3000–4000 investment for a 1 kWp rooftop plant. It is interesting to calculate how the OPV technology, with its lower efficiency, lower lifetime, and lower costs does perform in this metrics. Table 20.4 presents the costs (in €Wp^{-1}) of a 1 kWp OPV plant for various combinations in efficiency and lifetime. Table 20.4a presents the data for a BOM of €30 m^{-2} and a BOS of €40 m^{-2}, while Table 20.4b shows the data for a BOM of €50 m^{-2} and a BOS of €70 m^{-2}. Even for the second more conservative assumption, Table 20.4b shows that OPV will be competitive with an efficiency of around 7% and 7 years of lifetime. Depending on the parameters, a lowest cost technology such as OPV will attain a distinct competitive edge.

It is by far more interesting and relevant to answer the question whether a low-cost and low-performance technology can provide a sustainable solution for the TW challenge. For this question, one has to calculate the costs of electricity in €cents kWh^{-1}. The conversion from €Wp^{-1} into €cents kWh^{-1} is done easily when the sun

Table 20.4 Energy cost calculations in €Wp^{-1} for the presented model of 1 kWp grid connected rooftop plant under the following assumptions: BOM = €30 m^{-2} and BOS = €40 m^{-2}; BOM = €50 m^{-2} and BOS = €70 m^{-2}.

(a)

€Wp^{-1}	3 years	4 years	5 years	6 years	7 years	8 years	9 years
3%	7.3	5.4	4.4	3.6	3.1	2.7	2.4
4%	6.4	4.8	3.8	3.2	2.7	2.4	2.1
5%	5.7	4.3	3.4	2.9	2.5	2.1	1.9
6%	5.2	3.9	3.1	2.6	2.2	2.0	1.7
7%	4.4	3.3	2.6	2.2	1.9	1.6	1.5
8%	4.2	3.1	2.5	2.1	1.8	1.6	1.4
9%	3.8	2.8	2.3	1.9	1.6	1.4	1.3
10%	3.7	2.8	2.2	1.9	1.6	1.4	1.2

(b)

€Wp^{-1}	3 years	4 years	5 years	6 years	7 years	8 years	9 years
3%	12.2	9.2	7.3	6.1	5.2	4.6	4.1
4%	10.8	8.1	6.5	5.4	4.6	4.0	3.6
5%	9.7	7.3	5.8	4.8	4.1	3.6	3.2
6%	8.8	6.6	5.3	4.4	3.8	3.3	2.9
7%	7.4	5.5	4.4	3.7	3.2	2.8	2.5
8%	7.1	5.3	4.3	3.5	3.0	2.7	2.4
9%	6.5	4.8	3.9	3.2	2.8	2.4	2.2
10%	6.3	4.7	3.8	3.1	2.7	2.4	2.1

hours year^{-1} are known. In our calculation, we assumed a 1000 h of sun year^{-1}, a value that is typical for regions like middle Europe (e.g., Germany) or the northern East Coast of America (e.g., Massachusetts). We have investigated the power production costs for three different scenarios. Table 20.5 and Figure 20.18 show the costs for a BOS of €100 m^{-2}, Table 20.6 and Figure 20.19 show the costs in the case of €70 m^{-2} BOS, and Table 20.7 and Figure 20.20 analyze the scenario for a BOS of €40 m^{-2}, an optimistic but certainly achievable value in the long run. The calculations were run to identify the performance parameters to meet energy production costs of 50, 25, 10, and 5 cents kWh^{-1}, respectively. Figures [18–20] plot the efficiency of BOM costs for three different scenarios, whereas the symbol always indicates the value at 5 years of lifetime. The error bars and the guided lines around the symbols show the parameter variation in the case of a 3- and 10-year product, respectively.

It is surprising how significantly the BOS costs impact the cost calculations for a low-cost technology. It takes a BOS of less than €40 m^{-2} to enable power production at 5 cents kWh^{-1} for a PV product with ~10% efficiency, 5 years of lifetime, and a BOM of €30 m^{-2}. Another clear relation is seen by the trendiness connecting the efficiency error bars. An increase in lifetime flattens out the dependence between costs and efficiency. Modules with a longer lifetime are much less susceptible to cost reduction upon efficiency increase (in absolute numbers).

Table 20.5 Energy cost calculations in €cents kWh^{-1} for the presented model of 1 kWp grid connected rooftop plant under the following set of assumptions:

BOS €100 m^{-2}
BOM: varied from €10 to 100 m^{-2}.
Lifetime: varied from 3 to 10 years.
Efficiency: varied from 3 to 10%.

Efficiency plot BOS @ €100 m^{-2}

Energy costs (cents kWh^{-1})

BOM (€ m^{-2})	50		25			10			5	
	Efficiency (%)	Lifetime (years)	Efficiency (%)	Lifetime (years)	Efficiency (%)	Lifetime (years)	Efficiency (%)	Lifetime (years)	Efficiency (%)	Lifetime (years)
10			2.8	3	5.5–7	10–3				
30			3–4.6	10–3	7–10	10–5				
50	3.15	3	3.6–6.5	10–3	9–10	10–7	10	10		
70	3.3–4.1	5–3	4.3–8.3	10–3						
100	3–5.5	7–3	5.2–9	10–5						

Efficiency plot BOS @ €100 m^{-2}, lifetime @ 5 years

Energy costs (cents kWh^{-1})

BOM (€ m^{-2})	50		25			10			5	
	Efficiency (%)	Lifetime (years)	Efficiency (%)	Lifetime (years)	Efficiency (%)	Lifetime (years)	Efficiency (%)	Lifetime (years)	Efficiency (%)	Lifetime (years)
10			3.9	5	6	5				
30			5.25	5	10	5				
50			6.5	5						
70	3.3	5	8.6	5						
100	4.3	5								

The table is graphically summarized in Figure 20.18.

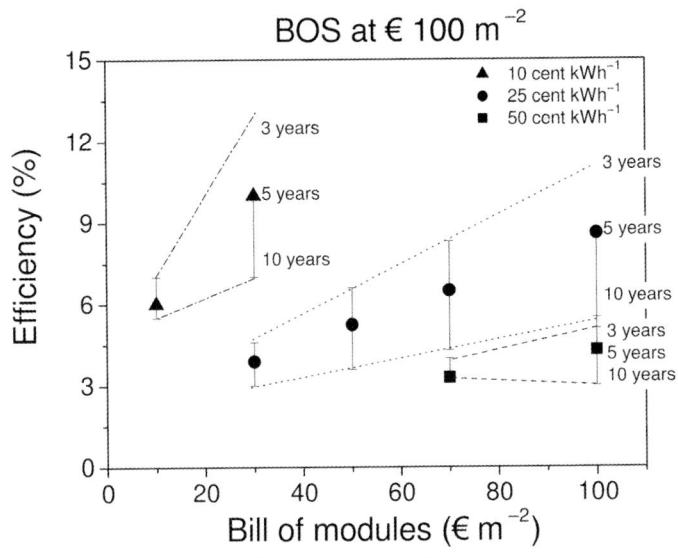

20.3 *Organic Photovoltaics and its Potential as a Low-Cost PV Technology* | 561

Figure 20.18 Dependence of energy costs on efficiency, BOM costs and lifetime of PV modules for a BOS of 100 € m^{-2}. Symbols indicate required efficiency to produce energy at costs of 10, 25 & 50 cents/kWh with modules of 5 years of lifetime. Trendlines indicate required efficiencies for 3 yrs and 10 yrs module lifetime.

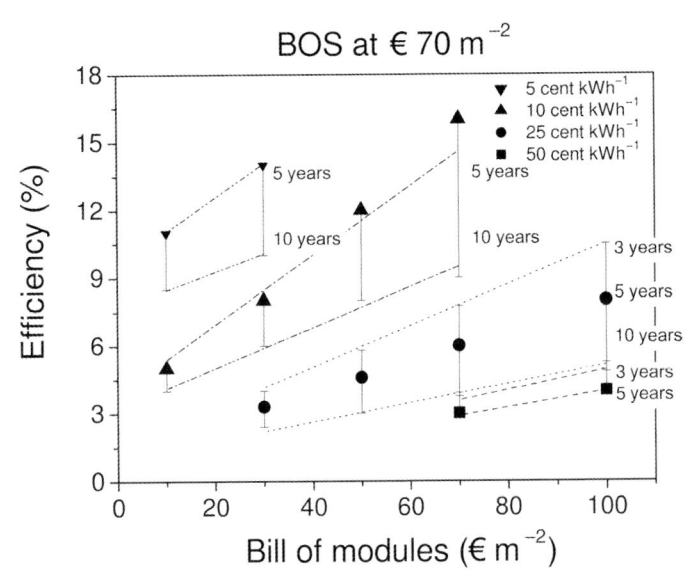

Figure 20.19 Dependence of energy costs on efficiency, BOM costs and lifetime of PV modules for a BOS of 70 € m^{-2}. Symbols indicate required efficiency to produce energy at costs of 5, 10, 25 & 50 cents/kWh with modules of 5 years of lifetime. Trendlines indicate required efficiencies for 3 yrs and 10 yrs module lifetime.

Table 20.6 Energy cost calculations in €cents kWh^{-1} for the presented model of 1 kWp grid connected rooftop plant under the following set of assumptions:

BOS €70 m^{-2}
BOM: varied from €10 to 100 m^{-2}.
Lifetime: varied from 3 to 10 years.
Efficiency: varied from 3 to 10%.

Efficiency plot BOS @ €70 m^{-2}

Energy costs (cents kWh^{-1})

BOM (€ m^{-2})	50 Efficiency (%)	50 Lifetime (years)	25 Efficiency (%)	25 Lifetime (years)	10 Efficiency (%)	10 Lifetime (years)	5 Efficiency (%)	5 Lifetime (years)
10					4–5.5	10–3		
30	3	3	3.3–4	5–3	6–10	10–3	8.5–11	10–5
50	3	3	3–5.8	10–3	7.5–12	10–5	10	10
70	3–3.9	5–3	3.75–7.75	10–3	9–16	10–5		
100	4–5.25	5–3	4.15–10.5	10–5				

Efficiency plot BOS @ €70 m^{-2}, lifetime @ 5 years

Energy costs (cents kWh^{-1})

BOM (€ m^{-2})	50 Efficiency (%)	50 Lifetime (years)	25 Efficiency (%)	25 Lifetime (years)	10 Efficiency (%)	10 Lifetime (years)	5 Efficiency (%)	5 Lifetime (years)
10					5	5		
30			3.3	5	8	5	8.5	5
50			4.6	5	12		14	5
70	3	5	6	5	16			
100	4	5	8	5				

The table is graphically summarized in Figure 20.19.

Table 20.7 Energy cost calculations in €cents kWh^{-1} for the presented model of 1 kWp grid connected rooftop plant under the following set of assumptions:

BOS €40 m^{-2}
BOM: varied from €10 to 100 m^{-2}.
Lifetime: varied from 3 to 10 years.
Efficiency: varied from 3 to 10%.

	\multicolumn{8}{c	}{Efficiency plot BOS @ €40 m^{-2}}
	\multicolumn{8}{c	}{Energy costs (cents kWh^{-1})}
BOM (€m^{-2})	\multicolumn{2}{c	}{50}	\multicolumn{2}{c	}{25}	\multicolumn{2}{c	}{10}	\multicolumn{2}{c	}{5}
	Efficiency (%)	Lifetime (years)	Efficiency (%)	Lifetime (years)	Efficiency (%)	Lifetime (years)	Efficiency (%)	Lifetime (years)
10					2.5–4	10–3	5.3–9	10–3
30			2.8–3.5	5–3	4.5–9	10–3	9–15	10–3
50			2.5–5.5	10–3	6–10	10–5	11	10
70	2.5–3.6	5–3	3.2–7.25	10–3	8–14	10–5		
100	3.8–5	5–3	4.2–10	10–3				

	\multicolumn{8}{c	}{Efficiency plot BOS @ €40 m^{-2}, lifetime @ 5 years}
	\multicolumn{8}{c	}{Energy costs (cents kWh^{-1})}
BOM (€m^{-2})	\multicolumn{2}{c	}{50}	\multicolumn{2}{c	}{25}	\multicolumn{2}{c	}{10}	\multicolumn{2}{c	}{5}
	Efficiency (%)	Lifetime (years)	Efficiency (%)	Lifetime (years)	Efficiency (%)	Lifetime (years)	Efficiency (%)	Lifetime (years)
10					3.6	5	7	5
30			2.8	5	7	5	12	5
50			4.15	5	10			
70	2.5	5	5.5	5	14			
100	3.8	5	7.5	5				

The table is graphically summarized in Figure 20.20.

20.3.4
Summary

The outcome of the cost calculation can be summarized best in the following way: a low-cost and lower performance PV technology such as OPV does meet all the cost requirements for power production. Lowest cost modules (i.e., BOM €30–50 m^{-2}) with a lifetime between 5 and 10 years and an efficiency between 10 and 5% can produce electricity at 10 €cents kWh^{-1}, even at a BOS of €70 m^{-2}. Lifetime and costs are directly related. Lower lifetime and lower efficiency can be easily compensated by lower module costs.

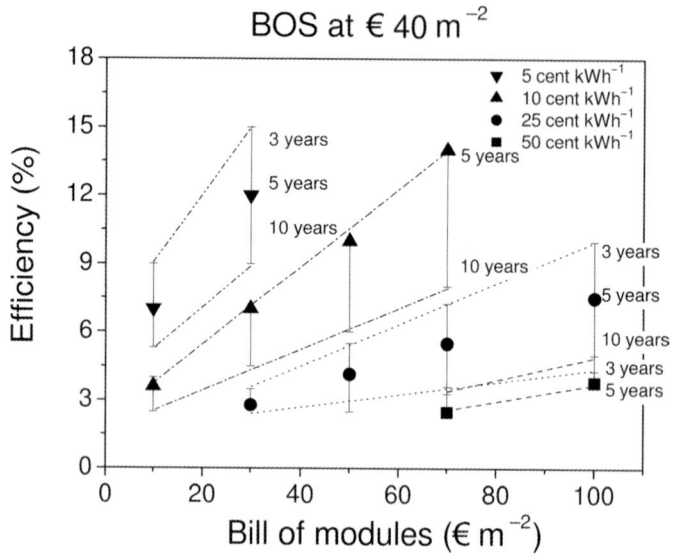

Figure 20.20 Dependence of energy costs on efficiency, BOM costs and lifetime of PV modules for a BOS of 40 € m^{-2}. Symbols indicate required efficiency to produce energy at costs of 5, 10, 25 & 50 cents/kWh with modules of 5 years of lifetime. Trendlines indicate required efficiencies for 3 yrs and 10 yrs module lifetime.

A second outcome of the cost calculations is the understanding how significantly a high BOS impacts cheap power production by PV. The current worldwide focus in cost reduction is on the BOM of the individual technologies. With the first low-cost technology, this trend will be reversed and reduction of BOS costs will become as important as reduction of the BOM costs.

References

1 UNDP, UNDESA, WEC (2001) World Energy Assessment Overview 2001 (United National Development Program, United Nation Department of Economy, and Social Affairs, World Energy Council 2001). Available at www.undp.org.
2 IEA (2006) Key World Energy Statistics (International Energy Agency 2006). Available at www.iea.org.
3 UNDP, UNDESA, WEC (2001) World Energy Assessment Overview 2004 Update (United National Development Program, United Nation Department of Economy, and Social Affairs, World Energy Council 2001). Available at www.undp.org.
4 Van Dielo, J., Maggetto, G. and Lataire, P. (2006) *Energy Conversion and Management*, 47, 2748.
5 Environmentally Sustainable Transport (EST) (2000) Synthesis Report of the OECD Projects.
6 Campbell, C.J. (2005) The end of the first half of the age of oil. 4th International Workshop on Oil and Gas Depletion, Lisbon, Portugal.
7 Hubbert, M.K. (1956) Nuclear energy and the fossil fuels. Presented before the

Spring Meeting of the Southern District, March 7–9, American Petroleum Institute, Plaza Hotel, San Antonio, Texas.

8 Laherrere, J. (2005) Forecasting production from discovery. 4th International Workshop on Oil and Gas Depletion, Lisbon, Portugal.

9 Bentley, R.W. (2002) Global oil & gas depletion: an overview. *Energy Policy*, **30**, 189–205.

10 *Climate Change 2007 – The Physical Science Basis*, Contribution of Working Group I to the Fourth Assessment Report of the International Panel on Climate Change, Cambridge University Press, New York, USA.

11 Grubb, M.J. and Meyer, N.I. (1993) Wind energy: resources, systems, and regional strategies, in *Renewable Energy: Sources for Fuels and Electricity* (eds T.B. Johansson *et al.*), Island Press, Washington, DC.

12 WEC (World Energy Council). (1994) *New Renewable Energy Resources: A Guide to the Future*, Kogan Page Limited, London.

13 Renewables in Global Energy Supply: An IEA Factsheet, International Energy Agency, January 2007. Available at www.iea.org.

14 Kammen, D.M. (2006) The rise of renewables. *Scientific American*, September.

15 Palmerini, C.G. (1993) Geothermal energy, in *Renewable Energy: Sources for Fuels and Electricity* (eds T.B. Johansson *et al.*), Island Press, Washington, DC.

16 The future of geothermal energy, Massachusetts Institute of Technology, 2006. Available at http://geothermal.inel.gov.

17 Nakicenovic, N., Grübler, N.A. and McDonald, A.(eds) (1998) *Global Energy Perspectives*, Cambridge University Press, Cambridge.

18 IEA (International Energy Agency) (1998) Biomass Energy: Data, Analysis and Trends. World Energy Outlook, Paris.

19 Concentrating solar power now, German Federal Ministry for the Environment, Nature Conservation and Nuclear Safety, German Federal Ministry for Economic Cooperation and Development, 2003.

20 Philibert, C. (2004) International Technology Collaboration and Climate Change Mitigation Case Study 1: Concentrating Solar Power Technologies, Environment Directorate, International Energy Agency.

21 U.S. Climate Change Technology Program – Technology Options for the Near and Long Term, August 2005.

22 Jäger-Waldau, A. (2006) PV Status Report 2006, Institute for Environment and Sustainability, European Commission.

23 Trends in Photovoltaic Application: Survey Report of Selected IEA Countries Between 1992 and 2005, International Energy Agency, Report IEA-PVPS T1–15:2006.

24 www.solarbuzz.com.

25 www.firstsolar.com.

26 Photon magazine, October 2005.

27 Energy for the Future: Renewable Sources of Energy, White paper for a Community Strategy and Action Plan, COM(1997)599 (21/11/97).

28 Towards a European Strategy for the Security of Energy Supply, COM (200) 769 Final.

29 Deutsche Bundesverband Solarwirtschaft, Statistische Zahlen der deutschen Solarwirtschaft, June 2006.

30 PVGIS Solar Irradiance Data, IES – Institute for Environment and Sustainability. http://ies.jrc.cec.eu.int/.

31 Jahn, U., Niemann, M., Blaesser, G., Dahl, R., Castello, S., Clavadetscher, L., Faiman, D., Mayer, D., van Otterdijk, K., Sachau, J., Sakuta, K., Yamaguchi, M. and Zoglauer, M. (1998) Proceedings of the 15th European Photovoltaic Solar Energy Conference, July, Vienna, Austria, Paper No. VD6.17, page 1 of 4.

32 Ransome, S. and Wohlgemuth, J. (2002) 28th PVSC, New Orleans, 5O3.1.

33 Energizing the Millennium Development Goals, United Nation Development Program, UNDP, 2005.

34 Hoffert, M., Caldiera, K., Jain, A.K., Haites, E.F., DanneyHarvey, L.D., Potter, S.D., Schlesinger, M.E., Schneider, S.H., Watts, R.G., Wrigley, T.M.L. and Wuebbles, D.J. (1998) *Nature*, **395**, 881.

35 Lewis, N. (2004) A global energy perspective. http://www.its.caltech.edu/~mmrc/nsl/energy.html.

36 Zweibel, K. (2006) *Thin Film Solar Cells* (eds J. Poortmans and V. Arkhipov), Wiley-VCH Verlag GmbH, Weinheim, Germany.

Index

a
ab initio methods 214
absorption coefficient 170, 330
AC signal 522
acceptor blend 166
– supersaturation 166
acceptor-type conjugated polymers 95
acrylate monomers 499
aerosols 537
air-stable polymer devices 272
– achievement 272
alkyl chains 5, 345
alkyl-substituted polycarbazole polymers 98
alternating polyfluorene copolymers (APFOs) 76
alumina films 495
alumina layers 499
aluminosilicate barrier 497
amorphous films 477, 480
anionic semiconductor species 233
annealing spin-cast thin films 29
anodic alumina template 339
antistatic coating 213
– applications 213
Arrhenius-like thermal activation 206
arylene vinylene based low-bandgap polymers 140
atomic force microscopy (AFM) 302, 391
– surface topography images 71
atomic layer deposition 495
Auger recombination process 193
Auger spectroscopy 27

b
backbone melting 10
balance of system (BOS) 543
ball-to-ball distances 163
Barix-coated plastic film 500
Barix multilayer technology 498
Barix process 504, 505
Barix resin system 499
benzo-ethylene-dioxythiophene 214
BHJ-based devices 174
bicontinuous interpenetrating network 59
black rollers 435
blending polystyrene (PS) 526
Boltzmanns constant 284
balance of modules (BOM) 554
bright-field TEM
– images 316, 320
– micrographs 309, 311
built-in intramolecular charge transfer 109
bulk-heterojunction (BHJ) 57, 59, 155, 165, 193, 284, 300, 319, 368, 348, 544
– architecture 347
– layers 375
– phase-separated 266
– photoactive layer 302
– photovoltaic devices 63, 94, 380
– solar cells 134, 138, 263, 269, 376, 382, 383, 544
– systems 309

c
cadmium-indium-gallium-selenide (CIGS) 543
cadmium telluride (CdTe) 543
Cadogan ring closure reaction 97
carbazole-based monomers 104
carbazole-based polymers 96, 111, 115, 121
– ladder-type 96, 100, 103, 111, 121, 124, 125
– structure 110
carbazole moiety 95
CdSe nanoparticles 181, 186, 199
– concentration 199
– synthesis 181

Organic Photovoltaics: Materials, Device Physics, and Manufacturing Technologies.
Edited by Christoph Brabec, Vladimir Dyakonov, and Ullrich Scherf
Copyright © 2008 WILEY-VCH Verlag GmbH & Co. KGaA, Weinheim
ISBN: 978-3-527-31675-5

Index

charge-carrier mobility 494
charge-transfer absorptions 18
charge-transport properties 179
chemical vapor deposition (CVD) 497
classic silicon-based solar cells 245
close-packed self-organized layers 109
close-packed titania mesopores 340
coefficient of linear thermal expansion (CLTE) 474
coffee stain effect 76
common cleaning techniques 219
– plasma cleaning 219
– UV ozonizing 219
– wet cleaning 219
concentration-dependent optical behavior 102
conductive polymers 245, 251
– largest shortcoming 251
conjugated polymers 368, 375, 385, 494
conjugated triblock copolymers 85
consumer electronics 555
conventional oil and gas 536
copolymers backbone 109
copper-metallized PET foils 523
copper phthalocyanine (CuPc) 230
corona device 521
cost-effective production process 425
cost-intensive printing 520
Coulomb interaction 180, 206
– long-range 205
crystal-packing effects 169
crystalline films 474
crystalline silicon modules 553
cubic zinc blend structure 184
current-voltage
– behaviors 192
– characteristics 186, 287
– curve 23, 202, 460, 463
cyclic voltammetry (CV)
– data 266
– experiments 121
mobile phones, see consumer electronics

d

Debye-Scherrer analysis 29
Debye-Scherrer diffraction rings 19
deposition techniques 375
– large-scale 375
– liquid-state 513
device film 173
– processes 185
– structures 332
dimethylformamide (DMF) 225
dimethylsulfoxide (DMSO) 220
dip coating 342
dispersion polymerization 247
– approach 255
donor-acceptor composites 179
– photovoltaic cells 266
donor-acceptor-donor segments 140
donor-acceptor interface 160, 300, 346, 348, 336, 337
donor-acceptor heterojunctions 57
donor-acceptor layer 206
donor-acceptor material 197, 202
donor-acceptor system 198, 451
– hybrid 368
donor blend 166
drift-diffusion motion 202
dye-sensitized photovoltaic cells 229
dye-sensitized solar cells (DSSC) 26, 58, 66, 181, 189, 244, 340, 370, 544

e

EDOT 215
– oxidative polymerization 215
electric field 346
electrical doping 226
electricity European exchange (EEX) 547
electrochemical oxidation 217
– approach 145
– characterization 144
– techniques 206
electrodes 332, 371, 426, 437, 441, 522
– structure 458
electroluminescence images 70
electron acceptors 189
electron-deficient heterocycles 130
electron-deficient phenyloxadiazoles 62
electron-donating monomers 130
electron diffraction patterns 311
electron donor 323
electron-hole pairs 272, 293, 331
– distance 288
electron-hole transfer 197
electron injection barriers 28
electron spectroscopy for chemical analysis (ESCA) 27
electron-transport layer 21
electron transfer 333, 347
electron transport materials (ETMs) 59, 63, 204
– domains 22
– mechanisms 205
electronic circuitry 472
electronic printing process 519
electronic relaxation process 199
electrophilic substitution reaction 100
electropolymerization 96, 146, 343, 347
electrostatic force microscopy (EFM) 72

ellipsometric data 435
ellipsometry 448
encapsulation process 504
– technology 506
energy conversion efficiency (ECE) 58
energy-filtered TEM (EFTEM) 307
energy storage systems 551, 552
energy transfer 350
enthalpic interaction 338
environmental scanning electron microscopy (ESEM) 75
EPIA, see European Photovoltaic Industry Association
equatorial wide-angle reflections 112
equatorial X-ray scattering intensity distribution 123
ethylene vinyl acetate copolymer (EVA) 494
European Photovoltaic Industry Association (EPIA) 544
evaporation processes 454
exciton diffusion 347
– length 332, 350
exciton harvesting 344, 347
exciton transport 197
external quantum efficiencies (EQEs) 58, 330, 334, 363
– curves 122
– spectrum 338

f

Förster process 197
Fermi level alignment 232
Fermi level pinning 233, 331
field-dependent measurements 205
field effect hole mobilities 256
field effect transistor (FET) 129, 144, 255
flexographic printing 518, 525
Flory–Huggins parameter 301
fluorene alkyl substituents 78
fluorenone containing copolymers 86
fluorine-doped tin oxide (FTO) 331, 364
fossil fuel 531, 537
fullerene-based electron transport 164
fullerene derivative 165, 169, 173
– absorption 169
– domains 19
– materials 171
– stability 173
– variations 165

g

gas-phase 513
geminate recombination 202
generic polymer 180
– structure 180
glass-based displays 479
glass substrates 492
global warming 534, 537
gravure printing 519, 525
grazing emission X-ray fluorescence (GEXRF) 28
grazing incidence small-angle X-ray 29
greenhouse gases 537
gross domestic product (GDP) 533

h

healing damaged devices 30
heat transfer coefficient 434
high angular annular dark field detector scanning transmission electron microscopy (HAADF-STEM) 225
high-boiling solvents 227
high-density data storage 472
high-efficiency devices 86
high-efficiency polymeric solar cells 300
high electron affinity 332, 368
high-energy-absorbed photon 193
high open-circuit voltage 188
high-performance devices 257, 322
high-quality conjugated polymer brushes 195
high-quality inorganic films 498
high-resolution microscopy techniques 302
highest occupied molecular orbital (HOMO) 100, 160, 231, 291, 300, 331, 445
hole-collecting electrode 180
hole-extracting layers 229
hole injection barriers 28
hole transport materials (HTMs) 58, 64
holographic microprisms 455
HOMO-LUMO
– bandgap 213
– gap estimation 16
– levels 64, 78
– transition 226
homogeneous thin films 455
HT-PAT polymorphs 9
HT-PHT films 27
HT-PT electron transport 17
HT-regioregular polythiophenes 7, 9, 17, 18, 25, 30
HT-HT sequences 9
– architecture 5
HTPT films 26
Hubbert graph 536
Hubbert law 537
hybrid organic 26
– nanocrystal solar cells 229
hydrolysis-resistant automotive wiring 472

i

ideal photovoltaic device 206
ideal polymer 79
idealized single-layer device 59
IMI systems 435, 436, 437
in situ PEDOT films 216
in situ polymerization 341, 343
– layers 213
– polythiophene 371
incident photon to current efficiency (IPCE) 58
indium-tin oxide (ITO) 220, 331, 428, 435, 493
– coated foil 526
– electrode 136, 253, 291, 363, 441, 426, 462
– film 441
– replacement 250, 251
inherent surface smoothness 478
inkjet printing 75, 76, 173, 518
inline measurement system 434
infiltration techniques 347
inorganic crystalline semiconductors 221
inorganic material 371
inorganic nanocomposites, *see* Hybrid organic
inorganic nanowires 348
inorganic semiconductor 179, 385
– nanoparticles 179, 202
inorganic solar cells 93
insoluble Fullerene layers 169
Institute for Solar Energy Systems (ISE) 464
Institute of Microsystem Technology (IMTEK) 464
insulator-to-metal transition 249
insulator-metal-insulator 435
interdigital electrode geometry 447
internal quantum efficiency (IQE) 195, 450
International Panel on Climate Change (IPCC) 538
intersystem crossing process 199
intrinsically conducting polymer (ICP) 213
inverted tandem cell structure 274
ion beam milling 27
ion beam technique 28
ion-solvating functional groups 67

k

Kelvin probe 232, 233
– measurements 366
Knoevenagel polycondensation 141

l

Lawessons reagent 142
light-absorbing material 243
light-emitting diodes (LEDs) 93, 135, 137
light-induced degradation 230
light-scattering measurements 256
liquid crystalline mesophase 12
lithographic structuring 442
long-term stability 228
long-wavelength emission 269
low-bandgap
– analogues 78
– building blocks 130
– materials 76, 167
– nanoparticles 192, 197
– polymers 81, 82, 121, 130, 136, 143, 146, 147, 171, 349
– segment 130, 135
– sensitizers 62
low current densities 334
low-energy hole 22
low-energy ion bombardment 430
low induced stress 494
low-lying conduction band, *see* High electron affinity
low-mobility polymer 333
low-performance technology 557, 558
low-resistance domains 23
low-temperature transitions 10
lowest unoccupied molecular orbital (LUMO) 4, 264, 291, 300
– level 159, 231
– variation 170
luminescence decay 201
LUMO–LUMO gap 349
LUMO–LUMO drop 349

m

macromolecular chemical design 116
macroscopic single domains 14
MALDI-TOF mass spectroscopy 216
mapping techniques 71
maximum power point (MPP) 290
MDMO-PPV 313
– matrix film 313, 314
mean carrier drift length 285
measurement techniques 344
MEH-PPV 335
melting temperatures 347
mesoporous titania/polymer 337
metal-assisted cross couplings 14
– reactions 5
metal electrodes 445
metal-insulator-metal (MIM) model 230
metal-like conductivity 96
metal oxide-polythiophene 365
metal oxide nanoparticle 189, 376
– devices 197

metal oxide networks 385
metallic electrode 331
methylene dioxythiophene (MDOT) 214
microcrystalline domains 9
microelectronics industry 517
microreplication technologies 441
microstructure 452
microwave absorption studies 249
mobile ionic impurities 28
module efficiency 459
moisture barriers 496
molecular-scale structure 202
monochromatic illumination 267, 343
monomer vapors 499
monomeric units 99
Monte Carlo techniques 206
– approaches 207
– modeling 202
– simulations 499
morphology fixation 169
morphology mapping 70
multicomponent system 165
– phase diagram 165
multilayer polymer tandem solar cell 273
multiple electron-hole pairs 193
multiple exciton generation 193

n

nanocrystal device 181
– classes 181
nanocrystal materials systems 181
nanocrystal photovoltaic device, *see* Generic polymer
nanocrystalline ZnO 373, 378
nanoimprint lithography (NIL) 339, 517
nanoparticle 375
– devices 196
– films 335
– shape 184
nanopillars 341
nanoporous devices 335
nanoporous metal oxide 369
nanoporous TiO$_2$ 368
nanoporous titania films 329, 335, 342
nanoscale phase separation 76
nanoscale structure-property relationships 9
nanostructure 333, 341
– electrodes 442
– geometry 348
– ideal structure 195
– metal oxides 368
– photovoltaic device 195, 197
– photovoltaics 181
– polymeric devices 341

– pores 385
nanowires, *see* Nanopillars
National Renewable Energy Laboratory (NREL) 245, 491, 551
near-field scanning optical microscopy (NSOM) 72
near-field scanning photocurrent microscopy (NSPM) 72
near infrared (NIR) 170
– photons 301
neutral conjugated polymers 245
new renewable energies (NRE) 539
next-generation organic photovoltaics 30
nickel-catalyzed reaction 96
– Yamamoto polymerization 97
nitrogen-bridged ladder-type polymer 100, 108
nonamphiphilic dye 352
nonoptimum phase morphology 273
nonspace-charge-limited devices 289
nonuniform electric field 287
n-type semiconductors 26, 162
nuclear power 534

o

oil shortage 534
open-circuit energy level diagrams 65
open-circuit voltage (V$_{oc}$) 21, 57, 134, 191, 193, 206, 231, 364, 459, 503
optical absorption spectra 19, 369
optical simulations 448
optimal device thickness 289
optimized blend compositions 167
– cyclohexafullerenes 167
OPVC fabrication 526
organic composite photovoltaics (OCPV) 329, 347
– OCPV models 329
organic electronic devices 299, 443, 447
organic field effect transitors (OFETs) 514
– circuits 517, 522, 524
– devices 527
organic hole transport materials 251
organic light-emitting diodes (OLEDs) 58, 213, 447, 491, 514,
– displays 485, 500
– materials 484
organic light-emitting polymers 471
organic moieties 250
– low-cost processability 250
organic photovoltaic (OPV) 3, 552
– absorber 492
– cells 494
– devices 155, 165, 446
– effect 165

– modules 552
– technologies 552
organic photovoltaic cell (OSC) 58, 59, 202, 213, 246, 434, 491, 514, 549
– devices 93, 245, 443, 447
– technology 26
organic polymeric materials 129
organic semiconducting materials 93
organic semiconductors 5, 17, 471, 513
organic solar cell 218, 244, 455, 448, 457, 463, 491, 492, 552
– architecture 458
– devices 65
– electrode interface 441
– PEDOT-type materials 218
organic thin-film devices 220
Organization for Economic Cooperation and Development (OECD) 531
organometallic derivatives 215
– metal-catalyzed coupling 215
organometallic synthesis 217
Ormecon dispersion process 249
ortho-dichlorobenzene (ODCB) 294
oscillator circuits 524
oxidant sodium peroxodisulfate 215
– polymerization 215
oxidative polymerization mechanism 215
oxygen-doping effects 15
oxygen-scavenging role 269
oxygen plasma treatment 230

p

Pacific Northwest National Laboratories (PNNL) 497
particle size distribution (PSD) 222
phenyl-C_{61}-butyric acid methyl ester (PCBM) 18, 166, 310
– assembly 18
– clusters 316
– crystals 166, 316, 318
– domains 310, 311, 316, 319, 322
– films 312, 314
– heterojunctions 319
– molecules 314
– nanocrystals 311, 316, 320
– phase separation 20
– polymer solar cells 287, 315
PCNEPV domains 305, 308
PDI-based device 118
– chemical structure 214
PEDOT 217, 218
– buffer layers 218, 229, 453, 455
– polymeric salts 216
– preparation 218

– properties 218
– redox states 217
– role 229
Peoples Republic of China (PRC) 543
perylene-based dyes 116
perylene dicarboximide antennae moiety 169
perylene monodicarboxiimide (PI) 62
perylene tetracarboxydiimide (PDI) 117
phase-segregated mixture 252
phenylene vinylene (PPV) derivative 61
phosphoric acid groups 332
photoactive layer deposition 255
– productive tool 255
photocurrents 364
– density 289
– generation 17
photodegradation reactions 482
photoinduced absorption (PIA) 369
– measurements 199
photoinduced ESR experiments 369
photolithography spin-coated semiconductor 523
photoluminescence images 70
photoluminescence quenching 68, 199, 343
photoluminescence spectral data 338
photoreaction mechanisms 481
photoresist structures 442
photovoltaic 549
– applications 185
– cells 336, 343
– conversion efficiency 254, 264, 491
– conversion solar energy 283
– devices 26, 117, 122, 179, 181, 190, 243, 250, 368, 375, 376, 382, 391
– energy 551
– fabrication method 253
– main steps 284
– polymers 250
physical vapor deposited (PVD) 497
PIA spectroscopy 369, 378, 379
planar heterojunction photo-voltaic cell 364
planar metal oxide 363
planar zinc oxide (ZnO) 363
plasma enhanced chemical vapor deposition (PECVD) 497
polychlorotrifluoroethylene (PCTFE) 496
polyester films 471
polyethylene naphthalate (PEN) 426, 471, 472, 496
– films 472, 475, 477, 478, 479, 480, 497
polyethylene terephthalate (PET) 425, 471, 475
– films 472, 477, 478, 479, 480
– heat-stabilized 485

– substrates   251, 502
polyethylenedioxythiophene (PEDOT) dispersions   247
polymer   68, 75, 181, 196
– bandgap   349
– blends   380, 383
– chains   333, 338, 346, 472
– dopant system   257
– engineering   349
– films   473
– flexibility   22
– influence   257
– matrix   312
– metal oxide   368
– mobility   344, 346
– morphology   257
– nanowires   348
– optical spacer   263
– organic solar cells   255
– phase separation   21
– photovoltaic devices   57, 243, 252
– physical processes   196
– production   243
– solar cells   18, 125, 263, 271, 272, 299, 311, 322, 375
polymeric semiconductor-based solar cells (PSCs)   299
polymethylmethacrylate (PMMA)   339
polystyrene sulfonic acid (PSS)   215
polythiophenes   3, 4, 10, 226
– controlled crystallization   291
– derivatives   332
– optical properties   226
– orientation   291
– phase   23
– plasticizers   12
– polymer chains   291
power-law decay   205
power conversion efficiency (PCE)   23, 57, 80, 264, 334
precipitation kinetics   166
precipitation thermodynamics   166
printing techniques   517
– process   525
pump-down curve   433
pump-probe spectroscopy   199
– absorption measurement   193
PV module installation   548
pyrrole-containing polymers   78

q

quantitative structure activity relationship (QSAR) analysis   174
quantum-conned systems   180

quantum efficiency (QE)   263
quantum rods   184
– formation   184
quasi living process   5
quasi-steady state photoinduced absorption experiment   203
quasi-two-dimensional behavior   16
quenching effects   173

r

radiative force (RF)   537
Raman spectroscopy   70, 71
random copolymers   143
rapid charge-pair recombination process   192
rapid spin coating   7
red-absorbing fluorine copolymer   189
red-shifted absorption   293
red-shifted optical absorption   7
reel-to-reel process   85, 442
refractive index   346
renewable energies (RE)   539, 553
– source   546, 548
residual charge moieties   217
resonant energy transfer (RET)   350, 351
rigid film   480
rigorous coupled wave analysis (RCWA)   448
ring oscillators   522
roll-coating equipment   432
roll-coating machines   429, 431, 436
roll-to-roll capable photolithography   523
roll-to-roll fabrication   514
roll-to-roll printing techniques   491
roll-to-roll processes   243, 522, 526
root-mean-square roughness   229
ruthenium dye   335
Rutherford backscattering spectrometry (RBS)   16, 28

s

SAED analysis   318
sandwich-structure devices   204
scanning electron microscopy (SEM)   119, 302, 337, 341
scanning Kelvin probe microscopy (SKPM)   71
scanning probe microscopy (SPM)   302
scanning transmission X-ray microscopy (STXM)   74
Schottky-Mott framework   232
Schottky–Mott model   207, 232, 233
secondary ion mass spectrometry (SIMS)   29, 74
selected area electron diffraction (SAED) analysis   19, 302

selenium-trioctylphosphine complex 184
self-assembled monolayer (SAM) 291
semiconductor-dielectric interfaces 520
semiconductor interface 231
semiconductor nanoparticles 179
– advantages 179
semicrystalline polymers 344, 345, 475
semiladder-type polymer 104
short-circuit current 23, 24, 348, 349, 376, 452, 455, 503
– density 348, 382, 450, 461, 463
– scales 461
side chain melting 10, 12
silicon-based PV 493
silicon-based solar cell 243, 428
silicon hybrid solar cells 229
single-layer laminate sheets 496
single-wall carbon nanotube (SWCNT) 231
single-walled nanotubes (SWNT) 25
single solar cells 277
– characteristics 277
singlet oxygen (O_2) 173
– net generation 173
– quantum yield 173
small-angle neutron spectroscopy 224
small-angle reflections 112
sol–gel process 272
– titanium oxide 263
– solution-based 264, 269
solar cells 229, 332, 341, 367, 441, 442, 491, 544
– buffer layer 229
– industry 426
– market 425
solar simulating devices
– arc lamps 482
  xenon 482
solar spectrum 334
solid-state crystalline order 9
solid-state devices 67
solid-state spectra trend 15
solid-state structure 6, 24
solid-state transfer methods 513
solution-processed devices 189, 190
solvent-plasticized chains 13
solvent effect 225
Solvophobic interactions 8
source-drain current 256
– saturation regime 256
Soxhlet extraction 15, 28
space-charge-limited (SCL) 286, 289, 306
space-charge density 205
spin-cast films 256
– photoluminescence spectra 256

– process 168
spin-coated films 75, 311, 316, 516
spin-coated polymer layer 336
spray pyrolysis method 335, 342, 429
sputtering process 429
stable charge-transfer-like configuration 233
stable colloidal suspensions 7
standard operating procedure (SOP) 503
state-of-the-art inorganic solar cells 283
state-of-the-art nuclear power plants 542
state-of-the-art performing devices 155
– fullerene derivatives 155
state-of-the-art polymer-fullerene cells 336
streak camera measurements 199
structure-property relationship 129
substituent-dependent processes 12
sulfonated polyaniline (SPAN) 253
– layer-by-layer deposition 253
surface potential mapping 72
Suzuki-type polymerization 97
Suzuki coupling 105
Suzuki polycondensation 100

t
tandem cell architecture 263, 272, 273
tandem solar cells 143
tandem structure 273
– electronic absorption spectrum 273
temperature-accelerated degradation mechanisms 500
temperature-dependent generation rate 288
temperature-dependent measurements 205
terawatt (TW) energy 549
tetradecylphosphonic acid (TDPA) 184
tetrapods 184, 185
– growth 185
TFB-rich domains 69, 73
thermodynamic equilibrium systems 246
thermopower measurements 249
thickness-dependent maxima 24
thin blend film 301
thin-film absorption spectra 16
thin-film devices 95, 229, 448
thin-film electrodes 428
thin-film encapsulation 498
thin-film inorganic PV materials 492
thin-film inorganic solar modules 496
thin-film photovoltaic (PV) cells 197, 220
– optical design 197
thin-film PV technology 554
thin-film technology 543
thin polymer films 310
thin semiconductor films 516
thin transparent films 26

thiophenes 135
– electron-donating strength 135
three-dimensional (3D) device
 architecture 441
time-correlated single-photon counting
 199, 201
time-of-fight (TOF) measurements 294
time-resolved luminescence
 measurements 199
time-resolved microwave conductivity
 (TRMC) 363
titania-based OCPVs 329, 331, 347
titania films 347
– layer 331, 336
– nanostructure 337, 339
titania precursor 334
titania templates 329
titania/polymer solar cell 331
– bilayer 335
titania/polymer system 334, 352
titanium dioxide 264
– nanoparticles 189
– sintered films 189
TM-polarized light 452
total primary energy supply (TPES) growth 531
transient absorption spectroscopic
 techniques 78
transition metal oxides 231
– inorganic buffer layers 231
transmission electron microscopy (TEM)
 273, 302, 341
transparent conductive oxide (TCO) 331,
 441, 429
– electrode 251, 331
– material 428
transversal magnetic (TM) polarization 443
trisopropylsilylethyny-substituted pentacene
 (TIPSEP) 146
TRMC technique 368, 390
twist glass transition 12
two dimensional polycyclic aromatic
 hydrocarbons 116
two-layer system 169
two-phase model 12
two-phase transitions 246

u
ultrasonic nebulizer 335
ultraviolet photoelectron spectroscopy (UPS)
 133

UV light exposure 228
UV stability 481
UV–Vis absorption 77, 78, 109
– band edge 16
UV–Vis spectroscopy 73, 257, 133

v
vacuum system 432
vanadium (III) oxide (V_2O_3) 231
various coating methods 514
vertical phase morphology 73
vibronic structure 7
vinylene dioxythiophene (VDOT) 214
voltage-dependent photocurrent 330

w
water-soluble polymer 215
water vapor transport rate (WVTR) 496
WAXS experiments 123
weak interaction 316
well-formed thin films 68
well-ordered nanoscale morphology 109
whisker formation 168
wide-bandgap inorganic semiconductor
 363
winding system 433, 434
witness cells 504
Wittig condensation 144
wrap-through device 459, 461

x
X-ray contour maps 9
X-ray diffraction (XRD) 216, 264, 341
X-ray diffractometry 111
X-ray photoelectron spectroscopy (XPS)
 337, 386
xerox process 96
XPS depth proling technique 270

y
Yamamoto polycondensation reaction 98,
 103, 105
Youngs modulus 479, 480

z
ZAO 428, 435
zero-loss-ltered 308
ZnO films 388
ZnO nanoparticles 190, 191, 377, 379
ZnO nanorod 374